P9-CTP-249

PAGE	ALGORITHM	DESCRIPTION
365	Algorithm 7.1	Composite Trapezoidal Rule
365	Algorithm 7.2	Composite Simpson Rule
378	Algorithm 7.3	Recursive Trapezoidal Rule
379	Algorithm 7.4	Romberg Integration
389	Algorithm 7.5	Adaptive Quadrature Using Simpson's Rule
397	Algorithm 7.6	Gauss–Legendre Quadrature
413	Algorithm 8.1	Golden Search for a Minimum
414	Algorithm 8.2	Nelder–Mead's Minimization Method
416	Algorithm 8.3	Local Minimum Search Using Quadratic Interpolation
418	Algorithm 8.4	Steepest Descent or Gradient Method
435	Algorithm 9.1	Euler's Method
441	Algorithm 9.2	Heun's Method
448	Algorithm 9.3	Taylor's Method of Order 4
460	Algorithm 9.4	Runge–Kutta Method of Order 4
461	Algorithm 9.5	Runge–Kutta–Fehlberg Method (RKF45)
471	Algorithm 9.6	Adams–Bashforth–Moulton Method
472	Algorithm 9.7	Milne–Simpson Method
473	Algorithm 9.8	The Hamming Method
488	Algorithm 9.9	Linear Shooting Method
496	Algorithm 9.10	Finite-Difference Method
507	Algorithm 10.1	Finite-Difference Solution for the Wave Equation
516	Algorithm 10.2	Forward-Difference Method for the Heat Equation
517	Algorithm 10.3	Crank–Nicholson Method for the Heat Equation
531	Algorithm 10.4	Dirichlet Method for Laplace's Equation
557	Algorithm 11.1	Power Method
558	Algorithm 11.2	Shifted Inverse Power Method
571	Algorithm 11.3	Jacobi Iteration for Eigenvalues and Eigenvectors
581	Algorithm 11.4	Reduction to Tridiagonal Form
587	Algorithm 11.5	The QL Method with Shifts

Numerical Methods

for Mathematics,
Science,
and Engineering

SECOND EDITION

Numerical Methods
for Mathematics,
Science,
and Engineering

JOHN H. MATHEWS
California State University, Fullerton

Prentice Hall, Englewood Cliffs, New Jersey 07632

Library of Congress Cataloging-in-Publication Data

MATHEWS, JOHN H., (date)
 Numerical methods for mathematics, science, and engineering / John
H. Mathews. — 2nd ed.
  p. cm.
 Rev. ed. of: Numerical methods for computer science, engineering,
and mathematics. 1987.
 Includes bibliographical references and index.
 ISBN 0-13-624990-6
 1. Numerical analysis. I. Mathews, John H., Numerical
methods for computer science, engineering, and mathematics.
II. Title.
QA297.M39   1992   91-42722
519.4—dc20   CIP

Editorial/production supervision
  and interior design: *Kathleen M. Lafferty*
Cover design: *Lundgren Graphics, Ltd.*
Prepress buyer: *Paula Massenaro*
Manufacturing buyer: *Lori Bulwin*
Acquisitions editor: *Steven R. Comny*

The first edition of this work was published under
the title, *Numerical Methods for Computer Science,
Engineering, and Mathematics.*

 Prentice-Hall, Inc.
A Simon & Schuster Company
Englewood Cliffs, New Jersey 07632

Printed in the United States of America

10 9 8 7 6 5 4 3

ISBN   0-13-624990-6

Prentice-Hall International (UK) Limited, *London*
Prentice-Hall of Australia Pty. Limited, *Sydney*
Prentice-Hall Canada Inc., *Toronto*
Prentice-Hall Hispanoamericana, S.A., *Mexico*
Prentice-Hall of India Private Limited, *New Delhi*
Prentice-Hall of Japan, Inc., *Tokyo*
Simon & Schuster Asia Pte. Ltd., *Singapore*
Editora Prentice-Hall do Brasil, Ltda., *Rio de Janeiro*

Contents

3   The Solution of Linear Systems $A\mathbf{X} = \mathbf{B}$ 122

4   Interpolation and Polynomial Approximation   191

5   Curve Fitting   257

10 Solution of Partial Differential Equations   498

11 Eigenvalues and Eigenvectors   536

Preface

Numerical Methods for Mathematics, Science, and Engineering, Second Edition, provides a rudimentary introduction to numerical analysis for either a single course or a year-long sequence and is suitable for undergraduate students in mathematics, science, and engineering. Ample material is presented so that instructors will be able to select topics appropriate to their needs. It is assumed that the reader is familiar with calculus and has taken a structured programming language such as BASIC, C, FORTRAN, or Pascal.

Students of all backgrounds enjoy numerical methods and this is kept in mind throughout the book. A variety of examples and problems sharpen one's skill in both the theory and practice of numerical analysis. Computer calculations are presented in the form of tables and graphs whenever possible so that the resulting numerical approximations are easier to interpret. Many figures for this second edition were obtained by using the software Mathematica™. The algorithms for the various numerical processes are given in pseudo-code and are easy for students to translate into BASIC, C, FORTRAN, or Pascal. The structure of the algorithms makes them easy to adapt to a programming environment such as MAPLE, Mathematica™, or MATLAB™.

Emphasis is placed on understanding why numerical methods work and their limitations. This is not easy for a first course; it involves a balance between theory, error analysis, and readability by students. An error analysis for each method is presented in a fashion that is appropriate for the method at hand and yet does not turn off the reader. A mathematical derivation for each method is given that uses elementary results and builds the student's understanding of numerical analysis. Computer assignments implementing the algorithms give students an opportunity to practice their skills at scientific programming.

Shorter numerical exercises can be carried out with a pocket calculator/computer, but others can be done more efficiently by computer. I have tried to be flexible on this issue and do not specify the precise hardware that must be used to solve any given problem. It is left for instructors to guide their students regarding the pedagogical use of numerical computations. Instructors must make assignments that are appropriate to the availability of computing resources for their particular courses.

The use of numerical analysis hardware, software packages, and libraries is encouraged. Sometimes the phrase "use a computer" occurs in an exercise. This must be interpreted in view of a school's particular learning environment. Instructors have the flexibility to permit their students to use the automatic root-finding and numerical integration routines found on some pocket calculator/computers or to use other popular software such as MathCad™, MATLAB™, Mathematica™, and IMSL™. Also, algorithms in the text are available in MATLAB™, FORTRAN, and Pascal and Mathematica™ notebooks for both IBM PC-compatible computers and APPLE Macintosh computers. These materials can be used to assist students in performing their "numerical experiments."

Acknowledgments

I would like to express my gratitude to all the people whose efforts contributed to both the first and second editions of this book. I thank the students at California State University, Fullerton. I thank my colleagues Stephen Goode, Mathew Koshy, Edward Sabotka, Harris Shultz, and Soo Tang Tan for their support in the first edition and Russell Egbert, William Gearhart, Ronald Miller, and John Pierce for their suggestions for the second edition. I also thank James Friel, chairman of the Mathematics Department at CSUF, for his encouragement.

I also express my gratitude to the reviewers who made recommendations for the first edition: Kenneth P. Bube, University of California, Los Angeles; Michael A. Freedman, University of Alaska, Fairbanks; Peter J. Gingo, University of Akron; George B. Miller, Central Connecticut State University; and Walter M. Patterson III, Lander College. For the second edition, I thank Richard T. Bumby, Rutgers University; Robert L. Curry, U.S. Army; Bruce Edwards, University of Florida; and David R. Hill, Temple University.

Finally, I wish to express my appreciation to the staff at Prentice Hall, especially Steven Conmy, mathematics editor, and Kathleen Lafferty, production editor, for their assistance and encouragement.

Software disks for both IBM and Macintosh computers are available to instructors who adopt the textbook. They include Pascal, FORTRAN and Matlab source code and Mathematica notebooks for all the algorithms. Inquires about the availability can be made to Prentice Hall, Inc. or the author. Comments and suggestions for improvements to the book and supporting software are welcome and can be made directly to me at (714) 773-3631 or via E-mail:
MATHEWS@FULLERTON.EDU.

John H. Mathews

1

Preliminaries

Consider the function $f(x) = \cos(x)$, its derivative $f'(x) = -\sin(x)$, and the integral $F(x) = \sin(x)$. These formulas were studied in calculus. The former is used to determine the slope $m = f'(x_0)$ of the curve $y = f(x)$ at a point $(x_0, f(x_0))$, and the latter is used to compute the area under the curve for $a \le x \le b$.

The slope at the point $(\pi/2, 0)$ is $m = f'(\pi/2) = -1$ and can be used to find the tangent line at this point [Figure 1.1(a)]:

$$y_{\tan} = m\left(x - \frac{\pi}{2}\right) + 0 = f'\left(\frac{\pi}{2}\right)\left(x - \frac{\pi}{2}\right) = -x + \frac{\pi}{2}.$$

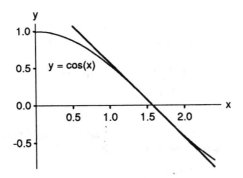

Figure 1.1 (a) The tangent line to the curve $y = \cos(x)$ at the point $(\pi/2, 0)$.

1

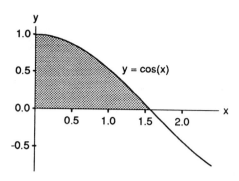

Figure 1.1 (b) The area under the curve $y = \cos(t)$ over the interval $[0, \pi/2]$.

The area under the curve for $0 \leq x \leq \pi/2$ is computed using an integral [Figure 1.1(b)]:

$$\text{area} = \int_0^{\pi/2} \cos(x)\, dx = F\!\left(\frac{\pi}{2}\right) - F(0) = \sin\!\left(\frac{\pi}{2}\right) - 0 = 1.$$

These are some of the results that we will need to use from calculus.

1.1 REVIEW OF CALCULUS

It is assumed that the reader is familiar with the notation and subject matter covered in the undergraduate calculus sequence. This included the topics real and complex numbers, continuity, limits, differentiation, integration, sequences, and series. Throughout the book we refer to the following results. They are illustrated with numerical examples that are characteristic of the study of numerical analysis.

Limits and Continuity

Definition 1.1. Assume that $f(x)$ is defined on the set S of real numbers. Then f is said to have the **limit** L at $x = x_0$, and we write

$$\lim_{x \to x_0} f(x) = L, \tag{1}$$

if given any $\epsilon > 0$, there exists a $\delta > 0$ such that whenever $x \in S$,

$$0 < |x - x_0| < \delta \quad \text{implies that} \quad |f(x) - L| < \epsilon.$$

When the h-increment notation $x = x_0 + h$ is used, equation (1) is equivalent to

$$\lim_{h \to 0} f(x_0 + h) = L. \tag{2}$$

Definition 1.2. Assume that $f(x)$ is defined on a set S of real numbers and let $x_0 \in S$. Then f is said to be **continuous** at $x = x_0$ if

$$\lim_{x \to x_0} f(x) = f(x_0). \tag{3}$$

The function f is said to be continuous on S if it is continuous at each point $x \in S$. The notation $C(S)$ stands for the set of all functions continuous on S. When S is an interval, the parentheses in this notation are omitted (e.g., the set of all functions continuous on the closed interval $[a, b]$ is denoted $C[a, b]$). When the h-increment notation $x = x_0 + h$ is used, equation (3) is equivalent to

$$\lim_{h \to 0} f(x_0 + h) = f(x_0). \tag{4}$$

For example, consider $f(x) = \cos(x) - \sqrt{2}/2$ over $[0, 0.6]$ and the value $x_0 = 0.4$ with the corresponding function value $y_0 = f(x_0) = f(0.4) = 0.21395$. For illustration, let us choose the tolerance as $\epsilon = 0.04$ and determine the corresponding δ. If x is restricted to lie in the $0.27998 < x < 0.49270$, the function value satisfy

$$f(x_0) - \epsilon = 0.17395 < f(x) < 0.25395 = f(x_0) + \epsilon.$$

Thus for $\epsilon = 0.04$ we choose $\delta = \min\{0.4 - 0.27998, 0.49270 - 0.4\} = \min\{0.12002, 0.09270\} = 0.09270$. Points on the portion of the graph $y = f(x)$ above the interval $[x_0 - \delta, x_0 + \delta] = [0.30730, 0.49270]$ will have y-coordinates that lie in $[y_0 - \epsilon, y_0 + \epsilon] = [0.17395, 0.25395]$. This portion of the graph is highlighted in Figure 1.2.

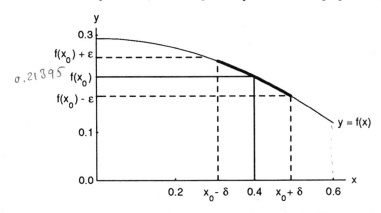

Figure 1.2 Investigating the continuity of $f(x) = \cos(x) - \sqrt{2}/2$ at $x_0 = 0.4$

Definition 1.3.   Suppose that $\{x_n\}_{n=1}^{\infty}$ is an infinite sequence. Then the sequence is said to have the limit L, and we write

$$\lim_{n \to \infty} x_n = L, \tag{5}$$

if given any $\epsilon > 0$, there exists a positive integer $N = N(\epsilon)$ such that

$$n > N \quad \text{implies that} \quad |x_n - L| < \epsilon.$$

When a sequence has a limit, we say that it is a **convergent sequence**. Another popular notation is that $x_n \to L$ as $n \to \infty$. Equation (5) is equivalent to

$$\lim_{n \to \infty} (L - x_n) = 0. \tag{6}$$

Thus we can view the sequence $\epsilon_n = L - x_n$ as an **error sequence**.

For example, if $x_n = [2n^3 + n \sin(n)]/(n^3 + 3n + 1)$, then $\lim\limits_{n \to \infty} x_n = 2$, so that $L = 2$. The error sequence $\epsilon_n = 2 - x_n$ tends to zero as $n \to \infty$. Figures 1.3(a) and (b) shows the behavior of $\{x_n\}$ and $\{\epsilon_n\}$.

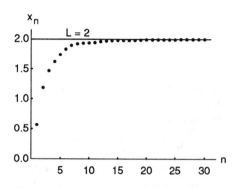

Figure 1.3   (a) A sequence $\{x_n\}$ where $L = 2 = \lim\limits_{n \to \infty} x_n$.

Figure 1.3   (b) The error sequence $\{\epsilon_n\} = \{L - x_n\}$ where $\lim\limits_{n \to \infty} \epsilon_n = \lim\limits_{n \to \infty} L - x_n = 0$.

Theorem 1.1.   Assume that $f(x)$ is defined on the set S and $x_0 \in S$. The following statements are equivalent:

$$\text{The function } f \text{ is continuous at } x_0. \tag{7}$$

$$\text{If } \lim_{n \to \infty} x_n = x_0, \text{ then } \lim_{n \to \infty} f(x_n) = f(x_0). \tag{8}$$

Theorem 1.2   (Intermediate Value Theorem).   Assume that $f \in C[a, b]$ and L is any number between $f(a)$ and $f(b)$. Then there exists a value c with $a < c < b$ such that $f(c) = L$.

For example, consider $f(x) = \cos(x - 1)$ over $[0, 1]$ and the constant $L = 0.8$. Then the solution to $f(x) = 0.8$ over $[0, 1]$ is $c_1 = 0.356499$. In the interval $[1, 2.5]$ the solution to $f(x) = 0.8$ is $c_2 = 1.643502$. These two cases are shown in Figure 1.4.

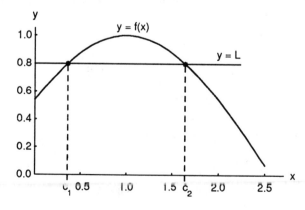

Figure 1.4   The intermediate value theorem applied to the function $f(x) = \cos(x - 1)$ over $[0, 1]$ and over the interval $[1, 2.5]$.

Theorem 1.3   (Extreme Value Theorem for a Continuous Function).   Assume that $f \in C[a, b]$. Then there exists a lower bound M_1 and an upper bound M_2 and two numbers $x_1, x_2 \in [a, b]$ such that

$$M_1 = f(x_1) \leq f(x) \leq f(x_2) = M_2 \qquad \text{whenever} \quad x \in [a, b]. \tag{9}$$

We sometimes express this by writing

$$M_1 = f(x_1) = \min_{a \leq x \leq b} \{f(x)\} \qquad \text{and} \qquad M_2 = f(x_2) = \max_{a \leq x \leq b} \{f(x)\}. \tag{10}$$

Differentiable Functions

Definition 1.4.   Assume that $f(x)$ is defined on an open interval containing x_0. Then f is said to be differentiable at x_0 if

$$\lim_{x \to x_0} \frac{f(x) - f(x_0)}{x - x_0} = f'(x_0) \tag{11}$$

exists. When this limit exists it is denoted by $f'(x_0)$ and is called the **derivative** of f at x_0. An equivalent way to express this limit is to use the h-increment notation:

$$\lim_{h \to 0} \frac{f(x_0 + h) - f(x_0)}{h} = f'(x_0). \tag{12}$$

A function that has a derivative at each point in S is said to be **differentiable** on S. The number $m = f'(x_0)$ is the slope of the tangent line to the curve $y = f(x)$ at $(x_0, f(x_0))$.

For example, let $f(x) = \ln(x)$, then $f'(x) = 1/x$. For $x_0 = 2$ and $h = 0.01$ we have the approximation

$$f'(x_0) = \frac{1}{2} \approx 0.4988 = \frac{0.698135 - 0.693147}{0.01} = \frac{f(2.01) - f(2.00)}{0.01}$$

$$= \frac{f(x_0 + h) - f(x_0)}{h}.$$

Theorem 1.4.   If $f(x)$ is differentiable at $x = x_0$, then $f(x)$ is continuous at $x = x_0$.

Theorem 1.5   (Rolle's Theorem).   Assume that $f \in C[a, b]$ and $f'(x)$ exists for all $a < x < b$. If $f(a) = f(b) = 0$, then there exists a value c, with $a < c < b$, such that $f'(c) = 0$.

Theorem 1.6   (Mean Value Theorem).   Assume that $f \in C[a, b]$ and $f'(x)$ exists for all $a < x < b$. Then there exists a number c, with $a < c < b$, such that

$$f'(c) = \frac{f(b) - f(a)}{b - a} = m. \tag{13}$$

For example, consider $f(x) = \sin(x)$ over $[a, b] = [0.1, 2.1]$. Then

$$m = \frac{f(b) - f(a)}{b - a} = \frac{f(2.1) - f(0.1)}{2.1 - 0.1} = \frac{0.863209 - 0.099833}{2.1 - 0.1} = 0.381688.$$

Using $f'(x) = \cos(x)$, the solution to $f'(c) = \cos(c) = 0.381688 = m$ is $c = 1.179174$. The line that goes through the points $(a, f(a))$ and $(b, f(b))$ is $y = 0.0998334 + 0.381688(x - 0.1) = 0.0616646 + 0.381688x$ and the line tangent to the curve at the point $(c, f(c))$ is $y = 0.924291 + 0.381688(x - 1.179174) = 0.474215 + 0.381688x$. The graphs of $f(x)$ and these two lines are shown in Figure 1.5.

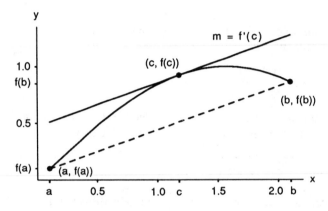

Figure 1.5   The mean value theorem applied to $f(x) = \sin(x)$ over the interval $[0.1, 2.1]$.

Theorem 1.7   (Extreme Value Theorem for a Diffentiable Function).   Assume that $f \in C[a, b]$ and $f'(x)$ exists for all $a < x < b$. Then there exists a lower bound M_1 and an upper bound M_2 and two numbers $x_1, x_2 \in [a, b]$ such that

$$M_1 = f(x_1) \le f(x) \le f(x_2) = M_2 \qquad \text{whenever } x \in [a, b]. \tag{14}$$

The numbers x_1 and x_2 occur either at endpoints of $[a, b]$ or where $f'(x) = 0$.

For example, consider $f(x) = 35 + 59.5x - 66.5x^2 + 15x^3$ over $[0, 3]$. Then $f'(x) = 59.5 - 133x + 45x^2$ and the solutions to $f'(x) = 0$ are $x_1 = 0.54955101$ and $x_2 = 2.4060045$. The minimum and maximum values of f over $[0, 3]$ are:

$$\min\{f(a), f(b), f(x_1), f(x_2)\} = \min\{35, 20, 50.104383, 2.118497\} = 2.118497$$

and

$$\max\{f(a), f(b), f(x_1), f(x_2)\} = \max\{35, 20, 50.104383, 2.118497\} = 50.104383,$$

respectively. The situation is shown in Figure 1.6.

Theorem 1.8   (Generalized Rolle's Theorem).   Assume that $f \in C[a, b]$ and that $f'(x), f''(x), \ldots, f^{(n)}(x)$ exist over (a, b) and $x_0, x_1, \ldots, x_n \in [a, b]$. If $f(x_j) = 0$ for $j = 0, 1, \ldots, n$, then there exists a value c, with $a < c < b$, such that

$$f^{(n)}(c) = 0. \tag{15}$$

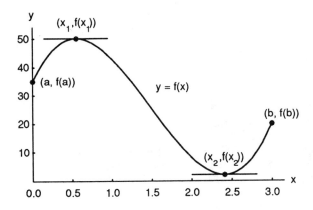

Figure 1.6 The extreme value theorem applied to the function $f(x) = 35 + 59.5x - 66.5x^2 + 15x^3$ over the interval $[0, 3]$.

Integrals

Theorem 1.9   (First Fundamental Theorem).   If f is continuous over $[a, b]$, then there exists a function F, called the **antiderivative** of f, such that

$$\int_a^b f(x)\,dx = F(b) - F(a) \qquad \text{where} \quad F'(x) = f(x). \tag{16}$$

Theorem 1.10   (Second Fundamental Theorem).   If f is continuous over $[a, b]$ and $a < x < b$, then

$$\frac{d}{dx} \int_a^x f(t)\,dt = f(x). \tag{17}$$

Theorem 1.11   (Mean Value Theorem for Integrals).   Assume that $f \in C[a, b]$ for $a \le x \le b$. Then there exists a number c with $a < c < b$ such that

$$\frac{1}{b-a} \int_a^b f(x)\,dx = f(c). \tag{18}$$

For example, consider $f(x) = \sin(x) + \frac{1}{3}\sin(3x)$ over the interval $[a, b] = [0, 2.5]$. The indefinite integral is $F(x) = -\cos(x) - \frac{1}{9}\cos(3x)$. The average value for the integral is:

$$\frac{1}{2.5 - 0} \int_0^{2.5} f(x)\,dx = \frac{F(2.5) - F(0.0)}{2.5} = \frac{0.762629 - (-1.111111)}{2.5} = \frac{1.873740}{2.5}$$

$$= 0.749496.$$

There are three solutions to the equation $f(c) = 0.749496$ over the interval $[0, 2.5]$: $c_1 = 0.440565$, $c_2 = 1.268010$, and $c_3 = 1.873583$. The area of the rectangle with base $b - a = 2.5$ and height $f(c_j) = 0.749496$ is $(b - a)f(c_j) = 1.873740$ and has the same

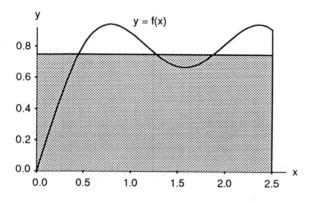

Figure 1.7 The mean value theorem for integrals applied to $f(x) = \sin(x) + \frac{1}{3}\sin(3x)$ over the interval $[0, 2.5]$.

numerical value of the integral of $f(x)$ taken over $[a, b]$. A comparison of the area under the curve $y = f(x)$ and that of the rectangle can be seen in Figure 1.7.

Theorem 1.12 (Weighted Integral Mean Value Theorem). Assume that $f, g \in C[a, b]$ and $g(x) \geq 0$ for $a \leq x \leq b$. Then there exists a number c with $a < c < b$ such that

$$\int_a^b f(x)g(x)\,dx = f(c)\int_a^b g(x)\,dx. \tag{19}$$

Definition 1.5 (Summation Definition of Integral). Assume that $f \in C[a, b]$, and suppose that $a = x_0 < x_1 < \cdots < x_n = b$ is a partition of $[a, b]$. For each $k = 1$, $2, \ldots, n$, select an arbitrary point t_k in the subinterval $[x_{k-1}, x_k]$ and introduce the difference notation $\Delta x_k = x_k - x_{k-1}$. Then the sum

$$\sum_{k=1}^{n} f(t_k)\,\Delta x_k \tag{20}$$

is called a **Riemann sum approximation** for the definite integral of $f(x)$ over $[a, b]$. We can speak of the definite integral

$$\int_a^b f(x)\,dx \tag{21}$$

as being the "limit" of the Riemann sums in (20), as the number of subintervals in the partition tend to infinity and the widths of the subintervals tend to zero.

For example, consider the function $f(x) = \sin(x)$ over the interval $[0, 5]$. Let $n = 5$ and choose t_k to be the midpoint of the each subinterval [i.e., $t_k = (2k - 1)/2$ for $k = 1, 2, 3, 4, 5$]. Then the Riemann sum for this partition is

$$S = \sin(0.5) + \sin(1.5) + \sin(2.5) + \sin(3.5) + \sin(4.5)$$
$$= 0.479426 + 0.997495 + 0.598472 - 0.350783 - 0.977530$$
$$= 0.747079.$$

This can be compared to the analytic solution

$$\int_0^5 \sin(x)\, dx = -\cos(5) + \cos(0) = -0.283662 + 1.000000 = 0.716338.$$

The error in this approximation is

$$\int_0^5 \sin(x)\, dx - S = 0.716338 - 0.747079 = -0.030741.$$

A graph of the curve $y = f(x)$ and the areas forming the Riemann sum is shown in Figure 1.8.

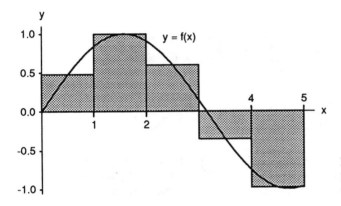

Figure 1.8 A Riemann sum with five partitions for the function $f(x) = \sin(x)$ over the interval $[0, 5]$.

Series

Definition 1.6. Let $\{a_n\}_{n=1}^{\infty}$ be a sequence. Then $\displaystyle\sum_{n=1}^{\infty} a_n$ is an infinite series. The nth partial sum is $S_n = \displaystyle\sum_{k=1}^{n} a_k$. The infinite series **converges** if and only if the sequence $\{S_n\}_{n=1}^{\infty}$ converges to a limit S, that is,

$$\lim_{n\to\infty} S_n = \lim_{n\to\infty} \sum_{k=1}^{n} a_k = S. \qquad (22)$$

If a series does not converge, we say that it **diverges**.

For example, consider $\{a_n\}_{n=1}^{\infty} = \left\{ \dfrac{1}{n(n+1)} \right\}_{n=1}^{\infty}$.

Then the nth partial sum is

$$S_n = \sum_{k=1}^{n} \frac{1}{k(k+1)} = \sum_{k=1}^{n} \left(\frac{1}{k} - \frac{1}{k+1}\right) = 1 - \frac{1}{n+1}.$$

Therefore, the sum of this infinite series is

$$S = \lim_{n\to\infty} S_n = \lim_{n\to\infty} \left(1 - \frac{1}{n+1}\right) = 1.$$

Theorem 1.13 (Taylor's Theorem). Assume that $f \in C^{n+1}[a, b]$ (i.e., f has continuous derivatives of order 1, 2, ... , $n + 1$). Suppose that $x_0 \in [a, b]$ is a fixed value; then for every $x \in [a, b]$, there exists a number $c = c(x)$ that lies between x_0 and x such that

$$f(x) = P_n(x) + R_n(x), \tag{23}$$

where

$$P_n(x) = \sum_{k=0}^{n} \frac{f^{(k)}(x_0)}{k!}(x - x_0)^k \tag{24}$$

and

$$R_n(x) = \frac{f^{(n+1)}(c)}{(n+1)!}(x - x_0)^{n+1}. \tag{25}$$

For example, let $f(x) = \sin(x)$; then $f'(x) = \cos(x)$, $f''(x) = -\sin(x)$, $f'''(x) = -\cos(x)$, ... The numerical values $f(0) = 0, f'(0) = 1, f''(0) = 0, f'''(0) = -1, \ldots$ must be substituted into formula (24) to obtain the Taylor polynomial $P(x)$ of degree $n = 9$ expanded about $x_0 = 0$:

$$P(x) = x - \frac{x^3}{3!} + \frac{x^5}{5!} - \frac{x^7}{7!} + \frac{x^9}{9!}.$$

A graph of both f and P over the interval $[0, 2\pi]$ is shown in Figure 1.9.

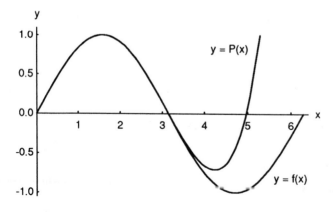

Figure 1.9 The graph of $f(x) = \sin(x)$ and the Taylor polynomial $P(x) = x - x^3/3! + x^5/5! - x^7/7! + x^9/9!$.

Corollary 1.1.   If $P_n(x)$ is the Taylor polynomial of degree n given in Theorem 1.13, then

$$P_n^{(k)}(x_0) = f^{(k)}(x_0) \qquad \text{for } k = 0, 1, \ldots, n. \tag{26}$$

Cramer's Rule

A practical method for solving 2×2 systems of linear equations is **Cramer's rule**, which we mention for the sake of completeness. Although Cramer's rule can be used to solve 3×3 systems, it is intractable when $N > 3$. In Section 3.4 we develop an efficient method for solving a large systems of equations. Consider the two linear equations

$$ax_1 + bx_2 = e, \tag{27}$$

$$cx_1 + dx_2 = f,$$

with the condition that $ad - bc \neq 0$. We can solve for the variable x_1 by eliminating the variable x_2. This is accomplished by multiplying the first equation by d and the second equation by b, and subtracting:

$$adx_1 + bdx_2 = ed$$
$$\underline{-(bcx_1 + bdx_2) = -bf}$$
$$adx_1 - bcx_1 = ed - bf$$

Hence $(ad - bc)x_1 = ed - bf$ and we can solve for x_1 and obtain

$$x_1 = \frac{ed - bf}{ad - bc}; \qquad \text{similarly,} \quad x_2 = \frac{af - ec}{ad - bc}. \tag{28}$$

The quotients in (28) can be expressed using determinants:

$$x_1 = \frac{\begin{vmatrix} e & b \\ f & d \end{vmatrix}}{\begin{vmatrix} a & b \\ c & d \end{vmatrix}} \quad \text{and} \quad x_2 = \frac{\begin{vmatrix} a & e \\ c & f \end{vmatrix}}{\begin{vmatrix} a & b \\ c & d \end{vmatrix}}. \tag{29}$$

Use Cramer's rule to solve the linear system

$$A\mathbf{X} = \begin{pmatrix} 3 & 1 \\ 7 & 4 \end{pmatrix}\begin{pmatrix} x_1 \\ x_2 \end{pmatrix} = \begin{pmatrix} 2 \\ 5 \end{pmatrix} = \mathbf{B}.$$

Solution. Using the equations in (29), we get

$$x_1 = \frac{\begin{vmatrix} 2 & 1 \\ 5 & 4 \end{vmatrix}}{\begin{vmatrix} 3 & 1 \\ 7 & 4 \end{vmatrix}} = \frac{8 - 5}{12 - 7} = 0.6, \qquad x_2 = \frac{\begin{vmatrix} 3 & 2 \\ 7 & 5 \end{vmatrix}}{\begin{vmatrix} 3 & 1 \\ 7 & 4 \end{vmatrix}} = \frac{15 - 14}{12 - 7} = 0.2.$$

Evaluation of a Polynomial

Let the polynomial $P(x)$ of degree n have the form

$$P(x) = a_n x^n + a_{n-1} x^{n-1} + \cdots + a_2 x^2 + a_1 x + a_0. \tag{30}$$

We now develop the algorithm known as **Horner's method** or **synthetic division** for evaluating a polynomial. It can be thought of as nested multiplication. For example, a fifth-degree polynomial $P_5(x)$ can be written in the nested multiplication form

$$P_5(x) = ((((a_5 x + a_4)x + a_3)x + a_2)x + a_1)x + a_0.$$

Since we do not want to enter the parentheses into lines of code in a computer program, we must create an algorithm that will carry out this intention.

Theorem 1.14   (Horner's Method for Polynomial Evaluation).   Assume that $P(x)$ is the polynomial given in equation (30) and $x = z$ is a number for which $P(z)$ is to be evaluated.

Set $b_n = a_n$ and compute

$$b_k = a_k + z b_{k+1} \text{ for } k = n - 1, n - 2, \ldots, 1, 0; \tag{31}$$

then $b_0 = P(z)$.

Moreover, if we define

$$Q_0(x) = b_n x^{n-1} + b_{n-1} x^{n-2} + \cdots + b_3 x^2 + b_2 x + b_1, \tag{32}$$

$$P(x) = (x - z)Q_0(x) + R_0, \tag{33}$$

where $Q_0(x)$ is the quotient polynomial of degree $n - 1$ and $R_0 = b_0 = P(z)$ is the remainder.

Proof. Substituting $Q_0(x)$ into the first term on the right side of (33) and substituting $R_0 = b_0$ for the second term yields

$$P(x) = (x-z)(b_n x^{n-1} + b_{n-1} x^{n-2} + \cdots + b_3 x^2 + b_2 x + b_1) + b_0. \tag{34}$$

The right side of (34) can be rewritten in decreasing powers of x:

$$P(x) = b_n x^n + (b_{n-1} - zb_n)x^{n-1} + \cdots + (b_2 - zb_3)x^2 + (b_1 - zb_2)x + (b_0 - zb_1). \tag{35}$$

The numbers b_k are determined by comparing the coefficients of x_k in equations (30) and (35), as shown in Table 1.1.

The value $P(z) = b_0$ is easily obtained by substituting $x = z$ in equation (32) and using the fact that $R_0 = b_0$:

$$P(z) = (z - z)Q_0(z) + R_0 = 0Q_0(z) + b_0 = b_0. \tag{36}$$

TABLE 1.1 Coefficients b_k for Horner's Method

x^k	Comparing (30) and (34)	Solving for b_k
x^n x^{n-1}	$a_n = b_n$ $a_{n-1} = b_{n-1} - zb_n$	$b_n = a_n$ $b_{n-1} = a_{n-1} + zb_n$
. . .	. . .	. . .
x^k	$a_k = b_k - zb_{k+1}$	$b_k = a_k + zb_{k+1}$
. . .	. . .	. . .
x^0	$a_0 = b_0 - zb_1$	$b_0 = a_0 + zb_1$

The recursive formulas for b_k given in (31) are easy to implement with a computer. A very simple algorithm is

```
       B(N) := A(N)
DO  FOR K=N−1 DOWNTO 0
   └──B(K) := A(K)+X*B(K+1)
```

The variable z was introduced in the equations above so that the term $(x - z)$ could be used. If the formulas in (36) are used with z replaced by x, the result will be $P(x) = b_0$. When hand calculations are done, it is easier to write the coefficients of $P(x)$ on a line and perform the calculation $b_k = a_k + xb_{k+1}$ below a_k in a column (see Table 1.2).

For example, evaluate $P(3)$ for the polynomial

$$P(x) = x^5 - 6x^4 + 8x^3 + 8x^2 \, 4x - 40.$$

TABLE 1.2 Horner's Table for the Synthetic Division Process

Input	a_n	a_{n-1}	a_{n-2}	. . .	a_k	. . .	a_2	a_1	a_0
x		xb_n	xb_{n-1}	. . .	xb_{k+1}	. . .	xb_3	xb_2	xb_1
	b_n	b_{n-1}	b_{n-2}	. . .	b_k	. . .	b_2	b_1	$b_0 = P(x)$ Output

Solution. The coefficients $a_5 = 1$, $a_4 = -6$, $a_3 = 8$, $a_2 = 8$, $a_1 = 4$, and $a_0 = -40$ and the value $x = 3$ are used in Table 1.2 to compute the coefficients b_k and the value $P(3) = b_0$.

	a_5	a_4	a_3	a_2	a_1	a_0
Input $x = 3$	1	-6 3	8 -9	8 -3	4 15	-40 57
	1 b_5	-3 b_4	-1 b_3	5 b_2	19 b_1	$17 = P(3) = b_0$ Output

Therefore, $P(3) = 17$.

EXERCISES FOR REVIEW OF CALCULUS

1. **(a)** Find

$$L = \lim_{n \to \infty} \frac{4n + 1}{2n + 1}.$$

Then determine $\{\epsilon_n\} = \{L - x_n\}$ and find $\lim_{n \to \infty} \epsilon_n$.

   (b) Find

$$L = \lim_{n \to \infty} \frac{2n^2 + 6n - 1}{4n^2 + 2n + 1}.$$

Then determine $\{\epsilon_n\} = \{L - x_n\}$ and find $\lim_{n \to \infty} \epsilon_n$.

2. Let $\{x_n\}_{n=1}^{\infty}$ be a sequence such that $\lim_{n \to \infty} x_n = 2$.
   (a) Find $\lim_{n \to \infty} \sin(x_n)$.   **(b)** Find $\lim_{n \to \infty} \ln(x_n^2)$.

3. Apply the intermediate value theorem to $f(x) = 3 + 2x - x^2$ using $L = 2$
   (a) over the interval $[-1, 0]$.   **(b)** over the interval $[0, 3]$.

4. Apply the extreme value theorem to $f(x) = x^3 - 3x + 1$ over $[-1, 2]$.

5. Apply Rolle's theorem to $f(x) = x^3 - x$ over $[0, 1]$.

6. Apply the mean value theorem to $f(x) = x^3 - 3x$ over $[1, 3]$.

7. Apply the generalized Rolle's theorem to $f(x) = x(x - 1)(x - 3)$ over $[0, 3]$.

8. Apply the first fundamental theorem of calculus to $f(x) = xe^x$ over $[0, 2]$.

9. Apply the second fundamental theorem of calculus to $\dfrac{d}{dx} \displaystyle\int_0^x t^2 \cos(t)\, dt$.

10. Apply the mean value theorem for integrals to $f(x) = \sin(x)$ over $[0, \pi/2]$.

11. Consider the function $f(x) = \cos(x)$ over the interval $[0, 5]$. Let $n = 5$ and choose t_k to be the midpoint of the each subinterval [i.e., $t_k = (2k - 1)/2$ for $k = 1, 2, 3, 4, 5$]. Find the Riemann sum for this partition.

12. Find the sum of the infinite geometric series $\displaystyle\sum_{n=1}^{\infty} \left(\frac{2}{3}\right)^n$.

13. Find the Taylor polynomial of degree $n = 8$ for $f(x) = \cos(x)$ expanded about $x_0 = 0$.

14. Given that $f(x) = \sin(x)$ and $P(x) = x - x^3/3! + x^5/5! - x^7/7! + x^9/9!$. Show that $P^{(k)}(0) = f^{(k)}(0)$ for $k = 1, 2, \ldots, 9$.

15. Use Cramer's rule to solve the following linear systems.

   (a) $\begin{pmatrix} 3 & 1 \\ 7 & 4 \end{pmatrix} \begin{pmatrix} x_1 \\ x_2 \end{pmatrix} = \begin{pmatrix} 3 \\ -2 \end{pmatrix}$ **(b)** $\begin{pmatrix} 3 & 4 \\ 1 & 3 \end{pmatrix} \begin{pmatrix} x_1 \\ x_2 \end{pmatrix} = \begin{pmatrix} 3 \\ -2 \end{pmatrix}$

   (c) $\begin{pmatrix} 3 & 4 \\ 1 & 3 \end{pmatrix} \begin{pmatrix} x_1 \\ x_2 \end{pmatrix} = \begin{pmatrix} 7 \\ 5 \end{pmatrix}$ **(d)** $\begin{pmatrix} 6 & 4 \\ 7 & 3 \end{pmatrix} \begin{pmatrix} x_1 \\ x_2 \end{pmatrix} = \begin{pmatrix} 3 \\ -2 \end{pmatrix}$

16. Use Horner's method to show that z_0 is a root of $P(x)$, and find the polynomial $Q_0(x)$ so that $P(x) = (x - z_0)Q_0(x)$.
   (a) $P(x) = x^4 + x^3 - 13x^2 - x + 12$, $z_0 = 3$.
   (b) $P(x) = x^4 + x^3 - 21x^2 - x + 20$, $z_0 = 4$.

17. Write a report on calculus and computers. See references [13, 18, 36, 55, 110, 111, 120, 122, 134, 162, 176, and 179].

1.2 BINARY NUMBERS

Introduction

We human beings do arithmetic using the decimal number system. Most computers do arithmetic using the binary number system. It may seem otherwise, because we communicate with the computer (input/output) in base 10 numbers. This transparency does not mean that the computer uses base 10. In fact, it converts our inputs to base 2 (or perhaps base 16), then performs base 2 arithmetic, and finally, translates the answer into base 10 before it prints it out to us. Some experimentation is required to verify this. One computer with nine decimal digits of accuracy gave the answer

$$\sum_{k=1}^{100,000} 0.1 = 9999.99447. \tag{1}$$

Here the intent was to add the number $\frac{1}{10}$ repeatedly 100,000 times. As everyone knows, the mathematical answer is exactly 10,000. One goal is to understand the reason for the computer's apparently flawed calculation. We shall see that something is lost when the computer translates the decimal fraction $\frac{1}{10}$ into a binary fraction.

Binary Numbers

For ordinary purposes we use base 10 numbers. For illustration, the number 1563 is expressible as

$$1563 = 1 \times 10^3 + 5 \times 10^2 + 6 \times 10^1 + 3 \times 10^0.$$

In general, let N denote a positive integer; then the digits $a_0, a_1, a_2, \ldots, a_K$ exist so that N has the base 10 expansion

$$N = a_K \times 10^K + a_{K-1} \times 10^{K-1} + \cdots + a_2 \times 10^2 + a_1 \times 10^1 + a_0 \times 10^0,$$

where the digits a_k are chosen from $\{0, 1, \ldots, 8, 9\}$. Thus N is expressed in decimal notation as

$$N = a_K a_{K-1} \cdots a_2 a_1 a_{0 \text{ ten}} \qquad \text{(decimal)} \tag{2}$$

If it is understood that 10 is the base, we write (2) as

$$N = a_K a_{K-1} \cdots a_2 a_1 a_0.$$

For example, we understand that $1563 = 1563_{\text{ten}}$.

Using powers of 2, the number 1563 can be written

$$1563 = 1 \times 2^{10} + 1 \times 2^9 + 0 \times 2^8 + 0 \times 2^7 + 0 \times 2^6 + 0 \times 2^5$$
$$+ 1 \times 2^4 + 1 \times 2^3 + 0 \times 2^2 + 1 \times 2^1 + 1 \times 2^0. \tag{3}$$

This can be verified by performing the calculation

$$1563 = 1024 + 512 + 16 + 8 + 2 + 1.$$

In general, let N denote a positive integer; the digits $b_0, b_1, b_2, \ldots, b_J$ exist so that N has the base 2 expansion

$$N = b_J \times 2^J + b_{J-1} \times 2^{J-1} + \cdots + b_2 \times 2^2 + b_1 \times 2^1 + b_0 \times 2^0, \tag{4}$$

where each digit b_j is either 0 or 1. Thus N is expressed in binary notation as

$$N = b_J b_{J-1} \cdots b_2 b_1 b_{0 \text{ two}} \qquad \text{(binary)}. \tag{5}$$

Using the notation (5) and the result in (3) yields

$$1563 = 11000011011_{\text{two}}.$$

Remarks.   We will always use the word "two" as a subscript at the end of a binary number. This will enable us to distinguish binary numbers from the ordinary base 10 usage. Thus 111 means one hundred eleven, whereas 111_{two} stands for seven.

It is usually the case that the binary representation for N will require more digits than the decimal representation. This is due to the fact that powers of 2 grow much more slowly than do powers of 10.

An efficient algorithm for finding the base 2 representation of the integer N can be derived from equation (4). Dividing both sides of (4) by 2, we obtain

$$\frac{N}{2} = b_J \times 2^{J-1} + b_{J-1} \times 2^{J-2} + \cdots + b_2 \times 2^1 + b_1 \times 2^0 + \frac{b_0}{2}.$$

Hence the remainder, upon dividing N by 2, is the digit b_0. Next we find b_1. If we write $N/2 = Q_0 + b_0/2$, then

$$Q_0 = b_J \times 2^{J-1} + b_{J-1} \times 2^{J-2} + \cdots + b_2 \times 2^1 + b_1 \times 2^0. \tag{6}$$

Now divide both sides of (6) by 2 to get

$$\frac{Q_0}{2} = b_J \times 2^{J-2} + b_{J-1} \times 2^{J-3} + \cdots + b_2 \times 2^0 + \frac{b_1}{2}.$$

Hence the remainder, upon dividing Q_0 by 2, is the digit b_1. This process is continued and generates sequences $\{Q_k\}$ and $\{b_k\}$ of quotients and remainders, respectively. The process is terminated when we find the first integer J such that $Q_J = 0$. The sequences obey the following formulas:

$$N = 2Q_0 + b_0$$
$$Q_0 = 2Q_1 + b_1$$
$$Q_1 = 2Q_2 + b_2$$
$$\cdot$$
$$\cdot \qquad\qquad\qquad (7)$$
$$\cdot$$
$$Q_{J-2} = 2Q_{J-1} + b_{J-1}$$
$$Q_{J-1} = 2Q_J + b_J \qquad (Q_J = 0).$$

Example 1.1

Show how to obtain $1563 = 11000011011_{two}$.

Solution.   Start with $N = 1563$ and construct the quotients and remainders according to the equations in (7):

$$1563 = 2 \times 781 + 1, \qquad b_0 = 1$$
$$781 = 2 \times 390 + 1, \qquad b_1 = 1$$
$$390 = 2 \times 195 + 0, \qquad b_2 = 0$$
$$195 = 2 \times 97 + 1, \qquad b_3 = 1$$
$$97 = 2 \times 48 + 1, \qquad b_4 = 1$$
$$48 = 2 \times 24 + 0, \qquad b_5 = 0$$
$$24 = 2 \times 12 + 0, \qquad b_6 = 0$$
$$12 = 2 \times 6 + 0, \qquad b_7 = 0$$
$$6 = 2 \times 3 + 0, \qquad b_8 = 0$$
$$3 = 2 \times 1 + 1, \qquad b_9 = 1$$
$$1 = 2 \times 0 + 1, \qquad b_{10} = 1.$$

Thus the binary representation for 1563 is

$$1563 = b_{10}b_9b_8 \ldots b_2b_1b_{0\ two} = 11000011011_{two}.$$

Sequences and Series

When fractions are expressed in decimal form, it is often the case that infinitely many digits are required. A familiar example is

$$\tfrac{1}{3} = 0.33333\overline{333} \ \ldots \ . \qquad\qquad (8)$$

Here the symbol $\overline{3}$ means that the digit 3 is repeated forever to form a decimal. It is understood that 10 is the base in (8). Moreover, it is the mathematical intent that (8) is the shorthand notation for the infinite series

$$S = 3 \times 10^{-1} + 3 \times 10^{-2} + 3 \times 10^{-3} + \cdots + 3 \times 10^{-n} + \cdots = \tfrac{1}{3}. \quad (9)$$

If only a finite number of digits are displayed, only an approximation is to $\frac{1}{3}$ is obtained. Suppose that we write

$$\frac{1}{3} \approx 0.333333 = \frac{333{,}333}{1{,}000{,}000}.$$

Then the error in this approximation is 1/3,000,000 and the reader can verify that

$$\frac{1}{3} = 0.333333 + \frac{1}{3{,}000{,}000}.$$

We should try to understand the nature of the expansion in (9). A naive approach is to multiply the right side by 10 and then subtract.

$$10S = 3 + 3 \times 10^{-1} + 3 \times 10^{-2} + 3 \times 10^{-3} + \cdots + 3 \times 10^{-n} + \cdots$$
$$-\quad S = \qquad\quad -3 \times 10^{-1} - 3 \times 10^{-2} - 3 \times 10^{-3} - \cdots - 3 \times 10^{-n} - \cdots$$
$$\overline{}$$
$$9S = 3 + 0 \times 10^{-1} + 0 \times 10^{-2} + 0 \times 10^{-3} + \cdots + 0 \times 10^{-n} + \cdots$$

Therefore, $S = \frac{3}{9} = \frac{1}{3}$. The theorems necessary to justify the calculation above can be found in most calculus books. We shall review a few of the concepts and the reader may want to refer to a standard reference on calculus to fill in all the details.

Definition 1.7   (Geometric Series).   The infinite series

$$\sum_{n=0}^{\infty} cr^n = c + cr + cr^2 + \cdots + cr^n + \cdots, \quad (10)$$

where $c \neq 0$ and $r \neq 0$, is called a **geometric series** with ratio r.

Theorem 1.15   (Geometric Series).   The geometric series has the following properties:

$$\text{If } |r| < 1, \text{ then } \sum_{n=0}^{\infty} cr^n = \frac{c}{1-r}. \quad (11)$$

$$\text{If } |r| \geq 1, \text{ then the series diverges.} \quad (12)$$

Proof. The summation formula for a finite geometric series is

$$S_n = c + cr + \cdots + cr^n = c \frac{1 - r^{n+1}}{1 - r} \qquad \text{for } r \neq 1. \quad (13)$$

To establish (11), we observe that

$$|r| < 1 \quad \text{implies that} \quad \lim_{n \to \infty} r^{n+1} = 0. \tag{14}$$

Taking the limit as $n \to \infty$, we use (13) and (14) to get

$$\lim_{n \to \infty} S_n = \frac{c}{1 - r}\left(1 - \lim_{n \to \infty} r^{n+1}\right) = \frac{c}{1 - r}(1 - 0).$$

By equation (22) of Section 1.1, the limit above establishes (11).

When $|r| \geqq 1$, the sequence $\{r^{n+1}\}$ does not converge. Hence the sequence $\{S_n\}$ in (13) does not tend to a limit. Therefore, (12) is established.

Example 1.2

Show that

$$1 + \tfrac{1}{4} + \tfrac{1}{16} + \tfrac{1}{64} + \tfrac{1}{256} + \cdots = \tfrac{4}{3}. \tag{15}$$

Solution.   We must first observe that $c = 1$ and $r = \tfrac{1}{4}$. Then using (11) we conclude that

$$1 + \frac{1}{4} + \left(\frac{1}{4}\right)^2 + \left(\frac{1}{4}\right)^3 + \cdots + \left(\frac{1}{4}\right)^n + \cdots = \frac{1}{1 - \tfrac{1}{4}} = \frac{4}{3}.$$

Binary Fractions

Binary fractions can be expressed as sums involving negative powers of 2. If R is a real number that lies in the range $0 < R < 1$, there exist digits $d_1, d_2, \ldots, d_n, \ldots$ so that

$$R = d_1 \times 2^{-1} + d_2 \times 2^{-2} + d_3 \times 2^{-3} + \cdots + d_n \times 2^{-n} + \cdots. \tag{16}$$

We usually express the quantity on the right side of (16) in the binary fraction notation

$$R = 0.d_1 d_2 d_3 \ldots d_n \ldots {}_{\text{two}}. \tag{17}$$

Since 2 is the base, the digits d_j are chosen from the set $\{0, 1\}$.

There are many real numbers whose binary representation requires infinitely many digits. The fraction $\tfrac{7}{10}$ can be expressed as 0.7_{ten}, yet its base 2 representation requires infinitely many digits:

$$\tfrac{7}{10} = 0.101100110011001100\overline{110011}\ldots {}_{\text{two}}. \tag{18}$$

The binary fraction in (18) is a repeating fraction where the group of four digits 0011 is repeated forever. To see what is happening, we could try to measure 0.7 inch on a ruler that is graduated in $\tfrac{1}{2}, \tfrac{1}{4}, \tfrac{1}{8},$ and $\tfrac{1}{16}$ inch. The result is shown in Figure 1.10. Thus 0.7 is located between $\tfrac{11}{16}$ and $\tfrac{3}{4}$ and satisfies the relation $\tfrac{11}{16} < \tfrac{7}{10} < \tfrac{3}{4}$. Using (16) and (17), it is easy to verify the expansions

$$\tfrac{11}{16} = \tfrac{1}{2} + \tfrac{0}{4} + \tfrac{1}{8} + \tfrac{1}{16} = 0.1011_{\text{two}} \quad \text{and} \quad \tfrac{3}{4} = \tfrac{1}{2} + \tfrac{1}{4} + \tfrac{0}{8} + \tfrac{0}{16} = 0.1100_{\text{two}}.$$

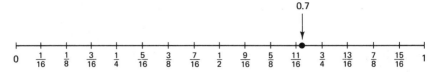

Figure 1.10   Location of the decimal number 0.7 on a binary scale.

Hence we have the binary relation

$$0.1011_{\text{two}} < \tfrac{7}{10} < 0.1100_{\text{two}}.$$

A ruler graduated in $\tfrac{1}{256}$ of an inch would reveal that

$$\frac{179}{256} < \frac{7}{10} < \frac{45}{64}.$$

The reader can verify that this leads to the binary relation

$$0.10110011_{\text{two}} < \tfrac{7}{10} < 0.10110100_{\text{two}}.$$

When the limit process is applied we get

$$\tfrac{7}{10} = 0.101100110011\ldots\ldots_{\text{two}}.$$

We now develop an algorithm for finding base 2 representations. If both sides of (16) are multipled by 2, the result is

$$2R = d_1 + (d_2 \times 2^{-1} + d_3 \times 2^{-2} + \cdots + d_n \times 2^{-n+1} + \cdots). \qquad (19)$$

The quantity in parentheses on the right side of (19) is a positive number and is less than 1. Therefore, we can take the integer part of both sides of (19) and obtain

$$d_1 = \text{int}(2R),$$

where $\text{int}(2R)$ is the integer part of the real number $2R$. To continue the process, take the fractional part of (19) and write

$$F_1 = \text{frac}(2R) = d_2 \times 2^{-1} + d_3 \times 2^{-2} + \cdots + d_n \times 2^{-n+1} + \cdots, \qquad (20)$$

where $\text{frac}(2R)$ is the fractional part of the real number $2R$. Multiplication of both sides of (20) by 2 results in

$$2F_1 = d_2 + (d_3 \times 2^{-1} + \cdots + d_n \times 2^{-n+2} + \cdots). \qquad (21)$$

Now take the integer part of (21) and obtain

$$d_2 = \text{int}(2F_1).$$

The process is continued, possibly ad infinitum, and two sequences $\{d_k\}$ and $\{F_k\}$ are recursively generated:

$$d_1 = \text{int}(2R), \qquad F_1 = \text{frac}(2R)$$

$$d_2 = \text{int}(2F_1), \qquad F_2 = \text{frac}(2F_1)$$

$$d_3 = \text{int}(2F_2), \qquad F_3 = \text{frac}(2F_2)$$

$$\begin{aligned} & \cdot \qquad\qquad\qquad\qquad \cdot \\ & \cdot \qquad\qquad\qquad\qquad \cdot \\ & \cdot \qquad\qquad\qquad\qquad \cdot \end{aligned}$$

$$d_n = \text{int}(2F_{n-1}), \qquad F_n = \text{frac}(2F_{n-1}).$$

$$\begin{aligned} & \cdot \qquad\qquad\qquad\qquad \cdot \\ & \cdot \qquad\qquad\qquad\qquad \cdot \\ & \cdot \qquad\qquad\qquad\qquad \cdot \end{aligned}$$

(22)

The binary representation of R is given by the convergent series

$$R = \sum_{j=1}^{\infty} d_j \times 2^{-j}.$$

Example 1.3

Show how to obtain the representation

$$\frac{7}{10} = 0.1011001100110\overline{0011}\ldots_{\text{two}}.$$

Solution.   Start with $R = 0.7$ and use the formulas in (22) to get

$$\begin{aligned}
2R &= 1.4, & d_1 &= 1 = \text{int}(1.4), & F_1 &= 0.4 = \text{frac}(1.4) \\
2F_1 &= 0.8, & d_2 &= 0 = \text{int}(0.8), & F_2 &= 0.8 = \text{frac}(0.8) \\
2F_2 &= 1.6, & d_3 &= 1 = \text{int}(1.6), & F_3 &= 0.6 = \text{frac}(1.6) \\
2F_3 &= 1.2, & d_4 &= 1 = \text{int}(1.2), & F_4 &= 0.2 = \text{frac}(1.2) \\
2F_4 &= 0.4, & d_5 &= 0 = \text{int}(0.4), & F_5 &= 0.4 = \text{frac}(0.4) \\
2F_5 &= 0.8, & d_6 &= 0 = \text{int}(0.8), & F_6 &= 0.8 = \text{frac}(0.8) \\
2F_6 &= 1.6, & d_7 &= 1 = \text{int}(1.6), & F_7 &= 0.6 = \text{frac}(1.6) \\
2F_7 &= 1.2, & d_8 &= 1 = \text{int}(1.2), & F_8 &= 0.2 = \text{frac}(1.2) \\
2F_8 &= 0.4, & d_9 &= 0 = \text{int}(0.4), & F_9 &= 0.4 = \text{frac}(0.4)
\end{aligned}$$

$$\begin{aligned} \cdot \qquad\qquad \cdot \qquad\qquad \cdot \\ \cdot \qquad\qquad \cdot \qquad\qquad \cdot \\ \cdot \qquad\qquad \cdot \qquad\qquad \cdot \end{aligned}$$

A repeating pattern is emerging: $d_k = d_{k+4}$ and $F_k = F_{k+4}$ for $k = 2, 3, \ldots$. Thus the binary representation for $\frac{7}{10}$ is

$$\frac{7}{10} = 0.d_1 d_2 d_3 d_4 d_5 d_6 d_7 d_8 \ldots_{\text{two}} = 0.101\overline{1\,0011}\ldots_{\text{two}}.$$

Geometric series can be used to determine the sum of a repeating binary fraction.

Example 1.4

Use geometric series to show that

$$\tfrac{1}{3} = 0.0101010101010\overline{1}\ldots \cdot_{\text{two}}.$$

Solution. If the result of Example 1.2 is used and 1 is subtracted from both sides of (15), we have

$$\tfrac{1}{3} = \tfrac{1}{4} + \left(\tfrac{1}{4}\right)^2 + \left(\tfrac{1}{4}\right)^3 + \cdots + \left(\tfrac{1}{4}\right)^j + \cdots .\tag{23}$$

Equation (23) can be rewritten in the form of (16):

$$\tfrac{1}{3} = 0 \times 2^{-1} + 1 \times 2^{-2} + 0 \times 2^{-3} + 1 \times 2^{-4} + \cdots + 0 \times 2^{-2j+1}$$
$$+ 1 \times 2^{-2j} + \cdots$$

or

$$\tfrac{1}{3} = 0.0101010101010\overline{1}\ldots \cdot_{\text{two}}.$$

Binary Shifting

If we are to find the rational number that is equivalent to an infinite repeating binary expansion, then a shift in the digits can be helpful. For example, let S be given by

$$S = 0.00011001100110011001100110\overline{0011}\ldots \cdot_{\text{two}}.\tag{24}$$

Multiplication of (24) by $32 = 2^5$ will shift the binary point five places to the right and $32S$ has the form

$$32S = 11.001100110011001100110\overline{0011}\ldots \cdot_{\text{two}}.\tag{25}$$

Similarly, when S is multiplied by 2, the result is

$$2S = 0.0011001100110011001100110\overline{0011}\ldots \cdot_{\text{two}}.\tag{26}$$

Since fractional parts in (25) and (26) agree, they will cancel out when we form the difference $32S - 2S$, that is,

$$32S - 2S = 11.00000000000000000000000000000\ldots \cdot_{\text{two}}.\tag{27}$$

Using $11_{\text{two}} = 3$ in (27), we obtain $30S = 3$. Therefore, $S = \tfrac{1}{10}$.

Scientific Notation

A standard way to present a real number, called **scientific notation**, is obtained by shifting the decimal point and supplying an appropriate power of 10. For example,

$$0.0000747 = 7.47 \times 10^{-5},$$

$$31.4159265 = 3.14159265 \times 10,$$

$$9{,}700{,}000{,}000 = 9.7 \times 10^9.$$

In chemistry, an important constant is Avogadro's number, which is 6.02252×10^{23}. It is the number of atoms in the gram atomic weight of an element. In computer science, $1K = 1.024 \times 10^3$.

Machine Numbers

Computers use a normalized floating-point binary representation for real numbers. This means that the mathematical quantity x is not actually stored in the computer. Instead, the computer stores a binary approximation to x:

$$x \approx \pm q \times 2^n. \tag{28}$$

The number q is the **mantissa** and it is a finite binary expression satisfying $\frac{1}{2} \le q < 1$. The integer n is called the **exponent**.

In a computer, only a small subset of the real number system is used. Typically, this subset contains only a portion of the binary numbers suggested by (38). The number of binary digits is restricted in both the numbers q and n. For example, consider the subset of positive numbers

$$0.d_1 d_2 d_3 d_4 \text{ two} \times 2^n, \tag{29}$$

where $d_1 = 1$ and d_2, d_3, d_4 are either 0 or 1 and n is chosen from the set $\{-3, -2, -1, 0, 1, 2, 3, 4\}$. There are eight choices for the mantissa and eight choices for the exponent in (29), and this produces a set of 64 numbers:

$$\{0.1000_{\text{two}} \times 2^{-3}, 0.1001_{\text{two}} \times 2^{-3}, \ldots, 0.1110_{\text{two}} \times 2^4, 0.1111_{\text{two}} \times 2^4\}. \tag{30}$$

The decimal form for these 64 numbers is given in Table 1.3. It is important to learn that when the mantissa and exponent in (28) are restricted, the computer has a limited number of values it chooses from to store as an approximation to the real number x.

Let us see what would happen if a computer had only a 4-bit mantissa and was required to perform the computation $\left(\frac{1}{10} + \frac{1}{5}\right) + \frac{1}{6}$. We assume that the computer rounds

TABLE 1.3 Decimal Equivalents for Set of Binary Numbers with a 4-Binary-Bit Mantissa and Exponent of $n = -3, -2, \ldots, 3, 4$

Mantissa	Exponent:							
	$n = -3$	$n = -2$	$n = -1$	$n = 0$	$n = 1$	$n = 2$	$n = 3$	$n = 4$
0.1000_{two}	0.0625	0.125	0.25	0.5	1	2	4	8
0.1001_{two}	0.0703125	0.140625	0.28125	0.5625	1.125	2.25	4.5	9
0.1010_{two}	0.078125	0.15625	0.3125	0.625	1.25	2.5	5	10
0.1011_{two}	0.0859375	0.171875	0.34375	0.6875	1.375	2.75	5.5	11
0.1100_{two}	0.09375	0.1875	0.375	0.75	1.5	3	6	12
0.1101_{two}	0.1015625	0.203125	0.40625	0.8125	1.625	3.25	6.5	13
0.1110_{two}	0.109375	0.21875	0.4375	0.875	1.75	3.5	7	14
0.1111_{two}	0.1171875	0.234375	0.46875	0.9375	1.875	3.75	7.5	15

all real numbers to the closest binary number in Table 1.3. At each step the reader can look in the table to see the best approximation is being used.

$$\begin{aligned} \tfrac{1}{10} &\approx 0.1101_{two} \times 2^{-3} = 0.01101_{two} \times 2^{-2} \\ \tfrac{1}{5} &\approx 0.1101_{two} \times 2^{-2} = \underline{0.1101_{two} \;\;\; \times 2^{-2}} \\ \tfrac{3}{10} & \qquad\qquad\qquad\qquad\qquad 1.00111_{two} \times 2^{-2}. \end{aligned} \tag{31}$$

The computer must now decide how to store the number $1.00111_{two} \times 2^{-2}$. We assume that it is rounded to $0.1010_{two} \times 2^{-1}$. The next step is

$$\begin{aligned} \tfrac{3}{10} &\approx 0.1010_{two} \times 2^{-1} = 0.1010_{two} \;\;\; \times 2^{-1} \\ \tfrac{1}{6} &\approx 0.1011_{two} \times 2^{-2} = \underline{0.01011_{two} \;\;\; \times 2^{-1}} \\ \tfrac{7}{15} & \qquad\qquad\qquad\qquad\qquad 0.11111_{two} \times 2^{-1}. \end{aligned} \tag{32}$$

The computer must decide how to store the number $0.11111_{two} \times 2^{-1}$. Since rounding is assumed to take place, it stores $0.1000_{two} \times 2^{0}$. Therefore, the computer's solution to the addition problem is

$$\tfrac{7}{15} \approx 0.1000_{two} \times 2^{0}. \tag{33}$$

The error in the computer's calculation is

$$\tfrac{7}{15} - 0.1000_{two} \approx 0.46666667 - 0.50000000 \approx -0.03333333. \tag{34}$$

Expressed as a percentage of $\tfrac{7}{15}$, this amounts to 7.14%.

Computer Accuracy

To store numbers accurately, computers must have floating-point binary numbers with at least 24 binary bits used for the mantissa; this translates to about seven decimal places. If a 32-binary-bit mantissa is used, numbers with nine decimal places can be stored. Let us return to the difficulty encountered in (1) when a computer added $\tfrac{1}{10}$ repeatedly.

Suppose that the mantissa q in (28) contains 32 binary bits. The condition $\tfrac{1}{2} \le q$ implies that the first digit is $d_1 = 1$. Hence q has the form

$$q = 0.1d_2d_3d_4\ldots d_{31}d_{32two}. \tag{35}$$

When fractions are represented in binary form, it is often the case that infinitely many digits are required. An example is

$$\tfrac{1}{10} = 0.0001100110011001100110011\overline{0011}\ldots{}_{two}. \tag{36}$$

When the 32-bit mantissa is used, truncation occurs and the computer uses the internal approximation

$$\tfrac{1}{10} \approx 2^{-3}(\underbrace{0.11001100110011001100110011001100_{two}}). \tag{37}$$

$$\text{32 significant digits}$$

The error in the approximation in (37) is

$$2^{-35}(0.1100\overline{110011}\ldots._{two}) \approx 2.328306437 \times 10^{-11}. \tag{38}$$

Because of (38), the computer must be in error when it sums the 100,000 addends of $\frac{1}{10}$ given in (1). The error must be greater than $100{,}000 \times 2.328306437 \times 10^{-11} = 2.328306437 \times 10^{-6}$. Indeed, there is a much larger error. Occasionally, the partial sum could be rounded up or down. Also, as the sum grows, the latter addends of $\frac{1}{10}$ are small compared to the current size of the sum, and their contribution is truncated more severely. The compounding effect of these errors actually produced the error $10{,}000 - 9999.99447 = 5.53 \times 10^{-3}$.

Computer Floating-Point Numbers

Computers have both an integer mode and a floating-point mode for representing numbers. The **integer mode** is used for performing calculations that are known to be integer value and have limited usage for numerical analysis. **Floating-point numbers** are used for scientific and engineering applications. It must be understood that any computer implementation of equation (28) places restrictions on the number of digits used in the mantissa q, and that the range of possible exponents n must be limited.

Computers that use 32 bits to represent single-precision real numbers use 8 bits for the exponent and 24 bits for the mantissa. They can represent real numbers whose magnitude is in the range

$$2.938736\text{E}-39 \quad \text{to} \quad 1.701412\text{E}+38$$

(i.e., 2^{-128} to 2^{127}) with six decimal digits of numerical precision (e.g., $2^{-23} = 1.2 \times 10^{-7}$).

Computers that use 48 bits to represent single-precision real numbers might use 8 bits for the exponent and 40 bits for the mantissa. They can represent real numbers in the range

$$2.9387358771\text{E}-39 \quad \text{to} \quad 1.7014118346\text{E}+38$$

(i.e., 2^{-128} to 2^{127}) with 11 decimal digits of precision (e.g., $2^{-39} = 1.8 \times 10^{-12}$).

If the computer has 64-bit double-precision real numbers, it might use 11 bits for the exponent and 53 bits for the mantissa. They can represent real numbers in the range

$$5.562684646268003 \times 10^{-309} \quad \text{to} \quad 8.988465674311580 \times 10^{307}$$

(i.e., 2^{-1024} to 2^{1023}) with about 16 decimal digits of precision (e.g., $2^{-52} = 2.2 \times 10^{-16}$).

EXERCISES FOR BINARY NUMBERS

1. Use a computer to accumulate the following sums.

(a) $10{,}000 - \displaystyle\sum_{k=1}^{100{,}000} 0.1$

(b) $10{,}000 - \displaystyle\sum_{k=1}^{80{,}000} 0.125$

Here the intent is to have the computer do repeated subtractions. Do **not** use the multiplication shortcut.

2. Use equations (4) and (5) and convert the following binary numbers to decimal form (base 10).
  (a) 10101_{two} (b) 111000_{two}
  (c) 1101101_{two} (d) 10110111_{two}

3. Use equations (16) and (17) and convert the following binary fractions to decimal form (base 10).
  (a) 0.11011_{two} (b) 0.10101_{two}
  (c) 0.1010101_{two} (d) 0.11011011_{two}

4. Convert the following binary numbers to decimal form (base 10).
  (a) 1.0110101_{two} (b) 11.0010010001_{two}

5. The fractions in Exercise 4 are approximately $\sqrt{2}$ and π. Find the error in these approximations, that is, find
  (a) $\sqrt{2} - 1.0110101_{two}$ (b) $\pi - 11.0010010001_{two}$
      Use $\sqrt{2} = 1.41421356237309. \ldots$ Use $\pi = 3.14159265358979. \ldots$

6. Follow Example 1.1 and convert the following to binary numbers.
  (a) 23   (b) 75   (c) 360   (d) 1766

7. Follow Example 1.3 and convert the following to a binary fraction of the form $0.d_1 d_2 \ldots d_n$ two.
  (a) $\frac{7}{16}$ (b) $\frac{13}{16}$ (c) $\frac{23}{32}$ (d) $\frac{75}{128}$

8. Follow Example 1.3 and obtain the expansions.
  (a) $\frac{1}{10} = 0.0001100110011. \ . \ ._{two}$ (b) $\frac{1}{3} = 0.010101010101. \ . \ ._{two}$
  (c) $\frac{1}{7} = 0.001001001\overline{001}. \ . \ ._{two}$
  Hint. For parts (b) and (c) use rational arithmetic, for example,

$$\text{int}\left(\tfrac{4}{3}\right) = 1 \quad \text{and} \quad \text{frac}\left(\tfrac{4}{3}\right) = \tfrac{1}{3}.$$

9. For the following seven-digit binary approximations, find the error in the approximation $R - 0.d_1 d_2 d_3 d_4 d_5 d_6 d_7$ two.
  (a) $\frac{1}{10} \approx 0.0001100_{two}$ (b) $\frac{1}{3} \approx 0.0101010_{two}$
  (c) $\frac{1}{7} \approx 0.0010010_{two}$
  Hint. First convert the binary expression to decimal.

10. For the expansion $\frac{1}{7} = 0.001001001\overline{001}. \ . \ ._{two}$, show that this is equivalent to $\frac{1}{7} = \frac{1}{8} + \frac{1}{64} + \frac{1}{512} + \cdots$. Then use Theorem 1.15 to establish the expansion.

11. For the expansion $\frac{1}{5} = 0.001100110011. \ . \ ._{two}$, show that this is equivalent to $\frac{1}{5} = \frac{3}{16} + \frac{3}{256} + \frac{3}{4096} + \cdots$. Then use Theorem 1.15 to establish the expansion.

12. Prove that any number 2^{-N}, where N is a positive integer, can be represented as a decimal number that has N digits, that is,

$$2^{-N} = 0.c_1 c_2 c_3 \ldots c_N$$

  Hint. $\frac{1}{2} = 0.5, \frac{1}{4} = 0.25.$

13. Write a report on hexadecimal numbers.

14. Use Table 1.3 and find what happens when a computer with a 4-bit mantissa performs the following calculations.
  (a) $\left(\frac{1}{3} + \frac{1}{5}\right) + \frac{1}{6}$ (b) $\left(\frac{1}{10} + \frac{1}{3}\right) + \frac{1}{5}$

(c) $\left(\frac{3}{17} + \frac{1}{9}\right) + \frac{1}{7}$ (d) $\left(\frac{7}{10} + \frac{1}{9}\right) + \frac{1}{7}$

15. (a) Prove that when 2 is replaced by 3 in all formulas in (7), the result is a method for finding the base 3 expansion of a positive integer.
    Express the following integers in base 3.
    (b) 10   (c) 23   (d) 49   (e) 123

16. (a) Prove that when 2 is replaced by 3 in all formulas in (22), the result is a method for finding the base 3 expansion of a positive number R that lines in the interval $0 < R < 1$.
    Express the following numbers in base 3.
    (b) $\frac{1}{3}$ (c) $\frac{1}{2}$ (d) $\frac{1}{10}$

17. (a) Prove that when 2 is replaced by 5 in all formulas in (7), the result is a method for finding the base 5 expansion of a positive integer.
    Express the following integers in base 5.
    (b) 10   (c) 32   (d) 49   (e) 144

18. (a) Prove that when 2 is replaced by 5 in all formulas in (22), the result is a method for finding the base 5 expansion of a positive number R that lies in the interval $0 < R < 1$.
    Express the following numbers in base 5.
    (b) $\frac{1}{3}$ (c) $\frac{1}{2}$ (d) $\frac{1}{10}$

19. Investigate floating-point real numbers on your computer.
    (a) How many digits in base 10 are used for the mantissa of a real floating-point number?
    (b) What is the range for the base 10 exponent of a real floating-point number?
    (c) Use functions in your computer to find approximations to the real numbers $2^{1/2}$, $2^{1/3}$, e, π, and $\ln(3)$. Formulas for BASIC, FORTRAN, and Pascal are given in the table.

Real number	BASIC formula	FORTRAN formula	Pascal formula
$2^{1/2} \approx 1.4142135623730950488$	SQR(2)	SQRT(2)	SQRT(2)
$2^{1/3} \approx 1.2599210498948731648$	2^(1/3)	2**(1/3)	EXP(LN(2)/3)
$e \approx 2.7182818284590452353$	EXP(1)	EXP(1)	EXP(1)
$\pi \approx 3.1415926535897932384$	4*ATN(1)	4*ATAN(1)	4*ARCTAN(1)
$\ln(3) \approx 1.0986122886681096913$	LOG(3)	ALOG(3)	LN(3)

20. Investigate how your computer stores real numbers.
    (a) How many bytes (and bits) are used for the mantissa of a real floating-point number?
    (b) What is the range for the base 2 exponent of a real floating-point number?
    (c) Write a program that will find the smallest positive integer N for which $1 + 2^{-N-1} = 1$ is a true expression for your computer.

21. In computer terminology, what are *exponent overflow* and *exponent underflow*?

22. When a computer translates a given real number x that is input to a stored binary machine number, what are the meanings of the phrases *computer rounding* and *computer chopping*?

23. In computer terminology, what is a *guard digit*?

Hexadecimal numbers. When the base of the number system is 16, the digits are 0, 1, 2, 3, 4, 5, 6, 7, 8, 9, A, B, C, D, E, F. Here A $= 10_{\text{ten}}$, B $= 11_{\text{ten}}$, C $= 12_{\text{ten}}$, D $= 13_{\text{ten}}$, E $= 14_{\text{ten}}$, and F $= 15_{\text{ten}}$.

24. Convert the following hexadecimal numbers to decimal (base 10) form.

(a) 213_{16} (b) $7C9_{16}$ (c) $1ABE_{16}$ (d) $F09C_{16}$

(e) 0.2_{16} (f) 0.99_{16} (g) $0.A4B_{16}$ (h) $0.F0B_{16}$

25. Adapt the formulas in (7) so that the base 16 expansion of a positive integer N can be found. Express the following integers in base 16.

(a) 512   (b) 2001   (c) 51,264   (d) 91,919

26. Adapt the formulas in (22) so that the base 16 expansion of real number R that lies in the interval $0 < R < 1$ can be found. Express the following real numbers in base 16.

(a) $\frac{1}{3}$ (b) $\frac{1}{2}$ (c) $\frac{1}{10}$ (d) $\frac{1}{15}$

27. Write a report on how to convert hexadecimal numbers to binary numbers, and vice versa.

28. Write a report on hexadecimal numbers. See References [8, 35, 51, 101, and 142].

29. Write a report on floating-point arithmetic. See References [8, 9, 35, 40, 41, 51, 57, 62, 90, 101, 103, 128, 129, 142, 153, 181, 184, and 208].

30. Write a report on scientific computing. See References [5, 71, 98, 103, 150, 151, 152, 158, 159, and 160].

31. Write a report on computer programming in numerical analysis. See References [12, 103, 119, 150, 151, and 152].

1.3 ERROR ANALYSIS

In the practice of numerical analysis it is important to be aware that computed solutions are not exact mathematical solutions. The precision of a numerical solution can be diminished in several subtle ways. Understanding these difficulties can often guide the practitioner in the proper implementation and/or development of numerical algorithms. We start with two important definitions.

Definition 1.8. Suppose that $\bar{p}$ is an approximation to p. The **error** is $E_p = p - \bar{p}$, and the **relative error** is $R_p = (p - \bar{p})/p$, provided that $p \neq 0$.

The error is simply the difference between the true value and the approximate value, whereas the relative is a portion of the true value.

Example 1.5

Find the error and relative error in the following three cases.

Let $x = 3.141592$ and $\bar{x} = 3.14$; then the error is

$$E_x = x - \bar{x} = 3.141592 - 3.140000 = 0.001592 \qquad (1a)$$

and the relative error is

$$R_x = \frac{x - \bar{x}}{x} = \frac{0.001592}{3.141592} = 0.000507.$$

Let $y = 1,000,000$ and $\bar{y} = 999,996$; then the error is

$$E_y = y - \bar{y} = 1,000,000 - 999,996 = 4 \qquad (1b)$$

and the relative error is

$$R_y = \frac{y - \bar{y}}{y} = \frac{4}{1,000,000} = 0.000004.$$

Let $z = 0.000012$ and $\bar{z} = 0.000009$; then the error is

$$E_z = z - \bar{z} = 0.000012 - 0.000009 = 0.000003 \tag{1c}$$

and the relative error is

$$R_z = \frac{z - \bar{z}}{z} = \frac{0.000003}{0.000012} = 0.25.$$

In case (1a), there is not too much difference between E_x and R_x and either could be used to determine the accuracy of $\bar{x}$. In case (1b), the value y is of magnitude 10^6, the error E_y is large, and the relative error R_y is small. We would most likely call $\bar{y}$ a good approximation to y. In case (1c), z is of magnitude 10^{-6} and the error E_z is the smallest of all three cases, but the relative error E_z is the largest. In terms of percentage, it amounts to 25%, and thus $\bar{z}$ is a bad approximation to z. Observe that as $|p|$ moves away from 1, either larger or smaller, the relative error R_p is a better indicator of the accuracy of the approximation than E_p. Relative error is preferred for floating-point representations since it deals directly with the mantissa.

Definition 1.9.   The number $\bar{p}$ is said to **approximate** p to d significant digits if d is the largest positive integer for which

$$\frac{|p - \bar{p}|}{|p|} < \frac{10^{-d}}{2}. \tag{2}$$

Example 1.6

Determine the number of significant digits for the approximations in Example 1.5.
If $x = 3.141592$ and $\bar{x} = 3.14$, then $|x - \bar{x}|/|x| = 0.000507 \approx 10^{-3}/2$. (3a)
Therefore, $\bar{x}$ approximates x to three significant digits (as expected).
If $y = 1,000,000$ and $\bar{y} = 999,996$, then $|y - \bar{y}|/|y| = 0.000004 < 10^{-5}/2$. (3b)
Therefore, $\bar{y}$ approximates y to five significant digits.
If $z = 0.000012$ and $\bar{z} = 0.000009$, then $|z - \bar{z}|/|z| = 0.25 < 10^{-0}/2$. (3c)
Therefore, $\bar{z}$ approximates z to no significant digits.

Truncation Error

The notion of truncation error usually refers to errors introduced when a more complicated mathematical expression is "replaced" with a more elementary formula. This terminology originates from the technique of replacing a complicated function with a truncated Taylor series. For example, the infinite Taylor series

$$e^{x^2} = 1 + x^2 + \frac{x^4}{2!} + \frac{x^6}{3!} + \frac{x^8}{4!} + \cdots + \frac{x^{2n}}{n!} + \cdots$$

might be replaced with just the first five terms $1 + x^2 + x^4/2! + x^6/3! + x^8/4!$. This might be done when approximating an integral numerically.

Example 1.7

Given that $\int_0^{1/2} e^{x^2} dx = 0.544987104184 = p$, determine the accuracy of the approximation obtained by replacing the integrand $f(x) = e^{x^2}$ with the truncated Taylor series $P_8(x) = 1 + x^2 + x^4/2! + x^6/3! + x^8/4!$.

Solution.   Term-by-term integration produces

$$\int_0^{1/2}\left(1 + x^2 + \frac{x^4}{2!} + \frac{x^6}{3!} + \frac{x^8}{4!}\right)dx = \left.\left(x + \frac{x^3}{3} + \frac{x^5}{5(2!)} + \frac{x^7}{7(3!)} + \frac{x^9}{9(4!)}\right)\right|_{x=0}^{x=1/2}$$

$$= \frac{1}{2} + \frac{1}{2^3 3} + \frac{1}{2^5 5(2!)} + \frac{1}{2^7 7(3!)} + \frac{1}{2^9 9(4!)}$$

$$= \frac{1}{2} + \frac{1}{24} + \frac{1}{320} + \frac{1}{5376} + \frac{1}{110,592}$$

$$= \frac{2,109,491}{3,870,720} = 0.544986720817 = \bar{p}.$$

Since $|p - \bar{p}|/|p| = 7.03442 \times 10^{-7} < 10^{-6}/2$, the approximation $\bar{p}$ agrees with the true answer $p = 0.544987104184$ to six significant digits. The graphs of $y = f(x)$ and $y = P_8(x)$ and the area under the curve for $0 \le x \le \frac{1}{2}$ are shown in Figure 1.11.

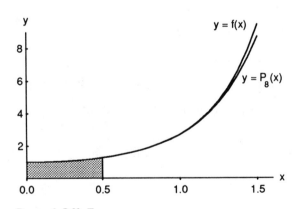

Figure 1.11   The graphs of $y = f(x) = e^{x^2}$, $y = P_8(x)$, and the area under the curve for $0 \le x \le \frac{1}{2}$.

Round-Off Error

A computer's representation of real numbers is limited to the fixed precision of the mantissa. True values are sometimes not stored exactly by a computer's representation. This is called **round-off error**. In the preceding section we saw that the real number $\frac{1}{10} = 0.000110011001100110011001100110011001100\ldots\,._{\text{two}}$ is truncated when it is stored in a computer. The actual number that is stored in the computer may undergo chopping or rounding of the last digit. Therefore, since the computer hardware works with only a limited number of digits in machine numbers, rounding errors are introduced and propagated in successive computations.

Chopping Off versus Rounding Off

Consider any real number p that is expressed in a **normalized decimal form**:

$$p = \pm 0.d_1 d_2 d_3 \ldots .d_k d_{k+1} \ldots \times 10^n,$$

$$\text{where} \quad 1 \le d_1 \le 9 \quad \text{and} \quad 0 \le d_j \le 9 \quad \text{for } j > 1. \tag{4}$$

Suppose that k is the maximum number of decimal digits carried in the floating-point computations of a computer; then the real number p is represented by $\text{fl}_{\text{chop}}(p)$, which is given by

$$\text{fl}_{\text{chop}}(p) = \pm 0.d_1 d_2 d_3 \ldots .d_k \times 10^n,$$

$$\text{where} \quad 1 \le d_1 \le 9 \quad \text{and} \quad 0 \le d_j \le 9 \quad \text{for } 1 < j \le k \tag{5}$$

and is called the **chopped floating point representation** of p. In this case the kth digit of $\text{fl}_{\text{chop}}(p)$ agrees with the kth digit of p. An alternative k-digit representation is the **rounded floating-point representation** $\text{fl}_{\text{round}}(p)$, which is given by

$$\text{fl}_{\text{round}}(p) = \pm 0.d_1 d_2 d_3 \ldots .r_k \times 10^n, \tag{6}$$

where $1 \le d_1 \le 9$ and $0 \le d_j \le 9$ for $1 < j < k$ and the last digit, r_k, is obtained by rounding the number $d_k.d_{k+1} d_{k+2} \ldots$ to the nearest integer. For example, the real number

$$p = \frac{22}{7} = 3.142857142857142857 \ldots$$

has the following six-digit representations:

$$\text{fl}_{\text{chop}}(p) = 0.314285 \times 10^1,$$

$$\text{fl}_{\text{round}}(p) = 0.314286 \times 10^1.$$

For common purposes we would probably write the chopping and rounding as 3.14285 and 3.14286, respectively.

Loss of Significance

Consider the two numbers $p = 3.1415926536$ and $q = 3.1415957341$, which are nearly equal and both carry 11 decimal digits of precision. Suppose that their difference is formed: $p - q = -0.0000030805$. Since the first six digits of p and q are the same, their difference $p - q$ contains only five decimal digits of precision. This phenomenon is called **loss of significance** or **subtractive cancellation**. This reduction in the precision of the final computed answer can creep in when it is not suspected.

Example 1.8

Compare the results of calculating $f(500)$ and $g(500)$ using six digits and rounding. The functions are $f(x) = x\left[\sqrt{x+1} - \sqrt{x}\right]$ and $g(x) = \dfrac{x}{\sqrt{x+1} + \sqrt{x}}$.

Solution. For the first function we compute

$$f(500) = 500\left[\sqrt{501} - \sqrt{500}\right] = 500[22.3830 - 22.3607] = 500[0.0223] = 11.1500.$$

For $g(x)$ we obtain

$$g(500) = \frac{500}{\sqrt{501} - \sqrt{500}} = \frac{500}{22.3830 + 22.3607} = \frac{500}{44.7437} = 11.1748.$$

The second function, $g(x)$, is algebraically equivalent to $f(x)$, as shown by the computation

$$f(x) = \frac{x\left[\sqrt{x+1} - \sqrt{x}\right]\left[\sqrt{x+1} + \sqrt{x}\right]}{\sqrt{x+1} + \sqrt{x}} = \frac{x\left[(\sqrt{x+1})^2 - (\sqrt{x})^2\right]}{\sqrt{x+1} + \sqrt{x}}$$

$$= \frac{x}{\sqrt{x+1} + \sqrt{x}}.$$

The answer, $g(500) = 11.1748$, involves less error and is the same as that obtained by rounding the true answer $11.174755300747198. . .$ to six digits.

The reader is encouraged to study Exercise 12 on how to avoid loss of significance in the quadratic formula. The next example shows that a truncated Taylor series will sometimes help avoid the loss of significance error.

Example 1.9

Compare the results of calculating $f(0.01)$ and $p(0.01)$ using six digits and rounding, where

$$f(x) = \frac{e^x - 1 - x}{x^2} \quad \text{and} \quad p(x) = \frac{1}{2} + \frac{x}{6} + \frac{x^2}{24}.$$

Solution. For the first function we compute

$$f(0.01) = \frac{e^{0.01} - 1 - 0.01}{(0.01)^2} = \frac{1.010050 - 1 - 0.01}{0.0001} = 0.5.$$

For the second function we obtain

$$p(0.01) = \frac{1}{2} + \frac{0.01}{6} + \frac{0.0001}{24} = 0.5 + 0.001667 + 0.000004 = 0.501671.$$

The answer, $p(0.01) = 0.501671$, contains less error and is the same as that obtained by rounding the true answer $0.50167084168057542. . .$ to six digits.

Remark. The function $p(x)$ is the Taylor polynomial of degree $n = 2$ for $f(x)$ expanded about $x = 0$.

For polynomial evaluation, the rearrangement of terms will sometimes produce a better result.

Example 1.10

Let $P(x) = ((x^3 - 3x^2) + 3x) - 1$ and $Q(x) = ((x - 3)x + 3)x - 1$. Use three-digit rounding arithmetic to compute approximations to $P(2.19)$ and $Q(2.19)$. Compare them with the true values, $P(2.19) = Q(2.19) = 1.685159$.

Solution.

$$P(2.19) = (((2.19)^3 - 3(2.19)^2) + 3(2.19)) - 1 \approx ((10.4 - 3(4.79)) + 6.57) - 1$$

$$\approx ((10.4 - 14.3) + 6.57) - 1 = (-3.9 + 6.57) - 1 = 2.67 - 1 = 1.67.$$

Hence $P(2.19) \approx 1.67$.

$$Q(2.19) = ((2.19 - 3)2.19 + 3)2.19 - 1 = ((-0.81)2.19 + 3)2.19 - 1$$

$$\approx (-1.77 + 3)2.19 - 1 = (1.23)2.19 - 1 \approx 2.69 - 1 = 1.69.$$

Hence $Q(2.19) \approx 1.69$. The errors are 0.015159 and -0.004841, respectively. Thus the approximation $Q(2.19) \approx 1.69$ has less error. Exercise 6 will explore the situation near the root of this polynomial.

$O(h^n)$ Order of Approximation

Often a function $f(h)$ is replaced with an approximation $p(h)$ and the error bound is known to be $M|h^n|$. This leads to the following definition.

Definition 1.10.   Assume that $f(h)$ is approximated by the function $p(h)$ and that there exists a real constant $M > 0$ and a positive integer n so that

$$\frac{|f(h) - p(h)|}{|h^n|} \le M \qquad \text{for sufficiently small } h. \tag{7}$$

We say that $p(h)$ approximates $f(h)$ with **order of approximation** $O(h^n)$ and write

$$f(h) = p(h) + O(h^n). \tag{8}$$

When relation (7) is rewritten in the form $|f(h) - p(h)| \le M|h^n|$ we see that the notation $O(h^n)$ stands in place of the error bound $M|h^n|$. The term $O(h^n)$ is pronounced "oh of h^n." The following result shows how to apply the definition to simple combinations of two functions.

Theorem 1.16.   Assume that $f(h) = p(h) + O(h^n)$, $g(h) = q(h) + O(h^m)$ and $r = \min\{m, n\}$. Then

$$f(h) + g(h) = p(h) + q(h) + O(h^r), \tag{9}$$

$$f(h)g(h) = p(h)q(h) + O(h^r), \tag{10}$$

$$\frac{f(h)}{g(h)} = \frac{p(h)}{q(h)} + O(h^r) \qquad \text{provided that } g(h) \ne 0 \text{ and } q(h) \ne 0. \tag{11}$$

It is instructive to consider $p(x)$ to be the Taylor polynomial approximation of $f(x)$ of degree n; then the remainder term is simply designated $O(h^{n+1})$, which stands for the presence of omitted terms starting with the power h^{n+1}. The remainder term converges to zero with the same rapidity that h^{n+1} converges to zero, as expressed in the relationship

$$O(h^{n+1}) \approx Mh^{n+1} \approx \frac{f^{(n+1)}(c)}{(n+1)!} h^{n+1} \qquad \text{for sufficiently small } h. \qquad (12)$$

Hence the notation $O(h^{n+1})$ stands in place of the quantity Mh^{n+1}, where M is a constant or "behaves like a constant."

Theorem 1.17 (Taylor's Theorem). Assume that $f \in C^{n+1}[a, b]$. If both x_0 and $x = x_0 + h$ lie in $[a, b]$, then

$$f(x_0 + h) = \sum_{k=0}^{n} \frac{f^{(k)}(x_0)}{k!} h^k + O(h^{n+1}). \qquad (13)$$

The following example illustrates the above theorems. The computations use the addition properties: (i) $O(h^p) + O(h^p) = O(h^p)$ and (ii) $O(h^p) + O(h^q) = O(h^r)$, where $r = \min\{p, q\}$, and the multiplicative property: (iii) $O(h^p)O(h^q) = O(h^s)$, where $s = p + q$.

Example 1.11

Consider the Taylor polynomial expansions:

$$e^h = 1 + h + \frac{h^2}{2!} + \frac{h^3}{3!} + O(h^4) \quad \text{and} \quad \cos(h) = 1 - \frac{h^2}{2!} + \frac{h^4}{4!} + O(h^6).$$

Determine the order of approximation for their sum and product.

Solution. For the sum we have

$$e^h + \cos(h) = 1 + h + \frac{h^2}{2!} + \frac{h^3}{3!} + O(h^4) + 1 - \frac{h^2}{2!} + \frac{h^4}{4!} + O(h^6)$$

$$= 2 + h + \frac{h^3}{3!} + O(h^4) + \frac{h^4}{4!} + O(h^6).$$

Since $O(h^4) + h^4/4! = O(h^4)$ and $O(h^4) + O(h^6) = O(h^4)$, this reduces to

$$e^h + \cos(h) = 2 + h + \frac{h^3}{3!} + O(h^4)$$

and the order of approximation is $O(h^4)$.

The product is treated similarly.

$$e^h \cos(h) = \left(1 + h + \frac{h^2}{2!} + \frac{h^3}{3!} + O(h^4) \right)\left(1 - \frac{h^2}{2!} + \frac{h^4}{4!} + O(h^6) \right)$$

$$= \left(1 + h + \frac{h^2}{2!} + \frac{h^3}{3!}\right)\left(1 - \frac{h^2}{2!} + \frac{h^4}{4!}\right) + \left(1 + h + \frac{h^2}{2!} + \frac{h^3}{3!}\right)O(h^6)$$

$$+ \left(1 - \frac{h^2}{2!} + \frac{h^4}{4!}\right)O(h^4) + O(h^4)O(h^6)$$

$$= 1 + h - \frac{h^3}{3} - \frac{5h^4}{24} - \frac{h^5}{24} + \frac{h^6}{48} + \frac{h^7}{144} + O(h^6) + O(h^4) + O(h^4)O(h^6).$$

Since

$$O(h^4)O(h^6) = O(h^{10}) \text{ and } -\frac{5h^4}{24} - \frac{h^5}{24} + \frac{h^6}{48} + \frac{h^7}{144} + O(h^6) + O(h^4) + O(h^{10}) = O(h^4),$$

the preceding equation is simplified to yield

$$e^h\cos(h) = 1 + h - \frac{h^3}{3} + O(h^4)$$

and the order of approximation is $O(h^4)$.

Order of Convergence of a Sequence

Numerical approximations are often arrived at by computing a sequence of approximations that get closer and closer to the desired answer. The order of convergence of a sequence is analogous to the order of an approximation.

Definition 1.11.   Suppose that $\lim_{n\to\infty} x_n = x$ and $\{r_n\}_{n=1}^{\infty}$ is a sequence with $\lim_{n\to\infty} r_n = 0$. We say that $\{x_n\}_{n=1}^{\infty}$ converges to x with the order of convergence $O(r_n)$ if there exists a constant $K > 0$ such that

$$\frac{|x_n - x|}{|r_n|} \le K \qquad \text{for } n \text{ sufficiently large.}$$

This is indicated by writing $x_n = x + O(r_n)$ or $x_n \to x$ with order of convergence $O(r_n)$.

Example 1.12

Let $x_n = \cos(n)/n^2$ and $r_n = 1/n^2$; then $\lim_{n\to\infty} x_n = 0$ with a rate of convergence $O(1/n^2)$.

Solution.   This follows immediately from the relation

$$\frac{|\cos(n)/n^2|}{|1/n^2|} = |\cos(n)| \le 1 \qquad \text{for all } n.$$

Propagation of Error

Let us investigate how error might be propagated in successive computations. Consider the addition of two numbers p and q (the true values) with the approximate values $\bar{p}$ and

$\bar{q}$, which contain errors ϵ_p and ϵ_q, respectively. Starting with $p = \bar{p} + \epsilon_p$ and $q = \bar{q} + \epsilon_q$, the sum is

$$p + q = (\bar{p} + \epsilon_p) + (\bar{q} + \epsilon_q) = (\bar{p} + \bar{q}) + (\epsilon_p + \epsilon_q). \tag{14}$$

Hence, for addition, the error in the sum is the sum of the errors of the addends.

The propagation of error in multiplication is more complicated. The product is

$$pq = (\bar{p} + \epsilon_p)(\bar{q} + \epsilon_q) = \bar{p}\bar{q} + \bar{p}\epsilon_q + \bar{q}\epsilon_p + \epsilon_p\epsilon_q. \tag{15}$$

Hence, if $\bar{p}$ and $\bar{q}$ are larger than 1 in absolute value, the terms $\bar{p}\epsilon_q$ and $\bar{q}\epsilon_p$ show that there is a possibility of magnification of the original errors ϵ_q and ϵ_p. Insights are gained if we look at the relative error. Rearrange the terms in (15) to get

$$pq - \bar{p}\bar{q} = \bar{p}\epsilon_q + \bar{q}\epsilon_p + \epsilon_p\epsilon_q. \tag{16}$$

Suppose that $p \neq 0$ and $q \neq 0$; then we can divide (16) by pq and obtain

$$\frac{pq - \bar{p}\bar{q}}{pq} = \frac{\bar{p}\epsilon_q + \bar{q}\epsilon_p + \epsilon_p\epsilon_q}{pq}$$

$$= \frac{\bar{p}\epsilon_q}{pq} + \frac{\bar{q}\epsilon_p}{pq} + \frac{\epsilon_p\epsilon_q}{pq}.$$

Furthermore, suppose that $\bar{p}/p \approx 1$, $\bar{q}/q \approx 1$, and $(\epsilon_p/p)(\epsilon_q/q) = R_p R_q \approx 0$. Then making these substitutions yields the simplified relationship

$$\frac{pq - \bar{p}\bar{q}}{pq} \approx \frac{\epsilon_q}{q} + \frac{\epsilon_p}{p} + 0 = R_q + R_p. \tag{17}$$

This shows that the relative error in the product pq is approximately the sum of the relative errors in the approximations $\bar{q}$ and $\bar{p}$.

Often an initial error will be propagated in a sequence of calculations. A quality which is desirable for any numerical process is that a small error in the initial conditions will produce small changes in the final result. An algorithm with this feature is called **stable**; otherwise, it is called **unstable**. Whenever possible we shall choose methods that are stable. The following definition is used to describe the propagation of error.

Definition 1.12.   Suppose that $E(n)$ represents the growth of error after n steps. If $|E(n)| \approx n\epsilon$, the growth of error is said to be **linear**. If $|E(n)| \approx K^n\epsilon$, the growth of error is called **exponential**. If $K > 1$, the exponential error grows without bound as $n \to \infty$, and if $0 < K < 1$, the exponential error diminishes to zero as $n \to \infty$.

The next two examples show how an initial error can propagate in either a stable or an unstable fashion.

Example 1.13

Show that the following three schemes can be used with infinite-precision rational arithmetic to recursively generate the terms in the sequence $\left\{1, \frac{1}{3}, \frac{1}{9}, \frac{1}{27}, \frac{1}{81}, \ldots\right\}$:

$$r_0 = 1 \quad \text{and} \quad r_n = \tfrac{1}{3}r_{n-1} \qquad\qquad\qquad \text{for } n = 1, 2, \ldots, \qquad (18a)$$

$$p_0 = 1, \quad p_1 = \tfrac{1}{3}, \quad \text{and} \quad p_n = \tfrac{4}{3}p_{n-1} - \tfrac{1}{3}p_{n-2} \qquad \text{for } n = 2, 3, \ldots, \qquad (18b)$$

$$q_0 = 1, \quad q_1 = \tfrac{1}{3}, \quad \text{and} \quad q_n = \tfrac{10}{3}q_{n-1} - q_{n-2} \qquad \text{for } n = 2, 3, \ldots, \qquad (18c)$$

Solution.   Formula (18a) is obvious. In (18b) the difference equation $p_n = \tfrac{4}{3}p_{n-1} - \tfrac{1}{3}p_{n-2}$ has the general solution $p_n = A(1/3^n) + B$. This can be verified by direct substitution:

$$\tfrac{4}{3}p_{n-1} - \tfrac{1}{3}p_{n-2} = \frac{4}{3}\left(A\frac{1}{3^{n-1}} + B\right) - \frac{1}{3}\left(A\frac{1}{3^{n-2}} + B\right) = \left(\frac{4}{3^n} - \frac{3}{3^n}\right)A - \left(\frac{4}{3} - \frac{1}{3}\right)B$$

$$= A\frac{1}{3^n} + B = p_n.$$

Setting $A = 1$ and $B = 0$ will generate the desired sequence. In (18c) the difference equation $q_n = \tfrac{10}{3}q_{n-1} - q_{n-2}$ has the general solution $q_n = A(1/3^n) + B3^n$. This, too, is verified by substitution:

$$\frac{10}{3}q_{n-1} - q_{n-2} = \frac{10}{3}\left(A\frac{1}{3^{n-1}} + B3^{n-1}\right) - \left(A\frac{1}{3^{n-2}} + B3^{n-2}\right)$$

$$= \left(\frac{10}{3^n} - \frac{9}{3^n}\right)A - (10 - 1)3^{n-2}B$$

$$= A\frac{1}{3^n} + B3^n = q_n.$$

Setting $A = 1$ and $B = 0$ generates the required sequence.

Example 1.14

Generate approximations to the sequence $\{x_n\} = \{1/3^n\}$ using the schemes

$$r_0 = 0.99996 \quad \text{and} \quad r_n = \tfrac{1}{3}r_{n-1} \qquad\qquad\qquad \text{for } n = 1, 2, \ldots, \qquad (19a)$$

$$p_0 = 1, \quad p_1 = 0.33332, \quad \text{and} \quad p_n = \tfrac{4}{3}p_{n-1} - \tfrac{1}{3}p_{n-2} \qquad \text{for } n = 2, 3, \ldots, \qquad (19b)$$

$$q_0 = 1, \quad q_1 = 0.33332, \quad \text{and} \quad q_n = \tfrac{4}{3}q_{n-1} - q_{n-2} \qquad \text{for } n = 2, 3, \ldots \qquad (19c)$$

In (19a) the initial error in r_0 is 0.00004, and in (19b) and (19c) the initial errors in p_1 and q_1 are 0.0000133333. Investigate the propagation of error for each scheme.

Solution.   Table 1.4 gives the first 10 numerical approximations for each sequence, and Table 1.5 gives the error in each formula. The error for $\{r_n\}$ is stable and decreases in an exponential manner. The error for $\{p_n\}$ is stable. The error for $\{q_n\}$ is unstable and grows at an exponential rate. Although the error for $\{p_n\}$ is stable, the terms $p_n \to 0$ as $n \to \infty$, so that the error eventually dominates and the terms past p_8 have no significant digits. Figures 1.12, 1.13, and 1.14 show the errors in $\{r_n\}$, $\{p_n\}$, and $\{q_n\}$, respectively.

TABLE 1.4   The Sequence $\{x_n\} = \{1/3^n\}$ and the Approximations $\{r_n\}$, $\{p_n\}$, and $\{q_n\}$

n	x_n	r_n	p_n	q_n
0	$1 = 1.0000000000$	0.9999600000	1.0000000000	1.0000000000
1	$\frac{1}{3} = 0.3333333333$	0.3333200000	0.3333200000	0.3333200000
2	$\frac{1}{9} = 0.1111111111$	0.1111066667	0.1110933330	0.1110666667
3	$\frac{1}{27} = 0.0370370370$	0.0370355556	0.0370177778	0.0369022222
4	$\frac{1}{81} = 0.0123456790$	0.0123451852	0.0123259259	0.0119407407
5	$\frac{1}{243} = 0.0041152263$	0.0041150617	0.0040953086	0.0029002469
6	$\frac{1}{729} = 0.0013717421$	0.0013716872	0.0013517695	-0.0022732510
7	$\frac{1}{2187} = 0.0004572474$	0.0004572291	0.0004372565	-0.0104777503
8	$\frac{1}{6561} = 0.0001524158$	0.0001524097	0.0001324188	-0.0326525834
9	$\frac{1}{19,683} = 0.0000508053$	0.0000508032	0.0000308063	-0.0983641945
10	$\frac{1}{59,049} = 0.0000169351$	0.0000169344	-0.0000030646	-0.2952280648

TABLE 1.5   The Error Sequences $\{x_n - r_n\}$, $\{x_n - p_n\}$, and $\{x_n - q_n\}$

n	$x_n - r_n$	$x_n - p_n$	$x_n - q_n$
0	0.0000400000	0.0000000000	0.0000000000
1	0.0000133333	0.0000133333	0.0000133333
2	0.0000044444	0.0000177778	0.0000444444
3	0.0000014815	0.0000192593	0.0001348148
4	0.0000004938	0.0000197531	0.0004049383
5	0.0000001646	0.0000199177	0.0012149794
6	0.0000000549	0.0000199726	0.0036449931
7	0.0000000183	0.0000199909	0.0109349977
8	0.0000000061	0.0000199970	0.0328049992
9	0.0000000020	0.0000199990	0.0984149998
10	0.0000000007	0.0000199997	0.2952449999

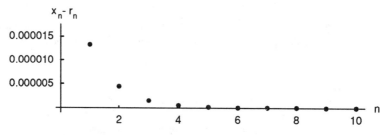

Figure 1.12   A stable decreasing error sequence $\{x_n - r_n\}$.

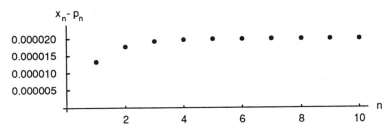

Figure 1.13   A stable error sequence $\{x_n - p_n\}$.

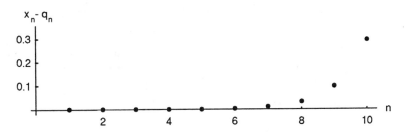

Figure 1.14   An unstable increasing error sequence $\{x_n - q_n\}$.

Uncertainty in Data

Data from real-world problems contain uncertainty or error. This type of error is referred to as noise. It will affect the accuracy of any numerical computation that is based on the data. An improvement of precision is not accomplished by performing successive computations on noisy data. Hence, if you start with data with d significant digits of accuracy, the result of a computation should be reported in d significant digits of accuracy. For example, suppose that the data $p_1 = 4.152$ and $p_2 = 0.07931$ both have four significant digits of accuracy. Then it is tempting to report all the digits that appear on your calculator (i.e., $p_1 + p_2 = 4.23131$). This is an oversight, because you should not report conclusions from noisy data that have more significant digits than the original data. The proper answer in this situation is $p_1 + p_2 = 4.231$.

EXERCISES FOR ERROR ANALYSIS

1. Find the error E_x and relative error R_x. Also, determine the number of significant digits in the approximation.
   (a) Let $x = 2.71828182$ and $\bar{x} = 2.7182$.
   (b) Let $y = 98,350$ and $\bar{y} = 98,000$.
   (c) Let $z = 0.000068$ and $\bar{z} = 0.00006$.

2. Complete the following computation: $\int_0^{1/4} e^{x^2}\,dx \approx \int_0^{1/4} (1 + x^2 + x^4/2! + x^6/3!)\,dx = \bar{p}.$
   State what type of error is present in this situation. Compare your answer with the true value
   $p = 0.2553074606.$

3. (a) Consider the data $p_1 = 1.414$ and $p_2 = 0.09125$, which have four significant digits of
       accuracy. Determine the proper answer for the sum $p_1 + p_2$ and the product $p_1 p_2$.
   (b) Consider the data $p_1 = 31.415$ and $p_2 = 0.027182$, which have five significant digits of
       accuracy. Determine the proper answer for the sum $p_1 + p_2$ and the product $p_1 p_2$.

4. Complete the following computation and state what type of error is present in this situation.

   (a) $\dfrac{\sin(\pi/4 + 0.00001) - \sin(\pi/4)}{0.00001} = \dfrac{0.70711385222 - 0.70710678119}{0.00001} = \cdots .$

   (b) $\dfrac{\ln(2 + 0.00005) - \ln(2)}{0.00005} = \dfrac{0.69317218025 - 0.69314718056}{0.00005} = \cdots .$

5. Sometimes the loss of significance error can be avoided by rearranging terms in the function or
   using a known identity from trigonometry or algebra. Find an equivalent formula for the
   following functions that avoids a loss of significance.

   (a) $\ln(x + 1) - \ln(x)$ for large x (b) $\sqrt{x^2 + 1} - x$ for large x

   (c) $\cos^2(x) - \sin^2(x)$ for $x \approx \dfrac{\pi}{4}$ (d) $\sqrt{\dfrac{1 + \cos(x)}{2}}$ for $x \approx \pi$

6. *Polynomial evaluation.* Let $P(x) = ((x^3 - 3x^2) + 3x) - 1$, $Q(x) = ((x - 3)x + 3) - 1$, and
   $R(x) = (x - 1)^3$.
   (a) Use four-digit rounding arithmetic and compute $P(2.72)$, $Q(2.72)$, and $R(2.72)$. In the
       computation of $P(x)$, assume that $(2.72)^2 = 7.398$ and $(2.72)^3 = 20.12$.
   (b) Use four-digit rounding arithmetic and compute $P(0.975)$, $Q(0.975)$, and $R(0.975)$. In the
       computation of $P(x)$, assume that $(0.975)^2 = 0.9506$ and $(0.975)^3 = 0.9268$.

7. Use three-digit rounding arithmetic to compute the following sums:

   (a) $\displaystyle\sum_{k=1}^{6} \frac{1}{3^k} = \frac{1}{3} + \frac{1}{9} + \frac{1}{27} + \frac{1}{81} + \frac{1}{243} + \frac{1}{729}$

   (b) $\displaystyle\sum_{k=1}^{6} \frac{1}{3^{7-k}} = \frac{1}{729} + \frac{1}{243} + \frac{1}{81} + \frac{1}{27} + \frac{1}{9} + \frac{1}{3}$

8. Discuss the propagation of error for the following:
   (a) The sum of three numbers $p + q + r = (\bar{p} + \epsilon_p) + (\bar{q} + \epsilon_q) + (\bar{r} + \epsilon_r)$.
   (b) The quotient of two numbers $p/q = (\bar{p} + \epsilon_p)/(\bar{q} + \epsilon_q)$.
   (c) The product of three numbers $pqr = (\bar{p} + \epsilon_p)(\bar{q} + \epsilon_q)(\bar{r} + \epsilon_r)$.

9. Given the Taylor polynomial expansions

   $$\frac{1}{1 - h} = 1 + h + h^2 + h^3 + O(h^4) \quad \text{and} \quad \cos(h) = 1 - \frac{h^2}{2!} + \frac{h^4}{4!} + O(h^6).$$

   Determine the order of approximation for their sum and product.

10. Given the Taylor polynomial expansions

$$e^h = 1 + h + \frac{h^2}{2!} + \frac{h^3}{3!} + \frac{h^4}{4!} + O(h^5) \quad \text{and} \quad \sin(h) = h - \frac{h^3}{3!} + O(h^5).$$

Determine the order of approximation for their sum and product.

11. Given the Taylor polynomial expansions

$$\cos(h) = 1 - \frac{h^2}{2!} + \frac{h^4}{4!} + O(h^6) \quad \text{and} \quad \sin(h) = h - \frac{h^3}{3!} + \frac{h^5}{5!} + O(h^7).$$

Determine the order of approximation for their sum and product.

12. *Improving the quadratic formula.* Assume that $a \neq 0$ and $b^2 - 4ac > 0$ and consider the equation $ax^2 + bx + c = 0$. The roots can be computed with the quadratic formulas

$$x_1 = \frac{-b + \sqrt{b^2 - 4ac}}{2a} \quad \text{and} \quad x_2 = \frac{-b - \sqrt{b^2 - 4ac}}{2a}. \tag{i}$$

Show that these roots can be calculated with the equivalent formulas

$$x_1 = \frac{-2c}{b + \sqrt{b^2 - 4ac}} \quad \text{and} \quad x_2 = \frac{-2c}{b - \sqrt{b^2 - 4ac}}. \tag{ii}$$

> *Remark.* In the cases when $|b| \approx \sqrt{b^2 - 4ac}$, one must proceed with caution to avoid loss of precision due to a catastrophic cancellation. If $b > 0$, then x_1 should be computed with formula (ii) and x_2 should be computed with formula (i). However, if $b < 0$, then x_1 should be computed using (i) and x_2 should be computed using (ii).

13. Use the appropriate formula for computing x_1 and x_2 mentioned in Exercise 12 to find the roots of the following quadratic equations.
 (a) $x^2 - 1,000.001x + 1 = 0$
 (b) $x^2 - 10,000.0001x + 1 = 0$
 (c) $x^2 - 100,000.00001x + 1 = 0$
 (d) $x^2 - 1,000,000.000001x + 1 = 0$

14. Study Exercises 12 and 13 and write an algorithm and/or computer program that will accurately compute the roots of a quadratic equation in all situations, including the troublesome ones when $|b| \approx \sqrt{b^2 - 4ac}$.

15. Generate approximations to the sequence $\{x_n\} = \{1/2^n\}$ (Table 1.6) using the following schemes:
 (a) $r_0 = 0.994$ and $r_n = \frac{1}{2}r_{n-1}$ for $n = 1, 2, \ldots$,

 (b) $p_0 = 1$, $p_1 = 0.497$, and $p_n = \frac{3}{2}p_{n-1} - \frac{1}{2}p_{n-2}$ for $n = 2, 3, \ldots$,

 (c) $q_0 = 1$, $q_1 = 0.497$, and $q_n = \frac{5}{2}q_{n-1} - q_{n-2}$ for $n = 2, 3, \ldots$,

Consider that the initial error in r_0 is 0.006 and the initial errors in p_1 and p_2 are 0.003. Investigate the error propagation for each scheme.

TABLE 1.6   The Sequence $\{x_n\} = \{1/2^n\}$ and the Approximations $\{r_n\}$, $\{p_n\}$, and $\{q_n\}$

n	x_n	r_n	p_n	q_n
0	1	0.994	1.	1.
1	0.5	0.497	0.497	0.497
2	0.25	0.2485	0.2455	0.2425
3	0.125	0.12425	0.11975	0.10925
4	0.0625	0.062125	0.056875	0.030625
5	0.03125	0.0310625	0.0254375	−0.0326875
6	0.015625	0.01553125	0.0097187	−0.11234375
7	0.0078125	0.007765625	0.001859375	−0.248171875
8	0.00390625	0.003882812	−0.002070312	−0.508085937

16. Write a report on round-off errors. See References [4, 9, 29, 35, 41, 49, 51, 76, 79, 81, 90, 94, 101, 117, 128, 146, 153, 160, 181, 184, 186, and 204].

17. Write a report on loss of significance (cancellation error). See References [3, 8, 35, 40, 79, and 142].

18. Write a report on error propagation. See References [4, 9, 40, 41, 49, 51, 78, 79, 81, 133, 142, 145, 153, and 204].

2

The Solution of Nonlinear Equations $f(x) = 0$

Consider the physical problem that involves a spherical ball of radius r which is submerged to a depth d in water (see Figure 2.1). Assume that the ball is constructed from a variety of longleaf pine that has a density of $\rho = 0.638$ and that its radius measures $r = 10$ cm. How much of the ball will be submerged when it is placed in water?

The mass M_w of water displaced when a sphere is submerged to a depth d is

$$M_w = \int_0^d \pi(r^2 - (x - r)^2)\, dx = \frac{\pi d^2(3r - d)}{3},$$

and the mass of the ball is $M_b = 4\pi r^3 \rho / 3$. Applying Archimedes' law, $M_w = M_b$, produces the following equation that must be solved:

$$\frac{\pi(d^3 - 3d^2 r + 4r^3\rho)}{3} = 0.$$

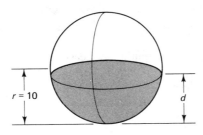

$r = 10$

d

Figure 2.1   The portion of a sphere of radius r that is to be submerged to a depth d.

In our case (with $r = 10$ and $\rho = 0.638$) this equation becomes

$$\frac{\pi(2552 - 30d^2 + d^3)}{3} = 0.$$

The graph of cubic polynomial $y = 2552 - 30d^2 + d^3$ is shown in Figure 2.2 and from it one can see that the solution lies near the value $d = 12$.

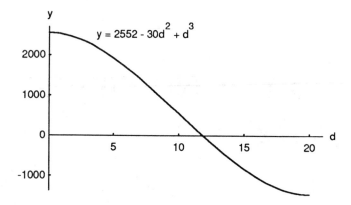

Figure 2.2   The cubic $y = 2552 - 30d^2 + d^3$.

The goal of this chapter is to develop a variety of methods for finding numerical approximations for the roots of an equation. For example, the bisection method could be applied to obtain the three roots $d_1 = -8.17607212$, $d_2 = 11.86150151$, and $d_3 = 26.31457061$. The first root, d_1, is not a feasible solution for this problem, because d cannot be negative. The third root, d_3, is larger than the diameter of the sphere and it is not the desired solution. The root $d_2 = 11.86150151$ lies in the interval [0, 20] and is the proper solution. Its magnitude is reasonable because a little more than one-half of the sphere must be submerged.

2.1 ITERATION FOR SOLVING x = g(x)

A fundamental principle in computer science is *iteration*. As the name suggests, it means that a process is repeated until an answer is achieved. Iterative techniques are used to find roots of equations, solutions of linear and nonlinear systems of equations, and solutions of differential equations. In this section we study the process of iteration using repeated substitution.

A rule or function $g(x)$ for computing successive terms is needed together with a starting value p_0. Then a sequence of values $\{p_k\}$ is obtained using the iterative rule $p_{k+1} = g(p_k)$. The sequence has the pattern

$$p_0 \quad \text{(starting value)}$$

$$p_1 = g(p_0)$$

$$p_2 = g(p_1)$$

$$\cdot$$
$$\cdot$$
$$\cdot$$

(1)

$$p_k = g(p_{k-1})$$

$$p_{k+1} = g(p_k)$$

$$\cdot$$
$$\cdot$$
$$\cdot$$

What can we learn from an unending sequence of numbers? If the numbers tend to a limit, we feel that something has been achieved. But what if the numbers diverge to infinity? The next example addresses this situation.

Example 2.1

Discuss the divergent iteration

$$p_0 = 1 \quad \text{and} \quad p_{k+1} = 1.001p_k \quad \text{for } k = 0, 1, \ldots .$$

Solution.   The first 100 terms look as follows:

$$p_1 = 1.001p_0 = 1.001 \times 1.000000 = 1.001000,$$

$$p_2 = 1.001p_1 = 1.001 \times 1.001000 = 1.002001,$$

$$p_3 = 1.001p_2 = 1.001 \times 1.002001 = 1.003003,$$

$$\cdot \qquad\qquad \cdot \qquad\qquad\qquad \cdot$$
$$\cdot \qquad\qquad \cdot \qquad\qquad\qquad \cdot$$
$$\cdot \qquad\qquad \cdot \qquad\qquad\qquad \cdot$$

$$p_{100} = 1.001p_{99} = 1.001 \times 1.104012 = 1.105116.$$

The process can be continued indefinitely and it is easily shown that $\lim_{n \to \infty} p_n = +\infty$. In Chapter 9 we will see that the sequence $\{p_k\}$ is a numerical solution to the differential equation $y' = 0.001y$. The solution is known to be $y(x) = \exp(0.001x)$. Indeed, if we compare the 100th term in the sequence with $y(100)$, we see that $p_{100} = 1.105116 \approx 1.105171 = \exp(0.1) = y(100)$.

In this section we are concerned with the type of functions $g(x)$ that produce convergent sequences $\{p_k\}$.

Finding Fixed Points

Definition 2.1   (Fixed Point).   A **fixed point** of a function $g(x)$ is a real number P such that $P = g(P)$.

Definition 2.2   (Fixed-Point Iteration).   The iteration $p_{n+1} = g(p_n)$ for $n = 0, 1, \ldots$ is called **fixed-point iteration**.

Theorem 2.1. Assume that g is a continuous function and that $\{p_n\}_{n=0}^{\infty}$ is a sequence generated by fixed-point iteration. If $\lim\limits_{n \to \infty} p_n = P$, then P is a fixed point of $g(x)$.

Proof. Suppose that

$$\lim_{n \to \infty} p_n = P; \quad \text{then} \quad \lim_{n \to \infty} p_{n+1} = P. \tag{2}$$

Now use the relation $p_{n+1} = g(p_n)$ and the continuity of $g(x)$ and (2) to obtain

$$P = \lim_{n \to \infty} p_{n+1} = \lim_{n \to \infty} g(p_n) = g(\lim_{n \to \infty} p_n) = g(P). \tag{3}$$

Therefore, $P = g(P)$, and P is a fixed point of $g(x)$.

Example 2.2

Discuss the convergent iteration

$$p_0 = 0.5 \quad \text{and} \quad p_{k+1} = \exp(-p_k) \quad \text{for } k = 0, 1, \ldots .$$

Solution. The first 10 terms are obtained by the calculations

$$p_1 = \exp(-0.500000) = 0.606531, \qquad p_2 = \exp(-0.606531) = 0.545239$$

$$p_3 = \exp(-0.545239) = 0.579703, \qquad p_4 = \exp(-0.579703) = 0.560065$$

$$\begin{matrix} . \\ . \\ . \end{matrix} \qquad\qquad\qquad\qquad \begin{matrix} . \\ . \\ . \end{matrix}$$

$$p_9 = \exp(-0.566409) = 0.567560, \qquad p_{10} = \exp(-0.567560) = 0.566907.$$

The sequence is converging and further calculations reveal that

$$\lim_{n \to \infty} p_n = 0.567143. \ldots$$

Thus we have found a fixed point of the function $\exp(-x)$, that is

$$0.567143. \ldots = \exp(-0.567143. \ldots).$$

Theorem 2.2. Assume that $g \in C[a, b]$.

If the range of the mapping $y = g(x)$ satisfies $a \le y \le b$ for all $a < x < b$, then g has a fixed point in $[a, b]$. $\tag{4}$

Furthermore, suppose that $g'(x)$ is defined over (a, b) and that a positive constant $K < 1$ exists with $\tag{5}$

$$|g'(x)| \le K < 1 \quad \text{for all } x \in (a, b),$$

then g has a unique fixed point P in $[a, b]$.

Proof of (4). If $g(a) = a$ or $g(b) = b$, the assertion is true. Otherwise, the range values $g(a)$ and $g(b)$ must satisfy $a < g(a) \le b$ and $a \le g(b) < b$. The function $f(x) \equiv x - g(x)$ has the property that

$$f(a) = a - g(a) < 0 \quad \text{and} \quad f(b) = b - g(b) > 0.$$

Now apply Theorem 1.2, the intermediate value theorem, to $f(x)$, with the constant $L = 0$, and conclude that there exists a number P with $a < P < b$ so that $f(P) = 0$. Therefore, $x = P$ is the desired fixed point of $g(x)$.

Proof of (5). Now we must show that this solution is unique. Let us make the additional assumption that there exist two fixed points P_1 and P_2 and see where it leads us. Apply Theorem 1.6, the mean value theorem for derivatives, to conclude that there exists a number d with $a < d < b$ so that $[g(P_2) - g(P_1)]/(P_2 - P_1) = g'(d)$. Next, use the facts that $g(P_1) = P_1$ and $g(P_2) = P_2$ to simplify the left side of this equation and obtain

$$1 = \frac{P_2 - P_1}{P_2 - P_1} g'(d).$$

But this contradicts the hypothesis that $|g'(x)| < 1$ over $[a, b]$, so that it is not possible for two fixed points to exist. Therefore, $g(x)$ has a unique fixed point P in $[a, b]$.

We now state a theorem that can be used to determine whether the iteration (1) will produce a convergent or divergent sequence.

Theorem 2.3   (Fixed-Point Theorem).   Assume that $g(x)$ and $g'(x)$ are continuous on a balanced interval $(a, b) = (P - \delta, P + \delta)$ that contains the unique fixed point P and that the starting value p_0 is chosen in this interval.

If $|g'(x)| \leq K < 1$ for all $a \leq x \leq b$, then the iteration $p_n = g(p_{n-1})$ will converge to P. In this case P is an attractive fixed point.   (6)

If $|g'(x)| > 1$ for all $a \leq x \leq b$, then the iteration $p_n = g(p_{n-1})$ will not converge to P. In this case P is a repulsive fixed point and the iteration exhibits local divergence.   (7)

Remark 1.   It is assumed that $p_0 \neq P$ in statement (7) and that "lucky guess" $g(p_0) = P$ is ruled out.

Remark 2.   Because $g(x)$ is continuous on an interval about $x = P$ it is permissible to use the simpler criterion $|g'(P)| \leq K < 1$ and $|g'(P)| > 1$ in (6) and (7), respectively.

Proof. We first show that the points $\{p_n\}_{n=0}^{\infty}$ all lie in (a, b). Starting with p_0, we apply Theorem 1.6, the mean value theorem. There exists a value $c_0 \in (a, b)$ so that

$$|P - p_1| = |g(P) - g(p_0)| = |g'(c_0)[P - p_0]| = |g'(c_0)| \, |P - p_0|$$
$$\leq K|P - p_0| < |P - p_0| < \delta.$$
                                                                                (8)

Therefore, p_1 is no farther from $x = P$ then p_0 was, and it follows that $p_1 \in (a, b)$ (see Figure 2.3). In general, suppose that $p_{n-1} \in (a, b)$; then

$$|P - p_n| = |g(P) - g(p_{n-1})| = |g'(c_{n-1})[P - p_{n-1}]| = |g'(c_{n-1})| \, |P - p_{n-1}|$$
$$\leq K|P - p_{n-1}| < |P - p_{n-1}| < \delta.$$
                                                                                (9)

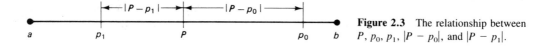

Figure 2.3 The relationship between P, p_0, p_1, $|P - p_0|$, and $|P - p_1|$.

Therefore, $p_n \in (a, b)$ and hence, by induction, all the points $\{p_n\}_{n=0}^{\infty}$ lie in (a, b).

Using the relationships above, it is easy to show that $\lim_{n \to \infty} p_n = P$ by proving that

$$\lim_{n \to \infty} |P - p_n| = 0. \tag{10}$$

A proof by induction will establish the inequality

$$|P - p_n| \le K^n |P - p_0|. \tag{11}$$

The case $n = 1$ follows from the details in relation (8). In general, suppose that $|P - p_{n-1}| \le K^{n-1}|P - p_0|$; then using this and the ideas in (9) we obtain

$$|P - p_n| \le K|P - p_{n-1}| \le KK^{n-1}|P - p_0| = K^n|P - p_0|.$$

Thus, by induction, inequality (11) holds for all n. Since $0 < K < 1$, the term K^n goes to zero as n goes to infinity. Hence

$$0 \le \lim_{n \to \infty} |P - p_n| \le \lim_{n \to \infty} K^n |P - p_0| = 0. \tag{12}$$

The limit of $|P - p_n|$ is squeezed between zero on the left and zero on the right, so we can conclude that $\lim_{n \to \infty} P - p_n = 0$. Therefore, statement (6) of Theorem 2.3 is proven. We leave statement (7) for the reader to investigate.

Corollary 2.1. Assume that g satisfies the hypothesis given in (6) of Theorem 2.3. Bounds for the error involved when using p_n to approximate P are given by

$$|P - p_n| \le K^n |P - p_0| \qquad \text{for all } n \ge 1 \tag{13}$$

and

$$|P - p_n| \le \frac{K^n |p_1 - p_0|}{1 - K} \qquad \text{for all } n \ge 1. \tag{14}$$

Graphical Interpretation of Fixed-Point Iteration

Since we seek a fixed point P to $g(x)$, it is necessary that the graph of the curve $y = g(x)$ and the line $y = x$ intersect at (P, P). Two simple types of convergent iteration, monotone and oscillating, are illustrated in Figure 2.4(a) and (b), respectively.

To visualize the process, start at p_0 on the x-axis and move vertically to the point $(p_0, p_1) = (p_0, g(p_0))$ on the curve $y = g(x)$. Then move horizontally from (p_0, p_1) to the point (p_1, p_1) on the line $y = x$. Finally, move vertically downward to p_1 on the x-axis. The recursion $p_{n+1} = g(p_n)$ is used to construct the point (p_n, p_{n+1}) on the graph, then

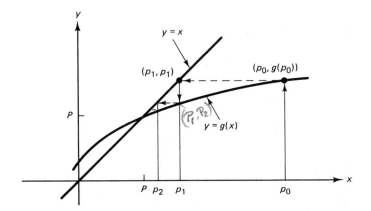

Figure 2.4   (a) Monotone convergence when $0 < g'(P) < 1$.

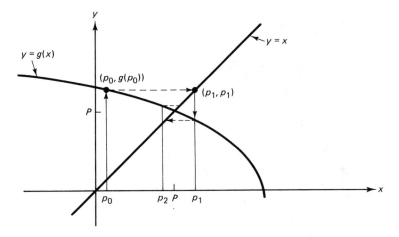

Figure 2.4   (b) Oscillating convergence when $-1 < g'(P) < 0$.

a horizontal motion locates (p_{n+1}, p_{n+1}) on the line $y = x$, and vertical movement ends up at p_{n+1} on the x-axis. The situation is shown in Figure 2.4.

If $|g'(P)| > 1$, then the iteration (1) produces a sequence that diverges away from P. The two simple types of divergent iteration, monotone and oscillating, are illustrated in Figure 2.5(a) and (b), respectively.

Example 2.3

Investigate the nature of the iteration (1) when the function $g(x) = 1 + x - x^2/4$ is used.

Solution.   The fixed points can be found by solving the equation $P = 1 + P - P^2/4$ and two solutions $P = -2$ and $P = 2$ are found. The derivative of the function is $g'(x) = 1 - x/2$, and there are two cases to consider.

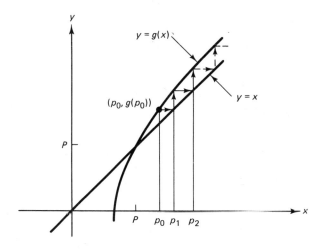

Figure 2.5 (a) Monotone divergence when $1 < g'(P)$.

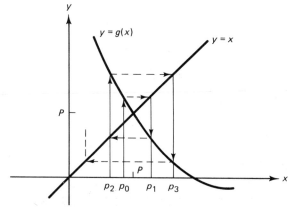

Figure 2.5 (b) Divergent oscillation when $g'(P) < -1$.

Case (i): $P = -2$
Start with $p_0 = -2.05$
then get $p_1 = -2.100625$
$p_2 = -2.20378135$
$p_3 = -2.41794441$

$$\lim_{n \to \infty} p_n = -\infty.$$

Since $|g'(x)| > \frac{3}{2}$ on $[-3, -1]$ by Theorem 2.3, the sequence will not converge to $P = -2$.

Case (ii): $P = 2$
Start with $p_0 = 1.6$
then get $p_1 = 1.96$
$p_2 = 1.9996$
$p_3 = 1.99999996$

$$\lim_{n \to \infty} p_n = 2.$$

Since $|g'(x)| < \frac{1}{2}$ on $[1, 3]$, Theorem 2.3 says that the sequence will converge to $P = 2$.

Theorem 2.3 does not state what will happen when $g'(P) = 1$. The next example has been specially constructed so that the sequence $\{p_k\}$ converges whenever $p_0 > P$ and it diverges if we choose $p_0 < P$.

Example 2.4

Investigate the nature of the iteration (1) when

$$g(x) = 2(x - 1)^{1/2} \qquad \text{for } x \geq 1.$$

Solution. Only one fixed point $P = 2$ exists. The derivative is $g'(x) = 1/(x - 1)^{1/2}$ and $g'(2) = 1$, so that Theorem 2.3 does not apply. There are two cases to consider, when the starting value lies to the left or right of $P = 2$.

Case (i): Start with $p_0 = 1.5$,
then get $p_1 = 1.41421356$
$\qquad p_2 = 1.28718851$
$\qquad p_3 = 1.07179943$
$\qquad p_4 = 0.53590832$

$\qquad \cdot$
$\qquad \cdot$
$\qquad \cdot$

$\qquad p_5 = 2(-0.46409168)^{1/2}.$

Since p_4 lies outisde the domain of $g(x)$, the term p_5 cannot be computed.

Case (ii): Start with $p_0 = 2.5$,
then get $p_1 = 2.44948974$
$\qquad p_2 = 2.40789513$
$\qquad p_3 = 2.37309514$
$\qquad p_4 = 2.34358284$

$\qquad \cdot$
$\qquad \cdot$
$\qquad \cdot$

$\qquad \lim_{n \to \infty} p_n = 2.$

This sequence is converging slowly to the value $P = 2$; indeed $p_{1000} = 2.00398714.$

Absolute and Relative Error Considerations

In Example 2.4, case (ii), the sequence converges slowly and after 1000 iterations the three consecutive terms are

$$p_{1000} = 2.00398714, \qquad p_{1001} = 2.00398317, \qquad p_{1002} = 2.00397921.$$

This should not be too disturbing; after all, we could compute a few more thousand terms and find a better approximation! But what about a criterion for stopping the iteration? Notice that if we use the difference between consecutive terms,

$$|p_{1001} - p_{1000}| = |2.00398317 - 2.00398714| = 0.00000397.$$

Yet the absolute error in the approximation p_{1000} is known to be

$$|P - p_{1000}| = |2.00000000 - 2.00398714| = 0.00398714.$$

This is about 1000 times larger than $|p_{1001} - p_{1000}|$ and it shows that closeness of consecutive terms does not guarantee that accuracy has been achieved. But it is usually the only criterion available and is often used to terminate an iterative procedure.

Algorithm 2.1   (Fixed-Point Iteration).   To find a solution to the equation $x = g(x)$ by starting with p_0 and iterating $p_{n+1} = g(p_n)$.

Tol := 0.000001	{Termination criterion}
Max := 200	{Maximum number of iterations}
Small := 0.000001	{Initialize the variable}
K := 0	{Initialize the counter}
RelErr := 1	{Initialize the variable}
INPUT Pterm	{The intial approximation}
Pnew := g(Pterm)	{The first iteration}

WHILE RelErr ≥ Tol and K ≤ Max DO	
K := K + 1	{Increment the counter}
Pold := Pterm	{Previous iterate p_{k-1}}
Pterm := Pnew	{Current iterate p_k}
Pnew := g(Pterm)	{Compute new iterate p_{k+1}}
Dg := Pnew − Pterm	{Difference in g(x)}
Delta := |Dg|	{Absolute error}
RelErr := 2∗Delta/[|Pnew|+Small]	{Relative Error}
Dx := Pterm − Pold	{Difference in x}
Slope := Dg/Dx	{$g'(p_k)$}

PRINT 'The computed fixed point of g(x) is' Pnew   {Output}
PRINT 'Consecutive iterates are within' Delta
IF |Slope| > 1 THEN
    PRINT "The sequence appears to be diverging."
ELSE
    PRINT "The sequence appears to be converging."
ENDIF

EXERCISES FOR ITERATION FOR SOLVING $x = g(x)$

When it is asked to implement the iteration (1), either a calculator or computer should be used.

1. Investigate the nature of the iteration (1) when

$$g(x) = -4 + 4x - \frac{x^2}{2}.$$

(a) Show that $P = 2$ and $P = 4$ are fixed points.
(b) Use starting value $p_0 = 1.9$ and compute p_1, p_2, and p_3.
(c) Use starting value $p_0 = 3.8$ and compute p_1, p_2, and p_3.
(d) Find the errors E_k and relative errors R_k for the values p_k in parts (b) and (c).
(e) What conclusions can be drawn from Theorem 2.3?

2. Graph the curve $y = g(x) = (6 + x)^{1/2}$ and the line $y = x$, and plot the fixed point (3, 3). Start with $p_0 = 7$ and construct $p_1 \approx 3.61$ and $p_2 \approx 3.10$ as indicated in Figure 2.4(a). Will fixed-point iteration converge?

3. Graph the curve $y = g(x) = 1 + 2/x$ and the line $y = x$, plot the fixed point $(2, 2)$. Start with $p_0 = 4$ and construct $p_1 \approx 1.5$ and $p_2 \approx 2.33$ as indicated in Figure 2.4(b). Will fixed-point iteration converge?

4. Let $g(x) = 0.4 + x - 0.1x^2$ and consider the iteration (1).
   (a) Start with $p_0 = 1.9$ and find $p_1, p_2, \ldots, p_5$. Will this sequence converge to the fixed-point $P = 2$? Why?
   (b) Start with $p_0 = -1.9$ and find $p_1, p_2, \ldots, p_5$. Will this sequence converge to the fixed point $P = -2$? Why?
   (c) Find the errors E_k and relative errors R_k for the values p_k in parts (a) and (b).

5. Let $g(x) = x^2 + x - 4$. Can fixed-point iteration (1) be used to find the solution to the equation $x = g(x)$? Why?

6. Suppose that $g(x)$ and $g'(x)$ are defined and continuous on (a, b) and that p_0, p_1, p_2 lie in this interval and $p_1 = g(p_0)$, $p_2 = g(p_1)$. Also, assume that there exists a constant K so that $|g'(x)| < K$. Show that $|p_2 - p_1| < K|p_1 - p_0|$. *Hint.* Use the mean value theorem.

7. Graph the curve $y = g(x) = x^2/3$ and the line $y = x$, and plot the fixed point $(3, 3)$. Start with $p_0 = 3.5$ and construct $p_1 \approx 4.08$ and $p_2 \approx 5.56$ as indicated in Figure 2.5(a). Will fixed-point iteration converge?

8. Graph the curve $y = g(x) = 2 + 2x - x^2$ and the line $y = x$, and plot the fixed point $(2, 2)$. Start with $p_0 = 2.5$ and construct $p_1 = 0.75$ and $p_2 \approx 2.94$ as indicated in Figure 2.5(b). Will fixed-point iteration converge?

9. Suppose that $g(x)$ and $g'(x)$ are continuous on (a, b) and that $|g'(x)| > 1$ on this interval. If the fixed point P and the initial approximations p_0 and p_1 lie in the interval, then show that $p_1 = g(p_0)$ results in $|E_1| = |P - p_1| > |P - p_0| = |E_0|$. Hence statement (7) of Theorem 2.3 is established. {local divergence}

10. Let $g(x) = x - 0.0001x^2$ and $p_0 = 1$ and consider fixed-point iteration.
    (a) Show that $p_0 > p_1 > \cdots > p_n > p_{n+1} \cdots$.
    (b) Show that $p_n > 0$ for all n.
    (c) Since the sequence $\{p_n\}$ is decreasing and bounded below, it has a limit. What is the limit?
    (d) Use a calculator or computer and find $p_1, p_2, \ldots, p_5$.

11. Let $g(x) = 1.5 + 0.5x$ and $p_0 = 4$ and consider fixed-point iteration.
    (a) Show that the fixed point is $P = 3$.
    (b) Show that $|P - p_n| = |P - p_{n-1}|/2$ for $n = 1, 2, \ldots$.
    (c) Show that $|P - p_n| = |P - p_0|/2^n$ for $n = 1, 2, \ldots$.

12. Suppose that $g(x) = x/2$ and $p_k = 2^{-k}$.
    (a) Find the quantity $|p_{k+1} - p_k|/|p_{k+1}|$.
    (b) Discuss what will happen if the relative error stopping criterion above is used in Algorithm 2.1.

13. Investigate the nature of the iteration (1) when $g(x) = 3(x - 2.25)^{1/2}$.
    (a) Show that $P = 4.5$ is the only fixed point.
    (b) Use starting value $p_0 = 4.4$ and comute p_1, p_2, p_3, and p_4. What do you conjecture about the limit?
    (c) Use starting value $p_0 = 4.6$ and compute p_1, p_2, p_3, and p_4. What do you conjecture about the limit?

14. For the fixed-point iteration (1), discuss why it is an advantage to have $g'(P) \approx 0$.

15. Verify that $P = 3$ is a solution to the equation $P = g(P)$ and find the other real solutions for
   (a) $g(x) = (x + 9/x)/2$
   (b) $g(x) = 18x/(x^2 + 9)$
   (c) $g(x) = x^3 - 24$
   (d) $g(x) = 2x^3/(3x^2 - 9)$
   (e) $g(x) = 81/(x^2 + 18)$

16. Find $g'(x)$ and $g'(3)$ for the functions $g(x)$ in Exercise 15.

17. Start with $p_0 = 3.1$ and use the iteration (1) to find p_1 and p_2 for the functions $g(x)$ in Exercise 15. Will the iteration converge to the fixed point $P = 3$? Why?

18. Write a report on the contraction mapping theorem.

2.2 BRACKETING METHODS FOR LOCATING A ROOT

Consider the familiar topic of interest. Suppose that you save money by making regular monthly deposits P and the annual interest rate is I; then the total amount A after N deposits is

$$A = P + P\left(1 + \frac{I}{12}\right) + P\left(1 + \frac{I}{12}\right)^2 + \cdots + P\left(1 + \frac{I}{12}\right)^{N-1}. \tag{1}$$

The first term on the right side of equation (1) is the last payment. Then the next-to-last payment, which has earned one period of interest, contributes $P(1 + I/12)$. The second-from-last payment has earned two periods of interest and contributes $P(1 + I/12)^2$, and so on. Finally, the first payment, which has earned interest for $N - 1$ periods, contributes $P(1 + I/12)^{N-1}$ toward the total. Recall that the formula for the sum of the first N terms of a geometric series is

$$1 + r + r^2 + r^3 + \cdots + r^{N-1} = \frac{1 - r^N}{1 - r}. \tag{2}$$

We can write (1) in the form

$$A = P\left[1 + \left(1 + \frac{I}{12}\right) + \left(1 + \frac{I}{12}\right)^2 + \cdots + \left(1 + \frac{I}{12}\right)^{N-1}\right],$$

and use the substitution $r = (1 + I/12)$ in (2) to obtain

$$A = P\frac{1 - (1 + I/12)^N}{1 - (1 + I/12)}.$$

This can be simplified to obtain the annuity-due equation,

$$A = \frac{P}{I/12}\left[\left(1 + \frac{I}{12}\right)^N - 1\right]. \tag{3}$$

The following example uses the annuity-due equation and requires a sequence of repeated calculations to find an answer.

Example 2.5

You save $250 per month for 20 years and desire that the total value of all payments and interest is $250,000 at the end of the 20 years. What interest rate I is needed to achieve your goal?

Solution.   If we hold $N = 240$ fixed, then A is a function of I alone; that is, $A = A(I)$. We will start with two guesses, $I_0 = 0.12$ and $I_1 = 0.13$, and perform a sequence of calculations to narrow down the final answer. Starting with $I_0 = 0.12$ yields

$$A(0.12) = \frac{250}{0.12/12} \left[\left(1 + \frac{0.12}{12} \right)^{240} - 1 \right] = 247{,}314.$$

Since this value is a little short of the goal, we next try $I_1 = 0.13$:

$$A(0.13) = \frac{250}{0.13/12} \left[\left(1 + \frac{0.13}{12} \right)^{240} - 1 \right] = 283{,}311.$$

This is a little high, so we try the value in the middle, $I_2 = 0.125$:

$$A(0.125) = \frac{250}{0.125/12} \left[\left(1 + \frac{0.125}{12} \right)^{240} - 1 \right] = 264{,}623.$$

This is again high and we conclude that the desired rate lies in the interval $[0.12, 0.125]$. The next guess is the midpoint $I_3 = 0.1225$:

$$A(0.1225) = \frac{250}{0.1225/12} \left[\left(1 + \frac{0.1225}{12} \right)^{240} - 1 \right] = 255{,}803.$$

This is high and the interval is now narrowed down to $[0.12, 0.1225]$. Our last calculation uses the midpoint approximation, $I_4 = 0.12125$:

$$A(0.12125) = \frac{250}{0.12125/12} \left[\left(1 + \frac{0.12125}{12} \right)^{240} - 1 \right] = 251{,}518.$$

Further iterations can be done to obtain as many significant digits as required. The purpose of this example was to find the value of I that produced a specified level L of the function value, that is to find a solution to $A(I) = L$. It is standard practice to place the constant L on the left and solve the equation $A(I) - L = 0$.

Definition 2.3   (Root of an Equation, Zero of a Function).   Assume that $f(x)$ is a continuous function. Any number r for which $f(r) = 0$ is called a **root of the equation** $f(x) = 0$. Also, we say that r is a **zero of the function** $f(x)$.

For example, the equation $2x^2 + 5x - 3 = 0$ has two real roots $r_1 = 0.5$ and $r_2 = -3$, whereas the corresponding function $f(x) = 2x^2 + 5x - 3 = (2x - 1)(x + 3)$ has two real zeros, $r_1 = 0.5$ and $r_2 = -3$.

The Bisection Method of Bolzano

In this section we develop our first bracketing method for finding a zero of a continuous function. We must start with an initial interval $[a, b]$, where $f(a)$ and $f(b)$ have opposite

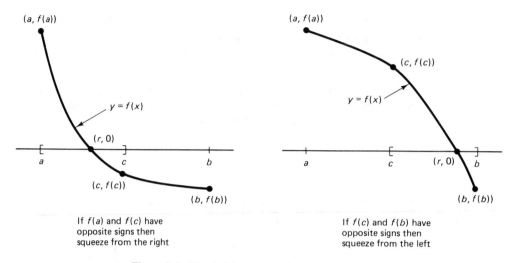

If $f(a)$ and $f(c)$ have opposite signs then squeeze from the right

If $f(c)$ and $f(b)$ have opposite signs then squeeze from the left

Figure 2.6   The decision process for the bisection method.

signs. Since the graph $y = f(x)$ of a continuous function is unbroken, it will cross the x-axis at a root $x = r$ that lies somewhere in the interval (see Figure 2.6). The bisection method systematically moves the endpoints of the interval closer and closer together until we obtain an interval of arbitrarily small width that brackets the root. The decision step for this process of interval halving is first to choose the midpoint $c = (a + b)/2$, and then analyze the three possibilities that might arise:

$$\text{If } f(a) \text{ and } f(c) \text{ have opposite signs, a root lies in } [a, c]. \tag{4}$$

$$\text{If } f(c) \text{ and } f(b) \text{ have opposite signs, a root lies in } [c, b]. \tag{5}$$

$$\text{If } f(c) = 0, \text{ we found a root at } x = c. \tag{6}$$

If either of cases (4) or (5) occur, we have found an interval half as wide as the original interval that contains the root, and we are "squeezing down on it" (see Figure 2.6). To continue the process, relabel the new smaller interval $[a, b]$ and repeat the process until the interval width is as small as desired. Since the bisection process involves sequences of nested intervals and their midpoints, we will use the following notation to keep track of the details in this process:

$[a_0, b_0]$ is the starting interval and c_0 is the midpoint, $c_0 = \dfrac{a_0 + b_0}{2}$.

$[a_1, b_1]$ is the second interval which brackets the root r and c_1 is its midpoint; the interval $[a_1, b_1]$ is half as wide as $[a_0, b_0]$. $\qquad(7)$

After arriving at the kth interval, $[a_n, b_n]$, which brackets r and has midpoint c_n, the interval $[a_{n+1}, b_{n+1}]$ is constructed, which also brackets r and is half as wide as $[a_n, b_n]$.

It is left for an exercise for the reader to show that the sequence of left endpoints is increasing and the sequence of right endpoints is decreasing; that is,

$$a_0 \le a_1 \le \cdots \le a_n \le a_{n+1} \le \cdots \le r$$
$$r \le \cdots \le b_{n+1} \le b_n \le \cdots \le b_1 \le b_0, \tag{8}$$

where

$$c_n = \frac{a_n + b_n}{2} \quad \text{and either} \tag{9}$$

$$[a_{n+1}, b_{n+1}] = [a_n, c_n] \quad \text{or} \quad [a_{n+1}, b_{n+1}] = [c_n, b_n] \quad \text{for all } n.$$

Theorem 2.4 (Bisection Theorem).   Assume that $f \in C[a, b]$ and that there exists a number $r \in [a, b]$ such that $f(r) = 0$. If $f(a)$ and $f(b)$ have opposite signs, and $\{c_n\}_{n=0}^{\infty}$ represents the sequence midpoints generated by the bisection process of (8) and (9), then

$$|r - c_n| \le \frac{b - a}{2^{n+1}} \quad \text{for } n = 0, 1, \ldots, \tag{10}$$

and therefore the sequence $\{c_n\}_{n=0}^{\infty}$ converges to the root $x = r$, that is,

$$\lim_{n \to \infty} c_n = r. \tag{11}$$

Proof. Since both the root r and the midpoint c_n lie in the interval $[a_n, b_n]$, the distance from c_n and r cannot be greater that half the width of this interval (see Figure 2.7). Thus

$$|r - c_n| \le \frac{|b_n - a_n|}{2} \quad \text{for all } n. \tag{12}$$

Observe that the successive interval widths form the pattern

$$|b_1 - a_1| = \frac{|b_0 - a_0|}{2^1},$$

$$|b_2 - a_2| = \frac{|b_1 - a_1|}{2} = \frac{|b_0 - a_0|}{2^2}.$$

It is left as an exercise to use mathematical induction and show that

$$|b_n - a_n| = \frac{|b_0 - a_0|}{2^n}. \tag{13}$$

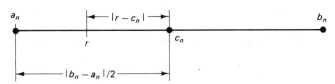

Figure 2.7   The root r and midpoint c_n of $[a_n, b_n]$ for the bisection method.

Combining (12) and (13) results in

$$|r - c_n| \le \frac{|b_0 - a_0|}{2^{n+1}} \quad \text{for all } n. \tag{14}$$

Now an argument similar to the one given in Theorem 2.3 can be used to show that (14) implies that the sequence $\{c_n\}_{n=0}^{\infty}$ converges to r and the proof of the theorem is complete.

Example 2.6

The function $h(x) = x \sin(x)$ occurs in the study of undamped forced oscillation. Find the value of x that lies in the interval $[0, 2]$, where the function takes on the value $h(x) = 1$. [The function $\sin(x)$ is evaluated in radians.]

Solution.   We use the bisection method to find a zero of the function $f(x) = x \sin(x) - 1$. Starting with $a_0 = 0$ and $b_0 = 2$, we compute

$$f(0) = -1.000000 \quad \text{and} \quad f(2) = 0.818595,$$

so that a root of $f(x) = 0$ lies in the interval $[0, 2]$. At the midpoint $c_0 = 1$, we find that $f(1) = -0.158529$. Hence the function changes sign on $[c_0, b_0] = [1, 2]$.

To continue, we squeeze from the left and set $a_1 = c_0$ and $b_1 = b_0$. The midpoint is $c_1 = 1.5$ and $f(c_1) = 0.496242$. Now, $f(1) = -0.158529$ and $f(1.5) = 0.496242$ imply that the root lies in the interval $[b_1, c_1] = [1.0, 1.5]$. The next decision is to squeeze from the right and set $a_2 = a_1$ and $b_2 = c_1$. In this manner we obtain a sequence $\{c_k\}$ that converges to $r \approx 1.114157141$. A sample of the calculations is given in Table 2.1.

TABLE 2.1   Bisection Method Solution of $x \sin(x) - 1 = 0$

k	Left endpoint, a_k	Midpoint, c_k	Right endpoint, b_k	Function value, $f(c_k)$
0	0.	1.	2.	-0.158529
1	1.0	1.5	2.0	0.496242
2	1.00	1.25	1.50	0.186231
3	1.000	1.125	1.250	0.015051
4	1.0000	1.0625	1.1250	-0.071827
5	1.06250	1.09375	1.12500	-0.028362
6	1.093750	1.109375	1.125000	-0.006643
7	1.1093750	1.1171875	1.1250000	0.004208
8	1.10937500	1.11328125	1.11718750	-0.001216
.	.	.	.	.
.	.	.	.	.
.	.	.	.	.

A virtue of the bisection method is that formula (7) provides a predetermined estimate for the accuracy of the computed solution. In Example 2.6 the width of the starting interval was $b_0 - a_0 = 2$. Suppose that Table 2.1 were continued to the thirty-first iterate; then by (7) the error bound would be $|e_{31}| \le (2 - 0)/2^{32} \approx 4.656612 \times 10^{-10}$. Hence c_{31} would be an approximation to r with nine decimal places of accuracy.

The number N of repreated bisections needed to guarantee that the Nth midpoint c_N is an approximation to a root and has an error less than the preassigned value Delta is

$$N = \text{int}\left(\frac{\ln(B - A) - \ln(\text{Delta})}{\ln(2)}\right). \qquad (15)$$

The proof of this formula is left as an exercise (see Exercise 19).

Another popular algorithm is the **method of false position** or the **regula falsi method**. It was developed because the bisection method converges at a fairly slow speed. As before, we assume that $f(a)$ and $f(b)$ have opposite signs. The bisection method used the midpoint of the interval $[a, b]$ as the next iterate. A better approximation is obtained if we find the point $(c, 0)$ where the straight line L joining the points $(a, f(a))$ and $(b, f(b))$ crosses the x-axis (Figure 2.8). To find the value c, we write down two versions of the slope m of the line L:

$$m = \frac{f(b) - f(a)}{b - a}, \qquad (16)$$

where the points $(a, f(a))$, $(b, f(b))$ are used, and

$$m = \frac{0 - f(b)}{c - b}, \qquad (17)$$

where the points $(c, 0)$ and $(b, f(b))$ are used.

Equating the slopes in (16) and (17) we have

$$\frac{f(b) - f(a)}{b - a} = \frac{0 - f(b)}{c - b},$$

which is easily solved for c to get

$$c = b - \frac{f(b)(b - a)}{f(b) - f(a)}. \qquad (18)$$

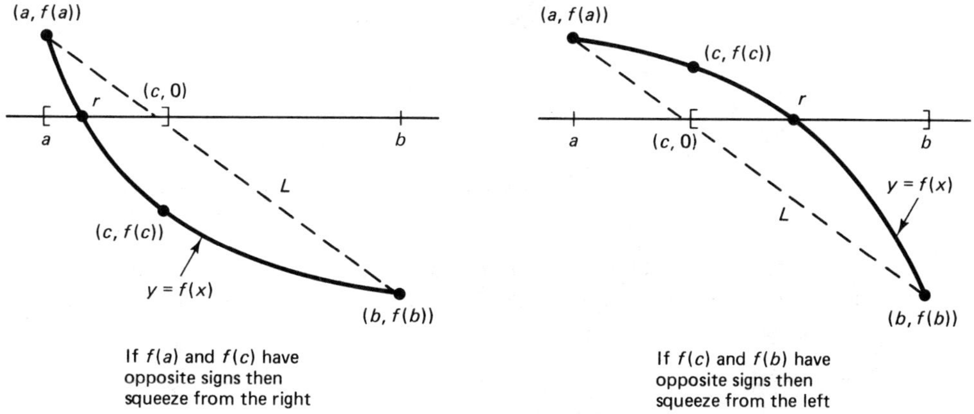

If $f(a)$ and $f(c)$ have opposite signs then squeeze from the right

If $f(c)$ and $f(b)$ have opposite signs then squeeze from the left

Figure 2.8 The decision process for the false position method.

The three possibilities are the same as before:

$f(a)$ and $f(c)$ have opposite signs and a root lies in $[a, c]$,

or   (19)

$f(c)$ and $f(b)$ have opposite signs and a root lies in $[c, b]$,

or

$f(c) = 0$ and you found a root.   (20)

Convergence of the False Position Method

The decision process implied by (19) together with (18) is used to construct a sequence of intervals $\{[a_n, b_n]\}$ each of which brackets the root. At each step the approximation to the root r is

$$c_n = b_n - \frac{f(b_n)(b_n - a_n)}{f(b_n) - f(a_n)},$$ (21)

and it can be proven that the sequence $\{c_n\}$ will converge to r. But beware, although the interval width $b_n - a_n$ is getting smaller, it is possible that it may not go to zero. If the graph $y = f(x)$ is concave near $(r, 0)$, one of the endpoints becomes fixed and the other one marches into the solution (Figure 2.9).

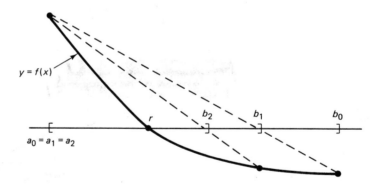

Figure 2.9   The stationary endpoint for the false position method.

Now we rework the solution to $x \sin(x) - 1 = 0$ using the method of false position and observe that it converges faster than the bisection method. Also, notice that $b_n - a_n$ does not go to zero.

Example 2.7

Use the false position method to find the root of $x \sin(x) - 1 = 0$ that is located in the interval $[0, 2]$. [The functions $\sin(x)$ is evaluated in radians.]

Solution.   Starting with $a_0 = 0$ and $b_0 = 2$, we have $f(0) = -1.00000000$ and $f(2) = 0.81859485$, so that a root lies in the interval $[0, 2]$. Using formula (21) , we get

$$c_0 = 2 - \frac{0.81859485(2 - 0)}{0.81859485 - (-1)} = 1.09975017 \text{ and } f(c_0) = -0.02001921.$$

The function changes sign on the interval $[c_0, b_0] = [1.09975017, 2]$, so we squeeze from the left and set $a_1 = c_0$ and $b_1 = b_0$. Formula (21) produces the next approximation:

$$c_1 = 2 - \frac{0.81859485(2 - 1.09975017)}{0.81859485 - (-0.02001921)} = 1.12124074 \text{ and } f(c_1) = 0.00983461.$$

Now $f(x)$ changes sign on $[a_1, c_1] = [1.09975017, 1.12124074]$, and the next decision is to squeeze from the right and set $a_2 = a_1$, $b_2 = c_1$. A summary of the calculations is given in Table 2.2.

TABLE 2.2   False Position Method Solution of $x \sin(x) - 1 = 0$

k	Left endpoint, a_k	Point of intersection, c_k	Right endpoint, b_k	Function value, $f(c_k)$
0	0.00000000	1.09975017	2.00000000	−0.02001921
1	1.09975017	1.12124074	2.00000000	0.00983461
2	1.09975017	1.11416120	1.12124074	0.00000563
3	1.09975017	1.11415714	1.11416120	0.00000000

The termination criterion used in the bisection method is not useful for the false position method and may result in an infinite loop. The closeness of consecutive iterates and the size of $|f(c_n)|$ are both used in the termination criterion for Algorithm 2.3. In Section 2.3 we discuss the reasons for this choice.

Algorithm 2.2   (Bisection Method).   To find a root of the equation $f(x) = 0$ in the interval $[a, b]$. Proceed with the method only if $f(x)$ is continuous and $f(a)$ and $f(b)$ have opposite signs.

Delta := 10^{-6} {Tolerance for width of interval}
Satisfied := "False"   {Condition for loop termination}

INPUT A, B   {Input endpoints of interval}
YA := F(A), YB := F(B)   {Compute function values}
Max := $1 + \text{INT}([\ln(B-A) - \ln(\text{Delta})]/\ln(2))$ {Calculation of the
                                                                                   maximum number of iterations}

IF   SIGN(YA) = SIGN(YB)   THEN   {Check to see if
 | PRINT   'The values f(a) and f(b)'   the bisection
 | PRINT   'do not differ in sign.'   method
 |__ TERMINATE THE ALGORITHM   applies}

```
DO  FOR  K = 1  TO  Max  UNTIL  Satisfied = "True"
    C := (A+B)/2                                          {Midpoint of interval}
    YC := F(C)                                            {Function value at midpoint}

        IF YC = 0  THEN
            A := C  and  B := C                           {Exact root is found}
            ELSEIF SIGN(YB)=SIGN(YC)  THEN
                B := C                                    {Squeeze from the right}
                YB := YC
            ELSE
                A := C                                    {Squeeze from the left}
                YA := YC
        ENDIF
    IF  B − A < Delta  THEN                               {Check for early
        Satisfied := "True"                                convergence}
END

    PRINT 'The computed root of f(x) = 0 is' C           {Output}
    PRINT 'The accuracy is +−' B−A
    PRINT 'The value of the function f(C) is' YC
```

Algorithm 2.3   (False Position or Regula Falsi Method).   To find a root of the equation $f(x) = 0$ in the interval $[a, b]$. Proceed with the method only if $f(x)$ is continuous and $f(a)$ and $f(b)$ have opposite signs.

```
    Delta := 10⁻⁶                        {Closeness for consecutive iterates}
    Epsilon := 10⁻⁶                      {Tolerance for the size of f(C)}
    Max := 199                           {Maximum number of iterations}
    Satisfied := "False"                 {Condition for loop termination}

    INPUT  A, B                          {Input the endpoints of the interval}
    YA := F(A), YB := F(B)               {Compute the function value}
```

```
DO  FOR K = 1 TO Max UNTIL Satisfied = "True"
      DX := YB*[B − A]/[YB − YA]                        {Change in iterate}
      C := B − DX                                        {New iterate}
      YC := F(C)                                         {Function value}
        IF YC = 0   THEN
            Satisfied = "True"                           {Exact root is found}
          ELSEIF SIGN(YB)=SIGN(YC)   THEN
              B := C, YB := YC                           {Squeeze from the right}
          ELSE
              A := C, YA := YC                           {Squeeze from the left}
        ENDIF
      IF |DX|<Delta  AND  |YC|<Epsilon  THEN             {Check for
        Satisfied = "True"                                convergence}
END
```

PRINT 'The computed root of f(x) = 0 is' C   {Output}
PRINT 'Consecutive iterates differ by' DX
PRINT 'The value of the function f(C) is' YC

EXERCISES FOR BRACKETING METHODS

In Exercises 1–3, find an approximation for the interest rate I that will yield the total annuity value A if 240 monthly payments P are made. Use the two starting values for I and compute the next three approximations using the bisection method.

1. Monthly payment $P = \$275$, annuity value $A = \$250,000$.
   Use the starting values $I = 0.11$ and 0.12.

2. Monthly payment $P = \$325$, annuity value $A = \$400,000$.
   Use the starting values $I = 0.13$ and 0.14.

3. Monthly payment $P = \$300$, annuity values $A = \$500,000$.
   Use the starting values $I = 0.15$ and 0.16.

4. Find an interval $a \le x \le b$ so that $f(a)$ and $f(b)$ have opposite signs for the following functions.
   (a) $f(x) = \exp(x) - 2 - x$
   (b) $f(x) = \cos(x) + 1 - x$ (x is in radians)
   (c) $f(x) = \ln(x) - 5 + x$
   (d) $f(x) = x^2 - 10x + 23$

In Exercises 5–8, start with the interval $[a_0, b_0]$ and use the bisection method to find an interval of width 0.05 that contains a solution of the given equation.

5. (a) $0 = \exp(x) - 2 - x$, $[a_0, b_0] = [1.0, 1.8]$
   (b) $0 = \exp(x) - 2 - x$, $[a_0, b_0] = [-2.4, -1.6]$

6. $0 = \cos(x) + 1 - x$, $[a_0, b_0] = [0.8, 1.6]$ (x is in radians)

7. $0 = \ln(x) - 5 + x$, $[a_0, b_0] = [3.2, 4.0]$

8. (a) $0 = x^2 - 10x + 23$, $[a_0, b_0] = [3.2, 4.0]$
   (b) $0 = x^2 - 10x + 23$, $[a_0, b_0] = [6.0, 6.8]$

9. Use a computer program and find the roots of $f(x) = 0$ accurate to 5×10^{-6} using the bisection method. Test your program using the functions in Exercise 4.

In Exercises 10–13, start with $[a_0, b_0]$ and use the false position method to compute $c_0, c_1, c_2,$ and c_3. Be sure to compute the function carefully.

10. $\exp(x) - 2 - x = 0$, $[a_0, b_0] = [-2.4, -1.6]$

11. $\cos(x) + 1 - x = 0$, $[a_0, b_0] = [0.8, 1.6]$ (x is in radians)

12. $\ln(x) - 5 + x = 0$, $[a_0, b_0] = [3.2, 4.0]$

13. $x^2 - 10x + 23 = 0$, $[a_0, b_0] = [6.0, 6.8]$

14. Use a computer program and find the roots of $f(x) = 0$ accurate to 5×10^{-6} using the false position method. Test your program using the functions in Exercise 4.

15. Denote the intervals that arise in the bisection method by $[a_0, b_0], [a_1, b_1], \ldots ,$ $[a_n, b_n], \ldots$.
   (a) Show that $a_0 \le a_1 \le \ldots \le a_n \le a_{n+1}$ and that $b_{n+1} \le b_n \le \ldots \le b_1 \le b_0$.
   (b) Show that $b_n - a_n = (b_0 - a_0)/2^n$.
   (c) Let the midpoint of each interval be $c_n = (a_n + b_n)/2$. Show that

$$\lim_{n \to \infty} a_n = \lim_{n \to \infty} c_n = \lim_{n \to \infty} b_n.$$

   Hint. Review convergence of monotone sequences in your calculus book.

16. What will happen if the bisection method is used with the function $f(x) = 1/(x - 2)$ and
   (a) the interval is [3, 7]?   **(b)** the interval is [1, 7]?

17. What will happen if the bisection method is used with the function $f(x) = \tan(x)$ (x is in radians) and
   (a) the interval is [3, 4]?   **(b)** the interval is [1, 3]?

18. Suppose that the bisection method is used to find a zero of $f(x)$ in the interval [2, 7]. How many times must this interval be bisected to guarantee that the approximation c_N has an accuracy of 5×10^{-9}?

19. Establish formula (15) for determining the number of iterations required in the bisection method. *Hint.* Use $|B - A|/2^{n+1} < \text{Delta}$ and take logarithms.

20. Show that formula (21) for the false position method is algebraically equivalent to

$$c_n = \frac{a_n f(b_n) - b_n f(a_n)}{f(b_n) - f(a_n)}.$$

21. What are the differences between the bisection method and the regula falsi method?

22. Write a report on the modified regula falsi method (sometimes referred to as the "modified false position" method).

23. Consider a spherical ball of radius $r = 15$ cm that is constructed from a variety of white oak which has a density of $\rho = 0.710$. How much of the ball will be submerged when it is placed in water? *Hint.* Use the formula $\pi(d^3 - 3d^2 r + 4r^3 \rho)/3 = 0$.

2.3 INITIAL APPROXIMATIONS AND CONVERGENCE CRITERIA

The bracketing methods depend on finding an interval $[a, b]$ so that $f(a)$ and $f(b)$ have opposite signs. Once the interval has been found, no matter how large, the iteration will proceed until a root is found. Hence these methods are called **globally convergent**. However, if $f(x) = 0$ has several roots in $[a, b]$, then a different starting interval must be used to find each root. It is not easy to locate these smaller intervals on which $f(x)$ changes sign.

In Section 2.4 we develop the Newton–Raphson method and the secant method for solving $f(x) = 0$. Both of these methods require that a close approximation to the root be given to guarantee convergence. Hence these methods are called **locally convergent**. They usually converge more rapidly than do global ones. Some hybrid algorithms start with a global method and switch to a local method when the iteration gets close to a root.

If the computation of roots is one part of a larger project, then a leisurely pace is suggested and the first thing to do is graph the function. If this is done by hand, it may be easier to rearrange the equation $f(x) = 0$ into an equivalent form $g(x) = h(x)$, where the graphs of $y = g(x)$ and $y = h(x)$ are easy to sketch.

For example, the curve $y = f(x) = x^2 - \exp(x)$ is difficult to sketch [see Figure 2.10(a)], but the functions $g(x) = x^2$ and $h(x) = \exp(x)$ are easy to graph and the abscissa of their point of intersection is the desired root p [see Figure 2.10(b)].

Suppose that a computer is used to graph $y = f(x)$ on $[a, b]$. The interval must be partitioned $a = x_0 < x_1 < \cdots < x_N = b$ and the function values $y_k = f(x_k)$ computed. Then the line segments between consecutive points (x_{k-1}, y_{k-1}) and (x_k, y_k) are plotted for $k = 1, 2, \ldots, N$. This process could take a significant amount of time, but there is something to gain. We can view the graph $y = f(x)$ and make decisions based on what it looks like. But more important, if the coordinates of the points (x_k, y_k) are stored, then they can be analyzed and the approximate location of roots determined.

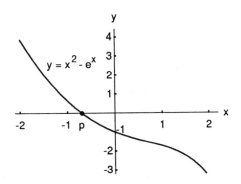

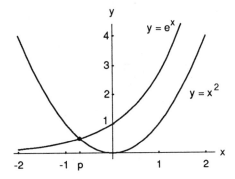

Figure 2.10   (a) The graph of $f(x) = x^2 - e^x$ and the root $p \approx -0.7$.

Figure 2.10   (b) The graphs of $g(x) = x^2$ and $h(x) = e^x$.

We must proceed carefully. There must be enough points so that we do not miss a root in a portion of the curve where the function is changing rapidly. If $f(x)$ is continuous and two adjacent points (x_{k-1}, y_{k-1}) and (x_k, y_k) lie on opposite sides of the x-axis, then the intermediate value theorem implies that at least one root lies in the interval $[x_{k-1}, x_k]$ and $(x_{k-1} + x_k)/2$ is an approximate root.

Near two closely spaced roots or near a double root, the line segment between (x_{k-1}, y_{k-1}) and (x_k, y_k) may fail to cross the axis. If $|f(x_k)|$ is smaller than a preassigned value ϵ, then x_k is a tentative approximate root. But the graph may be close to zero over a wide range of values near a double root, and we may get too many root approximations. Hence we add the requirement that the slope changes sign near (x_k, y_k), that is,

$$m_{k-1} = \frac{y_k - y_{k-1}}{x_k - x_{k-1}} \quad \text{and} \quad m_k = \frac{y_{k+1} - y_k}{x_{k+1} - x_k}$$

must have opposite signs. Since $x_k - x_{k-1} > 0$ and $x_{k+1} - x_k > 0$, it is not necessary to use the difference quotients and it will suffice to check to see if the differences $y_k - y_{k-1}$ and $y_{k+1} - y_k$ change sign. In this case x_k is the approximate root. Unfortunately, we cannot guarantee that this starting value will produce a convergent sequence. If the graph $y = f(x)$ has a local minimum (or maximum) that is extremely close to zero, then it is possible that x_k is reported as an approximate root when $f(x_k) \approx 0$, although x_k may not be close to a root.

Example 2.8

Find the approximate location of the roots of $x^3 - x^2 - x + 1 = 0$ on the interval $[-1.2, 1.2]$.

Solution.   For illustration, choose $N = 8$ and look at Table 2.3.

TABLE 2.3 Finding Approximate Locations for Roots

x_k	Function values		Differences in y		Significant changes in $f(x)$ or $f'(x)$
	y_{k-1}	y_k	$y_k - y_{k-1}$	$y_{k+1} - y_k$	
-1.2	-3.125	-0.968	2.157	1.329	
-0.9	-0.968	0.361	1.329	0.663	f changes sign in $[x_{k-1}, x_k]$
-0.6	0.361	1.024	0.663	0.159	
-0.3	1.024	1.183	0.159	-0.183	f' changes sign near x_k
0.0	1.183	1.000	-0.183	-0.363	
0.3	1.000	0.637	-0.363	-0.381	
0.6	0.637	0.256	-0.381	-0.237	
0.9	0.256	0.019	-0.237	0.069	f' changes sign near x_k
1.2	0.019	0.088	0.069	0.537	

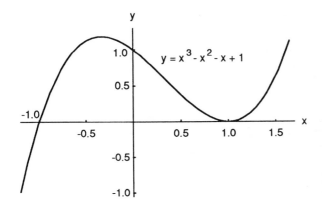

Figure 2.11 The graph of the cubic polynomial $y = x^3 - x^2 - x + 1$.

The three abscissas for consideration are -1.05, -0.3, and 0.9. Since $f(x)$ changes sign on the interval $[-1.2, -0.9]$, the value -1.05 is an approximate root; indeed, $f(-1.05) \approx -0.210$. Although the slope changes sign near -0.3, we find that $f(-0.3) = 1.183$; hence -0.3 is *not* near a root. Finally, the slope changes sign near 0.9 and $f(0.9) = 0.019$, so that 0.9 is an approximate root (Figure 2.11).

Checking for Convergence

A graph can be used to see the approximate location of a root, but an algorithm must be used to compute a value p_n that is an acceptable computer solution. Iteration is often used to produce a sequence $\{p_k\}$ that converges to a root p, and a termination criterion or strategy must be designed ahead of time so that the computer will stop when an accurate approximation is reached. Since the goal is to solve $f(x) = 0$, the final value p_n should have the property that $|f(p_n)| < \epsilon$.

The user can supply a tolerance value ϵ for the size of $|f(p_n)|$ and then an iterative process produces points $P_k = (p_k, f(p_k))$ until the last point P_n lies in the horizontal band bounded by the lines $y = +\epsilon$ and $y = -\epsilon$, as shown in Figure 2.12(a). This criterion is

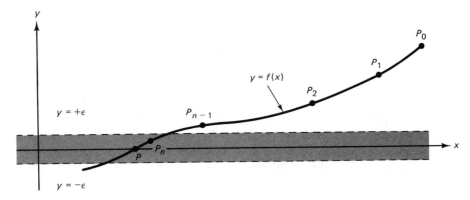

Figure 2.12   (a) The horizontal convergence band for locating a solution to $f(x) = 0$.

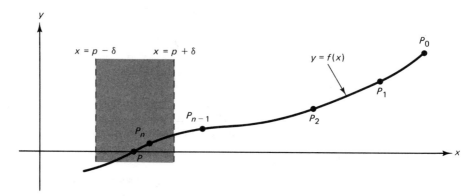

Figure 2.12   (b) The vertical convergence band for locating a solution to $f(x) = 0$.

useful if the user is trying to solve $h(x) = L$ by applying a root-finding algorithm to the function $f(x) = h(x) - L$.

Another termination criterion involves the abscissas, and we can try to determine if the sequence $\{p_k\}$ is converging. If we draw the vertical lines $x = p + \delta$ and $x = p - \delta$ on each side of $x = p$, we could decide to stop the iteration when the point P_n lies between these vertical lines, as shown in Figure 2.12(b).

The latter criterion is often desired, but it is difficult to implement because it involves the unknown solution p. We adapt this idea, and terminate further calculations when the consecutive iterates p_n and p_{n-1} are sufficiently close or if they agree within M significant digits.

Sometimes the user of an algorithm will be satisfied if $p_n \approx p_{n-1}$ and other times when $f(p_n) \approx 0$. Correct logical thinking is required to understand the consequences. If we require that $|p_n - p| < \delta$ and $|f(p_n)| < \epsilon$, the point P_n will be located in the rectangular region about the solution $(p, 0)$, as is shown in Figure 2.13(a). If we stipulate that

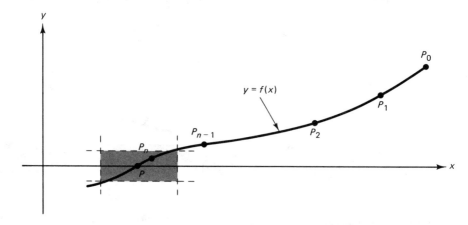

Figure 2.13   (a) The rectangular region defined by $|x - p| < \delta$ AND $|y| < \epsilon$.

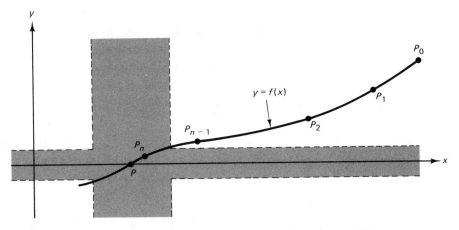

Figure 2.13 (b) The unbounded region defined by $|x - p| < \delta$ OR $|y| < \epsilon$.

$|p_n - p| < \delta$ or $|f(p)| < \epsilon$, the point P_n could be located anywhere in the region formed by the union of the horizontal and vertical strips, as shown in Figure 2.13(b).

The size of the tolerances δ and ϵ are crucial. If the tolerances are chosen too small, iteration may continue forever. They should be chosen about 100 times larger than 10^{-M}, where M is the number of decimal digits in the computer's floating-point numbers. The closeness of the abscissas is checked with one of the criteria

$$|p_n - p_{n-1}| < \delta \quad \text{(estimate for the absolute error)}$$

or

$$\frac{2|p_n - p_{n-1}|}{|p_n| + |p_{n-1}|} < \delta \quad \text{(estimate for the relative error).}$$

The closeness of the ordinate is usually checked by

$$|f(p_n)| < \epsilon.$$

Troublesome Functions

A computer solution to $f(x) = 0$ will almost always be in error due to round-off and/or instability in the calculations. If the graph $y = f(x)$ is steep near the root $(p, 0)$, then the root-finding problem is well conditioned (i.e., a solution with several significant digits is easy to obtain). If the graph $y = f(x)$ is shallow near $(p, 0)$, then the root-finding problem is ill-conditioned (i.e., the computed root may have only a few significant digits). This occurs quite often when $f(x)$ has a multiple root at p. This is discussed further in the next section.

Algorithm 2.4 (Approximate Location of Roots). To roughly estimate locations of the roots of the equation $f(x) = 0$ over the interval $[a, b]$, by using the equally spaced sample points $(x_k, f(x_k))$ and the following criteria:

(i) $y_{k-1}y_k < 0$, or
(ii) $|y_k| < \epsilon$ and $(y_k - y_{k-1})(y_{k+1} - y_k) < 0$.

That is, either $f(x_{k-1})$ and $f(x_k)$ have opposite signs, or $|f(x_k)|$ is small and the slope of the curve $y = f(x)$ changes sign near $(x_k, f(x_k))$.

```
Epsilon := 10⁻²                          {Tolerance for |f(xₖ)|}
INPUT  N                                 {Number of subintervals}
INPUT A, B                               {Endpoints of interval}
H := (B−A)/N                             {Subinterval width}

FOR K = 0  TO  N   DO
    X(K) := A + H∗K                      {Compute the abscissas}
    Y(K) := F(X(K))                      {Compute the ordinates}
M := 0                                   {Counter for the number of roots}
Y(N+1) := Y(N)                           {This permits one loop}

FOR  K = 1  TO  N  DO
  IF  Y(K−1)∗Y(K) ≦ 0  THEN              {Check for a change in
      M := M+1                            sign or finds a root}
      R(M) := [X(K−1) + X(K)]/2          {Approximate root found}
  S := [Y(K)−Y(K−1)]∗[Y(K+1)−Y(K)]
  IF  |Y(K)| < Epsilon AND S < 0 THEN    {Small function value
      M := M+1                            and slope changes sign}
      R(M) := X(K)                       {Tentative root found}

PRINT 'The approximate location of the roots are'
FOR  K = 1  TO  M  DO
    PRINT  R(K)
```

EXERCISES FOR INITIAL APPROXIMATIONS AND CONVERGENCE CRITERIA

In Exercises 1–6, determine functions $y = g(x)$ and $y = h(x)$ so that $g(x) = h(x)$ is equivalent to $0 = f(x)$; sketch the two curves $y = g(x)$ and $y = h(x)$ and determine the approximate location of the roots of $0 = f(x)$ graphically. If a computer plotter is available, then plot $y = f(x)$ directly.

1. $f(x) = x^2 - \exp(x)$ for $-2 \le x \le 2$.

2. $f(x) = x - \cos(x)$ for $-2 \leq x \leq 2$.
3. $f(x) = \sin(x) - 2\cos(2x)$ for $-2 \leq x \leq 2$.
4. $f(x) = \cos(x) + (1 + x^2)^{-1}$ for $-2 \leq x \leq 2$.
5. $f(x) = (x - 2)^2 - \ln(x)$ for $0.5 \leq x \leq 4.5$.
6. $f(x) = 2x - \tan(x)$ for $-1.4 \leq x \leq 1.4$.

7. A computer program that plots the graph of $y = f(x)$ over the interval $[a, b]$ using the points $(x_0, y_0), (x_1, y_1), \ldots , (x_N, y_N)$ usually scales the vertical height of the graph and a procedure must be written to determine the minimum and maximum values.

(a) Write an algorithm that will find the values

$$Y_{max} = \max_{k}\{y_k\} \quad \text{and} \quad Y_{min} = \min_{k}\{y_k\}.$$

(b) Write an algorithm that will find the approximate location of the extreme values of $f(x)$ on the interval $[a, b]$.

2.4 NEWTON–RAPHSON AND SECANT METHODS

Slope Methods for Finding Roots

If $f(x), f'(x)$, and $f''(x)$ are continuous near a root p, then this extra information regarding the nature of $f(x)$ can be used to develop algorithms that will produce sequences $\{p_k\}$ that converge faster to p than either the bisection or false position methods. The Newton–Raphson (or simply Newton's) method is one of the most useful and best known algorithms that rely on the continuity of $f'(x)$ and $f''(x)$. We shall introduce it graphically and then give a more rigorous treatment based on the Taylor polynomial.

Assume that the initial approximation p_0 is near the root p. Then the graph $y = f(x)$ intersects the x-axis at the point $(p, 0)$ and the point $(p_0, f(p_0))$ lies on the curve near the point $(p, 0)$ (see Figure 2.14). Define p_1 to be the point of intersection of the line tangent to the curve at the point $(p_0, f(p_0))$ and the x-axis. Then Figure 2.14 shows that p_1 will

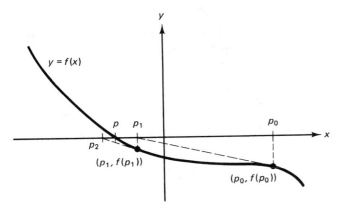

Figure 2.14   The geometric construction of p_1 and p_2 for the Newton–Raphson method.

be closer to p than p_0 in this case. An equation relating p_1 and p_0 can be found if we write down two versions for the slope of the tangent line L;

$$m = \frac{0 - f(p_0)}{p_1 - p_0}, \tag{1}$$

which is the slope of the line through $(p_1, 0)$ and $(p_0, f(p_0))$, and

$$m = f'(p_0), \tag{2}$$

which is the slope at the point $(p_0, f(p_0))$. Equating the values of the slope m in equations (1) and (2) and solving for p_1 results in

$$p_1 = p_0 - \frac{f(p_0)}{f'(p_0)}. \tag{3}$$

The process above can be repeated to obtain a sequence $\{p_k\}$ that converges to p. We now make these ideas more precise.

Theorem 2.5   (Newton–Raphson Theorem).   Assume that $f \in C^2[a, b]$ and there exists a number $p \in [a, b]$ where $f(p) = 0$. If $f'(p) \neq 0$, then there exists a $\delta > 0$ such that the sequence $\{p_k\}_{k=0}^{\infty}$ defined by the iteration

$$p_k = g(p_{k-1}) = p_{k-1} - \frac{f(p_{k-1})}{f'(p_{k-1})} \quad \text{for } k = 1, 2, \ldots \tag{4}$$

will converges to p for any initial approximation $p_0 \in [p - \delta, p + \delta]$.

Remark.   The function $g(x)$ defined by the formula

$$g(x) = x - \frac{f(x)}{f'(x)} \tag{5}$$

is called the **Newton–Raphson iteration function**. Since $f(p) = 0$ it is easy to see that $g(p) = p$. Thus the Newton–Raphson iteration for finding the root of the equation $f(x) = 0$ is accomplished by finding a fixed point of the equation $g(x) = x$.

Proof. The geometric construction of p_1 shown in Figure 2.14 does not help in understanding why p_0 needs to be close to p or why the continuity of $f''(x)$ is essential. Our analysis starts the Taylor polynomial of degree $n = 1$ with its remainder term:

$$f(x) = f(p_0) + f'(p_0)(x - p_0) + \frac{f''(c)(x - p_0)^2}{2!}, \tag{6}$$

where c lies somewhere between p_0 and x. Substituting $x = p$ into equation (6) and using the fact that $f(p) = 0$ produces

$$0 = f(p_0) + f'(p_0)(p - p_0) + \frac{f''(c)(p - p_0)^2}{2!}. \tag{7}$$

If p_0 is close enough to p, the last term on the right side of (7) will be small compared to the sum of the first two terms. Hence it can be neglected and we can use the approximation

$$0 \approx f(p_0) + f'(p_0)(p - p_0). \tag{8}$$

Solving for p in equation (8), we get $p \approx p_0 - f(p_0)/f'(p_0)$. This is used to define the next approximation p_1 to the root

$$p_1 = p_0 - \frac{f(p_0)}{f'(p_0)}. \tag{9}$$

When p_{k-1} is used in place of p_0 in equation (9), the general rule (4) is established. For most applications this is all that needs to be understood. However, to fully comprehend what is happening we need to consider the fixed-point iteration function, and apply Theorem 2.2 in our situation. The key is in the analysis of $g'(x)$:

$$g'(x) = 1 - \frac{f'(x)f'(x) - f(x)f''(x)}{[f'(x)]^2} = \frac{f(x)f''(x)}{[f'(x)]^2}.$$

Since $g'(p) = f(p)f''(p)/[f'(p)]^2 = 0 \, f''(p)/[f'(p)]^2 = 0$, and $g(x)$ is continuous, it is possible to find a $\delta > 0$ so that the hypothesis $|g'(x)| \leq 1$ of Theorem 2.2 is satisfied on $[p - \delta, p + \delta]$. Therefore, a sufficient condition for p_0 to initialize a convergent sequence $\{p_k\}_{k=0}^{\infty}$ which converges to a root of $f(x) = 0$ is that $p_0 \in [p - \delta, p + \delta]$ and that δ be chosen so that

$$\frac{|f(x)f''(x)|}{|f'(x)|^2} < 1 \qquad \text{for all } x \in [p - \delta, p + \delta]. \tag{10}$$

Corollary 2.2   (Newton's Iteration for Finding Square Roots).   Assume that $A > 0$ is a real number and let $p_0 > 0$ be an initial approximation to $\sqrt{A}$. Define the sequence $\{p_k\}_{k=0}^{\infty}$ using the recursive rule

$$p_k = \frac{p_{k-1} + A/p_{k-1}}{2} \qquad \text{for } k = 1, 2, \ldots . \tag{11}$$

Then the sequence $\{p_k\}_{k=0}^{\infty}$ converges to $\sqrt{A}$, that is, $\lim_{k \to \infty} p_k = \sqrt{A}$.

Outline of the Proof. Start with the function $f(x) = x^2 - A$, and notice that the roots of the equation $x^2 - A = 0$ are $\pm\sqrt{A}$. Now use $f(x)$ and the derivative $f'(x)$ in formula (4) and write down the Newton–Raphson iteration function

$$g(x) = x - \frac{f(x)}{f'(x)} = x - \frac{x^2 - A}{2x}. \tag{12}$$

This formula can be simplified to obtain

$$g(x) = \frac{x + A/x}{2}. \tag{13}$$

When $g(x)$ in (13) is used to define the recursive iteration in (5), the result is formula (11). It can be proven that the sequence that is generated by (11) will converge for any starting value $p_0 > 0$. The details are left for the exercises.

An important point of Corollary 2.2 is the fact that the iteration function $g(x)$ involved only the arithmetic operations $+$, $-$, $\times$, and $/$. If $g(x)$ had involved the calculation of a square root, we would be caught in the circular reasoning that being able to calculate the square root would permit you to recursively define a sequence that will converge to $\sqrt{A}$. For this reason, $f(x) = x^2 - A$ was chosen, because it involved only the arithmetic operations.

Example 2.9

Use Newton's square-root algorithm to find $5^{1/2}$.

Solution.   Starting with $p_0 = 2$ and using formula (13), we compute

$$p_1 = \frac{2 + 5/2}{2} = 2.25$$

$$p_2 = \frac{2.25 + 5/2.25}{2} = 2.236111111$$

$$p_3 = \frac{2.236111111 + 5/2.236111111}{2} = 2.236067978$$

$$p_4 = \frac{2.236067978 + 5/2.236067978}{2} = 2.236067978.$$

Further iterations produce $p_k \approx 2.236067978$ for $k > 4$, so we see that convergence accurate to nine significant digits has been achieved.

Now let us turn to a familiar problem from elementary physics and see why the location of a root is an important task. Suppose that a projectile is fired from the origin with an angle of elevation b_0 and initial velocity v_0. In elementary courses air resistance is neglected and we learn that the height $y = y(t)$ and the distance traveled $x = x(t)$ measured in feet obey the rules

$$y = v_y t - 16t^2 \quad \text{and} \quad x = v_x t, \tag{14}$$

where the horizontal and vertical components of the initial velocity are $v_x = v_0 \cos(b_0)$ and $v_y = v_0 \sin(b_0)$, respectively. The mathematical model expressed by the rules (14) are easy to work with but tend to give too high an altitude and too long a range for the projectile's path. If we make the additional assumption that the air resistance is proportional to the velocity, the equations of motion become

$$y = f(t) = (Cv_y + 32C^2)\left[1 - \exp\left(-\frac{t}{C}\right)\right] - 32Ct \tag{15}$$

and

$$x = r(t) = Cv_x\left[1 - \exp\left(-\frac{t}{C}\right)\right], \tag{16}$$

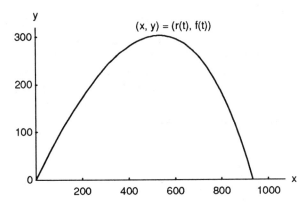

Figure 2.15 Path of a projectile with air resistance considered.

where $C = m/k$ and k is the coefficient of air resistance and m is the mass of the projectile. A larger value of C will result in a higher maximum altitude and a longer range for the projectile. The graph of a flight path of a projectile when air resistance is considered is shown in Figure 2.15. This improved model is more realistic but requires the use of a root-finding algorithm for solving $f(t) = 0$ to determine the elapsed time until the projectile hits the ground. The elementary model in (14) does not require a sophisticated procedure to find the elapsed time.

Example 2.10

A projectile is fired with an angle of elevation $b_0 = 45°$ and $v_y = v_x = 160$ ft/sec and $C = 10$. Find the elapsed time until impact and find the range.

Solution. Using formulas (15) and (16), the equations of motion are $y = f(t) = 4800[1 - \exp(-t/10)] - 320t$ and $x = r(t) = 1600[1 - \exp(-t/10)]$. Since $f(8) = 83.220972$ and $f(9) = -31.534367$, we will use the initial guess $p_0 = 8$. The derivative is $f'(t) = 480 \exp(-t/10) - 320$, and its value $f'(p_0) = f'(8) = -104.3220972$ is used in formula (5) to get

$$p_1 = 8 - \frac{83.22097200}{-104.3220972} = 8.797731010.$$

A summary of the calculations is given in Table 2.4.

The value p_4 has eight decimal places of accuracy and the time until impact is $t \approx 8.74217466$ sec. The range can now be computed using $r(t)$ and we get

$$r(8.74217466) = 1600[1 - \exp(-0.874217466)] = 932.4986302 \text{ ft.}$$

TABLE 2.4. Finding the Time When the Height $f(t)$ Is Zero

k	Time, p_k	$p_{k+1} - p_k$	Height, $f(p_k)$
0	8.00000000	0.79773101	83.22097200
1	8.79773101	−0.05530160	−6.68369700
2	8.74242941	−0.00025475	−0.03050700
3	8.74217467	−0.00000001	−0.00000100
4	8.74217466	0.00000000	0.00000000

The Division-by-Zero Error

One obvious pitfall of the Newton–Raphson method is the possibility of division by zero in formula (5), which would occur if $f'(p_{k-1}) = 0$. Algorithm 2.5 has a procedure to check for this situation, but what use is the final value p_{k-1} in this case? It is quite possible that $f(p_{k-1})$ is sufficiently close to zero and that p_{k-1} is an acceptable approximation to the root. We now investigate this situation and will uncover an interesting fact, namely, how fast the iteration converges.

Definition 2.4   (Order of a Root).   Assume that $f(x)$ and its derivatives $f'(x)$, $\ldots, f^{(M)}(x)$ are defined and continuous on an interval about $x = p$. We say that $f(x) = 0$ has a root of order M at $x = p$ if and only if

$$f(p) = 0, \quad f'(p) = 0, \quad f''(p) = 0, \quad \ldots, \quad f^{(M-1)}(p) = 0 \quad \text{and} \quad f^{(M)}(p) \neq 0. \tag{17}$$

A root of order $M = 1$ is often called a **simple root**, and if $M > 1$ it is called a **multiple root**. A root of order $M = 2$ is sometimes called a **double root**, and so on. The next result will illuminate these concepts.

Lemma 2.1.   If the equation $f(x) = 0$ has a root of order M at $x = p$, then there exists a continuous function $h(x)$ so that $f(x)$ can be expressed as the product

$$f(x) = (x - p)^M h(x), \quad \text{where} \quad h(p) \neq 0. \tag{18}$$

As an example, consider the function $f(x) = x^3 - 3x + 2$, which has a simple root at $p = -2$ and a double root at $p = 1$. This can be verified by considering the derivatives

$$f'(x) = 3x^2 - 3 \quad \text{and} \quad f''(x) = 6x.$$

At the value $p = -2$ we have $f(-2) = 0$ and $f'(-2) = 9$, so that $M = 1$ in Definition 2.4; hence $p = -2$ is a simple root. For the value $p = 1$ we have $f(1) = 0, f'(1) = 0$, and $f''(1) = 6$, so that $M = 2$ in Definition 2.4; hence $p = 1$ is a double root. Also, notice that $f(x)$ has the factorization

$$f(x) = (x + 2)(x - 1)^2.$$

Speed of Convergence

The distinguishing property we seek is the following. If p is a simple root of $f(x) = 0$, Newton's method will converge rapidly and the number of accurate decimal places (roughly) doubles with each iteration. On the other hand, if p is a multiple root, the error in each successive approximation is a fraction of the previous error. To make this precise, we define the order of convergence. This is a measure of how rapidly a sequence converges.

Definition 2.5   (Order of Convergence).   Assume that $\{p_n\}_{n=0}^{\infty}$ converges to p, and set $e_n = p - p_n$ for $n \geq 0$. If two positive constants $A \neq 0$ and $R > 0$ exist, and

$$\lim_{n \to \infty} \frac{|p - p_{n+1}|}{|p - p_n|^R} = \lim_{n \to \infty} \frac{|e_{n+1}|}{|e_n|^R} = A, \tag{19}$$

then the sequence is said to converge to p with **order of convergence** R. The number A is called the asymptotic error constant. The cases $R = 1, 2$ are given special consideration.

If $R = 1$, the convergence of $\{p_n\}_{n=0}^{\infty}$ is called **linear**.   (20)

If $R = 2$, the convergence of $\{p_n\}_{n=0}^{\infty}$ is called **quadratic**.   (21)

If R is large, the sequence $\{p_k\}$ converges rapidly to p; that is, relation (19) implies that for large values of n we have the approximation $|e_{n+1}| \approx A|e_n|^R$. For example, suppose that $R = 2$ and $|e_n| \approx 10^{-2}$; then we could expect that $|e_{n+1}| \approx A \times 10^{-4}$.

Some sequences converge at a rate that is not an integer, and we will see that the order of convergence of the secant method is $R = (1 + \sqrt{5})/2 \approx 1.618033989$.

Example 2.11

(Quadratic Convergence at a Simple Root) Start with $p_0 = -2.4$ and use Newton–Raphson iteration to find the root $p = -2$ of the polynomial $f(x) = x^3 - 3x + 2$.

Solution.   The iteration formula for computing $\{p_k\}$ is

$$p_k = g(p_{k-1}) = \frac{2p_{k-1}^3 - 2}{3p_{k-1}^2 - 3}. \tag{22}$$

Using formula (21) to check for quadratic convergence, we get the values in Table 2.5.

TABLE 2.5   Newton's Method Converges Quadratically at a Simple Root

k	p_k	$p_{k+1} - p_k$	$e_k = p - p_k$	$\dfrac{\|e_{k+1}\|}{\|e_k\|^2}$
0	-2.400000000	0.323809524	0.400000000	0.476190475
1	-2.076190476	0.072594465	0.076190476	0.619469086
2	-2.003596011	0.003587422	0.003596011	0.664202613
3	-2.000008589	0.000008589	0.000008589	
4	-2.000000000	0.000000000	0.000000000	

A detailed look at the rate of convergence in Example 2.11 will reveal that the error in each successive iteration is proportional to the square of the error in the previous iteration. That is,

$$|p - p_{k+1}| \approx A|p - p_k|^2,$$

where $A \approx \frac{2}{3}$. To check this, we use

$$|p - p_3| = 0.000008589 \quad \text{and} \quad |p - p_2|^2 = |0.003596011|^2 = 0.000012931$$

and it is easy to see that

$$|p - p_3| = 0.000008589 \approx 0.000008621 = \tfrac{2}{3}|p - p_2|^2.$$

Remark.   Theorem 2.6 will show that the constant A can be found by the calculation

$$A = \frac{1}{2}\frac{|f''(-2)|}{|f'(-2)|} = \frac{1}{2}\frac{|-12|}{|9|} = \frac{2}{3}.$$

Example 2.12

(Linear Convergence at a Double Root) Start with $p_0 = 1.2$ and use Newton–Raphson iteration to find the double root $p = 1$ of the polynomial $f(x) = x^3 - 3x + 2$.

Solution.   Using formula (20) to check for linear convergence, we get the values in Table 2.6.

TABLE 2.6   Newton's Method Converges Linear at a Double Root

k	p_k	$p_{k+1} - p_k$	$e_k = p - p_k$	$\dfrac{|e_{k+1}|}{|e_k|}$
0	1.200000000	−0.096969697	−0.200000000	0.515151515
1	1.103030303	−0.050673883	−0.103030303	0.508165253
2	1.052356420	−0.025955609	−0.052356420	0.496751115
3	1.026400811	−0.013143081	−0.026400811	0.509753688
4	1.013257730	−0.006614311	−0.013257730	0.501097775
5	1.006643419	−0.003318055	−0.006643419	0.500550093
.	.	.	.	.
.	.	.	.	.
.	.	.	.	.

Notice that the Newton–Raphson method is converging to the double root, but at a slow rate. The values $f(p_k)$ in Example 2.12 go to zero faster than the values of $f'(p_k)$, so that the quotient $f(p_k)/f'(p_k)$ in formula (5) is defined when $p_k \neq p$. The sequence is converging linearly and the error is decreasing by a factor of approximately $\frac{1}{2}$ with each successive iteration.

Theorem 2.6   (Convergence Rate for Newton–Raphson Iteration).   Suppose that Newton–Raphson iteration produces a sequence $\{p_n\}_{n=0}^{\infty}$ that converges to the root p of the function $f(x)$.

If p is a simple root, convergence is quadratic and

$$|e_{n+1}| \approx \frac{|f''(p)|}{2\,|f'(p)|}\,|e_n|^2 \qquad \text{for } n \text{ sufficiently large.} \tag{23}$$

If p is a multiple root of order M, convergence is linear and

$$|e_{n+1}| \approx \frac{M-1}{M} |e_n| \qquad \text{for } n \text{ sufficiently large.} \tag{24}$$

Pitfalls

The division-by-zero error was easy to anticipate, but there are other difficulties that are not so easy to spot. Suppose that the function is $f(x) = x^2 - 4x + 5$; then the sequence $\{p_k\}$ of real numbers generated by formula (5) will wander back and forth from left to right and **not** converge. A simple analysis of the situation reveals that $f(x) > 0$, and that the roots are complex numbers.

Sometimes the initial approximation p_0 is too far away from the desired root and the sequence $\{p_k\}$ converges to some other root. This usually happens when the slope $f'(p_0)$ is small and the tangent line to the curve $y = f(x)$ is nearly horizontal. For example, if $f(x) = \cos(x)$ and we seek the root $p = \pi/2$ and start with $p_0 = 3$, calculation reveals that $p_1 = -4.01525255$, $p_2 = -4.85265757$, . . . and $\{p_k\}$ will converge to a different root $-3\pi/2 \approx -4.71238898$.

Suppose that $f(x)$ is positive and monotone decreasing on the unbounded interval $[a, \infty)$ and $p_0 > a$; then the sequence $\{p_k\}$ might diverge to $+\infty$. For example, if $f(x) = x \exp(-x)$ and $p_0 = 2.0$, then

$$p_1 = 4.0, \quad p_2 \approx 5.333333333, \ldots, \quad p_{15} \approx 19.723549434, \ldots,$$

and $\{p_k\}$ diverges slowly to $+\infty$ [see Figure 2.16(a)]. This particular function has another surprising problem. The value of $f(x)$ goes to zero rapidly as x gets large, for example, $f(p_{15}) = 0.0000000536$, and it is possible that p_{15} could be mistaken for a root. For this reason we designed the stopping criterion in Algorithm 2.5 to involve the relative error $2|p_{k+1} - p_k|/(|p_k| + 10^{-6})$, and when $k = 15$ this value is 0.106817, so that the tolerance $\delta = 10^{-6}$ will help guard against reporting a false root.

Another phenomenon, **cycling**, occurs when the terms in the sequence $\{p_k\}$ tend to repeat, or almost repeat. For example, if $f(x) = x^3 - x - 3$ and the initial approximation is $p_0 = 0$, then the sequence is

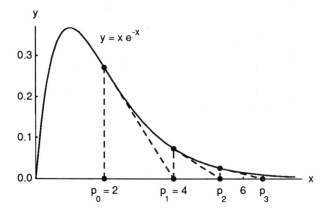

Figure 2.16   (a) Newton–Raphson iteration for $f(x) = xe^{-x}$ can produce a divergent sequence.

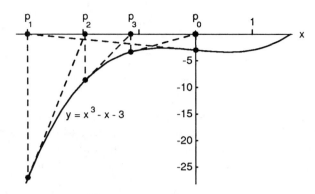

Figure 2.16   (b) Newton–Raphson iteration for $f(x) = x^3 - x - 3$ can produce a cyclic sequence.

$$p_1 = -3.000000, \quad p_2 = -1.961538, \quad p_3 = -1.147176, \quad p_4 = -0.006579,$$

$$p_5 = -3.000389, \quad p_6 = -1.961818, \quad p_7 = -1.147430, \ldots$$

and we are stuck in a cycle where $p_{k+4} \approx p_k$ for $k = 0, 1, 2, \ldots$ [see Figure 2.16(b)]. But if the starting value p_0 is sufficiently close to the root $p \approx 1.671699881$, then $\{p_k\}$ converges. If $p_0 = 2$, the sequence converges, $p_1 = 1.727272727$, $p_2 = 1.673691173$, $p_3 = 1.671702570$, $p_4 = 1.671699881$.

When $|g'(x)| \nless 1$ on an interval containing the root p, there is a chance of divergent oscillation. For example, let $f(x) = \arctan(x)$; then we have $g(x) = x - (1 + x^2) \arctan(x)$ and $g'(x) = -2x \arctan(x)$. If the starting value $p_0 = 1.45$ is chosen, then

$$p_1 \approx -1.550263297, \quad p_2 \approx 1.845931751, \quad p_3 \approx -2.889109054,$$

etc. [see Figure 2.16(c)]. But if the starting value is sufficiently close to the root $p = 0$, a convergent sequence results. If $p_0 = 0.5$, then

$$p_1 = -0.079559511, \quad p_2 = 0.000335302, \quad p_3 \approx 0.000000000.$$

The situations above point to the fact that we must be honest in reporting an answer. Sometimes the sequence does not converge. It is not always the case that after N iterations a solution is found. The user of a root-finding algorithm needs to be warned of the

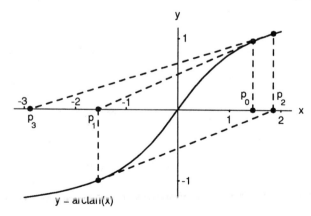

Figure 2.16   (c) Newton–Raphson iteration for $f(x) = \arctan(x)$ can produce a divergent oscillating sequence.

situation when a root is not found. If there is other information concerning the context of the problem, then it is less likely that an erroneous root will be found. Sometimes $f(x)$ has a definite interval in which a root is meaningful. If knowledge of the behavior of the function or a graph is available, then it is easier to choose p_0.

The Secant Method

The Newton–Raphson algorithm requires the evaluation of two functions per iteration, $f(p_{k-1})$ and $f'(p_{k-1})$. If they are not complicated expressions, the method is desirable. In some cases it may require a considerable amount of effort to use the rules of calculus and derive the formula for $f'(x)$ from $f(x)$. Hence it is desirable to have a method that converges almost as fast as Newton's method yet involves only evaluations of $f(x)$. The secant method will require only one evaluation of $f(x)$ per step and at a simple root has an order of convergence $R \approx 1.618033989$. It is almost as fast as Newton's method, which has order 2.

The formula involved in the secant method is the same one that was used in the regula falsi method, except that the logical decisions regarding how to define each succeeding term are different. Two initial points $(p_0, f(p_0))$ and $(p_1, f(p_1))$ near the point $(p, 0)$ are needed, as shown in Figure 2.17. Define p_2 to be the point of intersection of

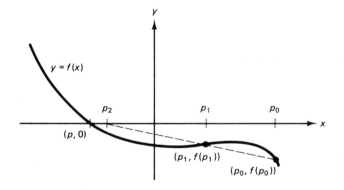

Figure 2.17   The geometric construction of p_2 for the secant method.

the line through these two points; then Figure 2.17 shows that p_2 will be closer to p than to either p_0 or p_1. The equation relating p_2, p_1, and p_0 is found by considering the slope

$$m = \frac{f(p_1) - f(p_0)}{p_1 - p_0} = \frac{0 - f(p_1)}{p_2 - p_1}. \qquad (25)$$

The values of m in (25) are the slope of the secant line through the first two approximations and the slope of the line through $(p_1, f(p_1))$ and $(p_2, 0)$, respectively. Solve for $p_2 = g(p_1, p_0)$ and get

$$p_2 = g(p_1, p_0) = p_1 - \frac{f(p_1)(p_1 - p_0)}{f(p_1) - f(p_0)}. \qquad (26)$$

The general term is given by the two-point iteration formula

$$p_{k+1} = g(p_k, p_{k-1}) = p_k - \frac{f(p_k)(p_k - p_{k-1})}{f(p_k) - f(p_{k-1})}. \tag{27}$$

Example 2.13

(Secant Method at a Simple Root) Start with $p_0 = -2.6$ and $p_1 = -2.4$ and use the secant method to find the root $p = -2$ of the polynomial function $f(x) = x^3 - 3x + 2$.

Solution.   In this case the iteration formula (27) is

$$p_{k+1} = g(p_k, p_{k-1}) = p_k - \frac{(p_k^3 - 3p_k + 2)(p_k - p_{k-1})}{p_k^3 - p_{k-1}^3 - 3p_k + 3p_{k-1}}. \tag{28}$$

This can be algebraically manipulated to obtain

$$p_{k+1} = g(p_k, p_{k-1}) = \frac{p_k^2 p_{k-1} + p_k p_{k-1}^2 - 2}{p_k^2 + p_k p_{k-1} + p_{k-1}^2 - 3}. \tag{29}$$

and the sequence of iterates is given in Table 2.7.

TABLE 2.7   Convergence of the Secant Method at a Simple Root

k	p_k	$p_{k+1} - p_k$	$e_k = p - p_k$	$\dfrac{\|e_{k+1}\|}{\|e_k\|^{1.618}}$
0	−2.600000000	0.200000000	0.600000000	0.914152831
1	−2.400000000	0.293401015	0.400000000	0.469497765
2	−2.106598985	0.083957573	0.106598985	0.847290012
3	−2.022641412	0.021130314	0.022641412	0.693608922
4	−2.001511098	0.001488561	0.001511098	0.825841116
5	−2.000022537	0.000022515	0.000022537	0.727100987
6	−2.000000022	0.000000022	0.000000022	
7	−2.000000000	0.000000000	0.000000000	

There is a relationship between the secant method and Newton's method. For a polynomial function $f(x)$, the secant method two-point formula $g(p_k, p_{k-1})$ will reduce to Newton's one-point formula $g(p_k)$ if p_k is replaced by p_{k-1}. Indeed, if we replace p_k by p_{k-1} in (29), then the right side becomes the same as the right side of (22).

Proofs about the rate of convergence of the secant method can be found in advanced texts on numerical analysis. Let us state that the error terms satisfy the relationship

$$|e_{k+1}| = |e_k|^{1.618} \left| \frac{f''(p)}{2f'(p)} \right|^{0.618} \tag{30}$$

where the order of convergence is $R = (1 + 5^{1/2})/2 \approx 1.618$ and the relation in (30) is valid only at simple roots.

To check this out we use Example 2.13 and the specific values $|p - p_5| = 0.000022537$, $|p - p_4|^{1.618} = 0.001511098^{1.618} = 0.000027296$, and $A = |f''(-2)/2f'(-2)|^{0.618} = (2/3)^{0.618} = 0.778351205$. Combine these and it is easy to see that

$$|p - p_5| = 0.000022537 \approx 0.000021246 = A|p - p_4|^{1.618}.$$

Accelerated Convergence

We would hope that there are root-finding techniques which converge faster than linearly when p is a root of order M. Our final result shows how a modification can be made to Newton's method so that convergence becomes quadratic at a multiple root.

Theorem 2.7   (Acceleration of Newton–Raphson Iteration).   Suppose that the Newton–Raphson algorithm produces a sequence that converges linearly to the root $x = p$ of order $M > 1$. Then the modified Newton–Raphson formula

$$p_n = p_{n-1} - \frac{Mf(p_{n-1})}{f'(p_{n-1})} \tag{31}$$

will produce a sequence $\{p_n\}_{n=0}^{\infty}$ that converges quadratically to p.

Example 2.14

(Acceleration of Convergence at a Double Root) Start with $p_0 = 1.2$ and use accelerated Newton–Raphson iteration to find the double root $p = 1$ of $f(x) = x^3 - 3x + 2$.

Solution.   Since $M = 2$, the acceleration formula (30) becomes

$$p_k = p_{k-1} - 2\frac{f(p_{k-1})}{f'(p_{k-1})} = \frac{p_{k-1}^3 + 3p_{k-1} - 4}{3p_{k-1}^2 - 3},$$

and we obtain the values in Table 2.8.

TABLE 2.8   Acceleration of Convergence at a Double Root

k	p_k	$p_{k+1} - p_k$	$e_k = p - p_k$	$\dfrac{|e_{k+1}|}{|e_k|^2}$
0	1.200000000	−0.193939394	−0.200000000	0.151515150
1	1.006060606	−0.006054519	−0.006060606	0.165718578
2	1.000006087	−0.000006087	−0.000006087	
3	1.000000000	0.000000000	0.000000000	

Table 2.9 compares speed of convergence of the various root-finding methods that we have studied so far. The value of the constant A is different for each method.

TABLE 2.9 Comparison of the Speed of Convergence

Method	Special considerations	Relation between successive error terms
Bisection		$\|e_{k+1}\| \approx \frac{1}{2}\|e_k\|$
Regula falsi		$\|e_{k+1}\| \approx A\|e_k\|$
Secant method	Multiple root	$\|e_{k+1}\| \approx A\|e_k\|$
Newton–Raphson	Multiple root	$\|e_{k+1}\| \approx A\|e_k\|$
Secant method	Simple root	$\|e_{k+1}\| \approx A\|e_k\|^{1.618}$
Newton–Raphson	Simple root	$\|e_{k+1}\| \approx A\|e_k\|^2$
Accelerated Newton–Raphson	Multiple root	$\|e_{k+1}\| \approx A\|e_k\|^2$

Algorithm 2.5   (Newton–Raphson Iteration).   To find a root of $f(x) = 0$ given one initial approximation p_0 and using the iteration

$$p_k = p_{k-1} - \frac{f(p_{k-1})}{f'(p_{k-1})} \qquad \text{for } k = 1, 2, \ldots .$$

```
Delta := 10⁻⁶, Epsilon := 10⁻⁶, Small := 10⁻⁶        {Tolerances}
Max := 99                                             {Maximum number of iterations}
Cond := 0                                             {Condition for loop termination}
INPUT  P0                                             {P0 must be close to the root}
Y0 := F(P0)                                           {Compute the function value}

DO   FOR K := 1 TO Max UNTIL Cond ≠ 0
     Df := F'(P0)                                     {Compute the derivative}
        IF  Df = 0  THEN                              {Check division
            Cond := 1
            Dp := 0                                       by zero}
        ELSE
            Dp := Y0/Df
     ENDIF
     P1 := P0 − Dp                                    {New iterate}
     Y1 := F(P1)                                      {New function value}
     RelErr := 2*|Dp|/[|P1|+Small]                    {Relative error}
     IF   RelErr<Delta AND |Y1|<Epsilon   THEN        {Check for
          IF Cond ≠ 1 THEN Cond := 2                   convergence}
     P0 := P1, Y0 := Y1                               {Update values}

     PRINT 'The current k-th iterate is' P1           {Output}
```

PRINT 'Consecutive iterates differ by' Dp
PRINT 'The value of f(x) is' Y1
IF   Cond = 0   THEN
PRINT 'The maximum number of iterations was exceeded.'
IF   Cond = 1   THEN
PRINT 'Division by zero was encountered.'
IF   Cond = 2   THEN
PRINT 'The root was found with the desired tolerances.'

Algorithm 2.6   (Secant Method).   To find a root of $f(x) = 0$ given two initial approximations p_0 and p_1 and using the iteration

$$p_{k+1} = p_k - \frac{f(p_k)[p_k - p_{k-1}]}{f(p_k) - f(p_{k-1})} \qquad \text{for } k = 1, 2, \ldots .$$

Delta := 10^{-6}, Epsilon := 10^{-6}, Small := 10^{-6} {Tolerances}
Max := 149   {Maximum number of iterations}
Cond := 0   {Condition for loop termination}

INPUT P0, P1   {These values must be close to the root}
Y0 := F(P0), Y1 := F(P1)   {Compute the function values}

DO   FOR K := 1 TO Max UNTIL Cond ≠ 0
          Df := [Y1 − Y0]/[P1 − P0]   {Compute the slope}
              IF   Df = 0   THEN   {Check division}
                      Cond := 1   by zero}
                      Dp := 0
                  ELSE
                      Dp := Y1/Df
              ENDIF
          P2 := P1 − Dp   {New iterate}
          Y2 := F(P2)   {New function value}
          RelErr := 2∗|Dp|/[|P2|+Small]   {Relative error}
          IF   RelErr<Delta AND |Y2|<Epsilon   THEN   {Check for
              IF Cond ≠ 1 THEN Cond := 2   convergence}
          P0 := P1, P1 := P2, Y0 := Y1, Y1 := Y2   {Update values}

PRINT 'The current k-th iterate is' P2
PRINT 'Consecutive iterates differ by' Dp
PRINT 'The value of f(x) is' Y2

```
IF   Cond = 0   THEN
     PRINT 'The maximum number of iterations was exceeded.'
IF   Cond = 1   THEN
     PRINT 'Division by zero was encountered.'
IF   Cond = 2   THEN
     PRINT 'The root was found with the desired tolerances'
```

EXERCISES FOR NEWTON–RAPHSON AND SECANT METHOD

For problems involving calculations, you can use either a *calculator* or a *computer*.

1. Use Newton's square-root algorithm.
   (a) Start with $p_0 = 3$ and find $8^{1/2}$.
   (b) Start with $p_0 = 7$ and find $50^{1/2}$.
   (c) Start with $p_0 = 10$ and find $91^{1/2}$.
   (d) Start with $p_0 = -3$ and find $-(8)^{1/2}$.

2. *Cube-root algorithm.* Start with $f(x) = x^3 - A$, where A is any real number, and derive the recursive formula

$$p_k = \frac{2p_{k-1} + A/p_{k-1}^2}{3} \qquad \text{for } k = 1, 2, \ldots$$

   for finding the cube root of A.

3. Use the iteration in Exercise 2 for finding cube roots.
   (a) Start with $p_0 = 2$ and find $7^{1/3}$.
   (b) Start with $p_0 = 3$ and find $30^{1/3}$.
   (c) Start with $p_0 = 6$ and find $200^{1/3}$.
   (d) Start with $p_0 = -2$ and find $(-7)^{1/3}$.

4. Consider $f(x) = x^N - A$, where N is a positive integer.
   (a) What real values are the solution to $f(x) = 0$ for the various choices of N and A that can arise?
   (b) Derive the recursive formula

$$p_k = \frac{(N - 1)p_{k-1} + A/p_{k-1}^{N-1}}{N} \qquad \text{for } k = 1, 2, \ldots$$

   for finding the nth root of A.

5. Establish the limit of the sequence in (11).

6. Write an algorithm that will use Newton's method to find $A^{1/N}$. Input the real number A and integer N. Return the value 0 for the case of $0^{1/N}$. Have the algorithm print an error message in the case when A is negative and N is an even integer. For the other cases start with $p_0 = A$ and use the stopping criterion

$$|p_k - p_{k-1}| < |p_k| \times 10^{-6}.$$

7. Let $f(x) = x^2 - 2x - 1$.
   (a) Find the Newton–Raphson formula $g(p_{k-1})$.

    (b) Start with $p_0 = 2.5$ and find, p_1, p_2, and p_3.
    (c) Start with $p_0 = -0.5$ and find p_1, p_2, and p_3.

8. Let $f(x) = x^3 - x - 3$.
    (a) Find the Newton–Raphson formula $g(p_{k-1})$.
    (b) Start with $p_0 = 1.6$ and find p_1, p_2, and p_3.
    (c) Start with $p_0 = 0.0$ and find p_1, p_2, p_3, p_4, and p_5. What do you conjecture about this sequence?

9. Let $f(x) = x^3 - x + 2$.
    (a) Find the Newton Raphson formula $g(p_{k-1})$.
    (b) Start with $p_0 = -1.5$ and find p_1, p_2, and p_3.

10. Suppose that the equations of motion for a projectile are

$$y = f(t) = 1600 \left[1 - \exp\left(-\frac{t}{5}\right) \right] - 160t,$$

$$x = r(t) = 800 \left[1 - \exp\left(-\frac{t}{5}\right) \right].$$

    (a) Start with $p_0 = 8$ and find the elapsed time until impact.
    (b) Find the range.

11. Suppose that the equations of motion for a projectile are

$$y = f(t) = 9600 \left[1 - \exp\left(-\frac{t}{15}\right) \right] - 480t,$$

$$x = r(t) = 2400 \left[1 - \exp\left(-\frac{t}{15}\right) \right].$$

    (a) Start with $p_0 = 9$ and find the elapsed time until impact.
    (b) Find the range.

12. Let $f(x) = (x - 2)^4$.
    (a) Find the Newton–Raphson formula $g(p_{k-1})$.
    (b) Start with $p_0 = 2.1$ and compute p_1, p_2, p_3, and p_4.
    (c) Is the sequence converging quadratically or linearly?

13. Let $f(x) = x^3 - 3x - 2$.
    (a) Find the Newton–Raphson formula $g(p_{k-1})$.
    (b) Start with $p_0 = 2.1$ and compute p_1, p_2, p_3, and p_4.
    (c) Is the sequence converging quadratically or linearly?

14. Consider the function $f(x) = \cos(x)$.
    (a) Find the Newton–Raphson formula $g(p_{k-1})$.
    (b) We want to find the root $p = 3\pi/2$. Can we use $p_0 = 3$? Why?
    (c) We want to find the root $p = 3\pi/2$. Can we use $p_0 = 5$? Why?

15. Find the point on the parabola $y = x^2$ that is closest to the point $(3, 1)$. Use the following steps.
    (a) Show that $d(x) = (x - 3)^2 + (x^2 - 1)^2$ is the distance squared between $(3, 1)$ and the point (x, y) on the parabola.
    (b) Show that when the derivative of $d(x)$ is set equal to zero we obtain the equation $f(x) = 4x^3 - 2x - 6 = 0$.
    (c) Start with $p_0 = 1.0$ and find the root of $f(x) = 0$.

16. Find the point on the parabola $y = x^2$ closest to $(1, 3)$.
   (a) Show that $d(x) = (x - 1)^2 + (x^2 - 3)^2$.
   (b) Start with $p_0 = 1.5$ and find the root of $d'(x) = f(x) = 0$.

17. Can Newton–Raphson iteration be used to solve $f(x) = 0$ if $f(x) = x^2 - 14x + 50$? Why?

18. Can Newton–Raphson iteration be used to solve $f(x) = 0$ if $f(x) = x^{1/3}$? Why?

19. Can Newton–Raphson iteration be used to solve $f(x) = 0$ if $f(x) = (x - 3)^{1/2}$ and the starting value is $p_0 = 4$?

20. An open-top box is constructed from a rectangular piece of sheet metal measuring 10 by 16 inches. Squares of what size should be cut from the corners if the volume of the box is 100 cubic inches? *Hint.* If the removed squares measure x inches on a side, the volume of the box will be height $\times$ width $\times$ length $= x(10 - 2x)(16 - 2x)$, so we need to find the roots of $f(x) = 4x^3 - 52x^2 + 160x - 100$.

21. An open-top box is constructed from a rectangular piece of sheet metal measuring 10 by 14 inches. Squares of what size should be cut from the corners if the volume of the box is 100 cubic inches?

22. Consider the function $f(x) = x \exp(-x)$.
   (a) Find the Newton–Raphson formula $g(p_{k-1})$.
   (b) If $p_0 = 0.2$, then find $p_1, \ldots, p_4$. What is $\lim p_k$?
   (c) If $p_0 = 20$, then find $p_1, \ldots, p_4$. What is $\lim p_k$?
   (d) What is the value of $f(p_4)$ in part (c)?

23. Consider the function $f(x) = \arctan(x)$.
   (a) Find the Newton–Raphson formula $g(p_{k-1})$.
   (b) If $p_0 = 1.0$, then find $p_1, \ldots, p_4$. What is $\lim p_k$?
   (c) If $p_0 = 2.0$, then find $p_1, \ldots, p_4$. What is $\lim p_k$?

24. In celestial mechanics, Kepler's equation is $y = x - e \sin(x)$ where y is the planet's mean anomaly, x its eccentric anomaly, and e the eccentricity of its orbit. Let $y = 1$ and $e = 0.5$ and find the root of $2x - 2 - \sin(x) = 0$ in the interval $[0, \pi]$.

25. The natural frequency P for an elastic beam of length L that is fixed at one end and pinned at the other end obeys the relation $\tan(KL) = \tanh(KL)$, where $P = K^2 C$ (C depends on certain measurements of the beam).
   (a) Find the solution to $\tan(x) = \tanh(x)$ near $p_0 = 4$.
   (b) Find the solution to $\tan(x) = \tanh(x)$ near $p_0 = 7$.

26. Prove that the sequence $\{p_k\}$ in equation (4) of Theorem 2.5 converges to p. Use the following steps.
   (a) Show that if p is a fixed point of $g(x)$ in equation (5), p is a zero of $f(x)$.
   (b) If p is a zero of $f(x)$ and $f'(p) \neq 0$, show that $g'(p) = 0$.
   (c) Use part (b) and Theorem 2.3 to show that the sequence $\{p_k\}$ in equation (4) converges to p.

27. If $h'(x)$ exists at x_0, prove that $h(x)$ is continuous at x_0.

28. Prove equation (23) of Theorem 2.6. Use the following steps.
   By Theorem 1.13 we can expand $f(x)$ about $x = p_k$ to get

$$f(x) = f(p_k) + f'(p_k)(x - p_k) + \tfrac{1}{2}f''(c_k)(x - p_k)^2.$$

Since p is a zero of $f(x)$, we set $x = p$ and obtain

$$0 = f(p_k) + f'(p_k)(p - p_k) + \tfrac{1}{2}f''(c_k)(p - p_k)^2.$$

(a) Now assume that $f'(x) \neq 0$ for all x near the root p.
Use the facts given above and $f'(p_k) \neq 0$ to show that

$$p - p_k + \frac{f(p_k)}{f'(p_k)} = \frac{-f''(c_k)}{2f'(p_k)} (p - p_k)^2.$$

(b) Assume that $f'(x)$ and $f''(x)$ do not change too rapidly, so that we can use the approximations $f'(p_k) \approx f'(p)$ and $f''(c_k) \approx f''(p)$. Now use part (a) to get

$$e_{k+1} \approx \frac{-f''(p)}{2f'(p)} (e_k)^2.$$

29. Suppose that A is a positive real number.
   (a) Show that A has the representation

$$A = q \times 2^{2m}, \qquad \text{where } \tfrac{1}{4} \leq q < 1 \quad \text{and} \quad m \text{ is an integer.}$$

   (b) Use part (a) to show that the square root is

$$A^{1/2} = q^{1/2} \times 2^m.$$

   Remark. Let $p_0 = (2q + 1)/3$, where $\tfrac{1}{4} \leq q < 1$, and use Newton's formula (11). After three iterations p_3 will be an approximation to $q^{1/2}$ with a precision of 24 binary digits. This is the algorithm that is often used in the computer's hardware to compute square roots.

30. **(a)** Show that formula (27) for the secant method is algebraically equivalent to

$$p_{k+1} = \frac{p_{k-1}f(p_k) - p_k f(p_{k-1})}{f(p_k) - f(p_{k-1})}.$$

   (b) Explain why loss of significance in subtraction makes this formula inferior for computational purposes to the one given in formula (27).

In Exercises 31–33, use the secant method and formula (27) and compute the next two iterates p_2 and p_3.

31. Let $f(x) = x^2 - 2x - 1$. Start with $p_0 = 2.6$, $p_1 = 2.5$.
32. Let $f(x) = x^3 - x - 3$. Start with $p_0 = 1.7$, $p_1 = 1.67$.
33. Let $f(x) = x^3 - x + 2$. Start with $p_0 = -1.5$, $p_1 = -1.52$.
34. Use the accelerated Newton–Raphson iteration in Theorem 2.3 to find the root p of order M for the following;
   (a) $f(x) = (x - 2)^5$, $M = 5$, $p = 2$; start with $p_0 = 1$.
   (b) $f(x) = \sin(x^3)$, $M = 3$, $p = 0$; start with $p_0 = 1$.
   (c) $f(x) = (x^2 - 3)^4$, $M = 4$, $p = 3^{1/2}$; start with $p_0 = 2$.
   (d) $f(x) = (x - 1) \ln(x)$, $M = 2$, $p = 1$; start with $p_0 = 2$.
35. Suppose that p is a root of order $M = 2$ for $f(x) = 0$. Prove that the accelerated Newton–Raphson iteration

$$p_k = p_{k-1} - 2\frac{f(p_{k-1})}{f'(p_{k-1})}$$

converges quadratically (cf. Exercise 28).

36. Halley's method is another way to speed up convergence of Newton's method. The Halley iteration formula is

$$g(x) = x - \frac{f(x)}{f'(x)} \left[1 - \frac{f(x)f''(x)}{2[f'(x)]^2} \right]^{-1}.$$

The term in brackets is the modification of the Newton–Raphson formula. Halley's method will yield cubic convergence $(R = 3)$ at simple zeros of $f(x)$.

(a) Start with $f(x) = x^2 - a$ and find Halley's iteration formula $g(x)$ for finding $\sqrt{a}$. Use $a = 5$ and $p_0 = 2$ and compute $p_1, p_2,$ and p_3.

(b) Start with $f(x) = x^3 - 3x + 2$ and find Halley's iteration formula $g(x)$. Use $p_0 = -2.4$ and compute $p_1, p_2,$ and p_3.

37. A modified Newton–Raphson method for multiple roots.

Fact. If P is a root of multiplicity M, then $f(x) = (x - p)^M q(x)$, where $q(P) \neq 0$.

(a) Show that $h(x) = f(x)/f'(x)$ has a simple root at $x = P$.

(b) Show that when the Newton–Raphson method is applied to finding the simple root P of $h(x)$, we get $g(x) = x - h(x)/h'(x)$, which becomes

$$g(x) = x - \frac{f(x)f'(x)}{[f'(x)]^2 - f(x)f''(x)}.$$

(c) The iteration using $g(x)$ in part (b) converges quadratically to P. Tell why this happens.

38. A catenary is the curve formed by a hanging cable. Assume that the lowest point is $(0, 0)$, then the formula for the catenary is $y = C \cosh(x/C) - C$. To determine the catenary that goes through $(\pm a, b)$ we must solve the equation $b = C \cosh(a/C) - C$ for C.

(a) Show that the catenary through $(\pm 10, 6)$ is $y = 9.1889 \cosh(x/9.1889) - 9.1889$ [see Figure 2.18(a)].

(b) Find the catenary in Figure 2.18(b) that passes through $(\pm 12, 5)$.

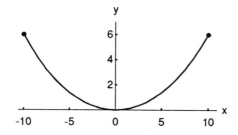

Figure 2.18   (a) The catenary through $(\pm 10, 6)$.

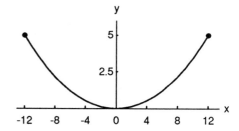

Figure 2.18   (b) The catenary through $(\pm 12, 5)$.

39. Write a report on the convergence rate of the secant method. See References [9, 35, 40, 41, 153, and 160].

40. Write a report on the quotient difference algorithm. See References [3, 29, 62, 78, 79, 86, 112, 152, and 200].

2.5 AITKEN'S PROCESS AND STEFFENSEN'S AND MULLER'S METHODS (OPTIONAL)

In Section 2.4 we saw that Newton's method converged slowly at a multiple root and the sequence of iterates $\{p_n\}$ exhibited linear convergence. Theorem 2.7 showed how to speed up convergence, but it depends on knowing the order of the root in advance.

Aitken's Process

A technique called **Aitken's Δ^2 process** can be used to speed up convergence of any sequence that is linearly convergent. In order to proceed we will need a definition.

Definition 2.6.   Given the sequence $\{p_n\}_{n=0}^{\infty}$, define the **forward difference** Δp_n by

$$\Delta p_n = p_{n+1} - p_n \qquad \text{for } n \geq 0. \tag{1}$$

Higher powers $\Delta^k p_n$ are defined recursively by

$$\Delta^k p_n = \Delta^{k-1}(\Delta p_n) \qquad \text{for } k \geq 2. \tag{2}$$

Theorem 2.8 (Aitken's Acceleration).   Assume that the sequence $\{p_n\}_{n=0}^{\infty}$ converges linearly to the limit p and that $p - p_n \neq 0$ for all $n \geq 0$. If there exists a real number A with $|A| < 1$ such that

$$\lim_{n \to \infty} \frac{p - p_{n+1}}{p - p_n} = A, \tag{3}$$

then the sequence $\{q_n\}_{n=0}^{\infty}$ defined by

$$q_n = p_n - \frac{(\Delta p_n)^2}{\Delta^2 p_n} = p_n - \frac{(p_{n+1} - p_n)^2}{p_{n+2} - 2p_{n+1} + p_n} \tag{4}$$

converges to p faster than $\{p_n\}_{n=0}^{\infty}$ in the sense that

$$\lim_{n \to \infty} \frac{|p - q_n|}{|p - p_n|} = 0. \tag{5}$$

Proof. We will show how to derive formula (4) and will leave the proof of (5) as an exercise. Since the terms in (3) are approaching a limit we can write

$$\frac{p - p_{n+1}}{p - p_n} \approx A \quad \text{and} \quad \frac{p - p_{n+2}}{p - p_{n+1}} \approx A \qquad \text{when } n \text{ is large.} \tag{6}$$

The relations in (6) imply that

$$(p - p_{n+1})^2 \approx (p - p_{n+2})(p - p_n). \tag{7}$$

When both sides of (7) are expanded and the terms p^2 canceled, the result is

$$p \approx \frac{p_{n+2}p_n - p_{n+1}^2}{p_{n+2} - 2p_{n+1} + p_n} = q_n \qquad \text{for } n = 0, 1, \ldots . \qquad (8)$$

The formula in (8) is used to define the term q_n. It can be rearranged algebraically to obtain formula (4), which has less error propagation when computer calculations are made.

Example 2.15

Show that the sequence $\{p_n\}$ in Example 2.2 exhibits linear convergence, and show that the sequence $\{q_n\}$ obtained by Aitken's Δ^2 process converges faster.

Solution. The sequence $\{p_n\}$ was obtained by fixed-point iteration using the function $g(x) = \exp(-x)$ and starting with $p_0 = 0.5$. After convergence has been achieved, the limit is $P \approx 0.567143290$. The values of p_n and q_n are given in Tables 2.10, and 2.11. For illustration, the value of q_1 is given by the calculation

$$q_1 = p_1 - \frac{(p_2 - p_1)^2}{p_3 - 2p_2 + p_1}$$

$$= 0.606530660 - \frac{(-0.061291448)^2}{0.095755331} = 0.567298989.$$

TABLE 2.10   Linearly Convergent Sequence $\{p_n\}$

n	p_n	$e_n = p_n - p$	$A_n = \dfrac{e_n}{e_{n-1}}$
1	0.606530660	0.039387369	−0.586616609
2	0.545239212	−0.021904079	−0.556119357
3	0.579703095	0.012559805	−0.573400269
4	0.560064628	−0.007078663	−0.563596551
5	0.571172149	0.004028859	−0.569155345
6	0.564862947	−0.002280343	−0.566002341

TABLE 2.11   Derived Sequence $\{q_n\}$ Using Aitken's Process

n	q_n	$q_n - p$
0	0.567298989	0.000155699
1	0.567193142	0.000049852
2	0.567159364	0.000016074
3	0.567148453	0.000005163
4	0.567144952	0.000001662
5	0.567143825	0.000000534

Although the sequence $\{q_n\}$ in Table 2.11 converges linearly, it converges faster than $\{p_n\}$ in the sense of Theorem 2.10, and usually Aitken's method gives a better improvement than this. When Aitken's process is combined with fixed-point iteration, the result is called **Steffensen's acceleration**. The details are given in Algorithm 2.7 and in the exercises.

Muller's Method

Muller's method is a generalization of the secant method, in the sense that it does not require the derivative of the function. It is an iterative method that requires three starting points: $(p_0, f(p_0))$, $(p_1, f(p_1))$, and $(p_2, f(p_2))$. A parabola is constructed that passes through the three points; then the quadratic formula is used to find a root of the quadratic for the next approximation. It has been proven that near a simple root Muller's method converges faster than the secant method and almost as fast as Newton's method. The method can be used to find real or complex zeros of a function and can be programmed to use complex arithmetic.

Without loss of generality, we assume that p_2 is the best approximation to the root and consider the parabola through the three starting values, as shown in Figure 2.19.

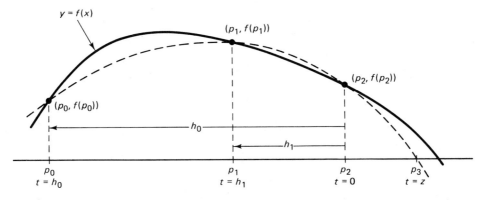

Figure 2.19   The starting approximations p_0, p_1, and p_2 for Muller's method, and the differences h_0 and h_1.

Make the change of variable

$$t = x - p_2, \tag{9}$$

and use the differences

$$h_0 = p_0 - p_2 \quad \text{and} \quad h_1 = p_1 - p_2. \tag{10}$$

Consider the quadratic polynomial, involving the variable t,

$$y = at^2 + bt + c. \tag{11}$$

Each point is used to obtain an equation involving a, b, and c:

$$\text{At } t = h_0:\ ah_0^2 + bh_0 + c = f_0.$$
$$\text{At } t = h_1:\ ah_1^2 + bh_1 + c = f_1. \tag{12}$$
$$\text{At } t = 0:\quad a0^2 + b0\ + c = f_2.$$

From the third equation in (12) we see that

$$c = f_2. \tag{13}$$

Substituting $c = f_2$ into the first two equations in (12) and using the definitions $e_0 = f_0 - c$ and $e_1 = f_1 - c$ results in the system

$$
\begin{aligned}
ah_0^2 + bh_0 &= f_0 - c = e_0, \\
ah_1^2 + bh_1 &= f_1 - c = e_1.
\end{aligned}
\tag{14}
$$

Cramer's rule can be used to solve the linear system in (14) (see Section 1.1). The required determinants are

$$
D = \begin{vmatrix} h_0^2 & h_0 \\ h_1^2 & h_1 \end{vmatrix} = h_0 h_1 (h_0 - h_1),
$$

$$
D_1 = \begin{vmatrix} e_0 & h_0 \\ e_1 & h_1 \end{vmatrix} = e_0 h_1 - e_1 h_0, \tag{15}
$$

$$
D_2 = \begin{vmatrix} h_0^2 & e_0 \\ h_1^2 & e_1 \end{vmatrix} = e_1 h_0^2 - e_0 h_1^2.
$$

The coefficients a and b are given by

$$a = \frac{D_1}{D} \quad \text{and} \quad b = \frac{D_2}{D}. \tag{16}$$

The quadratic formula is used to find the roots $t = z_1, z_2$ of (11).

$$z = \frac{-2c}{b \pm (b^2 - 4ac)^{1/2}} \tag{17}$$

Formula (17) is equivalent to the standard formula for the roots of a quadratic and is better in this case because we know that $c = f_2$.

To ensure stability of the method, we choose the root in (17) that has the smallest absolute value. If $b > 0$, use the positive sign with the square root, and if $b < 0$, use the negative sign. Then p_3 is shown in Figure 2.19 and is given by

$$p_3 = p_2 + z. \tag{18}$$

To update the iterates, choose p_0 and p_1 to be the two values selected from $\{p_0, p_1, p_2\}$ that lie closest to p_3 (i.e., throw out the one that is farthest away). Then replace p_2 with p_3. Although a lot of auxiliary calculations are done in Muller's method, it only requires one function evaluation per iteration.

If Muller's method is used to find the real roots of $f(x)$, it is possible that one may encounter complex approximations because the roots of the quadratic in (17) might be complex. In these cases the complex component will have a small magnitude and can be set equal to zero so that the calculations proceed with real numbers.

Comparison of Methods

Steffensen's method can be used with the Newton–Raphson fixed-point function $g(x) = x - f(x)/f'(x)$. In the next two examples we look at the roots of the polynomial $f(x) = x^3 - 3x + 2$. The Newton–Raphson function is $g(x) = (2x^3 - 2)/(3x^2 - 3)$. When this function is used in Algorithm 2.7, we get the calculations under the heading Steffensen with Newton in Tables 2.12 and 2.13. For example, starting with $p_0 = -2.4$, we would compute

$$p_1 = g(p_0) = -2.076190476,$$

$$p_2 = g(p_1) = -2.003596011.$$

Then Aitken's improvement will give $p_3 = -1.982618143$.

Example 2.16

(Convergence Near a Simple Root) This is a comparison of methods for the function $f(x) = x^3 - 3x + 2$ near the simple root at $p = -2$.

Solution. Newton's method and the secant method for this function were given in Examples 2.11 and 2.13, respectively. The summary of calculations for the methods is shown in Table 2.12.

TABLE 2.12   Comparison of Convergences Near a Simple Root

k	Secant method	Muller's method	Newton's method	Steffensen with Newton
0	−2.600000000	−2.600000000	−2.400000000	−2.400000000
1	−2.400000000	−2.500000000	−2.076190476	−2.076190476
2	−2.106598985	−2.400000000	−2.003596011	−2.003596011
3	−2.022641412	−1.985275287	−2.000008589	−1.982618143
4	−2.001511098	−2.000334062	−2.000000000	−2.000204982
5	−2.000022537	−2.000000218		−2.000000028
6	−2.000000022	−2.000000000		−2.000002389
7	−2.000000000			−2.000000000

Example 2.17

(Convergence Near a Double Root) This is a comparison of methods for the function $f(x) = x^3 - 3x + 2$ near the double root at $p = 1$.

Solution.   The summary of calculations is shown in Table 2.13.

TABLE 2.13   Comparison of Convergence Near a Double Root

k	Secant method	Muller's method	Newton's method	Steffensen with Newton
0	1.400000000	1.400000000	1.200000000	1.200000000
1	1.200000000	1.300000000	1.103030303	1.103030303
2	1.138461538	1.200000000	1.052356417	1.052356417
3	1.083873738	1.003076923	1.026400814	0.996890433
4	1.053093854	1.003838922	1.013257734	0.998446023
5	1.032853156	1.000027140	1.006643418	0.999223213
6	1.020429426	0.999997914	1.003325375	0.999999193
7	1.012648627	0.999999747	1.001663607	0.999999597
8	1.007832124	1.000000000	1.000832034	0.999999798
9	1.004844757		1.000416075	0.999999999
.			.	
.			.	
.			.	

Newton's method is the best choice for finding a simple root (see Table 2.12). At a double root, either Muller's method or Steffensen's method with the Newton–Raphson formula are good choices (see Table 2.13).

Algorithm 2.7   (Steffensen's Acceleration).   To quickly find a solution of the fixed-point equation $x = g(x)$ given an initial approximation p_0, where it is assumed that both $g(x)$ and $g'(x)$ are continuous and $|g'(p)| < 1$ and that ordinary fixed-point iteration converges slowly (linearly) to p.

Delta := 10^{-6}, Small := 10^{-6} {Tolerances}
Cond := 0   {Condition for loop termination}

INPUT   P0   {Input initial approximation}

```
DO    FOR   K = 1   TO   99   UNTIL   Cond ≠ 0
         P1 := g(P0)                                              {First new iterate}
         P2 := g(P1)                                              {Second new iterate}
         D1 := [P1 − P0]²                                         {Form the
         D2 := P2 − 2∗P1 + P0                                     differences}
            IF   D2 = 0   THEN                                    {Check division
                    Cond := 1                                     by zero}
                    DP := P2 − P1
                    P3 := P2
            ELSE
                DP := D1/D2
                P3 := P0 − DP                                     {Aitken's improvement}
            ENDIF
         RelErr := 2∗|DP|/[|P3|+Small]                            {Relative error}
         IF  RelErr < Delta   THEN                                {Check for
            IF Cond ≠ 1   THEN Cond := 2                          convergence}
         P0 := P3                                                 {Update iterate}
```

PRINT 'The computed fixed point of g(x) is' P0   {Output}
PRINT 'Consecutive iterates are closer than' DP
IF   Cond = 0   THEN
PRINT 'The maximum number of iterations was exceeded.'
IF   Cond = 1   THEN
PRINT 'Division by zero was encountered.'
IF   Cond = 2   THEN
PRINT 'The solution was found with the desired tolerance.'

Algorithm 2.8 (Muller's Method).   To find a root of the equation $f(x) = 0$ given three distinct initial approximations p_0, p_1, and p_2. (Eliminate the step that suppresses complex roots if complex variables are not used.)

Delta := 10^{-6}, Epsilon := 10^{-6}, Small := 10^{-6} {Tolerances}
Max := 99   {Maximum number of iterations}
Satisfied := "False"   {Condition for loop termination}

INPUT P0, P1, P2   {Input initial approximations}
Y0 := F(P0), Y1 := F(P1), Y2 := F(P2)   {Function values}

```
DO  FOR K = 1 TO Max UNTIL Satisfied = "True"
     H0 := P0 − P2, H1 := P1 − P2                          {Form differences}
     C := Y2, E0 := Y0 − C, E1 := Y1 − C
     Det := H0*H1*[H0 − H1]                                {Compute determinants
     A := [E0*H1 − H0*E1]/Det                                  and solve the
     B := [H0*H0*E1 − H1*H1*E0]/Det                            linear system}
     IF   B*B>4*A*C THEN Disc := SQRT(B*B−4*A*C)
     └── ELSE Disc := 0                              {This suppresses complex roots}
     IF   B<0   THEN   Disc := −Disc                        {Find the smallest
     Z := −2*C/[B + Disc]                                 root of the quadratic}
     P3 := P2 + Z                                         {Newest approximation}
     IF  |P3−P1| < |P3−P0|   THEN                               {Sort to make
     └── U := P1, P1 := P0, P0 := U, V := Y1, Y1 := Y0, Y0 := V    the values
     IF  |P3−P2|<|P3−P1|   THEN                                 P0, P1 closest
     └── U := P2, P2 := P1, P1 := U, V := Y2, Y2 := Y1, Y1 := V    to P3}
     P2 := P3, Y2 := F(P2)                                 {Update iterate}
     RelErr := 2*|Z|/[|P2|+Small]                          {Relative error}
     IF   RelErr < Delta   AND   |Y2| < Epsilon   THEN         {Check for
           Satisfied := "True"                                convergence}
END
```

```
IF   Satisfied = "True"   THEN                               {Output}
     PRINT "The computed root of f(x) is" P2
     PRINT "Consecutive iterates are closer than" |Z|
     └── PRINT "The current function value is" Y2
IF   Satisfied ≠ "True"   THEN
     PRINT "Muller's method did not find a root of f(x) = 0."
```

EXERCISES FOR AITKEN'S, STEFFENSEN'S, AND MULLER'S METHODS

1. Let $p_n = 1/2^n$. Show that $q_n = 0$ for all n, where q_n is given by formula (4).

2. Let $p_n = 1/n$. Show that $q_n = 1/(2n + 2)$ for all n, hence there is little acceleration of convergence. Does $\{p_n\}$ converge to 0 linearly? Why?

3. Let $p_n = 1/(2^n - 1)$. Show that $q_n = 1/(4^{n+1} - 1)$ for all n.

4. The sequence $p_n = 1/(4^n + 4^{-n})$ converges linearly to 0. Use Aitken's formula (4) to find q_1, q_2, and q_3 and hence speed up the convergence.

n	p_n	q_n
0	0.5	−0.26437542
1	0.23529412	_____
2	0.06225681	_____
3	0.01562119	_____
4	0.00390619	
5	0.00097656	

5. The sequence $\{p_n\}$ generated by fixed-point iteration, starting with $p_0 = 2.5$, and using the function $g(x) = (6 + x)^{1/2}$ converges linearly to $p = 3$. Use Aitken's formula (4) to find q_1, q_2, and q_3 and hence speed up the convergence.

n	p_n	q_n
0	2.5	3.00024351
1	2.91547595	_____
2	2.98587943	_____
3	2.99764565	_____
4	2.99960758	
5	2.99993460	

6. The sequence $\{p_n\}$ generated by fixed-point iteration, starting with $p_0 = 3.14$, and using the function $g(x) = \ln(x) + 2$ converges linearly to $p \approx 3.14619322$. Use Aitken's formula (4) to find q_1, q_2, and q_3 and hence speed up the convergence.

n	p_n	q_n
0	3.14	3.14619413
1	3.14422280	_____
2	3.14556674	_____
3	3.14599408	_____
4	3.14612992	
5	3.14617310	

7. For the equation $\cos(x) - 1 = 0$, the Newton–Raphson iteration function is $g(x) = x - [1 - \cos(x)]/\sin(x) = x - \tan(x/2)$. Use Steffensen's algorithm with $g(x)$ and start with $p_0 = 0.5$ and find p_1, p_2, and p_3, then find the next three values.

8. For the equation $x - \sin(x) = 0$, the Newton–Raphson iteration function is $g(x) = x - [x - \sin(x)]/[1 - \cos(x)]$. Use Steffensen's algorithm with $g(x)$ and start with $p_0 = 0.5$ and find p_1, p_2, and p_3, then find the next three values.

9. For the equation $\sin(x^3) = 0$, the Newton–Raphson iteration function is $g(x) = x - \sin(x^3)/[3x^2 \cos(x^3)] = x - \tan(x^3)/[3x^2]$. Use Steffensen's algorithm with $g(x)$ and start with $p_0 = 0.5$ and find p_1, p_2, and p_3, then find the next three values.

10. *Convergence of series.* Aitken's method can be used to speed up the convergence of a series. If the nth term of the series is

$$S_n = \sum_{k=1}^{n} A_k,$$

show that the derived series using Aitken's method is

$$T_n = S_n + \frac{A_{n+1}^2}{A_{n+1} - A_{n+2}}.$$

In Exercises 11–14, apply Aitken's method and the result of Exercise 10 to speed up convergence of the series.

11. $S_n = \displaystyle\sum_{k=1}^{n} (0.99)^k$

12. $S_n = \displaystyle\sum_{k=1}^{n} \dfrac{1}{4^k + 4^{-k}}$

13. $S_n = \displaystyle\sum_{k=1}^{n} \dfrac{k}{2^{k-1}}$

14. $S_n = \displaystyle\sum_{k=1}^{n} \dfrac{1}{k \times 2^k}$

15. Use Muller's method to find the root of $f(x) = x^3 - x - 2$. Start with $p_0 = 1.0$, $p_1 = 1.2$, and $p_2 = 1.4$ and find p_3, p_4, and p_5.

16. Use Muller's method to find a root of $f(x) = 4x^2 - \exp(x)$. Start with $p_0 = 4.0$, $p_1 = 4.1$, and $p_2 = 4.2$ and find p_3, p_4, and p_5.

17. Use Muller's method to find a root of $f(x) = 1 + 2x - \tan(x)$. Start with $p_0 = 1.5$, $p_1 = 1.4$, and $p_2 = 1.3$ and find p_3, p_4, and p_5.

18. Use Muller's method to find a root of $f(x) = 3\cos(x) + 2\sin(x)$. Start with $p_0 = 2.4$, $p_1 = 2.3$, and $p_2 = 2.2$ and find p_3, p_4, and p_5.

19. Start with formula (8) and add the terms p_{n+2} and $-p_{n+2}$ to the right side and show that an equivalent formula is

$$p \approx p_{n+2} - \frac{(p_{n+2} - p_{n+1})^2}{p_{n+2} - 2p_{n+1} + p_n} = q_n.$$

20. Prove statement (5) of Theorem 2.8.

21. Assume that the error in an iteration process satisfies the relation $e_{n+1} = Ke_n$ for some constant K and $|K| < 1$.
    (a) Find an expression for e_n that involves e_0, K, and n.
    (b) Find an expression for the smallest integer N so that

$$|e_N| < 10^{-8}.$$

2.6 ITERATION FOR NONLINEAR SYSTEMS

Iterative techniques will now be discussed that extend the previous methods to the case of nonlinear functions. Consider the functions

$$f_1(x, y) = x^2 - 2x - y + 0.5,$$
$$f_2(x, y) = x^2 + 4y^2 - 4. \tag{1}$$

We seek a method of solution for the system of nonlinear equations

$$f_1(x, y) = 0 \quad \text{and} \quad f_2(x, y) = 0. \tag{2}$$

The equations $f_1(x, y) = 0$ and $f_2(x, y) = 0$ implicitly define curves in the xy-plane. Hence a solution of the system (2) is a point (p, q) where the two curves cross [i.e., both $f_1(p, q) = 0$ and $f_2(p, q) = 0$]. The curves for the system in (1) are well known:

$$x^2 - 2x - y + 0.5 = 0 \qquad \text{is the graph of a parabola.}$$
$$x^2 + 4y^2 - 4 = 0 \qquad \text{is the graph of an ellipse.} \tag{3}$$

The graphs in Figure 2.20 show that there are two solution points, and that they are in the vicinity of $(-0.2, 1.0)$ and $(1.9, 0.3)$.

The first technique is fixed-point iteration. A method must be devised for generating a sequence $\{(p_k, q_k)\}$ that converges to the solution (p, q). The first equation in (3) can be used to solve directly for x. However, a multiple of y can be added to each side of the second equation to get $x^2 + 4y^2 - 8y - 4 = -8y$. The choice of adding $-8y$ is crucial and will be explained later. We now have an equivalent system of equations:

$$x = \frac{x^2 - y + 0.5}{2}, \tag{4}$$

$$y = \frac{-x^2 - 4y^2 + 8y + 4}{8}.$$

These two equations are used to write down the recursive formulas. Start with an initial point (p_0, q_0), then compute the sequence $\{(p_{k+1}, q_{k+1})\}$ using

$$p_{k+1} = g_1(p_k, q_k) = \frac{p_k^2 - q_k + 0.5}{2}, \tag{5}$$

$$q_{k+1} = g_2(p_k, q_k) = \frac{-p_k^2 - 4q_k^2 + 8q_k + 4}{8}.$$

Case (i): If we use the starting value $(p_0, q_0) = (0, 1)$, then

$$p_1 = \frac{0^2 - 1 + 0.5}{2} = -0.25 \quad q_1 = \frac{-0^2 - 4 \times 1^2 + 8 \times 1 + 4}{8} = 1.0.$$

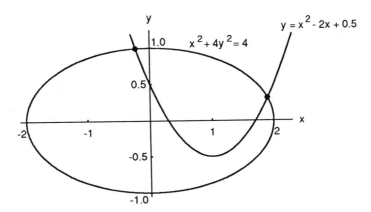

Figure 2.20 Graphs for the nonlinear system $y = x^2 - 2x + 0.5$ and $x^2 + 4y^2 = 4$.

Iteration will generate the sequence in case (i) of Table 2.14. In this case the sequence converges to the solution that lies near the starting value $(0, 1)$.

Case (ii): If we use the starting value $(p_0, q_0) = (2, 0)$, then

$$p_1 = \frac{2^2 - 0 + 0.5}{2} = 2.25 \quad q_1 = \frac{-2^2 - 4 \times 0^2 + 8 \times 0 + 4}{8} = 0.0.$$

Iteration will generate the sequence in case (ii) of Table 2.14. In this case the sequence diverges away from the solution.

TABLE 2.14 Fixed-Point Iteration Using the Formulas in (5)

	Case (i): Start with (0, 1)			Case (ii): Start with (2, 0)	
k	p_k	q_k	k	p_k	q_k
0	0.00	1.00	0	2.00	0.00
1	−0.25	1.00	1	2.25	0.00
2	−0.21875	0.9921875	2	2.78125	−0.1328125
3	−0.2221680	0.9939880	3	4.184082	−0.6085510
4	−0.2223147	0.9938121	4	9.307547	−2.4820360
5	−0.2221941	0.9938029	5	44.80623	−15.891091
6	−0.2222163	0.9938095	6	1,011.995	−392.60426
7	−0.2222147	0.9938083	7	512,263.2	−205,477.82
8	−0.2222145	0.9938084		This sequence is diverging.	
9	−0.2222146	0.9938084			

Iteration using formulas (5) cannot be used to find the second solution $(1.900677, 0.3112186)$. To find this point a different pair of iteration formulas are needed. Start with equations (3) and add $-2x$ to the first equation and $-11y$ to the second equation and get

$$x^2 - 4x - y + 0.5 = -2x \quad \text{and} \quad x^2 + 4y^2 - 11y - 4 = -11y.$$

These equations can then be used to obtain the iteration formulas

$$p_{k+1} = g_1(p_k, q_k) = \frac{-p_k^2 + 4p_k + q_k - 0.5}{2},$$

$$q_{k+1} = g_2(p_k, q_k) = \frac{-p_k^2 - 4q_k^2 + 11q_k + 4}{11}. \tag{6}$$

Table 2.15 shows how to use (6) to find the second solution.

TABLE 2.15   Fixed-Point
Iteration Using the
Formulas in (6)

k	p_k	q_k
0	2.00	0.00
1	1.75	0.0
2	1.71875	0.0852273
3	1.753063	0.1776676
4	1.808345	0.2504410
8	1.903595	0.3160782
12	1.900924	0.3112267
16	1.900652	0.3111994
20	1.900677	0.3112196
24	1.900677	0.3112186

Theory

We want to determine why equations (6) were suitable for finding the solution near (1.9, 0.3) and equations (5) were not. In Section 2.1 the size of the derivative at the fixed point was the necessary idea. When functions of several variables are used, the partial derivatives must be used. The generalization of "the derivative" for systems of functions of several variables is the Jacobian matrix. We will make only a few introductory ideas regarding this topic. More details can be found in any textbook on advanced calculus.

Definition 2.7   (Jacobian Matrix).   Assume that $f_1(x, y)$ and $f_2(x, y)$ are functions of the independent variables x and y; then their Jacobian matrix $J(x, y)$ is

$$\begin{pmatrix} \dfrac{\partial f_1}{\partial x} & \dfrac{\partial f_1}{\partial y} \\[2ex] \dfrac{\partial f_2}{\partial x} & \dfrac{\partial f_2}{\partial y} \end{pmatrix}. \tag{7}$$

Similarly, if $f_1(x, y, z)$, $f_2(x, y, z)$, and $f_3(x, y, z)$ are functions of the independent variables, x, y, and z, their 3×3 Jacobian matrix is defined as follows:

$$\begin{pmatrix} \dfrac{\partial f_1}{\partial x} & \dfrac{\partial f_1}{\partial y} & \dfrac{\partial f_1}{\partial z} \\[2ex] \dfrac{\partial f_2}{\partial x} & \dfrac{\partial f_2}{\partial y} & \dfrac{\partial f_2}{\partial z} \\[2ex] \dfrac{\partial f_3}{\partial x} & \dfrac{\partial f_3}{\partial y} & \dfrac{\partial f_3}{\partial z} \end{pmatrix}. \tag{8}$$

Example 2.18

Find the Jacobian matrix $J(x, y, z)$ of order 3×3, at the points $(1, 3, 2)$ and $(3, 2, 1)$, for the three functions

$$f_1(x, y, z) = x^3 - y^2 + y - z^4 + z^2,$$

$$f_2(x, y, z) = xy + yz + xz,$$

$$f_3(x, y, z) = \frac{y}{xz}.$$

Solution. The nine partial derivatives are

$$\frac{\partial f_1}{\partial x} = 3x^2, \qquad \frac{\partial f_1}{\partial y} = -2y + 1, \qquad \frac{\partial f_1}{\partial z} = -4z^3 + 2z,$$

$$\frac{\partial f_2}{\partial x} = y + z, \qquad \frac{\partial f_2}{\partial y} = x + z, \qquad \frac{\partial f_2}{\partial z} = y + x,$$

$$\frac{\partial f_3}{\partial x} = \frac{-y}{x^2 z}, \qquad \frac{\partial f_3}{\partial y} = \frac{1}{xz}, \qquad \frac{\partial f_3}{\partial z} = \frac{-y}{xz^2}.$$

These functions form the nine elements in the Jacobian matrix

$$J(x, y, z) = \begin{pmatrix} 3x^2 & -2y + 1 & -4z^3 + 2z \\ y + z & x + z & y + x \\ \dfrac{-y}{x^2 z} & \dfrac{1}{xz} & \dfrac{-y}{xz^2} \end{pmatrix}.$$

The values of $J(1, 3, 2)$ and $J(3, 2, 1)$ are

$$J(1, 3, 2) = \begin{pmatrix} 3 & -5 & -28 \\ 5 & 3 & 4 \\ -\frac{3}{2} & \frac{1}{2} & -\frac{3}{4} \end{pmatrix}, \quad J(3, 2, 1) = \begin{pmatrix} 27 & -3 & -2 \\ 3 & 4 & 5 \\ -\frac{2}{9} & \frac{1}{3} & -\frac{2}{3} \end{pmatrix}.$$

Generalized Differential

For a function of several variables, the differential is used to show how changes of the independent variable affect the change in the dependent variable. Suppose that we have three functions, each of which is a function of three variables,

$$u = f_1(x, y, z), \qquad v = f_2(x, y, z), \qquad w = f_3(x, y, z). \tag{9}$$

Suppose that the values of the functions in (9) are known at the point (x_0, y_0, z_0) and we wish to predict their value at a nearby point (x, y, z). Let du, dv, and dw denote differential changes in the dependent variables and dx, dy, and dz denote differential changes in the independent variables. These changes obey the relationships

$$du = \frac{\partial f_1}{\partial x}(x_0, y_0, z_0)\, dx + \frac{\partial f_1}{\partial y}(x_0, y_0, z_0)\, dy + \frac{\partial f_1}{\partial z}(x_0, y_0, z_0)\, dz,$$

$$dv = \frac{\partial f_2}{\partial x}(x_0, y_0, z_0)\, dx + \frac{\partial f_2}{\partial y}(x_0, y_0, z_0)\, dy + \frac{\partial f_2}{\partial z}(x_0, y_0, z_0)\, dz, \qquad (10)$$

$$dw = \frac{\partial f_3}{\partial x}(x_0, y_0, z_0)\, dx + \frac{\partial f_3}{\partial y}(x_0, y_0, z_0)\, dy + \frac{\partial f_3}{\partial z}(x_0, y_0, z_0)\, dz.$$

If vector notation is used, (10) can be compactly written by using the Jacobian matrix. The function changes are $d\mathbf{F}$ and the changes in the variables are denoted $d\mathbf{X}$

$$d\mathbf{F} = \begin{pmatrix} du \\ dv \\ dw \end{pmatrix} = J(x_0, y_0, z_0) \begin{pmatrix} dx \\ dy \\ dz \end{pmatrix} = J(x_0, y_0, z_0)\, d\mathbf{X}. \qquad (11)$$

Example 2.19

Use the Jacobian matrix to find the differential changes (du, dv, dw) when the independent variables change from $(1, 3, 2)$ to $(1.02, 2.97, 2.01)$ for the system of functions

$$u = f_1(x, y, z) = x^3 - y^2 + y - z^4 + z^2,$$

$$v = f_2(x, y, z) = xy + yz + xz,$$

$$w = f_3(x, y, z) = \frac{y}{xz}.$$

Solution. Use equation (11) with $J(1, 3, 2)$ of Example 2.18 and the differential changes $(dx, dy, dz) = (0.02, -0.03, 0.01)$ to obtain

$$\begin{pmatrix} du \\ dv \\ dw \end{pmatrix} = \begin{pmatrix} 3 & -5 & -28 \\ 5 & 3 & 4 \\ -\frac{3}{2} & \frac{1}{2} & -\frac{3}{4} \end{pmatrix} \begin{pmatrix} 0.02 \\ -0.03 \\ 0.01 \end{pmatrix} = \begin{pmatrix} -0.07 \\ 0.05 \\ -0.0525 \end{pmatrix}.$$

Notice that the function values at $(1.02, 2.97, 2.01)$ are close to the linear approximations obtained by adding the differentials $du = -0.07$, $dv = 0.05$, and $dw = -0.0525$ to the corresponding function values $f_1(1, 3, 2) = -17$, $f_2(1, 3, 2) = 11$, and $f_3(1, 3, 2) = 1.5$; that is,

$$f_1(1.02, 2.97, 2.01) = -17.072 \approx -17.07 = f_1(1, 3, 2) + du,$$

$$f_2(1.02, 2.97, 2.01) = 11.0493 \approx 11.05 = f_2(1, 3, 2) + dv,$$

$$f_3(1.02, 2.97, 2.01) = 1.44863916 \approx 1.4475 = f_3(1, 3, 2) + dw.$$

Convergence Near Fixed Points

The extension of the definitions and theorem in Section 2.1 to the case of two and three dimensions are now given. The notation for N-dimensional functions has not been used. The reader can easily find these extensions in many books on numerical analysis.

Definition 2.8.   A **fixed point** for the system of two equations

$$x = g_1(x, y) \quad \text{and} \quad y = g_2(x, y) \tag{12}$$

is a point (p, q) such that $p = g_1(p, q)$ and $q = g_2(p, q)$. Similarly, in three dimensions, a **fixed point** for the system

$$x = g_1(x, y, z), \quad y = g_2(x, y, z), \quad \text{and} \quad z = g_3(x, y, z) \tag{13}$$

is a point (p, q, r) such that $p = g_1(p, q, r)$, $q = g_2(p, q, r)$, and $r = g_3(p, q, r)$.

Definition 2.9.   For the functions (12), **fixed-point iteration** is

$$p_{k+1} = g_1(p_k, q_k) \quad \text{and} \quad q_{k+1} = g_2(p_k, q_k) \quad \text{for } k = 0, 1, \ldots. \tag{14}$$

Similarly, for the functions (13), **fixed-point iteration** is

$$p_{k+1} = g_1(p_k, q_k, r_k), \quad q_{k+1} = g_2(p_k, q_k, r_k), \quad \text{and} \quad r_{k+1} = g_3(p_k, q_k, r_k). \tag{15}$$

Theorem 2.9   (Fixed-Point Iteration).   Assume that the functions in (12) and (13) and their first partial derivatives are continuous on a region that contains the fixed point (p, q) or (p, q, r), respectively. If the starting point is chosen sufficiently close to the fixed point, then one of the following cases applies.

Case (i): Two dimensions. If (p_0, q_0) is sufficiently close to (p, q) and if

$$\left| \frac{\partial g_1}{\partial x}(p, q) \right| + \left| \frac{\partial g_1}{\partial y}(p, q) \right| < 1,$$
$$\left| \frac{\partial g_2}{\partial x}(p, q) \right| + \left| \frac{\partial g_2}{\partial y}(p, q) \right| < 1, \tag{16}$$

then the iteration in (14) converges to the fixed point (p, q).

Case (ii): Three dimensions. If (p_0, q_0, r_0) is sufficiently close to (p, q, r) and if

$$\left| \frac{\partial g_1}{\partial x}(p, q, r) \right| + \left| \frac{\partial g_1}{\partial y}(p, q, r) \right| + \left| \frac{\partial g_1}{\partial z}(p, q, r) \right| < 1,$$
$$\left| \frac{\partial g_2}{\partial x}(p, q, r) \right| + \left| \frac{\partial g_2}{\partial y}(p, q, r) \right| + \left| \frac{\partial g_2}{\partial z}(p, q, r) \right| < 1, \tag{17}$$
$$\left| \frac{\partial g_3}{\partial x}(p, q, r) \right| + \left| \frac{\partial g_3}{\partial y}(p, q, r) \right| + \left| \frac{\partial g_3}{\partial z}(p, q, r) \right| < 1,$$

then the iteration in (15) converges to the fixed point (p, q, r).

If conditions (16) or (17) are not met, the iteration might diverge. This will usually be the case if the sum of the magnitude of the partial derivatives is much larger than 1. Theorem 2.11 can be used to show why the iteration (5) converged to the fixed point near $(-0.2, 1.0)$. The partial derivatives are:

$$\frac{\partial}{\partial x} g_1(x, y) = x, \qquad \frac{\partial}{\partial y} g_1(x, y) = \frac{-1}{2},$$

$$\frac{\partial}{\partial x} g_2(x, y) = \frac{-x}{4}, \qquad \frac{\partial}{\partial y} g_2(x, y) = -y + 1,$$

Indeed, for all (x, y) satisfying $-0.5 < x < 0.5$ and $0.5 < y < 1.5$, the partial derivatives satisfy

$$\left| \frac{\partial}{\partial x} g_1(x, y) \right| + \left| \frac{\partial}{\partial y} g_1(x, y) \right| = |x| + |-0.5| < 1,$$

$$\left| \frac{\partial}{\partial x} g_2(x, y) \right| + \left| \frac{\partial}{\partial y} g_2(x, y) \right| = \frac{|-x|}{4} + |-y + 1| < 0.625 < 1.$$

Therefore, the partial derivative conditions in (16) are met and Theorem 2.11 implies that fixed-point iteration will converge to $(p, q) \approx (-0.2222146, 0.9938084)$. Notice that near the other fixed point $(1.90068, 0.31122)$ the partial derivatives do not meet the conditions in (16); hence convergence is not guaranteed. That is,

$$\left| \frac{\partial g_1}{\partial x} (1.90068, 0.31122) \right| + \left| \frac{\partial g_1}{\partial y} (1.90068, 0.31122) \right| = 2.40068 > 1,$$

$$\left| \frac{\partial g_2}{\partial x} (1.90068, 0.31122) \right| + \left| \frac{\partial g_2}{\partial y} (1.90068, 0.31122) \right| = 1.16395 > 1.$$

Seidel Iteration

An improvement of fixed-point iteration can be made. Suppose that p_{k+1} is used in the calculation of q_k (in three dimensions both p_{k+1} and q_{k+1} are used to compute r_k). When these modifications are incorporated in formulas (14) and (15), the method is called **Seidel iteration**:

$$p_{k+1} = g_1(p_k, q_k), \quad q_{k+1} = g_2(p_{k+1}, q_k), \tag{18}$$

and

$$p_{k+1} = g_1(p_k, q_k, r_k), \quad q_{k+1} = g_2(p_{k+1}, q_k, r_k), \quad r_{k+1} = g_3(p_{k+1}, q_{k+1}, r_k). \tag{19}$$

Algorithm 2.9 (Nonlinear Seidel Iteration). To solve the nonlinear fixed-point system

$$x = g_1(x, y, z),$$

$$y = g_2(x, y, z),$$

$$z = g_3(x, y, z),$$

given one initial approximation $\mathbf{P}_0 = (p_0, q_0, r_0)$, and generating a sequence $\{\mathbf{P}_k\} = \{(p_k, q_k, r_k)\}$ that converges to the solution $\mathbf{P} = (p, q, r)$ [i.e., $p = g_1(p, q, r)$, $q = g_2(p, q, r)$, and $r = g_3(p, q, r)$].

```
Tol  := 10⁻⁶                                      {Tolerance}
Sep := 1 , K := 0 , Max := 199                    {Initialize}
INPUT   P(0), Q(0), R(0)                           {Input}

WHILE   K < Max AND Sep > Tol DO
        K := K+1
        P(K) := G1(P(K−1), Q(K−1), R(K−1))         {Perform
        Q(K) := G2(P(K), Q(K−1), R(K−1))            Seidel
        R(K) := G3(P(K), Q(K), R(K−1))              iteration}
        Sep := |P(K)−P(K−1)| + |Q(K)−Q(K−1)| + |R(K)−R(K−1)|

IF      Sep < Tol   THEN                           {Output}
            PRINT "After ",K," iterations Seidel's method"
            PRINT "was successful and found the solution:"
        ELSE
            PRINT "Seidel iteration did not converge,"
            PRINT "After ",K," iterations the status is:"
ENDIF

        PRINT "P(",K,") = ", P(K)
        PRINT "Q(",K,") = ", Q(K)
        PRINT "R(",K,") = ", R(K)
```

EXERCISES FOR ITERATION FOR NONLINEAR SYSTEMS

In Exercises 1–6, for each nonlinear system,

$$x = g_1(x, y) \quad \text{and} \quad y = g_2(x, y)$$

(a) Find the Jacobian matrix $J(x, y)$ of partial derivatives

$$J(x,\ y) = \begin{vmatrix} \dfrac{\partial}{\partial x} g_1(x,\ y) & \dfrac{\partial}{\partial y} g_1(x,\ y) \\[2ex] \dfrac{\partial}{\partial x} g_2(x,\ y) & \dfrac{\partial}{\partial y} g_2(x,\ y) \end{vmatrix}.$$

Also find the matrix $J(p_0,\ q_0)$.

   (b) Compute $(p_1,\ q_1)$ and $(p_2,\ q_2)$ using fixed-point iteration and equations (14).
   (c) Compute $(p_1,\ q_1)$ and $(p_2,\ q_2)$ using Seidel iteration and equations (18).
   (d) Use a computer and find the solution $(p,\ q) = \lim\limits_{k \to \infty} (p_k,\ q_k)$.

1. $x = g_1(x,\ y) = \dfrac{8x - 4x^2 + y^2 + 1}{8}$ (hyperbola),

   $y = g_2(x,\ y) = \dfrac{2x - x^2 + 4y - y^2 + 3}{4}$ (circle),

   $(p_0,\ q_0) = (1.1,\ 2.0)$. See Figure 2.21.

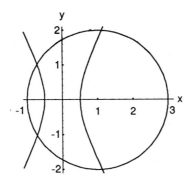

Figure 2.21   The hyperbola and circle for Exercise 1.

2. $x = g_1(x,\ y) = \dfrac{2x + x^2 - y}{2}$ (parabola),

   $y = g_2(x,\ y) = \dfrac{2x - x^2 + 8}{9} + \dfrac{4y - y^2}{4}$ (ellipse),

   $(p_0,\ q_0) = (-1.2,\ 1.4)$. See Figure 2.22.

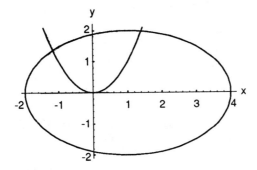

Figure 2.22   The parabola and ellipse for Exercises 2 and 3.

3. $x = g_1(x, y) = \dfrac{2x - x^2 + y}{2}$ (parabola),

$y = g_2(x, y) = \dfrac{2x - x^2 + 8}{9} + \dfrac{4y - y^2}{4}$ (ellipse),

$(p_0, q_0) = (1.4, 2.0)$. See Figure 2.22.

4. $x = g_1(x, y) = \dfrac{4x - x^3 + y}{4}$ (cubic),

$y = g_2(x, y) = -\dfrac{x^2}{9} + \dfrac{4y - y^2}{4} + 1$ (ellipse),

$(p_0, q_0) = (1.2, 1.8)$. See Figure 2.23.

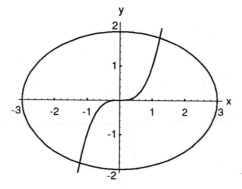

Figure 2.23 The cubic and ellipse for Exercise 4.

5. $x = g_1(x, y) = \dfrac{4x - x^2 + y + 3}{4}$ (parabola),

$y = g_2(x, y) = \dfrac{3 - xy + 2y}{2}$ (hyperbola),

$(p_0, q_0) = (2.1, 1.4)$. See Figure 2.24.

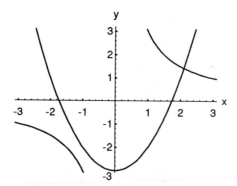

Figure 2.24 The parabola and hyperbola for Exercise 5.

6. $x = g_1(x, y) = \dfrac{y - x^3 + 3x^2 + 3x}{7}$ (cubic),

$y = g_2(x, y) = \dfrac{y^2 + 2y - x - 2}{2}$ (parabola),

$(p_0, q_0) = (-0.3, -1.3)$. See Figure 2.25.

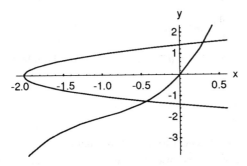

Figure 2.25   The cubic and parabola for Exercise 6.

7. Show that Jacobi iteration for 2×2 linear systems is a special case of fixed-point iteration (14). If A is diagonally dominant, then condition (16) is satisfied.

8. Show that Jacobi iteration for 3×3 linear systems is a special case of fixed-point iteration (15). If A is diagonally dominant, then condition (17) is satisfied.

9. We wish to solve the nonlinear system:

$$0 = 7x^3 - 10x - y - 1,$$

$$0 = 8y^3 - 11y + x - 1.$$

Sketch the graphs $y = 7x^3 - 10x - 1$ and $x = -8y^3 + 11y + 1$ and verify that there are nine points where the curves cross. Suppose that we use fixed-point iteration with the formulas

$$x = g_1(x, y) = \frac{7x^3 - y - 1}{10},$$

$$y = g_2(x, y) = \frac{8y^3 + x - 1}{11}.$$

Do some computer experimentation. Discover that no matter what starting value is used, only one of the nine solutions can be found using fixed-point iteration based on these two choices of $g_1(x, y)$ and $g_2(x, y)$. Which solution did you find? In Section 2.7 we will see that Newton's method is able to find all nine solutions.

10. Fixed point-iteration is used to solve the nonlinear system (12). Use the following steps to prove that conditions in (16) are sufficient to guarantee that $\{(p_k, q_k)\}$ converges to (p, q). Assume that there is a constant K with $0 < K < 1$, so that

$$\left| \frac{\partial}{\partial x} g_1(x, y) \right| + \left| \frac{\partial}{\partial y} g_1(x, y) \right| < K \quad \text{and} \quad \left| \frac{\partial}{\partial x} g_2(x, y) \right| + \left| \frac{\partial}{\partial y} g_2(x, y) \right| < K$$

for all (x, y) in the rectangle $a < x < b$, $c < y < d$. Also assume that $a < p_0 < b$ and $c < q_0 < d$. Define

$$e_k = p - p_k, \qquad E_k = q - q_k, \qquad r_k = \max\{|e_k|, |E_k|\}.$$

Use the following form of the mean value theorem applied to functions of two variables:

$$e_{k+1} = \frac{\partial}{\partial x} g_1(a_k^*, q_k)e_k + \frac{\partial}{\partial y} g_1(p, c_k^*)E_k,$$

$$E_{k+1} = \frac{\partial}{\partial x} g_2(b_k^*, q_k)e_k + \frac{\partial}{\partial y} g_2(p, d_k^*)E_k,$$

where a_k^* and b_k^* lie in $[a, b]$ and c_k^* and d_k^* lie in $[c, d]$. Prove the following things:
(a) $|e_1| \le Kr_0$ and $|E_1| \le Kr_0$
(b) $|e_2| \le Kr_1 \le K^2 r_0$ and $|E_2| \le Kr_1 \le K^2 r_0$
(c) $|e_k| \le Kr_{k-1} \le K^k r_0$ and $|E_k| \le Kr_{k-1} \le K^k r_0$
(d) $\lim_{k \to \infty} p_k = p$ and $\lim_{k \to \infty} p_k = q$

11. Write a report on the contraction-mapping fixed-point theorem.
12. Rewrite Algorithm 2.9 so that it does not use arrays to store all the calculations.

2.7 NEWTON'S METHOD FOR SYSTEMS

Transformations in Two Dimensions

Consider the system of nonlinear equations

$$u = f_1(x, y),$$
$$v = f_2(x, y),$$
(1)

which can be considered a transformation from the xy-plane into the uv-plane. We are interested in the behavior of this transformation near the point (x_0, y_0) whose image is the point (u_0, v_0). If the two functions have continuous partial derivatives, then the differential can be used to write a system of linear approximations that are valid near the point (x_0, y_0):

$$u - u_0 \approx \frac{\partial}{\partial x} f_1(x_0, y_0)(x - x_0) + \frac{\partial}{\partial y} f_1(x_0, y_0)(y - y_0),$$

$$v - v_0 \approx \frac{\partial}{\partial x} f_2(x_0, y_0)(x - x_0) + \frac{\partial}{\partial y} f_2(x_0, y_0)(y - y_0).$$
(2)

The system (2) is a local linear transformation that relates small changes in the independent variables to small changes in the dependent variable. When the Jacobian matrix $J(x_0, y_0)$ is used this relationship is easier to visualize:

$$\begin{pmatrix} u - u_0 \\ v - v_0 \end{pmatrix} \approx \begin{pmatrix} \dfrac{\partial}{\partial x} f_1(x_0, y_0) & \dfrac{\partial}{\partial y} f_1(x_0, y_0) \\ \dfrac{\partial}{\partial x} f_2(x_0, y_0) & \dfrac{\partial}{\partial y} f_2(x_0, y_0) \end{pmatrix} \begin{pmatrix} x - x_0 \\ y - y_0 \end{pmatrix}. \tag{3}$$

If the system in (1) is written as a vector function $\mathbf{V} = \mathbf{F}(\mathbf{X})$, the Jacobian $J(x, y)$ is the two-dimensional analog of the derivative, because (3) can be written

$$\Delta \mathbf{F} \approx J(x_0, y_0) \Delta \mathbf{X}. \tag{4}$$

Newton's Method in Two Dimensions

Consider the system (1) with u and v set equal to zero:

$$\begin{aligned} 0 &= f_1(x, y), \\ 0 &= f_2(x, y). \end{aligned} \tag{5}$$

Suppose that (p, q) is a solution of (5); that is,

$$\begin{aligned} 0 &= f_1(p, q), \\ 0 &= f_2(p, q). \end{aligned} \tag{6}$$

To develop Newton's method for solving (5), we need to consider small changes in the functions near the point (p_0, q_0):

$$\begin{aligned} \Delta u &= u - u_0, \quad \Delta p = x - p_0. \\ \Delta v &= v - v_0, \quad \Delta q = y - q_0. \end{aligned} \tag{7}$$

Set $(x, y) = (p, q)$ in (1), and use (6) to see that $(u, v) = (0, 0)$. Hence changes in the dependent variables are

$$\begin{aligned} u - u_0 &= f_1(p, q) - f_1(p_0, q_0) = 0 - f_1(p_0, q_0), \\ v - v_0 &= f_2(p, q) - f_2(p_0, q_0) = 0 - f_2(p_0, q_0). \end{aligned} \tag{8}$$

Use the result of (8) in (3) to get the linear approximation

$$\begin{pmatrix} \dfrac{\partial}{\partial x} f_1(p_0, q_0) & \dfrac{\partial}{\partial y} f_1(p_0, q_0) \\ \dfrac{\partial}{\partial x} f_2(p_0, q_0) & \dfrac{\partial}{\partial y} f_2(p_0, q_0) \end{pmatrix} \begin{pmatrix} \Delta p \\ \Delta q \end{pmatrix} \approx - \begin{pmatrix} f_1(p_0, q_0) \\ f_2(p_0, q_0) \end{pmatrix}. \tag{9}$$

If the Jacobian $J(p_0, q_0)$ in (9) is nonsingular, we can solve for $\Delta \mathbf{P} = (\Delta p, \Delta q) = (p, q) - (p_0, q_0)$ as follows:

$$\Delta \mathbf{P} \approx -J(p_0, q_0)^{-1} \mathbf{F}(p_0, q_0). \tag{10}$$

The next approximation $\mathbf{P}_1$ to the solution $\mathbf{P}$ is

$$\mathbf{P}_1 = \mathbf{P}_0 + \Delta\mathbf{P} = \mathbf{P}_0 - J(p_0, q_0)^{-1}\mathbf{F}(p_0, q_0). \tag{11}$$

Notice that (11) is the generalization of Newton's method for the one-variable case; that is, $p_1 = p_0 - [f'(p_0)]^{-1}f(p_0)$.

Outline of Newton's Method

Suppose that $\mathbf{P}_k$ has been obtained.

Step 1. Evaluate the function

$$\mathbf{F}(\mathbf{P}_k) = \begin{pmatrix} f_1(p_k, q_k) \\ f_2(p_k, q_k) \end{pmatrix}.$$

Step 2. Evaluate the Jacobian

$$J(\mathbf{P}_k) = \begin{vmatrix} \dfrac{\partial}{\partial x}f_1(p_k, q_k) & \dfrac{\partial}{\partial y}f_1(p_k, q_k) \\ \dfrac{\partial}{\partial x}f_2(p_k, q_k) & \dfrac{\partial}{\partial y}f_2(p_k, q_k) \end{vmatrix}.$$

Step 3. Solve the linear system

$$J(\mathbf{P}_k)\Delta\mathbf{P} = -\mathbf{F}(\mathbf{P}_k) \quad \text{for } \Delta\mathbf{P}.$$

Step 4. Compute the next point:

$$\mathbf{P}_{k+1} = \mathbf{P}_k + \Delta\mathbf{P}.$$

Now, repeat the process.

Example 2.20

Consider the nonlinear system:

$$0 = x^2 - 2x - y + 0.5,$$
$$0 = x^2 + 4y^2 - 4.$$

Use Newton's method with the starting value $(p_0, q_0) = (2.00, 0.25)$ and compute (p_1, q_1), (p_2, q_2), and (p_3, q_3).

Solution. The function vector and Jacobian matrix are

$$\mathbf{F}(x, y) = \begin{pmatrix} x^2 - 2x - y + 0.5 \\ x^2 + 4y^2 - 4 \end{pmatrix}, \quad J(x, y) = \begin{pmatrix} 2x - 2 & -1 \\ 2x & 8y \end{pmatrix}.$$

At the point $(2.00, 0.25)$ they take on the values

$$\mathbf{F}(2.00, 0.25) = \begin{pmatrix} 0.25 \\ 0.25 \end{pmatrix}, \quad J(2.00, 0.25) = \begin{pmatrix} 2.0 & -1.0 \\ 4.0 & 2.0 \end{pmatrix}.$$

The differentials Δp and Δq are solutions of the linear system

$$\begin{pmatrix} 2.0 & -1.0 \\ 4.0 & 2.0 \end{pmatrix} \begin{pmatrix} \Delta p \\ \Delta q \end{pmatrix} = -\begin{pmatrix} 0.25 \\ 0.25 \end{pmatrix}.$$

A straightforward calculation reveals that

$$\Delta \mathbf{P} = \begin{pmatrix} \Delta p \\ \Delta q \end{pmatrix} = \begin{pmatrix} -0.09375 \\ 0.0625 \end{pmatrix}.$$

The next point in the iteration is

$$\mathbf{P}_1 = \mathbf{P}_0 + \Delta \mathbf{P} = \begin{pmatrix} 2.00 \\ 0.25 \end{pmatrix} + \begin{pmatrix} -0.09375 \\ 0.0625 \end{pmatrix} = \begin{pmatrix} 1.90625 \\ 0.3125 \end{pmatrix}.$$

Similarly, the next two points are

$$\mathbf{P}_2 = \begin{pmatrix} 1.900691 \\ 0.311213 \end{pmatrix} \quad \text{and} \quad \mathbf{P}_3 = \begin{pmatrix} 1.900677 \\ 0.311219 \end{pmatrix}.$$

The coordinates of $\mathbf{P}_3$ are accurate to six decimal places. Calculations for finding $\mathbf{P}_2$ and $\mathbf{P}_3$ are summarized in Table 2.16.

TABLE 2.16 Function Values, Jacobian Matrices, and Differentials Required for Each Iteration in Newton's Solution to Example 2.20

$\mathbf{P}_k$	Solution of the linear system $\quad J(\mathbf{P}_k)\Delta \mathbf{P} = -\mathbf{F}(\mathbf{P}_k)$		$\mathbf{P}_k + \Delta \mathbf{P}$
$\begin{pmatrix} 2.00 \\ 0.25 \end{pmatrix}$	$\begin{pmatrix} 2.0 & -1.0 \\ 4.0 & 2.0 \end{pmatrix}$	$\begin{pmatrix} -0.09375 \\ 0.0625 \end{pmatrix} = -\begin{pmatrix} 0.25 \\ 0.25 \end{pmatrix}$	$\begin{pmatrix} 1.90625 \\ 0.3125 \end{pmatrix}$
$\begin{pmatrix} 1.90625 \\ 0.3125 \end{pmatrix}$	$\begin{pmatrix} 1.8125 & -1.0 \\ 3.8125 & 2.5 \end{pmatrix}$	$\begin{pmatrix} -0.005559 \\ -0.001287 \end{pmatrix} = -\begin{pmatrix} 0.008789 \\ 0.024414 \end{pmatrix}$	$\begin{pmatrix} 1.900691 \\ 0.311213 \end{pmatrix}$
$\begin{pmatrix} 1.900691 \\ 0.311213 \end{pmatrix}$	$\begin{pmatrix} 1.801381 & -1.000000 \\ 3.801381 & 2.489700 \end{pmatrix}$	$\begin{pmatrix} -0.000014 \\ 0.000006 \end{pmatrix} = -\begin{pmatrix} 0.000031 \\ 0.000038 \end{pmatrix}$	$\begin{pmatrix} 1.900677 \\ 0.311219 \end{pmatrix}$

Implementation of Newton's method can require the determination of several partial derivatives. It is permissible to use numerical approximations for the values of these partial derivatives, but care must be taken to determine the proper step size. In higher dimensions it is necessary to use the vector form for the functions and the general methods in Chapter 3 for solving the system (10) for $\Delta \mathbf{P}$.

Algorithm 2.10 (Newton–Raphson Method in Two Dimensions). To solve the nonlinear system

$$0 = f_1(x, y)$$

$$0 = f_2(x, y)$$

given one initial approximation (p_0, q_0) and using Newton–Raphson iteration.

Cond := 0, Max := 99, Delta := 10^{-5}, Epsilon := 10^{-5}, Small := 10^{-5}

INPUT P0, Q0   {Point must be close to the solution}
U0 := F₁(P0, Q0), V0 := F₂(P0, Q0)   {Compute the function values}

DO FOR K := 1 TO Max UNTIL Cond ≠ 0

    D(1, 1) := $\frac{\partial}{\partial x}$F₁(P0, Q0), D(1, 2) := $\frac{\partial}{\partial y}$F₁(P0, Q0)   {Compute the partial

    D(2, 1) := $\frac{\partial}{\partial x}$F₂(P0, Q0), D(2, 2) := $\frac{\partial}{\partial y}$F₂(P0, Q0)   derivatives}

    Det := D(1, 1)*D(2, 2) − D(1, 2)*D(2, 1)   {Compute the determinant}

    IF Det = 0 THEN   {Check division
        DP:= 0, DQ := 0, Cond := 1   by zero}
    ELSE
        DP := [U0*D(2, 2)−V0*D(1, 2)]/Det   {Solve the
        DQ := [V0*D(1, 1)−U0*D(2, 1)]/Det   linear system}

    P1 := P0 − DP, Q1 := Q0 − DQ   {New iterates and
    U1 := F₁(P1, Q1), V1 := F₂(P1, Q1)   function values}

    RelErr := [|DP|+|DQ|]/[|P1|+|Q1|+Small]   {Relative error}
    FnZero := |U1|+|V1|
    IF RelErr<Delta AND FnZero<Epsilon THEN   {Check for
        IF Cond ≠ 1 THEN Cond := 2   convergence}

    P0 := P1, Q0 := Q1, U0 := U1, V0 := V1   {Update values}

PRINT 'The current k-th iterate is' P1, Q1
PRINT 'The function values are' U1, V1
IF Cond=0 THEN PRINT 'The max. number of iterations was exceeded.'
IF Cond=1 THEN PRINT 'Division by zero was encounted.'
IF Cond=2 THEN PRINT 'The solution was found with the desired tolerances.'

EXERCISES FOR NEWTON'S METHOD FOR SYSTEMS

1. Consider the nonlinear system

$$0 = f_1(x, y) = x^2 - y - 0.2,$$

$$0 = f_2(x, y) = y^2 - x - 0.3.$$

These parabolas intersect in two points as shown in Figure 2.26.

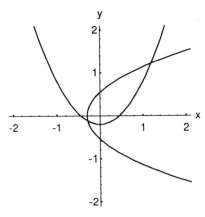

Figure 2.26   The parabolas for Exercise 1.

  (a) Start with $(p_0, q_0) = (1.2, 1.2)$ and use Newton's method to compute (p_1, q_1) and (p_2, q_2).
  (b) Start with $(p_0, q_0) = (-0.2, -0.2)$ and use Newton's method to compute (p_1, q_1) and (p_2, q_2).

2. Consider the nonlinear system

$$0 = f_1(x, y) = x^2 + y^2 - 2,$$

$$0 = f_2(x, y) = x^2 - y - 0.5x + 0.1.$$

The circle and parabola intersect in two points as shown in Figure 2.27.

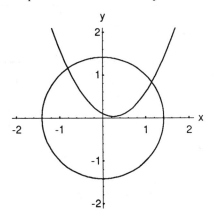

Figure 2.27   The circle and parabola for Exercise 2.

  (a) Start with $(p_0, q_0) = (1.2, 0.8)$ and use Newton's method to compute (p_1, q_1) and (p_2, q_2).

(b) Start with $(p_0, q_0) = (-0.8, 1.2)$ and use Newton's method to compute (p_1, q_1) and (p_2, q_2).

3. Consider the nonlinear system shown in Figure 2.28.

$$0 = f_1(x, y) = x^2 + y^2 - 2,$$

$$0 = f_2(x, y) = xy - 1.$$

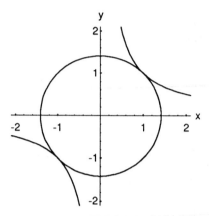

Figure 2.28   The circle and hyperbola for Exercise 3.

(a) Verify that the solutions are $(1, 1)$ and $(-1, -1)$.

(b) What difficulties might arise if we try to use Newton's method to find the solutions?

4. Show that Newton's method for two equations can be written in fixed-point iteration form

$$x = g_1(x, y), \qquad y = g_2(x, y).$$

where $g_1(x, y)$ and $g_2(x, y)$ are given by

$$g_1(x, y) = x - \frac{f_1(x, y) \dfrac{\partial}{\partial y} f_2(x, y) - f_2(x, y) \dfrac{\partial}{\partial y} f_1(x, y)}{D(x, y)},$$

$$g_2(x, y) = y - \frac{f_2(x, y) \dfrac{\partial}{\partial x} f_1(x, y) - f_1(x, y) \dfrac{\partial}{\partial x} f_2(x, y)}{D(x, y)},$$

and $D(x, y)$ is the determinant of the Jacobian

$$D(x, y) = |J(x, y)| = \frac{\partial}{\partial x} f_1(x, y) \frac{\partial}{\partial y} f_2(x, y) - \frac{\partial}{\partial x} f_2(x, y) \frac{\partial}{\partial y} f_1(x, y).$$

5. Write a report on the modified Newton method for systems of nonlinear equations.

In Exercises 6–11, solve the system of equations $0 = f_1(x, y)$ and $0 = f_2(x, y)$.

(a) Start with the given point (p_0, q_0) and use Newton's method to find (p_1, q_1) and (p_2, q_2).

(b) Use a computer program and find all the solutions

6. $0 = 2xy - 3$, $0 = x^2 - y - 2$. See Figure 2.29.
   (a) Start with $(p_0, q_0) = (1.5, 0.9)$.
7. $0 = x^2 + 4y^2 - 4$, $0 = x^2 - 2x - y + 1$. See Figure 2.30.
   (a) I. Start with $(p_0, q_0) = (1.5, 0.5)$.
       II. Start with $(p_0, q_0) = (-0.25, 1.1)$.

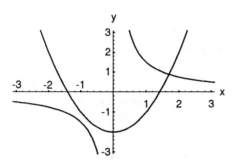

Figure 2.29 The hyperbola and parabola for Exercise 6.

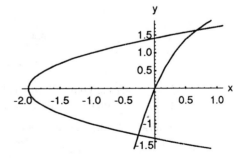

Figure 2.30 The ellipse and parabola for Exercise 7.

8. $0 = 3x^2 - 2y^2 - 1$, $0 = x^2 - 2x + y^2 + 2y - 8$
   (a) I. Start with $(p_0, q_0) = (-1.0, 1.0)$.
       II. Start with $(p_0, q_0) = (3.0, -3.4)$. See Figure 2.31.
9. $0 = -x + y^2 - 2$, $0 = x^3 - 3x^2 + 4x - y$
   (a) I. Start with $(p_0, q_0) = (0.5, 1.2)$.
       II. Start with $(p_0, q_0) = (-0.25, -1.30)$. See Figure 2.32.

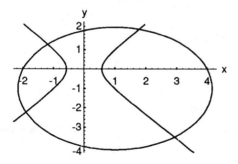

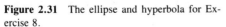

Figure 2.31 The ellipse and hyperbola for Exercise 8.

Figure 2.32 The parabola and cubic for Exercise 9.

10. $0 = 2x^3 - 12x - y - 1$, $0 = 3y^2 - 6y - x - 3$. See Figure 2.33.

   (a)   I. $(p_0, q_0) = (2.5, 2.5)$.   II. $(p_0, q_0) = (2.5, -1.0)$.
         III. $(p_0, q_0) = (0.0, 0.0)$.   IV. $(p_0, q_0) = (-2.5, 2.5)$.
         V. $(p_0, q_0) = (-2.5, 0.0)$.   VI. $(p_0, q_0) = (0.0, 2.5)$.

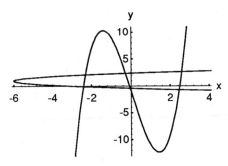

Figure 2.33   The cubic and parabola for Exercise 10.

11. $0 = 3x^2 - 2y^2 - 1$, $0 = x^2 - 2x + 2y - 8$. See Figure 2.34.

   (a)   I. $(p_0, q_0) = (2.5, 3.0)$.   II. $(p_0, q_0) = (-1.6, 1.6)$.
         III. $(p_0, q_0) = (5.6, -7.0)$.   IV. $(p_0, q_0) = (-3.0, -3.6)$.

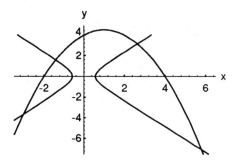

Figure 2.34   The hyperbola and parabola for Exercise 11.

12. Use Newton's method to find all nine solutions to

$$0 = 7x^3 - 10x - y - 1, \qquad 0 = 8y^3 - 11y + x - 1.$$

Use the starting points $(0, 0)$, $(1, 0)$, $(0, 1)$, $(-1, 0)$, $(0, -1)$, $(1, 1)$, $(-1, 1)$, $(1, -1)$, and $(-1, -1)$. Compare with Exercise 11 of Section 2.6.

In Exercises 13–18, extend Newton's method to three dimensions and find all solutions of the system.

13. $0 = x^2 - x + y^2 + z^2 - 5$, $0 = x^2 + y^2 - y + z^2 - 4$, $0 = x^2 + y^2 + z^2 + z - 6$.

   I. Start with $(p_0, q_0, r_0) = (-0.8, 0.2, 1.8)$.
   II. Start with $(p_0, q_0, r_0) = (1.2, 2.2, -0.2)$.

14. $0 = x^2 - x + 2y^2 + yz - 10, 0 = 5x - 6y + z, 0 = z - x^2 - y^2$.

   Start with $(p_0, q_0, r_0) = (1.1, 1.5, 3.5)$.

15. $0 = x^2 + y^2 - z^2 - 1, 0 = \dfrac{x^2}{4} + \dfrac{y^2}{9} + \dfrac{z^2}{4} - 1, 0 = x^2 + (y - 1)^2 - 1$.

   I. Start with $(p_0, q_0, r_0) = (0.8, 1.6, 1.5)$.
   II. Start with $(p_0, q_0, r_0) = (0.8, 1.6, -1.5)$.
   III. Start with $(p_0, q_0, r_0) = (-0.8, 1.6, 1.5)$.
   IV. Start with $(p_0, q_0, r_0) = (-0.8, 1.6, -1.5)$.

16. $0 = (x + 1)^2 + (y + 1)^2 - z, 0 = (x - 1)^2 + y^2 - z, 0 = 4x^2 + 2y^2 + z^2 - 16$.

17. $0 = 2x^2 + y^2 - 4z - 4, 0 = 3x^2 - 2y^2 - 5z, 0 = 4x^2 - 8x + 2y^2 + 4z^2$.

18. $0 = 9x^2 + 36y^2 + 4z^2 - 36, 0 = x^2 - 2y^2 - 20z, 0 = 16x - x^3 - 2y^2 - 16z^2$.

3

The Solution
of Linear Systems $A\mathbf{X} = \mathbf{B}$

Three planes form the boundary faces of a solid in the first octant which is shown in Figure 3.1. Suppose that the equations for these planes are

$$5x + y + z = 5, \quad x + 4y + z = 4 \quad \text{and} \quad x + y + 3z = 3.$$

What are the coordinates of the point of intersection of the three planes? Gaussian elimination can be used to find the solution of the linear system

$$x = 0.76, \quad y = 0.68, \quad \text{and} \quad z = 0.52.$$

In this chapter we develop numerical methods for solving systems of linear equations.

3.1 INTRODUCTION TO VECTORS AND MATRICES

Vectors

A real N-dimensional vector $\mathbf{X}$ is an ordered set of N real numbers and is usually written in the coordinate form

$$\mathbf{X} = (x_1, x_2, \ldots, x_N). \tag{1}$$

Here the numbers $x_1, x_2, \ldots, x_N$ are called the **components** of $\mathbf{X}$. The set consisting of all N-dimensional vectors is called **N-dimensional space**. When a vector is used to denote a point or position in space it is called a **position vector**. When it is used to denote a movement between two points in space it is called a **displacement vector**.

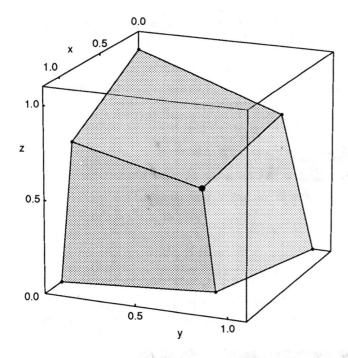

Figure 3.1 The intersection of three planes.

Let another vector be $\mathbf{Y} = (y_1, y_2, \ldots, y_N)$. The two vectors $\mathbf{X}$ and $\mathbf{Y}$ are said to be equal if and only if each corresponding coordinate is the same, that is,

$$\mathbf{X} = \mathbf{Y} \quad \text{if and only if} \quad x_j = y_j \qquad \text{for } j = 1, 2, \ldots, N. \tag{2}$$

The sum of the vectors $\mathbf{X}$ and $\mathbf{Y}$ is computed component by component, using the definition

$$\mathbf{X} + \mathbf{Y} = (x_1 + y_1, x_2 + y_2, \ldots, x_N + y_N). \tag{3}$$

The opposite of the vector $\mathbf{X}$ is obtained by replacing each coordinate with its opposite:

$$-\mathbf{X} = (-x_1, -x_2, \ldots, -x_N). \tag{4}$$

The difference $\mathbf{Y} - \mathbf{X}$ is formed by taking the difference in each coordinate:

$$\mathbf{Y} - \mathbf{X} = (y_1 - x_1, y_2 - x_2, \ldots, y_N - x_N). \tag{5}$$

Vectors in N-dimensional space obey the algebraic property

$$\mathbf{Y} - \mathbf{X} = \mathbf{Y} + (-\mathbf{X}). \tag{6}$$

If c is a real number (**scalar**), we define **scalar multiplication** $c\mathbf{X}$ as follows:

$$c\mathbf{X} = (cx_1, cx_2, \ldots, cx_N). \tag{7}$$

If c and d are scalars, then the weighted sum $c\mathbf{X} + d\mathbf{Y}$ is called a **linear combination** of $\mathbf{X}$ and $\mathbf{Y}$, and we write

$$c\mathbf{X} + d\mathbf{Y} = (cx_1 + dy_1, cx_2 + dy_2, \ldots, cx_N + dy_N). \tag{8}$$

The **dot product** of the two vectors $\mathbf{X}$ and $\mathbf{Y}$ is a scalar quantity (real number) defined by the equation

$$\mathbf{X} \cdot \mathbf{Y} = x_1 y_1 + x_2 y_2 + \cdots + x_N y_N. \tag{9}$$

The **length** (or **norm**) of the vector $\mathbf{X}$ is defined by

$$\|\mathbf{X}\| = (x_1^2 + x_2^2 + \cdots + x_N^2)^{1/2}. \tag{10}$$

Formula (10) is referred to as the **Euclidean norm** (or **length**) of the vector $\mathbf{X}$.

Scalar multiplication $c\mathbf{X}$ stretches the vector $\mathbf{X}$ when $|c| > 1$ and shrinks the vector when $|c| < 1$. This is shown by using equation (10):

$$\begin{aligned}
\|c\mathbf{X}\| &= (c^2 x_1^2 + c^2 x_2^2 + \cdots + c^2 x_N^2)^{1/2} \\
&= |c|(x_1^2 + x_2^2 + \cdots + x_N^2)^{1/2} = |c|\,\|\mathbf{X}\|.
\end{aligned} \tag{11}$$

An important relationship exists between the dot product and length of a vector. If both sides of equation (10) are squared and equation (9) is used with $\mathbf{Y}$ being replaced with $\mathbf{X}$, we have

$$\|\mathbf{X}\|^2 = x_1^2 + x_2^2 + \cdots + x_N^2 = \mathbf{X} \cdot \mathbf{X}. \tag{12}$$

If $\mathbf{X}$ and $\mathbf{Y}$ are position vectors that locate the points $(x_1, x_2, \ldots, x_N)$ and $(y_1, y_2, \ldots, y_N)$ in N-dimensional space, then the **displacement vector** from $\mathbf{X}$ to $\mathbf{Y}$ is given by the difference

$$\mathbf{Y} - \mathbf{X} \qquad \text{(displacement from position } \mathbf{X} \text{ to position } \mathbf{Y}\text{).} \tag{13}$$

Notice that if a particle starts at the position $\mathbf{X}$ and moves through the displacement $\mathbf{Y} - \mathbf{X}$, its new position is $\mathbf{Y}$. This can be obtained by the following vector sum:

$$\mathbf{Y} = \mathbf{X} + (\mathbf{Y} - \mathbf{X}). \tag{14}$$

Using equations (10) and (13), we can write down the formula for the distance between two points in N-space,

$$\|\mathbf{Y} - \mathbf{X}\| = [(y_1 - x_1)^2 + (y_2 - x_2)^2 + \cdots + (y_N - x_N)^2]^{1/2}. \tag{15}$$

When the distance between points is computed using formula (15), we say that the points lie in N-**dimensional Euclidean space**.

Example 3.1

Let $\mathbf{X} = (2, -3, 5, -1)$ and $\mathbf{Y} = (6, 1, 2, -4)$. The concepts mentioned above are now illustrated for vectors in 4-space.

Sum	$\mathbf{X} + \mathbf{Y} = (8, -2, 7, -5)$
Difference	$\mathbf{X} - \mathbf{Y} = (-4, -4, 3, 3)$

Scalar multiple	$3\mathbf{X} = (6, -9, 15, -3)$
Length	$\|\mathbf{X}\| = (4 + 9 + 25 + 1)^{1/2} = 39^{1/2}$
Dot product	$\mathbf{X} \cdot \mathbf{Y} = 12 - 3 + 10 + 4 = 23$
Displacement from $\mathbf{X}$ to $\mathbf{Y}$	$\mathbf{Y} - \mathbf{X} = (4, 4, -3, -3)$
Distance from $\mathbf{X}$ to $\mathbf{Y}$	$\|\mathbf{Y} - \mathbf{X}\| = (16 + 16 + 9 + 9)^{1/2} = 50^{1/2}$

It is sometimes useful to write vectors as columns instead of rows. For example,

$$\mathbf{X} = \begin{pmatrix} x_1 \\ x_2 \\ \cdot \\ \cdot \\ \cdot \\ x_N \end{pmatrix} \quad \text{and} \quad \mathbf{Y} = \begin{pmatrix} y_1 \\ y_2 \\ \cdot \\ \cdot \\ \cdot \\ y_N \end{pmatrix}. \tag{16}$$

Then the linear combination $c\mathbf{X} + d\mathbf{Y}$ is

$$c\mathbf{X} + d\mathbf{Y} = \begin{pmatrix} cx_1 + dy_1 \\ cx_2 + dy_2 \\ \cdot \quad \cdot \\ \cdot \quad \cdot \\ \cdot \quad \cdot \\ cx_N + dy_N \end{pmatrix}. \tag{17}$$

By choosing c and d appropriately in equation (17) we have the sum $1\mathbf{X} + 1\mathbf{Y}$, the difference $1\mathbf{X} - 1\mathbf{Y}$, and scalar multiple $c\mathbf{X} + 0\mathbf{Y}$. We use the superscript T, for transpose, to indicate that a row vector should be converted to a column vector, and vice versa.

$$(x_1, x_2, \ldots, x_N)^T = \begin{pmatrix} x_1 \\ x_2 \\ \cdot \\ \cdot \\ \cdot \\ x_N \end{pmatrix} \quad \text{and} \quad \begin{pmatrix} x_1 \\ x_2 \\ \cdot \\ \cdot \\ \cdot \\ x_N \end{pmatrix}^T = (x_1, x_2, \ldots, x_N). \tag{18}$$

The set of vectors has a zero element $\mathbf{0}$ which is defined by

$$\mathbf{0} = (0, 0, \ldots, 0). \tag{19}$$

Theorem 3.1   (Vector Algebra).   Suppose that $\mathbf{X}$, $\mathbf{Y}$, and $\mathbf{Z}$ are N-dimensional vectors and a and b are scalars (real numbers). The following properties of vector addition and scalar multiplication hold:

$\mathbf{Y} + \mathbf{X} = \mathbf{X} + \mathbf{Y}$	commutative property,	(20)
$\mathbf{O} + \mathbf{X} = \mathbf{X} + \mathbf{O} = \mathbf{X}$	zero vector,	(21)
$\mathbf{X} - \mathbf{X} = \mathbf{X} + (-\mathbf{X}) = \mathbf{O}$	the opposite vector,	(22)

$$(\mathbf{X} + \mathbf{Y}) + \mathbf{Z} = \mathbf{X} + (\mathbf{Y} + \mathbf{Z}) \qquad \text{associative property,} \qquad (23)$$

$$(a + b)\mathbf{X} = a\mathbf{X} + b\mathbf{X} \qquad \text{distributive property for scalars,} \qquad (24)$$

$$a(\mathbf{X} + \mathbf{Y}) = a\mathbf{X} + a\mathbf{Y} \qquad \text{distributive property for vectors,} \qquad (25)$$

$$a(b\mathbf{X}) = (ab)\mathbf{X} \qquad \text{associative property for scalars.} \qquad (26)$$

Matrices and Two-Dimensional Arrays

A **matrix** is a rectangular array of numbers that is arranged systematically in rows and columns. A matrix having M rows and N columns is called an $M \times N$ (read "M by N") matrix. The capital letter A denotes a matrix and the small subscripted letter $a_{i,j}$ denotes one of the numbers forming the matrix. We write

$$A = (a_{i,j})_{M \times N} \qquad \text{for } 1 \le i \le M, \quad 1 \le j \le N, \qquad (27)$$

where $a_{i,j}$ is the number in location (i, j) (i.e., stored in the ith row and jth column of the array). We refer to $a_{i,j}$ as the **element in location** (i, j). In expanded form we write

$$
\text{row } i \rightarrow
\begin{pmatrix}
a_{1,1} & a_{1,2} & \cdots & a_{1,j} & \cdots & a_{1,N} \\
a_{2,1} & a_{2,2} & \cdots & a_{2,j} & \cdots & a_{2,N} \\
\vdots & \vdots & & \vdots & & \vdots \\
a_{i,1} & a_{i,2} & \cdots & a_{i,j} & \cdots & a_{i,N} \\
\vdots & \vdots & & \vdots & & \vdots \\
a_{M,1} & a_{M,2} & \cdots & a_{M,j} & \cdots & a_{M,N}
\end{pmatrix} = A. \qquad (28)
$$

$$\uparrow$$
$$\text{column } j$$

The rows of the $M \times N$ matrix A are N-dimensional vectors:

$$\mathbf{V}_i = (a_{i,1}, a_{i,2}, \ldots, a_{i,j}, \ldots, a_{i,N}) \qquad \text{for } i = 1, 2, \ldots, M. \qquad (29)$$

The row vectors in (29) can also be viewed as $1 \times N$ matrices. Here we have sliced the $M \times N$ matrix A into M pieces which are $1 \times N$ matrices.

In this case we could express A as a column vector consisting of the row vectors $\mathbf{V}_i$, that is,

$$
A =
\begin{pmatrix}
\mathbf{V}_1 \\
\mathbf{V}_2 \\
\vdots \\
\mathbf{V}_i \\
\vdots \\
\mathbf{V}_M
\end{pmatrix} = [\mathbf{V}_1, \mathbf{V}_2, \ldots, \mathbf{V}_i, \ldots, \mathbf{V}_M]^T. \qquad (30)
$$

The columns of the $M \times N$ matrix A are M-dimensional vectors:

$$\mathbf{C}_1 = \begin{pmatrix} a_{1,1} \\ a_{2,1} \\ \vdots \\ a_{i,1} \\ \vdots \\ a_{M,1} \end{pmatrix}, \ldots, \mathbf{C}_j = \begin{pmatrix} a_{1,j} \\ a_{2,j} \\ \vdots \\ a_{i,j} \\ \vdots \\ a_{M,j} \end{pmatrix}, \ldots, \mathbf{C}_N = \begin{pmatrix} a_{1,N} \\ a_{2,N} \\ \vdots \\ a_{i,N} \\ \vdots \\ a_{M,N} \end{pmatrix}. \tag{31}$$

Each column vector in (31) can be viewed as an $M \times 1$ matrix. In this case we could express A as a row vector consisting of the column vectors $\mathbf{C}_j$:

$$A = [\mathbf{C}_1, \mathbf{C}_2, \ldots, \mathbf{C}_j, \ldots, \mathbf{C}_N]. \tag{32}$$

Example 3.2

Identify the row and column vectors associated with the 4×3 matrix

$$A = \begin{pmatrix} -2 & 4 & 9 \\ 5 & -7 & 1 \\ 0 & -3 & 8 \\ -4 & 6 & -5 \end{pmatrix}.$$

Solution.   The four row vectors are $\mathbf{V}_1 = (-2, 4, 9)$, $\mathbf{V}_2 = (5, -7, 1)$, $\mathbf{V}_3 = (0, -3, 8)$, and $\mathbf{V}_4 = (-4, 6, -5)$. The three column vectors are

$$\mathbf{C}_1 = \begin{pmatrix} -2 \\ 5 \\ 0 \\ -4 \end{pmatrix}, \quad \mathbf{C}_2 = \begin{pmatrix} 4 \\ -7 \\ -3 \\ 6 \end{pmatrix}, \quad \mathbf{C}_3 = \begin{pmatrix} 9 \\ 1 \\ 8 \\ -5 \end{pmatrix}.$$

Notice how A can be represented with these vectors:

$$A = \begin{pmatrix} \mathbf{V}_1 \\ \mathbf{V}_2 \\ \mathbf{V}_3 \\ \mathbf{V}_4 \end{pmatrix} = (\mathbf{C}_1, \mathbf{C}_2, \mathbf{C}_3).$$

Let $A = (a_{i,j})_{M \times N}$ and $B = (b_{i,j})_{M \times N}$ be two matrices of the same dimensions $M \times N$. The two matrices A and B are said to be equal if and only if each corresponding element is the same, that is,

$$A = B \quad \text{if and only if } a_{i,j} = b_{i,j} \quad \text{for } 1 \le i < M, \quad 1 \le j \le N. \tag{33}$$

The sum of the two $M \times N$ matrices A and B is computed element by element, using the defintion

$$A + B = (a_{i,j} + b_{i,j})_{M \times N} \qquad \text{for } 1 \le i \le M, \quad 1 \le j \le N. \tag{34}$$

The opposite of the matrix A is obtained by replacing each element with its opposite:

$$-A = (-a_{i,j})_{M \times N} \qquad \text{for } 1 \le i \le M, \quad 1 \le j \le N. \tag{35}$$

The difference $A - B$ is formed by taking the difference in each coordinate:

$$A - B = (a_{i,j} - b_{i,j})_{M \times N} \qquad \text{for } 1 \le i \le M, \quad 1 \le j \le N. \tag{36}$$

If c is a real number (scalar), we define scalar multiplication cA as follows:

$$cA = (ca_{i,j})_{M \times N} \qquad \text{for } 1 \le i \le M, \quad 1 \le j \le N. \tag{37}$$

If p and q are scalars, the weighted sum $pA + qB$ is called a linear combination of the matrices A and B, and we write

$$pA + qB = (pa_{i,j} + qb_{i,j})_{M \times N} \qquad \text{for } 1 \le i \le M, \quad 1 \le j \le N. \tag{38}$$

The zero matrix of order $M \times N$ consists of all zeros:

$$O = (0)_{M \times N}. \tag{39}$$

Example 3.3

Find the scalar multiples $2A$ and $3B$ and the linear combination $2A - 3B$ for the matrices

$$A = \begin{pmatrix} -1 & 2 \\ 7 & 5 \\ 3 & -4 \end{pmatrix}, \qquad B = \begin{pmatrix} -2 & 3 \\ 1 & -4 \\ -9 & 7 \end{pmatrix}.$$

Solution. Using formula (37), we obtain

$$2A = \begin{pmatrix} -2 & 4 \\ 14 & 10 \\ 6 & -8 \end{pmatrix}, \qquad 3B = \begin{pmatrix} -6 & 9 \\ 3 & -12 \\ -27 & 21 \end{pmatrix}.$$

The linear combination $2A - 3B$ is now found:

$$2A - 3B = \begin{pmatrix} -2 + 6 & 4 - 9 \\ 14 - 3 & 10 + 12 \\ 6 + 27 & -8 - 21 \end{pmatrix} = \begin{pmatrix} 4 & -5 \\ 11 & 22 \\ 33 & -29 \end{pmatrix}.$$

Theorem 3.2 (Matrix Arithmetic). Suppose that A, B, and C are $M \times N$ matrices and p and q are scalars (real numbers). The following properties of matrix addition and scalar multiplication hold:

$$B + A = A + B \qquad\qquad \text{commutative property,} \tag{40}$$

$$O + A = A + O = A \qquad \text{identity for addition,} \qquad (41)$$

$$A - A = A + (-A) = O \qquad \text{the opposite matrix,} \qquad (42)$$

$$(A + B) + C = A + (B + C) \qquad \text{associative property,} \qquad (43)$$

$$(p + q)A = pA + qA \qquad \text{distributive property for scalars,} \qquad (44)$$

$$p(A + B) = pA + pB \qquad \text{distributive property for matrices,} \qquad (45)$$

$$p(qA) = (pq)A \qquad \text{associative property for scalars.} \qquad (46)$$

EXERCISES FOR INTRODUCTION TO VECTORS AND MATRICES

1. Given the vectors **X** and **Y**, find (a) **X** + **Y**, (b) **X** − **Y**, (c) 3**X**, (d) $\|\mathbf{X}\|$, (e) **X** • **Y**, (f) **Y** − **X**, and (g) $\|\mathbf{Y} - \mathbf{X}\|$.
   (i) **X** = (3, −4) and **Y** = (−2, 8)
   (ii) **X** = (−6, 3, 2) and **Y** = (−8, 5, 1)
   (iii) **X** = (4, −8, 1) and **Y** = (1, −12, −11)
   (iv) **X** = (1, −2, 4, 2) and **Y** = (3, −5, −4, 0)

2. The angle θ between two vectors **X** and **Y** is given by the relation

$$\cos(\theta) = \frac{\mathbf{X} \cdot \mathbf{Y}}{\|\mathbf{X}\| \, \|\mathbf{Y}\|}.$$

   Find the angle between the following vectors.
   (a) **X** = (−6, 3, 2)   and   **Y** = (2, −2, 1)
   (b) **X** = (4, −8, 1)   and   **Y** = (3, 4, 12)

3. Two vectors **X** and **Y** are said to be orthogonal (perpendicular) if the angle between them is $\pi/2$.
   (a) Show that **X** and **Y** are orthogonal if and only if

$$\mathbf{X} \cdot \mathbf{Y} = 0.$$

   Use part (a) to determine if the following vectors are orthogonal.
   (b) **X** = (−6, 4, 2)   and   **Y** = (6, 5, 8)
   (c) **X** = (−4, 8, 3)   and   **Y** = (2, −5, 16)
   (d) **X** = (−5, 7, 2)   and   **Y** = (4, 1, 6)

4. Find (a) $A + B$, (b) $A - B$, and (c) $3A - 2B$ for the matrices

$$A = \begin{pmatrix} -1 & 9 & 4 \\ 2 & -3 & -6 \\ 0 & 5 & 7 \end{pmatrix}, \qquad B = \begin{pmatrix} -4 & 9 & 2 \\ 3 & -5 & 7 \\ 8 & 1 & -6 \end{pmatrix}.$$

5. The **transpose** of an $M \times N$ matrix A, denoted A^T, is an $N \times M$ matrix obtained from A by converting the rows of A to columns of A^T. That is, if $A = (a_{i,j})_{M \times N}$ and $A^T = (b_{j,i})_{N \times M}$, the elements satisfy the relation

$$b_{j,i} = a_{i,j} \qquad \text{for } 1 \le i \le M, \qquad 1 < j \le N.$$

Find the transpose of the following matrices.

(a) $\begin{pmatrix} -2 & 5 & 12 \\ 1 & 4 & -1 \\ 7 & 0 & 6 \\ 11 & -3 & 8 \end{pmatrix}$

(b) $\begin{pmatrix} 4 & 9 & 2 \\ 3 & 5 & 7 \\ 8 & 1 & 6 \end{pmatrix}$

6. The square matrix A of dimension $N \times N$ is said to be **symmetric** if $A = A^T$. (See Exercise 5 for the definition of A^T.) Determine whether the following square matrices are symmetric.

(a) $\begin{pmatrix} 1 & -7 & 4 \\ -7 & 2 & 0 \\ 4 & 0 & 3 \end{pmatrix}$

(b) $\begin{pmatrix} 4 & -7 & 1 \\ 0 & 2 & -7 \\ 3 & 0 & 4 \end{pmatrix}$

7. Write a report on norms of vectors and matrices. See References [9, 19, 29, 40, 49, 62, 90, 94, 96, 101, 117, 128, 145, 153, and 192].

3.2 PROPERTIES OF VECTORS AND MATRICES

A linear combination of the variables $x_1, x_2, \ldots, x_N$ is a sum

$$a_1 x_1 + a_2 x_2 + \cdots + a_N x_N \tag{1}$$

where a_k is the coefficient of x_k (for $k = 1, 2, \ldots, N$).

A linear equation in $x_1, x_2, \ldots, x_N$ is obtained by requiring the linear combination in (1) to take on a prescribed value b, that is,

$$a_1 x_1 + a_2 x_2 + \cdots + a_N x_N = b. \tag{2}$$

Systems of linear equations arise frequently and if M equations in N unknowns are given, we write

$$
\begin{aligned}
\text{Eq. } \langle 1 \rangle:\ & a_{1,1} x_1 + a_{1,2} x_2 + a_{1,3} x_3 + \cdots + a_{1,N} x_N = b_1 \\
\text{Eq. } \langle 2 \rangle:\ & a_{2,1} x_1 + a_{2,2} x_2 + a_{2,3} x_3 + \cdots + a_{2;N} x_N = b_2 \\
& \qquad \vdots \qquad\qquad \vdots \qquad\qquad \vdots \qquad\qquad\qquad \vdots \qquad\quad \vdots \\
\text{Eq. } \langle k \rangle:\ & a_{k,1} x_1 + a_{k,2} x_2 + a_{k,3} x_3 + \cdots + a_{k,N} x_N = b_k \\
& \qquad \vdots \qquad\qquad \vdots \qquad\qquad \vdots \qquad\qquad\qquad \vdots \qquad\quad \vdots \\
\text{Eq. } \langle M \rangle:\ & a_{M,1} x_1 + a_{M,2} x_2 + a_{M,3} x_3 + \cdots + a_{M,N} x_N = b_M.
\end{aligned}
\tag{3}
$$

To keep track of the different coefficients in each equation, it was necessary to use the two subscripts (k, j). The first subscript locates equation k and the second subscript locates the variable x_j.

A solution to (3) is a set of numerical values $x_1, x_2, \ldots, x_N$ that satisfies all the equations Eq. $\langle 1 \rangle$, Eq. $\langle 2 \rangle, \ldots$, Eq. $\langle M \rangle$. Hence a solution can be viewed as an N-dimensional vector:

$$\mathbf{X} = (x_1, x_2, \ldots, x_N). \tag{4}$$

Example 3.4

Concrete (used for sidewalks, etc.) is a mixture of portland cement, sand, and gravel. A distributor has three batches available for contractors. Batch 1 contains cement, sand, and gravel mixed in the proportions $\frac{1}{8} : \frac{3}{8} : \frac{4}{8}$. Batch 2 has the proportions $\frac{2}{10} : \frac{5}{10} : \frac{3}{10}$, and batch 3 has the proportions $\frac{2}{5} : \frac{3}{5} : \frac{0}{5}$.

Let $x_1, x_2,$ and x_3 denote the amount (in cubic yards) to be used from each batch to form a mixture of 10 cubic yards. Also, suppose that the mixture is to contain $b_1 = 2.3$, $b_2 = 4.8$, and $b_3 = 2.9$ cubic yards of portland cement, sand, and gravel, respectively. Then the system of linear equations for the ingredients is

$$\langle \text{cement} \rangle \quad 0.125x_1 + 0.200x_2 + 0.400x_3 = 2.3$$

$$\langle \text{sand} \rangle \quad 0.375x_1 + 0.500x_2 + 0.600x_3 = 4.8 \tag{5}$$

$$\langle \text{gravel} \rangle \quad 0.500x_1 + 0.300x_2 + 0.000x_3 = 2.9.$$

The solution to the linear system (5) is $x_1 = 4$, $x_2 = 3$, and $x_3 = 3$, which can be verified by direct substitution into the equations:

$$\langle \text{cement} \rangle \quad 0.125 \times 4 + 0.200 \times 3 + 0.400 \times 3 = 2.3$$

$$\langle \text{sand} \rangle \quad 0.375 \times 4 + 0.500 \times 3 + 0.600 \times 3 = 4.8$$

$$\langle \text{gravel} \rangle \quad 0.500 \times 4 + 0.300 \times 3 + 0.000 \times 3 = 2.9.$$

We now discuss how matrices and vectors are used to represent a linear system of equations. Each equation in (3) can be written using the dot-product notation for vectors:

Eq. $\langle 1 \rangle$: $\quad (a_{1,1}, a_{1,2}, \ldots, a_{1,j}, \ldots, a_{1,N}) \cdot (x_1, x_2, \ldots, x_j, \ldots, x_N) = b_1$

Eq. $\langle 2 \rangle$: $\quad (a_{2,1}, a_{2,2}, \ldots, a_{2,j}, \ldots, a_{2,N}) \cdot (x_1, x_2, \ldots, x_j, \ldots, x_N) = b_2$

$\qquad\qquad\qquad\qquad\qquad\qquad\qquad\qquad\qquad\qquad\qquad\qquad\qquad \vdots$

Eq. $\langle k \rangle$: $\quad (a_{k,1}, a_{k,2}, \ldots, a_{k,j}, \ldots, a_{k,N}) \cdot (x_1, x_2, \ldots, x_j, \ldots, x_N) = b_k$

$\qquad\qquad\qquad\qquad\qquad\qquad\qquad\qquad\qquad\qquad\qquad\qquad\qquad \vdots$

Eq. $\langle M \rangle$: $\; (a_{M,1}, a_{M,2}, \ldots, a_{M,j}, \ldots, a_{M,N}) \cdot (x_1, x_2, \ldots, x_j, \ldots, x_N) = b_M.$

One usually stores the coefficients $a_{k,j}$ in a matrix A of dimension $M \times N$ and the unknowns x_j are stored in an N-dimensional vector $\mathbf{X}$. The constants b_k are stored in an M-dimensional vector $\mathbf{B}$. It is conventional to use column vectors for both $\mathbf{X}$ and $\mathbf{B}$ and write

$$A\mathbf{X} = \begin{pmatrix} a_{1,1} & a_{1,2} & \cdots & a_{1,j} & \cdots & a_{1,N} \\ a_{2,1} & a_{2,2} & \cdots & a_{2,j} & \cdots & a_{2,N} \\ \vdots & \vdots & & \vdots & & \vdots \\ a_{k,1} & a_{k,2} & \cdots & a_{k,j} & \cdots & a_{k,N} \\ \vdots & \vdots & & \vdots & & \vdots \\ a_{M,1} & a_{M,2} & \cdots & a_{M,j} & \cdots & a_{M,N} \end{pmatrix} \begin{pmatrix} x_1 \\ x_2 \\ \vdots \\ x_j \\ \vdots \\ x_N \end{pmatrix} = \begin{pmatrix} b_1 \\ b_2 \\ \vdots \\ b_k \\ \vdots \\ b_M \end{pmatrix} = \mathbf{B}. \tag{6}$$

The matrix multiplication $A\mathbf{X} = \mathbf{B}$ in (6) is reminiscent of the dot-product multiplication for ordinary vectors, because each element b_k in $\mathbf{B}$ is the result obtained by taking the dot product of row k in matrix A with the column vector $\mathbf{X}$. We will use $A\mathbf{X}$ to denote the product of A times $\mathbf{X}$.

Example 3.5

Express the results of Example 3.4 with matrix notation.

Solution.   Equations (5) can be written in the matrix form

$$\begin{pmatrix} 0.125 & 0.200 & 0.400 \\ 0.375 & 0.500 & 0.600 \\ 0.500 & 0.300 & 0.000 \end{pmatrix} \begin{pmatrix} x_1 \\ x_2 \\ x_3 \end{pmatrix} = \begin{pmatrix} 2.3 \\ 4.8 \\ 2.9 \end{pmatrix}. \tag{7}$$

In vector notation, the column vector $(4, 3, 3)^T$ is shown to be the solution by performing the matrix calculation

$$\begin{pmatrix} 0.125 & 0.200 & 0.400 \\ 0.375 & 0.500 & 0.600 \\ 0.500 & 0.300 & 0.000 \end{pmatrix} \begin{pmatrix} 4 \\ 3 \\ 3 \end{pmatrix} = \begin{pmatrix} 0.5 + 0.6 + 1.2 \\ 1.5 + 1.5 + 1.8 \\ 2.0 + 0.9 + 0.0 \end{pmatrix} = \begin{pmatrix} 2.3 \\ 4.8 \\ 2.9 \end{pmatrix}.$$

Matrix Multiplication

Definition 3.1.   If $A = (a_{i,k})_{M \times N}$ and $B = (b_{k,j})_{N \times P}$ are two matrices with the property that A has as many columns as B has rows, the matrix product AB is defined to be the matrix C of dimension $M \times P$:

$$AB = C = (c_{i,j})_{M \times P} \tag{8}$$

where the element $c_{i,j}$ of C is given by the dot product of the ith row of A and the jth column of B:

$$c_{i,j} = (a_{i,1}, a_{i,2}, \ldots, a_{i,N}) \bullet (b_{1,j}, b_{2,j}, \ldots, b_{N,j})$$

$$= a_{i,1}b_{1,j} + a_{i,2}b_{2,j} + \cdots + a_{i,N}b_{N,j} = \sum_{k=1}^{N} a_{i,k}b_{k,j}, \tag{9}$$

for $i = 1, 2, \ldots, M$ and $j = 1, 2, \ldots, P$.

Using the notation $c_{i,j} = (\text{row}_i\, A) \bullet (\text{col}_j\, B)$ in equation (9), the matrix product AB can be viewed as follows:

$$AB = \begin{pmatrix} \text{row}_1\, A \bullet \text{col}_1\, B & \text{row}_1\, A \bullet \text{col}_2\, B & \cdots & \text{row}_1\, A \bullet \text{col}_j\, B & \cdots & \text{row}_1\, A \bullet \text{col}_P\, B \\ \text{row}_2\, A \bullet \text{col}_1\, B & \text{row}_2\, A \bullet \text{col}_2\, B & \cdots & \text{row}_2\, A \bullet \text{col}_j\, B & \cdots & \text{row}_2\, A \bullet \text{col}_P\, B \\ \vdots & \vdots & & \vdots & & \vdots \\ \text{row}_i\, A \bullet \text{col}_1\, B & \text{row}_i\, A \bullet \text{col}_2\, B & \cdots & \text{row}_i\, A \bullet \text{col}_j\, B & \cdots & \text{row}_i\, A \bullet \text{col}_P\, B \\ \vdots & \vdots & & \vdots & & \vdots \\ \text{row}_M\, A \bullet \text{col}_1\, B & \text{row}_M\, A \bullet \text{col}_2\, B & \cdots & \text{row}_M\, A \bullet \text{col}_j\, B & \cdots & \text{row}_M\, A \bullet \text{col}_P\, B \end{pmatrix}$$

Example 3.6

Find the product $C = AB$ for the following matrices, and tell why BA is not defined.

$$A = \begin{pmatrix} 2 & 3 \\ -1 & 4 \end{pmatrix}, \qquad B = \begin{pmatrix} 5 & -2 & 1 \\ 3 & 8 & -6 \end{pmatrix}.$$

Solution.   The matrix A has two columns and B has two rows, so that the matrix product AB is defined. The product of a 2×2 matrix and a 2×3 matrix is a 2×3 matrix. Computation reveals that

$$\begin{pmatrix} 2 & 3 \\ -1 & 4 \end{pmatrix} \begin{pmatrix} 5 & -2 & 1 \\ 3 & 8 & -6 \end{pmatrix} = \begin{pmatrix} 10 + 9 & -4 + 24 & 2 - 18 \\ -5 + 12 & 2 + 32 & -1 - 24 \end{pmatrix}.$$

Thus

$$C = \begin{pmatrix} 19 & 20 & -16 \\ 7 & 34 & -25 \end{pmatrix}.$$

When an attempt is made to form the product BA we discover that the dimensions are not compatible in this order because $(\text{row}_i\, B)$ is a three-dimensional vector and $(\text{col}_j\, A)$ is a two-dimensional vector. Hence the dot product $(\text{row}_i\, B) \bullet (\text{col}_j\, A)$ is not defined, that is,

$$\begin{pmatrix} 5 & -2 & 1 \\ 3 & 8 & -6 \end{pmatrix} \begin{pmatrix} 2 & 3 \\ -1 & 4 \end{pmatrix} = ?$$

If it happens that $AB = BA$, we say that A and B **commute**. Most often, even when AB and BA are both defined, the products are not necessarily the same.

Some Special Matrices

The $M \times N$ matrix whose elements are all zero is called the **zero matrix of dimension** $M \times N$ and is denoted by

$$O = (0)_{M \times N}. \tag{10}$$

When the dimension is clear, we use O to denote the zero matrix.

The **identity matrix of order** N is the square matrix given by

$$I_N = (\delta_{i,j})_{N \times N}, \qquad \text{where } \delta_{i,j} = \begin{cases} 1 & \text{when } i = j \\ 0 & \text{when } i \neq j. \end{cases} \tag{11}$$

It is the multiplicative identity, as illustrated in the next example.

Example 3.7

Let A be a 2×3 matrix. Then $I_2 A = A I_3 = A$.

Solution.   Multiplication of A on the left by I_2 results in

$$\begin{pmatrix} 1 & 0 \\ 0 & 1 \end{pmatrix} \begin{pmatrix} a_{11} & a_{12} & a_{13} \\ a_{21} & a_{22} & a_{23} \end{pmatrix} = \begin{pmatrix} a_{11} + 0 & a_{12} + 0 & a_{13} + 0 \\ 0 + a_{21} & 0 + a_{22} & 0 + a_{23} \end{pmatrix} = A.$$

Multiplication of A on the right by I_3 results in

$$\begin{pmatrix} a_{11} & a_{12} & a_{13} \\ a_{21} & a_{22} & a_{23} \end{pmatrix} \begin{pmatrix} 1 & 0 & 0 \\ 0 & 1 & 0 \\ 0 & 0 & 1 \end{pmatrix} = \begin{pmatrix} a_{11} + 0 + 0 & 0 + a_{12} + 0 & 0 + 0 + a_{13} \\ a_{21} + 0 + 0 & 0 + a_{22} + 0 & 0 + 0 + a_{23} \end{pmatrix} = A.$$

Some properties of matrix multiplication are given in the following theorem.

Theorem 3.3   (Matrix Multiplication).   Suppose that c is a scalar and that $A, B, C,$ and D are matrices such that the indicated sums and products are defined, then:

$A(BC) = (AB)C$	associativity of matrix multiplication,	(12)
$IA = A$ and $AI = A$	identity matrix,	(13)
$A(B + E) = AB + AE$	left distributive property,	(14)
$(A + D)B = AB + DB$	right distributive property,	(15)
$c(AB) = (cA)B = A(cB)$	scalar associative property.	(16)

The Inverse of a Nonsingular Matrix

The concept of inverse applies to matrices, but special attention must be given. An $N \times N$ matrix A is called **nonsingular** or **invertible** if there exists an $N \times N$ matrix B such that

$$AB = BA = I. \tag{17}$$

If no such matrix B can be found, A is said to be **singular**. When B can be found and (17) holds, we usually write $B = A^{-1}$ and use the familiar relation

$$AA^{-1} = A^{-1}A = I \qquad \text{if } A \text{ is nonsingular.} \tag{18}$$

It is easy to show that at most one matrix B can be found that satisfies relation (17). For suppose that C is also an inverse (i.e., $AC = CA = I$). Then properties (12) and (13) can be used to obtain

$$C = IC = (BA)C = B(AC) = BI = B.$$

Theorem 3.4   (Inverse of a 2 $\times$ 2 Matrix).   A necessary and sufficient condition for the matrix

$$A = \begin{pmatrix} a & b \\ c & d \end{pmatrix} \tag{19}$$

to have an inverse is that $ad - bc \neq 0$. If $ad - bc \neq 0$, then

$$A^{-1} = \frac{1}{ad - bc} \begin{pmatrix} d & -b \\ -c & a \end{pmatrix}. \tag{20}$$

Proof. If $ad - bc \neq 0$, the following calculation is valid:

$$\frac{1}{ad - bc} \begin{pmatrix} a & b \\ c & d \end{pmatrix} \begin{pmatrix} d & -b \\ -c & a \end{pmatrix} = \frac{1}{ad - bc} \begin{pmatrix} ad - bc & -ab + ba \\ cd - dc & -cb + da \end{pmatrix} = \begin{pmatrix} 1 & 0 \\ 0 & 1 \end{pmatrix}.$$

Hence A^{-1} given in formula (20) is the inverse of A.

Suppose that $ad - bc = 0$. If both $a = 0$ and $b = 0$, the first row of the product AB, for any matrix B, will contain only zero elements, so $AB \neq I$. Hence A does not have an inverse. On the other hand, if either $a \neq 0$ or $b \neq 0$, then

$$B \begin{pmatrix} a & b \\ c & d \end{pmatrix} \begin{pmatrix} b \\ -a \end{pmatrix} = B \begin{pmatrix} ab - ba \\ cb - da \end{pmatrix} = B \begin{pmatrix} 0 \\ 0 \end{pmatrix} = \begin{pmatrix} 0 \\ 0 \end{pmatrix}.$$

Thus for any matrix B, BA cannot be I, because

$$(BA) \begin{pmatrix} b \\ -a \end{pmatrix} = \begin{pmatrix} 0 \\ 0 \end{pmatrix} \neq \begin{pmatrix} b \\ -a \end{pmatrix}.$$

Hence A does not have an inverse.

We remark that if $ad - bc = 0$, then in Sections 3.4 and 3.5 we will see that the algorithm for constructing A^{-1} will encounter a division-by-zero error.

Example 3.8

Find the inverse of the matrix

$$A = \begin{pmatrix} 3 & 1 \\ 7 & 4 \end{pmatrix}.$$

Solution. Using formula (20), we find that

$$A^{-1} = \frac{1}{12 - 7} \begin{pmatrix} 4 & -1 \\ -7 & 3 \end{pmatrix} = \begin{pmatrix} 0.8 & -0.2 \\ -1.4 & 0.6 \end{pmatrix}.$$

Determinants

The determinant of a square matrix A is a scalar quantity (real number) and is denoted by det A. If A is an $N \times N$ matrix

$$A = \begin{pmatrix} a_{1,1} & a_{1,2} & \cdots & a_{1,N} \\ a_{2,1} & a_{2,2} & \cdots & a_{2,N} \\ \vdots & \vdots & & \vdots \\ a_{N,1} & a_{N,2} & \cdots & a_{N,N} \end{pmatrix},$$

it is customary to write

$$\det A = \begin{vmatrix} a_{1,1} & a_{1,2} & \cdots & a_{1,N} \\ a_{2,1} & a_{2,2} & \cdots & a_{2,N} \\ \vdots & \vdots & & \vdots \\ a_{N,1} & a_{N,2} & \cdots & a_{N,N} \end{vmatrix}.$$

Although the notation for a determinant may look like a matrix, its properties are completely different. For one, the determinant is a scalar quantity (real number). One goal of numerical methods is to develop an efficient algorithm for computing the determinant. The definition of det A found in most linear algebra textbooks is not tractable for computation when $N > 3$. We will review how to compute determinants of the first three orders. Evaluation of higher-order determinants is done using Gaussian elimination and is mentioned in the body of Algorithm 3.3.

If $A = (a_{1,1})$ is a 1×1 matrix, we define det $A = a_{1,1}$. If A is a 2×2 matrix,

$$A = \begin{pmatrix} a_{1,1} & a_{1,2} \\ a_{2,1} & a_{2,2} \end{pmatrix},$$

we define

$$\det A = a_{1,1}a_{2,2} - a_{1,2}a_{2,1}. \tag{21}$$

If B is a 3×3 matrix,

$$B = \begin{pmatrix} b_{1,1} & b_{1,2} & b_{1,3} \\ b_{2,1} & b_{2,2} & b_{2,3} \\ b_{3,1} & b_{3,2} & b_{3,3} \end{pmatrix},$$

we define

$$\det B = b_{1,1}b_{2,2}b_{3,3} + b_{1,2}b_{2,3}b_{3,1} + b_{1,3}b_{2,1}b_{3,2}$$
$$- b_{1,3}b_{2,2}b_{3,1} - b_{1,2}b_{2,1}b_{3,3} - b_{1,1}b_{2,3}b_{3,2}. \tag{22}$$

Before defining the determinant of the square matrix A of order N, we must introduce some preliminary concepts. Consider the set of integers $S = \{1, 2, \ldots, N\}$. An ordering of the elements of S is called a **permutation.** For example, if $S = \{1, 2, 3\}$, there are six permutations:

$$(1, 2, 3), \quad (1, 3, 2), \quad (2, 1, 3), \quad (2, 3, 1), \quad (3, 1, 2), \quad (3, 2, 1).$$

In general, the number of permutations of a set of N integers is $N!$.

Let $P = (j_1, j_2, \ldots, j_N)$ be a permutation of S. Let α_k be the number of integers following j_k that are smaller than j_k for $k = 1, 2, \ldots, N - 1$. The sum $\alpha_1 + \alpha_2 + \cdots + \alpha_{N-1}$ is called the **number of inversions** in the permutation. A permutation P is said to be **even (odd)** if it has an even (odd) number of inversions. For example, if $P = (3, 2, 1)$, then $\alpha_1 = 2$ and $\alpha_2 = 1$ and $\alpha_1 + \alpha_2 = 3$, so that P has an odd number of inversions. For each permutation P of S, we define

$$\delta(P) = \begin{cases} 0 & \text{if } P \text{ has an even number of inversions,} \\ 1 & \text{if } P \text{ has an odd number of inversions.} \end{cases}$$

For example, $\delta(3, 2, 1) = 1$ and $\delta(3, 1, 2) = 0$.

Making use of the definitions above, we define the **determinant** of the $N \times N$ matrix A as follows:

$$\det A = \sum (-1)^{\delta(P)} a_{1,j_1} a_{2,j_2} \cdots a_{N,j_N},$$

where the sum is taken over all permutations $P = (j_1, j_2, \ldots, j_N)$ of S. This definition is not tractable to implement when N is large. For $N = 10$ it would require $10 \times 10! \approx 3.6 \times 10^7$ multiplications.

Example 3.9

Find the determinant of the matrices

$$A = \begin{pmatrix} 4 & 9 \\ 1 & 3 \end{pmatrix}, \qquad B = \begin{pmatrix} 2 & 3 & 8 \\ -4 & 5 & -1 \\ 7 & -6 & 9 \end{pmatrix}.$$

Solution.   Using formula (21), we obtain

$$\det A = (4)(3) - (9)(1) = 3.$$

Using formula (22), $\det B$ is computed as follows:

$$\det B = (2)(5)(9) + (3)(-1)(7) + (8)(-4)(-6) - (8)(5)(7) - (3)(-4)(9) - (2)(-1)(-6)$$
$$= (90) + (-21) + (192) - (280) - (-108) - (12) = 77.$$

The following theorem gives sufficient conditions for the existence and uniqueness of solutions $A\mathbf{X} = \mathbf{B}$ for square matrices.

Theorem 3.5.   Assume that A is an $N \times N$ matrix. The following are equivalent:

Given any N-dimensional column vector $\mathbf{B}$, the linear system $A\mathbf{X} = \mathbf{B}$ has a unique solution.   (23)

Given any N-dimensional column vector $\mathbf{B}$, Gaussian elimination with row interchanges can be successfully performed to solve $A\mathbf{X} = \mathbf{B}$.   (24)

The matrix A is nonsingular (i.e., A^{-1} exists).   (25)

The equation $A\mathbf{X} = \mathbf{0}$ has the unique solution $\mathbf{X} = \mathbf{0}$.   (26)

$\det A \neq 0$.   (27)

Theorems 3.3 and 3.5 help relate matrix algebra to ordinary algebra. If statement (23) is true, then statement (25) together with properties (12) and (13) give the following line of reasoning:

$A\mathbf{X} = \mathbf{B}$ implies that $A^{-1}A\mathbf{X} = A^{-1}\mathbf{B}$ which implies that $\mathbf{X} = A^{-1}\mathbf{B}$.   (28)

Example 3.10

Use the inverse matrix A^{-1} in Example 3.8 and the reasoning in (28) to solve the linear system of equations $A\mathbf{X} = \mathbf{B}$:

$$AX = \begin{pmatrix} 3 & 1 \\ 7 & 4 \end{pmatrix} \begin{pmatrix} x_1 \\ x_2 \end{pmatrix} = \begin{pmatrix} 2 \\ 5 \end{pmatrix} = B.$$

Solution.   Using (28) we get

$$\mathbf{X} = A^{-1}\mathbf{B} = \begin{pmatrix} 0.8 & -0.2 \\ -1.4 & 0.6 \end{pmatrix} \begin{pmatrix} 2 \\ 5 \end{pmatrix} = \begin{pmatrix} 0.6 \\ 0.2 \end{pmatrix}.$$

Plane Rotations

Suppose that A is a 3×3 matrix and $\mathbf{U} = (x, y, z)$ is a three-dimensional vector; then the product $\mathbf{V} = A\mathbf{U}$ is another three-dimensional vector. This is an example of a linear transformation, and applications are found in the area of computer graphics. Consider the three special matrices:

$$R_x(\alpha) = \begin{pmatrix} 1 & 0 & 0 \\ 0 & \cos\alpha & -\sin\alpha \\ 0 & \sin\alpha & \cos\alpha \end{pmatrix}, \qquad R_y(\beta) = \begin{pmatrix} \cos\beta & 0 & \sin\beta \\ 0 & 1 & 0 \\ -\sin\beta & 0 & \cos\beta \end{pmatrix},$$

$$R_z(\gamma) = \begin{pmatrix} \cos\gamma & -\sin\gamma & 0 \\ \sin\gamma & \cos\gamma & 0 \\ 0 & 0 & 1 \end{pmatrix}.$$

These matrices $R_x(\alpha)$, $R_y(\beta)$, and $R_z(\gamma)$ are used to rotate points about the x, y, and z axes through an angle α, β, and γ, respectively. The inverses are $R_x(-\alpha)$, $R_y(-\beta)$, and $R_z(-\gamma)$ and they rotate space about the x, y, and z axes through the angles $-\alpha$, $-\beta$, and $-\gamma$, respectively. The next example illustrates the situation, and further investigations are left for the reader.

Example 3.11

A unit cube is situated in the first octant with one vertex at the origin. First, rotate the cube through an angle $\pi/4$ about the z-axis, then rotate this image through an angle $\pi/6$ about the y-axis. Find the images of all eight vertices of the cube.

Solution. The first rotation is given by the transformation

$$\mathbf{V} = R_z\left(\frac{\pi}{4}\right)\mathbf{U} = \begin{pmatrix} \cos\left(\frac{\pi}{4}\right) & -\sin\left(\frac{\pi}{4}\right) & 0 \\ \sin\left(\frac{\pi}{4}\right) & \cos\left(\frac{\pi}{4}\right) & 0 \\ 0 & 0 & 1 \end{pmatrix}\begin{pmatrix} x \\ y \\ z \end{pmatrix}$$

$$= \begin{pmatrix} 0.707107 & -0.707107 & 0.000000 \\ 0.707107 & 0.707107 & 0.000000 \\ 0.000000 & 0.000000 & 1.000000 \end{pmatrix}\begin{pmatrix} x \\ y \\ z \end{pmatrix}$$

Then the second rotation is given by

$$\mathbf{W} = R_y\left(\frac{\pi}{6}\right)\mathbf{V} = \begin{pmatrix} \cos\left(\frac{\pi}{6}\right) & 0 & \sin\left(\frac{\pi}{6}\right) \\ 0 & 1 & 0 \\ -\sin\left(\frac{\pi}{6}\right) & 0 & \cos\left(\frac{\pi}{6}\right) \end{pmatrix}$$

$$\mathbf{W} = \begin{pmatrix} 0.866025 & 0.000000 & 0.500000 \\ 0.000000 & 1.000000 & 0.000000 \\ -0.500000 & 0.000000 & 0.866025 \end{pmatrix}\mathbf{V}.$$

The composition of the two rotations is

$$\mathbf{W} = R_y\left(\frac{\pi}{6}\right)R_z\left(\frac{\pi}{4}\right)\mathbf{U} = \begin{pmatrix} 0.612372 & -0.612372 & 0.500000 \\ 0.707107 & 0.707107 & 0.000000 \\ -0.353553 & 0.353553 & 0.866025 \end{pmatrix}\begin{pmatrix} x \\ y \\ z \end{pmatrix}.$$

Numerical computations for the coordinates of the vertices of the starting cube are given in Table 3.1 and the images of these cubes are shown in Figure 3.2(a) to (c).

TABLE 3.1 The Coordinates of the Vertices of a Cube
Under Successive Rotations

U	$\mathbf{V} = R_z\left(\frac{\pi}{4}\right)\mathbf{U}$	$\mathbf{W} = R_y\left(\frac{\pi}{6}\right)R_z\left(\frac{\pi}{4}\right)\mathbf{U}$
$(0, 0, 0)^T$	$(0.000000, 0.000000, 0)^T$	$(0.000000, 0.000000, 0.000000)^T$
$(1, 0, 0)^T$	$(0.707107, 0.707107, 0)^T$	$(0.612372, 0.707107, -0.353553)^T$
$(0, 1, 0)^T$	$(-0.707107, 0.707107, 0)^T$	$(-0.612372, 0.707107, 0.353553)^T$
$(0, 0, 1)^T$	$(0.000000, 0.000000, 1)^T$	$(0.500000, 0.000000, 0.866025)^T$
$(1, 1, 0)^T$	$(0.000000, 1.414214, 0)^T$	$(0.000000, 1.414214, 0.000000)^T$
$(1, 0, 1)^T$	$(0.707107, 0.707107, 1)^T$	$(1.112372, 0.707107, 0.512472)^T$
$(0, 1, 1)^T$	$(-0.707107, 0.707107, 1)^T$	$(-0.112372, 0.707107, 1.219579)^T$
$(1, 1, 1)^T$	$(0.000000, 1.414214, 1)^T$	$(0.500000, 1.414214, 0.866025)^T$

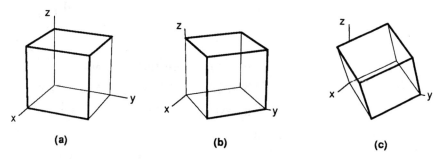

(a)	**(b)**	**(c)**

Figure 3.2   (a) The original starting cube. (b) $\mathbf{V} = R_z(\pi/4)\ \mathbf{U}$. Rotation about the z-axis. (c) $\mathbf{W} = R_y(\pi/6)\ \mathbf{V}$. Rotation about the y-axis.

EXERCISES FOR PROPERTIES OF VECTORS AND MATRICES

1. Find AB and BA for the following matrices.

$$A = \begin{pmatrix} -3 & 2 \\ 1 & 4 \end{pmatrix}, \qquad B = \begin{pmatrix} 5 & 0 \\ 2 & -6 \end{pmatrix}.$$

2. Find AB and BA for the following matrices.

$$A = \begin{pmatrix} 1 & -2 & 3 \\ 2 & 0 & 5 \end{pmatrix}, \qquad B = \begin{pmatrix} 3 & 0 \\ -1 & 5 \\ 3 & -2 \end{pmatrix}.$$

3. Let A, B, and C be given by

$$A = \begin{pmatrix} 3 & 1 \\ 0 & 4 \end{pmatrix}, \qquad B = \begin{pmatrix} 1 & 2 \\ -2 & -6 \end{pmatrix}, \qquad C = \begin{pmatrix} 2 & -5 \\ 3 & 4 \end{pmatrix}.$$

(a) Find $(AB)C$ and $A(BC)$.
(b) Find $A(B + C)$ and $AB + AC$.
(c) Find $(A + B)C$ and $AC + BC$.

4. We use the notation convention $A^2 = AA$.
Find A^2 and B^2 for the following matrices.

$$A = \begin{pmatrix} -1 & -7 \\ 5 & 2 \end{pmatrix}, \qquad B = \begin{pmatrix} 2 & 0 & 6 \\ -1 & 5 & -4 \\ 3 & -5 & 2 \end{pmatrix}.$$

5. Find the determinant of the following matrices, if it exists.

(a) $\begin{pmatrix} -1 & -7 \\ 5 & 2 \end{pmatrix}$ (b) $\begin{pmatrix} 2 & 0 & 1 \\ -3 & 0 & 5 \end{pmatrix}$

(c) $\begin{pmatrix} 1 & 2 \\ 3 & 4 \\ 0 & 0 \end{pmatrix}$ (d) $\begin{pmatrix} 2 & 0 & 6 \\ -1 & 5 & -4 \\ 3 & -5 & 2 \end{pmatrix}$

6. A unit cube is situated in the first octant with one vertex at the origin. First, rotate the cube through an angle $\pi/6$ about the y-axis, then rotate this image through an angle $\pi/4$ about the z-axis. Find the images of all eight vertices of the starting cube. Compare this result with the result of Example 3.11. Why is it different? Explain your answer with the noncommutative property of matrix multiplication [see Figure 3.3(a) to (c)].

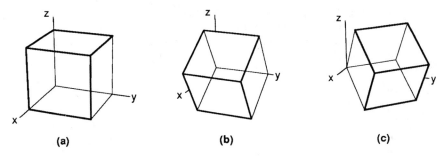

(a)   (b)   (c)

Figure 3.3   (a) The original starting cube. (b) $\mathbf{V} = R_y(\pi/6)\,\mathbf{U}$. Rotation about the y-axis. (c) $\mathbf{W} = R_z(\pi/4)\,\mathbf{V}$. Rotation about the z-axis.

7. A unit cube is situated in the first octant with one vertex at the origin. First, rotate the cube through an angle $\pi/12$ about the x-axis, then rotate this image through an angle $\pi/4$ about the z-axis. Find the images of all eight vertices of the starting cube [see Figure 3.4(a) to (c)].

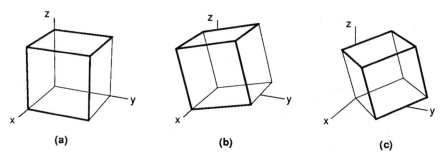

(a)   **(b)**   **(c)**

Figure 3.4 (a) The original starting cube. (b) $\mathbf{V} = R_x(\pi/12)\ \mathbf{U}$. Rotation about the x-axis. (c) $\mathbf{W} = R_z(\pi/4)\ \mathbf{V}$. Rotation about the z-axis.

8. Show that $R_x(\alpha)R_x(-\alpha) = \mathbf{I}$ by direct multiplication of the matrices $R_x(\alpha)$ and $R_x(-\alpha)$.

9. (a) Show that $R_x(\alpha)R_y(\beta) = \begin{pmatrix} \cos\beta & 0 & \sin\beta \\ \sin\beta\sin\alpha & \cos\alpha & -\cos\beta\sin\alpha \\ -\cos\alpha\sin\beta & \sin\alpha & \cos\beta\cos\alpha \end{pmatrix}$.

(b) Show that $R_y(\beta)R_x(\alpha) = \begin{pmatrix} \cos\beta & \sin\beta\sin\alpha & \cos\alpha\sin\beta \\ 0 & \cos\alpha & -\sin\alpha \\ -\sin\beta & \cos\beta\sin\alpha & \cos\beta\cos\alpha \end{pmatrix}$.

10. If A and B are nonsingular $N \times N$ matrices and $C = AB$, show that $C^{-1} = B^{-1}A^{-1}$. *Hint.* You must use the associative property of matrix multiplication.

11. Let A be an $M \times N$ matrix and $\mathbf{X}$ an N-dimensional vector.
   (a) How many multiplications are needed to calculate $A\mathbf{X}$?
   (b) How many additions are needed to calculate $A\mathbf{X}$?

12. Let A be an $M \times N$ matrix and B be an $N \times P$ matrix.
   (a) How many multiplications are needed to calculate AB?
   (b) How many additions are needed to calculate AB?

13. Let A be an $M \times N$ matrix, and let B and C be $N \times P$ matrices. Prove the left distributive law for matrix multiplication $A(B + C) = AB + AC$.

14. Let A and B be $M \times N$ matrices, and let C be an $N \times P$ matrix. Prove the right distributive law for matrix multiplication $(A + B)C = AC + BC$.

15. Find XX^T and X^TX, where $X = (1, -1, 2)$. *Note:* X^T is the transpose of X.

16. Write a report on band systems of equations. See References [29, 35, 41, 128, 160, and 192].

3.3 UPPER-TRIANGULAR LINEAR SYSTEMS

We will now develop the **back-substitution algorithm,** which is useful for solving a linear system of equations that has an upper-triangular coefficient matrix. This algorithm will be incorporated in the algorithm for solving a general system in Section 3.4.

Definition 3.2.   An $N \times N$ matrix $A = (a_{i,j})$ is called **upper-triangular** provided that the elements satisfy $a_{i,j} = 0$ whenever $i > j$. The $N \times N$ matrix $A = (a_{i,j})$ is called **lower-triangular** provided that $a_{ij} = 0$ whenever $i < j$.

We will develop a method for constructing the solution to upper-triangular systems of equations and leave the investigations of lower-triangular systems to the reader (see Exercise 8). If A is an upper-triangular matrix, $A\mathbf{X} = \mathbf{B}$ is an **upper-triangular system** of equations and has the form

$$
\begin{aligned}
a_{1,1}x_1 + a_{1,2}x_2 + a_{1,3}x_3 + \cdots + a_{1,N-1}x_{N-1} + \quad a_{1,N}x_N &= b_1 \\
a_{2,2}x_2 + a_{2,3}x_3 + \cdots + a_{2,N-1}x_{N-1} + \quad a_{2,N}x_N &= b_2 \\
a_{3,3}x_3 + \cdots + a_{3,N-1}x_{N-1} + \quad a_{3,N}x_N &= b_3 \\
\vdots \\
a_{N-1,N-1}x_{N-1} + a_{N-1,N}x_N &= b_{N-1} \\
a_{N,N}x_N &= b_N.
\end{aligned}
\tag{1}
$$

Theorem 3.6 (Back Substitution).   Suppose that $A\mathbf{X} = \mathbf{B}$ is an upper-triangular system with form given in (1). If

$$
a_{k,k} \neq 0 \qquad \text{for } k = 1, 2, \ldots, N,
\tag{2}
$$

then there exists a unique solution to (1).

Constructive Proof.   The solution is easy to find. The last equation involves only x_N, so we solve it first:

$$
x_N = \frac{b_N}{a_{N,N}}.
\tag{3}
$$

Now x_N is known and it can be used in the next-to-last equation:

$$
x_{N-1} = \frac{b_{N-1} - a_{N-1,N}x_N}{a_{N-1,N-1}}.
\tag{4}
$$

Now x_N and x_{N-1} are used to find x_{N-2}.

$$
x_{N-2} = \frac{b_{N-2} - a_{N-2,N-1}x_{N-1} - a_{N-2,N}x_N}{a_{N-2,N-2}}.
\tag{5}
$$

Once the values $x_N, x_{N-1}, \ldots, x_{k+1}$ are known, the general step is

$$
x_k = \frac{b_k - \sum_{j=k+1}^{N} a_{k,j}x_j}{a_{k,k}} \qquad \text{for } k = N-1, N-2, \ldots, 1.
\tag{6}
$$

The uniqueness of the solution is easy to see. The Nth equation implies that $b_N/a_{N,N}$ is the only possible value of x_N. Then finite induction is used to establish that $x_{N-1}, x_{N-2}, \ldots, x_1$ are unique.

Example 3.12

Use back substitution to solve the linear system

$$
\begin{aligned}
4x_1 - x_2 + 2x_3 + 3x_4 &= 20 \\
-2x_2 + 7x_3 - 4x_4 &= -7 \\
6x_3 + 5x_4 &= 4 \\
3x_4 &= 6.
\end{aligned}
$$

Solution.   Solving for x_4 in the last equation yields

$$x_4 = \frac{6}{3} = 2.$$

Using x_4 to solve for x_3 in the third equation, we obtain

$$x_3 = \frac{4 - 5(2)}{6} = -1.$$

Now x_3 and x_4 are used to find x_2 in the second equation:

$$x_2 = \frac{-7 - 7(-1) + 4(2)}{-2} = -4.$$

Finally, x_1 is obtained using the first equation:

$$x_1 = \frac{20 + 1(-4) - 2(-1) - 3(2)}{4} = 3.$$

The condition that $a_{k,k} \neq 0$ is essential because equation (6) involves division by $a_{k,k}$. If this requirement is not fulfilled, either no solution exists or infinitely many solutions exist.

Example 3.13

Show that there is no solution to the linear system

$$
\begin{aligned}
4x_1 - x_2 + 2x_3 + 3x_4 &= 20 \\
0x_2 + 7x_3 - 4x_4 &= -7 \\
6x_3 + 5x_4 &= 4 \\
3x_4 &= 6.
\end{aligned}
\tag{7}
$$

Solution.   Using the last equation in (7), we must have $x_4 = 2$, which is substituted into the second and third equations to obtain

$$
\begin{aligned}
7x_3 - 8 &= -7 \\
6x_3 + 10 &= 4.
\end{aligned}
\tag{8}
$$

The first equation in (8) implies that $x_3 = \frac{1}{7}$ and the second equation implies that $x_3 = -1$. This inconsistency leads to the conclusion that there is no solution to the linear system (7).

Example 3.14

Show that there are infinitely many solutions to

$$
\begin{aligned}
4x_1 - x_2 + 2x_3 + 3x_4 &= 20 \\
0x_2 + 7x_3 + 0x_4 &= -7 \\
6x_3 + 5x_4 &= 4 \\
3x_4 &= 6.
\end{aligned}
\tag{9}
$$

Solution.   Using the last equation in (9), we must have $x_4 = 2$, which is substituted into the second and third equations to get $x_3 = -1$, which checks out in both equations. But only the two values x_3 and x_4 have been obtained from the second through fourth equations, and when they are substituted into the first equation of (9) the result is

$$
x_2 = 4x_1 - 16,
\tag{10}
$$

which has infinitely many solutions; hence (9) has infinitely many solutions. If we choose a value of x_1 in (10), then the value of x_2 is uniquely determined. For example, if we include the equation $x_1 = 2$ in the system (9), then from (10) we compute $x_2 = -8$.

Theorem 3.7.   If $A = (a_{i,j})_{N \times N}$ is either upper-triangular or lower-triangular (or diagonal), then

$$
\det (A) = a_{1,1}a_{2,2} \cdots a_{N,N} = \prod_{i=1}^{N} a_{i,i}.
\tag{11}
$$

Product (Multiplication)

The value of the determinant for the matrix in Example 3.12 is

$$
\det (A) = 4(-2)(6)(3) = -144.
$$

Algorithm 3.1   (Back Substitution).   To solve the upper-triangular system,

$$
\begin{aligned}
a_{1,1}x_1 + a_{1,2}x_2 + \cdots + a_{1,N-1}x_{N-1} + a_{1,N}x_N &= b_1 \\
a_{2,2}x_2 + \cdots + a_{2,N-1}x_{N-1} + a_{2,N}x_N &= b_2 \\
&\;\; \vdots \\
a_{N-1,N-1}x_{N-1} + a_{N-1,N}x_N &= b_{N-1} \\
a_{N,N}x_N &= b_N.
\end{aligned}
$$

Proceed with the method only if all the diagonal elements are nonzero. First compute $x_N = b_N/a_{N,N}$ and then use the rule

$$x_r = \frac{b_r - \sum_{j=r+1}^{N} a_{r,j}x_j}{a_{r,r}} \qquad \text{for} \quad r = N-1, N-2, \ldots, 1.$$

Remark. Additional steps have been placed in the algorithm which perform the side calculations of finding the determinant of A.

```
INPUT N                                              {Number of equations}
VARiable declaration                                 {Dimension the arrays}
    REAL  A[1 .. N, 1 .. N] , B[1 .. N] , X[1 .. N]
INPUT  A[1 .. N, 1 .. N] , B[1 .. N]

DET := A(N, N)                                        {Initialize the variable}
X(N) := B(N)/A(N, N)                                  {Start the back substitution}

FOR    R = N - 1  DOWNTO  1  DO
       DET := DET*A(R, R)                             {Multiply the diagonal elements}
       SUM := 0                                       {Solve for X_r in row r}
           FOR   J = R + 1  TO  N  DO
                 SUM := SUM + A(R, J)*X(J)
       X(R) := [B(R) − SUM]/A(R, R)                   {End back substitution}

PRINT  "The value of  Det(A)  is "  DET               {Output}
PRINT  "The solution to the upper-triangular system is:"
       FOR    R = 1  TO  N  DO
             PRINT   "X(";R;") = "; X(R)
```

EXERCISES FOR UPPER-TRIANGULAR LINEAR SYSTEMS

In Exercises 1–3, solve the upper-triangular system and find the value of the determinant of the matrix of coefficients.

1.
$$3x_1 - 2x_2 + x_3 - x_4 = 8$$
$$4x_2 - x_3 + 2x_4 = -3$$
$$2x_3 + 3x_4 = 11$$
$$5x_4 = 15$$

2.
$$5x_1 - 3x_2 - 7x_3 + x_4 = -14$$
$$11x_2 + 9x_3 + 5x_4 = 22$$
$$3x_3 - 13x_4 = -11$$
$$7x_4 = 14$$

3. $4x_1 - x_2 + 2x_3 + 2x_4 - x_5 = 4$

$\qquad -2x_2 + 6x_3 + 2x_4 + 7x_5 = 0$

$\qquad\qquad x_3 - x_4 - 2x_5 = 3$

$\qquad\qquad\qquad -2x_4 - x_5 = 10$

$\qquad\qquad\qquad\qquad 3x_5 = 6$

4. Write a computer program that uses Algorithm 3.1. Use some of Exercises 1–3 as test cases.

5. Consider the two upper-triangular matrices

$$A = \begin{pmatrix} a_{11} & a_{12} & a_{13} \\ 0 & a_{22} & a_{23} \\ 0 & 0 & a_{33} \end{pmatrix} \quad \text{and} \quad B = \begin{pmatrix} b_{11} & b_{12} & b_{13} \\ 0 & b_{22} & b_{23} \\ 0 & 0 & b_{33} \end{pmatrix}.$$

Show that their product $C = AB$ is also upper-triangular.

6. Solve the lower-triangular system $A\mathbf{X} = \mathbf{B}$ and find det (A).

$$2x_1 \qquad\qquad\qquad\qquad = 6$$

$$-x_1 + 4x_2 \qquad\qquad\qquad = 5$$

$$3x_1 - 2x_2 - x_3 \qquad\qquad = 4$$

$$x_1 - 2x_2 + 6x_3 + 3x_4 = 2$$

7. Solve the lower-triangular system $A\mathbf{X} = \mathbf{B}$ and find det (A).

$$5x_1 \qquad\qquad\qquad\qquad = -10$$

$$x_1 + 3x_2 \qquad\qquad\qquad = 4$$

$$3x_1 + 4x_2 + 2x_3 \qquad\qquad = 2$$

$$-x_1 + 3x_2 - 6x_3 - x_4 = 5$$

8. *Forward-substitution algorithm.* A linear system is called lower-triangular provided that $a_{i,j} = 0$ when $j > i$. Given

$$a_{1,1}x_1 \qquad\qquad\qquad\qquad\qquad\qquad\qquad\qquad = b_1$$

$$a_{2,1}x_1 + a_{2,2}x_2 \qquad\qquad\qquad\qquad\qquad\qquad = b_2$$

$$a_{3,1}x_1 + a_{3,2}x_2 + a_{3,3}x_3 \qquad\qquad\qquad\qquad = b_3$$

$$\vdots \qquad\quad \vdots \qquad\quad \vdots \qquad\qquad\qquad\qquad\qquad \vdots$$

$$a_{N-1,1}x_1 + a_{N-1,2}x_2 + a_{N-1,3}x_3 + \cdots + a_{N-1,N-1}x_{N-1} \qquad\qquad = b_{N-1}$$

$$a_{N,1}x_1 + a_{N,2}x_2 + a_{N,3}x_3 + \cdots + a_{N,N-1}x_{N-1} + a_{N,N}x_N = b_N,$$

develop an algorithm for solving a lower-triangular system when the diagonal elements are nonzero (i.e., $a_{k,k} \neq 0$ for $k = 1, 2, \ldots, N$).

9. Show that back-substitution requires N divisions, $(N^2 - N)/2$ multiplications, and $(N^2 - N)/2$ additions or subtractions. *Hint.* You can use the formula

$$\sum_{k=1}^{M} k = (M^2 + M)/2.$$

3.4 GAUSSIAN ELIMINATION AND PIVOTING

In this section we develop an efficient scheme for solving a general system $A\mathbf{X} = \mathbf{B}$ of N equations and N unknowns. The crucial step is to construct an equivalent upper-triangular system $U\mathbf{X} = \mathbf{Y}$ that can be solved by the method of Section 3.3.

Two linear systems of dimension N by N are said to be *equivalent* provided that their solutions sets are the same. Theorems from linear algebra show that when certain transformations are applied to a given system, the solution sets do not change.

Definition 3.3   (Elementary Transformations).   The following operations applied to a linear system yield an equivalent system:

Interchanges: The order of two equations can be changed.   (1)

Scaling: Multiplying an equation by a nonzero constant.   (2)

Replacement: An equation can be replaced by the sum of that equation and a   (3)
multiple of any other equation.

It is common to use (3) by replacing an equation with the difference of that equation and a multiple of another equation. These concepts are illustrated in the next example.

Example 3.15

Find the parabola $y = A + Bx + Cx^2$ that passes through the three points $(1, 1)$, $(2, -1)$, and $(3, 1)$.

Solution.   For each point we obtain an equation relating the value of x to the value of y. The result is the system

$$\begin{aligned}
A + B + C &= 1 \quad \text{at } (1, 1) \\
A + 2B + 4C &= -1 \quad \text{at } (2, -1) \\
A + 3B + 9C &= 1 \quad \text{at } (3, 1).
\end{aligned} \tag{4}$$

The variable A is eliminated from the second and third equations by subtracting the first equation from them. This is an application of the replacement transformation (3) and the resulting system is

$$\begin{aligned}
A + B + C &= 1 \\
B + 3C &= -2 \\
2B + 8C &= 0.
\end{aligned} \tag{5}$$

The variable B is eliminated from the third equation in (5) by subtracting from it two times the second equation. We arrive at the equivalent upper-triangular system

$$
\begin{aligned}
A + B + \ C &= \quad 1 \\
B + 3C &= -2 \\
2C &= \quad 4.
\end{aligned}
\tag{6}
$$

The back-substitution algorithm is now used to find the coefficients $C = 4/2 = 2$, $B = -2 - 3(2) = -8$, and $A = 1 - (-8) - 2 = 7$, and the equation of the parabola is $y = 7 - 8x + 2x^2$.

It is efficient to store all the coefficients of the linear system $AX = \mathbf{B}$ in an array of dimension N by $N + 1$. The coefficients of $\mathbf{B}$ are stored in column $N + 1$ of the array (i.e., $a_{k,N+1} = b_k$). Each row contains all the coefficients necessary to represent one equation in the linear system. The **augmented matrix** is denoted by $[A, B]$ and the linear system is represented as follows:

$$
[A, B] =
\begin{pmatrix}
a_{1,1} & a_{1,2} & \cdots & a_{1,N} & b_1 \\
a_{2,1} & a_{2,2} & \cdots & a_{2,N} & b_2 \\
\vdots & \vdots & & \vdots & \vdots \\
a_{N,1} & a_{N,2} & \cdots & a_{N,N} & b_N
\end{pmatrix}.
\tag{7}
$$

The system $AX = \mathbf{B}$, with augmented matrix given in (7), can be solved by performing row operations on the augmented matrix $[A, B]$. The variables x_k are placeholders for the coefficients and can be omitted until the end of the calculation.

Definition 3.4 (Row Operations).   The following operations applied to the augmented matrix (7) yield an equivalent system.

Interchanges: The order of two rows can be changed.   (8)

Scaling: Multiplying a row by a nonzero constant.   (9)

Replacement: A row can be replaced by the sum of that row and a multiple
              of any other row; that is   (10)
$$\text{row}_r := \text{row}_r - m_{r,p}\text{row}_p.$$

It is common to use (10) by replacing a row with the difference of that row and a multiple of another row.

Definition 3.5 (Pivot).   The number $a_{r,r}$ in position (r, r) that is used to eliminate x_r in rows $r + 1, r + 2, \ldots, N$ is called the rth **pivotal element** and the rth row is called the **pivotal row**.

Example 3.16

Express the following system in augmented matrix form and find an equivalent upper-triangular system and the solution.

$$x_1 + 2x_2 + x_3 + 4x_4 = 13$$
$$2x_1 + 0x_2 + 4x_3 + 3x_4 = 28$$
$$4x_1 + 2x_2 + 2x_3 + x_4 = 20$$
$$-3x_1 + x_2 + 3x_3 + 2x_4 = 6.$$

Solution.   The augmented matrix is

$$
\begin{array}{l}
\text{pivot} \longrightarrow \\
m_{2,1} = 2 \\
m_{3,1} = 4 \\
m_{4,1} = -3
\end{array}
\left(
\begin{array}{cccc|c}
\underline{1} & 2 & 1 & 4 & 13 \\
2 & 0 & 4 & 3 & 28 \\
4 & 2 & 2 & 1 & 20 \\
-3 & 1 & 3 & 2 & 6
\end{array}
\right).
$$

The first row is used to eliminate elements in the first column below the diagonal. We refer to the first row as the pivotal row and the element $a_{1,1}$ is called the pivotal element. The values $m_{k,1}$ are the multiples of row 1 that are to be subtracted from row k for $k = 2, 3, 4$. The result after elimination is

$$
\begin{array}{l}
\\
\text{pivot} \longrightarrow \\
m_{3,2} = 1.5 \\
m_{4,2} = -1.75
\end{array}
\left(
\begin{array}{cccc|c}
1 & 2 & 1 & 4 & 13 \\
0 & \underline{-4} & 2 & -5 & 2 \\
0 & -6 & -2 & -15 & -32 \\
0 & 7 & 6 & 14 & 45
\end{array}
\right).
$$

The second row is used to eliminate elements in the second column that lie below the diagonal. The second row is the pivotal row and the values $m_{k,2}$ are the multiples of row 2 that are to be subtracted from row k for $k = 3, 4$. The result after elimination is

$$
\begin{array}{l}
\\
\\
\text{pivot} \longrightarrow \\
m_{4,3} = -1.9
\end{array}
\left(
\begin{array}{cccc|c}
1 & 2 & 1 & 4 & 13 \\
0 & -4 & 2 & -5 & 2 \\
0 & 0 & \underline{-5} & -7.5 & -35 \\
0 & 0 & 9.5 & 5.25 & 48.5
\end{array}
\right).
$$

Finally, the multiple $m_{4,3} = -1.9$ of the third row is subtracted from the fourth row and the result is the upper-triangular system

$$
\left(
\begin{array}{cccc|c}
1 & 2 & 1 & 4 & 13 \\
0 & -4 & 2 & -5 & 2 \\
0 & 0 & -5 & -7.5 & -35 \\
0 & 0 & 0 & -9 & -18
\end{array}
\right).
\tag{11}
$$

The back-substitution algorithm can be used to solve (11) and we get

$$x_4 = 2, \quad x_3 = 4, \quad x_2 = -1, \quad x_1 = 3.$$

The process described above is called **Gaussian elimination** and must be modified so that it can be used in most circumstances. If $a_{k,k} = 0$, row k cannot be used to eliminate the elements in column k, and row k must be changed with some row below the diagonal to obtain a nonzero pivot element. If this cannot be done, the system of equations does not have a unique solution.

Theorem 3.8   (Gaussian Elimination with Back Substitution).   Suppose that A is an $N \times N$ nonsingular matrix; there exists an equivalent system $U\mathbf{X} = \mathbf{Y}$ where U is an upper-triangular matrix with $u_{k,k} \neq 0$. After U and $\mathbf{Y}$ are constructed, back substitution can be used to solve $U\mathbf{X} = \mathbf{Y}$ for $\mathbf{X}$.

Proof. We will use the augmented matrix with $\mathbf{B}$ stored in column $N + 1$:

$$A\mathbf{X} = \begin{pmatrix} a_{1,1}^{(1)} & a_{1,2}^{(1)} & a_{1,3}^{(1)} & \cdots & a_{1,n}^{(1)} \\ a_{2,1}^{(1)} & a_{2,2}^{(1)} & a_{2,3}^{(1)} & \cdots & a_{2,n}^{(1)} \\ a_{3,1}^{(1)} & a_{3,2}^{(1)} & a_{3,3}^{(1)} & \cdots & a_{3,n}^{(1)} \\ \cdot & \cdot & \cdot & & \cdot \\ \cdot & \cdot & \cdot & & \cdot \\ \cdot & \cdot & \cdot & & \cdot \\ a_{n,1}^{(1)} & a_{n,2}^{(1)} & a_{n,3}^{(1)} & \cdots & a_{n,n}^{(1)} \end{pmatrix} \begin{pmatrix} x_1 \\ x_2 \\ x_3 \\ \cdot \\ \cdot \\ \cdot \\ x_n \end{pmatrix} = \begin{pmatrix} a_{1,n+1}^{(1)} \\ a_{2,n+1}^{(1)} \\ a_{3,n+1}^{(1)} \\ \cdot \\ \cdot \\ \cdot \\ a_{n,n+1}^{(1)} \end{pmatrix} = \mathbf{B}.$$

Then we will construct an equivalent upper-triangular system $U\mathbf{X} = \mathbf{Y}$:

$$U\mathbf{X} = \begin{pmatrix} a_{1,1}^{(1)} & a_{1,2}^{(1)} & a_{1,3}^{(1)} & \cdots & a_{1,n}^{(1)} \\ 0 & a_{2,2}^{(2)} & a_{2,3}^{(2)} & \cdots & a_{2,n}^{(2)} \\ 0 & 0 & a_{3,3}^{(3)} & \cdots & a_{3,n}^{(3)} \\ \cdot & \cdot & \cdot & & \cdot \\ \cdot & \cdot & \cdot & & \cdot \\ \cdot & \cdot & \cdot & & \cdot \\ 0 & 0 & 0 & \cdots & a_{n,n}^{(n)} \end{pmatrix} \begin{pmatrix} x_1 \\ x_2 \\ x_3 \\ \cdot \\ \cdot \\ \cdot \\ x_n \end{pmatrix} = \begin{pmatrix} a_{1,n+1}^{(1)} \\ a_{2,n+1}^{(2)} \\ a_{3,n+1}^{(3)} \\ \cdot \\ \cdot \\ \cdot \\ a_{n,n+1}^{(n)} \end{pmatrix} = \mathbf{Y}.$$

Step 1. Store the coefficients in the array. The superscript on $a_{r,c}^{(1)}$ means that this is the first time a number is stored in location (r, c).

$$\begin{matrix} a_{1,1}^{(1)} & a_{1,2}^{(1)} & a_{1,3}^{(1)} & \cdots & a_{1,n}^{(1)} & a_{1,n+1}^{(1)} \\ a_{2,1}^{(1)} & a_{2,2}^{(1)} & a_{2,3}^{(1)} & \cdots & a_{2,n}^{(1)} & a_{2,n+1}^{(1)} \\ a_{3,1}^{(1)} & a_{3,2}^{(1)} & a_{3,3}^{(1)} & \cdots & a_{3,n}^{(1)} & a_{3,n+1}^{(1)} \\ \cdot & \cdot & \cdot & & \cdot & \cdot \\ \cdot & \cdot & \cdot & & \cdot & \cdot \\ a_{n,1}^{(1)} & a_{n,2}^{(1)} & a_{n,3}^{(1)} & \cdots & a_{n,n}^{(1)} & a_{n,n+1}^{(1)}. \end{matrix}$$

Step 2. If necessary, switch rows so that $a_{1,1}^{(1)} \neq 0$; then eliminate x_1 in rows 2 through N. In this process $m_{r,1}$ is the multiple of row 1 that is subtracted from row r.

```
FOR     r = 2  TO  N  DO
 │   Set   m_r,1 := a_r,1^(1)/a_1,1^(1) and set a_r,1^(2) := 0
 │   FOR     c = 2  TO  N+1  DO
 └──── └──── a_r,c^(2) := a_r,c^(1) − m_r,1*a_1,c^(1)
```

The new elements are written $a_{r,c}^{(2)}$ to indicate that this is the second time that a number has been stored in the array at location (r, c). The result after step 2 is

$$
\begin{array}{ccccccc}
a_{1,1}^{(1)} & a_{1,2}^{(1)} & a_{1,3}^{(1)} & \cdots & a_{1,n}^{(1)} & a_{1,n+1}^{(1)} \\
0 & a_{2,2}^{(2)} & a_{2,3}^{(2)} & \cdots & a_{2,n}^{(2)} & a_{2,n+1}^{(2)} \\
0 & a_{3,2}^{(2)} & a_{3,3}^{(2)} & \cdots & a_{3,n}^{(2)} & a_{3,n+1}^{(2)} \\
\vdots & \vdots & \vdots & & \vdots & \vdots \\
0 & a_{n,2}^{(2)} & a_{n,3}^{(2)} & \cdots & a_{n,n}^{(2)} & a_{n,n+1}^{(2)}.
\end{array}
$$

Step 3. If necessary, switch the second row with some row below it so that $a_{2,2}^{(2)} \neq 0$, then eliminate x_2 in rows 3 through N. In this process $m_{r,2}$ is the multiple of row 2 that is subtracted from row r.

```
FOR     r = 3  TO  N  DO
 │   Set   m_r,2 := a_r,2^(2)/a_2,2^(2) and set a_r,2^(3) := 0
 │   FOR     c = 3  TO  N+1  DO
 └──── └──── a_r,c^(3) := a_r,c^(2) − m_r,2*a_2,c^(2)
```

The new elements are written $a_{r,c}^{(3)}$ to indicate that this is the third time that a number has been stored in the array at location (r, c). The result after step 3 is

$$
\begin{array}{ccccccc}
a_{1,1}^{(1)} & a_{1,2}^{(1)} & a_{1,3}^{(1)} & \cdots & a_{1,n}^{(1)} & a_{1,n+1}^{(1)} \\
0 & a_{2,2}^{(2)} & a_{2,3}^{(2)} & \cdots & a_{2,n}^{(2)} & a_{2,n+1}^{(2)} \\
0 & 0 & a_{3,3}^{(3)} & \cdots & a_{3,n}^{(3)} & a_{3,n+1}^{(3)} \\
\vdots & \vdots & \vdots & & \vdots & \vdots \\
0 & 0 & a_{n,3}^{(3)} & \cdots & a_{n,n}^{(3)} & a_{n,n+1}^{(3)}.
\end{array}
$$

Step $p + 1$. *This is the general step.* If necessary, switch row p with some row beneath it so that $a_{p,p}^{(p)} \neq 0$, then eliminate x_p in rows $p + 1$ through N. Here $m_{r,p}$ is the multiple of row p that is subtracted from row r.

```
FOR     r = p+1  TO  N  DO
 │   Set   m_r,p := a_r,p^(p)/a_p,p^(p) and set a_r,p^(p+1) := 0
 │   FOR     c = p+1  TO  N+1  DO
 └──── └──── a_r,c^(p+1) := a_r,c^(p) − m_r,p*a_p,c^(p)
```

The final result after x_{N-1} has been eliminated from row N is

$$
\begin{array}{ccccccc}
a_{1,1}^{(1)} & a_{1,2}^{(1)} & a_{1,3}^{(1)} & \cdots & a_{1,n}^{(1)} & a_{1,n+1}^{(1)} \\
0 & a_{2,2}^{(2)} & a_{2,3}^{(2)} & \cdots & a_{2,n}^{(2)} & a_{2,n+1}^{(2)} \\
0 & 0 & a_{3,3}^{(3)} & \cdots & a_{3,n}^{(3)} & a_{3,n+1}^{(3)} \\
\vdots & \vdots & \vdots & & \vdots & \vdots \\
0 & 0 & 0 & \cdots & a_{n,n}^{(n)} & a_{n,n+1}^{(n)}.
\end{array}
$$

The upper-triangularization process is now complete.

Since A is nonsingular, when row operations are performed the successive matrices are also nonsingular. This guarantees that $a_{k,k}^{(k)} \neq 0$ for all k in the construction process. Hence back substitution can be used to solve $U\mathbf{X} = \mathbf{Y}$ for $\mathbf{X}$, and the theorem is proven.

Pivoting to Avoid $a_{p,p}^{(p)} = 0$

If $a_{p,p}^{(p)} = 0$, row p cannot be used to eliminate the elements in column p below the diagonal. It is necessary to find row k, where $a_{k,p}^{(p)} \neq 0$ and $k > p$ and then interchange row p and row k so that a nonzero pivot element is obtained. This process is called **pivoting**, and the criterion for deciding which row to choose is called a **pivoting strategy**. The trivial pivoting strategy is as follows. If $a_{p,p}^{(p)} \neq 0$, do not switch rows. If $a_{p,p}^{(p)} = 0$, locate the first row below row p in which $a_{k,p}^{(p)} \neq 0$ and switch rows k and p. This will result in a new element $a_{p,p}^{(p)} \neq 0$, which is a nonzero pivot element.

Pivoting to Reduce Error

If there is more than one nonzero element in column p that lies on or below the diagonal, there is a choice to determine which rows to interchange. We must determine ahead of time a good pivoting strategy. The partial pivoting strategy is the most common one and is used in Algorithm 3.2. Because the computer uses fixed-precision arithmetic, it is possible that a small error is introduced each time an arithmetic operation is performed. To reduce the propagation of error, it is suggested that one check the magnitude of all the elements in column p that lie on or below the diagonal, and locate row k in which the element has the largest absolute value, that is,

$$|a_{k,p}| = \max \{|a_{p,p}|, |a_{p+1,p}|, \ldots, |a_{N-1,p}|, |a_{N,p}|\},$$

and then switch row p with row k if $k > p$. Usually, the larger pivot element will result in a smaller error being propagated.

In Sections 3.5 and 3.6 we will find that it takes a total of $(4N^3 + 9N^2 - 7N)/6$ arithmetic operations to solve an $N \times N$ system. When $N = 20$ the total number of arithmetic operations that must be performed is 5910, and the propagation of error in the computations could result in an erroneous answer. The technique of scaled partial pivoting or equilibrating can be used to further reduce the effect of error propagation. The reader can find an excellent discussion about these pivoting refinements in Reference [41].

Example 3.17

The values $x_1 = x_2 = 1.000$ are the solution to

$$1.133x_1 + 5.281x_2 = 6.414$$

$$24.14x_1 - 1.210x_2 = 22.93.$$

Use four-digit arithmetic and Gaussian elimination with trivial pivoting to find a computed approximate solution to the system.

Solution.   The multiple $m_{2,1} = 24.14/1.133 = 21.31$ of row 1 is to be subtracted from row 2 to obtain the upper-triangular system. Using four digits in the calculations we obtain the new coefficients

$$a_{2,2}^{(2)} = -1.210 - 21.31 \times 5.281 = -1.210 - 112.5 = -113.7$$

$$a_{2,3}^{(2)} = 22.93 - 21.31 \times 6.414 = 22.93 - 136.7 = -113.8.$$

The computer upper-triangular system is

$$1.133x_1 + 5.281x_2 = 6.414$$

$$-113.7x_2 = -113.8.$$

Back substitution is used to compute $x_2 = -113.8/(-113.7) = 1.001$, and $x_1 = (6.414 - 5.281 \times 1.001)/1.133 = (6.414 - 5.286)/1.133 = 0.9956$.

Example 3.18

Use four-digit arithmetic and Gaussian elimination with partial pivoting to solve the linear system in Example 3.17.

Solution.   Interchanging the rows, we have

$$24.14x_1 - 1.210x_2 = 22.93$$

$$1.133x_1 + 5.281x_2 = 6.414.$$

This time $m_{2,1} = 1.133/24.14 = 0.04693$ is the multiple of row 1 that is to be subtracted from row 2. The new coefficients are

$$a_{2,2}^{(2)} = 5.281 - 0.04693(-1.210) = 5.281 + 0.05679 = 5.338$$

$$a_{2,3}^{(2)} = 6.414 - 0.04693(22.93) = 6.414 - 1.076 = 5.338.$$

The computed upper-triangular system is

$$24.14x_1 - 1.210x_2 = 22.93$$

$$5.338x_2 = 5.338.$$

Back substitution is used to compute $x_2 = 5.338/5.338 = 1.000$, and $x_1 = (22.93 + 1.210 \times 1.000)/(24.14) = 24.14/24.14 = 1.000$.

Ill-Conditioning

A matrix A is called **ill-conditioned** if there exists a vector **B** for which small perturbations in the coefficients of A or **B** will produce large changes in $\mathbf{X} = A^{-1}\mathbf{B}$. The system $A\mathbf{X} = \mathbf{B}$ is ill-conditioned when A is ill-conditioned. In this case, numerical methods for computing an approximate solution are prone to have more error.

   One circumstance involving ill-conditioning occurs when A is "nearly singular" and the determinant of A is close to zero. Ill-conditioning can also occur in systems of two equations when two lines are nearly parallel (or in three equations when three planes are nearly parallel). A consequence of ill-conditioning is that substitution of erroneous values may appear to be genuine solutions. For example, consider the two equations:

$$x + 2y - 2.0 = 0,$$
$$2x + 3y - 3.4 = 0. \tag{12}$$

Substitution of $x_0 = 1.00$ and $y_0 = 0.48$ into these equations "almost produces zeros":

$$1 + 2(0.48) - 2.00 = 1.96 - 2.00 = -0.04 \approx 0,$$
$$2 + 3(0.48) - 3.40 = 3.44 - 3.40 = 0.04 \approx 0.$$

Here the discrepancy from 0 is only $\pm\, 0.04$. However, the true solution to this linear system is $x = 0.8$ and $y = 0.6$, so that the errors in the approximate solution are $x - x_0 = 0.80 - 1.00 = -0.20$ and $y - y_0 = 0.60 - 0.48 = -0.12$. Thus, merely substituting values into a set of equations is not a reliable test for accuracy. The rhombus-shaped region R in Figure 3.5 represents a set where both equations in (12) are "almost satisfied":

$$R = \{(x, y): |x + 2y - 2.0| < 0.1 \quad \text{and} \quad |2x + 3y - 3.4| < 0.2\}.$$

There are points in R that are far away from the solution point $(0.8, 0.6)$ and yet produce small values when substituted into the equations in (12). If it is suspected that a linear system is ill-conditioned, computations should be carried out in multiple-precision

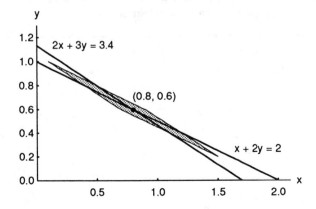

Figure 3.5   A region where two equations are "almost satisfied."

arithmetic. The interested reader should reach the topic of condition number of a matrix to get more information on this phenomenon.

Ill-conditioning has more drastic consequences when several equations are involved. Consider the problem of finding the cubic polynomial $y = c_1 x^3 + c_2 x^2 + c_3 x + c_4$ that passes through the four points (2, 8), (3, 27), (4, 64), and (5, 125). In Chapter 5 we will introduce the method of "least squares." Applying the method of least squares to find the coefficients requires that the following linear system be solved:

$$\begin{pmatrix} 20{,}514 & 4{,}424 & 978 & 224 \\ 4{,}424 & 978 & 224 & 54 \\ 978 & 224 & 54 & 14 \\ 224 & 54 & 14 & 4 \end{pmatrix} \begin{pmatrix} c_1 \\ c_2 \\ c_3 \\ c_4 \end{pmatrix} = \begin{pmatrix} 20{,}514 \\ 4{,}424 \\ 978 \\ 224 \end{pmatrix}.$$

A computer that carried about nine digits of precision was used to compute the coefficients and obtained $c_1 = 1.000004$, $c_2 = -0.000038$, $c_3 = 0.000126$, and $c_4 = -0.000131$. Although this computation is close to the true solution $c_1 = 1$ and $c_2 = c_3 = c_4 = 0$, it shows how easy it is for error to creep into the solution. Furthermore, suppose that the coefficient $a_{1,1} = 20{,}514$ in the upper left corner of the matrix is changed to the value 20,515 and the perturbed system is solved. Values obtained with the same computer were $c_1 = 0.642857$, $c_2 = 3.75000$, $c_3 = -12.3928$, and $c_4 = 12.7500$, which is a worthless answer. Ill-conditioning is not easy to detect. If the system is solved a second time with slightly perturbed coefficients and an answer that differs significantly from the first one is discovered, then it is realized that ill-conditioning is present. Sensitivity analysis is a topic in advanced numerical analysis and the reader is referred to Reference [66].

Algorithm 3.2 (Upper Triangularization Followed by Back Substitution).
To construct the solution to $A\mathbf{X} = \mathbf{B}$, by first reducing the augmented matrix $[A, B]$ to upper-triangular form and then performing back substitution.

```
           INPUT N , A[1 .. N, 1 .. N+1]                          {Input augmented matrix}
           FOR    J = 1  TO  N  DO                                {Initialize the pointer vector}
                     Row(J) := J
FOR        P = 1  TO  N−1  DO                                     {Start upper-triangularization}
           FOR    K = P+1  TO  N  DO                                  {Find pivot element}
                  IF   |A(Row(K), P)| > |A(Row(P), P)|      THEN
                             T := Row(P)                            {Switch the index for
                          Row(P) := Row(K)                             the p-th pivot row
                          Row(K) := T                                      if necessary}
                  ENDIF                                      {End of simulated row interchange}
                  IF    A(Row(P), P) = 0   THEN
                        PRINT "The matrix is singular."
                        TERMINATE ALGORITHM
                  ENDIF
           FOR    K = P+1  TO  N  DO
                        M := A(Row(K), P)/A(Row(P), P)                  {Form multiplier}
                        FOR   C = P+1  TO  N+1  DO                      {Eliminate Xp}
                              A(Row(K), C) := A(Row(K), C) − M∗A(Row(P), C)
END                                                 {End of the upper-triangularization routine}
                  IF    A(Row(N), N) = 0   THEN
                        PRINT "The matrix is singular."
                        TERMINATE ALGORITHM
                  ENDIF

           X(N) := A(Row(N), N+1)/A(Row(N), N)                   {Start the back substitution}
FOR        K = N−1  DOWNTO  1  DO
           SUM := 0
                        FOR    C = K+1  TO  N  DO
                              SUM := SUM + A(Row(K), C)∗X(C)
                  X(K) := [A(Row(K), N+1)−SUM]/A(Row(K), K)      {End back substitution}

FOR        K = 1  TO  N  DO
                  PRINT X(K)                                              {Output}
```

EXERCISES FOR GAUSSIAN ELIMINATION AND PIVOTING

In Exercises 1–4, show that $AX = B$ is equivalent to the upper-triangular system $UX = Y$ and find the solution.

1. $2x_1 + 4x_2 - 6x_3 = -4$ $2x_1 + 4x_2 - 6x_3 = -4$

    $x_1 + 5x_2 + 3x_3 = 10$ $3x_2 + 6x_3 = 12$

    $x_1 + 3x_2 + 2x_3 = 5$ $3x_3 = 3$

2. $x_1 + x_2 + 6x_3 = 7$ $x_1 + x_2 + 6x_3 = 7$

$-x_1 + 2x_2 + 9x_3 = 2$ $3x_2 + 15x_3 = 9$

$x_1 - 2x_2 + 3x_3 = 10$ $12x_3 = 12$

3. $2x_1 - 2x_2 + 5x_3 = 6$ $2x_1 - 2x_2 + 5x_3 = 6$

$2x_1 + 3x_2 + x_3 = 13$ $5x_2 - 4x_3 = 7$

$-x_1 + 4x_2 - 4x_3 = 3$ $0.9x_3 = 1.8$

4. $-5x_1 + 2x_2 - x_3 = -1$ $-5x_1 + 2x_2 - x_3 = -1$

$x_1 + 0x_2 + 3x_3 = 5$ $0.4x_2 + 2.8x_3 = 4.8$

$3x_1 + x_2 + 6x_3 = 17$ $-10x_3 = -10$

5. Find the parabola $y = A + Bx + Cx^2$ that passes through $(1, 4)$, $(2, 7)$, and $(3, 14)$.

6. Find the parabola $y = A + Bx + Cx^2$ that passes through $(1, 6)$, $(2, 5)$, and $(3, 2)$.

7. Find the parabola $y = A + Bx + Cx^2$ that passes through $(1, 2)$, $(2, 2)$, and $(4, 8)$.

In Exercises 8–10, show that $A\mathbf{X} = \mathbf{B}$ is equivalent to the upper-triangular system $U\mathbf{X} = \mathbf{Y}$ and find the solution.

8. $4x_1 + 8x_2 + 4x_3 + 0x_4 = 8$ $4x_1 + 8x_2 + 4x_3 + 0x_4 = 8$

$x_1 + 5x_2 + 4x_3 - 3x_4 = -4$ $3x_2 + 3x_3 - 3x_4 = -6$

$x_1 + 4x_2 + 7x_3 + 2x_4 = 10$ $4x_3 + 4x_4 = 12$

$x_1 + 3x_2 + 0x_3 - 2x_4 = -4$ $x_4 = 2$

9. $2x_1 + 4x_2 - 4x_3 + 0x_4 = 12$ $2x_1 + 4x_2 - 4x_3 + 0x_4 = 12$

$x_1 + 5x_2 - 5x_3 - 3x_4 = 18$ $3x_2 - 3x_3 - 3x_4 = 12$

$2x_1 + 3x_2 + x_3 + 3x_4 = 8$ $4x_3 + 2x_4 = 0$

$x_1 + 4x_2 - 2x_3 + 2x_4 = 8$ $3x_4 = -6$

10. $x_1 + 2x_2 + 0x_3 - x_4 = 9$ $x_1 + 2x_2 + 0x_3 - x_4 = 9$

$2x_1 + 3x_2 - x_3 + 0x_4 = 9$ $-x_2 - x_3 + 2x_4 = -9$

$0x_1 + 4x_2 + 2x_3 - 5x_4 = 26$ $-2x_3 + 3x_4 = -10$

$5x_1 + 5x_2 + 2x_3 - 4x_4 = 32$ $1.5x_4 = -3$

11. Find the solution to the following linear system.

$$x_1 + x_2 + 0x_3 + 4x_4 = 3$$

$$2x_1 - x_2 + 5x_3 + 0x_4 = 2$$

$$5x_1 + 2x_2 + x_3 + 2x_4 = 5$$

$$-3x_1 + 0x_2 + 2x_3 + 6x_4 = -2$$

12. **(a)** Write a computer program that uses Algorithm 3.2. Use some of Exercises 8–11 as test cases.
    (b) Extend the program so that it will compute $\mathbf{R} = \mathbf{B} - A\mathbf{X}$. This quantity is called the *residual* and can be used for a check to see that the correct solution was obtained.

13. Find the solution to the following linear system.

$$
\begin{aligned}
x_1 + 2x_2 \qquad\qquad\quad &= 7 \\
2x_1 + 3x_2 - x_3 \qquad\quad &= 9 \\
4x_2 + 2x_3 + 3x_4 &= 10 \\
2x_3 - 4x_4 &= 12
\end{aligned}
$$

14. Find the solution to the following linear system.

$$
\begin{aligned}
x_1 + x_2 \qquad\qquad\quad &= 5 \\
2x_1 - x_2 + 5x_3 \qquad\quad &= -9 \\
3x_2 - 4x_3 + 2x_4 &= 19 \\
2x_3 + 6x_4 &= 2
\end{aligned}
$$

15. Many applications involve matrices with many zeros. Of practical importance are **tridiagonal systems** of the form

$$
\begin{aligned}
d_1x_1 + c_1x_2 \qquad\qquad\qquad\qquad\qquad\qquad &= b_1 \\
a_1x_1 + d_2x_2 + c_2x_3 \qquad\qquad\qquad\qquad &= b_2 \\
a_2x_2 + d_3x_3 + c_3x_4 \qquad\qquad\quad &= b_3 \\
\vdots \qquad\quad \vdots \qquad\quad \vdots \qquad\qquad\quad \vdots \\
a_{N-2}x_{N-2} + d_{N-1}x_{N-1} + c_{N-1}x_N &= b_{N-1} \\
a_{N-1}x_{N-1} + d_Nx_N &= b_N.
\end{aligned}
$$

Write an algorithm that will solve a tridiagonal system. You may assume that row interchanges are not needed and that row k can be used to eliminate x_k in row $k + 1$.

16. How could you modify Algorithm 3.2 so that it will efficiently solve M linear systems with the same matrix A but different column vectors $\mathbf{B}$? The M linear systems look like

$$
A\mathbf{X}_1 = \mathbf{B}_1, \quad A\mathbf{X}_2 = \mathbf{B}_2, \quad \ldots, \quad A\mathbf{X}_M = \mathbf{B}_M.
$$

17. Write a report on the Gauss–Jordan method. See References [29, 44, 51, 62, 79, 85, 90, 117, and 152].

18. Write a report on pivoting strategies. See References [9, 29, 35, 40, 41, 58, 79, 96, 101, 117, 128, 145, 146, 152, 153, and 160].

19. Write a report on the condition number of a matrix. See References [9, 19, 29, 40, 41, 57, 62, 74, 94, 96, 98, 101, 117, 128, 145, 152, 153, 160, and 192].

20. Write a report on ill-conditioned matrices. See References [9, 19, 29, 40, 41, 47, 49, 62, 94, 101, 128, 145, 153, 192, and 197].

21. Write a report on iterative improvement (residual correction). See References [8, 9, 19, 29, 40, 41, 49, 51, 58, 72, 90, 94, 96, 97, 117, 137, 152, 153, and 160].

22. The Hilbert matrix is a classical ill-conditioned matrix, and small changes in its coefficients will produce a large change in the solution to the perturbed system.
    (a) Solve $A\mathbf{X} = \mathbf{B}$ using the Hilbert matrix of dimension 5×5:

$$A = \begin{pmatrix} \frac{1}{1} & \frac{1}{2} & \frac{1}{3} & \frac{1}{4} & \frac{1}{5} \\ \frac{1}{2} & \frac{1}{3} & \frac{1}{4} & \frac{1}{5} & \frac{1}{6} \\ \frac{1}{3} & \frac{1}{4} & \frac{1}{5} & \frac{1}{6} & \frac{1}{7} \\ \frac{1}{4} & \frac{1}{5} & \frac{1}{6} & \frac{1}{7} & \frac{1}{8} \\ \frac{1}{5} & \frac{1}{6} & \frac{1}{7} & \frac{1}{8} & \frac{1}{9} \end{pmatrix} \qquad \mathbf{B} = \begin{pmatrix} 1 \\ 0 \\ 0 \\ 0 \\ 0 \end{pmatrix}.$$

(b) Solve $A\mathbf{X} = \mathbf{B}$ using

$$A = \begin{pmatrix} 1.0 & 0.5 & 0.33333 & 0.25 & 0.2 \\ 0.5 & 0.33333 & 0.25 & 0.2 & 0.16667 \\ 0.33333 & 0.25 & 0.2 & 0.16667 & 0.14286 \\ 0.25 & 0.2 & 0.16667 & 0.14286 & 0.125 \\ 0.2 & 0.16667 & 0.14286 & 0.125 & 0.11111 \end{pmatrix} \qquad \mathbf{B} = \begin{pmatrix} 1 \\ 0 \\ 0 \\ 0 \\ 0 \end{pmatrix}.$$

Note: The matrices in parts (a) and (b) are different.

23. The Rockmore Corp. is considering the purchase of a new computer, and will choose either the DoGood 174 or the MightDo 11. They test both computers' ability to solve the linear system

$$34x + 55y - 21 = 0,$$
$$55x + 89y - 34 = 0.$$

The DoGood 174 computer gives $x = -0.11$ and $y = 0.45$ and its check for accuracy is found by substitution:

$$34(-0.11) + 55(0.45) - 21 = 0.01,$$
$$55(-0.11) + 89(0.45) - 34 = 0.00.$$

The MightDo 11 computer gives $x = -0.99$ and $y = 1.01$ and its check for accuracy is found by substitution:

$$34(-0.99) + 55(1.01) - 21 = 0.89,$$
$$55(-0.99) + 89(1.01) - 34 = 1.44.$$

Which computer gave the better answer? Why?

3.5 MATRIX INVERSION

The Vector Decomposition of A^{-1} الايجاد المعكوس لـ A^{-1}

The following discussion is presented for matrices of dimension 3×3, but the concepts apply to matrices of dimension $N \times N$. Consider the three linear systems $AC_j = E_j$ for $j = 1, 2, 3$, where E_1, E_2, and E_3 are the standard base vectors

$$AC_1 = \begin{pmatrix} 5 & 6 & -3 \\ 1 & 3 & 1 \\ 4 & 2 & -6 \end{pmatrix} \begin{pmatrix} 2.0 \\ -1.0 \\ 1.0 \end{pmatrix} = \begin{pmatrix} 1 \\ 0 \\ 0 \end{pmatrix} = E_1,$$

$$AC_2 = \begin{pmatrix} 5 & 6 & -3 \\ 1 & 3 & 1 \\ 4 & 2 & -6 \end{pmatrix} \begin{pmatrix} -3.0 \\ 1.8 \\ -1.4 \end{pmatrix} = \begin{pmatrix} 0 \\ 1 \\ 0 \end{pmatrix} = E_2,$$

$$AC_3 = \begin{pmatrix} 5 & 6 & -3 \\ 1 & 3 & 1 \\ 4 & 2 & -6 \end{pmatrix} \begin{pmatrix} -1.5 \\ 0.8 \\ -0.9 \end{pmatrix} = \begin{pmatrix} 0 \\ 0 \\ 1 \end{pmatrix} = E_3.$$

This can be compactly written with the matrix notation

$$AC_1 = E_1, \quad AC_2 = E_2, \quad AC_3 = E_3.$$

The column vectors C_1, C_2, C_3 and E_1, E_2, E_3 are used to form the matrices $C = [C_1, C_2, C_3]$ and $I = [E_1, E_2, E_3]$. Using the row-by-column rule for matrix multiplication, we obtain

$$A[C_1, C_2, C_3] = [E_1, E_2, E_3] = I.$$

Hence, we have found that $A^{-1} = [C_1, C_2, C_3]$. When the matrices are displayed this is easy to see:

$$AC = \begin{pmatrix} 5 & 6 & -3 \\ 1 & 3 & 1 \\ 4 & 2 & -6 \end{pmatrix} \begin{pmatrix} 2.0 & -3.0 & -1.5 \\ -1.0 & 1.8 & 0.8 \\ 1.0 & -1.4 & -0.9 \end{pmatrix} = \begin{pmatrix} 1 & 0 & 0 \\ 0 & 1 & 0 \\ 0 & 0 & 1 \end{pmatrix} = I.$$

Conversely, if we want to find the inverse matrix, then this is accomplished by first representing A^{-1} by its three columns

$$A^{-1} = [C_1, C_2, C_3]. \tag{1}$$

Then the identity matrix $I = [E_1, E_2, E_3]$ is used to write

$$A[C_1, C_2, C_3] = [E_1, E_2, E_3] = I. \tag{2}$$

Hence the solution of (2) is equivalent to

$$A\mathbf{C}_1 = \mathbf{E}_1, \quad A\mathbf{C}_2 = \mathbf{E}_2, \quad A\mathbf{C}_3 = \mathbf{E}_3. \tag{3}$$

When displayed, we see that we must solve three linear systems,

$$A\mathbf{C}_1 = \begin{pmatrix} a_{1,1} & a_{1,2} & a_{1,3} \\ a_{2,1} & a_{2,2} & a_{2,3} \\ a_{3,1} & a_{3,2} & a_{3,3} \end{pmatrix} \begin{pmatrix} c_{1,1} \\ c_{2,1} \\ c_{3,1} \end{pmatrix} = \begin{pmatrix} 1 \\ 0 \\ 0 \end{pmatrix} = \mathbf{E}_1,$$

$$A\mathbf{C}_2 = \begin{pmatrix} a_{1,1} & a_{1,2} & a_{1,3} \\ a_{2,1} & a_{2,2} & a_{2,3} \\ a_{3,1} & a_{3,2} & a_{3,3} \end{pmatrix} \begin{pmatrix} c_{1,2} \\ c_{2,2} \\ c_{3,2} \end{pmatrix} = \begin{pmatrix} 0 \\ 1 \\ 0 \end{pmatrix} = \mathbf{E}_2, \tag{4}$$

$$A\mathbf{C}_3 = \begin{pmatrix} a_{1,1} & a_{1,2} & a_{1,3} \\ a_{2,1} & a_{2,2} & a_{2,3} \\ a_{3,1} & a_{3,2} & a_{3,3} \end{pmatrix} \begin{pmatrix} c_{1,3} \\ c_{2,3} \\ c_{3,3} \end{pmatrix} = \begin{pmatrix} 0 \\ 0 \\ 1 \end{pmatrix} = \mathbf{E}_3.$$

When the solution vectors in (4) are combined to form $C = [\mathbf{C}_1, \mathbf{C}_2, \mathbf{C}_3]$, then it is easy to see that C is the inverse of A:

$$AC = \begin{pmatrix} a_{1,1} & a_{1,2} & a_{1,3} \\ a_{2,1} & a_{2,2} & a_{2,3} \\ a_{3,1} & a_{3,2} & a_{3,3} \end{pmatrix} \begin{pmatrix} c_{1,1} & c_{1,2} & c_{1,3} \\ c_{2,1} & c_{2,2} & c_{2,3} \\ c_{3,1} & c_{3,2} & c_{3,3} \end{pmatrix} = \begin{pmatrix} 1 & 0 & 0 \\ 0 & 1 & 0 \\ 0 & 0 & 1 \end{pmatrix}.$$

A slight modification of the Gaussian elimination process will give an algorithm for finding the inverse matrix. In Section 3.4 the augmented matrix contained one extra column. In the case of matrix inversion we add three column vectors $\mathbf{E}_1, \mathbf{E}_2, \mathbf{E}_3$ to the right of A to form the augmented matrix $[A, \mathbf{E}_1, \mathbf{E}_2, \mathbf{E}_3] = [A, I]$. A succession of row operations are used to eliminate the elements above as well as below the diagonal in the augmented matrix so that the identity matrix appears on the left side and the three solution vectors appear on the right side, that is, we reduce the augmented matrix to the form $[I, \mathbf{C}_1, \mathbf{C}_2, \mathbf{C}_3] = [I, C]$. Since the elements in column P that lie above and below the diagonal are eliminated, it is not necessary to use back substitution.

Example 3.19

Find the inverse of the matrix

$$\begin{pmatrix} 2 & 0 & 1 \\ 3 & 2 & 5 \\ 1 & -1 & 0 \end{pmatrix}.$$

Solution.   Start with the augmented matrix $[A, I]$.

$$\begin{pmatrix} 2 & 0 & 1 & | & 1 & 0 & 0 \\ 3 & 2 & 5 & | & 0 & 1 & 0 \\ 1 & -1 & 0 & | & 0 & 0 & 1 \end{pmatrix}.$$

Interchange rows 1 and 3 so that $a_{1,1} = 1$.

$$\begin{pmatrix} 1 & -1 & 0 & | & 0 & 0 & 1 \\ 3 & 2 & 5 & | & 0 & 1 & 0 \\ 2 & 0 & 1 & | & 1 & 0 & 0 \end{pmatrix}.$$

Eliminate elements in column 1 that lie below the diagonal.

$$\begin{pmatrix} 1 & -1 & 0 & | & 0 & 0 & 1 \\ 0 & 5 & 5 & | & 0 & 1 & -3 \\ 0 & 2 & 1 & | & 1 & 0 & -2 \end{pmatrix}.$$

Divide row 2 by 5 so that $a_{2,2} = 1$.

$$\begin{pmatrix} 1 & -1 & 0 & | & 0 & 0 & 1 \\ 0 & 1 & 1 & | & 0 & 0.2 & -0.6 \\ 0 & 2 & 1 & | & 1 & 0 & -2 \end{pmatrix}.$$

Eliminate elements in column 2 that lie above and below the diagonal.

$$\begin{pmatrix} 1 & 0 & 1 & | & 0 & 0.2 & 0.4 \\ 0 & 1 & 1 & | & 0 & 0.2 & -0.6 \\ 0 & 0 & -1 & | & 1 & -0.4 & -0.8 \end{pmatrix}.$$

Change the sign in row 3 so that $a_{3,3} = 1$ and eliminate the elements in column 3 that lie above the diagonal.

$$\begin{pmatrix} 1 & 0 & 0 & | & 1 & -0.2 & -0.4 \\ 0 & 1 & 0 & | & 1 & -0.2 & -1.4 \\ 0 & 0 & 1 & | & -1 & 0.4 & 0.8 \end{pmatrix}.$$

The identity matrix now appears on the left side of the augmented matrix and the inverse on the right side. Thus

$$A^{-1} = \begin{pmatrix} 1 & -0.2 & -0.4 \\ 1 & -0.2 & -1.4 \\ -1 & 0.4 & 0.8 \end{pmatrix}.$$

The solution of an $N \times N$ system of linear equations can be accomplished by inverting the matrix A. The solution to $A\mathbf{X} = \mathbf{B}$ is given by $\mathbf{X} = A^{-1}\mathbf{B}$. However, this is not a computationally efficient way to solve the system and the reader is encouraged to work out Exercise 8 to determine how inefficient it is. Sometimes the inverse is useful in its own right. For example, in the statistical treatment of the fitting of a function to observed data by the method of least squares, the entries of A^{-1} give information about the magnitude of errors in the data.

EXERCISES FOR MATRIX INVERSION

In Exercises 1–7:
(a) Find the inverse of the given matrix.
(b) Check your answer by computing the product AA^{-1}.

1. $\begin{pmatrix} 1 & 1 & 2 \\ 1 & 2 & 4 \\ 2 & 4 & 7 \end{pmatrix}$

2. $\begin{pmatrix} 1 & -3 & 3 \\ -2 & 4 & -5 \\ 1 & -5 & 3 \end{pmatrix}$

3. $\begin{pmatrix} 1 & -2 & 3 \\ -2 & 4 & -5 \\ 1 & -5 & 3 \end{pmatrix}$

4. $\begin{pmatrix} 1 & 3 & -2 \\ -2 & -4 & 6 \\ 1 & 5 & 2 \end{pmatrix}$

5. $\begin{pmatrix} 2 & -3 & -5 & 2 \\ 1 & -4 & 7 & 4 \\ 0 & 2 & 0 & -1 \\ 2 & 1 & 4 & 1 \end{pmatrix}$

6. $\begin{pmatrix} 3 & -9 & 27 & -81 \\ -4 & 16 & -64 & 256 \\ 5 & -25 & 125 & -625 \\ -6 & 36 & -216 & 1296 \end{pmatrix}$

7. $\begin{pmatrix} 16 & -120 & 240 & -140 \\ -120 & 1200 & -2700 & 1680 \\ 240 & -2700 & 6480 & -4200 \\ -140 & 1680 & -4200 & 2800 \end{pmatrix}$

8. Formulas for the arithmetic operations count for solving $A\mathbf{X} = \mathbf{B}$, for finding A^{-1} and for computing the product $A^{-1}\mathbf{B}$ are readily available and are given in the table below.

TABLE 3.2 Operations Count for Solving a Linear System

N	Operations needed to solve $A\mathbf{X} = \mathbf{B}$: $(4N^3 + 9N^2 - 7N)/6$ (see Algorithm 3.2)	Operations needed to find A^{-1}: $(16N^3 - 9N^2 - N)/6$ (see Problem 11)	Operations needed to multiply $A^{-1}\mathbf{B}$: $2N^2 - N$
3	28	58	15
5	—	295	45
8	428	—	120
10	—	—	190
15	—	—	—
20	—	—	—

(a) Complete Table 3.2.

(b) Discuss the amount of work required to solve $A\mathbf{X} = \mathbf{B}$ directly using Gaussian elimination and that required to find A^{-1} and then compute the product $\mathbf{X} = A^{-1}\mathbf{B}$.

(c) If the goal is to solve $A\mathbf{X} = \mathbf{B}$, which way is best?

9. Let $A = (a_{i,j})_{N \times N}$ be a nonsingular upper-triangular matrix, $a_{i,j} = 0$ when $i > j$. Then $C = A^{-1}$ is also upper-triangular. Let $I_N = (d_{i,j})_{N \times N}$, where $d_{i,i} = 1$ and $d_{i,j} = 0$ when $i \neq j$. An extension of the back-substitution algorithm can be made to find $A^{-1} = (c_{i,j})$. Show that the elements of A^{-1} can be computed using the following algorithm:

$$
\begin{aligned}
&\text{For} \quad j = N \;\; \text{downto} \;\; 1 \;\; \text{do} \\
&\qquad \text{For} \quad i = j \;\; \text{downto} \;\; 1 \;\; \text{do} \\
&\qquad\qquad c_{i,j} = \Big[d_{i,j} - \sum_{k=i+1}^{j} a_{i,k} c_{k,j} \Big] / a_{i,i}
\end{aligned}
$$

Remark. When $i = j$ the lower index for the summation is larger than the upper index. This is interpreted to mean that no terms are to be added, that is,

$$
0 = \sum_{k=j+1}^{j} a_{j,k} c_{k,j}.
$$

10. Let $A = (a_{i,j})_{N \times N}$ be a nonsingular lower-triangular matrix, $a_{i,j} = 0$ when $j > i$. Then $C = A^{-1}$ is also lower-triangular. Let $I_N = (d_{i,j})_{N \times N}$, where $d_{i,i} = 1$ and $d_{i,j} = 0$ when $i \neq j$. An extension of the forward-substitution algorithm in Exercise 8 of Section 3.3 can be made to find $A^{-1} = (c_{i,j})$. Show that the elements of A^{-1} can be computed using the following algorithm:

$$
\begin{aligned}
&\text{For} \quad j = 1 \;\; \text{to} \;\; N \;\; \text{do} \\
&\qquad \text{For} \quad i = j \;\; \text{to} \;\; N \;\; \text{do} \\
&\qquad\qquad c_{i,j} = \Big[d_{i,j} - \sum_{k=j}^{i-1} a_{i,k} c_{k,j} \Big] / a_{i,i}
\end{aligned}
$$

Remark. When $i = j$ the lower index for the summation is larger than the upper index. This is interpreted to mean that no terms are to be added, that is,

$$
0 = \sum_{k=j}^{i-1} a_{j,k} c_{k,j}.
$$

11. Modify Algorithm 3.2 to find A^{-1}. Use the special augmented matrix $[A, \mathbf{E}_1, \mathbf{E}_2, \ldots, \mathbf{E}_N]$, where $\mathbf{E}_J$ is the standard basis vector and is stored in column $N + J$ of the augmented matrix.

12. Use a computer to find the inverse of the Hilbert matrix of dimension 5×5 given in Exercise 22 of Section 3.4.

3.6 TRIANGULAR FACTORIZATION

In Section 3.3 we saw how easy it is to solve an upper-triangular system. Now we introduce the concept of factorization of a given matrix A into the product of an upper-triangular matrix U with nonzero diagonal elements and a lower-triangular matrix L that has 1's along the diagonal. For ease of notation we illustrate the concepts with matrices of dimension 4×4, but they apply to an arbitrary system of dimension $N \times N$.

Definition 3.6.   The nonsingular matrix A has a **triangular factorization** if it can be expressed as the product of a lower-triangular matrix L and an upper-triangular matrix U:

$$A = LU. \tag{1}$$

In matrix form this is written as

$$
\begin{pmatrix}
a_{1,1} & a_{1,2} & a_{1,3} & a_{1,4} \\
a_{2,1} & a_{2,2} & a_{2,3} & a_{2,4} \\
a_{3,1} & a_{3,2} & a_{3,3} & a_{3,4} \\
a_{4,1} & a_{4,2} & a_{4,3} & a_{4,4}
\end{pmatrix}
=
\begin{pmatrix}
1 & 0 & 0 & 0 \\
m_{2,1} & 1 & 0 & 0 \\
m_{3,1} & m_{3,2} & 1 & 0 \\
m_{4,1} & m_{4,2} & m_{4,3} & 1
\end{pmatrix}
\begin{pmatrix}
u_{1,1} & u_{1,2} & u_{1,3} & u_{1,4} \\
0 & u_{2,2} & u_{2,3} & u_{2,4} \\
0 & 0 & u_{3,3} & u_{3,4} \\
0 & 0 & 0 & u_{4,4}
\end{pmatrix}.
$$

The condition that A is nonsingular implies that $u_{k,k} \neq 0$ for all k. The notation for the entries in L are $m_{i,j}$, and the reason for the choice of $m_{i,j}$ instead of $l_{i,j}$ will be pointed out soon.

Solution of a Linear System

Suppose that the coefficient matrix A for the linear system $A\mathbf{X} = \mathbf{B}$ has a triangular factorization (1); then the solution to

$$LU\mathbf{X} = \mathbf{B} \tag{2}$$

can be obtained by defining $\mathbf{Y} = U\mathbf{X}$ and then solving two systems

first solve $L\mathbf{Y} = \mathbf{B}$ for $\mathbf{Y}$, then solve $U\mathbf{X} = \mathbf{Y}$ for $\mathbf{X}$. $\tag{3}$

In equation form, we must first solve the lower-triangular system

$$
\begin{aligned}
y_1 &= b_1 \\
m_{2,1}y_1 + y_2 &= b_2 \\
m_{3,1}y_1 + m_{3,2}y_2 + y_3 &= b_3 \\
m_{4,1}y_1 + m_{4,2}y_2 + m_{4,3}y_3 + y_4 &= b_4
\end{aligned}
\tag{4}
$$

to obtain y_1, y_2, y_3, and y_4 and use them in solving the system

$$
\begin{aligned}
u_{1,1}x_1 + u_{1,2}x_2 + u_{1,3}x_3 + u_{1,4}x_4 &= y_1 \\
u_{2,2}x_2 + u_{2,3}x_3 + u_{2,4}x_4 &= y_2 \\
u_{3,3}x_3 + u_{3,4}x_4 &= y_3 \\
u_{4,4}x_4 &= y_4.
\end{aligned}
\tag{5}
$$

Example 3.20

Solve

$$
\begin{aligned}
4x_1 + 3x_2 - x_3 &= -2 \\
-2x_1 - 4x_2 + 5x_3 &= 20 \\
x_1 + 2x_2 + 6x_3 &= 7.
\end{aligned}
$$

Use the triangular factorization method and the fact that

$$
\begin{pmatrix} 4 & 3 & -1 \\ -2 & -4 & 5 \\ 1 & 2 & 6 \end{pmatrix} = \begin{pmatrix} 1 & 0 & 0 \\ -0.5 & 1 & 0 \\ 0.25 & -0.5 & 1 \end{pmatrix} \begin{pmatrix} 4 & 3 & -1 \\ 0 & -2.5 & 4.5 \\ 0 & 0 & 8.5 \end{pmatrix}.
$$

Solution.   Use the forward-substitution method to solve

$$
\begin{aligned}
y_1 &= -2 \\
-0.5y_1 + y_2 &= 20 \\
0.25y_1 - 0.5y_2 + y_3 &= 7.
\end{aligned}
\tag{6}
$$

Compute the values $y_1 = -2$, $y_2 = 20 + 0.5(-2) = 19$, and $y_3 = 7 - 0.25(-2) + 0.5(19) = 17$. Next write the system $U\mathbf{X} = \mathbf{Y}$:

$$
\begin{aligned}
4x_1 + 3x_2 - x_3 &= -2 \\
-2.5x_2 + 4.5x_3 &= 19 \\
8.5x_3 &= 17.
\end{aligned}
\tag{7}
$$

Now use back substitution and compute the solution $x_3 = 17/8.5 = 2$, $x_2 = [19 - 4.5(2)]/(-2.5) = -4$, and $x_1 = [-2 - 3(-4) + 2]/4 = 3$.

Triangular Factorization

We now discuss how to obtain the triangular factorization. If row interchanges are not necessary when using Gaussian elimination, the multipliers $m_{i,j}$ are the subdiagonal entries in L.

Example 3.21

Use Gaussian elimination to construct the triangular factorization of the matrix

$$A = \begin{pmatrix} 4 & 3 & -1 \\ -2 & -4 & 5 \\ 1 & 2 & 6 \end{pmatrix}.$$

Solution.   The matrix L will be constructed from an identity matrix placed at the left. For each row operation used to construct the upper-triangular matrix, the multipliers $m_{i,j}$ will be put in their proper place at the left. Start with

$$A = \begin{pmatrix} 1 & 0 & 0 \\ 0 & 1 & 0 \\ 0 & 0 & 1 \end{pmatrix} \begin{pmatrix} 4 & 3 & -1 \\ -2 & -4 & 5 \\ 1 & 2 & 6 \end{pmatrix}.$$

Row 1 is used to eliminate the elements of A in column 1 below $a_{1,1}$. The multiples $m_{2,1} = -0.5$ and $m_{3,1} = 0.25$ of row 1 are subtracted from rows 2 and 3, respectively. These multipliers are put in the matrix at the left and the result is

$$A = \begin{pmatrix} 1 & 0 & 0 \\ -0.5 & 1 & 0 \\ 0.25 & 0 & 1 \end{pmatrix} \begin{pmatrix} 4 & 3 & -1 \\ 0 & -2.5 & 4.5 \\ 0 & 1.25 & 6.25 \end{pmatrix}.$$

Row 2 is used to eliminate the elements of A in column 2 below $a_{2,2}$. The multiple $m_{3,2} = -0.5$ of the second row is subtracted from row 3, and the multiplier is entered in the matrix at the left and we have the desired triangular factorization of A:

$$A = \begin{pmatrix} 1 & 0 & 0 \\ -0.5 & 1 & 0 \\ 0.25 & -0.5 & 1 \end{pmatrix} \begin{pmatrix} 4 & 3 & -1 \\ 0 & -2.5 & 4.5 \\ 0 & 0 & 8.5 \end{pmatrix}. \tag{8}$$

Theorem 3.9   (Direct Factorization $A = LU$, No Row Interchanges).   Suppose that Gaussian elimination, without row interchanges, can be successfully performed to solve the general linear system $A\mathbf{X} = \mathbf{B}$. Then the matrix A can be factored as the product of a lower-triangular matrix L and an upper-triangular matrix U:

$$A = LU.$$

Furthermore, L can be constructed to have 1's on its diagonal and U will have nonzero diagonal elements. After finding L and U, the solution $\mathbf{X}$ is computed in two steps;

1. Solve $L\mathbf{Y} = \mathbf{B}$ for $\mathbf{Y}$ using forward substitution.
2. Solve $U\mathbf{X} = \mathbf{Y}$ for $\mathbf{X}$ using back substitution.

Proof.   We will show that when the Gaussian elimination process is followed and $\mathbf{B}$ is stored in column $N + 1$ of the augmented matrix, the result after the upper-triangularization step is the equivalent upper-triangular system $U\mathbf{X} = \mathbf{Y}$. The matrices L, U and vectors $\mathbf{B}$, $\mathbf{Y}$ will have the form

$$
L = \begin{pmatrix}
1 & 0 & 0 & & 0 \\
m_{2,1} & 1 & 0 & \cdots & 0 \\
m_{3,1} & m_{3,2} & 1 & \cdots & 0 \\
\cdot & \cdot & \cdot & & \cdot \\
\cdot & \cdot & \cdot & & \cdot \\
\cdot & \cdot & \cdot & & \cdot \\
m_{n,1} & m_{n,2} & m_{n,3} & \cdots & 1
\end{pmatrix},
\qquad
\mathbf{B} = \begin{pmatrix}
a^{(1)}_{1,n+1} \\
a^{(1)}_{2,n+1} \\
a^{(1)}_{3,n+1} \\
\cdot \\
\cdot \\
\cdot \\
a^{(1)}_{n,n+1}
\end{pmatrix}
$$

$$
U = \begin{pmatrix}
a^{(1)}_{1,1} & a^{(1)}_{1,2} & a^{(1)}_{1,3} & \cdots & a^{(1)}_{1,n} \\
0 & a^{(2)}_{2,2} & a^{(2)}_{2,3} & \cdots & a^{(2)}_{2,n} \\
0 & 0 & a^{(3)}_{3,3} & \cdots & a^{(3)}_{3,n} \\
\cdot & \cdot & \cdot & & \cdot \\
\cdot & \cdot & \cdot & & \cdot \\
\cdot & \cdot & \cdot & & \cdot \\
0 & 0 & 0 & \cdots & a^{(n)}_{n,n}
\end{pmatrix},
\qquad
\mathbf{Y} = \begin{pmatrix}
a^{(1)}_{1,n+1} \\
a^{(2)}_{2,n+1} \\
a^{(3)}_{3,n+1} \\
\cdot \\
\cdot \\
\cdot \\
a^{(n)}_{n,n+1}
\end{pmatrix}.
$$

Remark.   To find just L and U, the $(n + 1)$st column is not needed.

Step 1. Store the coefficients in the array. The superscript on $a^{(1)}_{r,c}$ means that this is the first time a number is stored in location (r, c).

$$
\begin{matrix}
a^{(1)}_{1,1} & a^{(1)}_{1,2} & a^{(1)}_{1,3} & \cdots & a^{(1)}_{1,n} & a^{(1)}_{1,n+1} \\
a^{(1)}_{2,1} & a^{(1)}_{2,2} & a^{(1)}_{2,3} & \cdots & a^{(1)}_{2,n} & a^{(1)}_{2,n+1} \\
a^{(1)}_{3,1} & a^{(1)}_{3,2} & a^{(1)}_{3,3} & \cdots & a^{(1)}_{3,n} & a^{(1)}_{3,n+1} \\
\cdot & \cdot & \cdot & & \cdot & \cdot \\
\cdot & \cdot & \cdot & & \cdot & \cdot \\
\cdot & \cdot & \cdot & & \cdot & \cdot \\
a^{(1)}_{n,1} & a^{(1)}_{n,2} & a^{(1)}_{n,3} & \cdots & a^{(1)}_{n,n} & a^{(1)}_{n,n+1}.
\end{matrix}
$$

Step 2. Eliminate x_1 in rows 2 through N and store the multiplier $m_{r,1}$ used to eliminate x_1 in row r in the array at location $(r, 1)$.

```
FOR    r = 2  TO  N  DO
         m_{r,1} := a^{(1)}_{r,1}/a^{(1)}_{1,1}  and  a_{r,1} := m_{r,1}
         FOR    c = 2  TO  N+1  DO
                  a^{(2)}_{r,c} := a^{(1)}_{r,c} − m_{r,1}*a^{(1)}_{1,c}
```

The new elements are written $a^{(2)}_{r,c}$ to indicate that this is the second time that a number has been stored in the array at location (r, c). The result after step 2 is

$$\begin{array}{cccccc}
a_{1,1}^{(1)} & a_{1,2}^{(1)} & a_{1,3}^{(1)} & \cdots & a_{1,n}^{(1)} & a_{1,n+1}^{(1)} \\
m_{2,1} & a_{2,2}^{(2)} & a_{2,3}^{(2)} & \cdots & a_{2,n}^{(2)} & a_{2,n+1}^{(2)} \\
m_{3,1} & a_{3,2}^{(2)} & a_{3,3}^{(2)} & \cdots & a_{3,n}^{(2)} & a_{3,n+1}^{(2)} \\
\vdots & \vdots & \vdots & & \vdots & \vdots \\
m_{n,1} & a_{n,2}^{(2)} & a_{n,3}^{(2)} & \cdots & a_{n,n}^{(2)} & a_{n,n+1}^{(2)}.
\end{array}$$

Step 3. Eliminate x_2 in rows 3 through N and store the multiplier $m_{r,2}$ used to eliminate x_2 in row r in the array at location $(r, 2)$.

```
FOR    r = 3  TO  N  DO
       m_{r,2} := a_{r,2}^{(2)}/a_{2,2}^{(2)}  and  a_{r,2} := m_{r,2}
       FOR    c = 3  TO  N+1  DO
              a_{r,c}^{(3)} := a_{r,c}^{(2)} − m_{r,2}*a_{2,c}^{(2)}
```

The new elements are written $a_{r,c}^{(3)}$ to indicate that this is the third time that a number has been stored in the array at location (r, c). The result after step 3 is

$$\begin{array}{cccccc}
a_{1,1}^{(1)} & a_{1,2}^{(1)} & a_{1,3}^{(1)} & \cdots & a_{1,n}^{(1)} & a_{1,n+1}^{(1)} \\
m_{2,1} & a_{2,2}^{(2)} & a_{2,3}^{(2)} & \cdots & a_{2,n}^{(2)} & a_{2,n+1}^{(2)} \\
m_{3,1} & m_{3,2} & a_{3,3}^{(3)} & \cdots & a_{3,n}^{(3)} & a_{3,n+1}^{(3)} \\
\vdots & \vdots & \vdots & & \vdots & \vdots \\
m_{n,1} & m_{n,2} & a_{n,3}^{(3)} & \cdots & a_{n,n}^{(3)} & a_{n,n+1}^{(3)}.
\end{array}$$

Step $p + 1$. This is the general step. Eliminate x_p in rows $p + 1$ through N and store the multipliers $m_{r,p}$ at locations (r, p).

```
FOR    r = p+1  TO  N  DO
       m_{r,p} := a_{r,p}^{(p)}/a_{p,p}^{(p)}  and  a_{r,p} := m_{r,p}
       FOR    c = p+1  TO  N+1  DO
              a_{r,c}^{(p+1)} := a_{r,c}^{(p)} − m_{r,p}*a_{p,c}^{(p)}
```

The final result after x_{N-1} has been eliminated from row N is

$$\begin{array}{cccccc}
a_{1,1}^{(1)} & a_{1,2}^{(1)} & a_{1,3}^{(1)} & \cdots & a_{1,n}^{(1)} & a_{1,n+1}^{(1)} \\
m_{2,1} & a_{2,2}^{(2)} & a_{2,3}^{(2)} & \cdots & a_{2,n}^{(2)} & a_{2,n+1}^{(2)} \\
m_{3,1} & m_{3,2} & a_{3,3}^{(3)} & \cdots & a_{3,n}^{(3)} & a_{3,n+1}^{(3)} \\
\vdots & \vdots & \vdots & & \vdots & \vdots \\
m_{n,1} & m_{n,2} & m_{n,3} & \cdots & a_{n,n}^{(n)} & a_{n,n+1}^{(n)}.
\end{array}$$

The upper-triangularization process is now complete. Notice that one array is used to store the elements of both L and U. The 1's of L are not stored, nor are the 0's of L and U that lie above and below the diagonal, respectively. Only the essential coefficients needed to reconstruct L and U are stored!

We must now verify that the product is $LU = A$. Suppose that $C = LU$; and consider the case when $r \leq c$. Then $c_{r,c}$ is

$$c_{r,c} = m_{r,1}a_{1,c}^{(1)} + m_{r,2}a_{2,c}^{(2)} + m_{r,r-1}a_{r-1,c}^{(r-1)} + a_{r,c}^{(r)}. \tag{9}$$

Using the replacement equations in steps 1 through $p + 1 = r$, we obtain the following substitutions:

$$
\begin{aligned}
m_{r,1}a_{1,c}^{(1)} &= a_{r,c}^{(1)} - a_{r,c}^{(2)}, \\
m_{r,2}a_{2,c}^{(2)} &= a_{r,c}^{(2)} - a_{r,c}^{(3)}, \\
&\ \ \vdots \\
m_{r,r-1}a_{r-1,c}^{(r-1)} &= a_{r,c}^{(r-1)} - a_{r,c}^{(r)}.
\end{aligned}
\tag{10}
$$

When the substitutions in (10) are used in (9), the result is

$$c_{r,c} = a_{r,c}^{(1)} - a_{r,c}^{(2)} + a_{r,c}^{(2)} - a_{r,c}^{(3)} + \cdots + a_{r,c}^{(r-1)} - a_{r,c}^{(r)} + a_{r,c}^{(r)} = a_{r,c}^{(1)}.$$

The other case $r > c$ is similar to prove.

Counting Arithmetic Operations

The process for triangularizing is the same for both the Gaussian elimination and triangular factorization methods. We can count the operations if we look at the first N columns of the augmented matrix in Theorem 3.9. The outer loop of step $p + 1$ requires $N - p = N - (p + 1) + 1$ divisions to compute the multipliers $m_{r,p}$. Inside the loops, but for the first N columns only, a total of $(N - p)(N - p)$ multiplications and the same number of subtractions are required to compute the new row elements $a_{r,c}^{(p+1)}$. This process is carried out for $p = 1, 2, \ldots, N - 1$. Thus the triangular factorization portion $A = LU$ requires

$$\sum_{p=1}^{N-1} (N - p)(N - p + 1) = \frac{N^3 - N}{3} \text{ multiplications and divisions} \tag{11}$$

and

$$\sum_{p=1}^{N-1} (N - p)(N - p) = \frac{2N^3 - 3N^2 + N}{6} \text{ subtractions.} \tag{12}$$

To establish (11) we use the summation formulas

$$\sum_{k=1}^{M} k = \frac{M(M + 1)}{2} \quad \text{and} \quad \sum_{k=1}^{M} k^2 = \frac{M(M + 1)(2M + 1)}{6}$$

Using the change of variable $k = N - p$, we rewrite (11) as

$$\sum_{p=1}^{N-1} (N - p)(N - p + 1) = \sum_{p=1}^{N-1} (N - p) + \sum_{p=1}^{N-1} (N - p)^2$$

$$= \sum_{k=1}^{N-1} k + \sum_{k=1}^{N-1} k^2$$

$$= \frac{(N - 1)(N)}{2} + \frac{(N - 1)(N)(2N - 1)}{6}$$

$$= \frac{N^3 - N}{3}$$

Once the triangular factorization $A = LU$ has been obtained, the solution to the lower-triangular system $L\mathbf{Y} = \mathbf{B}$ will require $0 + 1 + \cdots + N - 1 = (N^2 - N)/2$ multiplications and subtractions; no divisions are required because the diagonal elements of L are 1's. Then the solution of the upper-triangular system $U\mathbf{X} = \mathbf{Y}$ requires $1 + 2 + \cdots + N = (N^2 + N)/2$ multiplications and divisions and $(N^2 - N)/2$ subtractions. Therefore, finding the solution to $LU\mathbf{X} = \mathbf{B}$ requires

$$N^2 \quad \begin{matrix} \text{multiplications} \\ \text{and divisions} \end{matrix} \quad \text{and} \quad N^2 - N \quad \text{subtractions.}$$

We see that the bulk of the calculations lie in the triangularization portion of the solution. If the linear system is to be solved many times, with the same coefficient matrix A but with different column vectors $\mathbf{B}$, it is not necessary to triangularize the matrix each time if the factors are saved. This is the reason the triangular factorization method is usually chosen over the elimination method. However, if only one linear system is solved, the two methods are the same, except that the triangular factorization method stores the multipliers.

Permutation Matrices

The $A = LU$ factorization in Theorem 3.9 assumes that there are no row interchanges. It is possible that a nonsingular matrix A cannot be directly factored as $A = LU$.

Example 3.22

Show that the following matrix cannot be directly factored as $A = LU$.

$$A = \begin{pmatrix} 1 & 2 & 6 \\ 4 & 8 & -1 \\ -2 & 3 & 5 \end{pmatrix}.$$

Solution. Suppose that A has a direct factorization LU; then

$$\begin{pmatrix} 1 & 2 & 6 \\ 4 & 8 & -1 \\ -2 & 3 & 5 \end{pmatrix} = \begin{pmatrix} 1 & 0 & 0 \\ m_{21} & 1 & 0 \\ m_{31} & m_{32} & 1 \end{pmatrix} \begin{pmatrix} u_{11} & u_{12} & u_{13} \\ 0 & u_{22} & u_{23} \\ 0 & 0 & u_{33} \end{pmatrix}. \tag{13}$$

The matrices L and U on the right-hand side of (13) can be multiplied and each element of the product compared with the corresponding element of matrix A. In the first column $1 = 1u_{11}$, then $4 = m_{21}u_{11} = m_{21}$, and finally $-2 = m_{31}u_{11} = m_{31}$. In the second column $2 = 1u_{12}$, then $8 = m_{21}u_{12} + 1u_{22} = 4(2) + u_{22}$ implies that $u_{22} = 0$, and finally $3 = m_{31}u_{12} + m_{32}u_{22} = (-2)(2) + m_{32}(0) = -4$, which is a contradiction. Therefore, A does not have an LU factorization.

A permutation of the first N positive integers $1, 2, \ldots, N$ is an arrangement k_1, $k_2, \ldots, k_N$ of these integers in a definite order. For example, $1, 4, 2, 3, 5$ is a permutation of the five integers $1, 2, 3, 4, 5$. The standard base vectors $\mathbf{E}_i = (0, 0, \ldots, 1_i, \ldots, 0)$, for $i = 1, 2, \ldots, N$, are used in the next definition.

Definition 3.7. An $N \times N$ **permutation matrix** P is a matrix with precisely one entry whose value is 1 in each column and each row, and all of whose other entries are 0. The rows of P are a permutation of the rows of the identity matrix, and can be written as

$$P = [\mathbf{E}_{k_1}, \mathbf{E}_{k_2}, \ldots, \mathbf{E}_{k_N}]^T. \tag{14}$$

The elements of $P = (p_{i,j})$ have the form

$$p_{i,j} = \begin{cases} 1 & \text{when } j = k_i, \\ 0 & \text{otherwise.} \end{cases}$$

For example, the following 4×4 matrix is a permutation matrix.

$$P = \begin{pmatrix} 0 & 1 & 0 & 0 \\ 1 & 0 & 0 & 0 \\ 0 & 0 & 0 & 1 \\ 0 & 0 & 1 & 0 \end{pmatrix} = [\mathbf{E}_2, \mathbf{E}_1, \mathbf{E}_4, \mathbf{E}_3]^T. \tag{15}$$

Theorem 3.10. Suppose that $P = [\mathbf{E}_{k_1}, \mathbf{E}_{k_2}, \ldots, \mathbf{E}_{k_n}]^T$ is a permutation matrix. The product PA is a new matrix whose rows consist of the rows of A rearranged in the order $\text{row}_{k_1} A, \text{row}_{k_2} A, \ldots, \text{row}_{k_n} A$.

Example 3.23

Let A be a 4×4 matrix and let P be the permutation matrix given in (15); then PA is the matrix whose rows consist of the rows of A rearranged in the order $\text{row}_2 A, \text{row}_1 A, \text{row}_4 A, \text{row}_3 A$.

Solution. Computing the product, we have

$$
\begin{pmatrix} 0 & 1 & 0 & 0 \\ 1 & 0 & 0 & 0 \\ 0 & 0 & 0 & 1 \\ 0 & 0 & 1 & 0 \end{pmatrix}
\begin{pmatrix} a_{1,1} & a_{1,2} & a_{1,3} & a_{1,4} \\ a_{2,1} & a_{2,2} & a_{2,3} & a_{2,4} \\ a_{3,1} & a_{3,2} & a_{3,3} & a_{3,4} \\ a_{4,1} & a_{4,2} & a_{4,3} & a_{4,4} \end{pmatrix}
=
\begin{pmatrix} a_{2,1} & a_{2,2} & a_{2,3} & a_{2,4} \\ a_{1,1} & a_{1,2} & a_{1,3} & a_{1,4} \\ a_{4,1} & a_{4,2} & a_{4,3} & a_{4,4} \\ a_{3,1} & a_{3,2} & a_{3,3} & a_{3,4} \end{pmatrix}.
$$

Theorem 3.11.   If P is a permutation matrix, then it is nonsingular and $P^{-1} = P^{T}$.

Theorem 3.12.   If A is a nonsingular matrix, then there exists a permutation matrix P so that PA has a triangular factorization

$$
PA = LU. \tag{16}
$$

The proof can be found in advanced texts; see Reference [128].

Example 3.24

If rows 2 and 3 of the matrix in Example 3.22 are interchanged, the resulting matrix PA can be factored.

Solution.   The permutation matrix that switches rows 2 and 3 is $P = [\mathbf{E}_1, \mathbf{E}_3, \mathbf{E}_2]^{T}$. Computing the product PA, we obtain

$$
PA =
\begin{pmatrix} 1 & 0 & 0 \\ 0 & 0 & 1 \\ 0 & 1 & 0 \end{pmatrix}
\begin{pmatrix} 1 & 2 & 6 \\ 4 & 8 & -1 \\ -2 & 3 & 5 \end{pmatrix}
=
\begin{pmatrix} 1 & 2 & 6 \\ -2 & 3 & 5 \\ 4 & 8 & -1 \end{pmatrix}.
$$

Now Gaussian elimination without row interchanges can be used:

$$
\begin{array}{c}
\text{pivot} \longrightarrow \\
m_{2,1} = -2 \\
m_{3,1} = 4
\end{array}
\begin{pmatrix} \underline{1} & 2 & 6 \\ -2 & 3 & 5 \\ 4 & 8 & -1 \end{pmatrix}.
$$

After x_2 has been eliminated from column 2, row 3, we have

$$
\begin{array}{c}
\\
\text{pivot} \longrightarrow \\
m_{3,2} = 0
\end{array}
\begin{pmatrix} 1 & 2 & 6 \\ 0 & \underline{7} & 17 \\ 0 & 0 & -25 \end{pmatrix} = U.
$$

Extending the Gaussian Elimination Process

The following theorem is an extension of Theorem 3.9, which includes the cases when row interchanges are required. Thus the solution to any linear system $A\mathbf{X} = \mathbf{B}$ where A is nonsingular can be found.

Theorem 3.13   (Indirect Factorization $PA = LU$).   Let A be a given $N \times N$ matrix. Assume that Gaussian elimination can be performed successfully to solve the general linear system $AX = \mathbf{B}$ but that row interchanges are required. Then there exists a permutation matrix P so that the matrix product PA can be factored as the product of a lower-triangular matrix L and an upper-triangular matrix U:

$$PA = LU.$$

Furthermore, L can be constructed to have 1's on its diagonal and U will have nonzero diagonal elements. After finding L and U, the solution $\mathbf{X}$ is found in four steps:

1. Construct the matrices L, U, and P.
2. Compute the column vector $P\mathbf{B}$.
3. Solve $LY = P\mathbf{B}$ for $\mathbf{Y}$ using forward substitution.
4. Solve $UX = \mathbf{Y}$ for $\mathbf{X}$ using back substitution.

Remark.   Suppose that $AX = \mathbf{B}$ is to be solved for A fixed and several different vectors $\mathbf{B}$. Then step 1 is performed only once and steps 2 to 4 are used to find the solution $\mathbf{X}$ that corresponds to $\mathbf{B}$. Steps 2 to 4 are a computationally efficient way to construct the solution $\mathbf{X}$ and require $O(n^2)$ operations instead of the $O(n^3)$ operations required by Gaussian elimination.

The only difference between Gaussian elimination and triangular factorization is the delay in performing calculations involving the column vector $\mathbf{B}$ until the factorization of A is accomplished. If row interchanges are required, then we will keep track of them in the pointer vector Row(P). The elements of $\mathbf{B}$ are not changed during the triangularization portion of the algorithm, and they are rearranged correctly using Row(P) during the forward-substitution step. The value of the determinant det (A) is the product of the diagonal elements of U multiplied by $(-1)^Q$, where Q is the number of row interchanges that are required. By Theorem 3.13 we can find a factorization of a PA. Since the permutation matrix keeps track of row interchanges, they can be remembered in a pointer vector Row(P), and this will save space.

Algorithm 3.3   ($PA = LU$ Factorization with Pivoting).   To construct the solution to the linear system $AX = \mathbf{B}$ by performing the steps:

1. Find a permutation matrix P, lower-triangular matrix L, and upper-triangular matrix U that satisfy
$$PA = LU.$$
2. Compute $P\mathbf{B}$ and form the equivalent linear system
$$LUX = P\mathbf{B}.$$
3. Solve the lower-triangular system
$$LY = P\mathbf{B} \qquad \text{for } \mathbf{Y}.$$
4. Solve the upper-triangular system
$$UX = \mathbf{Y} \qquad \text{for } \mathbf{X}.$$

Remarks. This algorithm is an extension of Algorithm 3.2. Since the diagonal elements of L are all 1's, these values do not need to be stored. The coefficients of L below the main diagonal and the nonzero coefficients of U overwrite the matrix A.

VARiable declaration

```
        REAL   A[1 . . N, 1 . . N], X[1 . . N], Y[1 . . N], B[1 . . N]
        INTEGER   Row[1 . . N]
        INPUT N , A[1 . . N, 1 . . N]                {Input the matrix A at this time}
        DET := 1                                     {Initialize the variable}


        FOR    J = 1  TO  N  DO                       {Initialize the pointer vector}
             └────────Row(J) := J


FOR     P = 1   TO   N−1   DO                         {Start LU factorization}

        FOR    K = P+1  TO  N  DO                        {Find the pivot element}
               IF  |A(Row(K), P)| > |A(Row(P), P)|   THEN
                       T := Row(P)                   {Switch the index for
                   Row(P) := Row(K)                   the p-th pivot row
                   Row(K) := T                        if necessary}
                       DET := − DET                 {Change the sign of DET}
              └──────── ENDIF                  {End of simulated row interchange}


               IF   A(Row(P), P) = 0   THEN
               │   Print "The matrix is singular."
               │   TERMINATE ALGORITHM
               ENDIF


        DET := DET*A(Row(P), P)                       {Multiply the diagonal elements}


        FOR    K = P+1   TO   N   DO
               A(Row(K), P) := A(Row(K), P)/A(Row(P), P)        {Find the multipliers}
               FOR  C = P+1   TO   N   DO                        {Eliminate X_p}
              └───────└────A(Row(K), C) := A(Row(K), C) − A(Row(K), P)*A(Row(P), C)


END                                               {End of the L*U factorization routine}
        DET := DET*A(Row(N), N)                        {Multiply the diagonal elements}


        INPUT   B[1 . . N]                            {Input the column vector now}


        Y(1) := B(Row(1))                             {Start the forward substitution}
```

```
FOR     K = 2  TO  N  DO
    SUM := 0
    FOR  C = 1  TO  K−1  DO
        SUM := SUM + A(Row(K), C)*Y(C)          {End forward substitution}
    Y(K)  := B(Row(K)) − SUM


    IF     A(Row(N), N) = 0  THEN
        PRINT "The matrix is singular."
        TERMINATE ALGORITHM
    ENDIF


    X(N) := Y(N)/A(Row(N), N)                   {Start the back substitution}


FOR     K = N−1  DOWNTO  1  DO
    SUM := 0
        FOR    C = K+1  TO  N  DO
            SUM := SUM+A(Row(K), C)*X(C)
    X(K) := [Y(K)−SUM]/A(Row(K), K)             {End back substitution}


FOR     K = 1  TO  N  DO
    PRINT  X(K)                                 {Output}
```

EXERCISES FOR TRIANGULAR FACTORIZATION

1. Solve $LY = B$, $UX = Y$ and verify that $B = AX$ for **(a)** $B^T = (-4, 10, 5)$ and **(b)** $B^T = (20, 49, 32)$, where $A = LU$ is

$$\begin{pmatrix} 2 & 4 & -6 \\ 1 & 5 & 3 \\ 1 & 3 & 2 \end{pmatrix} = \begin{pmatrix} 1 & 0 & 0 \\ \frac{1}{2} & 1 & 0 \\ \frac{1}{2} & \frac{1}{3} & 1 \end{pmatrix} \begin{pmatrix} 2 & 4 & -6 \\ 0 & 3 & 6 \\ 0 & 0 & 3 \end{pmatrix}.$$

2. Solve $LY = B$, $UX = Y$ and verify that $B = AX$ for **(a)** $B^T = (7, 2, 10)$ and **(b)** $B^T = (25, 35, 7)$, where $A = LU$ is

$$\begin{pmatrix} 1 & 1 & 6 \\ -1 & 2 & 9 \\ 1 & -2 & 3 \end{pmatrix} = \begin{pmatrix} 1 & 0 & 0 \\ -1 & 1 & 0 \\ 1 & -1 & 1 \end{pmatrix} \begin{pmatrix} 1 & 1 & 6 \\ 0 & 3 & 15 \\ 0 & 0 & 12 \end{pmatrix}.$$

3. Solve $LY = B$, $UX = Y$ and verify that $B = AX$ for **(a)** $B^T = (6, 13, 3)$ and **(b)** $B^T = (3.0, 1.5, -1.5)$, where $A = LU$ is

$$\begin{pmatrix} 2 & -2 & 5 \\ 2 & 3 & 1 \\ -1 & 4 & -4 \end{pmatrix} = \begin{pmatrix} 1 & 0 & 0 \\ 1 & 1 & 0 \\ -0.5 & 0.6 & 1 \end{pmatrix} \begin{pmatrix} 2 & -2 & 5 \\ 0 & 5 & -4 \\ 0 & 0 & 0.9 \end{pmatrix}.$$

4. Find the triangular factorization $A = LU$ for the matrices

(a) $\begin{pmatrix} -5 & 2 & -1 \\ 1 & 0 & 3 \\ 3 & 1 & 6 \end{pmatrix}$
(b) $\begin{pmatrix} 1 & 0 & 3 \\ 3 & 1 & 6 \\ -5 & 2 & -1 \end{pmatrix}$

5. Find the triangular factorization $A = LU$ for the matrices

(a) $\begin{pmatrix} 4 & 2 & 1 \\ 2 & 5 & -2 \\ 1 & -2 & 7 \end{pmatrix}$
(b) $\begin{pmatrix} 1 & -2 & 7 \\ 4 & 2 & 1 \\ 2 & 5 & -2 \end{pmatrix}$

6. Solve $L\mathbf{Y} = \mathbf{B}$, $U\mathbf{X} = \mathbf{Y}$ and verify that $\mathbf{B} = A\mathbf{X}$ for (a) $\mathbf{B}^T = (8, -4, 10, -4)$ and (b) $\mathbf{B}^T = (28, 13, 23, 4)$, where $A = LU$ is

$$\begin{pmatrix} 4 & 8 & 4 & 0 \\ 1 & 5 & 4 & -3 \\ 1 & 4 & 7 & 2 \\ 1 & 3 & 0 & -2 \end{pmatrix} = \begin{pmatrix} 1 & 0 & 0 & 0 \\ \frac{1}{4} & 1 & 0 & 0 \\ \frac{1}{4} & \frac{2}{3} & 1 & 0 \\ \frac{1}{4} & \frac{1}{3} & -\frac{1}{2} & 1 \end{pmatrix} \begin{pmatrix} 4 & 8 & 4 & 0 \\ 0 & 3 & 3 & -3 \\ 0 & 0 & 4 & 4 \\ 0 & 0 & 0 & 1 \end{pmatrix}.$$

7. Solve $L\mathbf{Y} = \mathbf{B}$, $U\mathbf{X} = \mathbf{Y}$ and verify that $\mathbf{B} = A\mathbf{X}$ for (a) $\mathbf{B}^T = (12, 18, 8, 8)$ and (b) $\mathbf{B}^T = (-2, -10, 11, 4)$, where $A = LU$ is

$$\begin{pmatrix} 2 & 4 & -4 & 0 \\ 1 & 5 & -5 & -3 \\ 2 & 3 & 1 & 3 \\ 1 & 4 & -2 & 2 \end{pmatrix} = \begin{pmatrix} 1 & 0 & 0 & 0 \\ \frac{1}{2} & 1 & 0 & 0 \\ 1 & -\frac{1}{3} & 1 & 0 \\ \frac{1}{2} & \frac{2}{3} & \frac{1}{2} & 1 \end{pmatrix} \begin{pmatrix} 2 & 4 & -4 & 0 \\ 0 & 3 & -3 & -3 \\ 0 & 0 & 4 & 2 \\ 0 & 0 & 0 & 3 \end{pmatrix}.$$

8. Solve $L\mathbf{Y} = \mathbf{B}$, $U\mathbf{X} = \mathbf{Y}$ and verify that $\mathbf{B} = A\mathbf{X}$ for (a) $\mathbf{B}^T = (9, 9, 26, 32)$ and (b) $\mathbf{B}^T = (2.0, 7.5, -7.0, 7.5)$, where $A = LU$ is

$$\begin{pmatrix} 1 & 2 & 0 & -1 \\ 2 & 3 & -1 & 0 \\ 0 & 4 & 2 & -5 \\ 5 & 5 & 2 & -4 \end{pmatrix} = \begin{pmatrix} 1 & 0 & 0 & 0 \\ 2 & 1 & 0 & 0 \\ 0 & -4 & 1 & 0 \\ 5 & 5 & -3.5 & 1 \end{pmatrix} \begin{pmatrix} 1 & 2 & 0 & -1 \\ 0 & -1 & -1 & 2 \\ 0 & 0 & -2 & 3 \\ 0 & 0 & 0 & 1.5 \end{pmatrix}.$$

9. Find the triangular factorization $A = LU$ for the matrix

$$\begin{pmatrix} 1 & 1 & 0 & 4 \\ 2 & -1 & 5 & 0 \\ 5 & 2 & 1 & 2 \\ -3 & 0 & 2 & 6 \end{pmatrix}.$$

10. Establish the formula in (12).

11. Write a computer program that uses Algorithm 3.3 to solve a linear system of equations $A\mathbf{X} = \mathbf{B}$. Use some of Exercises 1–3 and 6–8 as test cases.

12. Write an output routine for Algorithm 3.3 so that it will print out the matrices A, P, L, and U. Use some of the exercises above as test cases; switch rows of your test case to see what happens.

13. Modify Algorithm 3.3 so that it will compute A^{-1} by repeatedly solving N linear systems

$$A\mathbf{C}_J = \mathbf{E}_J, \quad \text{for} \quad J = 1, 2, \ldots, N.$$

Then

$$A[\mathbf{C}_1, \mathbf{C}_2, \ldots, \mathbf{C}_N] = [\mathbf{E}_1, \mathbf{E}_2, \ldots, \mathbf{E}_N] = I, \quad \text{and} \quad A^{-1} = [\mathbf{C}_1, \mathbf{C}_2, \ldots, \mathbf{C}_N].$$

Make sure that you compute the LU factorization only once!

14. Prove that a triangular factorization is unique in the following sense: If A is nonsingular and $L_1 U_1 = A = L_2 U_2$, then $L_1 = L_2$ and $U_1 = U_2$.

15. Prove that the product of two $N \times N$ upper-triangular matrices is an upper-triangular matrix.

16. Prove that the inverse of a nonsingular $N \times N$ upper-triangular matrix is an upper-triangular matrix.

17. Kirchhoff's voltage law says that the sum of the voltage drops around any closed path in the network in a given direction is zero. When this principle is applied to the circuit shown in Figure 3.6, we obtain the following linear system of equations:

$$
\begin{aligned}
(R_1 + R_3 + R_4)I_1 \; + & \; R_3 I_2 \; + & R_4 I_3 &= E_1 \\
R_3 I_1 \; + \; (R_2 + R_3 + R_5)I_2 \; - & & R_5 I_3 &= E_2 \\
R_4 I_1 \; - & \; R_5 I_2 \; + \; (R_4 + R_5 + R_6)I_3 &= 0.
\end{aligned}
$$

Solve for the currents I_1, I_2, and I_3 if
(a) $R_1 = 1$, $R_2 = 1$, $R_3 = 2$, $R_4 = 1$, $R_5 = 2$, $R_6 = 4$, and $E_1 = 23$, $E_2 = 29$
(b) $R_1 = 1$, $R_2 = 0.75$, $R_3 = 1$, $R_4 = 2$, $R_5 = 1$, $R_6 = 4$, and $E_1 = 12$, $E_2 = 21.5$
(c) $R_1 = 1$, $R_2 = 2$, $R_3 = 4$, $R_4 = 3$, $R_5 = 1$, $R_6 = 5$, and $E_1 = 41$, $E_2 = 38$

18. Write a report on Choleski's factorization. See References [9, 29, 40, 41, 51, 90, 97, 152, 153, and 160].

19. Write a report on linear programming (the simplex method). See References [19, 27, 35, 37, 41, 44, 50, 53, 79, 83, 94, 104, 115, 135, 152, 153, 154, 165, and 169].

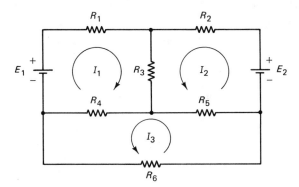

Figure 3.6   The electrical network for Exercise 17.

3.7 ITERATIVE METHODS FOR LINEAR SYSTEMS

The goal of this section is to extend some of the iterative methods introduced in Chapter 2 to higher dimensions. We consider an extension of fixed-point iteration that applies to systems of linear equations.

Jacobi Iteration

Example 3.25

Consider the system of equations

$$
\begin{aligned}
4x - y + z &= 7 \\
4x - 8y + z &= -21 \\
-2x + y + 5z &= 15.
\end{aligned}
\tag{1}
$$

These equations can be written in the form

$$
x = \frac{7 + y - z}{4}
$$

$$
y = \frac{21 + 4x + z}{8}
\tag{2}
$$

$$
z = \frac{15 + 2x - y}{5}.
$$

This suggests the following Jacobi iterative process:

$$
x_{k+1} = \frac{7 + y_k - z_k}{4}
$$

$$
y_{k+1} = \frac{21 + 4x_k + z_k}{8}
\tag{3}
$$

$$
z_{k+1} = \frac{15 + 2x_k - y_k}{5}.
$$

Let us show that if we start with $(x_0, y_0, z_0) = (1, 2, 2)$, the iteration in (3) appears to converge to the solution $(2, 4, 3)$.

Solution. Substitute $x_0 = 1$, $y_0 = 2$, $z_0 = 2$ into the right-hand side of each equation in (3) to obtain the new values

$$
x_1 = \frac{7 + 2 - 2}{4} = 1.75,
$$

$$
y_1 = \frac{21 + 4 + 2}{8} = 3.375,
$$

$$
z_1 = \frac{15 + 2 - 2}{5} = 3.00.
$$

The new point $\mathbf{P}_1 = (1.75, 3.375, 3.00)$ is closer to $(2, 4, 3)$ than $\mathbf{P}_0$. Iteration using (3) generates a sequence of points $\{\mathbf{P}_k\}$ that converges to the solution $(2, 4, 3)$ (see Table 3.3).

TABLE 3.3 Convergent Jacobi Iteration for the Linear System (1)

k	x_k	y_k	z_k
0	1.0	2.0	2.0
1	1.75	3.375	3.0
2	1.84375	3.875	3.025
3	1.9625	3.925	2.9625
4	1.99062500	3.97656250	3.00000000
5	1.99414063	3.99531250	3.00093750
⋮	⋮	⋮	⋮
15	1.99999993	3.99999985	2.99999993
⋮	⋮	⋮	⋮
19	2.00000000	4.00000000	3.00000000

This process is called **Jacobi iteration** and can be used to solve certain types of linear systems. After 19 steps, the iteration has converged to the nine-digit machine approximation $(2.00000000, 4.00000000, 3.00000000)$. This requires more computational effort than Gaussian elimination.

Linear systems with as many as 100,000 variables often arise in the solution of partial differential equations. The coefficient matrix for these systems is sparse; that is, the nonzero entries form a pattern. An iterative process provides an efficient method for solving these large systems.

Sometimes, the Jacobi method does not work. Let us experiment and see that a rearrangement of the original linear system can result in a system of iteration equations that will produce a divergent sequence of points.

Example 3.26

Let the linear system (1) be rearranged as follows:

$$\begin{aligned} -2x + y + 5z &= 15 \\ 4x - 8y + z &= -21 \\ 4x - y + z &= 7. \end{aligned} \tag{4}$$

These equations can be written in the form

$$x = \frac{-15 + y + 5z}{2},$$

$$y = \frac{21 + 4x + z}{8}, \tag{5}$$

$$z = 7 - 4x + y.$$

This suggests the following Jacobi iterative process:

$$x_{k+1} = \frac{-15 + y_k + 5z_k}{2},$$

$$y_{k+1} = \frac{21 + 4x_k + z_k}{8}, \tag{6}$$

$$z_{k+1} = 7 - 4x_k + y_k.$$

See that if we start with $(x_0, y_0, z_0) = (1, 2, 2)$, then iteration using (6) will diverge away from the solution $(2, 4, 3)$.

Solution.   Substitute $x_0 = 1$, $y_0 = 2$, and $z_0 = 2$ into the right-hand side of each equation in (6) to obtain the new values x_1, y_1, and z_1:

$$x_1 = \frac{-15 + 2 + 10}{2} = -1.5,$$

$$y_1 = \frac{21 + 4 + 2}{8} = 3.375,$$

$$z_1 = 7 - 4 + 2 = 5.00.$$

The new point $\mathbf{P}_1 = (-1.5, 3.375, 5.00)$ is farther away from the solution $(2, 4, 3)$ than $\mathbf{P}_0$. Iteration using equations (6) produces a divergent sequence (see Table 3.4).

TABLE 3.4   Divergent Jacobi Iteration for the Linear System (4)

k	x_k	y_k	z_k
0	1.0	2.0	2.0
1	−1.5	3.375	5.0
2	6.6875	2.5	16.375
3	34.6875	8.015625	−17.25
4	−46.617188	17.8125	−123.73438
5	−307.929688	−36.150391	211.28125
6	502.62793	−124.929688	1202.56836
:	:	:	:

Gauss–Seidel Iteration

Sometimes the convergence can be speeded up. Observe that $\{x_k\}$, $\{y_k\}$, and $\{z_k\}$ converge to 2, 4, and 3, respectively. It seems reasonable that x_{k+1} could be used in place of x_k in the computation of y_{k+1}. Similarly, the values x_{k+1} and y_{k+1} might be used in the computation of z_{k+1}. The next example shows what happens when this is applied to the equations in Example 3.25.

Example 3.27

Consider the system of equations given in (1), and the Gauss–Seidel iterative process suggested by (2):

$$x_{k+1} = \frac{7 + y_k - z_k}{4},$$

$$y_{k+1} = \frac{21 + 4x_{k+1} + z_k}{8}, \tag{7}$$

$$z_{k+1} = \frac{15 + 2x_{k+1} - y_{k+1}}{5}.$$

See that if we start with $\mathbf{P}_0 = (x_0, y_0, z_0) = (1, 2, 2)$, iteration using (7) will converge to the solution $(2, 4, 3)$.

Solution.   Substitute $y_0 = 2$, $z_0 = 2$ into the first equation and obtain

$$x_1 = \frac{7 + 2 - 2}{4} = 1.75.$$

Then substitute $x_1 = 1.75$, $z_0 = 2$ into the second equation and get

$$y_1 = \frac{21 + 4 \times 1.75 + 2}{8} = 3.75.$$

Finally, substitute $x_1 = 1.75$, $y_1 = 3.75$ into the third equation and get

$$z_1 = \frac{15 + 2 \times 1.75 - 3.75}{5} = 2.95.$$

The new point $\mathbf{P}_1 = (1.75, 3.75, 2.95)$ is closer to $(2, 4, 3)$ than $\mathbf{P}_0$ and is better than the value given in Example 3.25. Iteration using (7) generates a sequence $\{\mathbf{P}_k\}$ converges to $(2, 4, 3)$ (see Table 3.5).

TABLE 3.5 Convergent Gauss–Seidel
Iteration for the System (1)

k	x_k	y_k	z_k
0	1.0	2.0	2.0
1	1.75	3.75	2.95
2	1.95	3.96875	2.98625
3	1.995625	3.99609375	2.99903125
⋮	⋮	⋮	⋮
8	1.99999983	3.99999988	2.99999996
9	1.99999998	3.99999999	3.00000000
10	2.00000000	4.00000000	3.00000000

In view of Examples 3.25 and 3.26 it is necessary to have some criterion to determine whether the Jacobi iteration will converge. Hence we make the following definition.

Definition 3.8.   A matrix A of dimension $N \times N$ is said to be **strictly diagonally dominant** provided that

$$|a_{k,k}| > |a_{k,1}| + \cdots + |a_{k,k-1}| + |a_{k,k+1}| + \cdots + |a_{k,N}| \qquad (8)$$
$$\text{for } k = 1, 2, \ldots, N.$$

This means that in each row of the matrix, the magnitude of the diagonal coefficient must exceed the sum of the magnitudes of the other coefficients in the row. The matrix in Example 3.25 is diagonally dominant because

$$\text{In row 1:} \quad |4| > |-1| + |1|.$$
$$\text{In row 2:} \quad |-8| > |4| + |1|.$$
$$\text{In row 3:} \quad |5| > |-2| + |1|.$$

All the rows satisfy relation (8) in Definition 3.8; therefore, the matrix A for the linear system (1) is diagonally dominant.

The matrix in Example 3.26 is not diagonally dominant because

$$\text{In row 1:} \quad |-2| < |1| + |5|.$$
$$\text{In row 2:} \quad |-8| > |4| + |1|.$$
$$\text{In row 3:} \quad |1| < |4| + |-1|.$$

Rows 1 and 3 do not satisfy relation (8) in Definition 3.8; therefore, the matrix A for the linear system (4) is not diagonally dominant.

Suppose that the given linear system is

$$
\begin{aligned}
a_{1,1}x_1 + a_{1,2}x_2 + \cdots + a_{1,j}x_j + \cdots + a_{1,N}x_N &= b_1 \\
a_{2,1}x_1 + a_{2,2}x_2 + \cdots + a_{2,j}x_j + \cdots + a_{2,N}x_N &= b_2 \\
&\ \vdots \\
a_{j,1}x_1 + a_{j,2}x_2 + \cdots + a_{j,j}x_j + \cdots + a_{j,N}x_N &= b_j \qquad (9)\\
&\ \vdots \\
a_{N,1}x_1 + a_{N,2}x_2 + \cdots + a_{N,j}x_j + \cdots + a_{N,N}x_N &= b_N.
\end{aligned}
$$

Let the kth point be $\mathbf{P}_k = (x_1^{(k)}, x_2^{(k)}, \ldots, x_j^{(k)}, \ldots, x_N^{(k)})$; then the next point is $\mathbf{P}_{k+1} = (x_1^{(k+1)}, x_2^{(k+1)}, \ldots, x_j^{(k+1)}, \ldots, x_N^{(k+1)})$. The superscript (k) on the coordinates of $\mathbf{P}_k$ enables us to identify the coordinates that belong to this point. The iteration formulas use row j to solve for $x_j^{(k+1)}$ in terms of a linear combination of the previous values $x_1^{(k)}, x_2^{(k)}, \ldots, x_{j-1}^{(k)}, x_{j+1}^{(k)}, \ldots, x_N^{(k)}$:

Jacobi iteration:

$$x_j^{(k+1)} = \frac{b_j - a_{j,1}x_1^{(k)} - \cdots - a_{j,j-1}x_{j-1}^{(k)} - a_{j,j+1}x_{j+1}^{(k)} - \cdots - a_{j,N}x_N^{(k)}}{a_{j,j}}, \qquad (10)$$

for $j = 1, 2, \ldots, N$.

Jacobi iteration uses all old coordinates to generate all new coordinates, whereas Gauss–Seidel iteration uses the new coordinates as they become available:

Gauss–Seidel iteration:

$$x_j^{(k+1)} = \frac{b_j - a_{j,1}x_1^{(k+1)} - \cdots - a_{j,j-1}x_{j-1}^{(k+1)} - a_{j,j+1}x_{j+1}^{(k)} - \cdots - a_{j,N}x_N^{(k)}}{a_{j,j}}, \qquad (11)$$

for $j = 1, 2, \ldots, N$.

Theorem 3.14   (Jacobi Iteration).   Suppose that A is a diagonally dominant matrix. Then $A\mathbf{X} = \mathbf{B}$ has a unique solution $\mathbf{X} = \mathbf{P}$. Iteration using formula (10) will produce a sequence of vectors $\{\mathbf{P}_k\}$ that will converge to $\mathbf{P}$ for any choice of the starting vector $\mathbf{P}_0$.

Proof. See Reference [41].

It can be proven that the Gauss–Seidel method will converge when the matrix A is diagonally dominant. In many cases the Gauss–Seidel method will converge faster than the Jacobi method; hence it is usually preferred (compare Examples 3.25 and 3.27). It is important to understand the slight modification of formula (10) that has been made to obtain formula (11). In some cases the Jacobi method will converge even though the Gauss–Seidel method will not.

Convergence

A measure of the closeness between points is needed so that we can determine if $\{\mathbf{P}_k\}$ is converging to $\mathbf{P}$. The Euclidean distance between

$$\mathbf{P} = (x_1, x_2, \ldots, x_N) \quad \text{and} \quad \mathbf{Q} = (y_1, y_2, \ldots, y_N)$$

is

$$D(\mathbf{P}, \mathbf{Q}) = [(x_1 - y_1)^2 + (x_2 - y_2)^2 + \cdots + (x_N - y_N)^2]^{1/2}. \qquad (12)$$

Its disadvantage is that it requires considerable computing effort. Hence we introduce a different norm $\|\mathbf{X}\|_1$ that will be used to define the separation between points.

Definition 3.9.   Let $\mathbf{X}$ be an N-dimensional vector. We define the ℓ_1 **norm of X** to be the function $\|\mathbf{X}\|_1$:

$$\|\mathbf{X}\|_1 = \sum_{j=1}^{N} |x_j|. \qquad (13)$$

The following result ensures that $\|\mathbf{X}\|_1$ has the mathematical structure of a metric, hence is suitable to use as a generalized "distance formula."

Theorem 3.15.   Let $\mathbf{X}$ and $\mathbf{Y}$ be N-dimensional vectors and c be a scalar. Then the function $\|\mathbf{X}\|_1$ has the following properties:

$$\|\mathbf{X}\|_1 \geq 0, \tag{14}$$

$$\|\mathbf{X}\|_1 = 0 \quad \text{if and only if} \quad \mathbf{X} = \mathbf{0}, \tag{15}$$

$$\|c\mathbf{X}\|_1 = |c|\,\|\mathbf{X}\|_1, \tag{16}$$

$$\|\mathbf{X} + \mathbf{Y}\|_1 \leq \|\mathbf{X}\|_1 + \|\mathbf{Y}\|_1. \tag{17}$$

Proof. We prove (17) and leave the others as exercises. For each j, the triangle inequality for real numbers states that $|x_j + y_j| \leq |x_j| + |y_j|$. Summing these yields inequality (17):

$$\|\mathbf{X} + \mathbf{Y}\|_1 = \sum_{j=1}^{N} |x_j + y_j| \leq \sum_{j=1}^{N} |x_j| + \sum_{j=1}^{N} |y_j| = \|\mathbf{X}\|_1 + \|\mathbf{Y}\|_1.$$

The norm given by (13) can be used to define the "separation between" points.

Definition 3.10.   Suppose that $\mathbf{X}$ and $\mathbf{Y}$ are two points in N-dimensional space. We define the **separation** between $\mathbf{X}$ and $\mathbf{Y}$ in terms of the ℓ_1 distance between them:

$$\|\mathbf{X} - \mathbf{Y}\|_1 = \sum_{j=1}^{N} |x_j - y_j|.$$

Example 3.28

Determine the Euclidean distance and separation between the points $\mathbf{P} = (2, 4, 3)$ and $\mathbf{Q} = (1.75, 3.75, 2.95)$.

Solution.   The Euclidean distance is

$$D(\mathbf{P}, \mathbf{Q}) = [(2 - 1.75)^2 + (4 - 3.75)^2 + (3 - 2.95)^2]^{1/2} = 0.3570714214.$$

The separation is

$$\|\mathbf{P} - \mathbf{Q}\|_1 = |2 - 1.75| + |4 - 3.75| + |3 - 2.95| = 0.55.$$

The separation is easier to compute and use for determining convergence in N-dimensional space.

Algorithm 3.4   (Jacobi Iteration).   To solve the linear system $A\mathbf{X} = \mathbf{B}$ by starting with $\mathbf{P}_0 = \mathbf{0}$ and generating a sequence $\{\mathbf{P}_k\}$ that converges to the solution $\mathbf{P}$ (i.e., $A\mathbf{P} = \mathbf{B}$). A sufficient condition for the method to be applicable is that A is diagonally dominant.

```
          Tol := 10⁻⁶                                                        {Tolerance}
          Sep := 1, K := 1, Max := 99                                        {Initialize}
          Cond := 1                                                {Condition of matrix}

          INPUT   N, A[1 . . N, 1 . . N], B[1 . . N]                           {Input}

FOR   R = 1   TO   N   DO
    │     Row := 0                                                        {Check for
    │     FOR   C = 1   TO   N   DO  Row := Row+|A(R, C)|                  diagonal
    └─────IF    Row ≥ 2*|A(R, R)|   THEN   Cond := 0                      dominance}

IF     Cond = 0   THEN
    │       PRINT "The matrix is not diagonally dominant."
    │       TERMINATE ALGORITHM
ENDIF

          FOR   J = 0   TO   N   DO   P(J) := 0, Pnew(J) := 0              {Initialize}
WHILE   K < Max AND Sep > Tol   DO
    │     FOR   R = 1   TO   N   DO                                        {Perform
    │         │   Sum := B(R)                                              Jacobi
    │         │   FOR   C = 1   TO   N   DO                                iteration}
    │         │   └─────IF C ≠ R   THEN   Sum := Sum−A(R, C)*P(C)
    │         └───Pnew(R) := Sum/A(R, R)
    │     Sep := 0                                                        {Convergence
    │     FOR   J = 1   TO   N   DO   Sep := Sep+|Pnew(J)−P(J)|            criterion}
    │     FOR   J = 1   TO   N   DO   P(J) := Pnew(J)                     {Update values,
    └─────K := K+1                                                 increment the counter}

IF     Sep < Tol   THEN                                                     {Output}
    │       PRINT "The solution to the linear system is:"
    │       ELSE
    │       PRINT "Jacobi iteration did not converge:"
ENDIF

FOR J = 1   TO   N   DO   PRINT "P(" ; J ; ") = " ; P(J)
```

Algorithm 3.5   (Gauss–Seidel Iteration).   To solve the linear system $AX = B$ by starting with $\mathbf{P}_0 = \mathbf{0}$ and generating a sequence $\{\mathbf{P}_k\}$ that converges to the solution $\mathbf{P}$ (i.e., $A\mathbf{P} = \mathbf{B}$). A sufficient condition for the method to be applicable is that A is diagonally dominant.

```
        FOR  J = 1  TO  N  DO  P(J) := 0, Pold(J) := 0              {Initialize}
WHILE  K < Max AND Sep > Tol  DO
        FOR  R = 1  TO  N  DO                                        {Perform
            Sum := B(R)                                         Gauss-Seidel
            FOR  C = 1  TO  N  DO                                   iteration}
                IF C ≠ R  THEN  Sum := Sum−A(R, C)∗P(C)
            P(R) := Sum/A(R, R)
        Sep := 0                                               {Convergence
        FOR  J = 1  TO  N  DO  Sep := Sep+|P(J)−Pold(J)|           criterion}
        FOR  J = 1  TO  N  DO  Pold(J) := P(J)                 {Update values,
        K := K+1                                         increment the counter}

IF    Sep < Tol  THEN                                             {Output}
        PRINT "The solution to the linear system is:"
      ELSE
        PRINT "Gauss-Seidel iteration did not converge:"
ENDIF

FOR  J = 1  TO  N  DO  PRINT "P(" ; J ; ") = " ; P(J)
```

EXERCISES FOR ITERATIVE METHODS FOR LINEAR SYSTEMS

In Exercises 1–8:

(a) Start with $\mathbf{P}_0 = \mathbf{0}$ and use Jacobi iteration to find $\mathbf{P}_k$ ($k = 1, 2, 3$). Will Jacobi iteration converge to the solution?

(b) Start with $\mathbf{P}_0 = \mathbf{0}$ and use Gauss–Seidel iteration to find $\mathbf{P}_k$ ($k = 1, 2, 3$). Will Gauss–Seidel iteration converge to the solution?

(c) Write a computer program that uses iteration, and solve the linear system.

1. $4x - y = 15$
    $x + 5y = 9$

2. $8x - 3y = 10$
    $-x + 4y = 6$

3. $-x + 3y = 1$
    $6x - 2y = 2$

4. $2x + 3y = 1$
    $7x - 2y = 1$

5. $5x - y + z = 10$
    $2x + 8y - z = 11$
    $-x + y + 4z = 3$

6. $2x + 8y - z = 11$
    $5x - y + z = 10$
    $-x + y + 4z = 3$

7. $x - 5y - z = -8$
    $4x + y - z = 13$
    $2x - y - 6z = -2$

8. $4x + y - z = 13$
    $x - 5y - z = -8$
    $2x - y - 6z = -2$

9. Consider the following linear system:

$$5x + 3y = 6$$
$$4x - 2y = 8.$$

Can either Jacobi or Gauss–Seidel iteration be used to solve this linear system? Why?

10. Can Jacobi iteration ever be used to find the solution to the following system?

$$2x + y - 5z = 9$$
$$x - 5y - z = 14$$
$$7x - y - 3z = 26.$$

11. Consider the following tridiagonal linear system, and assume that the coefficient matrix is diagonally dominant.

$$d_1 x_1 + c_1 x_2 \qquad\qquad\qquad\qquad = b_1$$
$$a_1 x_1 + d_2 x_2 + c_2 x_3 \qquad\qquad\qquad = b_2$$
$$a_2 x_2 + d_3 x_3 + c_3 x_4 \qquad\qquad = b_3$$
$$\vdots \qquad\qquad \vdots$$
$$a_{N-2} x_{N-2} + d_{N-1} x_{N-1} + c_{N-1} x_N = b_{N-1}$$
$$a_{N-1} x_{N-1} + d_N x_N = b_N.$$

Write an iterative algorithm that will solve this system.

12. Tridiagonal matrices were involved in the construction of a cubic spline. Find the solution to the following systems.

(a)
$$4m_1 + m_2 = 3$$
$$m_1 + 4m_2 + m_3 = 3$$
$$m_2 + 4m_3 + m_4 = 3$$
$$m_3 + 4m_4 + m_5 = 3$$
$$\vdots$$
$$m_{48} + 4m_{49} + m_{50} = 3$$
$$m_{49} + 4m_{50} = 3$$

(b)
$$4m_1 + m_2 = 1$$
$$m_1 + 4m_2 + m_3 = 2$$
$$m_2 + 4m_3 + m_4 = 1$$
$$m_3 + 4m_4 + m_5 = 2$$
$$\vdots$$
$$m_{48} + 4m_{49} + m_{50} = 1$$
$$m_{49} + 4m_{50} = 2$$

13. Use Gauss–Seidel iteration to solve the following band system.

$$
\begin{aligned}
12x_1 - 2x_2 + x_3 &= 5 \\
-2x_1 + 12x_2 - 2x_3 + x_4 &= 5 \\
x_1 - 2x_2 + 12x_3 - 2x_4 + x_5 &= 5 \\
x_2 - 2x_3 + 12x_4 - 2x_5 + x_6 &= 5 \\
\vdots \qquad \vdots \qquad \vdots \qquad \vdots \qquad \vdots \qquad \vdots \\
x_{46} - 2x_{47} + 12x_{48} - 2x_{49} + x_{50} &= 5 \\
x_{47} - 2x_{48} + 12x_{49} - 2x_{50} &= 5 \\
x_{48} - 2x_{49} + 12x_{50} &= 5.
\end{aligned}
$$

14. Let $\mathbf{X} = (x_1, x_2, \dots, x_N)$. Prove that the Euclidean length

$$
\|\mathbf{X}\|_2 = \left(\sum_{k=1}^{N} x_k^2 \right)^{1/2}
$$

satisfies the four properties given in (14)–(17).

15. Let $\mathbf{X} = (x_1, x_2, \dots, x_N)$. Prove that the function

$$
\|\mathbf{X}\|_\infty = \max_{1 \le k \le N} |x_k|
$$

satisfies the four properties given in (14)–(17).

16. Write a report on the SOR method. See References [10, 29, 40, 41, 49, 137, 139, 152, 160, 175, 199, 207].

4

Interpolation and Polynomial Approximation

The computational procedure used in computer software for the evaluation of a library function such as $\sin(x)$, $\cos(x)$, or e^x involves polynomial approximation. The state-of-the-art methods use rational functions (which are the quotient of polynomials). However, the theory of polynomial approximation is suitable for a first course in numerical analysis, and we will mainly consider them in this chapter. Suppose that the function $f(x) = e^x$ is to be approximated by a polynomial of degree $n = 2$ over the interval $[-1, 1]$. The Taylor polynomial is shown in Figure 4.1(a) and can be contrasted with the Chebyshev approximation in Figure 4.1(b). The maximum error for the Taylor approximation is 0.218282, whereas the maximum error for the Chebyshev polynomial is 0.056468. In this chapter we develop the basic theory needed to investigate these matters.

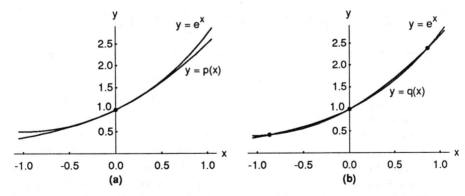

Figure 4.1   (a) The Taylor polynomial $p(x) = 1.000000 + 1.000000x + 0.500000x^2$ which approximates $f(x) = e^x$ over $[-1, 1]$. (b) The Chebyshev approximation $q(x) = 1.000000 + 1.129772x + 0.532042x^2$ for $f(x) = e^x$ over $[-1, 1]$.

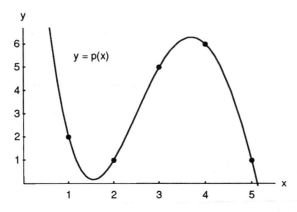

Figure 4.2 The graph of the collocation polynomial that passes through $(1, 2)$, $(2, 1)$, $(3, 5)$, $(4, 6)$, and $(5, 1)$.

An associated problem involves the construction of the collocation polynomial. Given $n + 1$ points in the plane (no two of which are aligned vertically), the collocation polynomial is the unique polynomial of degree $\leq n$ which passes through the points. In cases where data are known to a high degree of precision, the collocation polynomial is sometimes used to find a polynomial that passes through the given data points. There are a variety of methods that can be used to construct the collocation polynomial: solving a linear system for its coefficients, the use of Lagrange coefficient polynomials, and the construction of a divided differences table and the coefficients of the Newton polynomial. All three techniques are important for a practitioner of numerical analysis to know. For example, the collocation polynomial of degree $n = 4$ that passes through the five points $(1, 2)$, $(2, 1)$ $(3, 5)$ $(4, 6)$, and $(5, 1)$ is

$$p(x) = \frac{504 - 806x + 427x^2 - 82x^3 + 5x^4}{24},$$

and a graph showing both the points and the polynomial is given in Figure 4.2.

4.1 TAYLOR SERIES AND CALCULATION OF FUNCTIONS

Limit processes are the basis of calculus. For example, the derivative

$$f'(x) = \lim_{h \to 0} \frac{f(x + h) - f(x)}{h}$$

is the limit of the difference quotient where both numerator and denominator go to zero. Taylor series illustrate another type of limit process. In this case an infinite number of terms are added together by taking the limit of certain partial sums. An important application is their use to represent the elementary functions $\sin(x)$, $\cos(x)$, $\exp(x)$, $\ln(x)$. Table 4.1 gives several of the common Taylor series expansions. The partial sums can be accumulated until an approximation to the function is obtained which has the accuracy that is specified. Series solutions are used in the areas of engineering and physics.

We want to learn how a finite sum can be used to obtain a good approximation to

TABLE 4.1 Taylor Series Expansions for Some Common Functions

$\sin(x) = x - \dfrac{x^3}{3!} + \dfrac{x^5}{5!} - \dfrac{x^7}{7!} + \cdots$	for all x
$\cos(x) = 1 - \dfrac{x^2}{2!} + \dfrac{x^4}{4!} - \dfrac{x^6}{6!} + \cdots$	for all x
$\exp(x) = 1 + x + \dfrac{x^2}{2!} + \dfrac{x^3}{3!} + \dfrac{x^4}{4!} + \cdots$	for all x
$\ln(1 + x) = x - \dfrac{x^2}{2} + \dfrac{x^3}{3} - \dfrac{x^4}{4} + \cdots$	$-1 < x \le 1$
$\arctan(x) = x - \dfrac{x^3}{3} + \dfrac{x^5}{5} - \dfrac{x^7}{7} + \cdots$	$-1 \le x \le 1$
$(1 + x)^p = 1 + px + \dfrac{p(p-1)}{2!}x^2 + \dfrac{p(p-1)(p-2)}{3!}x^3 + \cdots$	for $\lvert x \rvert < 1$

an infinite sum. For illustration we shall use the exponential series in Table 4.1 to compute the number $e = \exp(1)$, which is the base of the natural logarithm and exponential functions. Here we choose $x = 1$ and use the series

$$\exp(1) = 1 + \frac{1}{1!} + \frac{1^2}{2!} + \frac{1^3}{3!} + \frac{1^4}{4!} + \cdots + \frac{1^k}{k!} + \cdots .$$

The definition for the sum of an infinite series in Section 1.1 requires that the partial sums S_n tend to a limit. The values for several of these sums are given in Table 4.2.

TABLE 4.2 Partial Sums S_n Used to Determine e

n	$S_n = 1 + \dfrac{1}{1!} + \dfrac{1}{2!} + \cdots + \dfrac{1}{n!}$
0	1.0
1	2.0
2	2.5
3	2.666666666666. . .
4	2.708333333333. . .
5	2.716666666666. . .
6	2.718055555555. . .
7	2.718253968254. . .
8	2.718278769841. . .
9	2.718281525573. . .
10	2.718281801146. . .
11	2.718281826199. . .
12	2.718281828286. . .
13	2.718281828447. . .
14	2.718281828458. . .
15	2.718281828459. . .

A natural way to think about the power series representation of a function is to view the expansion as the limiting case of polynomials of increasing degree. If enough terms are added, then an accurate approximation will be obtained. This needs to be made precise. What degree should be chosen for the polynomial, and how do we calculate the coefficients for the powers of x in the polynomial? Theorem 4.1 answers these questions.

Theorem 4.1 (Taylor Polynomial Approximation).   Assume that $f \in C^{N+1}[a, b]$ and $x_0 \in [a, b]$ is a fixed value. If $x \in [a, b]$, then

$$f(x) = P_N(x) + E_N(x), \tag{1}$$

where $P_N(x)$ is a polynomial that can be used to approximate $f(x)$:

$$f(x) \approx P_N(x) = \sum_{k=0}^{N} \frac{f^{(k)}(x_0)}{k!} (x - x_0)^k. \tag{2}$$

The error term $E_N(x)$ has the form

$$E_N(x) = \frac{f^{(N+1)}(c)}{(N+1)!} (x - x_0)^{N+1} \tag{3}$$

for some value $c = c(x)$ that lies between x_0 and x.

Remark.   The polynomial in (2) can be expressed in the familiar form

$$P_N(x) = a_0 + a_1(x - x_0) + a_2(x - x_0)^2 + \cdots + a_N(x - x_0)^N, \tag{4}$$

$$\text{where} \quad a_k = \frac{f^{(k)}(x_0)}{k!}.$$

Proof. The proof is left as an exercise (see Exercises 20 and 21).

Relation (4) indicates how the coefficients a_k of the Taylor polynomial are calculated. Although the error term (3) involves a similar expression, notice that $f^{(N+1)}(c)$ is to be evaluated and that the number c may differ from the number x_0. For this reason we do *not* try to evaluate $E_N(x)$; it is used to determine a bound for the accuracy of the approximation.

Example 4.1

Show why 15 terms are all that are needed to obtain the 13-digit approximation $e \approx 2.718281828459$ in Table 4.1.

Solution.   Expand $f(x) = \exp(x)$ in a Taylor polynomial of degree 15 using the fixed value $x_0 = 0$ and involving the powers $(x - 0)^k = x^k$. The derivatives required are $f'(x) = \exp(x)$, $f''(x) = \exp(x), \ldots, f^{(16)}(x) = \exp(x)$. The first 15 derivatives are used to calculate the coefficients $a_k = \exp(0)/k! = 1/k!$ and are used to write

$$P_{15}(x) = 1 + x + \frac{x^2}{2!} + \frac{x^3}{3!} + \cdots + \frac{x^{15}}{15!}. \tag{5}$$

Setting $x = 1$ in (5) gives the partial sum $S_{15} = P_{15}(1)$. The remainder term is needed to show the accuracy of the approximation

$$E_{15}(x) = \frac{f^{(16)}(c)x^{16}}{16!} \tag{6}$$

Since we chose $x_0 = 0$ and $x = 1$, the value c lies between them (i.e., $0 < c < 1$), which implies that $e^c < e^1$. Notice that the partial sums in Table 4.2 are bounded above by 3; this permits us to write the crude inequality $e^1 < 3$. Combining these two inequalities yields $e^c < 3$, which is used in the following calculation

$$|E_{15}(1)| = \frac{|f^{(16)}(c)|}{16!} \leq \frac{e^c}{16!} < \frac{3}{16!} = 1.433843 \times 10^{-13}.$$

Therefore, all the digits in the approximation $e \approx 2.718281828459.\ .\ .$ are correct, because the actual error (whatever it is) must be less than 2 in the thirteenth decimal place.

Instead of giving a rigorous proof of Theorem 4.1, we shall discuss some of the features of the approximation; the reader can look in any standard reference on calculus for more details. For illustration, we again use the function $f(x) = \exp(x)$ and the value $x_0 = 0$. From elementary calculus, we know that the slope of the curve $y = f(x)$ at the point $(x, f(x))$ is $f'(x) = \exp(x)$. Hence the slope at the point $(0, 1)$ is $f'(0) = 1$. Therefore, the tangent line to the curve at the point $(0, 1)$ is $y = 1 + x$. This is the same formula that would be obtained if we use $N = 1$ in Theorem 4.1, that is, $P_1(x) = f(0) + f'(0)x/1! = 1 + x$. Therefore, $P_1(x)$ is the equation of the line tangent to the curve. The graphs are shown in Figure 4.3.

Observe that the approximation $\exp(x) \approx 1 + x$ is good near the center $x_0 = 0$ and that the distance between curves grows as x moves away from 0. Notice that the slopes of the curves agree at $(0, 1)$. In calculus we learned that the second derivative indicates whether a curve is concave up or concave down. The study of curvature[†] shows that if two

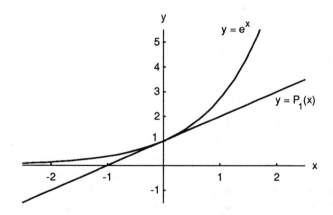

Figure 4.3 The graphs of $y = e^x$ and $y = P_1(x) = 1 + x$.

[†]The curvature K of a graph $y = f(x)$ at (x, y) is defined by

$$K = |f''(x_0)|(1 + [f'(x_0)]^2)^{-3/2}.$$

curves $y = f(x)$ and $y = g(x)$ have the property that $f(x_0) = g(x_0)$, $f'(x_0) = g'(x_0)$, and $f''(x_0) = g''(x_0)$, they have the same curvature at x_0. This property would be desirable for a polynomial function that approximates $f(x)$. Corollary 4.1 shows that the Taylor polynomial has this property for $N \geq 2$.

Corollary 4.1. If $P_N(x)$ is the Taylor polynomial of degree N given in Theorem 4.1, then

$$P_N^{(k)}(x_0) = f^{(k)}(x_0) \qquad \text{for } k = 0, 1, \ldots, N. \tag{7}$$

Proof. Set $x = x_0$ in equation (3) and the result is $P_N(x_0) = f(x_0)$. Differentiate both sides of (3) and get

$$P_N'(x) = a_1 + 2a_2(x - x_0) + 3a_3(x - x_0)^2 + \cdots + Na_N(x - x_0)^{N-1}. \tag{8}$$

By definition (4), $a_1 = f'(x_0)$. Use this in (8) with $x = x_0$ to obtain $P_N'(x_0) = f'(x_0)$. Successive differentiations of (8) with the appropriate use of definition (4) will establish the other identities in (7). The details are left as an exercise.

Applying Corollary 4.1, we see that $y = P_2(x)$ has the properties $f(x_0) = P_2(x_0)$, $f'(x_0) = P_2'(x_0)$, and $f''(x_0) = P_2''(x_0)$; hence the graphs have the same curvature at x_0. For example, consider $f(x) = \exp(x)$ and $P_2(x) = 1 + x + x^2/2$. The graphs are shown in Figure 4.4 and it is seen that they are curved up in the same fashion at $(0, 1)$.

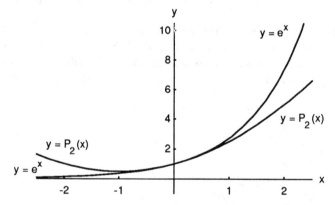

Figure 4.4 The graphs of $y = e^x$ and $y = P_2(x) = 1 + x + x^2/2$.

In the theory of approximation, one seeks to find an accurate polynomial approximation to the analytic function[†] $f(x)$ over $[a, b]$. This is one technique used in developing computer software. The accuracy of a Taylor polynomial is increased when we choose N large. The accuracy of any given polynomial will generally decrease as the value of x moves away from the center x_0. Hence we must choose N large enough and restrict the maximum value of $|x - x_0|$ so that the error does not exceed a specified bound.

[†]The function $f(x)$ is analytic at x_0 if it has continuous derivatives of all orders and can be expressed as a Taylor series in an interval about x_0.

If we choose the interval width to be $2R$ and x_0 in the center (i.e., $|x - x_0| < R$), the absolute value of the error satisfies the relation

$$|\text{error}| = |E_N(x)| \le \frac{MR^{N+1}}{(N+1)!}. \tag{9}$$

where $M \le \max\{|f^{(N+1)}(z)|: x_0 - R \le z \le x_0 + R\}$. If N is fixed, and the derivatives are uniformly bounded, the error bound in (9) is proportional to $R^{N+1}/(N+1)!$ and decreases if R goes to zero. For R fixed, the error bound is inversely proportional to $(N+1)!$ and it goes to zero as N gets large. Table 4.3 shows how the choices of these two parameters affect the accuracy of the approximation $\exp(x) \approx P_N(x)$ over the interval $|x| \le R$. The error is smallest when N is largest and R smallest. Graphs for P_2, P_3, and P_4 are given in Figure 4.5.

TABLE 4.3 Values for the Error Bound $|\text{error}| < e^R R^{N+1}/[(N+1)!]$ Using the Approximation $\exp(x) \approx P_N(x)$ for $|x| \le R$

	$R = 2.0$, $\|x\| \le 2.0$	$R = 1.5$, $\|x\| \le 1.5$	$R = 1.0$, $\|x\| \le 1.0$	$R = 0.5$, $\|x\| \le 0.5$
$\exp(x) \approx P_5(x)$	0.65680499	0.07090172	0.00377539	0.00003578
$\exp(x) \approx P_6(x)$	0.18765857	0.01519323	0.00053934	0.00000256
$\exp(x) \approx P_7(x)$	0.04691464	0.00284873	0.00006742	0.00000016
$\exp(x) \approx P_8(x)$	0.01042548	0.00047479	0.00000749	0.00000001

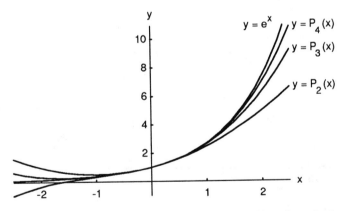

Figure 4.5 The graphs of $y = e^x$, $y = P_2(x)$, $y = P_3(x)$, and $y = P_4(x)$.

Example 4.2

Establish the error bounds for the approximation $\exp(x) \approx P_8(x)$ on the intervals $|x| \le 1.0$ and $|x| \le 0.5$.

Solution.   If $|x| \le 1.0$, then $|f^{(9)}(c)| = |e^c| \le e^{1.0}$ implies that

$$|\text{error}| = |E_8(x)| \le \frac{e^{1.0}(1.0)^9}{9!} \approx 0.00000749.$$

For the interval $|x| \leq 0.5$, $|f^{(9)}(c)| = |e^c| \leq e^{0.5}$ implies that

$$|\text{error}| = |E_8(x)| \leq \frac{e^{0.5}(0.5)^9}{9!} \approx 0.00000001.$$

Example 4.3

If $f(x) = \exp(x)$, show that $N = 9$ is the smallest integer, so that $|\text{error}| = |E_N(x)| \leq 0.0000005$ for x in $[-1, 1]$. Hence $P_9(x)$ can be used to compute approximate values of $\exp(x)$ that will be accurate in the sixth decimal place.

Solution. We need to find the smallest integer N so that

$$|\text{error}| = |E_N(x)| = \frac{e^c(1)^{N+1}}{(N+1)!} \leq 0.0000005.$$

In Example 4.2 we saw that $N = 8$ is too small, so we try $N = 9$ and discover that $|E_N(x)| \leq e^1(1)^{9+1}/[(9+1)!] \leq 0.000000749$. This value is slightly larger than desired; hence we would be likely to choose $N = 10$. But we used $e^c \leq e^1$ as a crude estimate in finding the error bound. Hence 0.000000749 is a little larger than the actual error. Figure 4.6 shows a graph of $E_9(x) = \exp(x) - P_9(x)$. Notice that the maximum vertical range is about 3×10^{-7} and occurs at the right endpoint $(1, E_9(1))$. Indeed, the maximum error on the interval is $E_9(1) = 2.718281828 - 2.718281526 \approx 3.024 \times 10^{-7}$. Therefore, $N = 9$ is justified.

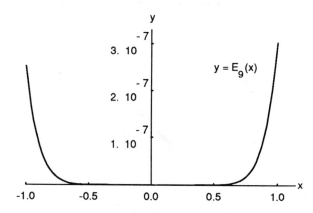

Figure 4.6   The graph of the error $y = E_9(x) = e^x - P_9(x)$.

When series are used to approximate the elementary functions it is worthwhile to take advantage of the special properties of the functions. Suppose that we wanted to compute $\exp(86.3)$ using series. Then we would have to be patient and use $N = 150$ to obtain

$$\exp(86.3) \approx P_{150}(86.3) \approx 3.01726735 \times 10^{37}. \tag{10}$$

This sum was accumulated using a computer that has floating-point numbers with nine digits of accuracy.

We could save a considerable amount of effort if we use the identity $e^{86.3} = e^{86}e^{0.3}$ and the fact that we have already found e^1 accurate to 13 decimal places. We need only to compute $\exp(0.3) \approx P_9(0.3) \approx 1.34985881$ and use the formula

$$e^{86}e^{0.3} \approx (2.718281828459)^{86}(1.34985881)$$

$$= 2.23524660 \times 10^{37}(1.34985881) = 3.01726732 \times 10^{37}. \qquad (11)$$

The products in (11) were computed in double-precision arithmetic. If only nine significant digits are carried, a larger error results:

$$2.71828183^{86}(1.34985881) = 2.23524670 \times 10^{37}(1.34985881)$$

$$= 3.01726744 \times 10^{37}.$$

This time the answer has seven significant digits of accuracy because the error in the approximation to e is propagated in the multiplications.

Theorem 4.2   (Alternating Series).   Suppose that $c_k > 0$ and that $c_k \geq c_{k+1} > 0$ for $k = J, J + 1, \ldots$ and that $\lim_{k \to \infty} c_k = 0$. Then

$$S = \sum_{k=J}^{\infty} (-1)^k c_k = \lim_{N \to \infty} \sum_{k=J}^{N} (-1)^k c_k \qquad (12)$$

is a convergent infinite series. If the partial sum $S_N = \sum_{k=J}^{N} (-1)^k c_k$ is used to approximate S, the truncation error $E_N = S - S_N$ is no larger than the magnitude of the next term (see Figure 4.7):

$$|E_N| = |S - S_N| \leq c_{N+1}. \qquad (13)$$

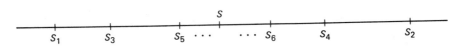

Figure 4.7   The partial sums $\{S_n\}$ in an alternating series.

Another way to state the result in (12) is to say that the partial sums converge to S, that is,

$$S = \lim_{N \to \infty} S_N = \lim_{N \to \infty} \sum_{k=J}^{N} (-1)^k c_k = \sum_{k=J}^{\infty} (-1)^k c_k. \qquad (14)$$

As a note of caution, we investigate what happens when an alternating series is used to compute $\exp(-11)$. Theorem 4.2 implies that the truncation error for $P_{50}(-11)$ is no larger than $11^{51}/51! = 8.324955 \times 10^{-14}$. A computer with nine decimal digits of accuracy was used to accumulate this sum and it gave the answer $P_{50}(-11) = 1.60363964 \times 10^{-5}$. The mathematical error analysis might lead us to believe that all nine digits were correct. However, the computer's built-in exponential function gives the answer $\exp(-11) = 1.67017008 \times 10^{-5}$. Therefore, $P_{50}(-11)$ has roughly two significant digits of accuracy! So far, this is the worst discrepancy that we have encountered.

The reason for the flawed calculation is that each term has its own round-off error, and the error in the total will be as big as the largest error in one of its addends. Theorem 4.2 discusses the accuracy of the mathematical approximation

$$P_{50}(-11) = 1 - 11 + \frac{11^2}{2} - \frac{11^3}{6} + \cdots - \frac{11^{49}}{49!} + \frac{11^{50}}{50!}$$

$$= \frac{34{,}694{,}835{,}499{,}645{,}991{,}522{,}891{,}035{,}426{,}510{,}983{,}460{,}551{,}595{,}052{,}804{,}990{,}261}{2{,}077{,}323{,}488{,}949{,}756{,}030{,}572{,}543{,}416{,}847{,}535{,}608{,}522{,}480{,}812{,}032{,}000{,}000{,}000{,}000}.$$

The computer actually calculated a different sum:

$$P_{50}(-11) \approx 1 - 11 + 60.5 - 221.833333 + \cdots + 0.00000000000385975186$$

$$\approx 0.00000160363964.$$

The maximum error occurs with the tenth and eleventh terms in the series when the computer uses $11^{10}/10! \approx 7147.65889$ and the round-off error is 5.7782×10^{-6}. Table 4.4 gives some of the terms, the computer's approximation, and the round-off error for the term.

If an accurate approximation for $\exp(-11)$ is needed, it can be found by using $x = 11$ and computing $P_{50}(11) = 59{,}874.1417$. The error for each term in the series is less than 0.0000057782, and the total error is less than $51 \times 5.7782 \times 10^{-6} = 0.00029469$.

TABLE 4.4   Errors for Some Terms in the Approximation $\exp(-11) \approx P_{50}(-11)$

k	$(-11)^k/k!$	Computer approximation	Computer's round-off error
0	1.	1.	0.0000000000000000
1	$-11.$	$-11.$	0.0000000000000000
2	60.5	60.5	0.0000000000000000
3	$-221.8333333333333333\ldots$	-221.833333	-0.0000003333333333
4	$610.0416666666666666\ldots$	610.041667	-0.0000003333333333
5	$-1{,}342.0916666666666666\ldots$	$-1{,}342.09167$	0.0000003333333333
10	$7{,}147.6588957782186948\ldots$	$7{,}147.65889$	0.0000057782186948
15	$-3{,}194.4100699966086664\ldots$	$-3{,}194.41007$	0.0000000033913335
20	$276.5216160254471009\ldots$	276.521616	0.0000000254471009
25	$-6.9850810562949810\ldots$	-6.98508106	0.0000000037050189
30	$0.0657840306839339\ldots$	0.0657840307	-0.0000000000160660
35	$-0.0002719639449612\ldots$	-0.000271963945	0.0000000000000387
40	$0.0000005547053290\ldots$	0.000000554705329	0.0000000000000000
50	$0.0000000000003859\ldots$	0.0000000000003859	0.0000000000000000

This is a small fraction of the total 59,874.1417, so that eight significant digits are guaranteed. Now the reciprocal can be taken to obtain

$$\exp(-11) \approx \frac{1}{59,874.1417} = 1.67017008 \times 10^{-5}.$$

Example 4.4

Show that $\ln(2) = 1 - \frac{1}{2} + \frac{1}{3} - \frac{1}{4} + \frac{1}{5} - \frac{1}{6} + \cdots$ is a convergent series. How many terms are needed so that the truncation error is less than 2×10^{-4}?

Solution.   In general, $c_k = 1/k$ and it is easy to verify that

$$c_k = \frac{1}{k} \geq \frac{1}{k+1} = c_{k+1} \quad \text{and} \quad \lim_{k \to \infty} c_k = \lim_{k \to \infty} \frac{1}{k} = 0.$$

By Theorem 4.2 we conclude that the series is convergent. To make $|E_N| < 2 \times 10^{-4}$ we seek the smallest integer N so that $c_{N+1} < 2 \times 10^{-4}$ and find $N = 5000$. Calculation reveals that

$$S_{4999} \approx 0.693247197 \quad \text{with} \quad E_{4999} \approx -0.000100016$$

$$S_{5000} \approx 0.693047197 \quad \text{with} \quad E_{5000} \approx \;\;\; 0.000099984$$

$$S_{5001} \approx 0.693247157 \quad \text{with} \quad E_{5001} \approx -0.000099976.$$

Notice that the series is slowly converging and that the accuracy of the partial sums S_{5000} and S_{5001} is not much better than S_{4999}. In the exercises there will be a series that will converge faster.

Methods for Evaluating a Polynomial

There are several mathematically equivalent ways to evaluate a polynomial. Consider, for example, the function

$$f(x) = (x - 1)^8. \tag{15}$$

The binomial formula can be used to expand $f(x)$ in powers of x, and the result is

$$g(x) = x^8 - 8x^7 + 28x^6 - 56x^5 + 70x^4 - 56x^3 + 28x^2 - 8x + 1. \tag{16}$$

A third way to evaluate $f(x)$ is **Horner's method**, which is also called **nested multiplication**. It is explained in detail in Section 1.1. When applied to formula (16), nested multiplication permits us to write

$$h(x) = (((((((x - 8)x + 28)x - 56)x + 70)x - 56)x + 28)x - 8)x + 1. \tag{17}$$

To evaluate $f(x)$ requires one subtraction and an eighth-power calculation. Notice that the eighth power can be computed by seven multiplications. The function $g(x)$ can be evaluated using powers of x in the calculations, but this could be time consuming. If the powers of x were stored, then $g(x)$ could be computed using seven multiplications and eight additions or subtractions. The function $h(x)$ is also seen to require seven multiplications and eight additions or subtractions.

Algorithm 4.1 uses the idea of storing the terms $(x - x_0)^k/k!$. If this is done in the sequential order

$$\frac{(x - x_0)}{1!},$$

$$\frac{(x - x_0)^2}{2!} = \frac{(x - x_0)(x - x_0)}{2},$$

$$\vdots$$

$$\frac{(x - x_0)^k}{k!} = \frac{[(x - x_0)^{k-1}/(k - 1)!](x - x_0)}{k},$$

a larger value of k can be used before the computer encounters an exponent overflow or underflow.

Theorem 4.3   (Taylor Series).   Assume that $f(x)$ is analytic and has continuous derivatives of all orders $N = 1, 2, \ldots$, on an interval (a, b) containing x_0. Suppose that the Taylor polynomials in (2) tend to a limit

$$S(x) = \lim_{N \to \infty} P_N(x) = \lim_{N \to \infty} \sum_{k=0}^{N} \frac{f^{(k)}(x_0)}{k!} (x - x_0)^k; \tag{18}$$

then $f(x)$ has the Taylor series expansion:

$$f(x) = \sum_{k=0}^{\infty} \frac{f^{(k)}(x_0)}{k!} (x - x_0)^k. \tag{19}$$

Proof. This follows directly from the definition of convergence of series in Section 1.1. The limit condition is often stated by saying that the error term must go to zero as N goes to infinity. Therefore, a necessary and sufficient condition for (19) to hold is that

$$\lim_{N \to \infty} E_N(x) = \lim_{N \to \infty} \frac{f^{(N+1)}(c)(x - x_0)^{N+1}}{(N + 1)!} = 0. \tag{20}$$

where c depends on N and x.

Theorem 4.4   (Ratio Test).   For every **power series**

$$f(x) = \sum_{k=0}^{\infty} a_k(x - x_0)^k \tag{21}$$

there exists a number R with $0 \le R \le +\infty$, called the **radius of convergence** of the series. It is used to determine regions where the series converges and diverges:

If $|x - x_0| < R$, then the series in (21) converges. (22)

If $|x - x_0| > R$, then the series in (21) diverges. (23)

The value R can often be found by using the **ratio test**:

$$R = \lim_{k \to \infty} \frac{|a_k|}{|a_{k+1}|} \qquad \text{when this limit exists.} \qquad (24)$$

Algorithm 4.1   (Evaluation of a Taylor Series).   To approximate

$$P(x) = \sum_{k=0}^{\infty} \frac{f^{(k)}(x_0)(x - x_0)^k}{k!} = \sum_{k=0}^{\infty} \frac{D(k)(x - x_0)^k}{k!}$$

by computing a partial sum with at most N terms.

```
            READ   Tol                          {Termination criterion, e.g., 10^-7}
            READ   N                             {Maximum degree, e.g., 100}
            READ   X0                            {Point of expansion}
            READ   D(0), . . . , D(N)            {Get the derivatives of f(x)}
            Close := 1                           {Closeness of consecutive partial sums}

            K := 0                               {Initialize the counter}
            Sum := D(0)                          {Initialize the variable}
            Prod := 1                            {Variable that holds (x − x0)^k/k!}
            INPUT   X                            {The independent variable}
            IF   X = X0   THEN   CLOSE := 0
WHILE   CLOSE ≥ Tol and K < N   DO
            K := K + 1
            Prod := Prod*(X−X0)/K                {Make the factor (x − x0)^k/k!}
            WHILE   D(K) = 0 and K < N   DO      {If D(K) = 0, then continue
                    K := K + 1
                    Prod := Prod*(X−X0)/K         to build up (x − x0)^k/k!}
            Term := D(K)*Prod
            IF   Term ≠ 0   THEN   Close := |Term|
            Sum := Sum + Term

IF      Close < Tol and K ≤ N   THEN                                  {Output}
            PRINT  'The sum of the Taylor polynomial is'   Sum
                   'Consecutive partial sums are closer than'   Close
    ELSE
            PRINT  'The current partial sum is'   Sum
                   'Convergence has NOT been achieved.'
ENDIF
```

EXERCISES FOR TAYLOR SERIES AND CALCULATION OF FUNCTIONS

1. Let $f(x) = \sin(x)$ and apply Theorem 4.1.

(a) Use $x_0 = 0$ and find $P_5(x)$, $P_7(x)$, and $P_9(x)$.

(b) Use the polynomials in part (a) and evaluate $P_5(0.5)$, $P_5(1.0)$, $P_7(0.5)$, $P_7(1.0)$, $P_9(0.5)$, and $P_9(1.0)$ and compare with $\sin(0.5)$ and $\sin(1.0)$. [Evaluate $\sin(x)$ with x in radians.]

(c) Show that if $|x| \leq 1$, the approximation

$$\sin(x) \approx x - \frac{x^3}{3!} + \frac{x^5}{5!} - \frac{x^7}{7!} + \frac{x^9}{9!}$$

has the error bound $|E_9| < 1/10! = 0.000000275573.\ \ldots$

(d) Use $x_0 = \pi/4$ and find $P_5(x)$, which involves powers of $(x - \pi/4)$.

(e) Use the polynomial in part (d) and evaluate $P_5(0.5)$ and $P_5(1.0)$ and compare with $\sin(0.5)$ and $\sin(1.0)$.

2. Let $f(x) = \cos(x)$ and apply Theorem 4.1.

(a) Use $x_0 = 0$ and find $P_4(x)$, $P_6(x)$, and $P_8(x)$.

(b) Use the polynomials in part (a) and evaluate $P_4(0.5)$, $P_4(1.0)$, $P_6(0.5)$, $P_6(1.0)$, $P_8(0.5)$, and $P_8(1.0)$ and compare with $\cos(0.5)$ and $\cos(1.0)$. [Evaluate $\cos(x)$ with x in radians.]

(c) Show that if $|x| \leq 1$, the approximation

$$\cos(x) \approx 1 - \frac{x^2}{2!} + \frac{x^4}{4!} - \frac{x^6}{6!} + \frac{x^8}{8!}$$

has the error bound $|E_8| < 1/9! = 0.00000275573.\ \ldots$

(d) Use $x_0 = \pi/4$ and find $P_4(x)$, which involves powers of $(x - \pi/4)$.

(e) Use the polynomial in part (d) and evaluate $P_4(0.5)$ and $P_4(1.0)$ and compare with $\cos(0.5)$ and $\cos(1.0)$.

3. Does $f(x) = x^{1/2}$ have a Taylor series expansion about $x_0 = 0$? Justify your answer.

4. (a) Find the Taylor polynomial of degree $N = 5$ for $f(x) = 1/(1 + x)$ expanded about $x_0 = 0$.

(b) Find the error term for the polynomial in part (a).

5. Find the Taylor polynomial of degree $N = 3$ for $f(x) = \exp(-x^2/2)$ expanded about $x_0 = 0$.

6. Use $f(x) = (2 + x)^{1/2}$ and apply Theorem 4.1.

(a) Find the Taylor polynomial $P_3(x)$ expanded about the value $x_0 = 2$ that involves powers of $(x - 2)$.

(b) Use $P_3(x)$ to find an approximation to $3^{1/2}$.

(c) Find the maximum value of $|f^{(4)}(c)|$ on the interval $1 \leq c \leq 3$ and find a bound for $|E_3(x)|$.

7. (a) Use the geometric series

$$\frac{1}{1 + t^2} = 1 - t^2 + t^4 - t^6 + t^8 - \cdots \qquad \text{or } |t| < 1,$$

and integrate both sides term by term to obtain

$$\arctan(x) = x - \frac{x^3}{3} + \frac{x^5}{5} - \frac{x^7}{7} + \cdots \qquad \text{for } |x| < 1.$$

(b) Use $\pi/6 = \arctan(3^{-1/2})$ and the series in part (a) to show that

$$\pi = 3^{1/2} \times 2\left(1 - \frac{3^{-1}}{3} + \frac{3^{-2}}{5} - \frac{3^{-3}}{7} + \frac{3^{-4}}{9} - \cdots\right).$$

(c) Use the series in part (b) to compute π accurate to eight digits.
Fact. $\pi \approx 3.1415926535897932384.$ . . . .

8. Use $f(x) = \ln(1 + x)$ and $x_0 = 0$, and apply Theorem 4.1.
   (a) Show that $f^{(k)}(x) = (-1)^{k-1}[(k-1)!]/(1+x)^k$.
   (b) Show that the Taylor polynomial of degree N is

$$P_N(x) = x - \frac{x^2}{2} + \frac{x^3}{3} - \frac{x^4}{4} + \cdots + \frac{(-1)^{N-1}x^N}{N}.$$

   (c) Show that the error term for $P_N(x)$ is

$$E_N(x) = \frac{(-1)^N x^{N+1}}{(N+1)(1+c)^{N+1}}.$$

   (d) Evalute $P_3(0.5)$, $P_6(0.5)$, and $P_9(0.5)$. Compare with $\ln(1.5)$.
   (e) Show that if $0.0 \le x \le 0.5$, the approximation

$$\ln(x) \approx x - \frac{x^2}{2} + \frac{x^3}{3} - \frac{x^4}{4} + \cdots + \frac{x^7}{7} - \frac{x^8}{8} + \frac{x^9}{9}$$

   has the error bound $|E_9| \le (0.5)^{10}/10 = 0.00009765.$ . . . .

9. (a) Let N go to infinity in Exercise 8(b) and obtain

$$\ln(1+x) = x - \frac{x^2}{2} + \frac{x^3}{3} - \cdots + \frac{(-1)^{n-1}x^n}{n} + \cdots .$$

   (b) Change the variable in part (a) and obtain the expansion

$$\ln(1-x) = -x - \frac{x^2}{2} - \frac{x^3}{3} - \cdots - \frac{x^n}{n} - \cdots .$$

   (c) Subtract the series in part (b) from the series in part (a) to obtain

$$\ln\left(\frac{1+x}{1-x}\right) = 2\left(x + \frac{x^3}{3} + \cdots + \frac{x^{2n-1}}{2n-1} + \cdots\right).$$

   (d) Set $x = \frac{1}{3}$ in part (c) and show that

$$\ln(2) = 2\left(3^{-1} + \frac{3^{-3}}{3} + \frac{3^{-5}}{5} + \cdots + \frac{3^{-2n+1}}{2n-1} + \cdots\right).$$

   (e) Use the ratio test and show that the series in parts (a) and (b) converge when $|x| < 1$.

10. *Binomial series.* Let $f(x) = (1+x)^p$ and $x_0 = 0$.
   (a) Show that $f^{(k)}(x) = p(p-1)\cdots(p-k+1)(1+x)^{p-k}$.
   (b) Show that the Taylor polynomial of degree N is

$$P_N(x) = 1 + px + \frac{p(p-1)x^2}{2!} + \cdots + \frac{p(p-1)\cdots(p-N+1)x^N}{N!}.$$

(c) Show that $E_N(x) = p(p - 1) \cdots (p - N)x^{N+1}/[(1 + c)^{N+1-p}(N + 1)!]$.

(d) Set $p = \frac{1}{2}$ and compute $P_2(0.5)$, $P_4(0.5)$, and $P_6(0.5)$. Compare with $(1.5)^{1/2}$.

(e) Show that if $0.0 \leq x \leq 0.5$, then the approximation

$$(1 + x)^{1/2} \approx 1 + \frac{x}{2} - \frac{x^2}{8} + \frac{x^3}{16} - \frac{5x^4}{128} + \frac{7x^5}{256}$$

has the error bound $|E_5| \leq (0.5)^6(21/1024) = 0.0003204. \ldots$

(f) Show that if $p = N$ is a positive integer,

$$P_N(x) = 1 + Nx + \frac{N(N - 1)x^2}{2!} + \cdots + Nx^{N-1} + x^N.$$

Notice that this is the familiar binomial expansion.

11. Let $f(x) = x^3 - 2x^2 + 2x$. Find the Taylor polynomial $P_3(x)$ expanded about $x_0 = 1$ that involves powers of $(x - 1)$.

12. Show that

$$f(x) = 1 + 3(x - 1) + 3(x - 1)^2 + (x - 1)^3$$

$$g(x) = -1 + 3(x + 1) - 3(x + 1)^2 + (x + 1)^3$$

are two representations for the polynomial $p(x) = x^3$.

13. Assume that the Taylor polynomial for $f(x)$ of degree $N = 3$ expanded about the value $x_0 = 2$ is given by $P_3(x) = 2 + 3(x - 2) - 5(x - 2)^2 + 4(x - 2)^3$.

(a) Find $P_3(2.1)$ and $P_3(1.9)$. **(b)** Find $[P_3(2.1) - P_3(1.9)]/0.2$.

(c) Show that $f'(2) = 3$. *Remark.* An approximation to $f'(2)$ is given in part (b).

14. Finish the proof of Corollary 4.1 by writing down the expression for $P_N^{(k)}(x)$ and showing that

$$P_N^{(k)}(x_0) = f^{(k)}(x_0) \qquad \text{for } k = 2, 3, \ldots, N.$$

15. Write a report on the proof of Theorem 4.2 (you can look it up in a calculus book). Be sure to include a discussion about the even partial sums $\{S_{2k}\}$ and the odd partial sums $\{S_{2k+1}\}$. In your own words, describe why the limit exists.

16. Modify Algorithm 4.1 so that it will find the sum of a power series.

17. In your own words, describe the similarities and differences between Taylor series and power series.

The Taylor polynomial of degree $N = 2$ for a function $f(x, y)$ of two variables x and y expanded about the point (a, b) is

$$P_2(x, y) = f(a, b) + f_x(a, b)(x - a) + f_y(a, b)(y - b)$$

$$+ \frac{f_{xx}(a, b)(x - a)^2}{2} + f_{xy}(a, b)(x - a)(y - b) + \frac{f_{yy}(a, b)(y - b)^2}{2}.$$

18. **(a)** Show that the Taylor polynomial of degree $N = 2$ for $f(x, y) = y/x$ expanded about $(1, 1)$ is

$$P_2(x, y) = 1 - (x - 1) + (y - 1) + 2(x - 1)^2 - (x - 1)(y - 1).$$

(b) Find $P_2(1.05, 1.1)$ and compare with $f(1.05, 1.1)$.

19. (a) Show that the Taylor polynomial of degree $N = 2$ for $f(x, y) = (1 + x - y)^{1/2}$ expanded about $(0, 0)$ is

$$P_2(x, y) = 1 + \frac{x}{2} - \frac{y}{2} - \frac{x^2}{8} + \frac{xy}{4} - \frac{y^2}{8}.$$

(b) Find $P_2(0.04, 0.08)$ and compare with $f(0.04, 0.08)$.

Exercises 20 and 21 form a proof of Taylor's theorem.

20. Let $g(t)$ and its derivatives $g^{(k)}(t)$, for $k = 1, 2, \ldots, N + 1$, be continuous on the interval (a, b), which contains x_0. Suppose that there exist two distinct points x and x_0 such that $g(x) = 0$, $g(x_0) = 0$, $g'(x_0) = 0$, $g''(x_0) = 0, \ldots, g^{(N)}(x_0) = 0$. Prove that there exists a value c that lies between x_0 and x such that $g^{(N+1)}(c) = 0$.

   Remark. $g(t)$ is a function of t and the values x and x_0 are to be treated as constants with respect to the variable t.

   Hint. Use Rolle's theorem[†] on the interval with endpoints x_0 and x to find the number c_1 such that $g'(c_1) = 0$. Then use Rolle's theorem applied to the function $g'(t)$ on the interval with endpoints x_0, c_1 to find the number c_2 such that $g''(c_2) = 0$. Inductively repeat the process until the number c_{N+1} is found such that $g^{(N+1)}(c_{N+1}) = 0$.

21. Use the result of Exercise 20 and the special function

$$g(t) = f(t) - P_N(t) - E_N(x) \frac{(t - x_0)^{N+1}}{(x - x_0)^{N+1}},$$

where $P_N(x)$ is the Taylor polynomial of degree N to prove that the error term $E_N(x) = f(x) - P_N(x)$ has the form

$$E_N(x) = f^{(N+1)}(c) \frac{(x - x_0)^{N+1}}{(N + 1)!}.$$

Hint. Find $g^{(N+1)}(t)$ and evaluate it at $t = c$.

4.2 INTRODUCTION TO INTERPOLATION

In Section 4.1 we saw how a Taylor polynomial can be used to approximate the function $f(x)$. The information needed to construct the Taylor polynomial is the value of f and its derivatives at x_0. A shortcoming is that the higher-order derivatives must be known, and often they are either not available or they are hard to compute.

Suppose that the function $y = f(x)$ is known at the $N + 1$ points $(x_0, y_0), \ldots,$ (x_N, y_N), where the values x_k are spread out over the interval $[a, b]$ and satisfy

$$a \le x_0 < x_1 < \cdots < x_N \le b \quad \text{and} \quad y_k = f(x_k).$$

A polynomial $P(x)$ of degree N shall be constructed which passes through these $N + 1$ points. In the construction, only the numerical values x_k and y_k are needed. Hence the

---

[†]*Rolle's theorem.* Let f be continuous on $[a, b]$ and differentiable on (a, b). If $f(a) = f(b)$, there exists a number c in (a, b) such that $f'(c) = 0$.

higher-order derivatives are not necessary. The polynomial $P(x)$ can be used to approximate $f(x)$ over the entire interval $[a, b]$. However, if the error function $E(x) = f(x) - P(x)$ is required, then we will need to know $f^{(N+1)}(x)$ and a bound for its magnitude, that is,

$$M = \max\{|f^{(N+1)}(x)|: \quad \text{for } a \le x \le b\}.$$

Situations in statistical and scientific analysis arise where the function $y = f(x)$ is available only at $N + 1$ tabulated points (x_k, y_k) and a method is needed to approximate $f(x)$ at nontabulated abscissas. If there is a significant amount of error in the tabulated values, then the methods of curve fitting in Chapter 5 should be considered. On the other hand, if the points (x_k, y_k) are known to a high degree of accuracy, then the polynomial curve $y = P(x)$ that passes through them can be considered. When $x_0 < x < x_N$ the approximation $P(x)$ is called an **interpolated value**. If either $x < x_0$ or $x_N < x$, then $P(x)$ is called an **extrapolated value**. Polynomials are used to design software algorithms to approximate functions, for numerical differentiation, for numerical integration, and for making computer-drawn curves that must pass through specified points.

Let us briefly mention how to evaluate the polynomial $P(x)$:

$$P(x) = a_N x^N + a_{N-1} x^{N-1} + \cdots + a_2 x^2 + a_1 x + a_0. \tag{1}$$

Horner's method of synthetic division is an efficient way to evaluate $P(x)$. The derivative $P'(x)$ is

$$P'(x) = N a_N x^{N-1} + (N - 1) a_{N-1} x^{N-2} + \cdots + 2 a_2 x + a_1 \tag{2}$$

and the indefinite integral $I(x)$, which satisfies $I'(x) = P(x)$, is

$$I(x) = \frac{a_N x^{N+1}}{N + 1} + \frac{a_{N-1} x^N}{N} + \cdots + \frac{a_2 x^3}{3} + \frac{a_1 x^2}{2} + a_0 x + C, \tag{3}$$

where C is the constant of integration. Algorithm 4.2 shows how to adapt Horner's method to evaluate $P'(x)$ and $I(x)$.

Example 4.5

The polynomial $P(x) = -0.02x^3 + 0.2x^2 - 0.4x + 1.28$ passes through the four points $(1, 1.06)$, $(2, 1.12)$, $(3, 1.34)$, and $(5, 1.78)$. Find (a) $P(4)$, (b) $P'(4)$, (c) $\int_1^4 P(x)\,dx$, and (d) $P(5.5)$.
(e) Show how to find the coefficients of $P(x)$.

Solution.   Use Algorithm 4.2(i)–(iii) with $x = 4$.

(a) $b_3 = a_3 = -0.02$

$\quad b_2 = a_2 + b_3 x = 0.2 + (-0.02)(4) = 0.12$

$\quad b_1 = a_1 + b_2 x = -0.4 + (0.12)(4) = 0.08$

$\quad b_0 = a_0 + b_1 x = 1.28 + (0.08)(4) = 1.60.$

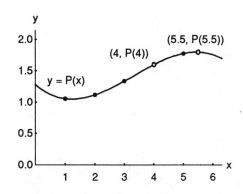

Figure 4.8 (a) The approximating polynomial $P(x)$ can be used for interpolation at the point $(4, P(4))$ and extrapolation at the point $(5.5, P(5.5))$.

Figure 4.8 (b) The approximating polynomial $P(x)$ is differentiated and $P'(x)$ is used to find the slope at the interpolation point $(4, P(4))$.

The interpolated value is $P(4) = 1.60$ [see Figure 4.8(a)].

(b) $d_2 = 3a_3 = -0.06$

$d_1 = 2a_2 + d_2x = 0.4 + (-0.06)(4) = 0.16$

$d_0 = a_1 + d_1x = -0.4 + (0.16)(4) = 0.24.$

The numerical derivative is $P'(4) = 0.24$ [see Figure 4.8(b)].

(c) $i_4 = \dfrac{a_3}{4} = -0.005$

$i_3 = \dfrac{a_2}{3} + i_4x = 0.06666667 + (-0.005)(4) = 0.04666667$

$i_2 = \dfrac{a_1}{2} + i_3x = -0.2 + (0.04666667)(4) = -0.01333333$

$i_1 = a_0 + i_2x = 1.28 + (-0.01333333)(4) = 1.22666667$

$i_0 = 0 + i_1x = 0 + (1.22666667)(4) = 4.90666667.$

Hence $I(4) = 4.90666667$. Similarly, $I(1) = 1.14166667$. Therefore, $\int_1^4 P(x)\, dx = I(4) - I(1) = 3.765$ [see Figure 4.8(c)].

(d) Use Algorithm 4.2(i) with $x = 5.5$

$b_3 = a_3 = -0.02$

$b_2 = a_2 + b_3x = 0.2 + (-0.02)(5.5) = 0.09$

$b_1 = a_1 + b_2x = -0.4 + (0.09)(5.5) = 0.095$

$b_0 = a_0 + b_1x = 1.28 + (0.095)(5.5) = 1.8025.$

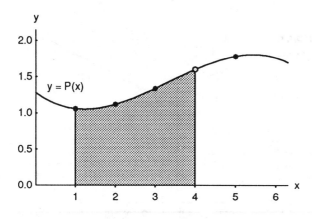

Figure 4.8 (c) The approximating polynomial $P(x)$ is integrated and its antiderivative is used to find the area under the curve for $1 \leq x \leq 4$.

The extrapolated value is $P(5.5) = 1.8025$ [see Figure 4.8(a)].

(e) The methods of Chapter 3 can be used to find the coefficients. Assume that $P(x) = A + Bx + Cx^2 + Dx^3$; then at each value $x = 1, 2, 3, 5$ we get a linear equation involving A, B, C, and D.

$$\text{At } x = 1: \quad A + 1B + 1C + 1D = 1.06$$
$$\text{At } x = 2: \quad A + 2B + 4C + 8D = 1.12$$
$$\text{At } x = 3: \quad A + 3B + 9C + 27D = 1.34 \tag{4}$$
$$\text{At } x = 5: \quad A + 5B + 25C + 125D = 1.78.$$

The solution to (4) is $A = 1.28$, $B = -0.4$, $C = 0.2$, and $D = -0.02$.

This method for finding the coefficients is mathematically sound, but sometimes the matrix is difficult to solve accurately. In this chapter we design algorithms specifically for polynomials.

Let us return to the topic of using a polynomial to calculate approximations to a known function. In Section 4.1 we saw that the fifth-degree Taylor polynomial for $f(x) = \ln(1 + x)$ is

$$T(x) = x - \frac{x^2}{2} + \frac{x^3}{3} - \frac{x^4}{4} + \frac{x^5}{5}. \tag{5}$$

If $T(x)$ is used to approximate $\ln(1 + x)$ on the interval $[0, 1]$, then the error is 0 at $x = 0$ and is largest when $x = 1$ (see Table 4.5). Indeed, the error between $T(1)$ and the correct value $\ln(2)$ is 13%. We seek a polynomial of degree 5 that will approximate $\ln(1 + x)$ better over the interval $[0, 1]$. The polynomial $P(x)$ in Example 4.6 is an interpolating polynomial and will approximate $\ln(1 + x)$ with an error no bigger than 0.00002385 over the interval $[0, 1]$.

TABLE 4.5 Values of the Taylor polynomial $T(x)$ of Degree 5, and the Function $\ln(1 + x)$ and the Error $\ln(1 + x) - T(x)$ on $[0, 1]$

x	Taylor polynomial, $T(x)$	Function, $\ln(1 + x)$	Error, $\ln(1 + x) - T(x)$
0.0	0.00000000	0.00000000	0.00000000
0.2	0.18233067	0.18232156	−0.00000911
0.4	0.33698133	0.33647224	−0.00050909
0.6	0.47515200	0.47000363	−0.00514837
0.8	0.61380267	0.58778666	−0.02601601
1.0	0.78333333	0.69314718	−0.09018615

Example 4.6

Consider the function $f(x) = \ln(1 + x)$ and the polynomial

$$P(x) = 0.02957206x^5 - 0.12895295x^4 + 0.28249626x^3 - 0.48907554x^2 + 0.99910735x$$

based on the six nodes $x_k = k/5$ for $k = 0, 1, \ldots, 5$. The following are empirical descriptions of the approximation $P(x) \approx \ln(1 + x)$.

1. $P(x_k) = f(x_k)$ at each node (see Table 4.6).
2. The maximum error on the interval $[-0.1, 1.1]$ occurs at $x = -0.1$ and $|\text{error}| \leq 0.00026334$ for $-0.1 \leq x \leq 1.1$ (see Figure 4.10). Hence the graph of $y = P(x)$ would appear identical to that of $y = \ln(1 + x)$ (see Figure 4.9).

TABLE 4.6 Values of the Approximating Polynomial $P(x)$ of Example 4.6, and the Function $f(x) = \ln(1 + x)$ and the Error $E(x)$ on $[-0.1, 1.1]$

x	Approximating polynomial $P(x)$	Function, $f(x) = \ln(1 + x)$	Error function, $E(x) = f(x) - P(x)$
−0.1	−0.10509718	−0.10536052	−0.00026334
0.0	0.00000000	0.00000000	0.00000000
0.1	0.09528988	0.09531018	0.00002030
0.2	0.18232156	0.18232156	0.00000000
0.3	0.26237015	0.26236426	−0.00000589
0.4	0.33647224	0.33647224	0.00000000
0.5	0.40546139	0.40546511	0.00000372
0.6	0.47000363	0.47000363	0.00000000
0.7	0.53063292	0.53062825	−0.00000467
0.8	0.58778666	0.58778666	0.00000000
0.9	0.64184118	0.64185389	0.00001271
1.0	0.69314718	0.69314718	0.00000000
1.1	0.74206529	0.74193734	−0.00012795

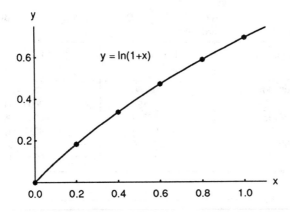

Figure 4.9   The graph of $y = P(x)$, which "lies on top" of the graph $y = \ln(1 + x)$.

3. The maximum error on the interval $[0, 1]$ occurs at $x = 0.06472456$ and $|\text{error}| \leq 0.00002385$ for $0 \leq x \leq 1$ (see Figure 4.10).

Remark. At a node x_k we have $f(x_k) = P(x_k)$. Hence $E(x_k) = 0$ at a node. The graph of $E(x) = f(x) - P(x)$ looks like a vibrating string, with the nodes being the abscissa where there is no displacement.

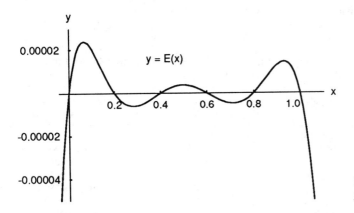

Figure 4.10   The graph of the error $y = E(x) = \ln(1 + x) - P(x)$.

Algorithm 4.2   (Polynomial Calculus).   To evaluate the polynomial $P(x)$, its derivative $P'(x)$, and its integral $\int P(x)\,dx$ by performing synthetic division.

```
INPUT   N                                    {Degree of P(x)}
INPUT   A(0), A(1), . . . , A(N)             {Coefficients of P(x)}
INPUT   C                                    {Constant of integration}
INPUT   X                                    {Independent variable}
```

(i) Algorithm to evaluate P(x)
B(N) := A(N)
FOR K = N−1 DOWNTO 0 DO
    B(K) := A(K) + B(K+1)*X
PRINT "The value P(x) is", B(0)
	Space-saving version:
	Poly := A(N)
	FOR K = N−1 DOWNTO 0 DO
	    Poly := A(K) + Poly*X
	"The value P(x) is", Poly

(ii) Algorithm to evaluate P′(x)
D(N−1) := N*A(N)
FOR K = N−1 DOWNTO 1 DO
    D(K−1) := K*A(K) + D(K)*X
PRINT "The value P′(x) is", D(0)
	Space-saving version:
	Deriv := N*A(N)
	FOR K = N−1 DOWNTO 1 DO
	    Deriv := K*A(K) + Deriv*X
	"The value P′(x) is", Deriv

(iii) Algorithm to evaluate I(x)
I(N+1) := A(N)/[N+1]
FOR K = N DOWNTO 1 DO
    I(K) := A(K−1)/K + I(K+1)*X
I(0) := C + I(1)*X
PRINT "The value of I(x) is", I(0)
	Space-saving version:
	Integ := A(N)/[N+1]
	FOR K = N DOWNTO 1 DO
	    Integ := A(K−1)/K + Integ*X
	Integ := C + Integ*X
	"The value of I(x) is", Integ

EXERCISES FOR INTRODUCTION TO INTERPOLATION

1. Consider $P(x) = -0.02x^3 + 0.1x^2 - 0.2x + 1.66$, which passes through the four points (1, 1.54), (2, 1.5), (3, 1.42), and (5, 0.66).
   (a) Find $P(4)$.
   (b) Find $P'(4)$.
   (c) Find the integral of $P(x)$ taken over [1, 4].
   (d) Find the extrapolated value $P(5.5)$.
   (e) Show how to find the coefficients of $P(x)$.

2. Consider $P(x) = -0.04x^3 + 0.14x^2 - 0.16x + 2.08$, which passes through the four points (0, 2.08), (1, 2.02), (2, 2.00), and (4, 1.12).
   (a) Find $P(3)$.
   (b) Find $P'(3)$.
   (c) Find the integral of $P(x)$ taken over [0, 3].
   (d) Find the extrapolated value $P(4.5)$.
   (e) Show how to find the coefficients of $P(x)$.

3. Consider $P(x) = -0.029166667x^3 + 0.275x^2 - 0.570833333x + 1.375$, which passes through (1, 1.05), (2, 1.10), (3, 1.35), and (5, 1.75).
   (a) Show that the ordinates 1.05, 1.1, 1.35, and 1.75 differ from those of Example 4.5 by less than 1.8%, yet the coefficients of x^3 and x differ by more than 42%.
   (b) Find $P(4)$ and compare with Example 4.5.
   (c) Find $P'(4)$ and compare with Example 4.5.
   (d) Find the integral of $P(x)$ taken over [1, 4] and compare with Example 4.5.
   (e) Find the extrapolated value $P(5.5)$ and compare with Example 4.5.
   Remark. Part (a) shows that the computation of the coefficients of an interpolating polynomial is an ill-conditioned problem.

In Exercises 4–6, for the given function $f(x)$, the fifth-degree polynomial $P(x)$ passes through the six points $(0, f(0))$, $(0.2, f(0.2))$, $(0.4, f(0.4))$, $(0.6, f(0.6))$, $(0.8, f(0.8))$, and $(1, f(1))$. The six coefficients of $P(x)$ are $a_0, a_1, \ldots, a_5$, where

$$P(x) = a_5 x^5 + a_4 x^4 + a_3 x^3 + a_2 x^2 + a_1 x + a_0.$$

(a) Use Algorithm 4.1(i) to compute the interpolated values $P(0.3)$, $P(0.4)$, and $P(0.5)$ and compare with $f(0.3)$, $f(0.4)$, and $f(0.5)$.
(b) Use Algorithm 4.1(i) to compute the extrapolated values $P(-0.1)$ and $P(1.1)$ and compare with $f(-0.1)$ and $f(1.1)$.
(c) Use a computer and make a table of values for $P(x_k)$, $f(x_k)$, and $E(x_k)$, where $x_k = k/100$ for $k = 0, 1, \ldots, 100$.
(d) Use a computer to solve the 6×6 system of linear equations

$$a_0 + a_1 x_j^1 + a_2 x_j^2 + a_3 x_j^3 + a_4 x_j^4 + a_5 x_j^5 = f(x_j)$$

using $x_j = (j - 1)/5$ and $j = 1, 2, 3, 4, 5, 6$ for the six unknowns $\{a_k\}_{k=0}^5$.

4. If $f(x) = \exp(x)$, then the six coefficients are

$$a_5 = 0.01385431, \qquad a_2 = 0.49906876,$$

$$a_4 = 0.03486637, \qquad a_1 = 1.00008255,$$

$$a_3 = 0.17040984, \qquad a_0 = 1.00000000.$$

5. If $f(x) = \sin(x)$, then the six coefficients are

$$a_5 = 0.00725244, \qquad a_2 = 0.00024394,$$

$$a_4 = 0.00161306, \qquad a_1 = 0.99997802,$$

$$a_3 = -0.16761647, \qquad a_0 = 0.00000000.$$

6. If $f(x) = \cos(x)$, then the six coefficients are

$$a_5 = -0.00396206, \qquad a_2 = -0.49944788,$$

$$a_4 = 0.04604659, \qquad a_1 = -0.00004812,$$

$$a_3 = -0.00228623, \qquad a_0 = 1.00000000.$$

7. Write a report on the approximation of functions. See References [34, 44, 114, 149, 157, 161, and 182.]

4.3 LAGRANGE APPROXIMATION

Interpolation means to estimate a missing function value by taking a weighted average of known function values at neighboring points. Linear interpolation uses a line segment that passes through two points. The slope between (x_0, y_0) and (x_1, y_1) is $m = (y_1 - y_0)/(x_1 - x_0)$, and the point-slope formula for the line $y = y_0 + m(x - x_0)$ can be rearranged as

$$y = P(x) = y_0 + (y_1 - y_0)\frac{x - x_0}{x_1 - x_0}. \tag{1}$$

When formula (1) is expanded, the result is a polynomial of degree ≤ 1. Evaluation of $P(x)$ at x_0 and x_1 produces y_0 and y_1, respectively,

$$P(x_0) = y_0 + (y_1 - y_0)0 = y_0 \quad \text{and} \quad P(x_1) = y_0 + (y_1 - y_0)1 = y_1. \tag{2}$$

The French mathematician Joseph Louis Lagrange used a slightly different method to find this polynomial. He noticed that it could be written as

$$y = P_1(x) = y_0 \frac{x - x_1}{x_0 - x_1} + y_1 \frac{x - x_0}{x_1 - x_0}. \tag{3}$$

Each term on the right side of (3) involves a linear factor; hence the sum is a polynomial of degree ≤ 1. The quotients in (3) are denoted by

$$L_{1,0}(x) = \frac{x - x_1}{x_0 - x_1} \quad \text{and} \quad L_{1,1}(x) = \frac{x - x_0}{x_1 - x_0}. \tag{4}$$

Computation reveals that $L_{1,0}(x_0) = 1, L_{1,0}(x_1) = 0, L_{1,1}(x_0) = 0,$ and $L_{1,1}(x_1) = 1$ so that the polynomial $P_1(x)$ in (3) also passes through the two given points:

$$P_1(x_0) = y_0 + y_1 0 = y_0 \quad \text{and} \quad P_1(x_1) = y_0 0 + y_1 = y_1. \tag{5}$$

The terms $L_{1,0}(x)$ and $L_{1,1}(x)$ in (4) are called the **Lagrange coefficient polynomials** based on the nodes x_0 and x_1. Using this notation (3) can be written in summation form:

$$P_1(x) = \sum_{k=0}^{1} y_k L_{1,k}(x). \tag{6}$$

Suppose that the ordinates y_k are computed with the formula $y_k = f(x_k)$. If $P_1(x)$ is used to approximate $f(x)$ over the interval $[x_0, x_1]$, we call the process **interpolation**. If $x < x_0$ (or $x_1 < x$), then using $P_1(x)$ to approximate $f(x)$ is called **extrapolation**. The next example illustrates these concepts.

Example 4.7

Consider the graph $y = f(x) = \cos(x)$ over $[0.0, 1.2]$.
(a) Use the nodes $x_0 = 0.0, x_1 = 1.2$ to construct a linear interpolation polynomial $P_1(x)$.
(b) Use the nodes $x_0 = 0.2, x_1 = 1.0$ to construct a linear approximating polynomial $Q_1(x)$.

Solution. Using (3) with the abscissas $x_0 = 0.0$, $x_1 = 1.2$ and the ordinates $y_0 = \cos(0.0) = 1.000000$, $y_1 = \cos(1.2) = 0.362358$ produces

$$P_1(x) = 1.000000 \frac{x - 1.2}{0.0 - 1.2} + 0.362358 \frac{x - 0.0}{1.2 - 0.0}$$

$$= -0.833333(x - 1.2) + 0.301965(x - 0.0).$$

When the nodes $x_0 = 0.2$, $x_1 = 1.0$ with $y_0 = \cos(0.2) = 0.980067$, $y_1 = \cos(1.0) = 0.540302$ are used, the result is

$$Q_1(x) = 0.980067 \frac{x - 1.0}{0.2 - 1.0} + 0.540302 \frac{x - 0.2}{1.0 - 0.2}$$

$$= -1.225083(x - 1.0) + 0.675378(x - 0.2).$$

Figure 4.11(a) and (b) shows the graph $y = \cos(x)$ and compares it with $y = P_1(x)$ and $y = Q_1(x)$, respectively. Numerical computations are given in Table 4.7 and reveal that $Q_1(x)$ has less error at the points x_k that satisfy $0.1 \leq x_k \leq 1.1$. The largest tabulated error, $f(0.6) - P_1(0.6) = 0.144157$, is reduced to $f(0.6) - Q_1(0.6) = 0.065151$ by using $Q_1(x)$.

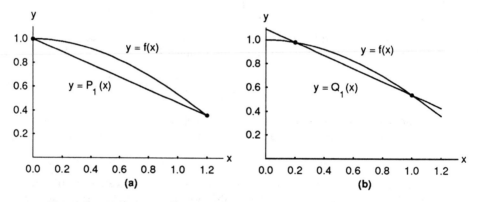

Figure 4.11 (a) The linear approximation of $y = P_1(x)$ where the nodes $x_0 = 0.0$ and $x_1 = 1.2$ are the endpoints of the interval $[a, b]$. (b) The linear approximation of $y = Q_1(x)$ where the nodes $x_0 = 0.2$ and $x_1 = 1.0$ lie inside the interval $[a, b]$.

The generalization of (6) is the construction of a polynomial $P_N(x)$ of degree at most N that passes through the $N + 1$ points (x_0, y_0), (x_1, y_1), . . . , (x_N, y_N) and has the form

$$P_N(x) = \sum_{k=0}^{N} y_k L_{N,k}(x), \tag{7}$$

where $L_{N,k}(x)$ is the Lagrange coefficient polynomial based on these nodes:

$$L_{N,k}(x) = \frac{(x - x_0) \cdots (x - x_{k-1})(x - x_{k+1}) \cdots (x - x_N)}{(x_k - x_0) \cdots (x_k - x_{k-1})(x_k - x_{k+1}) \cdots (x_k - x_N)}. \tag{8}$$

TABLE 4.7 Comparison of $f(x) = \cos(x)$ and the Linear Approximations $P_1(x)$ and $Q_1(x)$

x_k	$f(x_k) = \cos(x_k)$	$P_1(x_k)$	$f(x_k) - P_1(x_k)$	$Q_1(x_k)$	$f(x_k) - Q_1(x_k)$
0.0	1.000000	1.000000	0.000000	1.090008	−0.090008
0.1	0.995004	0.946863	0.048141	1.035037	−0.040033
0.2	0.980067	0.893726	0.086340	0.980067	0.000000
0.3	0.955336	0.840589	0.114747	0.925096	0.030240
0.4	0.921061	0.787453	0.133608	0.870126	0.050935
0.5	0.877583	0.734316	0.143267	0.815155	0.062428
0.6	0.825336	0.681179	0.144157	0.760184	0.065151
0.7	0.764842	0.628042	0.136800	0.705214	0.059628
0.8	0.696707	0.574905	0.121802	0.650243	0.046463
0.9	0.621610	0.521768	0.099842	0.595273	0.026337
1.0	0.540302	0.468631	0.071671	0.540302	0.000000
1.1	0.453596	0.415495	0.038102	0.485332	−0.031736
1.2	0.362358	0.362358	0.000000	0.430361	−0.068003

It is understood that the terms $(x - x_k)$ and $(x_k - x_k)$ do not appear on the right side of equation (8). It is appropriate to introduce the product notation for (8) and we write

$$L_{N,k}(x) = \frac{\displaystyle\prod_{\substack{j=0 \\ j \neq k}}^{N} (x - x_j)}{\displaystyle\prod_{\substack{j=0 \\ j \neq k}}^{N} (x_k - x_j)}. \tag{9}$$

Here the notation in (9) indicates that in the numerator, the product of the linear factors $(x - x_j)$ are to be formed, but the factor $(x_k - x_k)$ is to be left out (skipped). A similar construction occurs in the denominator.

A straightforward calculation shows that for each fixed k, the Lagrange coefficient polynomial $L_{N,k}(x)$ has the property

$$L_{N,k}(x_j) = 1 \quad \text{when } j = k \quad \text{and} \quad L_{N,k}(x_j) = 0 \quad \text{when } j \neq k. \tag{10}$$

Then direct substitution of these values into (7) is used to show that the polynomial curve $y = P_N(x)$ goes through (x_j, y_j):

$$P_N(x_j) = y_0 L_{N,0}(x_j) + \cdots + y_j L_{N,j}(x_j) + \cdots + y_N L_{N,N}(x_j) \tag{11}$$

$$= y_0 0 + \cdots + y_j 1 + \cdots + y_N 0 = y_j.$$

To show that $P_N(x)$ is unique, we invoke the fundamental theorem of algebra, which states that a polynomial $T(x)$ of degree $\leq N$ has at most N roots. In other words, if $T(x)$ is zero at $N + 1$ distinct abscissas, it is identically zero. Suppose that $P_N(x)$ is not unique and that there exists another polynomial of $Q_N(x)$ of degree $\leq N$ which passes through the $N + 1$ points. Form the difference polynomial $T(x) \equiv P_N(x) - Q_N(x)$. Observe that the polynomial $T(x)$ has degree $\leq N$ and that $T(x_j) = P_N(x_j) - Q_N(x_j) = y_j - y_j = 0$ for $j = 0, 1, \ldots, N$. Therefore, $T(x) \equiv 0$ and it follows that $Q_N(x) \equiv P_N(x)$.

When (7) is expanded the result is similar to (3). The Lagrange quadratic interpolating polynomial through the three points (x_0, y_0), (x_1, y_1), and (x_2, y_2) is

$$P_2(x) = y_0 \frac{(x - x_1)(x - x_2)}{(x_0 - x_1)(x_0 - x_2)} + y_1 \frac{(x - x_0)(x - x_2)}{(x_1 - x_0)(x_1 - x_2)} + y_2 \frac{(x - x_0)(x - x_1)}{(x_2 - x_0)(x_2 - x_1)}. \quad (12)$$

The Lagrange cubic interpolating polynomial through the four points (x_0, y_0), (x_1, y_1), (x_2, y_2), and (x_3, y_3) is

$$P_3(x) = y_0 \frac{(x - x_1)(x - x_2)(x - x_3)}{(x_0 - x_1)(x_0 - x_2)(x_0 - x_3)} + y_1 \frac{(x - x_0)(x - x_2)(x - x_3)}{(x_1 - x_0)(x_1 - x_2)(x_1 - x_3)}$$

$$+ y_2 \frac{(x - x_0)(x - x_1)(x - x_3)}{(x_2 - x_0)(x_2 - x_1)(x_2 - x_3)} + y_3 \frac{(x - x_0)(x - x_1)(x - x_2)}{(x_3 - x_0)(x_3 - x_1)(x_3 - x_2)}. \quad (13)$$

Example 4.8

Consider $y = f(x) = \cos(x)$ over $[0.0, 1.2]$. (a) Use the three nodes $x_0 = 0.0$, $x_1 = 0.6$, and $x_2 = 1.2$ to construct a quadratic approximation polynomial $P_2(x)$. (b) Use the four nodes $x_0 = 0.0$, $x_1 = 0.4$, $x_2 = 0.8$, and $x_3 = 1.2$ to construct a cubic polynomial $P_3(x)$.

Solution. Using $x_0 = 0.0$, $x_1 = 0.6$, $x_2 = 1.2$ and $y_0 = \cos(0.0) = 1.0$, $y_1 = \cos(0.6) = 0.825336$, $y_2 = \cos(1.2) = 0.362358$ in equation (12) produces

$$P_2(x) = 1.0 \frac{(x - 0.6)(x - 1.2)}{(0.0 - 0.6)(0.0 - 1.2)} + 0.825336 \frac{(x - 0.0)(x - 1.2)}{(0.6 - 0.0)(0.6 - 1.2)}$$

$$+ 0.362358 \frac{(x - 0.0)(x - 0.6)}{(1.2 - 0.0)(1.2 - 0.6)}$$

$$= 1.388889(x - 0.6)(x - 1.2) - 2.292599(x - 0.0)(x - 1.2)$$

$$+ 0.503275(x - 0.0)(x - 0.6).$$

Using $x_0 = 0.0$, $x_1 = 0.4$, $x_2 = 0.8$, $x_3 = 1.2$ and $y_0 = \cos(0.0) = 1.0$, $y_1 = \cos(0.4) = 0.921061$, $y_2 = \cos(0.8) = 0.696707$, $y_3 = \cos(1.2) = 0.362358$ in equation (13) produces

$$P_3(x) = 1.000000 \frac{(x - 0.4)(x - 0.8)(x - 1.2)}{(0.0 - 0.4)(0.0 - 0.8)(0.0 - 1.2)}$$

$$+ 0.921061 \frac{(x - 0.0)(x - 0.8)(x - 1.2)}{(0.4 - 0.0)(0.4 - 0.8)(0.4 - 1.2)}$$

$$+ 0.696707 \frac{(x - 0.0)(x - 0.4)(x - 1.2)}{(0.8 - 0.0)(0.8 - 0.4)(0.8 - 1.2)}$$

$$+ 0.362358 \frac{(x - 0.0)(x - 0.4)(x - 0.8)}{(1.2 - 0.0)(1.2 - 0.4)(1.2 - 0.8)}$$

$$= -2.604167(x - 0.4)(x - 0.8)(x - 1.2) + 7.195789(x - 0.0)(x - 0.8)(x - 1.2)$$

$$- 5.443021(x - 0.0)(x - 0.4)(x - 1.2) + 0.943640(x - 0.0)(x - 0.4)(x - 0.8).$$

The graphs of $y = \cos(x)$ and the polynomials $y = P_2(x)$ and $y = P_3(x)$ are shown in Figure 4.12(a) and (b), respectively.

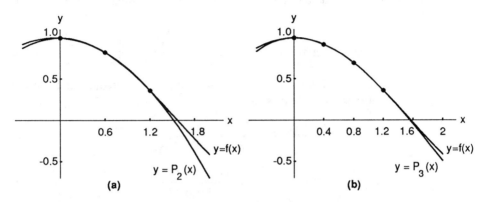

Figure 4.12   (a) The quadratic approximation polynomial $y = P_2(x)$ based on the nodes $x_0 = 0.0$, $x_1 = 0.6$, and $x_2 = 1.2$. (b) The cubic approximation polynomial $y = P_3(x)$ based on the nodes $x_0 = 0.0$, $x_1 = 0.4$, $x_2 = 0.8$, and $x_3 = 1.2$.

Error Terms and Error Bounds

It is important to understand the nature of the error term when the Lagrange polynomial is used to approximate a continuous function $f(x)$. It is similar to the error term for the Taylor polynomial, except that the factor $(x - x_0)^{N+1}$ is replaced with the product $(x - x_0)(x - x_1) \cdots (x - x_N)$. This is expected because interpolation is exact at each of the $N + 1$ nodes x_k, where we have $E_N(x_k) = f(x_k) - P_N(x_k) = y_k - y_k = 0$, for $k = 1$, $2, \ldots, N$.

Theorem 4.5 (Lagrange Polynomial Approximation).   Assume that $f \in C^{N+1}[a, b]$ and that $x_0, x_1, \ldots, x_N \in [a, b]$ are $N + 1$ nodes. If $x \in [a, b]$, then

$$f(x) = P_N(x) + E_N(x), \tag{14}$$

where $P_N(x)$ is a polynomial that can be used to approximate $f(x)$:

$$f(x) \approx P_N(x) = \sum_{k=0}^{N} f(x_k)L_{N,k}(x). \tag{15}$$

The error term $E_N(x)$ has the form

$$E_N(x) = \frac{(x - x_0)(x - x_1) \cdots (x - x_N) f^{(N+1)}(c)}{(N + 1)!}, \tag{16}$$

for some value $c = c(x)$ that lies in the interval $[a, b]$.

Proof. As an example of the general method, we establish (16) when $N = 1$. The general case is discussed in Exercise 13. Start by defining the special function $g(t)$ as follows:

$$g(t) = f(t) - P_1(t) - E_1(x) \frac{(t - x_0)(t - x_1)}{(x - x_0)(x - x_1)}. \tag{17}$$

Notice that x, x_0, and x_1 are constants with respect to the variable t and that $g(t)$ evaluates to be zero at these three values; that is,

$$g(x) = f(x) - P_1(x) - E_1(x) \frac{(x - x_0)(x - x_1)}{(x - x_0)(x - x_1)} = f(x) - P_1(x) - E_1(x) = 0,$$

$$g(x_0) = f(x_0) - P_1(x_0) - E_1(x) \frac{(x_0 - x_0)(x_0 - x_1)}{(x - x_0)(x - x_1)} = f(x_0) - P_1(x_0) = 0,$$

$$g(x_1) = f(x_1) - P_1(x_1) - E_1(x) \frac{(x_1 - x_0)(x_1 - x_1)}{(x - x_0)(x - x_1)} = f(x_1) - P_1(x_1) = 0.$$

Suppose that x lies in the open interval (x_0, x_1). Applying Rolle's theorem to $g(t)$ on the interval $[x_0, x]$ produces a value d_0 with $x_0 < d_0 < x$ such that

$$g'(d_0) = 0. \tag{18}$$

A second application of Rolle's theorem to $g(t)$ on $[x, x_1]$ will produce a value d_1 with $x < d_1 < x_1$ such that

$$g'(d_1) = 0. \tag{19}$$

Equations (18) and (19) show that the function $g'(t)$ is zero at $t = d_0$ and $t = d_1$. A third use of Rolle's theorem, but this time applied to $g'(t)$ over $[d_0, d_1]$, produces a value c for which

$$g^{(2)}(c) = 0. \tag{20}$$

Now go back to (17) and compute the derivatives $g'(t)$ and $g''(t)$:

$$g'(t) = f'(t) - P_1'(t) - E_1(x) \frac{(t - x_0) + (t - x_1)}{(x - x_0)(x - x_1)} \tag{21}$$

$$g''(t) = f''(t) - 0 - E_1(x) \frac{2}{(x - x_0)(x - x_1)}. \tag{22}$$

In (22) we have used the fact that $P_1(x)$ is a polynomial of degree $N = 1$; hence its second derivative is $P_1''(t) \equiv 0$. Evaluation of (22) at the point $t = c$, and using (20), yields

$$0 = f''(c) - E_1(x) \frac{2}{(x - x_0)(x - x_1)}. \tag{23}$$

Solving (23) for $E_1(x)$ results in the desired form (16) for the remainder:

$$E_1(x) = \frac{(x - x_0)(x - x_1)f^{(2)}(c)}{2!}, \tag{24}$$

and the proof is complete.

The next result addresses the special case when the nodes for the Lagrange polynomial are equally spaced $x_k = x_0 + hk$ for $k = 0, 1, \ldots, N$ and the polynomial $P_N(x)$ is used only for interpolation inside the interval $[x_0, x_N]$.

Theorem 4.6   (Error Bounds for Lagrange Interpolation, Equally Spaced Nodes).   Assume that $f(x)$ is defined on $[a, b]$ that contains the equally spaced nodes $x_k = x_0 + hk$. Suppose that $f(x)$ and the derivatives up to the order $N + 1$ are continuous and bounded on the special subintervals $[x_0, x_1]$, $[x_0, x_2]$, and $[x_0, x_3]$, respectively; that is,

$$\left| f^{(N+1)}(x) \right| \leq M_{N+1} \qquad \text{for } x_0 \leq x \leq x_N, \tag{25}$$

for $N = 1, 2, 3$. The error terms (16) corresponding to the cases $N = 1, 2, 3$ have the following useful bounds on their magnitude:

$$\left| E_1(x) \right| \leq \frac{h^2 M_2}{8} \qquad \text{valid for } x \in [x_0, x_1], \tag{26}$$

$$\left| E_2(x) \right| \leq \frac{h^3 M_3}{9\sqrt{3}} \qquad \text{valid for } x \in [x_0, x_2], \tag{27}$$

$$\left| E_3(x) \right| \leq \frac{h^4 M_4}{24} \qquad \text{valid for } x \in [x_0, x_3]. \tag{28}$$

Proof.   We establish (26) and leave the others for the reader. Using the change of variables $x - x_0 = t$ and $x - x_1 = t - h$, the error term $E_1(x)$ can be written

$$E_1(x) = E_1(x_0 + t) = \frac{(t^2 - ht)f^{(2)}(c)}{2!} \qquad \text{for } 0 \leq t \leq h. \tag{29}$$

The bound for the derivative for this case is

$$\left| f^{(2)}(c) \right| \leq M_2 \qquad \text{for } x_0 \leq c \leq x_1. \tag{30}$$

Now determine a bound for the expression $(t^2 - ht)$ in the numerator of (29); call this term $\phi(t) = t^2 - ht$. Since $\phi'(t) = 2t - h$, there is one critical point $t = h/2$ which is the solution to $\phi'(t) = 0$. The extreme values of $\phi(t)$ over $[0, h]$ occurs either at an endpoint, $\phi(0) = 0$, $\phi(h) = 0$ or at the critical point $\phi(h/2) = -h^2/4$. Since the latter value is the largest, we have established the bound:

$$|\phi(t)| = |t^2 - ht| \leq \frac{|-h^2|}{4} = \frac{h^2}{4} \qquad \text{for } 0 \leq t \leq h. \tag{31}$$

Using (30) and (31) to estimate the magnitude of the product in the numerator in (29) results in

$$|E_1(x)| = \frac{|\phi(t)|\,|f^{(2)}(c)|}{2!} \leq \frac{h^2 M_2}{8}, \tag{32}$$

and formula (26) is established.

Comparison of Accuracy and $O(h^{N+1})$

The significance of Theorem 4.6 is to understand a simple relationship between the size of the error terms for linear, quadratic, and cubic interpolation. In each case the error bound $|E_N(x)|$ depends on h in two ways. First, h^{N+1} is explicitly present, so that $|E_N(x)|$ is proportional to h^{N+1}. Second, the values M_{N+1} generally depend on h and tend to $|f^{(N+1)}(x_0)|$ as h goes to zero. Therefore, as h goes to zero $E_N(x)$ converges to zero with the same rapidity that h^{N+1} converges to zero. The notation $O(h^{N+1})$ is used when discussing this behavior. For example, the error bound (26) can be expressed as

$$|E_1(x)| = O(h^2) \qquad \text{valid for } x \in [x_0, x_1].$$

The notation $O(h^2)$ stands in place of $h^2 M_2/8$ in relation (26) and is meant to convey the idea that the bound for the error term is approximately a multiple of h^2; that is,

$$|E_1(x)| \leq Ch^2 \approx O(h^2).$$

As a consequence, if the derivatives of $f(x)$ are uniformly bounded on the interval and $|h| < 1$, then choosing N large will make h^{N+1} small, and the higher-degree approximating polynomial will have less error.

Example 4.9

Consider $y = f(x) = \cos(x)$ over $[0.0, 1.2]$. Use formulas (26) to (28) and determine the error bounds for the Lagrange polynomials $P_1(x)$, $P_2(x)$, and $P_3(x)$ that were constructed in Examples 4.7 and 4.8.

Solution. First determine the bounds M_2, M_3, and M_4 for the derivatives $|f^{(2)}(x)|$, $|f^{(3)}(x)|$, and $|f^{(4)}(x)|$, respectively, taken over the interval $[0.0, 1.2]$.

$$|f^{(2)}(x)| = |-\cos(x)| \leq |-\cos(0.0)| = 1.000000 \qquad \text{so that } M_2 = 1.000000,$$

$$|f^{(3)}(x)| = |\sin(x)| \leq |\sin(1.2)| = 0.932039 \qquad \text{so that } M_3 = 0.932039,$$

$$|f^{(4)}(x)| = |\cos(x)| \leq |\cos(0.0)| = 1.000000 \qquad \text{so that } M_4 = 1.000000.$$

For $P_1(x)$ the spacing of the nodes is $h = 1.2$, and its error bound is

$$|E_1(x)| \leq \frac{h^2 M_2}{8} \leq \frac{(1.2)^2 1.000000}{8} = 0.180000. \tag{33}$$

For $P_2(x)$ the spacing of the nodes is $h = 0.6$, and its error bound is

$$|E_2(x)| \leq \frac{h^3 M_3}{9\sqrt{3}} \leq \frac{(0.6)^3 0.932039}{9\sqrt{3}} = 0.012915. \tag{34}$$

For $P_3(x)$ the spacing of the nodes is $h = 0.4$, and its error bound is

$$|E_3(x)| \leq \frac{h^4 M_4}{24} \leq \frac{(0.4)^4 1.000000}{24} = 0.001067. \tag{35}$$

From Example 4.7 we saw that $|E_1(0.6)| = |\cos(0.6) - P_1(0.6)| = 0.144157$, so the bound 0.180000 in (33) is reasonable. The graphs of the error functions $E_2(x) = \cos(x) - P_2(x)$ and $E_3(x) = \cos(x) - P_3(x)$ are shown in Figure 4.13(a) and (b), respectively, and numerical computations are given in Table 4.8. Using values in the table we find that $|E_2(1.0)| = |\cos(1.0) - P_2(1.0)| = 0.008416$ and $|E_3(0.2)| = |\cos(0.2) - P_3(0.2)| = 0.000855$, which is in reasonable agreement with the bounds 0.012915 and 0.001067 given in (34) and (35), respectively.

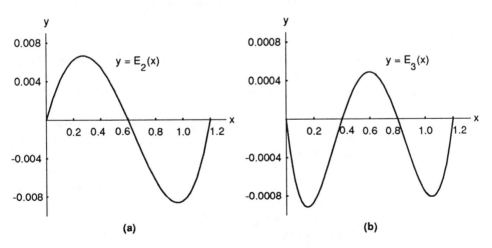

Figure 4.13   (a) The error function $E_2(x) = \cos(x) - P_2(x)$. (b) The error function $E_3(x) = \cos(x) - P_3(x)$.

TABLE 4.8   Comparison of $f(x) = \cos(x)$ and the Quadratic and Cubic Polynomial Approximations $P_2(x)$ and $P_3(x)$

x_k	$f(x_k) = \cos(x_k)$	$P_2(x_k)$	$E_2(x_k)$	$P_3(x_k)$	$E_3(x_k)$
0.0	1.000000	1.000000	0.0	1.000000	0.0
0.1	0.995004	0.990911	0.004093	0.995835	−0.000831
0.2	0.980067	0.973813	0.006253	0.980921	−0.000855
0.3	0.955336	0.948707	0.006629	0.955812	−0.000476
0.4	0.921061	0.915592	0.005469	0.921061	0.0
0.5	0.877583	0.874468	0.003114	0.877221	0.000361
0.6	0.825336	0.825336	0.0	0.824847	0.00089
0.7	0.764842	0.768194	−0.003352	0.764491	0.000351
0.8	0.696707	0.703044	−0.006338	0.696707	0.0
0.9	0.621610	0.629886	−0.008276	0.622048	−0.000438
1.0	0.540302	0.548719	−0.008416	0.541068	−0.000765
1.1	0.453596	0.459542	−0.005946	0.454320	−0.000724
1.2	0.362358	0.362358	0.0	0.362358	0.0

Algorithm 4.3   (Lagrange Approximation).   To evaluate the Lagrange polynomial $P(x) = \displaystyle\sum_{k=0}^{N} y_k L_{N,k}(x)$, of degree N, based on the $N + 1$ points (x_k, y_k) for $k = 0, 1, \ldots, N$.

```
INPUT N                              {Degree of the polynomial}
INPUT X₀, X₁, . . . , Xₙ             {The nodes of Pₙ(x)}
INPUT Y₀, Y₁, . . . , Yₙ             {The ordinates of the points}

INPUT T                              {The independent variable}
Sum := 0                             {Initialize variable}
FOR  K = 0  TO  N  DO
     Term := Yₖ                      {Initialize variable}
     FOR  J = 0  TO  N  DO           {Form terms to be added
          IF  J ≠ K  THEN              to avoid division by 0}
               Term := Term*[T−Xⱼ]/[Xₖ−Xⱼ]
     Sum := Sum + Term

PRINT 'The value of the Lagrange interpolating polynomial'
      'based on the nodes X₀, X₁, . . . , Xₙ is' Sum
```

EXERCISES FOR LAGRANGE APPROXIMATION

1. Find Lagrange polynomials that approximate $f(x) = x^3$.
   (a) Find the linear interpolation polynomial $P_1(x)$ using the nodes $x_0 = -1$ and $x_1 = 0$.
   (b) Find the quadratic interpolation polynomial $P_2(x)$ using the nodes $x_0 = -1$, $x_1 = 0$, and $x_2 = 1$.
   (c) Find the cubic interpolation polynomial $P_3(x)$ using the nodes $x_0 = -1$, $x_1 = 0$, $x_2 = 1$, and $x_3 = 2$.
   (d) Find the linear interpolation polynomial $P_1(x)$ using the nodes $x_0 = 1$ and $x_1 = 2$.
   (e) Find the quadratic interpolation polynomial $P_2(x)$ using the nodes $x_0 = 0$, $x_1 = 1$, and $x_2 = 2$.

2. Let $f(x) = x + 2/x$.
   (a) Use quadratic Lagrange interpolation based on the nodes $x_0 = 1$, $x_1 = 2$, and $x_2 = 2.5$ to approximate $f(1.5)$ and $f(1.2)$.
   (b) Use cubic Lagrange interpolation based on the nodes $x_0 = 0.5$, $x_1 = 1$, $x_2 = 2$, and $x_3 = 2.5$ to approximate $f(1.5)$ and $f(1.2)$.

3. Let $f(x) = 8x/2^x$.
   (a) Use quadratic Lagrange interpolation based on the nodes $x_0 = 0$, $x_1 = 1$, and $x_2 = 2$ to approximate $f(1.5)$ and $f(1.3)$.
   (b) Use cubic Lagrange interpolation based on the nodes $x_0 = 0$, $x_1 = 1$, $x_2 = 2$, and $x_3 = 3$ to approximate $f(1.5)$ and $f(1.3)$.

4. Let $f(x) = 2 \sin(x\pi/6)$, where x is in radians.
   (a) Use quadratic Lagrange interpolation based on the nodes $x_0 = 0$, $x_1 = 1$, and $x_2 = 3$ to approximate $f(2)$ and $f(2.4)$.
   (b) Use cubic Lagrange interpolation based on the nodes $x_0 = 0$, $x_1 = 1$, $x_2 = 3$, and $x_3 = 5$ to approximate $f(2)$ and $f(2.4)$.

5. Let $f(x) = 2 \sin(x\pi/6)$, where x is in radians.
   (a) Use quadratic Lagrange extrapolation based on the nodes $x_0 = 0$, $x_1 = 1$, and $x_2 = 3$ to approximate $f(4)$ and $f(3.5)$.
   (b) Use cubic Lagrange interpolation based on the nodes $x_0 = 0$, $x_1 = 1$, $x_2 = 3$, and $x_3 = 5$ to approximate $f(4)$ and $f(3.5)$.

6. Write down the error term $E_3(x)$ for cubic Lagrange interpolation to $f(x)$, where interpolation is to be exact at the four nodes $x_0 = -1$, $x_1 = 0$, $x_2 = 3$, and $x_3 = 4$ and $f(x)$ is given by
   (a) $f(x) = 4x^3 - 3x + 2$
   (b) $f(x) = x^4 - 2x^3$
   (c) $f(x) = x^5 - 5x^4$

7. Let $L_{N,0}(x)$, $L_{N,1}(x)$, . . . , $L_{N,N}(x)$ be the Lagrange coefficient polynomials based on the $N + 1$ nodes x_0, x_1, . . . , x_N. Show that $L_{N,k}(x_k) = 1$ for $k = 0, 1, . . . , N$ and that $L_{N,k}(x_j) = 0$ whenever $j \neq k$.

8. Consider the Lagrange coefficient polynomials $L_{2,k}(x)$ that are used for quadratic interpolation at the nodes x_0, x_1, x_2. Define $g(x) = L_{2,0}(x) + L_{2,1}(x) + L_{2,2}(x) - 1$.
   (a) Show that g is a polynomial of degree ≤ 2.
   (b) Show that $g(x_k) = 0$ for $k = 0, 1, 2$.
   (c) Show that $g(x) = 0$ for all x. *Hint.* Use the fundamental theorem of algebra.

9. Let $f(x)$ be a polynomial of degree $\leq N$. Let $P_N(x)$ be the Lagrange interpolating polynomial of degree $\leq N$ based on the $N + 1$ nodes $x_0, x_1, \ldots, x_N$. Show that $f(x) = P_N(x)$ for all x. *Hint.* Show that the error term $E_N(x)$ is identically zero.

10. Consider the function $f(x) = \sin(x)$ on the interval $[0, 1]$. Use Theorem 4.6 to determine the step size h so that
    (a) linear Lagrange interpolation has an accuracy of 10^{-6} [i.e., find h so that $|E_1(x)| < 5 \times 10^{-7}$].
    (b) quadratic Lagrange interpolation has an accuracy of 10^{-6} [i.e., find h so that $|E_2(x)| < 5 \times 10^{-7}$].
    (c) cubic Lagrange interpolation has an accuracy of 10^{-6}.

11. Start with equation (16) and $N = 2$, and prove inequality (27). Let $x_1 = x_0 + h$, $x_2 = x_0 + 2h$. Prove that if $x_0 \leq x \leq x_2$, then

$$|x - x_0|\,|x - x_1|\,|x - x_2| \leq \frac{2h^3}{3 \times 3^{1/2}}.$$

Hint. Use the substitutions $t = x - x_1$, $t + h = x - x_0$, and $t - h = x - x_2$ and the function $v(t) = t^3 - th^2$ on the interval $-h \leq t \leq h$. Set $v'(t) = 0$ and solve for t in terms of h.

12. *Linear interpolation in two dimensions.* Consider the polynomial $z = P(x, y) = A + Bx + Cy$ that passes through the three points (x_0, y_0, z_0), (x_1, y_1, z_1), and (x_2, y_2, z_2). Then A, B, and C are the solution values for the linear system of equations

$$A + x_0 B + y_0 C = z_0$$

$$A + x_1 B + y_1 C = z_1$$

$$A + x_2 B + y_2 C = z_2.$$

    (a) Find A, B, and C so that $z = P(x, y)$ passes through the points $(1, 1, 5)$, $(2, 1, 3)$, and $(1, 2, 9)$.
    (b) Find A, B, and C so that $z = P(x, y)$ passes through the points $(1, 1, 2.5)$, $(2, 1, 0)$, and $(1, 2, 4)$.
    (c) Find A, B, and C so that $z = P(x, y)$ passes through the points $(2, 1, 5)$, $(1, 3, 7)$, and $(3, 2, 4)$.
    (d) Can values A, B, and C be found so that $z = P(x, y)$ passes through the points $(1, 2, 5)$, $(3, 2, 7)$, and $(1, 2, 0)$? Why?

13. Use the generalized Rolle's theorem[†] and the special function

$$g(t) = f(t) - P_N(t) - E_N(x) \frac{(t - x_0)(t - x_1) \cdots (t - x_N)}{(x - x_0)(x - x_1) \cdots (x - x_N)},$$

where $P_N(x)$ is the Lagrange polynomial of degree N to prove that the error term $E_N(x) = f(x) - P_N(x)$ has the form

$$E_N(x) = (x - x_0)(x - x_1) \cdots (x - x_N) \frac{f^{(N+1)}(c)}{(N + 1)!}.$$

Hint. Find $g^{(N+1)}(t)$ and then evaluate it at $t = c$.

[†]*Generalized Rolle's theorem.* Let $g(t)$ be continuous on $[a, b]$ and suppose that its derivatives $g^{(k)}(t)$, for $k = 1, 2, \ldots, N + 1$, are continuous on (a, b). If there are $N + 2$ distinct points $x, x_0, x_1, \ldots, x_N$ such that $g(x) = 0$ and $g(x_k) = 0$ for $k = 0, 1, \ldots, N$, there exists a value c in (a, b) such that $g^{(N+1)}(c) = 0$.

4.4 NEWTON POLYNOMIALS

It is sometimes useful to find several approximating polynomials $P_1(x)$, $P_2(x)$, . . . , $P_N(x)$ and then choose the one that suits our needs. If the Lagrange polynomials are used, there is no constructive relationship between $P_{N-1}(x)$ and $P_N(x)$. Each polynomial has to be constructed individually, and the work required to compute the higher-degree polynomials involves many computations. We take a new approach and construct Newton polynomials that have the recursive pattern

$$P_1(x) = a_0 + a_1(x - x_0), \tag{1}$$

$$P_2(x) = a_0 + a_1(x - x_0) + a_2(x - x_0)(x - x_1), \tag{2}$$

$$P_3(x) = a_0 + a_1(x - x_0) + a_2(x - x_0)(x - x_1) + a_3(x - x_0)(x - x_1)(x - x_2), \tag{3}$$

$$\vdots$$

$$P_N(x) = a_0 + a_1(x - x_0) + a_2(x - x_0)(x - x_1) + a_3(x - x_0)(x - x_1)(x - x_2)$$
$$+ a_4(x - x_0)(x - x_1)(x - x_2)(x - x_3) + \cdots \tag{4}$$
$$+ a_N(x - x_0) \cdots (x - x_{N-1}).$$

Here the polynomial $P_N(x)$ is obtained from $P_{N-1}(x)$ using the recursive relationship

$$P_N(x) = P_{N-1}(x) + a_N(x - x_0)(x - x_1)(x - x_2) \cdots (x - x_{N-1}). \tag{5}$$

The polynomial (4) is said to be a Newton polynomial with the N **centers** x_0, x_1, . . . , x_{N-1}. It involves sums of products of linear factors up to

$$a_N(x - x_0)(x - x_1)\ (x - x_2) \cdots (x - x_{N-1})$$

so that $P_N(x)$ will simplify to be an ordinary polynomial of degree $\leq N$.

Example 4.10

Given the centers $x_0 = 1$, $x_1 = 3$, $x_2 = 4$, and $x_3 = 4.5$ and the coefficients $a_0 = 5$, $a_1 = -2$, $a_2 = 0.5$, $a_3 = -0.1$, and $a_4 = 0.003$, find $P_1(x)$, $P_2(x)$, $P_3(x)$, and $P_4(x)$ and evaluate $P_k(2.5)$ for $k = 1, 2, 3, 4$.

Solution.   Using formulas (1) to (4), we have

$$P_1(x) = 5 - 2(x - 1),$$

$$P_2(x) = 5 - 2(x - 1) + 0.5(x - 1)(x - 3),$$

$$P_3(x) = 5 - 2(x - 1) + 0.5(x - 1)(x - 3) - 0.1(x - 1)(x - 3)(x - 4),$$

$$P_4(x) = P_3(x) + 0.003(x - 1)(x - 3)(x - 4)(x - 4.5).$$

Evaluating the polynomials at $x = 2.5$ results in

$$P_1(2.5) = 5 - 2(1.5) = 2,$$

$$P_2(2.5) = P_1(2.5) + 0.5(1.5)(-0.5) = 1.625,$$

$$P_3(2.5) = P_2(2.5) - 0.1(1.5)(-0.5)(-1.5) = 1.5125,$$
$$P_4(2.5) = P_3(2.5) + 0.03(1.5)(-0.5)(-1.5)(-2.0) = 1.50575.$$

Nested Multiplication

If N is fixed and the polynomial $P_N(x)$ is evaluated many times, then nested multiplication should be used. The process is similar to nested multiplication for ordinary polynomials, except that the centers x_k must be subtracted from the independent variable x. The nested multiplication form for $P_3(x)$ is

$$P_3(x) = [[a_3(x - x_2) + a_2](x - x_1) + a_1](x - x_0) + a_0. \tag{6}$$

To evaluate $P_3(x)$ for a given value of x, start with the innermost grouping and form successively the quantities

$$\begin{aligned}
S_3 &= a_3, \\
S_2 &= S_3(x - x_2) + a_2, \\
S_1 &= S_2(x - x_1) + a_1, \\
S_0 &= S_1(x - x_0) + a_0.
\end{aligned} \tag{7}$$

The quantity S_0 is now $P_3(x)$.

Example 4.11

Compute $P_3(2.5)$ in Example 4.10 using nested multiplication.

Solution. Using (6), we write

$$P_3(x) = [[-0.1(x - 4) + 0.5](x - 3) - 2](x - 1) + 5.$$

The values in (7) are

$$\begin{aligned}
S_3 &= -0.1, \\
S_2 &= -0.1(2.5 - 4) + 0.5 = 0.65, \\
S_1 &= 0.65(2.5 - 3) - 2 = -2.325, \\
S_0 &= -2.325(2.5 - 1) + 5 = 1.5125.
\end{aligned}$$

Therefore, $P_3(2.5) = 1.5125$.

Polynomial Approximation, Nodes, and Centers

Suppose that we want to find the coefficients a_k for all the polynomials $P_1(x), \ldots, P_N(x)$ that approximate a given function $f(x)$. Then $P_k(x)$ will be based on the centers $x_0, x_1, \ldots, x_k$ and have the nodes $x_0, x_1, \ldots, x_{k+1}$. For the polynomial $P_1(x)$ the coefficients a_0 and a_1 have a familiar meaning. In this case

$$P_1(x_0) = f(x_0) \quad \text{and} \quad P_1(x_1) = f(x_1). \tag{8}$$

Using (1) and (8) to solve for a_0, we find that

$$f(x_0) = P_1(x_0) = a_0 + a_1(x_0 - x_0) = a_0. \tag{9}$$

Hence $a_0 = f(x_0)$. Next, using (1), (8), and (9), we have

$$f(x_1) = P_1(x_1) = a_0 + a_1(x_1 - x_0) = f(x_0) + a_1(x_1 - x_0),$$

which can be solved for a_1 and we get

$$a_1 = \frac{f(x_1) - f(x_0)}{x_1 - x_0}. \tag{10}$$

Hence a_1 is the slope of the secant line through the two points $(x_0, f(x_0))$ and $(x_1, f(x_1))$.

The coefficients a_0 and a_1 are the same for both $P_1(x)$ and $P_2(x)$. Evaluating (2) at the node x_2, we find that

$$f(x_2) = P_2(x_2) = a_0 + a_1(x_2 - x_0) + a_2(x_2 - x_0)(x_2 - x_1). \tag{11}$$

The values for a_0 and a_1 in (9) and (10) can be used in (11) to obtain

$$a_2 = \frac{f(x_2) - a_0 - a_1(x_2 - x_0)}{(x_2 - x_0)(x_2 - x_1)}$$

$$= \left[\frac{f(x_2) - f(x_0)}{x_2 - x_0} - \frac{f(x_1) - f(x_0)}{x_1 - x_0} \right] \Big/ (x_2 - x_1).$$

For computational purposes we prefer to write this last quantity as

$$a_2 = \left[\frac{f(x_2) - f(x_1)}{x_2 - x_1} - \frac{f(x_1) - f(x_0)}{x_1 - x_0} \right] \Big/ (x_2 - x_0). \tag{12}$$

The two formulas for a_2 can be shown to be equivalent by writing the quotients over the common denominator $(x_2 - x_1)(x_2 - x_0)(x_1 - x_0)$. The details are left for the reader. The numerator in (12) is the difference between first-order divided differences. In order to proceed, we need to introduce the idea of divided differences.

Definition 4.1 (Divided Differences).   The divided differences for a function $f(x)$ are defined as follows:

$$f[x_k] = f(x_k),$$

$$f[x_{k-1}, x_k] = \frac{f[x_k] - f[x_{k-1}]}{x_k - x_{k-1}},$$

$$f[x_{k-2}, x_{k-1}, x_k] = \frac{f[x_{k-1}, x_k] - f[x_{k-2}, x_{k-1}]}{x_k - x_{k-2}}, \tag{13}$$

$$f[x_{k-3}, x_{k-2}, x_{k-1}, x_k] = \frac{f[x_{k-2}, x_{k-1}, x_k] - f[x_{k-3}, x_{k-2}, x_{k-1}]}{x_k - x_{k-3}}.$$

The recursive rule for constructing higher-order divided differences is

$$f[x_{k-j}, x_{k-j+1}, \ldots, x_k] = \frac{f[x_{k-j+1}, \ldots, x_k] - f[x_{k-j}, \ldots, x_{k-1}]}{x_k - x_{k-j}} \tag{14}$$

and is used to construct the divided differences in Table 4.9.

TABLE 4.9 Divided-Difference Table for $y = f(x)$

x_k	$f[x_k]$	$f[\ ,\ ]$	$f[\ ,\ ,\ ]$	$f[\ ,\ ,\ ,\ ]$	$f[\ ,\ ,\ ,\ ,\ ]$
x_0	$f[x_0]$				
x_1	$f[x_1]$	$f[x_0, x_1]$			
x_2	$f[x_2]$	$f[x_1, x_2]$	$f[x_0, x_1, x_2]$		
x_3	$f[x_3]$	$f[x_2, x_3]$	$f[x_1, x_2, x_3]$	$f[x_0, x_1, x_2, x_3]$	
x_4	$f[x_4]$	$f[x_3, x_4]$	$f[x_2, x_3, x_4]$	$f[x_1, x_2, x_3, x_4]$	$f[x_0, x_1, x_2, x_3, x_4]$

The coefficient a_k of $P_N(x)$ depends on the values $f(x_j)$ for $j = 0, 1, \ldots, k$. The next theorem shows that a_k can be computed using divided difference:

$$a_k = f[x_0, x_1, \ldots, x_k]. \tag{15}$$

Theorem 4.7 (Newton Polynomial).   Suppose that $x_0, x_1, \ldots, x_N$ are $N + 1$ distinct numbers in $[a, b]$. There exists a unique polynomial $P_N(x)$ of degree at most N with the property that

$$f(x_j) = P_N(x_j) \qquad \text{for } j = 0, 1, \ldots, N.$$

The Newton form of this polynomial is

$$P_N(x) = a_0 + a_1(x - x_0) + \cdots + a_N(x - x_0)(x - x_1) \cdots (x - x_{N-1}), \tag{16}$$

where $a_k = f[x_0, x_1, \ldots, x_k]$ for $k = 0, 1, \ldots, N$.

Remark.   If $\{(x_j, y_j)\}_{j=0}^{N}$ is a set of points whose abscissas are distinct, the values $f(x_j) = y_j$ can be used to construct the unique polynomial of degree $\leq N$ that passes through the $N + 1$ points.

Corollary 4.2 (Newton Approximation).   Assume that $P_N(x)$ is the Newton polynomial given in Theorem 4.7 and is used to approximate the function $f(x)$, that is,

$$f(x) = P_N(x) + E_N(x). \tag{17}$$

If $f \in C^{N+1}[a, b]$, then for each $x \in [a, b]$, there corresponds a number $c = c(x)$ in (a, b), so that the error term has the form

$$E_N(x) = \frac{(x - x_0)(x - x_1) \cdots (x - x_N)f^{(N+1)}(c)}{(N + 1)!}. \tag{18}$$

Remark.   The error term $E_N(x)$ is the same as the one for Lagrange interpolation, which was introduced in equation (34) of Section 4.3.

It is of interest to start with a known function $f(x)$ that is a polynomial of degree N, and compute its divided-difference table. In this case we know that $f^{(N+1)}(x) = 0$ for all x, and calculation will reveal that the $(N + 1)$st divided difference is zero. This will happen because the divided difference (14) is proportional to a numerical approximation for the jth derivative. More details can be found in Reference [41].

Example 4.12

Let $f(x) = x^3 - 4x$. Construct the divided-difference table based on the nodes $x_0 = 1$, $x_1 = 2, \ldots, x_5 = 6$, and find the Newton polynomial $P_3(x)$ based on x_0, x_1, x_2, x_3.

Solution.   See Table 4.10.

TABLE 4.10

x_k	$f[x_k]$	First divided difference	Second divided difference	Third divided difference	Fourth divided difference	Fifth divided difference
$x_0 = 1$	-3					
$x_1 = 2$	0	3				
$x_2 = 3$	15	15	6			
$x_3 = 4$	48	33	9	1		
$x_4 = 5$	105	57	12	1	0	
$x_5 = 6$	192	87	15	1	0	0

The coefficients $a_0 = -3$, $a_1 = 3$, $a_2 = 6$, and $a_3 = 1$ of $P_3(x)$ appear on the diagonal of the divided-difference table. The centers $x_0 = 1$, $x_1 = 2$, and $x_2 = 3$ are the values in the first column. Using formula (3), we write

$$P_3(x) = -3 + 3(x - 1) + 6(x - 1)(x - 2) + (x - 1)(x - 2)(x - 3).$$

Example 4.13

Construct a divided-difference table for $f(x) = \cos(x)$ based on the five points $(k, \cos(k))$ for $k = 0, 1, \ldots, 4$. Use it to find the coefficients a_k and the four Newton interpolating polynomials $P_k(x)$ for $k = 1, 2, 3, 4$.

Solution.   For simplicity we round off the values to seven decimal places, which are displayed in Table 4.11.

TABLE 4.11 The Divided-Difference Table Used for Constructing the Newton Polynomials $P_k(x)$ in Example 4.13

x_k	$f[x_k]$	$f[\ ,\ ]$	$f[\ ,\ ,\ ]$	$f[\ ,\ ,\ ,\ ]$	$f[\ ,\ ,\ ,\ ,\ ]$
$x_0 = 0.0$	1.0000000				
$x_1 = 1.0$	0.5403023	−0.4596977			
$x_2 = 2.0$	−0.4161468	−0.9564491	−0.2483757		
$x_3 = 3.0$	−0.9899925	−0.5738457	0.1913017	0.1465592	
$x_4 = 4.0$	−0.6536436	0.3363489	0.4550973	0.0879318	−0.0146568

The nodes x_0, x_1, x_2, x_3 and the diagonal elements a_0, a_1, a_2, a_3, a_4 in Table 4.11 are used in formula (16) and we write down the first four Newton polynomials:

$$P_1(x) = 1.0000000 - 0.4596977(x - 0.0),$$

$$P_2(x) = 1.0000000 - 0.4596977(x - 0.0) - 0.2483757(x - 0.0)(x - 1.0),$$

$$P_3(x) = 1.0000000 - 0.4596977(x - 0.0) - 0.2483757(x - 0.0)(x - 1.0)$$
$$+ 0.1465592(x - 0.0)(x - 1.0)(x - 2.0),$$

$$P_4(x) = 1.0000000 - 0.4596977(x - 0.0) - 0.2483757(x - 0.0)(x - 1.0)$$
$$+ 0.1465592(x - 0.0)(x - 1.0)(x - 2.0)$$
$$- 0.0146568(x - 0.0)(x - 1.0)(x - 2.0)(x - 3.0).$$

The following sample calculation shows how to find the coefficient a_2.

$$f[x_0, x_1] = \frac{f[x_1] - f[x_0]}{x_1 - x_0} = \frac{0.5403023 - 1.0000000}{1.0 - 0.0} = -0.4596977,$$

$$f[x_1, x_2] = \frac{f[x_2] - f[x_1]}{x_2 - x_1} = \frac{-0.4161468 - 0.5403023}{2.0 - 1.0} = -0.9564491,$$

$$a_2 = f[x_0, x_1, x_2] = \frac{f[x_1, x_2] - f[x_0, x_1]}{x_2 - x_0} = \frac{-0.9564491 + 0.4596977}{2.0 - 0.0} = -0.2483757.$$

The graphs of $y = \cos(x)$ and $y = P_1(x)$, $y = P_2(x)$, $y = P_3(x)$ are shown in Figure 4.14(a), (b), and (c), respectively.

For computational purposes the divided differences in Table 4.9 need to be stored in an array which is chosen to be $D(k, j)$. Thus (15) becomes

$$D(k, j) = f[x_{k-j}, x_{k-j+1}, \ldots, x_k] \qquad \text{for } j \le k. \tag{19}$$

Relation (14) is used to obtain the formula to recursively compute the entries in the array:

$$D(k, j) = \frac{D(k, j - 1) - D(k - 1, j - 1)}{x_k - x_{k-j}}. \tag{20}$$

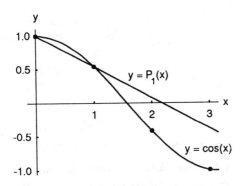

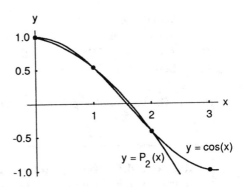

Figure 4.14 (a) Graphs of $y = \cos(x)$ and the linear Newton polynomial $y = P_1(x)$ based on the nodes $x_0 = 0.0$ and $x_1 = 1.0$.

Figure 4.14 (b) Graphs of $y = \cos(x)$ and the quadratic Newton polynomial $y = P_2(x)$ based on the nodes $x_0 = 0.0$, $x_1 = 1.0$, and $x_2 = 2.0$.

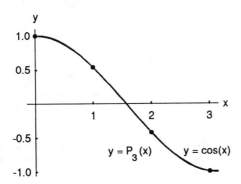

Figure 4.14 (c) Graphs of $y = \cos(x)$ and the cubic Newton polynomial $y = P_2(x)$ based on the nodes $x_0 = 0.0$, $x_1 = 1.0$, $x_2 = 2.0$, and $x_3 = 3.0$.

Notice that the value a_k in (15) is the diagonal element $a_k = D(k, k)$. The algorithm for computing the divided differences and evaluating $P_N(x)$ is now given. We remark that Exercise 13 investigates how to modify the algorithm so that the values $\{a_k\}$ are computed using a one-dimensional array.

Algorithm 4.4   (Nested Multiplication with Multiple Centers).   To evaluate the Newton form of a polynomial

$$P(x) = a_0 + a_1(x - x_0) + a_2(x - x_0)(x - x_1) + a_3(x - x_0)(x - x_1)(x - x_2)$$
$$+ \cdots + a_N(x - x_0)(x - x_1) \cdots (x - x_{N-1}),$$

with multiple centers $x_0, x_1, \ldots, x_{N-1}$.

```
INPUT   N                              {Degree of P(x)}
INPUT   A(0), A(1), . . . , A(N)       {Coefficients of P(x)}
INPUT   X(0), X(1), . . . , X(N−1)     {Centers for P(x)}
```

```
      INPUT  X
        Sum := A(N)
   FOR  K = N−1  DOWNTO  0  DO
     └──── Sum := Sum*[X − X(K)] + A(K)
```

{Independent variable}
{Initialize variable}
{Implement nested
multiplication}

PRINT 'The value of the polynomial P(X) is' Sum

Algorithm 4.5 (Newton Interpolation Polynomial). To construct and evaluate the Newton polynomial of degree $\leq N$ that passes through $(x_k, y_k) = (x_k, f(x_k))$ for $k = 0, 1, \ldots, N$:

$$P(x) = d_{0,0} + d_{1,1}(x - x_0) + d_{2,2}(x - x_0)(x - x_1)$$
$$+ \cdots + d_{N,N}(x - x_0)(x - x_1) \cdots (x - x_{N-1}),$$

where

$$d_{k,0} = y_k \quad \text{and} \quad d_{k,j} = \frac{d_{k,j-1} - d_{k-1,j-1}}{x_k - x_{k-j}}.$$

```
      INPUT  N
      INPUT  X(0), X(1), . . . , X(N)
      INPUT  Y(0), Y(1), . . . , Y(N)
```

{Degree of P(x)}
{Abscissas of the points}
{Ordinates of the points}

```
   FOR  K = 0  TO  N  DO
     └────D(K,0) := Y(K)
```

{Store ordinates in column 0
of the array D(K,J)}

```
   FOR  J = 1  TO  N  DO
     │    FOR  K = J  TO  N  DO
     └────└────── D(K,J) := [D(K,J−1) − D(K−1,J−1)]/[X(K) − X(K−J)]
```

{Compute the divided-
difference table}

```
      INPUT  X
```

{Independent variable}

```
        Sum := D(N,N)
   FOR  K = N−1  DOWNTO  0  DO
     └──── Sum := Sum*[X − X(K)] + D(K,K)
```

{Nested multiplication
is used to
evaluate P(x)}

PRINT 'The value of the Newton polynomial P(x) is' Sum
      'Based on the interpolation nodes' $x_0, x_1, \ldots, x_N$
      'and the ordinates' $y_0, y_1, \ldots, y_N$

EXERCISES FOR NEWTON POLYNOMIALS

In Exercises 1–4, use the centers x_0, x_1, x_2, and x_3 and coefficients a_0, a_1, a_2, a_3, and a_4 to find the Newton polynomials $P_1(x)$, $P_2(x)$, $P_3(x)$, and $P_4(x)$ and evaluate them at the value $x = c$. *Hint.* Use equations (1)–(4) and the techniques of Example 4.10.

1. $a_0 = 4$, $a_1 = -1$, $a_2 = 0.4$, $a_3 = 0.01$, $a_4 = -0.002$,
   $x_0 = 1$, $x_1 = 3$, $x_2 = 4$, $x_3 = 4.5$, $c = 2.5$

2. $a_0 = 5$, $a_1 = -2$, $a_2 = 0.5$, $a_3 = -0.1$, $a_4 = 0.003$,
   $x_0 = 0$, $x_1 = 1$, $x_2 = 2$, $x_3 = 3$, $c = 2.5$

3. $a_0 = 7$, $a_1 = 3$, $a_2 = 0.1$, $a_3 = 0.05$, $a_4 = -0.04$,
   $x_0 = -1$, $x_1 = 0$, $x_2 = 1$, $x_3 = 4$, $c = 3$

4. $a_0 = -2$, $a_1 = 4$, $a_2 = -0.04$, $a_3 = 0.06$, $a_4 = 0.005$,
   $x_0 = -3$, $x_1 = -1$, $x_2 = 1$, $x_3 = 4$, $c = 2$

In Exercises 5–12:
  (a) Compute the divided-difference table for the tabulated function.
  (b) Write down the Newton polynomials $P_1(x)$, $P_2(x)$, $P_3(x)$, and $P_4(x)$.
  (c) Evaluate the Newton polynomials in part (b) at the given values of x.
  (d) Compare the values in part (c) with function value $f(x)$.

5. $f(x) = 3 \times 2^x$
   $x = 1.5, 2.5$

k	x_k	$f(x_k)$
0	−1.0	1.5
1	0.0	3.0
2	1.0	6.0
3	2.0	12.0
4	3.0	24.0

6. $f(x) = 3 \times 2^x$
   $x = 1.5, 2.5$

k	x_k	$f(x_k)$
0	1.0	6.0
1	2.0	12.0
2	0.0	3.0
3	3.0	24.0
4	−1.0	1.5

7. $f(x) = 3.6/x$
   $x = 2.5, 3.5$

k	x_k	$f(x_k)$
0	1.0	3.60
1	2.0	1.80
2	3.0	1.20
3	4.0	0.90
4	5.0	0.72

8. $f(x) = 3.6/x$
   $x = 2.5, 3.5$

k	x_k	$f(x_k)$
0	3.0	1.20
1	2.0	1.80
2	4.0	0.90
3	1.0	3.60
4	5.0	0.72

9. $f(x) = 3 \sin^2(\pi x/6)$
  $x = 1.5, 3.5$

k	x_k	$f(x_k)$
0	0.0	0.00
1	1.0	0.75
2	2.0	2.25
3	3.0	3.00
4	4.0	2.25

10. $f(x) = 3x \times 2^{-x}$
  $x = 1.5, 3.5$

k	x_k	$f(x_k)$
0	0.0	0.000
1	1.0	1.500
2	2.0	1.500
3	3.0	1.125
4	4.0	0.750

11. $f(x) = x^{1/2}$
  $x = 4.5, 7.5$

k	x_k	$f(x_k)$
0	4.0	2.00000
1	5.0	2.23607
2	6.0	2.44949
3	7.0	2.64575
4	8.0	2.82843

12. $f(x) = \exp(-x)$
  $x = 0.5, 1.5$

k	x_k	$f(x_k)$
0	0.0	1.00000
1	1.0	0.36788
2	2.0	0.13534
3	3.0	0.04979
4	4.0	0.01832

13. Verify that the following modification of Algorithm 4.5 is an equivalent way to compute the Newton polynomial interpolation.

```
FOR  K = 0  TO  N  DO
    A(K) := Y(K)
FOR  J = 1  TO  N  DO
    FOR  K = N  DOWNTO  J  DO
        A(K) := [A(K) − A(K−1)]/[X(K) − X(K−J)]
    INPUT  X
    Sum := A(N)
FOR  K = N−1  DOWNTO  0  DO
    Sum := Sum*[X − X(K)] + A(K)
```

14. Consider the M points $(x_0, y_0), \ldots, (x_M, y_M)$.
  (a) If the $(N + 1)$st divided differences are zero, then show that the $(N + 2)$nd up to the Mth divided differences are zero.
  (b) If the $(N + 1)$st divided differences are zero, then show that there exists a polynomial $P_N(x)$ of degree N such that

$$P_N(x_k) = y_k \qquad \text{for } k = 0, 1, \ldots, M.$$

In Exercises 15–17, use the result of Problem 14 to find the polynomial $P_N(x)$ that goes through the $M + 1$ points ($N < M$).

15.

x_k	y_k
0	−2
1	2
2	4
3	4
4	2
5	−2

16.

x_k	y_k
1	8
2	17
3	24
4	29
5	32
6	33

17.

x_k	y_k
0	5
1	5
2	3
3	5
4	17
5	45
6	95

18. Write a report on iterated interpolation. See References [29, 78, 81, 90, 126, 128, 129, 181, 184, and 208].

19. Write a report on inverse interpolation. See References [9, 19, 29, 35, 41, 62, 81, 128, 153, 166, 181, and 191].

4.5 CHEBYSHEV POLYNOMIALS (OPTIONAL)

We now turn our attention to polynomial interpolation for $f(x)$ over $[-1, 1]$ based on the nodes $-1 \le x_0 < x_1 < \cdots < x_N \le 1$. Both the Lagrange and Newton polynomials satisfy

$$f(x) = P_N(x) + E_N(x),$$

where

$$E_N(x) = Q(x) \frac{f^{(N+1)}(c)}{(N+1)!} \qquad (1)$$

and $Q(x)$ is the polynomial of degree $N + 1$:

$$Q(x) = (x - x_0)(x - x_1) \cdots (x - x_N). \qquad (2)$$

Using the relationship

$$|E_N(x)| \le |Q(x)| \frac{\max_{-1 \le x \le 1}\{|f^{(N+1)}(x)|\}}{(N+1)!},$$

our task is to follow Chebyshev's derivation on how to select the set of nodes $\{x_k\}_{k=0}^{N}$ which minimize $\max_{-1 \le x \le 1}\{|Q(x)|\}$. This leads us to a discussion of Chebyshev polynomials and some of their properties. To begin, the first eight Chebyshev polynomials are listed in Table 4.12.

TABLE 4.12   Chebyshev Polynomials
$T_0(x)$ through $T_7(x)$

$T_0(x) = 1$
$T_1(x) = x$
$T_2(x) = 2x^2 - 1$
$T_3(x) = 4x^3 - 3x$
$T_4(x) = 8x^4 - 8x^2 + 1$
$T_5(x) = 16x^5 - 20x^3 + 5x$
$T_6(x) = 32x^6 - 48x^4 + 18x^2 - 1$
$T_7(x) = 64x^7 - 112x^5 + 56x^3 - 7x$

Properties of Chebyshev Polynomials

Several important properties about Chebyshev polynomials need to be discussed.

Property 1. Recurrence Relation

Chebyshev polynomials can be generated in the following way. Set $T_0(x) = 1$ and $T_1(x) = x$ and use the recurrence relation

$$T_k(x) = 2xT_{k-1}(x) - T_{k-2}(x) \qquad \text{for } k = 2, 3, \ldots . \tag{3}$$

Property 2. Leading Coefficient

The coefficient of x^N in $T_N(x)$ is 2^{N-1} when $N \geq 1$.

Property 3. Symmetry

When $N = 2M$, $T_{2M}(x)$ is an even function, that is,

$$T_{2M}(-x) = T_{2M}(x). \tag{4}$$

When $N = 2M + 1$, $T_{2M+1}(x)$ is an odd function, that is,

$$T_{2M+1}(-x) = -T_{2M+1}(x). \tag{5}$$

Property 4. Trigonometric Representation on $[-1, 1]$

$$T_N(x) = \cos(N \arccos(x)) \qquad \text{for } -1 \leq x \leq 1. \tag{6}$$

Property 5. Distinct Zeros in $[-1, 1]$

$T_N(x)$ has N distinct zeros x_k that lie in the interval $[-1, 1]$ (see Figure 4.15):

$$x_k = \cos\left(\frac{(2k + 1)\pi}{2N}\right) \qquad \text{for } k = 0, 1, \ldots, N - 1. \tag{7}$$

These values are called the **Chebyshev abscissas (nodes)**.

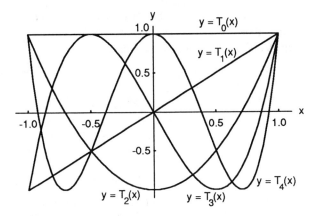

Figure 4.15 Graphs of the Chebyshev polynomials $T_0(x)$, $T_1(x)$, . . . , $T_4(x)$ over $[-1, 1]$.

Property 6. Extreme Values

$$|T_N(x)| \leq 1 \qquad \text{for } -1 \leq x \leq 1. \tag{8}$$

Property 1 is often used as the definition for higher-degree Chebyshev polynomials. Let us show that $T_3(x) = 2xT_2(x) - T_1(x)$. Using the expressions for $T_1(x)$ and $T_2(x)$ in Table 4.12, we obtain

$$2xT_2(x) - T_1(x) = 2x(2x^2 - 1) - x = 4x^3 - 3x = T_3(x).$$

Property 2 is proved by observing that the recurrence relation doubles the leading coefficient of $T_{N-1}(x)$ to get the leading coefficient of $T_N(x)$.

Property 3 is established by showing that $T_{2M}(x)$ involves only even powers of x and $T_{2M+1}(x)$ involves only odd powers of x. The details are left for the reader.

The proof of property 4 uses the trigonometric identity

$$\cos(k\theta) = \cos(2\theta)\cos((k-2)\theta) - \sin(2\theta)\sin((k-2)\theta).$$

Substitute $\cos(2\theta) = 2\cos^2(\theta) - 1$ and $\sin(2\theta) = 2\sin(\theta)\cos(\theta)$ and get

$$\cos(k\theta) = 2\cos(\theta)\,[\cos(\theta)\cos((k-2)\theta) - \sin(\theta)\sin((k-2)\theta)] - \cos((k-2)\theta),$$

which is simplified as

$$\cos(k\theta) = 2\cos(\theta)\cos((k-1)\theta) - \cos((k-2)\theta).$$

Finally, substitute $\theta = \arccos(x)$ and obtain

$$2x\cos((k-1)\arccos(x)) - \cos((k-2)\arccos(x)) = \cos(k\arccos(x)) \tag{9}$$

$$\text{for } -1 \leq x \leq 1.$$

The first two Chebyshev polynomials are $T_0(x) = \cos(0 \arccos(x)) = 1$ and $T_1(x) = \cos(1 \arccos(x)) = x$. Now assume that $T_k(x) = \cos(k \arccos(x))$ for $k = 2, 3, \ldots, N - 1$. Formula (3) is used with (9) to establish the general case:

$$T_N(x) = 2xT_{N-1}(x) - T_{N-2}(x) = 2x \cos((N - 1) \arccos(x)) - \cos((N - 2) \arccos(x))$$

$$= \cos(N \arccos(x)) \qquad \text{for } -1 \le x \le 1.$$

Properties 5 and 6 are consequences of property 4.

Minimax

The Russian mathematician Chebyshev studied how to minimize the upper bound for $|E_N(x)|$. One upper bound can be formed by taking the product of the maximum value of $|Q(x)|$ over all x in $[-1, 1]$ and the maximum value $|f^{(N+1)}(x)/[(N + 1)!]|$ over all x in $[-1, 1]$. To minimize the factor $\max \{|Q(x)|\}$, Chebyshev discovered that $x_0, x_1, \ldots, x_N$ should be chosen so that $Q(x) = (1/2^N)T_{N+1}(x)$.

Theorem 4.8. Assume that N is fixed. Among all possible choices for $Q(x)$ in equation (2), and thus among all possible choices for the distinct nodes $\{x_k\}_{k=0}^N$ in $[-1, 1]$, the polynomial $T(x) = T_{N+1}(x)/2^N$ is the unique choice that has the property

$$\max_{-1 \le x \le 1} \{|T(x)|\} \le \max_{-1 \le x \le 1} \{|Q(x)|\}.$$

Moreover,

$$\max_{-1 \le x \le 1} \{|T(x)|\} = \frac{1}{2^N}. \tag{10}$$

Proof. The proof can be found in Reference [29].

The consequence of this result can be stated by saying that for Lagrange interpolation $f(x) = P_N(x) + E_N(x)$ on $[-1, 1]$, the minimum value of the error bound

$$\max \{|Q(x)|\} \max \{|f^{(N+1)}(x)/(N + 1)!|\}$$

is achieved when the nodes $\{x_k\}$ are the Chebyshev abscissas of $T_{N+1}(x)$. As an illustration, we look at the Lagrange coefficient polynomials that are used in forming $P_3(x)$. First we use equally spaced nodes and then the Chebyshev nodes. Recall that the Lagrange polynomial of degree $N = 3$ has the form

$$P_3(x) = f(x_0)L_{3,0}(x) + f(x_1)L_{3,1}(x) + f(x_2)L_{3,2}(x) + f(x_3)L_{3,3}(x). \tag{11}$$

Equally Spaced Nodes

If $f(x)$ is approximated by a polynomial of degree at most $N = 3$ on $[-1, 1]$, the equally spaced nodes $x_0 = -1$, $x_1 = -\frac{1}{3}$, $x_2 = \frac{1}{3}$, and $x_3 = 1$ are easy to use for calculations. Substitution of these values into formula (8) of Section 4.3 and simplifying will produce the coefficient polynomials $L_{3,k}(x)$ in Table 4.13.

TABLE 4.13 Lagrange Coefficient Polynomials Used to Form
$P_3(x)$ Based on Equally Spaced Nodes $x_k = -1 + 2k/3$

$$L_{3,0}(x) = -0.06250000 + 0.06250000x + 0.56250000x^2 - 0.56250000x^3$$
$$L_{3,1}(x) = 0.56250000 - 1.68750000x - 0.56250000x^2 + 1.68750000x^3$$
$$L_{3,2}(x) = 0.56250000 + 1.68750000x - 0.56250000x^2 - 1.68750000x^3$$
$$L_{3,3}(x) = -0.06250000 - 0.06250000x + 0.56250000x^2 + 0.56250000x^3$$

Chebyshev Nodes

When $f(x)$ is to be approximated by a polynomial of degree at most 3, using the Chebyshev nodes $x_0 = \cos(7\pi/8)$, $x_1 = \cos(5\pi/8)$, $x_2 = \cos(3\pi/8)$, and $x_3 = \cos(\pi/8)$, the coefficients polynomials are tedious to find (but this can be done by a computer). The results after simplification are shown in Table 4.14.

TABLE 4.14 Coefficient Polynomials Used to Form $P_3(x)$ Based
on the Chebyshev Nodes $x_k = \cos((7 - 2k)\pi/8)$

$$C_0(x) = -0.10355339 + 0.11208538x + 0.70710678x^2 - 0.76536686x^3$$
$$C_1(x) = 0.60355339 - 1.57716102x - 0.70710678x^2 + 1.84775906x^3$$
$$C_2(x) = 0.60355339 + 1.57716102x - 0.70710678x^2 - 1.84775906x^3$$
$$C_3(x) = -0.10355339 - 0.11208538x + 0.70710678x^2 + 0.76536686x^3$$

Example 4.14

Compare the Lagrange interpolating polynomials of degree $N = 3$ for $f(x) = \exp(x)$ that are obtained by using the coefficient polynomials in Tables 4.13 and 4.14, respectively.

Solution. Using equally spaced nodes we get the polynomial

$$P(x) = 0.99519577 + 0.99904923x + 0.54788486x^2 + 0.17615196x^3.$$

This is obtained by finding the function values

$$f(x_0) = \exp(-1) = 0.36787944, \qquad f(x_1) = \exp\left(\frac{-1}{3}\right) = 0.71653131,$$

$$f(x_2) = \exp\left(\frac{1}{3}\right) = 1.39561243, \qquad f(x_3) = \exp(1) = 2.71828183,$$

and using the coefficient polynomials $L_{3,k}(x)$ in Table 4.13, and forming the linear combination

$$P(x) = 0.36787944L_{3,0}(x) + 0.71653131L_{3,1}(x) + 1.39561243L_{3,2}(x) + 2.71828183L_{3,3}(x).$$

Similarly, when the Chebyshev nodes are used, we obtain

$$V(x) = 0.99461532 + 0.99893323x + 0.54290072x^2 + 0.17517569x^3.$$

Notice that the coefficients are different from those of $P(x)$. This is a consequence of using different nodes and function values:

$$f(x_0) = \exp(-0.92387953) = 0.39697597,$$

$$f(x_1) = \exp(-0.38268343) = 0.68202877,$$

$$f(x_2) = \exp(0.38268343) = 1.46621380,$$

$$f(x_3) = \exp(0.92387953) = 2.51904417.$$

Then the alternative set of coefficient polynomials $C_k(x)$ in Table 4.14 are used to form the linear combination:

$$V(x) = 0.39697597C_0(x) + 0.68202877C_1(x) + 1.46621380C_2(x) + 2.51904417C_3(x).$$

For a comparison of the accuracy of $P(x)$ and $V(x)$, the error functions are graphed in Figure 4.16(a) and (b), respectively. The maximum error $|\exp(x) - P(x)|$ occurs at $x = 0.75490129$, and

$$|\exp(x) - P(x)| \leq 0.00998481 \qquad \text{for } -1 \leq x \leq 1.$$

The maximum error $|\exp(x) - V(x)|$ occurs at $x = 1$, and we get

$$|\exp(x) - V(x)| \leq 0.00665687 \qquad \text{for } -1 \leq x \leq 1.$$

Notice that the maximum error in $V(x)$ is about two-thirds the maximum error in $P(x)$. Also, the error is spread out more evenly over the interval.

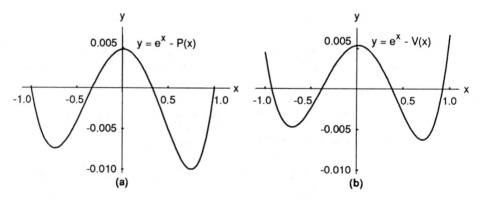

Figure 4.16   (a) The error function $y = e^x - P(x)$ for Lagrange approximation over $[-1, 1]$. (b) The error function $y = e^x - V(x)$ for Lagrange approximation over $[-1, 1]$.

Runge Phenomenon

We now look deeper to see the advantage of using the Chebyshev interpolation nodes. Consider Lagrange interpolating to $f(x)$ over the interval $[-1, 1]$ based on equally spaced nodes. Does the error $E_N(x) = f(x) - P_N(x)$ tend to zero as N increases? For functions like $\sin(x)$ or $\exp(x)$ where all the derivatives are bounded by the same constant M, the answer is yes. In general the answer to this question is no, and it is easy to find functions for which the sequence $\{P_N(x)\}$ does not converge. If $f(x) = 1/(1 + 12x^2)$, the maximum of the error term $E_N(x)$ grows when $N \to \infty$. This nonconvergence is called the **Runge phenomenon** (see Reference [90], pp. 275–278). The Lagrange polynomial of degree 10 based on 11 equally spaced nodes for this function is shown in Figure 4.17(a). Wild oscillations occur near the end of the interval. If the number of nodes is increased, then the oscillations become larger. This problem occurs because the nodes are equally spaced!

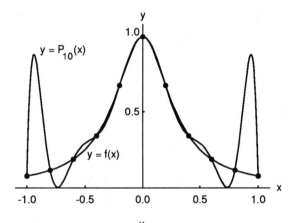

Figure 4.17   (a) The polynomial approximation to $y = 1/(1 + 12x^2)$ based on 11 equally spaced nodes over $[-1, 1]$.

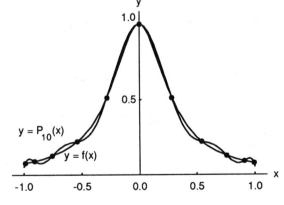

Figure 4.17   (b) The polynomial approximation to $y = 1/(1 + 12x^2)$ based on 11 Chebyshev nodes over $[-1, 1]$.

If the Chebyshev nodes are used to construct an interpolating polynomial of degree 10 to $f(x) = 1/(1 + 12x^2)$, the error is much smaller as seen in Figure 4.17(b). Under the condition that Chebyshev nodes are used, the error $E_N(x)$ will go to zero as $N \to \infty$. In general, if $f(x)$ and $f'(x)$ are continuous on $[-1, 1]$, then it can be proven that Chebyshev interpolation will produce a sequence of polynomials $\{P_N(x)\}$ that converges uniformly to $f(x)$ over $[-1, 1]$.

Transforming the Interval

Sometimes it is necessary to take a problem stated on an interval $[a, b]$ and reformulate the problem on the interval $[c, d]$ where the solution is known. If the approximation $P_N(x)$ to $f(x)$ is to be obtained on the interval $[a, b]$, then we change the variable so that the problem is reformulated on $[-1, 1]$:

$$x = \left(\frac{b - a}{2}\right)t + \frac{a + b}{2} \quad \text{or} \quad t = 2\frac{x - a}{b - a} - 1, \tag{12}$$

where $a \leq x \leq b$ and $-1 \leq t \leq 1$.

The required Chebyshev nodes of $T_{N+1}(t)$ on $[-1, 1]$ are

$$t_k = \cos\left((2N + 1 - 2k)\frac{\pi}{2N + 2}\right) \quad \text{for } k = 0, 1, \ldots, N. \tag{13}$$

and the interpolating nodes on $[a, b]$ are obtained by using (11):

$$x_k = t_k \frac{b - a}{2} + \frac{a + b}{2} \quad \text{for } k = 0, 1, \ldots, N. \tag{14}$$

The next result gives the error bound for this approximation on $[a, b]$.

Theorem 4.9   (Lagrange–Chebyshev Approximation Polynomial).   Assume that $P_N(x)$ is the Lagrange polynomial that is based on the Chebyshev nodes given in (14). If $f \in C^{N+1}[a, b]$, then

$$|f(x) - P_N(x)| \leq \frac{2(b - a)^{N+1}}{4^{N+1}(N + 1)!} \max_{a \leq x \leq b} \{|f^{(N+1)}(x)|\}. \tag{15}$$

Example 4.15

For $f(x) = \sin(x)$ on $[0, \pi/4]$, find the Chebyshev nodes and the error bound (14) for the Lagrange polynomial $P_5(x)$.

Solution.   Formulas (12) and (13) are used to find the nodes;

$$x_k = \cos\left(\frac{(11 - 2k)\pi}{12}\right)\frac{\pi}{8} + \frac{\pi}{8} \quad \text{for } k = 0, 1, \ldots, 5.$$

Using the bound $|f^{(6)}(x)| \leq |-\sin(\pi/4)| = 2^{-1/2} = M$ in (14), we get

$$|f(x) - P_N(x)| \leq \left(\frac{\pi}{8}\right)^6\left(\frac{2}{6!}\right)2^{-1/2} \leq 0.00000720.$$

Orthogonal Property

In Example 4.14, the Chebyshev nodes were used to find the Lagrange interpolating polynomial. In general this implies that the Chebyshev polynomial of degree N can be obtained by Lagrange interpolation based on the $N + 1$ nodes that are the $N + 1$ zeros of

$T_{N+1}(x)$. However, a direct approach to finding the approximation polynomial is to express $P_N(x)$ as a linear combination of the polynomials $T_k(x)$ which were given in Table 4.12. Therefore, the Chebyshev interpolating polynomial can be written in the form:

$$P_N(x) = \sum_{k=0}^{N} c_k T_k(x) = c_0 T_0(x) + c_1 T_1(x) + \cdots + c_N T_N(x). \tag{16}$$

The coefficients $\{c_k\}$ in (16) are easy to find. The technical proof requires the use of following orthogonality properties. Let

$$x_k = \cos\left(\pi \frac{2k + 1}{2N + 2}\right) \qquad \text{for } k = 0, 1, \ldots, N; \tag{17}$$

$$\sum_{k=0}^{N} T_i(x_k) T_j(x_k) = 0 \qquad \text{when } i \neq j, \tag{18}$$

$$\sum_{k=0}^{N} T_i(x_k) T_j(x_k) = \frac{N + 1}{2} \qquad \text{when } i = j \neq 0, \tag{19}$$

$$\sum_{k=0}^{N} T_0(x_k) T_0(x_k) = N + 1. \tag{20}$$

Property 4 and the identities (18) to (20) can be used to prove the following theorem.

Theorem 4.10   (Chebyshev Approximation).   The Chebyshev approximation polynomial $P_N(x)$ of degree $\leq N$ for $f(x)$ over $[-1, 1]$ can be written as a sum of $\{T_j(x)\}$:

$$f(x) \approx P_N(x) = \sum_{j=1}^{N} c_j T_j(x). \tag{21}$$

The coefficients $\{c_j\}$ are computed with the formulas

$$c_0 = \frac{1}{N + 1} \sum_{k=0}^{N} f(x_k) T_0(x_k) = \frac{1}{N + 1} \sum_{k=0}^{N} f(x_k) \tag{22}$$

and

$$c_j = \frac{2}{N + 1} \sum_{k=0}^{N} f(x_k) T_j(x_k)$$

$$= \frac{2}{N + 1} \sum_{k=0}^{N} f(x_k) \cos\left(\frac{j\pi(2k + 1)}{2N + 2}\right) \qquad \text{for } j = 1, 2, \ldots, N. \tag{23}$$

Example 4.16

Find the Chebyshev polynomial $P_3(x)$ that approximates the function $f(x) = \exp(x)$ over $[-1, 1]$.

Solution. The coefficients are calculated using (22) and (23), and the nodes $x_k = \cos(\pi(2k + 1)/8)$ for $k = 0, 1, 2, 3$.

$$c_0 = \frac{1}{4} \sum_{k=0}^{3} \exp(x_k)T_0(x_k) = \frac{1}{4} \sum_{k=0}^{3} \exp(x_k) = 1.26606568,$$

$$c_1 = \frac{1}{2} \sum_{k=0}^{3} \exp(x_k)T_1(x_k) = \frac{1}{2} \sum_{k=0}^{3} \exp(x_k)x_k = 1.13031500,$$

$$c_2 = \frac{1}{2} \sum_{k=0}^{3} \exp(x_k)T_2(x_k) = \frac{1}{2} \sum_{k=0}^{3} \exp(x_k) \cos\left(2\pi \frac{2k + 1}{8}\right) = 0.27145036,$$

$$c_3 = \frac{1}{2} \sum_{k=0}^{3} \exp(x_k)T_3(x_k) = \frac{1}{2} \sum_{k=0}^{3} \exp(x_k) \cos\left(3\pi \frac{2k + 1}{8}\right) = 0.04379392.$$

Therefore, the Chebyshev polynomial $P_3(x)$ for $\exp(x)$ is

$$P_3(x) = 1.26606568T_0(x) + 1.13031500T_1(x) + 0.27145036T_2(x) + 0.04379392T_3(x). \quad (24)$$

If the Chebyshev polynomial (24) is expanded in powers of x, the result is

$$P_3(x) = 0.99461532 + 0.99893324x + 0.54290072x^2 + 0.17517568x^3,$$

which is the same as the polynomial $V(x)$ in Example 4.14. If the goal is to find the Chebyshev polynomial, formulas (22) and (23) are preferred.

Algorithm 4.6 (Chebyshev Approximation). To construct and evaluate the Chebyshev interpolating polynomial of degree N over the interval $[-1, 1]$, where

$$P(x) = \sum_{j=0}^{N} c_j T_j(x)$$

is based on the nodes

$$x_k = \cos\left(\frac{(2k + 1)\pi}{2N + 2}\right).$$

INPUT N   {Degree of polynomial}
Pi := 3.1415926535
D := Pi/(2*N+2)

```
FOR   K = 0   TO   N   DO
      X(K) := COS((2*K+1)*D)                          {Make the nodes}
      Y(K) := F(X(K))                            {Store the function values}
      C(K) := 0                                   {Initialize the array}
```

```
FOR   K = 0   DO   N   DO                              {Calculate
      Z := (2*K+1)*D                                   the sums
      FOR   J = 0   TO   N   DO                         for the
            C(J) := C(J) + Y(K)*COS(J*Z)             coefficients}
```

```
      C(0) := C(0)/(N+1)                              {Multiple of
FOR   J = 1   TO   N   DO                              sum forms
      C(J) := 2*C(J)/(N+1)                           coefficient}
```

{Procedure for evaluating the Chebyshev polynomial approximation}

```
INPUT   X                                      {Input independent variable}
```

```
      T(0) := 1, T(1) := X                        (Recursively generate
IF    N>1   THEN                                     the Chebyshev
      FOR   J = 1   TO   N−1   DO                      coefficient
            T(J+1) := 2*X*T(J) − T(J−1)              polynomials}
```

```
      P := C(0)*T(0)                                  {Evaluate
FOR   J = 1   TO   N   DO                             Chebyshev
      P := P + C(J)*T(J)                             polynomial}
```

PRINT "The value of the Chebyshev poly. P(x) is" P   {Output}

EXERCISES FOR CHEBYSHEV POLYNOMIALS

1. Use property (1) and
   (a) construct $T_4(x)$ from $T_3(x)$ and $T_2(x)$.
   (b) construct $T_5(x)$ from $T_4(x)$ and $T_3(x)$.
2. Use property (1) and
   (a) construct $T_6(x)$ from $T_5(x)$ and $T_4(x)$.
   (b) construct $T_7(x)$ from $T_6(x)$ and $T_5(x)$.
3. Use mathematical induction to prove property 2.
4. Use mathematical induction to prove property 3.
5. Find the maximum and minimum values of $T_2(x)$.

6. Find the maximum and minimum values of $T_3(x)$. *Hint.* $T_3'\left(\frac{1}{2}\right) = 0$ and $T_3'\left(-\frac{1}{2}\right) = 0$.

7. Find the maximum and minimum values of $T_4(x)$. *Hint.* $T_4'(0) = 0$, $T_4'(2^{-1/2}) = 0$, and $T_4'(-2^{-1/2}) = 0$.

8. Let $f(x) = \sin(x)$ on $[-1, 1]$.
   (a) Use the coefficient polynomials in Table 4.14 to obtain the Lagrange–Chebyshev polynomial approximation $P_3(x)$.
   (b) Find the error bound for $|\sin(x) - P_3(x)|$.

9. Let $f(x) = \ln(x + 2)$ on $[-1, 1]$.
   (a) Use the coefficient polynomials in Table 4.14 to obtain the Lagrange–Chebyshev polynomial approximation $P_3(x)$.
   (b) Find the error bound for $|\ln(x + 2) - P_3(x)|$.

10. The Lagrange polynomial of degree $N = 2$ has the form

$$f(x) = f(x_0)L_{2,0}(x) + f(x_1)L_{2,1}(x) + f(x_2)L_{2,2}(x).$$

If the Chebyshev nodes $x_0 = \cos(5\pi/6)$, $x_1 = 0$, and $x_2 = \cos(\pi/6)$ are used, show that the coefficient polynomials are

$$L_{2,0}(x) = -0.57735027x + \frac{2x^2}{3} = \frac{-x}{\sqrt{3}} + \frac{2x^2}{3},$$

$$L_{2,1}(x) = 1.0000000 - \frac{4x^2}{3} = 1 - \frac{4x^2}{3},$$

$$L_{2,2}(x) = 0.57735027x + \frac{2x^2}{3} = \frac{x}{\sqrt{3}} + \frac{2x^2}{3}.$$

11. Let $f(x) = \cos(x)$ on $[-1, 1]$.
   (a) Use the coefficient polynomials in Exercise 10 to get the Lagrange–Chebyshev polynomial approximation $P_2(x)$.
   (b) Find the error bound for $|\cos(x) - P_2(x)|$.

12. Let $f(x) = \exp(x)$ on $[-1, 1]$.
   (a) Use the coefficient polynomials in Exercise 10 to get the Lagrange–Chebyshev polynomial approximation $P_2(x)$.
   (b) Find the error bound for $|\exp(x) - P_2(x)|$.

In Exercises 13–15, compare the Taylor polynomial and the Lagrange–Chebyshev approximation to $f(x)$ on $[-1, 1]$. Find their error bounds.

13. $f(x) = \sin(x)$ and $N = 7$, the Lagrange–Chebyshev polynomial is

$$\sin(x) \approx 0.99999998x - 0.16666599x^3 + 0.00832995x^5 - 0.00019297x^7.$$

14. $f(x) = \cos(x)$ and $N = 6$, the Lagrange–Chebyshev polynomial is

$$\cos(x) \approx 1 - 0.49999734x^2 + 0.04164535x^4 - 0.00134608x^6.$$

15. $f(x) = \exp(x)$ and $N = 7$, the Lagrange–Chebyshev polynomial is

$$\exp(x) \approx 0.99999980 + 0.99999998x + 0.50000634x^2 + 0.16666737x^3$$

$$+ 0.04163504x^4 + 0.00832984x^5 + 0.00143925x^6 + 0.00020399x^7.$$

In Exercises 16–20, use formulas (22) and (23) to compute the coefficients $\{c_k\}$ for the Chebyshev polynomial approximation $P_N(x)$ to $f(x)$ over $[-1, 1]$, when **(a)** $N = 3$, **(b)** $N = 4$, **(c)** $N = 5$, and **(d)** $N = 6$. **(e)** Use a computer program to solve parts (a) to (d).

16. $f(x) = \exp(x)$ **17.** $f(x) = \sin(x)$

18. $f(x) = \cos(x)$ **19.** $f(x) = \ln(x + 2)$

20. $f(x) = (x + 2)^{1/2}$

21. Write a report on the economization of power series. See References [3, 9, 29, 41, 51, 62, 76, 85, 88, 117, 153, and 184].

22. Write a report on Legendre polynomials. See References [9, 29, 40, 41, 75, 152, and 153].

23. Write a report on the Remes algorithm. See References [9, 19, 56, 88, 128, 149, 152, and 153].

24. Prove equation (18).   **25.** Prove equation (19).

4.6 PADÉ APPROXIMATIONS

In this section we introduce the notion of rational approximations for functions. The function $f(x)$ will be approximated over a small portion of its domain. For example, if $f(x) = \cos(x)$, it is sufficient to have a formula to generate approximations on the interval $[0, \pi/2]$. Then trigonometric identities can be used to compute $\cos(x)$ for any value x that lies outside $[0, \pi/2]$.

A rational approximation to $f(x)$ on $[a, b]$ is the quotient of two polynomials $P_N(x)$ and $Q_M(x)$ of degrees N and M, respectively. We use the notation $R_{N,M}(x)$ to denote this quotient:

$$R_{N,M}(x) = \frac{P_N(x)}{Q_M(x)} \qquad \text{for } a \le x \le b. \tag{1}$$

Our goal is to make the maximum error as small as possible. For a given amount of computational effort, one can usually construct a rational approximation that has a smaller overall error on $[a, b]$ than a polynomial approximation. Our development is an introduction and will be limited to Padé approximations.

The *method of Padé* requires that $f(x)$ and its derivatives are continuous at $x = 0$. There are two reasons for the arbitrary choice of $x = 0$. First, it makes the manipulations simpler. Second, a change of variable can be used to shift the calculations over to an interval that contains zero. The polynomials used in (1) are

$$P_N(x) = p_0 + p_1 x + p_2 x^2 + \cdots + p_N x^N \tag{2}$$

and

$$Q_M(x) = 1 + q_1 x + q_2 x^2 + \cdots + q_M x^M. \tag{3}$$

The polynomials in (2) and (3) are constructed so that $f(x)$ and $R_{N,M}(x)$ agree at $x = 0$ and their derivatives up to $N + M$ agree at $x = 0$. In the case $Q_0(x) = 1$, the approximation is just the Maclaurin expansion for $f(x)$. For a fixed value of $N + M$ the error is smallest when $P_N(x)$ and $Q_M(x)$ have the same degree or when $P_N(x)$ has degree one higher than $Q_M(x)$.

Notice that the constant coefficient of $Q_M(x)$ is $q_0 = 1$. This is permissible, because it cannot be 0 and $R_{N,M}(x)$ is not changed when both $P_N(x)$ and $Q_M(x)$ are divided by the same constant. Hence the rational function $R_{N,M}(x)$ has $N + M + 1$ unknown coefficients. Assume that $f(x)$ is analytic, and has the Maclaurin expansion

$$f(x) = a_0 + a_1 x + a_2 x^2 + \cdots + a_k x^k + \cdots, \tag{4}$$

and form the difference $f(x)Q_M(x) - P_N(x) = Z(x)$:

$$\left(\sum_{j=0}^{\infty} a_j x^j \right) \left(\sum_{j=0}^{M} q_j x^j \right) - \sum_{j=0}^{N} p_j x^j = \sum_{j=N+M+1}^{\infty} c_j x^j. \tag{5}$$

The lower index $j = M + N + 1$ in the summation on the right side of (5) is chosen because the first $N + M$ derivatives of $f(x)$ and $R_{N,M}(x)$ are to agree at $x = 0$.

When the left side of (5) is multiplied out and the coefficients of the powers of x^j are set equal to zero for $j = 0, 1, \ldots, N + M$, the result is a system of $N + M + 1$ linear equations:

$$a_0 - p_0 = 0$$

$$q_1 a_0 + a_1 - p_1 = 0$$

$$q_2 a_0 + q_1 a_1 + a_2 - p_2 = 0 \tag{6}$$

$$q_3 a_0 + q_2 a_1 + q_1 a_2 + a_3 - p_3 = 0$$

$$q_M a_{N-M} + q_{M-1} a_{N-M+1} \cdots + a_N - p_N = 0$$

and

$$q_M a_{N-M+1} + q_{M-1} a_{N-M+2} + \cdots + q_1 a_N \qquad + a_{N+1} = 0$$

$$q_M a_{N-M+2} + q_{M-1} a_{N-M+3} + \cdots + q_1 a_{N+1} \qquad + a_{N+2} = 0 \tag{7}$$

$$\vdots \qquad\qquad\qquad\qquad\qquad\qquad\qquad\qquad\qquad\qquad \vdots$$

$$q_M a_N \qquad + q_{M-1} a_{N+1} \qquad + \cdots + q_1 a_{N+M-1} + a_{N+M} = 0.$$

Notice that in each equation the sum of the subscripts on the factors of each product is the same, and this sum increases consecutively from 0 to $N + M$. The M equations in (7) involve only the unknowns $q_1, q_2, \ldots, q_M$ and must be solved first. Then the equations in (6) are used successively to find $p_0, p_1, \ldots p_N$.

Example 4.17

Establish the Padé approximation

$$\cos(x) \approx R_{4,4}(x) = \frac{15{,}120 - 6900 x^2 + 313 x^4}{15{,}120 + 660 x^2 + 13 x^4}. \tag{8}$$

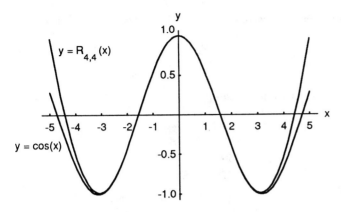

Figure 4.18 The graph of $y = \cos(x)$ and its Padé approximation $R_{4,4}(x)$.

See Figure 4.18 for the graphs of $\cos(x)$ and $R_{4,4}(x)$ over $[-5, 5]$.

Solution. If the Maclaurin expansion for $\cos(x)$ was used, we would obtain nine equations in nine unknowns. Instead, notice that both $\cos(x)$ and $R_{4,4}(x)$ are even functions and involve powers of (x^2). We can simplify the computations if we start with $f(x) = \cos(x^{1/2})$:

$$f(x) = 1 - \frac{1}{2}x + \frac{1}{24}x^2 - \frac{1}{720}x^3 + \frac{1}{40,320}x^4 - \cdots . \tag{9}$$

In this case equation (5) becomes

$$\left(1 - \frac{x}{2} + \frac{x^2}{24} - \frac{x^3}{720} + \frac{x^4}{40,320} + \ldots\right)(1 + q_1x + q_2x^2) - p_0 - p_1x - p_2x^2$$

$$= 0 + 0x + 0x^2 + 0x^3 + 0x^4 + c_5x^5 + c_6x^6 + \cdots .$$

When the coefficients of the first five powers of x are compared, we get the following system of linear equations:

$$1 - p_0 = 0$$

$$-\frac{1}{2} + q_1 - p_1 = 0$$

$$\frac{1}{24} - \frac{1}{2}q_1 + q_2 - p_2 = 0 \tag{10}$$

$$-\frac{1}{720} + \frac{1}{24}q_1 - \frac{1}{2}q_2 = 0$$

$$\frac{1}{40,320} - \frac{1}{720}q_1 + \frac{1}{24}q_2 = 0.$$

The last two equations in (10) must be solved first. They can be rewritten in a form that is easy to solve:

$$q_1 - 12q_2 = \frac{1}{30} \quad \text{and} \quad -q_1 + 30q_2 = \frac{-1}{56}.$$

First find q_2 by adding the equations, then find q_1:

$$q_2 = \frac{1}{18}\left(\frac{1}{30} - \frac{1}{56}\right) = \frac{13}{15,120},$$

$$q_1 = \frac{1}{30} + \frac{156}{15,120} = \frac{11}{252}. \tag{11}$$

Now the first three equations of (10) are used. It is obvious that $p_0 = 1$, and we can use q_1 and q_2 in (11) to solve for p_1 and p_2:

$$p_1 = -\frac{1}{2} + \frac{11}{252} = -\frac{115}{252},$$

$$p_2 = \frac{1}{24} - \frac{11}{504} + \frac{13}{15,120} = \frac{313}{15,120}. \tag{12}$$

Now use the coefficients in (11) and (12) to form the rational approximation to $f(x)$:

$$f(x) \approx \frac{1 - 115x/252 + 313x^2/15,120}{1 + 11x/252 + 13x^2/15,120}. \tag{13}$$

Since $\cos(x) = f(x^2)$, we can substitute x^2 for x in equation (13) and the result is the formula for $R_{4,4}(x)$ in (8).

Continued Fraction Form

The Padé approximation $R_{4,4}(x)$ in Example 4.13 requires a minimum of 12 arithmetic operations to perform an evaluation. It is possible to reduce this number to seven by the use of continued fractions. This is accomplished by starting with (8) and finding the quotient and its polynomial remainder.

$$R_{4,4}(x) = \frac{15,120/313 - (6900/313)x^2 + x^4}{15,120/13 + (660/13)x^2 + x^4}$$

$$= \frac{313}{13} - \frac{296,280}{169} \frac{12,600/823 + x^2}{15,120/13 + (660/13)x^2 + x^4}.$$

The process is carried out once more using the term in the previous remainder. The result is

$$R_{4,4}(x) = \frac{313}{13} - \frac{296,280/169}{\dfrac{15,120/13 + (660/13)x^2 + x^4}{12,600/823 + x^2}}$$

$$= \frac{313}{13} - \cfrac{\dfrac{296{,}280}{169}}{\dfrac{379{,}380}{10{,}699} + x^2 + \cfrac{420{,}078{,}960/677{,}329}{12{,}600/823 + x^2}}$$

The fractions are converted to decimal form for computational purposes and we obtain

$$R_{4,4}(x) = 24.07692308 - \cfrac{1753.13609467}{35.45938873 + x^2 + 620.19928277/(15.30984204 + x^2)}.$$

$$(14)$$

To evaluate (14), first compute and store x^2, then proceed from the bottom right term in the denominator and tally the operations: addition, division, addition, addition, division, and subtraction. Hence it takes a total of seven arithmetic operations to evalute $R_{4,4}(x)$ in continued fraction form in (14).

We can compare $R_{4,4}(x)$ with the Taylor polynomial $P_6(x)$ of degree $n = 6$, which requires seven arithmetic operations to evaluate when it is written in the nested form

$$P_6(x) = 1 + x^2\left(-\tfrac{1}{2} + x^2\left(\tfrac{1}{24} - \tfrac{1}{720}x^2\right)\right)$$
$$= 1 + x^2(-0.5 + x^2(0.0416666667 - 0.0013888889x^2)).$$

$$(15)$$

The graphs of $E_R(x) = \cos(x) - R_{4,4}(x)$ and $E_P(x) = \cos(x) - P_6(x)$ over $[-1, 1]$ are shown in Figure 4.19(a) and (b), respectively. The largest errors occur at the endpoints and are $E_R(1) = -0.0000003599$ and $E_T(1) = 0.0000245281$, respectively. The magnitude of the largest error for $R_{4,4}(x)$ is about 1.467% of the error for $P_6(x)$. The Padé approximation outperforms the Taylor approximation better on smaller intervals, and

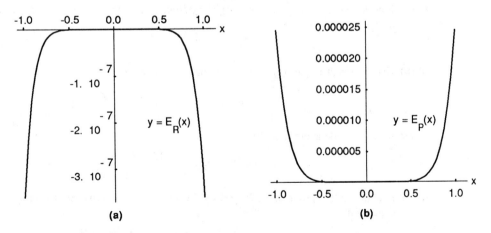

Figure 4.19 (a) Graph of the error $E_R(x) = \cos(x) - R_{4,4}(x)$ for the Padé approximation $R_{4,4}(x)$. (b) Graph of the error $E_P(x) = \cos(x) - P_6(x)$ for the Taylor approximation $P_6(x)$.

over $[-0.1, 0.1]$ we find that $E_R(1) = -0.0000000004$ and $E_T(1) = 0.0000000966$ so that the magnitude of the error for $R_{4,4}(x)$ is about 0.384% of the magnitude of the error for $P_6(x)$.

EXERCISES FOR PADÉ APPROXIMATIONS

1. Establish the Padé approximation:

$$\exp(x) \approx R_{1,1}(x) = \frac{2 + x}{2 - x}.$$

2. **(a)** Find the Padé approximation $R_{1,1}(x)$ for $f(x) = \ln(1 + x)/x$. *Hint.* Start with the Maclaurin expansion:

$$f(x) = 1 - \frac{x}{2} + \frac{x^2}{3} - \cdots .$$

   (b) Use the result in part (a) to establish the approximation

$$\ln(1 + x) \approx R_{2,1}(x) = \frac{6x + x^2}{6 + 4x}.$$

3. **(a)** Find $R_{1,1}(x)$ for $f(x) = \tan(x^{1/2})/x^{1/2}$. *Hint.* Start with the Maclaurin expansion:

$$f(x) = 1 + \frac{x}{3} + \frac{2x^2}{15} + \cdots .$$

   (b) Use the result in part (a) to establish the approximation

$$\tan(x) \approx R_{3,2}(x) = \frac{15x - x^3}{15 - 6x^2}.$$

4. **(a)** Find $R_{1,1}(x)$ for $f(x) = \arctan(x^{1/2})/x^{1/2}$. *Hint.* Start with the Maclaurin expansion:

$$f(x) = 1 - \frac{x}{3} + \frac{x^2}{5} - \cdots .$$

   (b) Use the result in part (a) to establish the approximation

$$\arctan(x) \approx R_{3,2}(x) = \frac{15x + 4x^3}{15 + 9x^2}.$$

5. Establish the Padé approximation:

$$\exp(x) \approx R_{2,2}(x) = \frac{12 + 6x + x^2}{12 - 6x + x^2}.$$

6. **(a)** Find the Padé approximation $R_{2,2}(x)$ for $f(x) = \ln(1 + x)/x$. *Hint.* Start with the Maclaurin expansion:

$$f(x) = 1 - \frac{x}{2} + \frac{x^2}{3} - \frac{x^3}{4} + \frac{x^4}{5} - \cdots .$$

(b) Use the result in part (a) to establish

$$\ln(1 + x) \approx R_{3,2}(x) = \frac{30x + 21x^2 + x^3}{30 + 36x + 9x^2}.$$

7. Find $R_{2,2}(x)$ for $f(x) = \tan(x^{1/2})/x^{1/2}$. *Hint*. Start with the Maclaurin expansion:

$$f(x) = 1 + \frac{x}{3} + \frac{2x^2}{15} + \frac{17x^3}{315} + \frac{62x^4}{2835} + \cdots .$$

(b) Use the result in part (a) to establish

$$\tan(x) \approx R_{5,4}(x) = \frac{945x - 105x^3 + x^5}{945 - 420x^2 + 15x^4}.$$

8. (a) Find $R_{2,2}(x)$ for $f(x) = \arctan(x^{1/2})/x^{1/2}$. *Hint*. Start with the Maclaurin expansion:

$$f(x) = 1 - \frac{x}{3} + \frac{x^2}{5} - \frac{x^3}{7} + \frac{x^4}{9} - \cdots .$$

(b) Use the result in part (a) to establish

$$\arctan(x) \approx R_{5,4}(x) = \frac{945x + 735x^3 + 64x^5}{945 + 1050x^2 + 225x^4}.$$

9. Establish the Padé approximation:

$$\exp(x) \approx R_{2,2}(x) = \frac{120 + 60x + 12x^2 + x^3}{120 - 60x + 12x^2 - x^3}.$$

10. Establish the Padé approximation:

$$\exp(x) \approx R_{4,4}(x) = \frac{1680 + 840x + 180x^2 + 20x^3 + x^4}{1680 - 840x + 180x^2 - 20x^3 + x^4}.$$

11. Compare the following approximations to $f(x) = \exp(x)$.

$$\text{Taylor:} \quad T_4(x) = 1 + x + \frac{x^2}{2} + \frac{x^3}{6} + \frac{x^4}{24}$$

$$\text{Padé:} \quad R_{2,2}(x) = \frac{12 + 6x + x^2}{12 - 6x + x^2}$$

Evaluate $T_4(x)$ and $R_{2,2}(x)$ for $x = -0.8, -0.4, 0.4,$ and 0.8.

12. Compare the following approximations to $f(x) = \ln(1 + x)$.

$$\text{Taylor:} \quad T_5(x) = x - \frac{x^2}{2} + \frac{x^3}{3} - \frac{x^4}{4} + \frac{x^5}{5}$$

$$\text{Padé:} \quad R_{3,2}(x) = \frac{30x + 21x^2 + x^3}{30 + 36x + 9x^2}$$

Evaluate $T_5(x)$ and $R_{3,2}(x)$ for $x = -0.6, -0.2, 0.2,$ and 0.6.

13. Compare the following approximations to $f(x) = \tan(x)$.

$$\text{Taylor:} \quad T_9(x) = x + \frac{x^3}{3} + \frac{2x^5}{15} + \frac{17x^7}{315} + \frac{62x^9}{2835}$$

$$\text{Padé:} \quad R_{5,4}(x) = \frac{945x - 105x^3 + x^5}{945 - 420x^2 + 15x^4}$$

Evaluate $T_9(x)$ and $R_{5,4}(x)$ for $x = 0.4, 0.8,$ and 1.2.

14. Compare the following Padé approximations to $f(x) = \sin(x)$ over the interval $[-1.2, 1.2]$.

(a) $R_{5,4}(x) = \dfrac{166320x - 22260x^3 + 551x^5}{15(11088 + 364x^2 + 5x^4)}$

(b) $R_{7,6}(x) = \dfrac{11511339840x - 1640635920x^3 + 52785432x^5 - 479249x^7}{7(1644477120 + 39702960x^2 + 453960x^4 + 2623x^6)}$

15. Compare the following Padé approximations to $f(x) = \cos(x)$ over the interval $[-1.2, 1.2]$.

$$R_{6,6}(x) = \frac{39251520 - 18471600x^2 + 1075032x^4 - 14615x^6}{39251520 + 1154160x^2 + 16632x^4 + 127x^6)}$$

5

Curve Fitting

Application of numerical techniques in science and engineering often involve "curve fitting" of experimental data. In 1601 the German astronomer Johannes Kepler formulated the third law of planetary motion, $T = Cx^{3/2}$, where x is the distance to the sun measured in millions of kilometers, T is the orbital period measured in days, and C is a constant. The observed data pairs (x, T) for the first four planets Mercury, Venus, Earth, and Mars are (58, 88), (108, 225), (150, 365), and (228, 687), and the coefficient C obtained from the method of "least squares" is $C = 0.199769$. The curve $T = 0.199769x^{3/2}$ and the data points are shown in Figure 5.1.

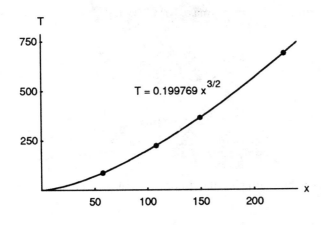

Figure 5.1 The least-squares fit $T = 0.199769x^{3/2}$ for the first four planets using Kepler's third law of planetary motion.

5.1 LEAST-SQUARES LINE

In science and engineering it is often the case that an experiment produces a set of data points $(x_1, y_1), \ldots, (x_N, y_N)$, where the abscissas $\{x_k\}$ are distinct. One goal of numerical methods is to determine a formula $y = f(x)$ that relates these variables. Usually, a class of allowable formulas is chosen and then coefficients must be determined. There are many possibilities for the type of function that can be used. Often there is an underlying mathematical model, based on the physical situation, that will determine the form of the function. In this section we emphasize the class of linear functions of the form

$$y = f(x) = Ax + B. \tag{1}$$

In Chapter 4 we saw how to construct a polynomial that passes through a set of points. If all the numerical values $\{x_k\}$, $\{y_k\}$ are known to several significant digits of accuracy, then polynomial interpolation can be used successfully, otherwise it cannot. Some experiments are devised using specialized equipment so that the data points will have at least five digits of accuracy. However, many experiments are done with equipment that is reliable only to three or fewer digits of accuracy. Often there is an experimental error in the measurements, and although three digits are recorded for the values $\{x_k\}$ and $\{y_k\}$, it is realized that the true value $f(x_k)$ satisfies

$$f(x_k) = y_k + e_k, \tag{2}$$

where e_k is the measurement error.

How do we find the best linear approximation of form (1) that goes near (not always through) the points? To answer this question we need to discuss the **errors** (also called **deviations** or **residuals**);

$$e_k = f(x_k) - y_k \quad \text{for } 1 \le k \le N. \tag{3}$$

There are several norms that can be used with the residuals in (3) to measure how far the curve $y = f(x)$ lies from the data.

$$\text{Maximum error:} \quad E_\infty(f) = \max_{1 \le k \le N} \{|f(x_k) - y_k|\}. \tag{4}$$

$$\text{Average error:} \quad E_1(f) = \frac{1}{N} \sum_{k=1}^{N} |f(x_k) - y_k|. \tag{5}$$

$$\text{Root-mean-square (RMS) error:} \quad E_2(f) = \left[\frac{1}{N} \sum_{k=1}^{N} |f(x_k) - y_k|^2 \right]^{1/2}. \tag{6}$$

The next example shows how to apply these norms when a function and a set of points are given.

Example 5.1

Compare the maximum error, average error, and rms error for the linear approximation $y = f(x) = 8.6 - 1.6x$ to the data points $(-1, 10)$, $(0, 9)$, $(1, 7)$, $(2, 5)$, $(3, 4)$, $(4, 3)$, $(5, 0)$, and $(6, -1)$.

Solution.   The errors are found using the values for $f(x_k)$ and e_k given in Table 5.1.

TABLE 5.1   Calculations for Finding $E_1(f)$ and $E_2(f)$ for Example 5.1

x_k	y_k	$f(x_k) = 8.6 - 1.6x_k$	$\|e_k\|$	e_k^2
-1	10.0	10.2	0.2	0.04
0	9.0	8.6	0.4	0.16
1	7.0	7.0	0.0	0.00
2	5.0	5.4	0.4	0.16
3	4.0	3.8	0.2	0.04
4	3.0	2.2	0.8	0.64
5	0.0	0.6	0.6	0.36
6	-1.0	-1.0	0.0	0.00
			2.6	1.40

$$E_\infty(f) = \max\{0.2, 0.4, 0.0, 0.4, 0.2, 0.8, 0.6, 0.0\} = 0.8, \tag{7}$$

$$E_1(f) = \frac{1}{8}(2.6) = 0.325, \tag{8}$$

$$E_2(f) = \left(\frac{1.4}{8}\right)^{1/2} \approx 0.41833. \tag{9}$$

We can see that the maximum error is largest, and if one point is badly in error, its value determines $E_\infty(f)$. The average error $E_1(f)$ simply averages the absolute value of the error at the various points. It is often used because it is easy to compute. The error $E_2(f)$ is often used when the statistical nature of the errors is considered.

A "best-fitting" line is found by minimizing one of the quantities in equations (4) to (6). Hence there are three best-fitting lines that we could find. The third norm $E_2(f)$ is the traditional choice because it is much easier to minimize $E_2(f)$ computationally.

Finding the Least-Squares Line

A best-fitting line can be found by minimizing one of the quantities $E_\infty(f)$, $E_1(f)$, or $E_2(f)$ given in equations (4) to (6), respectively. Hence there are three possibilities for best-fitting lines. The third norm $E_2(f)$ is preferred and is the traditional choice because it has applications in statistics and it is much easier to minimize computationally. Let $\{(x_k, y_k)\}_{k=1}^{N}$ be a set of N points, where the abscissas $\{x_k\}$ are distinct. The **least-squares line** $y = f(x) = Ax + B$ is the line that minimizes the root-mean-square error $E_2(f)$.

The quantity $E_2(f)$ will be a minimum if and only if the quantity $N[E_2(f)]^2 =$ $\sum_{k=1}^{N} (Ax_k + B - y_k)^2$ is a minimum. The latter is visualized geometrically by minimizing the sum of the squares of the vertical distances from the points to the line. The next result explains this process.

Theorem 5.1 (Least-Squares Line). Suppose that $\{(x_k, y_k)\}_{k=1}^{N}$ are N points where the abscissas $\{x_k\}_{k=1}^{N}$ are distinct. The coefficients of the least-squares line

$$y = Ax + B$$

are the solution to the following linear system, known as the **normal equations**:

$$\left(\sum_{k=1}^{N} x_k^2\right) A + \left(\sum_{k=1}^{N} x_k\right) B = \sum_{k=1}^{N} x_k y_k,$$

$$\left(\sum_{k=1}^{N} x_k\right) A + NB = \sum_{k=1}^{N} y_k. \tag{10}$$

Proof. Geometrically, we start with the line $y = Ax + B$. The vertical distance d_k from the data point (x_k, y_k) to the point $(x_k, mx_k + b)$ on the line is $d_k = Ax_k + B - y_k$ (see Figure 5.2). We must minimize the sum of the squares of the vertical distances d_k:

$$E(A, B) = \sum_{k=1}^{N} (Ax_k + B - y_k)^2 = \sum_{k=1}^{N} d_k^2. \tag{11}$$

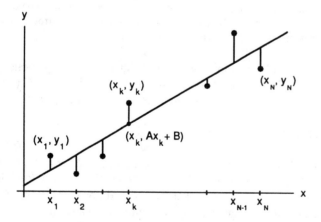

Figure 5.2 The vertical distances between the points $\{(x_k, y_k)\}$ and the least-squares line $y = Ax + B$.

The minimum value of $E(A, B)$ is determined by setting the partial derivatives $\partial E/\partial A$ and $\partial E/\partial B$ equal to zero and solving these equations for A and B. Notice that $\{x_k\}$ and $\{y_k\}$

are constants in equation (11) and that A and B are the variables! Hold B fixed and differentiate $E(A, B)$ with respect to A and get

$$\frac{\partial E(A, B)}{\partial A} = \sum_{k=1}^{N} 2(Ax_k + B - y_k)^1(x_k) = 2 \sum_{k=1}^{N} (Ax_k^2 + Bx_k - x_ky_k). \qquad (12)$$

Now hold A fixed and differentiate $E(A, B)$ with respect to B and get

$$\frac{\partial E(A, B)}{\partial B} = \sum_{k=1}^{N} 2(Ax_k + B - y_k)^1 = 2 \sum_{k=1}^{N} (Ax_k + B - y_k). \qquad (13)$$

Setting the partial derivatives equal to zero in (12) and (13), use the distributive properties of summation and obtain

$$0 = \sum_{k=1}^{N} (Ax_k^2 + Bx_k - x_ky_k) = A \sum_{k=1}^{N} x_k^2 + B \sum_{k=1}^{N} x_k - \sum_{k=1}^{N} x_ky_k, \qquad (14)$$

$$0 = \sum_{k=1}^{N} (Ax_k + B - y_k) = A \sum_{k=1}^{N} x_k + NB - \sum_{k=1}^{N} y_k. \qquad (15)$$

Equations (14) and (15) can be rearranged in the standard form for a system and results in the normal equations (10). The solution to this system can be obtained by applying Cramer's rule and is illustrated in the next example (also see Exercise 6). However, the method employed in Algorithm 5.1 translates the data points so that a well-conditioned matrix is employed (see Exercise 8).

Example 5.2

Find the least-squares line for the data points given in Example 5.1.

Solution.   The sums required for the normal equations (15) are easily obtained using the values in Table 5.2.

TABLE 5.2 Obtaining the Coefficients
for Normal Equations

x_k	y_k	x_k^2	x_ky_k
-1	10	1	-10
0	9	0	0
1	7	1	7
2	5	4	10
3	4	9	12
4	3	16	12
5	0	25	0
6	-1	36	-6
20	37	92	25

The linear system involving A and B is

$$92A + 20B = 25$$

$$20A + 8B = 37.$$

The solution is obtained by the calculations

$$A = \frac{\begin{vmatrix} 25 & 20 \\ 37 & 8 \end{vmatrix}}{\begin{vmatrix} 92 & 20 \\ 20 & 8 \end{vmatrix}} = \frac{-540}{336} \approx -1.6071429$$

and

$$B = \frac{\begin{vmatrix} 92 & 25 \\ 20 & 37 \end{vmatrix}}{\begin{vmatrix} 92 & 20 \\ 20 & 8 \end{vmatrix}} = \frac{2904}{336} \approx 8.6428571.$$

Therefore, the least-squares line is (see Figure 5.3)

$$y = -1.6071429x + 8.6428571.$$

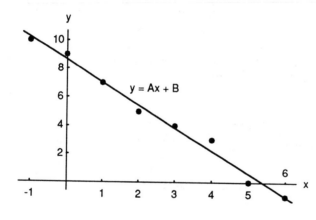

Figure 5.3   The least-squares line
$y = -1.6071429x + 8.6428571.$

The Power Fit $y = Ax^M$

Some situations involve $f(x) = Ax^M$, where M is a known constant. The example of planetary motion given in Figure 5.1 is an example. In these cases there is only one parameter A to be determined.

Theorem 5.2   (Power Fit).   Suppose that $\{(x_k, y_k)\}_{k=1}^{N}$ are N points where abscissas are distinct. The coefficient A of the least-squares power curve

$$y = Ax^M$$

is given by

$$A = \left(\sum_{k=1}^{N} x_k^M y_k\right) \Big/ \left(\sum_{k=1}^{N} x_k^{2M}\right). \tag{16}$$

Using the least-squares technique, we seek a minimum of the function $E(A)$:

$$E(A) = \sum_{k=1}^{N} (A x_k^M - y_k)^2. \tag{17}$$

In this case it will suffice to solve $E'(A) = 0$. The derivative is

$$E'(A) = 2 \sum_{k=1}^{N} (A x_k^M - y_k)^1 (x_k^M) = 2 \sum_{k=1}^{N} (A x_k^{2M} - x_k^M y_k). \tag{18}$$

Hence the coefficient A is the solution of the equation

$$0 = A \sum_{k=1}^{N} x_k^{2M} - \sum_{k=1}^{N} x_k^M y_k, \tag{19}$$

which reduces to the formula given in equation (16).

Example 5.3

Students collected the experimental data in Table 5.3. The relation is $d = \frac{1}{2} g t^2$, where d is distance in meters and t is time in seconds. Find the gravitational constant g.

Solution.   The values in Table 5.3 are used to find the summations required in formula (18), where the power used is $M = 2$.

TABLE 5.3 Obtaining the Coefficient for a Power Fit

Time, t_k	Distance, d_k	$d_k t_k^2$	t_k^4
0.200	0.1960	0.00784	0.0016
0.400	0.7850	0.12560	0.0256
0.600	1.7665	0.63594	0.1296
0.800	3.1405	2.00992	0.4096
1.000	4.9075	4.90750	1.0000
		7.68680	1.5664

The coefficient is $A = 7.68680/1.5664 = 4.9073$ and we get $d = 4.9073 t^2$ and $g = 2A = 9.8146$ m/sec^2.

Algorithm 5.1   (Least-Squares Line).   To construct the least-squares line $y = Ax + B$ that fits the N data points $(x_1, y_1), \ldots, (x_N, y_N)$.

Remark. The algorithm is computationally stable; it gives reliable results in cases when the linear system (15) is ill-conditioned (see Exercise 5).

```
        Read N                                    {Number of points}
FOR     K = 1  TO  N  DO
        READ X(K), Y(K)                           {Get the N points}

        Xmean := 0
FOR     K = 1  TO  N  DO                           {Find the mean x̄}
        Xmean := Xmean + X(K)
        Xmean := Xmean/N

        Ymean := 0
FOR     K = 1  TO  N  DO                           {Find the mean ȳ}
        Ymean := Ymean + Y(K)
        Ymean := Ymean/N

        SumX := 0                                  {Sum (xₖ − x̄)²}
FOR     K = 1  TO  N  DO
        SumX := SumX + [X(K) − Xmean]²

        SumXY := 0                                 {Sum (xₖ − x̄)(yₖ − ȳ)}
FOR     K = 1  TO  N  DO
        SumXY := SumXY + [X(K) − Xmean]*[Y(K) − Ymean]

        A := SumXY/SumX                            {Compute the slope}
        B := Ymean − A*Xmean                       {Compute the y-intercept}

PRINT "Least-squares line y = Ax+B"               {Output}
PRINT "The coefficients are"
PRINT "A = ", A
PRINT "B = ", B
```

EXERCISES FOR LEAST-SQUARES LINE

In Exercises 1–3, find the least-squares line $y = f(x) = Ax + B$ for the data and calculate $E_2(f)$.

1. (a)

x_k	y_k	$f(x_k)$
-2	1	1.2
-1	2	1.9
0	3	2.6
1	3	3.3
2	4	4.0

(b)

x_k	y_k	$f(x_k)$
-6	7	7.0
-2	5	4.6
0	3	3.4
2	2	2.2
6	0	-0.2

(c)

x_k	y_k	$f(x_k)$
-4	-3	-3.0
-1	-1	-0.9
0	0	-0.2
2	1	1.2
3	2	1.9

2. (a)

x_k	y_k	$f(x_k)$
-4	1.2	0.44
-2	2.8	3.34
0	6.2	6.24
2	7.8	9.14
4	13.2	12.04

(b)

x_k	y_k	$f(x_k)$
-6	-5.3	-6.00
-2	-3.5	-2.84
0	-1.7	-1.26
2	0.2	0.32
6	4.0	3.48

(c)

x_k	y_k	$f(x_k)$
-8	6.8	7.32
-2	5.0	3.81
0	2.2	2.64
4	0.5	0.30
6	-1.3	-0.87

3. (a)

x_k	y_k	$f(x_k)$
-2	1	0.4
0	3	3.3
2	6	6.2
4	8	9.1
6	13	12.0

(b)

x_k	y_k	$f(x_k)$
-2	7	7.6
1	5	3.7
2	2	2.4
4	0	-0.2
5	-2	-1.5

(c)

x_k	y_k	$f(x_k)$
-3	-6	-6.7
1	-4	-3.3
3	-2	-1.6
5	0	0.1
9	4	3.5

4. Find the *power fit* $y = Ax$, where $M = 1$ which is a line through the origin, for the data and calculate $E_2(f)$.

(a)

x_k	y_k	$f(x_k)$
1	1.6	1.58
2	2.8	3.16
3	4.7	4.74
4	6.4	6.32
5	8.0	7.90

(b)

x_k	y_k	$f(x_k)$
3	1.6	1.722
4	2.4	2.296
5	2.9	2.870
6	3.4	3.444
8	4.6	4.592

(c)

x_k	y_k	$f(x_k)$
-4	-3	-2.8
-1	-1	-0.7
0	0	0.0
2	1	1.4
3	2	2.1

5. Define the data means $\bar{x}$ and $\bar{y}$ by

$$\bar{x} = \frac{1}{N} \sum_{k=1}^{N} x_k \quad \text{and} \quad \bar{y} = \frac{1}{N} \sum_{k=1}^{N} y_k.$$

Show that the point $(\bar{x}, \bar{y})$ lies on the least-squares line.

6. Show that the solution of the system in (15) is given by

$$A = \frac{1}{D} \left(N \sum_{k=1}^{N} x_k y_k - \sum_{k=1}^{N} x_k \sum_{k=1}^{N} y_k \right),$$

$$B = \frac{1}{D} \left(\sum_{k=1}^{N} x_k^2 \sum_{k=1}^{N} y_k - \sum_{k=1}^{N} x_k \sum_{k=1}^{N} x_k y_k \right),$$

$$D = N \sum_{k=1}^{N} x_k^2 - \left(\sum_{k=1}^{N} x_k \right)^2.$$

7. Show that the value D in Exercise 6 is nonzero. *Hint.* Show that $D = N \sum_{k=1}^{N} (x_k - \bar{x})^2$.

8. Show that the coefficients A and B for the least squares line can be computed as follows. First compute the data means $\bar{x}$ and $\bar{y}$ in Exercise 5, then perform the calculations.

$$C = \sum_{k=1}^{N} (x_k - \bar{x})^2 \qquad A = \frac{1}{C} \sum_{k=1}^{N} (x_k - \bar{x})(y_k - \bar{y}) \qquad B = \bar{y} - A\bar{x}.$$

Hint. Use $X_k = x_k - \bar{x}$, $Y_k = y_k - \bar{y}$ and first find the line $Y = AX$.

9. Compute the missing entries in the table and determine which curve fits the data better. Use $E_2(f)$ to check the fits.

x_k	y_k	$y_k \approx 2.069x_k^2$	$y_k \approx 1.516x_k^3$
0.3	0.1	0.19	0.04
0.6	0.4	.	.
0.9	1.2	.	.
1.2	2.8	2.98	2.62
1.5	5.0	.	.

10. Write a report on linear correlation of data. Be sure to include a discussion of the coefficient of correlation.

11. Find the power fits $y = Ax^2$ and $y = Bx^3$ for the following data and use $E_2(f)$ to determine which curve fits best.

(a)

x_k	y_k
2.0	5.1
2.3	7.5
2.6	10.6
2.9	14.4
3.2	19.0

(b)

x_k	y_k
2.0	5.9
2.3	8.3
2.6	10.7
2.9	13.7
3.2	17.0

R^2

12. Find the power fits $y = A/x$ and $y = B/x^2$ for the following data and use $E_2(f)$ to determine which curve fits best.

(a)

x_k	y_k
0.5	7.1
0.8	4.4
1.1	3.2
1.8	1.9
4.0	0.9

(b)

x_k	y_k
0.7	8.1
0.9	4.9
1.1	3.3
1.6	1.6
3.0	0.5

R

13. Hooke's law states that $F = kx$, where F is the force (ounces) used to stretch a spring and x is the increase in its length (in inches). Find an approximation to the spring constant k for the following data.

(a)

x_k	F_k
0.2	3.6
0.4	7.3
0.6	10.9
0.8	14.5
1.0	18.2

(b)

x_k	F_k
0.2	5.3
0.4	10.6
0.6	15.9
0.8	21.2
1.0	26.4

14. Find the gravitational constant g for the following sets of data. Use the power fit that was shown in Example 5.3.

(a)

time, t_k	Distance, d_k
0.200	0.1960
0.400	0.7835
0.600	1.7630
0.800	3.1345
1.000	4.8975

(b)

Time, t_k	Distance, d_k
0.200	0.1965
0.400	0.7855
0.600	1.7675
0.800	3.1420
1.000	4.9095

15. The following data gives the distance of the nine planets from the sun and their sidereal period in days.

Planet	Distance from sun (km × 10^6)	Sidereal period (days)
Mercury	57.59	87.99
Venus	108.11	224.70
Earth	149.57	365.26
Mars	227.84	686.98
Jupiter	778.14	4332.4
Saturn	1427.0	10759
Uranus	2870.3	30684
Neptune	4499.9	60188
Pluto	5909.0	90710

(a) Use a power fit of the form $y = Cx^{3/2}$ for the first four planets.

(b) Use a power fit of the form $y = Cx^{3/2}$ for all nine planets.

16. Derive the normal equation for finding the least-squares linear fit through the origin $y = Ax$.

17. Derive the normal equation for finding the least-squares power fit $y = Ax^2$.

18. Derive the normal equations for finding the least-squares parabola $y = Ax^2 + B$.

5.2 CURVE FITTING

Data Linearization Method for y = Ce^Ax

Suppose that we are given the points (x_1, y_1), (x_2, y_2), . . . , (x_N, y_N) and want to fit an exponential curve of the form

$$y = Ce^{Ax}. \tag{1}$$

The first step is to take the logarithm of both sides:

$$\ln(y) = Ax + \ln(C). \tag{2}$$

Then introduce the change of variables:

$$Y = \ln(y), \quad X = x, \quad \text{and} \quad B = \ln(C). \tag{3}$$

This results in a linear relation between the new variables X and Y:

$$Y = AX + B. \tag{4}$$

The original points (x_k, y_k) in the xy-plane are transformed into $(X_k, Y_k) = (x_k, \ln(y_k))$ in the XY-plane. This process is called **data linearization**. Then the least-squares line (4) is fit to the points $\{(X_k, Y_k)\}$. The normal equations for finding A and B are

$$\left(\sum_{k=1}^{N} X_k^2\right)A + \left(\sum_{k=1}^{N} X_k\right)B = \sum_{k=1}^{N} X_k Y_k,$$

$$\left(\sum_{k=1}^{N} X_k\right)A + NB = \sum_{k=1}^{N} Y_k. \tag{5}$$

After A and B have been found, the parameter C in equation (1) is computed:

$$C = e^B. \tag{6}$$

Example 5.4

Use the data linearization method and find the exponential fit $y = Ce^{Ax}$, for the five data points (0, 1.5), (1, 2.5), (2, 3.5), (3, 5.0), and (4, 7.5).

Solution.   Apply the transformation (3) to the original points and obtain

$$\{(X_k, Y_k)\} = \{(0, \ln(1.5)), (1, \ln(2.5)), (2, \ln(3.5)), (3, \ln(5.0)), (4, \ln(7.5))\}$$
$$= \{(0, 0.40547), (1, 0.91629), (2, 1.25276), (3, 1.60944), (4, 2.01490)\}. \tag{7}$$

These transformed points are shown in Figure 5.4 and exhibit a linearized form. The equation of the least-squares line $Y = AX + B$ for the points (7) in Figure 5.4 is

$$Y = 0.391202X + 0.457367. \tag{8}$$

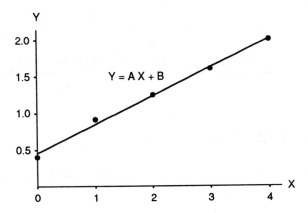

Figure 5.4   The transformed data points $\{(X_k, Y_k)\}$.

Calculation of the coefficients for the normal equations in (5) is shown in Table 5.4.

TABLE 5.4 Obtaining Coefficients of the Normal Equations for the Transformed Data Points $\{(X_k, Y_k)\}$

x_k	y_k	X_k	$Y_k = \ln(y_k)$	X_k^2	$X_k Y_k$
0.0	1.5	0.0	0.405465	0.0	0.000000
1.0	2.5	1.0	0.916291	1.0	0.916291
2.0	3.5	2.0	1.252763	4.0	2.505526
3.0	5.0	3.0	1.609438	9.0	4.828314
4.0	7.5	4.0	2.014903	16.0	8.059612
		10.0	6.198860	30.0	16.309743
		$= \sum X_k$	$= \sum Y_k$	$= \sum X_k^2$	$= \sum X_k Y_k$

The resulting linear system (5) for determining A and B is

$$30A + 10B = 16.309742$$
$$10A + 5B = 6.198860. \tag{9}$$

The solution is $A = 0.3912023$ and $B = 0.457367$. Then C is obtained with the calculation $C = \exp(0.457367) = 1.579910$, and these values A and C are substituted into equation (1) to obtain the exponential fit (see Figure 5.5):

$$y = 1.579910e^{0.3912023x} \qquad \text{(fit by data linearization)}. \tag{10}$$

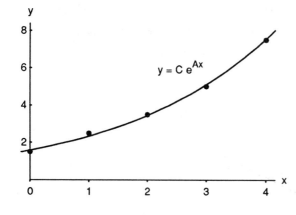

Figure 5.5 The exponential fit $y = 1.579910e^{0.3912023x}$ obtained by using the data linearization method.

Nonlinear Least-Squares Method for $y = Ce^{Ax}$

Suppose that we are given the points (x_1, y_1), (x_2, y_2), . . . , (x_N, y_N) and want to fit an exponential curve:

$$y = Ce^{Ax}. \tag{11}$$

The nonlinear least-squares procedure requires that we find a minimum of

$$E(A, C) = \sum_{k=1}^{N} (Ce^{Ax_k} - y_k)^2. \tag{12}$$

The partial derivatives of $E(A, C)$ with respect to A and C are

$$\frac{\partial E}{\partial A} = 2 \sum_{k=1}^{N} (Ce^{Ax_k} - y_k)(Cx_k e^{Ax_k}) \tag{13}$$

and

$$\frac{\partial E}{\partial C} = 2 \sum_{k=1}^{N} (Ce^{Ax_k} - y_k)(e^{Ax_k}). \tag{14}$$

When the partial derivatives in (13) and (14) are set equal to zero and then simplified, the resulting normal equations are

$$C \sum_{k=1}^{N} x_k e^{2Ax_k} - \sum_{k=1}^{N} x_k y_k e^{Ax_k} = 0,$$

$$C \sum_{k=1}^{N} e^{2Ax_k} - \sum_{k=1}^{N} y_k e^{Ax_k} = 0. \tag{15}$$

The equations in (15) are nonlinear in the unknowns A and C and can be solved using Newton's method. This is a time-consuming computation and the iteration involved requires good starting values for A and C. Many software packages have a built-in minimization subroutine for functions of several variables that can be used to minimize $E(A, C)$ directly. For example, the Nelder–Mead simplex algorithm can be used to minimize (12) directly and bypass the need for equations (13) to (15).

Example 5.5

Use the nonlinear least-squares method and determine the exponential fit $y = Ce^{Ax}$, for the five data points $(0, 1.5)$, $(1, 2.5)$, $(2, 3.5)$, $(3, 5.0)$, and $(4, 7.5)$.

Solution.   For this solution we must minimize the quantity $E(A, C)$, which is

$$E(A, C) = (C - 1.5)^2 + (Ce^a - 2.5)^2 + (Ce^{2a} - 3.5)^2 + (Ce^{3a} - 5.0)^2 + (Ce^{4a} - 7.5)^2. \tag{16}$$

An iterative can be used to determine that the minimum value of $E(A, C)$ is 0.040866 and the values of A and C that produce this minimum value are $A = 0.383575$ and $C = 1.610869$; hence the corresponding exponential fit is

$$y = 1.610869e^{0.383575x} \quad \text{(fit by nonlinear least squares).} \tag{17}$$

A comparison of the solutions using data linearization and nonlinear least squares is given in Table 5.5. There is a slight difference in the coefficients. For the purpose of

TABLE 5.5 Comparison of the Two Exponential Fits

x_k	y_k	$1.5799e^{0.39120x}$	$1.6109e^{0.38357x}$
0.0	1.5	1.5799	1.6109
1.0	2.5	2.3363	2.3640
2.0	3.5	3.4548	3.4692
3.0	5.0	5.1088	5.0911
4.0	7.5	7.5548	7.4713
5.0		11.1716	10.9644
6.0		16.5202	16.0904
7.0		24.4293	23.6130
8.0		36.1250	34.6527
9.0		53.4202	50.8535
10.0		78.9955	74.6287

interpolation it can be seen that the approximations differ by no more than 2% over the interval [0, 4] (see Table 5.5 and Figure 5.6). If there is a normal distribution of the errors in the data, (17) is usually the preferred choice. When extrapolation beyond the range of data is made, the two solutions will diverge and the discrepancy increases to about 6% when $x = 10$.

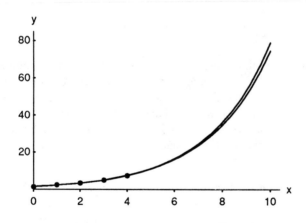

Figure 5.6 A graphical comparison of the two exponential curves.

Transformations for Data Linearization

The technique of data linearization has been used by scientists to fit curves such as $y = C\exp(Ax)$, $y = Cx^A$, $y = A\ln(x) + B$, and $y = A/x + B$. Once the curve has been chosen, a suitable transformation of the variables must be found so that a linear relation is obtained. For example, the reader can verify that $y = D/(x + C)$ is transformed into a linear problem $Y = AX + B$ by using the change of variables (and constants) $X = xy$, $Y = y$, $C = -1/A$, and $D = -B/A$. Graphs of several cases of the possibilities for the curves are shown in Figure 5.7, and other useful transformations are given in Table 5.6.

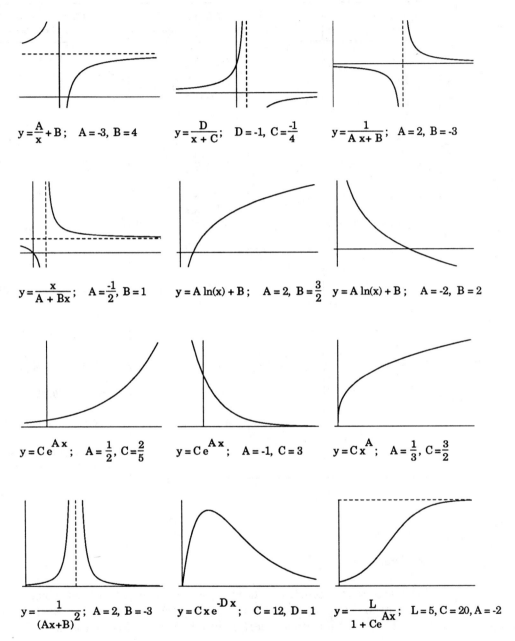

$y = \dfrac{A}{x} + B$; $A = -3, B = 4$ $y = \dfrac{D}{x + C}$; $D = -1, C = \dfrac{-1}{4}$ $y = \dfrac{1}{A\,x + B}$; $A = 2, B = -3$

$y = \dfrac{x}{A + Bx}$; $A = \dfrac{-1}{2}, B = 1$ $y = A\ln(x) + B$; $A = 2, B = \dfrac{3}{2}$ $y = A\ln(x) + B$; $A = -2, B = 2$

$y = C\,e^{A\,x}$; $A = \dfrac{1}{2}, C = \dfrac{2}{5}$ $y = C\,e^{A\,x}$; $A = -1, C = 3$ $y = C\,x^{A}$; $A = \dfrac{1}{3}, C = \dfrac{3}{2}$

$y = \dfrac{1}{(Ax+B)^{2}}$; $A = 2, B = -3$ $y = C\,x\,e^{-D\,x}$; $C = 12, D = 1$ $y = \dfrac{L}{1 + Ce^{Ax}}$; $L = 5, C = 20, A = -2$

Figure 5.7   Possibilities for the curves used in ''data linearization.''

TABLE 5.6   Change of Variable(s) for Data Linearization

Function, $y = f(x)$	Linearized form, $Y = AX + B$	Change of variable(s) and constants
$y = \dfrac{A}{x} + B$	$y = A\dfrac{1}{x} + B$	$X = \dfrac{1}{x},\ Y = y$
$y = \dfrac{D}{x + C}$	$y = \dfrac{-1}{C}(xy) + \dfrac{D}{C}$	$X = xy,\ Y = y$ $C = \dfrac{-1}{A},\ D = \dfrac{-B}{A}$
$y = \dfrac{1}{Ax + B}$	$\dfrac{1}{y} = Ax + B$	$X = x,\ Y = \dfrac{1}{y}$
$y = \dfrac{x}{A + Bx}$	$\dfrac{1}{y} = A\dfrac{1}{x} + B$	$X = \dfrac{1}{x},\ Y = \dfrac{1}{y}$
$y = A\ln(x) + B$	$y = A\ln(x) + B$	$X = \ln(x),\ Y = y$
$y = C\exp(Ax)$	$\ln(y) = Ax + \ln(C)$	$X = x,\ Y = \ln(y)$ $C = \exp(B)$
$y = Cx^A$	$\ln(y) = A\ln(x) + \ln(C)$	$X = \ln(x),\ Y = \ln(y)$ $C = \exp(B)$
$y = (Ax + B)^{-2}$	$y^{-1/2} = Ax + B$	$X = x,\ Y = y^{-1/2}$
$y = Cx\exp(-Dx)$	$\ln\left(\dfrac{y}{x}\right) = -Dx + \ln(C)$	$X = x,\ Y = \ln\left(\dfrac{y}{x}\right)$ $C = \exp(B),\ D = -A$
$y = \dfrac{L}{1 + C\exp(Ax)}$	$\ln\left(\dfrac{L}{y} - 1\right) = Ax + \ln(C)$	$X = x,\ Y = \ln\left(\dfrac{L}{y} - 1\right)$ $C = \exp(B)$ and L is a constant that must be given

Linear Least Squares

The linear least-squares problem is stated as follows. Suppose that N data points $\{(x_k, y_k)\}$ and a set of M linear independent functions $\{f_j(x)\}$ are given. We want to find M coefficients $\{c_j\}$ so that the function $f(x)$ given by the linear combination

$$f(x) = \sum_{j=1}^{M} c_j f_j(x) \tag{18}$$

will minimize the sum of the squares of the errors

$$E(c_1, c_2, \ldots, c_M) = \sum_{k=1}^{N} [f(x_k) - y_k]^2 = \sum_{k=1}^{N} [\sum_{j=1}^{M} c_j f_j(x_k) - y_k]^2. \qquad (19)$$

For E to be minimized, it is necessary that each partial derivative be zero (i.e., $\partial E/\partial c_i = 0$ for $i = 1, 2, \ldots, M$), and this results in the system of equations;

$$\sum_{k=1}^{N} [\sum_{j=1}^{M} c_j f_j(x_k) - y_k] [f_i(x_k)] = 0 \qquad \text{for } i = 1, 2, \ldots, M. \qquad (20)$$

Interchanging the order of the summations in (20) will produce an $M \times M$ system of linear equations where the unknowns are the coefficients $\{c_j\}$. They are called the normal equations:

$$\sum_{j=1}^{M} \left[\sum_{k=1}^{N} f_i(x_k) f_j(x_k) \right] c_j = \sum_{k=1}^{N} f_i(x_k) y_k \qquad \text{for } i = 1, 2, \ldots, M. \qquad (21)$$

The Matrix Formulation

Although (21) is easily recognized as a system of M equations in M unknowns, one must be clever so that wasted computations are not performed when writing the system in matrix notation. The key is to write down the matrices F and F^T as follows:

$$F = \begin{pmatrix} f_1(x_1) & f_2(x_1) & \cdots & f_M(x_1) \\ f_1(x_2) & f_2(x_2) & \cdots & f_M(x_2) \\ f_1(x_3) & f_2(x_3) & \cdots & f_M(x_3) \\ \vdots & \vdots & & \vdots \\ f_1(x_N) & f_2(x_N) & \cdots & f_M(x_N) \end{pmatrix}, \quad F^T = \begin{pmatrix} f_1(x_1) & f_1(x_2) & f_1(x_3) & \cdots & f_1(x_N) \\ f_2(x_1) & f_2(x_2) & f_2(x_3) & \cdots & f_2(x_N) \\ \vdots & \vdots & \vdots & & \vdots \\ f_M(x_1) & f_M(x_2) & f_M(x_3) & \cdots & f_M(x_N) \end{pmatrix}.$$

Consider the product of F^T and the column vector $\mathbf{Y}$:

$$F^T \mathbf{Y} = \begin{pmatrix} f_1(x_1) & f_1(x_2) & f_1(x_3) & \cdots & f_1(x_N) \\ f_2(x_1) & f_2(x_2) & f_2(x_3) & \cdots & f_2(x_N) \\ \vdots & \vdots & \vdots & & \vdots \\ f_M(x_1) & f_M(x_2) & f_M(x_3) & \cdots & f_M(x_N) \end{pmatrix} \begin{pmatrix} y_1 \\ y_2 \\ \vdots \\ y_N \end{pmatrix}. \qquad (22)$$

The element in the ith row of the product $F^T \mathbf{Y}$ in (22) is the same as the ith element in the column vector in equation (21), that is,

$$\sum_{k=1}^{N} f_i(x_k) y_k = \left(\text{row}_i \, F^T \cdot (y_1, y_2, \ldots, y_N)^T \right). \qquad (23)$$

Now consider the product $F^T F$, which is an $M \times M$ matrix:

$$
F^T F = \begin{pmatrix} f_1(x_1) & f_1(x_2) & f_1(x_3) & \cdots & f_1(x_N) \\ f_2(x_1) & f_2(x_2) & f_2(x_3) & \cdots & f_2(x_N) \\ \vdots & \vdots & \vdots & & \vdots \\ f_M(x_1) & f_M(x_2) & f_M(x_3) & \cdots & f_M(x_N) \end{pmatrix} \begin{pmatrix} f_1(x_1) & f_2(x_1) & \cdots & f_M(x_1) \\ f_1(x_2) & f_2(x_2) & \cdots & f_M(x_2) \\ f_1(x_3) & f_2(x_3) & \cdots & f_M(x_3) \\ \vdots & \vdots & & \vdots \\ f_1(x_N) & f_2(x_N) & \cdots & f_M(x_N) \end{pmatrix}.
$$

The element in the ith row and jth column of $F^T F$ is the coefficient of c_j in the ith row in equation (21), that is,

$$
\sum_{k=1}^{N} f_i(x_k) f_j(x_k) = f_i(x_1) f_j(x_1) + f_i(x_2) f_j(x_2) + \cdots + f_i(x_N) f_j(x_N). \tag{24}
$$

When M is small, a computationally efficient way to calculate the linear least-squares coefficients for (18) is to store the matrix F and compute $F^T F$ and $F^T \mathbf{Y}$ and then solve the linear system

$$
F^T F \mathbf{C} = F^T \mathbf{Y} \qquad \text{for the coefficient vector } \mathbf{C}. \tag{25}
$$

Polynomial Fitting

When the foregoing method is adapted to using the functions $\{f_j(x) = x^{j-1}\}$ and the index of summation ranges from $j = 1$ to $j = M + 1$, the function $f(x)$ will be a polynomial of degree M:

$$
f(x) = c_1 + c_2 x + c_3 x^2 + \cdots + c_{M+1} x^M. \tag{26}
$$

We now show how to find the **least-squares parabola** and the extension to a polynomial of higher degree is easily made and is left for the reader.

Theorem 5.3   (Least-Squares Parabola).   Suppose that $\{(x_k, y_k)\}_{k=1}^{N}$ are N points where the abscissas are distinct. The coefficients of the least-squares parabola

$$
y = f(x) = Ax^2 + Bx + C \tag{27}
$$

are the solution values A, B, and C of the linear system

$$
\left(\sum_{k=1}^{N} x_k^4 \right) A + \left(\sum_{k=1}^{N} x_k^3 \right) B + \left(\sum_{k=1}^{N} x_k^2 \right) C = \sum_{k=1}^{N} y_k x_k^2,
$$

$$
\left(\sum_{k=1}^{N} x_k^3 \right) A + \left(\sum_{k=1}^{N} x_k^2 \right) B + \left(\sum_{k=1}^{N} x_k \right) C = \sum_{k=1}^{N} y_k x_k, \tag{28}
$$

$$
\left(\sum_{k=1}^{N} x_k^2 \right) A + \left(\sum_{k=1}^{N} x_k \right) B + NC = \sum_{k=1}^{N} y_k.
$$

Proof. The coefficients A, B, and C will minimize the quantity:

$$E(A, B, C) = \sum_{k=1}^{N} (Ax_k^2 + Bx_k + C - y_k)^2. \tag{29}$$

The partial derivatives $\partial E/\partial A$, $\partial E/\partial B$, and $\partial E/\partial C$ must all be zero. This results in

$$0 = \frac{\partial E(A, B, C)}{\partial A} = 2 \sum_{k=1}^{N} (Ax_k^2 + Bx_k + C - y_k)^1(x_k^2),$$

$$0 = \frac{\partial E(A, B, C)}{\partial B} = 2 \sum_{k=1}^{N} (Ax_k^2 + Bx_k + C - y_k)^1(x_k), \tag{30}$$

$$0 = \frac{\partial E(A, B, C)}{\partial C} = 2 \sum_{k=1}^{N} (Ax_k^2 + Bx_k + C - y_k)^1(1).$$

Using the distributive property of addition, we can move the values A, B, and C outside the summations in (30) to obtain the normal equations that are given in (28).

Example 5.6

Find the least-squares parabola for the four points $(-3, 3)$, $(0, 1)$, $(2, 1)$, and $(4, 3)$.

Solution. The entries in Table 5.7 are used to compute the summations required in the linear system (30).

TABLE 5.7 Obtaining the Coefficients for the Least-Squares Parabola of Example 5.6

x_k	y_k	x_k^2	x_k^3	x_k^4	$x_k y_k$	$x_k^2 y_k$
-3	3	9	-27	81	-9	27
0	1	0	0	0	0	0
2	1	4	8	16	2	4
4	3	16	64	256	12	48
3	8	29	45	353	5	79

The linear system (30) for finding A, B, and C becomes

$$353A + 45B + 29C = 79$$

$$45A + 29B + 3C = 5.$$

$$29A + 3B + 4C = 8.$$

The solution to the linear system is $A = 585/3278$, $B = -631/3278$, and $C = 1394/1639$, and the desired parabola is (see Figure 5.8)

$$y = \frac{585}{3278} x^2 - \frac{631}{3278} x + \frac{1394}{1639} = 0.178462x^2 - 0.192495x + 0.850519.$$

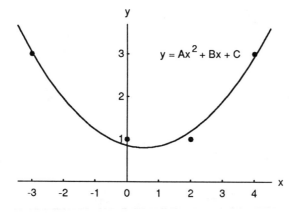

Figure 5.8 The least-squares parabola for Example 5.6.

Polynomial Wiggle

It is tempting to use a least-squares polynomial to fit data that is nonlinear. But if the data do not exhibit a "polynomial nature," the resulting curve may exhibit large oscillations. This phenomenon, called **polynomial wiggle**, becomes more pronounced with higher-degree polynomials. For this reason we seldom use a polynomial of degree 6 or above unless it is known that the true function we are working with is a polynomial.

For example, let $f(x) = 1.44/x^2 + 0.24x$ be used to generate the six data points (0.25, 23.1), (1.0, 1.68), (1.5, 1.0), (2.0, 0.84), (2.4, 0.826), and (5.0, 1.2576). The result of curve fitting with the least-squares polynomials

$$P_2(x) = 22.93 - 16.96x + 2.553x^2,$$

$$P_3(x) = 33.04 - 46.51x + 19.51x^2 - 2.296x^3,$$

$$P_4(x) = 39.92 - 80.93x + 58.39x^2 - 17.15x^3 + 1.680x^4,$$

and

$$P_5(x) = 46.02 - 118.1x + 119.4x^2 - 57.51x^3 + 13.03x^4 - 1.085x^5$$

is shown in Figure 5.9(a) to (d). Notice that $P_3(x)$, $P_4(x)$, and $P_5(x)$ exhibit a large wiggle in the interval [2, 5]. Even though $P_5(x)$ goes through the six points, it produces the worst fit. If we must use a polynomial to these data, $P_2(x)$ should be the choice.

Algorithm 5.2 (Least-Squares Polynomial). To construct the least-squares polynomial of degree M of the form

$$P_M(x) = c_1 + c_2x + c_3x^2 + c_4x^3 + \cdots + c_M x^{M-1} + c_{M+1}x^M$$

that fits the N data points $(x_1, y_1), \ldots, (x_N, y_N)$.

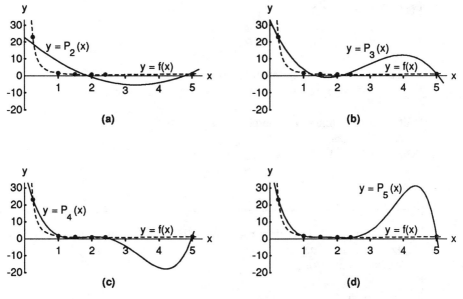

Figure 5.9   (a) Using $P_2(x)$ to fit data. (b) Using $P_3(x)$ to fit data. (c) Using $P_4(x)$ to fit data. (d) Using $P_5(x)$ to fit data.

```
READ N                                                    {Number of points}
READ M                                                    {Degree of polynomial}
DIMENSION X[1 . . N],Y[ 1 . . N],B[1 . . M+1],C[1 . . M+1],P[0 . . 2M]
DIMENSION A[1 . . M+1, 1 . . M+1]


FOR    K = 1  TO  N  DO                                    {Get the N points}
       READ X(K), Y(K)


FOR    R = 1 TO M+1 DO B(R) := 0                           {Zero the array}
FOR    K = 1 TO  N   DO                                    {Compute
       Y := Y(K), X := X(K), P := 1                          the
       FOR    R = 1  TO   M+1   DO                            column
              B(R) := B(R) + Y*P                              vector}
              P := P*X


FOR    J = 1 TO 2*M DO P(J) := 0                          {Zero the array}
       P(0) := N
FOR    K = 1  TO  N  DO                                   {Compute
       X := X(K), P := X(K)                                 the sum of
       FOR    J = 1  TO  2*M   DO                            powers
              P(J) := P(J) + P                               of x_k}
              P := P*X
```

```
FOR    R = 1  TO  M+1   DO              {Determine
       FOR    C = 1  TO  M+1    DO          the matrix
              A(R, C) = P(R+C−2)            entries}
```

USE a linear systems algorithm to solve the M+1 equations:

$$A*C = B \text{ for coefficient vector } C = (c_1, c_2, \ldots, c_M, c_{M+1}).$$

PRINT "Least squares polynomial of degree M"   {Output}
PRINT "The coefficients are"
PRINT C(1), C(2), . . . , C(M+1)

Algorithm 5.3   (Nonlinear Curve Fitting).   To construct a nonlinear fit $y = g(x)$ given the N data points $(x_1, y_1), \ldots, (x_N, y_N)$ by using "data linearization." Available curves are

$$y = \frac{A}{x} + B, \qquad y = \frac{D}{x + C}, \qquad y = \frac{1}{Ax + B}, \qquad y = \frac{x}{A + Bx}$$

$$y = A \ln(x) + B, \qquad y = C \exp(Ax), \qquad y = Cx^A, \qquad y = (Ax + B)^{-2}$$

$$y = Cx \exp(-Dx), \qquad y = \frac{L}{1 + C \exp(Ax)}.$$

```
READ N                                   {Number of points}
FOR J = 1 TO N DO READ X(J), Y(J)        {Get the N points}
```

CASES {Select the nonlinear form you wish to fit.}

(i) $y = A/x + B$	(ii) $y = D/[x + C]$
(iii) $y = 1/[Ax + B]$	(iv) $y = x/[A + Bx]$
(v) $y = A \ln (x) + B$	(vi) $y = C*\exp (Ax)$
(vii) $y = C*x^A$	(viiii) $y = [Ax + B]^{-2}$

END

CASE {Make transformation(s) to linearize the data j = 1, 2, . . . , N.}

(i) $X_j = 1/x_j$, $Y_j = y_j$	(ii) $X_j = x_j y_j$, $Y_j = y_j$
(iii) $X_j = x_j$, $Y_j = 1/y_j$	(iv) $X_j = 1/x_j$, $Y_j = 1/y_j$
(v) $X_j = \ln (x_j)$, $Y_j = y_j$	(vi) $X_j = x_j$, $Y_j = \ln (y_j)$
(vii) $X_j = \ln (x_j)$, $Y_j = \ln (y_j)$	(viii) $X_j = x_j$, $Y_j = y_j^{-1/2}$

END

USE Algorithm 5.1 to find the least-squares line   Y = A*X + B
for the transformed data points {X$_j$, Y$_j$}.

CASES {Make coefficient transformation(s) if necessary.}
| (ii) C = −1/A, D = −B/A
| (vi) C = exp (B)
| (vii) C = exp (B)
END

PRINT "The least-squares curve for the data is:"   {Output}
CASES {Print appropriate equation and coefficients}

(i) y = A/x + B	(ii) y = D/[x + C]
(iii) y = 1/[Ax + B]	(iv) y = x/[A + Bx]
(v) y = A ln (x) + B	(vi) y = C*exp (Ax)
(vii) y = C*x^A	(viii) y = [Ax + B]$^{-2}$

END

EXERCISES FOR CURVE FITTING

1. Find the least-squares parabolic fit $y = Ax^2 + Bx + C$.

(a)

x_k	y_k
−3	15
−1	5
1	1
3	5

(b)

x_k	y_k
−3	−1
−1	25
1	25
3	1

2. Find the least-squares parabolic fit $y = Ax^2 + Bx + C$.

(a)

x_k	y_k
−2	−5.8
−1	1.1
0	3.8
1	3.3
2	−1.5

(b)

x_k	y_k
−2	2.8
−1	2.1
0	3.25
1	6.0
2	11.5

(c)

x_k	y_k
−2	10
−1	1
0	0
1	2
2	9

3. Use $E_1(f)$ to determine which curve fits best.

(a)

x_k	y_k	$y_k \approx 0.5102x^2$	$y_k \approx 0.4304 \times 2^x$
1	0.7	0.51	0.86
2	2.0	.	.
3	4.2	.	.
4	8.0	.	.
5	13.0	12.76	13.77

(b) Change the last entry to (5, 15.0) and determine the best fit $y \approx 0.5613x^2$ or $y \approx 0.4773 \times 2^x$.

4. **(a)** Find the curve fit $y = C \exp(Ax)$ by using the change of variables $X = x$, $Y = \ln(y)$, $C = \exp(B)$ to linearize the data points.
   (b) Find the curve fit $y = Cx^A$ by using the change of variables $X = \ln(x)$, $Y = \ln(y)$, and $C = \exp(B)$ to linearize the data points.
   (c) Use $E_1(f)$ and determine which curve fit in part (a) or (b) is best.

x_k	y_k	$\ln(x_k)$	$\ln(y_k)$
1	0.6	0.0000	-0.5108
2	1.9	0.6931	0.6419
3	4.3	1.0986	1.4586
4	7.6	1.3863	2.0281
5	12.6	1.6094	2.5337

5. Follow the instructions for Exercise 4.

x_k	y_k	$\ln(x_k)$	$\ln(y_k)$
1	0.7	0.0000	-0.3567
2	1.7	0.6931	0.5306
3	3.4	1.0986	1.2238
4	6.7	1.3863	1.9021
5	12.7	1.6094	2.5416

6. **(a)** Find the curve fit $y = C \exp(Ax)$ by using the change of variables $X = x$, $Y = \ln(y)$, $C = \exp(B)$ to linearize the data points.
   (b) Find the curve fit $y = 1/(Ax + B)$ by using the change of variables $X = x$, $Y = 1/y$ to linearize the data points.
   (c) Use $E_1(f)$, and determine which curve fit in part (a) or (b) is best.

x_k	y_k	$\ln(y_k)$	$1/y_k$
-1	6.62	1.8901	0.1511
0	3.94	1.3712	0.2538
1	2.17	0.7747	0.4608
2	1.35	0.3001	0.7407
3	0.89	-0.1165	1.1236

7. Follow the instructions for Exercise 6.

x_k	y_k	$\ln(y_k)$	$1/y_k$
-1	6.62	1.8901	0.1511
0	2.78	1.0225	0.3597
1	1.51	0.4121	0.6623
2	1.23	0.2070	0.8130
3	0.89	-0.1165	1.1236

8. (a) Find the curve fit $y = C \exp(Ax)$ by using the transformations $X = x$, $Y = \ln(y)$, $C = \exp(B)$.

(b) Find the curve fit $y = (Ax + B)^{-2}$ by using the transformations $X = x$, $Y = y^{-1/2}$.

(c) Use $E_1(f)$ and determine which curve fit in part (a) or (b) is best.

(i)

x_k	y_k
-1	13.45
0	3.01
1	0.67
2	0.15

(ii)

x_k	y_k
-1	13.65
0	1.38
1	0.49
3	0.15

9. *Logistic population growth.* When the population $P(t)$ is bounded by the limiting value L, it follows a logistic curve and has the form $P(t) = L/[1 + C \exp(At)]$. Find A and C for the following data, where L is a known value.

(a) (0, 200), (1, 400), (2, 650), (3, 850), (4, 950), and $L = 1000$

(b) (0, 500), (1, 1000), (2, 1800), (3, 2800), (4, 3700), and $L = 5000$

10. Use the data for the U.S. population and find the logistic curve $P(t)$. Estimate the population in the year 2000.

(a) Assume that $L = 800$ (million).

Year	t_k	P_k
1800	-10	5.3
1850	-5	23.2
1900	0	76.1
1950	5	152.3

(b) Assume that $L = 800$ (million).

Year	t_k	P_k
1900	0	76.1
1920	2	106.5
1940	4	132.6
1960	6	180.7
1980	8	226.5

11. The least-squares plane $z = Ax + By + C$ for the N points $(x_1, y_1, z_1), \ldots, (x_N, y_N, z_N)$ is obtained by minimizing

$$E(A, B, C) = \sum_{k=1}^{N} (Ax_k + By_k + C - z_k)^2.$$

Derive the normal equations:

$$\left(\sum_{k=1}^{N} x_k^2\right) A + \left(\sum_{k=1}^{N} x_k y_k\right) B + \left(\sum_{k=1}^{N} x_k\right) C = \sum_{k=1}^{N} z_k x_k$$

$$\left(\sum_{k=1}^{N} x_k y_k\right) A + \left(\sum_{k=1}^{N} y_k^2\right) B + \left(\sum_{k=1}^{N} y_k\right) C = \sum_{k=1}^{N} z_k y_k$$

$$\left(\sum_{k=1}^{N} x_k\right) A + \left(\sum_{k=1}^{N} y_k\right) B + NC = \sum_{k=1}^{N} z_k.$$

12. Find the least-squares plane for the following data.
    (a) (1, 1, 7), (1, 2, 9), (2, 1, 10), (2, 2, 11), (2, 3, 12)
    (b) (1, 2, 6), (2, 3, 7), (1, 1, 8), (2, 2, 8), (2, 1, 9)
    (c) (3, 1, −3), (2, 1, −1), (2, 2, 0), (1, 1, 1), (1, 2, 3)

13. Consider the following table of data.

x_k	y_k
1.0	2.0
2.0	5.0
3.0	10.0
4.0	17.0
5.0	26.0

When the change of variables $X = xy$ and $Y = y$ are used with the function $y = D/(x + C)$, the transformed least-squares fit is

$$y = \frac{-17.719403}{x - 5.476617}.$$

When the change of variables $X = x$ and $Y = 1/y$ are used with the function $y = 1/(Ax + B)$, the transformed least-squares fit is

$$y = \frac{1}{-0.1064253x + 0.4987330}.$$

Determine which fit is best and why one of the solutions is completely absurd.

In Exercises 15–22, show how to linearize the given formula.

14. $y = \dfrac{A}{x} + B$

15. $y = \dfrac{D}{x + C}$

16. $y = \dfrac{1}{Ax + B}$

17. $y = \dfrac{x}{A + Bx}$

18. $y = A \ln(x) + B$

19. $y = Cx^A$

20. $y = (Ax + B)^{-2}$

21. $y = Cx \exp(-Dx)$

22. Write a report on orthogonal polynomials. See References [9, 19, 29, 34, 40, 41, 44, 76, 81, 90, 96, 126, 128, 143, 145, 149, 152, 153, and 169].

23. Write a report on least squares. See References [39, 92, 109, 112, and 152].

5.3 INTERPOLATION BY SPLINE FUNCTIONS

Polynomial interpolation for a set of $N + 1$ points $\{(x_k, y_k)\}$ is frequently unsatisfactory. As discussed in Section 5.2, a polynomial of degree N can have $N − 1$ relative maxima and minima and the graph can wiggle in order to pass through the points. Another method is to "piece together" the graphs of lower-degree polynomials $S_k(x)$ and interpolate between the successive nodes (x_k, y_k) and (x_{k+1}, y_{k+1}) (see Figure 5.10). The two

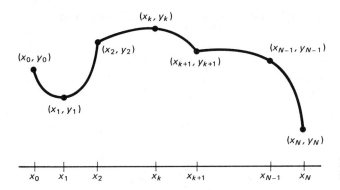

Figure 5.10 Piecewise polynomial interpolation.

adjacent portions of the curve $y = S_k(x)$ and $y = S_{k+1}(x)$, which lie above $[x_k, x_{k+1}]$ and $[x_{k+1}, x_{k+2}]$, respectively, pass through the common **knot** (x_{k+1}, y_{k+1}). The two portions of the graph are "tied together" at the knot (x_{k+1}, y_{k+1}) and the set of functions $\{S_k(x)\}$ form a piecewise polynomial curve which is denoted by $S(x)$.

Piecewise Linear Interpolation

The simplest polynomial to use, a polynomial of degree 1, produces a polygonal path that consists of line segments that pass through the points. The Lagrange polynomial from Section 4.3 is used to represent this piecewise linear curve:

$$S_k(x) = y_k \frac{x - x_{k+1}}{x_k - x_{k+1}} + y_{k+1} \frac{x - x_k}{x_{k+1} - x_k} \quad \text{for } x_k \leq x \leq x_{k+1}. \tag{1}$$

The resulting curve looks like a "broken line" (see Figure 5.11).

An equivalent expression can be obtained if we use the point-slope formula for a line segment

$$S_k(x) = y_k + d_k(x - x_k),$$

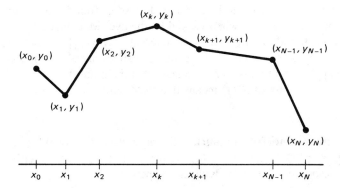

Figure 5.11 Piecewise linear interpolation (a linear spline).

where $d_k = (y_{k+1} - y_k)/(x_{k+1} - x_k)$. The resulting linear spline function can be written in the form

$$S(x) = \begin{cases} y_0 + d_0(x - x_0) & \text{for } x \text{ in } [x_0, x_1], \\ y_1 + d_1(x - x_1) & \text{for } x \text{ in } [x_1, x_2], \\ \vdots & \vdots \\ y_k + d_k(x - x_k) & \text{for } x \text{ in } [x_k, x_{k+1}], \\ \vdots & \vdots \\ y_{N-1} + d_{N-1}(x - x_{N-1}) & \text{for } x \text{ in } [x_{N-1}, x_N]. \end{cases} \qquad (2)$$

The form of equation (2) is better than equation (1) for the explicit calculation of $S(x)$. It is assumed that the abscissas are ordered $x_0 < x_1 < \cdots < x_{N-1} < x_N$. For a fixed value of x, the interval $[x_k, x_{k+1}]$ containing x can be found by successively computing the differences $x - x_1, \ldots, x - x_k, x - x_{k+1}$ until $k + 1$ is the smallest integer such that $x - x_{k+1} < 0$. Hence we have found k so that $x_k \le x \le x_{k+1}$, and the value of the spline function $S(x)$ is

$$S(x) = S_k(x) = y_k + d_k(x - x_k) \qquad \text{for } x_k \le x \le x_{k+1}. \qquad (3)$$

These techniques can be extended to higher-degree polynomials. For example, if an odd number of nodes $x_0, \ldots, x_1, \ldots, x_{2M}$ are given, then a piecewise quadratic polynomial can be constructed on each subinterval $[x_{2k}, x_{2k+2}]$, for $k = 0, 1, \ldots, M - 1$. A shortcoming of the resulting quadratic spline is that the curvature at the even nodes x_{2k} changes abruptly and this can cause an undesired bend or distortion in the graph. The second derivative of a quadratic spline is discontinuous at the even nodes. If we use piecewise cubic polynomials, then both the first and second derivatives can be made continuous.

Piecewise Cubic Splines

The fitting of a polynomial curve to a set of data points has applications in the areas of drafting and computer graphics. This drafter wants to draw a "smooth curve" through data points that are not subject to error. It is common to use a french curve and subjectively draw a curve that looks smooth when viewed by the eye. Mathematically, it is possible to construct cubic functions $S_k(x)$ on each interval $[x_k, x_{k+1}]$ so that the resulting piecewise curve $y = S(x)$ and its first and second derivatives are all continuous on the larger interval $[x_0, x_N]$. The continuity of $S'(x)$ means that the graph $Y = S(x)$ will not have sharp corners. The continuity of $S''(x)$ means that the "radius of curvature" is defined at each point.

Definition 5.1   (Cubic Spline Interpolant).   Suppose that $\{(x_k, y_k)\}_{k=0}^{N}$ are $N + 1$ points where $a = x_0 < x_1 < \cdots < x_N = b$. The function $S(x)$ is called a **cubic spline** if there exists N cubic polynomials $S_k(x)$ with the properties:

I. $S(x) = S_k(x) = s_{k,0} + s_{k,1}(x - x_k) + s_{k,2}(x - x_k)^2 + s_{k,3}(x - x_k)^3$
for $x \in [x_k, x_{k+1}]$ for each $k = 0, 1, \ldots, N - 1$.

II. $S(x_k) = y_k$ for $k = 0, 1, \ldots, N$.
The spline passes through each data point.

III. $S_k(x_{k+1}) = S_{k+1}(x_{k+1})$ for $k = 0, 1, \ldots, N - 2$.
The spline forms a continuous function.

IV. $S_k'(x_{k+1}) = S_{k+1}'(x_{k+1})$ for $k = 0, 1, \ldots, N - 2$.
The spline forms a smooth function.

V. $S_k''(x_{k+1}) = S_{k+1}''(x_{k+1})$ for $k = 0, 1, \ldots, N - 2$.
The second derivative is continuous.

Existence of Cubic Splines

Let us try to ascertain if it is possible to construct a cubic spline that satisfies properties I to V. Each cubic polynomial $S_k(x)$ has four unknown constants, hence there are $4N$ coefficients to be determined. Loosely speaking, we have $4N$ degrees of freedom or conditions that must be specified. The data points supply $N + 1$ conditions, and properties III, IV, and V each supply $N - 1$ conditions. Hence, $N + 1 + 3(N - 1) = 4N - 2$ conditions are specified. This leaves us two additional degrees of freedom. We will call them **endpoint constraints**; they will involve either $S'(x)$ or $S''(x)$ at x_0 and x_N and will be discussed later. We now proceed with the construction.

Since $S(x)$ is piecewise cubic, its second derivative $S''(x)$ is piecewise linear on $[x_0, x_N]$. The linear Lagrange interpolation formula gives the following representation for $S''(x) = S_k''(x)$:

$$S_k''(x) = S''(x_k) \frac{x - x_{k+1}}{x_k - x_{k+1}} + S''(x_{k+1}) \frac{x - x_k}{x_{k+1} - x_k}. \tag{4}$$

Use $m_k = S''(x_k)$, $m_{k+1} = S''(x_{k+1})$, and $h_k = x_{k+1} - x_k$ in (4) to get

$$S_k''(x) = \frac{m_k}{h_k}(x_{k+1} - x) + \frac{m_{k+1}}{h_k}(x - x_k) \tag{5}$$

for $x_k \leq x \leq x_{k+1}$ and $k = 0, 1, \ldots, N - 1$. Integrating (5) twice will introduce two constants of integration, and the result can be manipulated so that it has the form

$$S_k(x) = \frac{m_k}{6h_k}(x_{k+1} - x)^3 + \frac{m_{k+1}}{6h_k}(x - x_k)^3 + p_k(x_{k+1} - x) + q_k(x - x_k). \tag{6}$$

Substituting x_k and x_{k+1} into equation (6) and using the values $y_k = S_k(x_k)$ and $y_{k+1} = S_k(x_{k+1})$ yields the following equations that involve p_k and q_k, respectively:

$$y_k = \frac{m_k}{6}h_k^2 + p_k h_k \quad \text{and} \quad y_{k+1} = \frac{m_{k+1}}{6}h_k^2 + q_k h_k. \tag{7}$$

These two equations are easily solved for p_k and q_k, and when these values are substituted into equation (6), the result is the following expression for the cubic function $S_k(x)$:

$$S_k(x) = \frac{m_k}{6h_k}(x_{k+1} - x)^3 + \frac{m_{k+1}}{6h_k}(x - x_k)^3 + \left(\frac{y_k}{h_k} - \frac{m_k h_k}{6}\right)(x_{k+1} - x) \tag{8}$$

$$+ \left(\frac{y_{k+1}}{h_k} - \frac{m_{k+1} h_k}{6}\right)(x - x_k).$$

Notice that the representation (8) has been reduced to a form that involves only the unknown coefficients $\{m_k\}$. To find these values we must use the derivative of (8), which is

$$S_k'(x) = -\frac{m_k}{2h_k}(x_{k+1} - x)^2 + \frac{m_{k+1}}{2h_k}(x - x_k)^2 - \left(\frac{y_k}{h_k} - \frac{m_k h_k}{6}\right) \tag{9}$$

$$+ \frac{y_{k+1}}{h_k} - \frac{m_{k+1} h_k}{6}.$$

Evaluating (9) at x_k and simplifying the result yields

$$S_k'(x_k) = -\frac{m_k}{3}h_k - \frac{m_{k+1}}{6}h_k + d_k, \qquad \text{where } d_k = \frac{y_{k+1} - y_k}{h_k}. \tag{10}$$

Similarly, we can replace k by $k - 1$ in (9) to get the expression for $S_{k-1}'(x)$ and evaluate it at x_k to obtain

$$S_{k-1}'(x_k) = \frac{m_k}{3}h_{k-1} + \frac{m_{k-1}}{6}h_{k-1} + d_{k-1}. \tag{11}$$

Now use property IV and equations (10) and (11) to obtain an important relation involving m_{k-1}, m_k, and m_{k+1}:

$$h_{k-1}m_{k-1} + 2(h_{k-1} + h_k)m_k + h_k m_{k+1} = u_k, \qquad \text{where } u_k = 6(d_k - d_{k-1}) \tag{12}$$

$$\text{for } k = 1, 2, \ldots, N - 1.$$

Construction of Cubic Splines

Observe that the unknowns in (12) are the desired values $\{m_k\}$ and the other terms are constants obtained by performing simple arithmetic with the data points $\{(x_k, y_k)\}$. Therefore, in reality system (12) is an underdetermined system of $N - 1$ linear equations involving $N + 1$ unknowns. Hence two additions equations must be supplied. They are used to eliminate m_0 from equation 1 and m_N from equation $N - 1$, in system (12). The standard strategies for the endpoints constraints are summarized in Table 5.8.

TABLE 5.8 Endpoint Constraints for a Cubic Spline

Description of the strategy	Equations involving m_0 and m_N
(i) "Clamped cubic spline": specify $S'(x_0)$, $S'(x_N)$ (the "best choice" if the derivatives are known)	$m_0 = \dfrac{3}{h_0}[d_0 - S'(x_0)] - \dfrac{m_1}{2}$, $\quad m_N = \dfrac{3}{h_{N-1}}[S'(x_N) - d_{N-1}] - \dfrac{m_{N-1}}{2}$
(ii) "Natural cubic spline" (a "relaxed curve")	$m_0 = 0, \ m_N = 0$
(iii) Extrapolate $S''(x)$ to the endpoints.	$m_0 = m_1 - \dfrac{h_0(m_2 - m_1)}{h_1}$, $\quad m_N = m_{N-1} + \dfrac{h_{N-1}(m_{N-1} - m_{N-2})}{h_{n-2}}$
(iv) $S''(x)$ is constant near the endpoints.	$m_0 = m_1, \ m_N = m_{N-1}$
(v) Specify $S''(x)$ at each endpoint.	$m_0 = S''(x_0), \ m_N = S''(x_N)$

Consider strategy (v) in Table 5.8. If m_0 is given, then $h_0 m_0$ can be computed, and the first equation (when $k = 1$) of (12) is

$$2[h_0 + h_1]m_1 + h_1 m_2 = u_1 - h_0 m_0. \tag{13}$$

Similarly, if m_N is given, then $h_{N-1}m_N$ can be computed, and the last equation (when $k = N - 1$) of (12) is

$$h_{N-2}m_{N-2} + 2(h_{N-2} + h_{N-1})m_{N-1} = u_{N-1} - h_{N-1}m_N. \tag{14}$$

Equations (13) and (14) with (12) used for $k = 2, 3, \ldots, N - 2$ form $N - 1$ linear equations involving the coefficients $m_1, m_2, \ldots, m_{N-1}$.

Regardless of the particular strategy chosen in Table 5.8, we can rewrite equations 1 and $N - 1$ in (12) and obtain a tridiagonal linear system of the form $HM = V$, which involves $m_1, m_2, \ldots, m_{N-1}$:

$$\begin{pmatrix} b_1 & c_1 & & & & \\ a_1 & b_2 & c_2 & & & \\ & & \cdot & & & \\ & & & \cdot & & \\ & & & & \cdot & \\ & & a_{N-3} & b_{N-2} & c_{N-2} \\ & & & a_{N-2} & b_{N-1} \end{pmatrix} \begin{pmatrix} m_1 \\ m_2 \\ \vdots \\ \\ \vdots \\ m_{N-2} \\ m_{N-1} \end{pmatrix} = \begin{pmatrix} v_1 \\ v_2 \\ \vdots \\ \\ \vdots \\ v_{N-2} \\ v_{N-1} \end{pmatrix}. \tag{15}$$

The linear system in (15) is diagonally dominant and has a unique solution (see Chapter 3 for details). After the coefficients $\{m_k\}$ are determined, the spline coefficients $\{s_{k,j}\}$ for $S_k(x)$ are computed using the formulas

$$s_{k,0} = y_k, \qquad s_{k,1} = d_k - \frac{h_k(2m_k + m_{k+1})}{6},$$

$$s_{k,2} = \frac{m_k}{2}, \qquad s_{k,3} = \frac{m_{k+1} - m_k}{6h_k}. \tag{16}$$

Each cubic polynomial $S_k(x)$ can be written in nested multiplication form for efficient computation:

$$S_k(x) = [(s_{k,3}w + s_{k,2})w + s_{k,1}]w + y_k, \qquad \text{where } w = x - x_k \tag{17}$$

and $S_k(x)$ is used on the interval $x_k \le x \le x_{k+1}$.

Equations (12) together with a strategy from Table 5.8 can be used to construct a cubic spline with distinctive properties at the ends. Specifically, the values for m_0 and m_N in Table 5.8 are used to customize the first and last equations in (12) and form the system of $N - 1$ equations given in (15). Then the tridiagonal system is solved for the remaining coefficients $m_1, m_2, \ldots, m_{N-1}$. Finally, the formulas in (16) are used to determine the spline coefficients. For reference, we now state how the equations must be prepared for each different type of spline.

Lemma 5.1   (Clamped Spline).   There exists a unique cubic spline with the first derivative boundary conditions $S'(a) = d_0$ and $S'(b) = d_N$.

Proof. Solve the linear system

$$\left(\tfrac{3}{2}h_0 + 2h_1\right)m_1 + h_1 m_2 = u_1 - 3[d_0 - S'(x_0)],$$

$$h_{k-1}m_{k-1} + 2(h_{k-1} + h_k)m_k + h_k m_{k+1} = u_k \qquad \text{for } k = 2, 3, \ldots, N - 2,$$

$$h_{N-2}m_{N-2} + \left(2h_{N-2} + \tfrac{3}{2}h_{N-1}\right)m_{N-1} = u_{N-1} - 3[S'(x_N) - d_{N-1}].$$

Remark.   The clamped spline involves slope at the ends. This spline can be visualized as the curve obtained when a "flexible elastic rod" is forced to pass through the points, and the rod is clamped at each end with a fixed slope. This spline would be useful to a draftsman for drawing a "smooth" curve through several points.

Lemma 5.2   (Natural Spline).   There exists a unique cubic spline with the free boundary conditions $S''(a) = 0$ and $S''(b) = 0$.

Proof. Solve the linear system

$$2(h_0 + h_1)m_1 + h_1 m_2 = u_1,$$

$$h_{k-1}m_{k-1} + 2(h_{k-1} + h_k)m_k + h_k m_{k+1} = u_k \qquad \text{for } k = 2, 3, \ldots, N - 2,$$

$$h_{N-2}m_{N-2} + 2(h_{N-2} + h_{N-1})m_{N-1} = u_{N-1}.$$

Remark.   The natural spline is the curve obtained by forcing a flexible elastic rod through the points but letting the slope at the ends be free to equilibrate to the position that minimizes the oscillatory behavior of the curve. It is useful for fitting a curve to experimental data that are significant to several significant digits.

Lemma 5.3   (Extrapolated Spline).   There exists a unique cubic spline which uses extrapolation from the interior nodes at x_1 and x_2 to determine $S''(a)$ and extrapolation from the nodes at x_{N-1} and x_{N-2} to determine $S''(b)$.

Proof. Solve the linear system:

$$\left(3h_0 + 2h_1 + \frac{h_0^2}{h_1}\right) m_1 + \left(h_1 - \frac{h_0^2}{h_1}\right) m_2 = u_1,$$

$$h_{k-1}m_{k-1} + 2(h_{k-1} + h_k)m_k + h_k m_{k+1} = u_k \qquad \text{for } k = 2, 3, \ldots, N - 2,$$

$$\left(h_{N-2} - \frac{h_{N-1}^2}{h_{N-2}}\right) m_{N-2} + \left(2h_{N-2} + 3h_{N-1} + \frac{h_{N-1}^2}{h_{N-2}}\right) m_{N-1} = u_{N-1}.$$

Remark.   The extrapolated spline is equivalent to assuming that the end cubic is an extension of the adjacent cubic; that is, the spline forms a single cubic curve over the interval $[x_0, x_2]$ and another single cubic curve over the interval $[x_{N-2}, x_N]$.

Lemma 5.4   (Parabolically Terminated Spline).   There exists a unique cubic spline that uses $S''(x) \equiv 0$ on the interval $[x_0, x_1]$ and $S''(x) \equiv 0$ on $[x_{N-1}, x_N]$.

Proof. Solve the linear system

$$(3h_0 + 2h_1)m_1 + h_1 m_2 = u_1,$$

$$h_{k-1}m_{k-1} + 2(h_{k-1} + h_k)m_k + h_k m_{k+1} = u_k, \qquad \text{for } k = 2, 3, \ldots, N - 2,$$

$$h_{N-2}m_{N-2} + (2h_{N-2} + 3h_{N-1})m_{N-1} = u_{N-1}.$$

Remark.   The assumption that $S''(x) \equiv 0$ on the interval $[x_0, x_1]$ forces the cubic to degenerate to a quadratic over $[x_0, x_1]$ and a similar situation occurs over $[x_{N-1}, x_N]$.

Lemma 5.5   (Endpoint Curvature-Adjusted Spline).   There exists a unique cubic spline with the second derivative boundary conditions $S''(a)$ and $S''(b)$ specified.

Proof. Solve the linear system

$$2(h_0 + h_1)m_1 + h_1 m_2 = u_1 - h_0 S''(x_0),$$

$$h_{k-1}m_{k-1} + 2(h_{k-1} + h_k)m_k + h_k m_{k+1} = u_k \qquad \text{for } k = 2, 3, \ldots, N - 2,$$

$$h_{N-2}m_{N-2} + 2(h_{N-2} + h_{N-1})m_{N-1} = u_{N-1} - h_{N-1} S''(x_N).$$

Remark.   Imposing values for $S''(a)$ and $S''(b)$ permits the practitioner to adjust the curvature at each endpoint.

The next five examples illustrate the behavior of the various splines. It is possible to "mix" the end conditions to obtain and even wider variety of possibilities, but we leave these variations for the reader to investigate.

Example 5.7

Find the clamped cubic spline that passes through (0, 0.0), (1, 0.5), (2, 2.0), and (3, 1.5) with the first derivative boundary conditions: $S'(0) = 0.2$ and $S'(3) = -1$.

Solution. First, compute the quantities: $h_0 = h_1 = h_2 = 1$ and $d_0 = (y_1 - y_0)/h_0 = (0.5 - 0.0)/1 = 0.5$, $d_1 = (y_2 - y_1)/h_1 = (2.0 - 0.5)/1 = 1.5$, $d_2 = (y_3 - y_2)/h_2 = (1.5 - 2.0)/1 = -0.5$, $u_1 = 6(d_1 - d_0) = 6(1.5 - 0.5) = 6.0$, and $u_2 = 6(d_2 - d_1) = 6(-0.5 - 1.5) = -12.0$. Then use Lemma 5.1 and obtain the equations

$$\left(\tfrac{3}{2} + 2\right)m_1 + m_2 = 6.0 - 3[0.5 - 0.2] = 5.1,$$

$$m_1 + \left(2 + \tfrac{3}{2}\right)m_2 = -12.0 - 3[-1.0 - (-0.5)] = -10.5.$$

When these equations are simplified and put in matrix notation, we have

$$\begin{pmatrix} 3.5 & 1.0 \\ 1.0 & 3.5 \end{pmatrix} \begin{pmatrix} m_1 \\ m_2 \end{pmatrix} = \begin{pmatrix} 5.1 \\ -10.5 \end{pmatrix}.$$

It is a straightforward task to compute the solution $m_1 = 2.52$ and $m_2 = -3.72$. Now apply the equations in (i) of Table 5.8 to determine the coefficients m_0 and m_3:

$$m_0 = 3(0.5 - 0.2) - \frac{2.52}{2} = -0.36,$$

$$m_3 = 3(-1.0 + 0.5) - \frac{3.72}{2} = 0.36.$$

Next, the values $m_0 = -0.36$, $m_1 = 2.52$, $m_2 = -3.72$, and $m_3 = 0.36$ are substituted into the equations (16) to find the spline coefficients. The solution is

$$S_0(x) = 0.48x^3 - 0.18x^2 + 0.2x \qquad \text{for } 0 \le x \le 1,$$

$$S_1(x) = -1.04(x - 1)^3 + 1.26(x - 1)^2 + 1.28(x - 1) + 0.5 \qquad \text{for } 1 \le x \le 2, \qquad (18)$$

$$S_2(x) = 0.68(x - 2)^3 - 1.86(x - 2)^2 + 0.68(x - 2) + 2.0 \qquad \text{for } 2 \le x \le 3.$$

This clamped cubic spline is shown in Figure 5.12.

Example 5.8

Find the natural cubic spline that passes though (0, 0.0), (1, 0.5), (2, 2.0), and (3, 1.5) with the free boundary conditions: $S''(0) = 0$ and $S''(3) = 0$.

Solution. Use the same values $\{h_k\}$, $\{d_k\}$, and $\{u_k\}$ that were computed in Example 5.7. Then use Lemma 5.2 and obtain the equations

$$2(1 + 1)m_1 + m_2 = 6.0,$$

$$m_1 + 2(1 + 1)m_2 = -12.0.$$

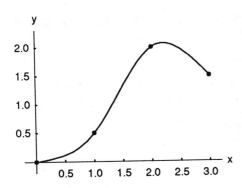

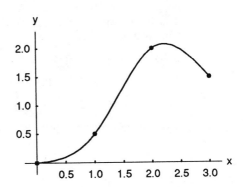

Figure 5.12 The clamped cubic spline with derivative boundary conditions: $S'(0) = 0.2$ and $S'(3) = -1$.

Figure 5.13 The natural cubic spline with $S''(0) = 0$ and $S''(3) = 0$.

The matrix form of this linear system is

$$\begin{pmatrix} 4.0 & 1.0 \\ 1.0 & 4.0 \end{pmatrix} \begin{pmatrix} m_1 \\ m_2 \end{pmatrix} = \begin{pmatrix} 6.0 \\ -12.0 \end{pmatrix}.$$

It is easy to find the solution $m_1 = 2.4$ and $m_2 = -3.6$. Since $m_0 = S''(0) = 0$ and $m_3 = S''(0) = 0$, when equations (16) are used to find the spline coefficients, the result is

$$S_0(x) = 0.4x^3 + 0.1x \qquad\qquad\qquad\text{for } 0 \le x \le 1,$$

$$S_1(x) = -(x - 1)^3 + 1.2(x - 1)^2 + 1.3(x - 1) + 0.5 \qquad \text{for } 1 \le x \le 2, \qquad (19)$$

$$S_2(x) = 0.6(x - 2)^3 - 1.8(x - 2)^2 + 0.7(x - 2) + 2.0 \qquad \text{for } 2 \le x \le 3.$$

This natural cubic spline is shown in Figure 5.13.

Example 5.9

Find the extrapolated cubic spline through $(0. 0.0)$, $(1, 0.5)$, $(2, 2.0)$, and $(3, 1.5)$.

Solution.   Use the values $\{h_k\}$, $\{d_k\}$, and $\{u_k\}$ from Example 5.7 with Lemma 5.3 and obtain the linear system:

$$(3 + 2 + 1)m_1 + (1 - 1)m_2 = 6.0,$$

$$(1 - 1)m_1 + (2 + 3 + 1)m_2 = -12.0.$$

The matrix form is

$$\begin{pmatrix} 6.0 & 0.0 \\ 0.0 & 6.0 \end{pmatrix} \begin{pmatrix} m_1 \\ m_2 \end{pmatrix} = \begin{pmatrix} 6.0 \\ -12.0 \end{pmatrix},$$

and it is trivial to obtain $m_1 = 1.0$ and $m_2 = -2.0$. Now apply the equations in (iii) of Table 5.8 to compute m_0 and m_3:

$$m_0 = 1.0 - (-2.0 - 1.0) = 4.0,$$

$$m_3 = -2.0 + (-2.0 - 1.0) = -5.0.$$

Finally, the values for $\{m_k\}$ are substituted in equations (16) to find the spline coefficients. The solution is

$$S_0(x) = -0.5x^3 + 2.0x^2 - x \qquad \text{for } 0 \le x \le 1,$$
$$S_1(x) = -0.5(x-1)^3 + 0.5(x-1)^2 + 1.5(x-1) + 0.5 \qquad \text{for } 1 \le x \le 2, \qquad (20)$$
$$S_2(x) = -0.5(x-2)^3 - (x-2)^2 + (x-2) + 2.0 \qquad \text{for } 2 \le x \le 3.$$

The extrapolated cubic spline is shown in Figure 5.14.

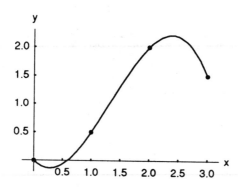

Figure 5.14 The extrapolated cubic spline.

Example 5.10

Find the parabolically terminated cubic spline though $(0, 0.0)$, $(1, 0.5)$, $(2, 2.0)$, and $(3, 1.5)$.

Solution.   Use $\{h_k\}$, $\{d_k\}$, and $\{u_k\}$ from Example 5.7 and then apply Lemma 5.4 to obtain

$$(3+2)m_1 + m_2 = 6.0,$$
$$m_1 + (2+3)m_2 = -12.0.$$

The matrix form is

$$\begin{pmatrix} 5.0 & 1.0 \\ 1.0 & 5.0 \end{pmatrix} \begin{pmatrix} m_1 \\ m_2 \end{pmatrix} = \begin{pmatrix} 6.0 \\ -12.0 \end{pmatrix},$$

and the solution is $m_1 = 1.75$ and $m_2 = -2.75$. Since $S''(x) \equiv 0$ on the subinterval at each end, formulas (iv) in Table 5.8 imply that we have $m_0 = m_1 = 1.75$ and $m_3 = m_2 = -2.75$. Then the values for $\{m_k\}$ are substituted in equations (16) to get the solution

$$S_0(x) = 0.875x^2 - 0.375x \qquad \text{for } 0 \le x \le 1,$$
$$S_1(x) = -0.75(x-1)^3 + 0.875(x-1)^2 + 1.375(x-1) + 0.5 \qquad \text{for } 1 \le x \le 2, \quad (21)$$
$$S_2(x) = -1.375(x-2)^2 + 0.875(x-2) + 2.0 \qquad \text{for } 2 \le x \le 3.$$

This parabolically terminated cubic spline is shown in Figure 5.15.

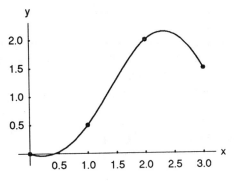

Figure 5.15   The parabolically terminated cubic spline.

Example 5.11

Find the curvature-adjusted cubic spline through $(0, 0.0)$, $(1, 0.5)$, $(2, 2.0)$, and $(3, 1.5)$ with the second derivative boundary conditions $S''(0) = -0.3$ and $S''(3) = 3.3$.

Solution.   Use $\{h_k\}$, $\{d_k\}$, and $\{u_k\}$ from Example 5.7 and then apply Lemma 5.5 to obtain

$$2(1 + 1)m_1 + m_2 = 6.0 - (-0.3) = 6.3,$$

$$m_1 + 2(1 + 1)m_2 = -12.0 - (3.3) = -15.3.$$

The matrix form is

$$\begin{pmatrix} 4.0 & 1.0 \\ 1.0 & 4.0 \end{pmatrix} \begin{pmatrix} m_1 \\ m_2 \end{pmatrix} = \begin{pmatrix} 6.3 \\ -15.3 \end{pmatrix},$$

and the solution is $m_1 = 2.7$ and $m_2 = -4.5$. The given boundary conditions are used to determine $m_0 = S''(0) = -0.3$ and $m_3 = S''(3) = 3.3$. Substitution of $\{m_k\}$ in equations (16) produces the solution

$$S_0(x) = 0.5x^3 - 0.15x^2 + 0.15x \qquad\qquad \text{for } 0 \le x \le 1,$$

$$S_1(x) = -1.2(x - 1)^3 + 1.35(x - 1)^2 + 1.35(x - 1) + 0.5 \qquad \text{for } 1 \le x \le 2, \qquad (22)$$

$$S_2(x) = 1.3(x - 2)^3 - 2.25(x - 2)^2 + 0.45(x - 2) + 2.0 \qquad \text{for } 2 \le x \le 3.$$

This curvature-adjusted cubic spline is shown in Figure 5.16.

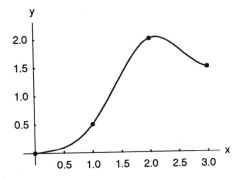

Figure 5.16   The curvature adjusted cubic spline with $S''(0) = -0.3$ and $S''(3) = 3.3$.

Suitability of Cubic Splines

A practical feature of splines is the minimum of the oscillatory behavior they possess. Consequently, among all functions $f(x)$ which are twice continuously differentiable on $[a, b]$ and interpolate a given set of data points $\{(x_k, y_k)\}_{k=0}^{N}$, the cubic spline has less wiggle. The next result explains this phenomenon.

Theorem 5.4 (Minimum Property of Cubic Splines). Assume that $f \in C^2[a, b]$ and $S(x)$ is the unique cubic spline interpolant for $f(x)$ which passes through $\{(x_k, f(x_k))\}_{k=0}^{N}$ and satisfies the clamped end conditions $S'(a) = f'(a)$ and $S'(b) = f'(b)$. Then

$$\int_a^b [S''(x)]^2 \, dx \le \int_a^b [f''(x)]^2 \, dx. \tag{23}$$

Proof. Use integration by parts and the end conditions to obtain

$$\int_a^b S''(x)[f''(x) - S''(x)] \, dx = S''(x)[f'(x) - S'(x)] \Big|_{x=a}^{x=b} - \int_a^b S'''(x)[f'(x) - S'(x)] \, dx$$

$$= 0 - 0 - \int_a^b S'''(x)[f'(x) - S'(x)] \, dx.$$

Since $S'''(x) \equiv 6s_{k,3}$ on the subinterval $[x_k, x_{k+1}]$, its follows that

$$\int_{x_k}^{x_{k+1}} S'''(x)[f'(x) - S'(x)] \, dx = 6s_{k,3}[f(x) - S(x)] \Big|_{x=x_k}^{x=x_{k+1}} = 0$$

$$\text{for } k = 0, 1, \ldots, N - 1.$$

Hence $\int_a^b S''(x)[f''(x) - S''(x)] \, dx = 0$ and it follows that

$$\int_a^b S''(x)f''(x) \, dx = \int_a^b [S''(x)]^2 \, dx. \tag{24}$$

Since $0 \le [f''(x) - S''(x)]^2$ we get the integral relationship

$$0 \le \int_a^b [f''(x) - S''(x)]^2 \, dx$$

$$= \int_a^b [f''(x)]^2 \, dx - 2 \int_a^b f''(x)S''(x) \, dx + \int_a^b [S''(x)]^2 \, dx. \tag{25}$$

Now the result in (24) is substituted into (25) and the result is

$$0 \le \int_a^b [f''(x)]^2 \, dx - \int_a^b [S''(x)]^2 \, dx,$$

and this is easily rewritten to obtain relation (23) and the result is proven.

Algorithm 5.4   (Cubic Splines).   To construct and evaluate a cubic spline interpolant $S(x)$ for $N + 1$ data points (x_0, y_0), (x_1, y_1), ... , (x_N, y_N). Provision is made for the following choices of endpoint constraints:

(i)   "Clamped cubic spline"; specify $S'(x_0)$ and $S'(x_N)$.

(ii)   "Natural cubic spline" (a "relaxed curve").

(iii)   Extrapolate $S''(x)$ to the endpoints.

(iv) $S''(x)$ is constant near the endpoints.

(v)   Specify $S''(x)$ at each endpoint.

H(0) := X(1) − X(0)   {Difference in abscissa}
D(0) := [Y(1) − Y(0)]/H(0)   {Difference quotient}

FOR   K = 1   TO   N−1   DO
⎮   H(K) := X(K+1) − X(K)   {Differences in abscissa}
⎮   D(K) := [Y(K+1) − Y(K)]/H(K)   {Difference quotients}
⎮   A(K) := H(K)   {Subdiagonal elements}
⎮   B(K) := 2*[H(K−1) + H(K)]   {Diagonal elements}
⎣   C(K) := H(K)   {Superdiagonal elements}

FOR   K = 1   TO   N−1   DO   {Determine the
⎣_____ V(K) := 6*[D(K) − D(K−1)]   column vector}

CASES   {Modify the matrix and/or column vector}

(i)   Set B(1) := B(1) − H(0)/2
          V(1) := V(1) − 3*[D(0) − S'(x₀)]   {Input S'(x₀)}
          B(N−1) := B(N−1) − H(N−1)/2
          V(N−1) := V(N−1) − 3*[S'(xₙ) − D(N−1)]   {Input S'(xₙ)}

(ii)   Set M(0) := 0 and M(N) := 0

(iii)   Set B(1) := B(1) + H(0) + H(0)*H(0)/H(1)
          C(1) := C(1) − H(0)*H(0)/H(1)
          B(N−1) := B(N−1) + H(N−1) + H(N−1)*H(N−1)/H(N−2)
          A(N−2) := A(N−2) − H(N−1)*H(N−1)/H(N−2)

(iv)   Set B(1) := B(1) + H(0) and B(N−1) := B(N−1) + H(N−1)

(v)   Set V(1) := V(1) − H(0)*S''(x₀)   {Input S''(x₀)}
          V(N−1) := V(N−1) − H(N−1)*S''(xₙ)   {Input S''(xₙ)}
END

```
FOR    K = 2  TO  N-1  DO              {Gaussian elimination is used
       T := A(K-1)/B(K-1)               to produce an upper-triangular
       B(K) := B(K) - T*C(K-1)          system with "two diagonals"}
       V(K) := V(K) - T*V(K-1)
```

```
       M(N-1) := V(N-1)/B(N-1)         {Back substitution
FOR    K = N-2  DOWNTO  1  DO           is used to find m_k}
       M(K) := [V(K) - C(K)*M(K+1)]/B(K)
```

CASES   {Determine the values M(0) and M(N)}

  (i)   Set M(0) := 3*[D(0) - S'(x_0)]/H(0) - M(1)/2
        M(N) := 3*[S'(x_N) - D(N-1)]/H(N-1) - M(N-1)/2

  (ii) Set M(0) := 0 and M(N) := 0

  (iii) Set M(0) := M(1) - H(0)*[M(2) - M(1)]/H(1)
        M(N) := M(N-1) + H(N-1)*[M(N-1) - M(N-2)]/H(N-2)

  (iv) Set M(0) := M(1) and M(N) := M(N-1)

  (v)   Set M(0) := S''(x_0) and M(N) := S''(x_N)
END

```
FOR    K = 0  TO  N-1  DO
       S(K, 0) := Y(K)                          {Compute and store
       S(K, 1) := D(K) - H(K)*[2*M(K) + M(K+1)]/6   the coefficients
       S(K, 2) := M(K)/2                            for each cubic
       S(K, 3) := [M(K+1) - M(K)]/[6*H(K)]      polynomial S_k(x)}
```

{Procedure for evaluating the cubic spline above on $[x_0, x_N]$}
```
INPUT  X                               {Input the independent variable}
FOR    J = 1  TO  N  DO                 {Find the interval
       IF  X(J-1) ≤ X ≤ X(J)   THEN      so that x lies
           Set  K := J-1  and  J := N    in [x_k, x_{k+1}]}
       IF  X = X(0)  THEN   Set K := 0
       W := X - X(K)                    {Evaluate the spline}
       Z := [[S(K, 3)*W + S(K, 2)]*W + S(K, 1)]*W + S(K, 0)
PRINT   "The value of the spline S(x) is" Z          {Output}
```

EXERCISES FOR INTERPOLATION BY SPLINE FUNCTIONS

1. Consider the cubic polynomial $S(x) = a_0 + a_1x + a_2x^2 + a_3x^3$.
    (a) Show that the conditions $S(1) = 1$, $S'(1) = 0$, $S(2) = 2$, and $S'(2) = 0$ produce the system of equations

$$a_0 + a_1 + a_2 + a_3 = 1$$
$$a_1 + 2a_2 + 3a_3 = 0$$
$$a_0 + 2a_1 + 4a_2 + 8a_3 = 2$$
$$a_1 + 4a_2 + 12a_3 = 0.$$

(b) Solve the system in part (a) and obtain $S(x) = 6 - 12x + 9x^2 - 2x^3$. The graph of this cubic polynomial is shown in Figure 5.17.

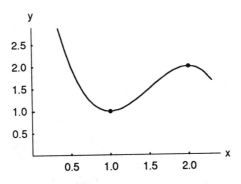

Figure 5.17   Figure for Exercise 1.

2. Consider the cubic polynomial $S(x) = a_0 + a_1x + a_2x^2 + a_3x^3$.
   (a) Show that the conditions $S(1) = 3$, $S'(1) = -4$, $S(2) = 1$, and $S'(2) = 2$ produce the system of equations

$$a_0 + a_1 + a_2 + a_3 = 3$$
$$a_1 + 2a_2 + 3a_3 = -4$$
$$a_0 + 2a_1 + 4a_2 + 8a_3 = 1$$
$$a_1 + 4a_2 + 12a_3 = 2.$$

   (b) Solve the system in part (a) and obtain $S(x) = 5 + 2x - 6x^2 + 2x^3$. The graph of this cubic polynomial is shown in Figure 5.18.

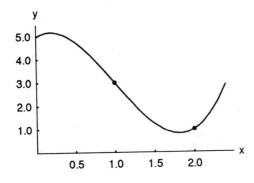

Figure 5.18   Figure for Exercise 2.

3. Consider the function

$$f(x) = \begin{cases} \dfrac{19}{2} - \dfrac{81x}{4} + 15x^2 - \dfrac{13x^3}{4} & \text{for } 1 \le x \le 2, \\[3mm] \dfrac{-77}{2} + \dfrac{207x}{4} - 21x^2 + \dfrac{11x^3}{4} & \text{for } 2 \le x \le 3. \end{cases}$$

Show that $f(x)$ is a cubic spline by verifying that both formulas and their first two derivatives agree at $x = 2$. The graph $y = f(x)$ is shown in Figure 5.19.

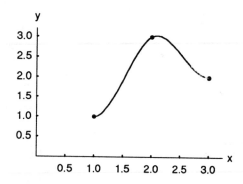

Figure 5.19 Figure for Exercise 3.

4. Consider the function

$$f(x) = \begin{cases} 18 - \dfrac{75x}{2} + 26x^2 - \dfrac{11x^3}{2} & \text{for } 1 \le x \le 2, \\[3mm] -70 + \dfrac{189x}{2} - 40x^2 + \dfrac{11x^3}{2} & \text{for } 2 \le x \le 3. \end{cases}$$

Show that $f(x)$ is a cubic spline by verifying that both formulas and their first two derivatives agree at $x = 2$. The graph $y = f(x)$ is shown in Figure 5.20.

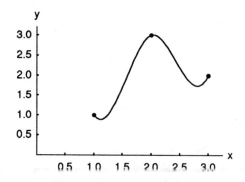

Figure 5.20 Figure for Exercise 4.

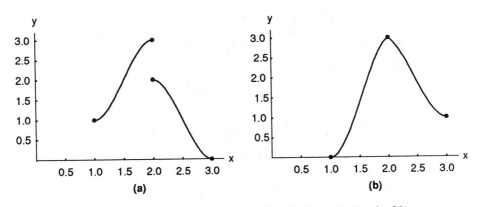

Figure 5.21   (a) Figure for Exercise 5(a). (b) Figure for Exercise 5(b).

5. Show that the following functions are not cubic splines.

   (a) $f(x) = \begin{cases} 11 - 24x + 18x^2 - 4x^3 & \text{for } 1 \le x \le 2 \\ -54 + 72x - 30x^2 + 4x^3 & \text{for } 2 < x \le 3 \end{cases}$ [see Figure 5.21(a)]

   (b) $f(x) = \begin{cases} 13 - 31x + 23x^2 - 5x^3 & \text{for } 1 \le x \le 2 \\ -35 + 51x - 22x^2 + 3x^3 & \text{for } 2 \le x \le 3 \end{cases}$ [see Figure 5.21(b)]

In Exercises 6 and 7, consider five different splines. Show that the linear system in (i) to (v) arises when Lemmas 1 to 5 are applied, respectively. After you have found $\{m_k\}$, use formula (16) to find the coefficients $\{s_{k,j}\}$ of $\{S_k(x)\}$. For part (i) use $S'(x_0)$ and $S'(x_3)$, and for (v) use $S''(x_0)$ and $S''(x_3)$.

6. Use the points $(0, 1)$, $(1, 4)$, $(2, 0)$, and $(3, -2)$, and use

$$S'(0) = 2, \quad S'(3) = 2 \quad \text{and} \quad S''(0) = -1.5, \quad S''(3) = 3.$$

   (i) $3.5m_1 + m_2 = -45, \quad m_1 + 3.5m_2 = 0$

   (ii) $4m_1 + m_2 = -42, \quad m_1 + 4m_2 = 12$

   (iii) $6m_1 = -42, \quad 6m_2 = 12$

   (iv) $5m_1 + m_2 = -42, \quad m_1 + 5m_2 = 12$

   (v) $4m_1 + m_2 = -40.5, \quad m_1 + 4m_2 = 9$

7. Use the function $f(x) = x + 2/x$ for $x_0 = 1/2$, $x_1 = 1$, $x_2 = \frac{3}{2}$, and $x_3 = 2$, and

$$S'\left(\tfrac{1}{2}\right) = -7, \quad S'(2) = \tfrac{1}{2} \quad \text{and} \quad S''\left(\tfrac{1}{2}\right) = 32, \quad S''(2) = \tfrac{1}{2}.$$

   (i) $1.75m_1 + 0.5m_2 = 4, \quad 0.5m_1 + 1.75m_2 = 3.5$

   (ii) $2m_1 + 0.5m_2 = 16, \quad 0.5m_1 + 2m_2 = 4$

   (iii) $3m_1 = 16, \quad 3m_2 = 4$

(iv) $2.5m_1 + 0.5m_2 = 16$, $0.5m_1 + 2.5m_2 = 4$

(v) $2m_1 + 0.5m_2 = 0$, $0.5m_1 + 2m_2 = 3.75$

8. Use the substitutions

$$x_{k+1} - x = h_k + (x_k - x)$$

and

$$(x_{k+1} - x)^3 = h_k^3 + 3h_k^2(x_k - x) + 3h_k(x_k - x)^2 + (x_k - x)^3$$

to show that when equation (8) is expanded into powers of $(x - x_k)$, the coefficients are those given in equations (16).

9. Consider each cubic function $S_k(x)$ over $[x_k, x_{k+1}]$.

(a) Give a formula for $\int_{x_k}^{x_{k+1}} S_k(x)\, dx$.

Then evaluate $\int_{x_0}^{x_3} S(x)\, dx$ for case (ii), the natural spline, in part (ii) of

(b) Exercise 6   (c) Exercise 7

In Exercises 10 and 11, consider five different splines. Show that the linear system in (i)–(v) arises when Lemmas 1 to 5 are applied, respectively. Find $\{m_k\}$ and $\{S_k(x)\}$. For (i) use $S'(x_0)$, $S'(x_4)$ and for (v) use $S''(x_0)$, $S''(x_4)$.

10. Use the points $(0, 5)$, $(1, 2)$, $(2, 1)$, $(3, 3)$, and $(4, 1)$, and use

$$S'(0) = -2, \quad S'(4) = -1 \quad \text{and} \quad S''(0) = 0.5, \quad S''(4) = -1.9$$

(i) $3.5m_1 + m_2 = 15$, $m_1 + 4m_2 + m_3 = 18$, $m_2 + 3.5m_3 = -27$

(ii) $4m_1 + m_2 = 12$, $m_1 + 4m_2 + m_3 = 18$, $m_2 + 4m_3 = -24$

(iii) $6m_1 = 12$, $m_1 + 4m_2 + m_3 = 18$, $6m_3 = -24$

(iv) $5m_1 + m_2 = 12$, $m_1 + 4m_2 + m_3 = 18$, $m_2 + 5m_3 = -24$

(v) $4m_1 + m_2 = 11.5$, $m_1 + 4m_2 + m_3 = 18$, $m_2 + 4m_3 = -22.1$

11. Use the points $(0, 2)$, $(1, 3)$, $(2, 2)$, $(3, 2)$, and $(4, 1)$, and use

$$S'(0) = 1.6, \quad S'(4) = -1 \quad \text{and} \quad S''(0) = -1.4, \quad S''(4) = -1.4.$$

(i) $3.5m_1 + m_2 = -10.2$, $m_1 + 4m_2 + m_3 = 6$, $m_2 + 3.5m_3 = -6$

(ii) $4m_1 + m_2 = -12$, $m_1 + 4m_2 + m_3 = 6$, $m_2 + 4m_3 = -6$

(iii) $6m_1 = -12$, $m_1 + 4m_2 + m_3 = 6$, $6m_3 = -6$

(iv) $5m_1 + m_2 = -12$, $m_1 + 4m_2 + m_3 = 6$, $m_2 + 5m_3 = -6$

(v) $4m_1 + m_2 = -10.6$, $m_1 + 4m_2 + m_3 = 6$, $m_2 + 4m_3 = -4.6$

12. Show how strategy (i) in Table 5.8 and system (12) are combined to obtain the equations in Lemma 5.1.

13. Show how strategy (iii) in Table 5.8 and system (12) are combined to obtain the equations in Lemma 5.3.

14. The distance d_k that a car has traveled at time t_k is given in the table below. Use the values $S'(0) = 0$ and $S'(8) = 98$, and find the clamped spline for the points.

Time, t_k	0	2	4	6	8
Distance, d_k	0	40	160	300	480

In Exercises 15–17, use a computer to find the five different splines in cases (i) to (v) for the given points.

15. Use the points (0, 1), (1, 0), (2, 0), (3, 1), (4, 2), (5, 2), and (6, 1) and

$$S'(0) = -0.6, \quad S'(6) = -1.8 \quad \text{and} \quad S''(0) = 1, \quad S''(6) = -1.$$

16. Use the points (0, 0), (1, 4), (2, 8), (3, 9), (4, 9), (5, 8), and (6, 6) and

$$S'(0) = 1, \quad S'(6) = -2 \quad \text{and} \quad S''(0) = 1, \quad S''(6) = -1.$$

17. Use the points (0, 0), (1, 2), (2, 3), (3, 2), (4, 2), (5, 1), and (6, 0) and

$$S'(0) = 1.5, \quad S'(6) = -0.3 \quad \text{and} \quad S''(0) = -1, \quad S''(6) = 1.$$

18. Write a report on Hermite interpolation. See References [9, 29, 40, 41, 79, 81, 90, 92, 128, 153, 191, 193, and 208].

19. Write a report on basic splines (*B*-splines). See References [35, 96, 101, 149, and 160].

5.4 FOURIER SERIES AND TRIGONOMETRIC POLYNOMIALS

Scientists and engineers often study physical phenomena, such as light and sound, which have a periodic character. They are described by functions $f(x)$ which are periodic, that is,

$$g(x + P) = g(x) \qquad \text{for all } x. \tag{1}$$

The number P is called a **period** of the function.

It will suffice to consider functions that have period 2π. If $g(x)$ has period P, then $f(x) = g(Px/2\pi)$ will be periodic with period 2π. This is verified by the observation

$$f(x + 2\pi) = g\left(\frac{Px}{2\pi} + P\right) = g\left(\frac{Px}{2\pi}\right) = f(x). \tag{2}$$

Henceforth in this section we shall assume that $f(x)$ is a function that is periodic with period 2π, that is,

$$f(x + 2\pi) = f(x) \qquad \text{for all } x. \tag{3}$$

The graph $y = f(x)$ is obtained by repeating the portion of the graph in any interval of length 2π, as shown in Figure 5.22.

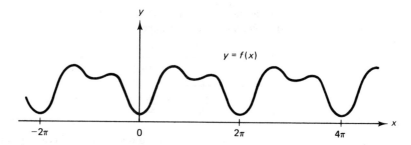

Figure 5.22 A continuous function $f(x)$ with period 2π.

Examples of functions with period 2π are $\sin(jx)$ and $\cos(jx)$, where j is an integer. This raises the question: Can a periodic function be represented by the sum of terms involving $a_j \cos(jx)$ and $b_j \sin(jx)$? We will soon see that the answer is yes.

Definition 5.2 (Piecewise Continuous). The function $f(x)$ is said to be **piecewise continuous** on $[a, b]$ if there exists values $t_0, t_1, \ldots, t_K$ with $a = t_0 < t_1 < \cdots < t_K = b$ such that $f(x)$ is continuous on each open interval $t_{i-1} < x < t_i$ for $i = 1$, $2, \ldots, K$, and $f(x)$ has left- and right-hand limits at each of the points t_i. The situation is illustrated in Figure 5.23.

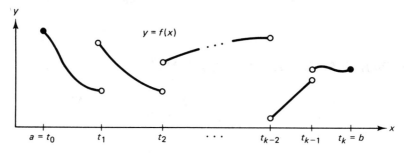

Figure 5.23 A piecewise continuous function over $[a, b]$.

Definition 5.3 (Fourier Series). Assume that $f(x)$ is periodic with period 2π and that $f(x)$ is piecewise continuous on $[-\pi, \pi]$. The **Fourier series** $S(x)$ for $f(x)$ is

$$S(x) = \frac{a_0}{2} + \sum_{j=1}^{\infty} (a_j \cos(jx) + b_j \sin(jx)), \tag{4}$$

where the coefficients a_j and b_j are computed with **Euler's formulas**:

$$a_j = \frac{1}{\pi} \int_{-\pi}^{\pi} f(x) \cos(jx)\, dx \qquad \text{for } j = 0, 1, \ldots \tag{5}$$

and

$$b_j = \frac{1}{\pi} \int_{-\pi}^{\pi} f(x) \sin(jx)\, dx \qquad \text{for } j = 1, 2, \ldots . \tag{6}$$

The factor $\frac{1}{2}$ in the constant term $a_0/2$ in the Fourier series (4) has been introduced for convenience so that a_0 could be obtained from the general formula (5) by setting $j = 0$. Convergence of the Fourier series is discussed in the next result.

Theorem 5.5   (Fourier Expansion).   Assume that $S(x)$ is the Fourier series for $f(x)$ over $[-\pi, \pi]$. If $f'(x)$ is piecewise continuous on $[-\pi, \pi]$ and has a left-hand and right-hand derivative at each point in this interval, then $S(x)$ is convergent for all $x \in [-\pi, \pi]$. The relation

$$S(x) = f(x)$$

holds at all points $x \in [-\pi, \pi]$ where $f(x)$ is continuous. If $x = a$ is a point of discontinuity of f, then

$$S(a) = \frac{f(a^-) + f(a^+)}{2},$$

where $f(a^-)$ and $f(a^+)$ denote the left- and right-hand limits, respectively. With this understanding, we obtain the Fourier expansion:

$$f(x) = \frac{a_0}{2} + \sum_{j=1}^{\infty} (a_j \cos(jx) + b_j \sin(jx)). \tag{7}$$

A brief outline of the derivation of formulas (5) and (6) is given at the end of the section.

Example 5.12

Show that the function $f(x) = x/2$ for $-\pi < x < \pi$, extended periodically by the equation $f(x + 2\pi) = f(x)$, has the Fourier series representation

$$f(x) = \sum_{j=1}^{\infty} \frac{(-1)^{j+1}}{j} \sin(jx) = \sin(x) - \frac{\sin(2x)}{2} + \frac{\sin(3x)}{3} - \cdots .$$

Solution.   Using Euler's formulas and integration by parts, we get

$$a_j = \frac{1}{\pi} \int_{-\pi}^{\pi} \frac{x}{2} \cos(jx)\, dx = \frac{x \sin(jx)}{2\pi j} + \frac{\cos(jx)}{2\pi j^2} \Bigg|_{-\pi}^{\pi} = 0$$

for $j = 1, 2, 3, \ldots$

and

$$b_j = \frac{1}{\pi} \int_{-\pi}^{\pi} \frac{x}{2} \sin(jx)\, dx = \frac{-x \cos(jx)}{2\pi j} + \frac{\sin(jx)}{2\pi j^2} \Bigg|_{-\pi}^{\pi} = \frac{(-1)^{j+1}}{j}$$

for $j = 1, 2, 3, \ldots$.

The coefficient a_0 is obtained by a separate calculation:

$$a_0 = \frac{1}{\pi} \int_{-\pi}^{\pi} \frac{x}{2} \, dx = \left. \frac{x^2}{4\pi} \right|_{-\pi}^{\pi} = 0.$$

The calculations above show that all the coefficients of the cosine functions are zero. The graphs of $f(x)$ and the partial sums

$$S_2(x) = \sin(x) - \frac{\sin(2x)}{2},$$

$$S_3(x) = \sin(x) - \frac{\sin(2x)}{2} + \frac{\sin(3x)}{3}$$

and

$$S_4(x) = \sin(x) - \frac{\sin(2x)}{2} + \frac{\sin(3x)}{3} - \frac{\sin(4x)}{4}$$

are shown in Figure 5.24.

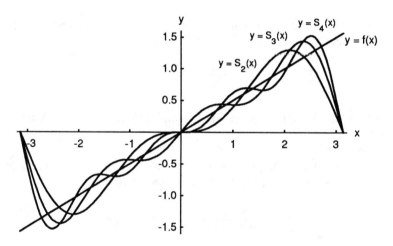

Figure 5.24   The function $f(x) = x/2$ over $[-\pi, \pi]$ and its trigonometric approximations $S_2(x)$, $S_3(x)$ and $S_4(x)$.

We now state some general properties of Fourier series. The proofs are left as exercises.

Theorem 5.6   (Cosine Series).   Suppose that $f(x)$ is an even function; that is, $f(-x) = f(x)$ holds for all x. If $f(x)$ has period 2π and if $f(x)$ and $f'(x)$ are piecewise continuous, then the Fourier series for $f(x)$ involves only cosine terms:

$$f(x) = \frac{a_0}{2} + \sum_{j=1}^{\infty} a_j \cos(jx), \qquad (8)$$

where

$$a_j = \frac{2}{\pi} \int_0^{\pi} f(x) \cos(jx)\, dx \qquad \text{for } j = 0, 1, \ldots \qquad (9)$$

Theorem 5.7 (Sine Series). Suppose that $f(x)$ is an odd function; that is, $f(-x) = -f(x)$ holds for all x. If $f(x)$ has period 2π and if $f(x)$ and $f'(x)$ are piecewise continuous, then the Fourier series for $f(x)$ involves only sine terms:

$$f(x) = \sum_{j=1}^{\infty} b_j \sin(jx), \qquad (10)$$

where

$$b_j = \frac{2}{\pi} \int_0^{\pi} f(x) \sin(jx)\, dx \qquad \text{for } j = 1, 2, \ldots \qquad (11)$$

Example 5.13

Show that the function $f(x) = |x|$ for $-\pi < x < \pi$, extended periodically by the equation $f(x + 2\pi) = f(x)$, has the Fourier cosine representation

$$
\begin{aligned}
f(x) &= \frac{\pi}{2} - \frac{4}{\pi} \sum_{j=1}^{\infty} \frac{\cos((2j-1)x)}{(2j-1)^2} \\
&= \frac{\pi}{2} - \frac{4}{\pi} \left[\cos(x) + \frac{\cos(3x)}{3^2} + \frac{\cos(5x)}{5^2} + \cdots \right].
\end{aligned}
\qquad (12)
$$

Solution. The function $f(x)$ is an even function, so we can use Theorem 5.4 and need only to compute the coefficients $\{a_j\}$:

$$a_j = \frac{2}{\pi} \int_0^{\pi} x \cos(jx)\, dx = \left. \frac{2x \sin(jx)}{\pi j} + \frac{2 \cos(jx)}{\pi j^2} \right|_0^{\pi}$$

$$= \frac{2 \cos(j\pi) - 2}{\pi j^2} = \frac{2[(-1)^j - 1]}{\pi j^2} \qquad \text{for } j = 1, 2, 3, \ldots$$

Since $[(-1)^j - 1] = 0$ when j is even, the cosine series will involve only the odd terms. The odd coefficients have the pattern

$$a_1 = \frac{-4}{\pi}, \quad a_3 = \frac{-4}{\pi 3^2}, \quad a_5 = \frac{-4}{\pi 5^2}, \ldots$$

The coefficient a_0 is obtained by the separate calculation

$$a_0 = \frac{2}{\pi} \int_0^\pi x \, dx = \left. \frac{x^2}{\pi} \right|_0^\pi = \pi.$$

Therefore, we have found the desired coefficients in (12).

Proof of Euler's Formulas for Theorem 5.2. The following heuristic argument assumes the existence and convergence of the Fourier series representation. To determine a_0, we can integrate both sides of (7) and get

$$\int_{-\pi}^\pi f(x) \, dx = \int_{-\pi}^\pi \left\{ \frac{a_0}{2} + \sum_{j=1}^\infty [a_j \cos(jx) + b_j \sin(jx)] \right\} dx$$

$$= \int_{-\pi}^\pi \frac{a_0}{2} \, dx + \sum_{j=1}^\infty a_j \int_{-\pi}^\pi \cos(jx) \, dx + \sum_{j=1}^\infty b_j \int_{-\pi}^\pi \sin(jx) \, dx$$

$$= a_0 + 0 + 0.$$

(13)

Justification for switching the order of integration and summation requires a detailed treatment of uniform convergence and can be found in advanced texts. Hence we have shown that

$$a_0 = \frac{1}{\pi} \int_{-\pi}^\pi f(x) \, dx. \tag{14}$$

To determine a_m, we let $m > 0$ be a fixed integer and multiply both sides of (7) by $\cos(mx)$ and integrate both sides to obtain

$$\int_{-\pi}^\pi f(x) \cos(mx) \, dx = \frac{a_0}{2} \int_{-\pi}^\pi \cos(mx) \, dx + \sum_{j=1}^\infty a_j \int_{-\pi}^\pi \cos(jx) \cos(mx) \, dx$$

$$+ \sum_{j=1}^\infty b_j \int_{-\pi}^\pi \sin(jx) \cos(mx) \, dx.$$

(15)

Equation (15) can be simplified by using the orthogonal properties of the trigonometric functions, which are now stated. The value of the first term on the right-hand side of (15) is

$$\frac{a_0}{2} \int_{-\pi}^\pi \cos(mx) \, dx = \left. \frac{a_0 \sin(mt)}{2m} \right|_{-\pi}^\pi = 0. \tag{16}$$

The value of the term involving $\cos(jx) \cos(mx)$ is found by using the trigonometric identity

$$\cos(jx) \cos(mx) = \tfrac{1}{2} \cos((j + m)x) + \tfrac{1}{2} \cos((j - m)x). \tag{17}$$

When $j \neq m$, then (17) is used to get

$$a_j \int_{-\pi}^{\pi} \cos(jx) \cos(mx)\, dx = \tfrac{1}{2} a_j \int_{-\pi}^{\pi} \cos((j + m)x)\, dx$$

$$+ \tfrac{1}{2} a_j \int_{-\pi}^{\pi} \cos((j - m)x)\, dx = 0 + 0 = 0. \tag{18}$$

When $j = m$, the value of the integral is

$$a_m \int_{-\pi}^{\pi} \cos(jx) \cos(mx)\, dx = a_m \pi. \tag{19}$$

The value of the term on the right side of (15) involving $\sin(jx) \cos(mx)$ is found by using the trigonometric identity

$$\sin(jx) \cos(mx) = \tfrac{1}{2}\sin((j + m)x) + \tfrac{1}{2}\sin((j - m)x). \tag{20}$$

For all values of j and m in (20) we obtain

$$b_j \int_{-\pi}^{\pi} \sin(jx) \cos(mx)\, dx = \tfrac{1}{2} b_j \int_{-\pi}^{\pi} \sin((j + m)x)\, dx$$

$$+ \tfrac{1}{2} b_j \int_{-\pi}^{\pi} \sin((j - m)x)\, dx = 0 + 0 = 0. \tag{21}$$

Therefore, using the results of (16), (18), (19), and (21) in equation (15), we conclude that

$$\pi a_m = \int_{-\pi}^{\pi} f(x) \cos(mx)\, dx, \qquad \text{for } m = 1, 2, \ldots . \tag{22}$$

Therefore, Euler's formula (5) is established.
    Euler's formula (6) is proven similarly.

Trigonometric Polynomial Approximation

Definition 5.4   (Trigonometric Polynomial).   A series of the form

$$T_M(x) = \frac{a_0}{2} + \sum_{j=1}^{M} (a_j \cos(jx) + b_j \sin(jx)) \tag{23}$$

is called a **trigonometric polynomial** of order M.

Theorem 5.8   (Discrete Fourier Series).   Assume that $\{(x_j, y_j)\}_{j=0}^{N}$, where $y_j = f(x_j)$ and that the $N + 1$ points have equally spaced abscissas

$$x_j = -\pi + \frac{2j\pi}{N} \qquad \text{for } j = 0, 1, \ldots, N. \tag{24}$$

If $f(x)$ is periodic with period 2π and $2M < N$, then there exists a trigonometric polynomial $T_M(x)$ of the form (23) that minimizes the quantity

$$\sum_{k=1}^{N} (f(x_k) - T_M(x_k))^2. \tag{25}$$

The coefficients a_j and b_j of this polynomial are computed with the formulas

$$a_j = \frac{2}{N} \sum_{k=1}^{N} f(x_k) \cos(jx_k) \qquad \text{for } j = 0, 1, \ldots, M, \tag{26}$$

and

$$b_j = \frac{2}{N} \sum_{k=1}^{N} f(x_k) \sin(jx_k) \qquad \text{for } j = 1, 2, \ldots, M. \tag{27}$$

Although formulas (26) and (27) are defined with the least-squares procedure, they can also be viewed as numerical approximations to the integrals in Euler's formulas (5) and (6). Euler's formulas give the coefficients for the Fourier series of a continuous function, whereas formulas (26) and (27) give the trigonometric polynomial coefficients for curve fitting to data points. The next example uses data points generated by the function $f(x) = x/2$ at discrete points. When more points are used, the trigonometric polynomial coefficients get closer to the Fourier series coefficients.

Example 5.14

Use the 12 equally spaced points $x_k = -\pi + k\pi/6$ for $k = 1, 2, \ldots, 12$ and find the trigonometric polynomial approximation for $M = 5$ to the 12 data points $\{(x_k, f(x_k))\}$, where $f(x) = x/2$. Also, compare the results when 60 and 360 points are used, and with the first five terms of the Fourier series expansion for $f(x)$ that is given in Example 5.12.

Solution. Since the periodic extension is assumed, at a point of discontinuity, the function value $f(\pi)$ must be computed using the formula

$$f(\pi) = \frac{f(\pi^-) + f(\pi^+)}{2} = \frac{\pi/2 - \pi/2}{2} = 0. \tag{28}$$

The function $f(x)$ is an odd function, hence the coefficients for the sine terms are all zero (i.e., $a_j = 0$ for all j). The trigonometric polynomial of degree $M = 5$ involves only the cosine terms and when formula (26) is used with (28), we get

$$T_5(x) = 0.9770486 \sin(x) - 0.4534498 \sin(2x) + 0.26179938 \sin(3x)$$
$$\tag{29}$$
$$-0.1511499 \sin(4x) + 0.0701489 \sin(5x).$$

The graph of $T_5(x)$ is shown in Figure 5.25.

The coefficients of the fifth-degree trigonometric polynomial change slightly when the number of interpolation points increases to 60 and 360. As the number of points increases, they get closer to the coefficients of the Fourier series expansion of $f(x)$. The results are compared in Table 5.9.

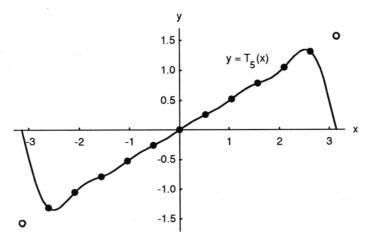

Figure 5.25 The trigonometric polynomial $T_5(x)$ of degree $M = 5$, based on 12 data points that lie on the line $y = x/2$.

TABLE 5.9 Comparison of Trigonometric Polynomial Coefficients for Approximations to $f(x) = x/2$ over $[-\pi, \pi]$

	Trigonometric polynomial coefficients			Fourier series coefficients
	12 Points	60 Points	360 Points	
b_1	0.97704862	0.99908598	0.99997462	1.0
b_2	−0.45344984	−0.49817096	−0.49994923	−0.5
b_3	0.26179939	0.33058726	0.33325718	0.33333333
b_4	−0.15114995	−0.24633386	−0.24989845	−0.25
b_5	0.07014893	0.19540972	0.19987306	0.2

Algorithm 5.5   (Trigonometric Polynomials).   To construct and evaluate the trigonometric polynomial of order M of the form

$$P(x) = \frac{a_0}{2} + \sum_{j=1}^{M} [a_j \cos(jx) + b_j \sin(jx)],$$

based on the N equally spaced values $x_k = -\pi + 2\pi k/N$. The construction is possible provided that $2M + 1 \leq N$.

```
INPUT N                                              {Number of points}
INPUT X(1), . . . , X(N), Y(1), Y(2), . . . , Y(N)      {Data points}
INPUT M                             {Degree of trigonometric polynomial}
Max := INT((N−1)/2)                        {Maximum degree possible}
IF  M > Max  THEN  M := Max


FOR    K = 1  TO  N  DO                              {Calculate sums
        A(0) := A(0) + Y(K)                             needed for
        FOR    J = 1  TO  M  DO                         coefficients}
              T := J*X(K)
              A(J) := A(J) + Y(K)*COS(T)
              B(J) := B(J) + Y(K)*SIN(T)


FOR    J = 0  TO  M  DO                             {Multiple of
        A(J) := 2*A(J)/N                               sum forms
        B(J) := 2*B(J)/N                               coefficient}
```

{Procedure for evaluating the trigonometric polynomial above}

```
INPUT X                                      {Input independent variable}


P := A(0)/2                                         {Sum terms of
FOR  J = 1  TO  M  DO                               trigonometric
        P := P + A(J)*COS(J*X) + B(J)*SIN(J*X)       polynomial}


PRINT "The value of the trig. poly. P(x) is" P              {Output}
```

EXERCISES FOR FOURIER SERIES

1. Find the Fourier sine series for the function

$$f(x) = \begin{cases} 1 & \text{for } 0 < x < \pi, \\ -1 & \text{for } -\pi < x < 0. \end{cases}$$

2. Find the Fourier cosine series for the function

$$f(x) = \begin{cases} \dfrac{\pi}{2} - x & \text{for } 0 \leq x < \pi, \\ \dfrac{\pi}{2} + x & \text{for } -\pi \leq x < 0. \end{cases}$$

3. In Exercise 1 set $x = \pi/2$ and conclude that

$$\frac{\pi}{4} = 1 - \frac{1}{3} + \frac{1}{5} - \frac{1}{7} + \cdots.$$

4. In Exercise 2 set $x = 0$ and conclude that

$$\frac{\pi^2}{8} = 1 + \frac{1}{3^2} + \frac{1}{5^2} + \frac{1}{7^2} + \cdots .$$

5. Find the Fourier series for the function

$$f(x) = \begin{cases} x & \text{for } 0 \leq x < \pi, \\ 0 & \text{for } -\pi < x < 0. \end{cases}$$

6. Find the Fourier cosine series for the function

$$f(x) = \begin{cases} -1 & \text{for } \dfrac{\pi}{2} < x < \pi, \\[2mm] 1 & \text{for } \dfrac{-\pi}{2} < x < \dfrac{\pi}{2}, \\[2mm] -1 & \text{for } -\pi < x < \dfrac{-\pi}{2}. \end{cases}$$

7. Find the Fourier sine series for the function

$$f(x) = \begin{cases} \pi - x & \text{for } \dfrac{\pi}{2} \leq x < \pi, \\[2mm] x & \text{for } \dfrac{-\pi}{2} \leq x < \dfrac{\pi}{2}, \\[2mm] -\pi - x & \text{for } -\pi \leq x < \dfrac{-\pi}{2}. \end{cases}$$

8. Find the Fourier cosine series for the function $f(x) = x^2/4$ extended periodically by the equation $f(x + 2\pi) = f(x)$.

9. Prove Theorem 5.6

10. Prove Theorem 5.7.

11. Modify Algorithm 5.5 so that it will find the trigonometric polynomial of period $P = B - A$ when the data points are equally spaced over the interval $[A, B]$.

12. Write a report on how to adapt the linear least-squares procedure to find the trigonometric polynomial when the data points are not equally spaced in the interval $[-\pi, \pi]$.

13. Use a numerical integration program and find approximations for the coefficients $a_0, a_1, a_2, a_3,$ $a_4,$ and a_5 in Exercise 8. At least four significant digits of accuracy are required.

14. Use a numerical integration program and find approximations for the coefficients $a_0, a_1, \ldots,$ a_5 and $b_1, b_2, \ldots, b_5$ in Exercise 5. At least four significant digits of accuracy are required.

15. Use Algorithm 5.5 with $N = 12$ points and follow Example 5.14 to find the trigonometric polynomial of degree $M = 5$ for the equally spaced points $\{(x_k, f(x_k)): k = 1, \ldots, 12\}$, where $f(x)$ is the function in **(a)** Exercise 1; **(b)** Exercise 2; **(c)** Exercise 5; **(d)** Exercise 6; **(e)** Exercise 7.

16. The temperature cycle in a suburb of Los Angeles on November 8 is given in the table below. There are 24 data points.
   (a) Find the trigonometric polynomial for the temperature that involves terms up to $\cos(2\pi x/24)$ and $\sin(2\pi x/24)$. *Hint.* Find a_0, a_1, a_2 and b_1, b_2.
   (b) Compare the values of the trigonometric polynomial in part (a) with the values in the table.

Time P.M.	Degrees Fahrenheit	Time A.M.	Degrees Fahrenheit
1	66	1	58
2	66	2	58
3	65	3	58
4	64	4	58
5	63	5	57
6	63	6	57
7	62	7	57
8	61	8	58
9	60	9	60
10	60	10	64
11	59	11	67
Midnight	58	Noon	68

   (c) Repeat parts (a) and (b) using temperatures from your locale.

17. The yearly temperature cycle for Fairbanks, Alaska, is given in the table below. There are 13 equally spaced data points, which corresponds to a measurement every 28 days.
   (a) Find the trigonometric polynomial for the temperature that involves terms up to $\cos(2\pi x/13)$ and $\sin(2\pi x/13)$. *Hint.* Find a_0, a_1, and b_0.
   (b) Compare the values of the trigonometric polynomial in part (a) with the values in the table.

Calendar date	Average degrees Fahrenheit
Jan. 1	-14
Jan. 29	-9
Feb. 26	2
Mar. 26	15
Apr. 23	35
May 21	52
June 18	62
July 16	63
Aug. 13	58
Sept. 10	50
Oct. 8	34
Nov. 5	12
Dec. 3	-5

18. Write a report on the fast Fourier transform. See References [25, 29, 33, 40, 51, 62, 79, 96, 98, 112, 136, 141, 145, 149, 150, 152, 153, 155, 169, and 210].

6

Numerical Differentiation

Formulas for numerical derivatives are important in developing algorithms for solving boundary value problems for ordinary differential equations and partial differential equations (see Chapters 9 and 10). Standard examples of numerical differentiation often use known functions so that the numerical approximation can be compared with the exact answer. For illustration we use the Bessel function $J_1(x)$, whose tabulated values can be found in standard reference books. Eight equally spaced points over $[0, 7]$ are $(0, 0.0000)$, $(1, 0.4400)$, $(2, 0.5767)$, $(3, 0.3391)$, $(4, -0.0660)$, $(5, -0.3276)$, $(6, -0.2767)$, and $(7, -0.004)$. The underlying principle is differentiation of an interpolation polynomial. Let us focus our attention on finding $J_1'(2)$. The interpolation polynomial $p_2(x) = -0.0710 + 0.6982x - 0.1872x^2$ passes through the three points $(1, 0.4400)$, $(2, 0.5767)$, and $(3, 0.3391)$ and is used to obtain $J_1'(2) \approx p_2'(2) = -0.0505$. This quadratic polynomial $p_2(x)$ and its tangent line at $(2, J_1(2))$ are shown in Figure 6.1(a). If five interpolation points are used, a better approximation can be determined. The polynomial $p_4(x) = 0.4986x + 0.011x^2 - 0.0813x^3 + 0.0116x^4$ passes through $(0, 0.0000)$, $(1, 0.4400)$, $(2, 0.5767)$, $(3, 0.3391)$, $(4, -0.0660)$ and is used to obtain $J_1'(2) \approx p_4'(2) = -0.0618$. The quartic polynomial $p_4(x)$ and its tangent line at $(2, J_1(2))$ are shown in Figure 6.1(b). The true value for the derivative is $J_1'(2) = -0.0645$ and the errors in $p_2(2)$ and $p_4(2)$ are -0.0140 and -0.0026, respectively. In this chapter we develop the introductory theory needed to investigate the accuracy of numerical differentiation.

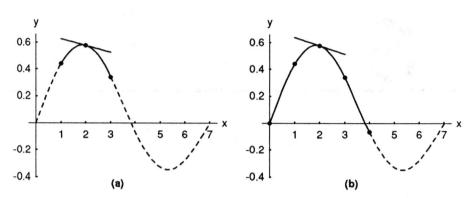

Figure 6.1   (a) The tangent to $p_2(x)$ at $(2, 0.5767)$ with slope $p_2'(2) = -0.0505$.
(b) The tangent to $p_4(x)$ at $(2, 0.5767)$ with slope $p_4'(2) = -0.0618$.

6.1 APPROXIMATING THE DERIVATIVE

The Limit of the Difference Quotient

We now turn our attention to the numerical process for approximating the derivative of $f(x)$:

$$f'(x) = \lim_{h \to 0} \frac{f(x + h) - f(x)}{h}. \tag{1}$$

The method seems straightforward; choose a sequence $\{h_k\}$ so that $h_k \to 0$ and compute the limit of the sequence:

$$D_k = \frac{f(x + h_k) - f(x)}{h_k} \qquad \text{for } k = 1, 2, \ldots, n, \ldots \tag{2}$$

The reader may notice that we will only compute a finite number of terms $D_1, D_2, \ldots$, D_N in the sequence (2), and it appears that we should use D_N for our answer. The following question is often posed: Why compute $D_1, D_2, \ldots, D_{N-1}$? Equivalently, we could ask: What value h_N should be chosen so that D_N is a good approximation to the derivative $f'(x)$? To answer this question, we must look at an example to see why there is no simple solution.

For example, consider the function $f(x) = e^x$ and use the step size $h = 1, \frac{1}{2}$, and $\frac{1}{4}$ to construct the secant lines between the points $(0, 1)$ and $(h, f(h))$, respectively. As h gets small, the secant line approaches the tangent line as shown in Figure 6.2. Although Figure 6.2 gives a good visualization of the process described in (1), we must make numerical computations with $h = 0.00001$ to get an acceptable numerical answer, and for this value of h the graph of the tangent line and secant line would be indistinguishable.

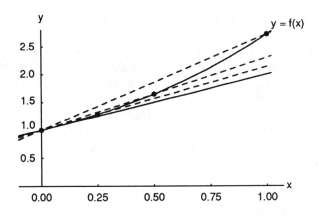

Figure 6.2   Several secant lines for $y = e^x$.

Example 6.1

Let $f(x) = \exp(x)$ and $x = 1$. Compute the difference quotients D_k using the step sizes $h_k = 10^{-k}$ for $k = 1, 2, \ldots , 10$. Carry nine decimal places in all the calculations.

Solution.   The table of values $f(1 + h_k)$ and $[f(1 + h_k) - f(1)]/h_k$ that are used in the computation of D_k are shown in Table 6.1.

TABLE 6.1   Finding the Difference Quotients $D_k = [\exp(1 + h_k) - e]/h_k$

h_k	$f_k = f(1 + h_k)$	$f_k - e$	$D_k = [f_k - e]/h_k$
$h_1 = 0.1$	3.004166024	0.285884196	2.858841960
$h_2 = 0.01$	2.745601015	0.027319187	2.731918700
$h_3 = 0.001$	2.721001470	0.002719642	2.719642000
$h_4 = 0.0001$	2.718553670	0.000271842	2.718420000
$h_5 = 0.00001$	2.718309011	0.000027183	2.718300000
$h_6 = 10^{-6}$	2.718284547	0.000002719	2.719000000
$h_7 = 10^{-7}$	2.718282100	0.000000272	2.720000000
$h_8 = 10^{-8}$	2.718281856	0.000000028	2.800000000
$h_9 = 10^{-9}$	2.718281831	0.000000003	3.000000000
$h_{10} = 10^{-10}$	2.718281828	0.000000000	0.000000000

The largest value $h_1 = 0.1$ does not produce a good approximation $D_1 \approx f'(1)$, because the step size h_1 is too large and the difference quotient is the slope of the secant line through two points that are not close enough to each other. When formula (2) is used with a fixed precision of nine decimal places, h_9 produced the approximation $D_9 = 3$, and h_{10} produced $D_{10} = 0$. If h_k is too small, then the computed function values $f(x + h_k)$ and $f(x)$ are very close together. The difference $f(x + h_k) - f(x)$ can exhibit the problem of loss of significance due to the subtraction of quantities that are nearly equal. The value $h_{10} = 10^{-10}$ is so small that the stored values of $f(x + h_{10})$ and $f(x)$ are the same, and hence the computed difference quotient is zero. In Example 6.1 the mathematical value for the limit is $f'(1) \approx 2.718281828$. Observe that the value $h_5 = 10^{-5}$ gives the best approximation, $D_5 = 2.7183$.

Example 6.1 shows that it is not easy to find numerically the limit in equation (2). The sequence starts to converge to e, and D_5 is the closest; then the terms move away from e. In Algorithm 6.1 it is suggested that terms in the sequence $\{D_k\}$ should be computed until $|D_{N+1} - D_N| \geq |D_N - D_{N-1}|$. This is an attempt to determine the best approximation before the terms start to move away from the limit. When this criterion is applied to Example 6.1, we have $0.0007 = |D_6 - D_5| > |D_5 - D_4| = 0.00012$; hence D_5 is the answer we choose. We now proceed to develop formulas that give a reasonable amount of accuracy for larger values of h.

The Central-Difference Formulas

If the function $f(x)$ can be evaluated at values that lie to the left and right of x, then the best two-point formula will involve abscissas that are chosen symmetrically on both sides of x.

Theorem 6.1   [Centered Formula of Order $O(h^2)$].   Assume that $f \in C^3[a, b]$ and that $x - h, x, x + h \in [a, b]$. Then

$$f'(x) \approx \frac{f(x + h) - f(x - h)}{2h}. \tag{3}$$

Furthermore, there exists a number $c = c(x) \in [a, b]$ so that

$$f'(x) = \frac{f(x + h) - f(x - h)}{2h} + E_{\text{trunc}}(f, h), \tag{4}$$

where

$$E_{\text{trunc}}(f, h) = -\frac{h^2 f^{(3)}(c)}{6} = O(h^2).$$

The term $E(f, h)$ is called the **truncation error**.

Proof. Start with the Taylor expansions for $f(x + h)$ and $f(x - h)$:

$$f(x + h) = f(x) + f'(x)h + \frac{f^{(2)}(x)h^2}{2!} + \frac{f^{(3)}(c_1)h^3}{3!} \tag{5}$$

and

$$f(x - h) = f(x) - f'(x)h + \frac{f^{(2)}(x)h^2}{2!} - \frac{f^{(3)}(c_2)h^3}{3!}. \tag{6}$$

After (6) is subtracted from (5), the result is

$$f(x + h) - f(x - h) = 2f'(x)h + \frac{[f^{(3)}(c_1) + f^{(3)}(c_2)]h^3}{3!}. \tag{7}$$

Since $f^{(3)}(x)$ is continuous, the intermediate value theorem can be used to find a value c so that

$$\frac{f^{(3)}(c_1) + f^{(3)}(c_2)}{2} = f^{(3)}(c). \tag{8}$$

This can be substituted into (7) and the terms rearranged to yield

$$f'(x) = \frac{f(x + h) - f(x - h)}{2h} - \frac{f^{(3)}(c)h^2}{3!}. \tag{9}$$

The first term on the right side of (9) is the central-difference formula (3) and the second term is the truncation error, and the proof is complete.

Suppose that the value of the third derivative $f^{(3)}(c)$ does not change too rapidly; then the truncation error in (4) goes to zero in the same manner as h^2, which is expressed by using the notation $O(h^2)$. When computer calculations are used it is not desirable to choose h too small. For this reason it is useful to have a formula for approximating $f'(x)$ which has a truncation error term of the order $O(h^4)$.

Theorem 6.2   [Centered Formula of Order $O(h^4)$].   Assume that $f \in C^5[a, b]$ and that $x - 2h, x - h, x, x + h, x + 2h \in [a, b]$. Then

$$f'(x) \approx \frac{-f(x + 2h) + 8f(x + h) - 8f(x - h) + f(x - 2h)}{12h}. \tag{10}$$

Furthermore, there exists a number $c = c(x) \in [a, b]$ so that

$$f'(x) = \frac{-f(x + 2h) + 8f(x + h) - 8f(x - h) + f(x - 2h)}{12h} + E_{\text{trunc}}(f, h),$$

where   (11)

$$E_{\text{trunc}}(f, h) = \frac{h^4 f^{(5)}(c)}{30} = O(h^4).$$

Proof. One way to derive formula (10) is as follows. Start with the fifth degree Taylor expansion

$$f(x + h) - f(x - h) = 2f'(x)h + \frac{2f^{(3)}(x)h^3}{3!} + \frac{2f^{(5)}(c_1)h^5}{5!}. \tag{12}$$

Then use the step size $2h$ instead of h and write down the following approximation:

$$f(x + 2h) - f(x - 2h) = 4f'(x)h + \frac{16f^{(3)}(x)h^3}{3!} + \frac{64f^{(5)}(c_2)h^5}{5!}. \tag{13}$$

Next multiply the terms in equation (12) by 8 and subtract (13) from it. The the terms involving $f^{(3)}(x)$ will be eliminated and we get

$$-f(x + 2h) + 8f(x + h) - 8f(x - h) + f(x - 2h) \qquad (14)$$

$$= 12f'(x)h + \frac{[16f^{(5)}(c_1) - 64f^{(5)}(c_2)]h^5}{120}.$$

If $f^{(5)}(x)$ has one sign and if its magnitude does not change rapidly, we can find a value c that lies in $[x - 2h, x + 2h]$ so that

$$16f^{(5)}(c_1) - 64f^{(5)}(c_2) = -48f^{(5)}(c). \qquad (15)$$

After (15) is substituted into (14) and the result is solved for $f'(x)$, we obtain

$$f'(x) \approx \frac{-f(x + 2h) + 8f(x + h) - 8f(x - h) + f(x - 2h)}{12h} + \frac{f^{(5)}(c)h^4}{30}. \qquad (16)$$

The first term on the right side of (16) is the central-difference formula (10) and the second term is the truncation error, and the theorem is proven.

Suppose that $|f^{(5)}(c)|$ is bounded for $c \in [a, b]$; then the truncation error in (11) goes to zero in the same manner as h^4, which is expressed with the notation $O(h^4)$. Now we can make a comparison of the two formulas (3) and (10). Suppose that $f(x)$ has five continuous derivatives and that $|f^{(3)}(c)|$ and $|f^{(5)}(c)|$ are about the same. Then the truncation error for the fourth-order formula (10) is $O(h^4)$ and will go to zero faster than the truncation error $O(h^2)$ for the second-order formula (3). This permits the use of a larger step size.

Example 6.2

Let $f(x) = \cos(x)$.

   (a) Use formulas (3) and (10) with step sizes $h = 0.1, 0.01, 0.001,$ and 0.0001 and calculate approximations for $f'(0.8)$. Carry nine decimal places in all the calculations.

   (b) Compare with the true value $f'(0.8) = -\sin(0.8)$.

Solution.   (a) Using formula (3) with $h = 0.01$, we get

$$f'(0.8) \approx \frac{f(0.81) - f(0.79)}{0.02} \approx \frac{0.689498433 - 0.703845316}{0.02} \approx -0.717344150.$$

Using formula (10) with $h = 0.01$, we get

$$f'(0.8) \approx \frac{-f(0.82) + 8f(0.81) - 8f(0.79) + f(0.78)}{0.12}$$

$$\approx \frac{-0.682221207 + 8 \times 0.689498433 - 8 \times 0.703845316 + 0.710913538}{0.12}$$

$$\approx -0.717356108.$$

   (b) The error in approximation for formulas (3) and (10) is -0.000011941 and 0.000000017, respectively. In this example formula (10) gives a better approximation to

$f'(0.8)$ than formula (3) when $h = 0.01$. The error analysis will illuminate this example and show why this happened. The other calculations are summarized in Table 6.2.

TABLE 6.2 Numerical Differentiation Using Formulas (3) and (10)

Step size	Approximation by formula (3)	Error using formula (3)	Approximation by formula (10)	Error using formula (10)
0.1	−0.716161095	−0.001194996	−0.717353703	−0.000002389
0.01	−0.717344150	−0.000011941	−0.717356108	0.000000017
0.001	−0.717356000	−0.000000091	−0.717356167	0.000000076
0.0001	−0.717360000	−0.000003909	−0.717360833	0.000004742

Error Analysis and Optimum Step Size

An important topic in the study of numerical differentiation is the effect of the computer's round-off error. Let us examine the formulas more closely. Assume that a computer is used to make numerical computations and that

$$f(x_0 - h) = y_{-1} + e_{-1} \quad \text{and} \quad f(x_0 + h) = y_1 + e_1,$$

where $f(x_0 - h)$ and $f(x_0 + h)$ are approximated by the numerical values y_{-1} and y_1 and e_{-1} and e_1 are the associated round-off errors, respectively. The following result indicates the complex nature of error analysis for numerical differentiation.

Corollary 6.1(a).   Assume that f satisfies the hypothesis of Theorem 6.1 and we use the **computational formula**

$$f'(x_0) \approx \frac{y_1 - y_{-1}}{2h}. \tag{17}$$

The error analysis is explained by the following equations:

$$f'(x_0) = \frac{y_1 - y_{-1}}{2h} + E(f, h) \tag{18}$$

and

$$E(f, h) = E_{\text{round}}(f, h) + E_{\text{trunc}}(f, h) = \frac{e_1 - e_{-1}}{2h} - \frac{h^2 f^{(3)}(c)}{6}, \tag{19}$$

where the **total error term** $E(f, h)$ has a part due to round-off error plus a part due to truncation error.

Corollary 6.1(b).   Assume that f satisfies the hypothesis of Theorem 6.1 and that numerical computations are made. If $|e_{-1}| \le \epsilon$, $|e_1| \le \epsilon$ and $M = \max_{a \le x \le b} \{|f^{(3)}(x)|\}$, then

$$|E(f, h)| \le \frac{\epsilon}{h} + \frac{Mh^2}{6}, \tag{20}$$

and the value of h that minimizes the right-hand side of (19) is

$$h = \left(\frac{3\epsilon}{M}\right)^{1/3}. \tag{21}$$

When h is small, the portion of (19) involving $(e_1 - e_{-1})/2h$ can be relatively large. In Example 6.2, when $h = 0.0001$, this difficulty was encountered. The round-off errors are

$$f(0.8001) = 0.696634970 + e_1, \qquad \text{where } e_1 \approx -0.0000000003,$$
$$f(0.7999) = 0.696778442 + e_{-1}, \qquad \text{where } e_{-1} \approx 0.0000000005.$$

The truncation error term is

$$\frac{-h^2 f^{(3)}(c)}{6} \approx -(0.0001)^2 \frac{\sin(0.8)}{6} \approx 0.000000001.$$

The error term $E(f, h)$ in (19) can now be estimated:

$$E(f, h) \approx \frac{-0.0000000003 - 0.0000000005}{0.0002} - 0.000000001 = -0.000004001.$$

Indeed, the computed numerical approximation for the derivative using $h = 0.0001$ is found by the calculation

$$f'(0.8) \approx \frac{f(0.8001) - f(0.7999)}{0.0002} = \frac{0.696634970 - 0.696778442}{0.0002} = -0.717360000$$

and a loss of about four significant digits is evident. The error is -0.000003909 and this is close to the predicted error, -0.000004001.

When formula (21) is applied to Example 6.2, we can use the bound $|f^{(3)}(x)| \le |\sin(x)| \le 1 = M$ and the value $e = 0.5 \times 10^{-9}$ for the magnitude of the round-off error. The optimal value for h is easily calculated: $h = (1.5 \times 10^{-9}/1)^{1/3} = 0.001144714$. The step size $h = 0.001$ was closest to the optimal value 0.001144714 and it gave the best

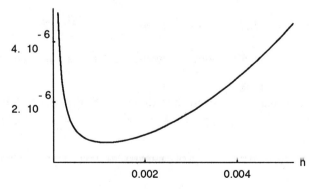

Error bound

4. 10^{-6}

2. 10^{-6}

0.002   0.004 h

Figure 6.3 Finding the optimum step size $h = 0.00114462$ when formula (21) is applied to $f(x) = \cos(x)$ in Example 6.2.

approximation to $f'(0.8)$ among the four choices involving formula (3) (see Table 6.2 and Figure 6.3).

An error analysis of formula (10) is similar. Assume that a computer is used to make numerical computations and that $f(x_0 + kh) = y_k + e_k$.

Corollary 6.2(a).   Assume that f satisfies the hypothesis of Theorem 6.2 and we use the **computational formula**

$$f'(x_0) \approx \frac{-y_2 + 8y_1 - 8y_{-1} + y_{-2}}{12h}. \tag{22}$$

The error analysis is explained by the following equations:

$$f'(x_0) = \frac{-y_2 + 8y_1 - 8y_{-1} + y_{-2}}{12h} + E(f, h) \tag{23}$$

and

$$E(f, h) = E_{\text{round}}(f, h) + E_{\text{trunc}}(f, h) = \frac{-e_2 + 8e_1 - 8e_{-1} + e_{-2}}{12h} + \frac{h^4 f^{(5)}(c)}{30}, \tag{24}$$

where the total error term $E(f, h)$ has a part due to round-off error plus a part due to truncation error.

Corollary 6.2(b).   Assume that f satisfies the hypothesis of Theorem 6.2 and that numerical computations are made. If $|e_k| \leq \epsilon$ and $M = \max\limits_{a \leq x \leq b} \{|f^{(5)}(x)|\}$, then

$$|E(f, h)| \leq \frac{3\epsilon}{2h} + \frac{Mh^4}{30}, \tag{25}$$

and the value of h that minimizes the right-hand side of (25) is

$$h = \left(\frac{45\epsilon}{4M}\right)^{1/5}. \tag{26}$$

When formula (25) is applied to Example 6.2, we can use the bound $|f^{(5)}(x)| \leq |\sin(x)| \leq 1 = M$ and the value $e = 0.5 \times 10^{-9}$. The optimal step size is $h = (22.5 \times 10^{-9}/4)^{1/5} = 0.022388475$. The step size $h = 0.01$ was closest to the optimal value, 0.022388475, and it gave the best approximation to $f'(0.8)$ among the four choices involving formula (10) (see Table 6.2 and Figure 6.4).

We should not end the discussion of Example 6.2 without mentioning that numerical differentiation formulas can be obtained by an alternative derivation. They can be derived by differentiation of an interpolation polynomial. For example, the Lagrange form of the quadratic polynomial $p_2(x)$ that passes through the three points $(0.7, \cos(0.7))$, $(0.8, \cos(0.8))$, and $(0.9, \cos(0.9))$ is

$$p_2(x) = 38.2421094(x - 0.8)(x - 0.9) - 69.6706709(x - 0.7)(x - 0.9)$$

$$+ 31.0804984(x - 0.7)(x - 0.8).$$

Error bound

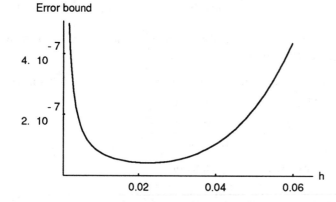

Figure 6.4   Finding the optimum step size $h = 0.02250556$ when formula (26) is applied to $f(x) = \cos(x)$ in Example 6.2.

This polynomial can be expanded to obtain the usual form:

$$p_2(x) = 1.046875165 - 0.159260044x - 0.348063157x^2.$$

A similar computation can be used to obtain the quartic polynomial $p_4(x)$ that passes through $(0.6, \cos(0.6))$, $(0.7, \cos(0.7))$, $(0.8, \cos(0.8))$, $(0.9, \cos(0.9))$ and $(1.0, \cos(1.0))$:

$$p_4(x) = 0.998452927 + 0.009638391x - 0.523291341x^2 + 0.026521229x^3$$

$$+ 0.028981100x^4.$$

When these polynomials are differentiated, they produce $p_2'(0.8) = -0.716161095$ and $p_4'(0.8) = -0.717353703$, which agree with the values listed under $h = 0.1$ in Table 6.2. The graphs of $p_2(x)$ and $p_4(x)$ and their tangent lines at $(0.8, \cos(0.8))$ are shown in Figure 6.5(a) and (b), respectively.

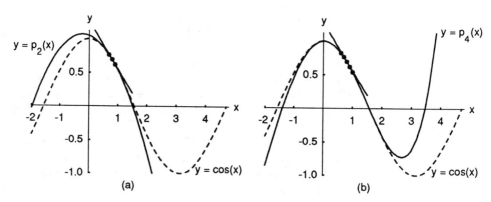

Figure 6.5   (a) The graph $y = \cos(x)$ and the interpolating polynomial $p_2(x)$ used to estimate $f'(0.8) \approx p_2'(0.8) = -0.716161095$. (b) The graph of $y = \cos(x)$ and the interpolating polynomial $p_4(x)$ used to estimate $f'(0.8) = p_4'(0.8) = 0.717353703$.

Richardson's Extrapolation

In this section we emphasize the relationship between formulas (3) and (10). Let $f_k = f(x_k)$ and use the notation $D_0(h)$ and $D_0(2h)$ to denote the approximations to $f'(x_0)$ that are obtained from (3) with step sizes h and $2h$, respectively:

$$f'(x_0) \approx D_0(h) + Ch^2 \tag{27}$$

and

$$f'(x_0) \approx D_0(2h) + 4Ch^2. \tag{28}$$

If we multiply relation (27) by 4 and subtract relation (28) from this product then the terms involving C cancel and the result is

$$3f'(x_0) \approx 4D_0(h) - D_0(2h) = \frac{4(f_1 - f_{-1})}{2h} - \frac{f_2 - f_{-2}}{4h}. \tag{29}$$

Next solve for $f'(x_0)$ in (29) and get

$$f'(x_0) \approx \frac{4D_0(h) - D_0(2h)}{3} = \frac{-f_2 + 8f_1 - 8f_{-1} + f_{-2}}{12h}. \tag{30}$$

The last expression in (30) is the central-difference formula (10).

Example 6.3

Let $f(x) = \cos(x)$. Use (27) and (28) with $h = 0.01$ and show how the linear combination $[4D_0(h) - D_0(2h)]/3$ in (30) can be used to obtain the approximation to $f'(0.8)$ given in (10). Carry nine decimal places in all the calculations.

Solution.   Use (27) and (28) with $h = 0.01$ to get

$$D_0(h) \approx \frac{f(0.81) - f(0.79)}{0.02} \approx \frac{0.689498433 - 0.703845316}{0.02} \approx -0.717344150$$

and

$$D_0(2h) \approx \frac{f(0.82) - f(0.78)}{0.04} \approx \frac{0.682221207 - 0.710913538}{0.04} \approx -0.717308275.$$

Now the linear combination in (30) is computed:

$$f'(0.8) \approx \frac{4D_0(h) - D_0(2h)}{3} \approx \frac{4(-0.717344150) - (-0.717308275)}{3}$$

$$\approx -0.717356108.$$

This is exactly the same as the solution in Example 6.2 that used (10) directly to approximate to $f'(0.8)$.

The method of obtaining a formula for $f'(x_0)$ of higher order from a formula of lower order is called **extrapolation**. The proof requires that the error term for (3) can be expanded in a series containing only even powers of h. We have already seen how to use step sizes h and $2h$ to remove the term involving h^2. To see how h^4 is removed, let $D_1(h)$ and $D_1(2h)$ denote the approximations to $f'(x_0)$ of order $O(h^4)$ obtained with formula (16) using step sizes h and $2h$, respectively. Then

$$f'(x_0) = \frac{-f_2 + 8f_1 - 8f_{-1} + f_{-2}}{12h} + \frac{h^4 f^{(5)}(c_1)}{30} \approx D_1(h) + Ch^4 \tag{31}$$

and

$$f'(x_0) = \frac{-f_4 + 8f_2 - 8f_{-2} + f_{-4}}{12h} + \frac{16h^4 f^{(5)}(c_2)}{30} \approx D_1(2h) + 16Ch^4. \tag{32}$$

Suppose that $f^{(5)}(x)$ has one sign and does not change too rapidly; then the assumption that $f^{(5)}(c_1) \approx f^{(5)}(c_2)$ can be used to eliminate the terms involving h^4 in (31) and (32) and the result is

$$f'(x_0) \approx \frac{16D_1(h) - D_1(2h)}{15}. \tag{33}$$

The general pattern for improving calculations is stated in the next result.

Theorem 6.3 (Richardson's Extrapolation). Suppose that two approximations of order $O(h^{2k})$ for $f'(x_0)$ are $D_{k-1}(h)$ and $D_{k-1}(2h)$ and that they satisfy

$$f'(x_0) = D_{k-1}(h) + c_1 h^{2k} + c_2 h^{2k+2} + \cdots \tag{34}$$

and

$$f'(x_0) = D_{k-1}(2h) + 4^k c_1 h^{2k} + 4^{k+1} c_2 h^{2k+2} + \cdots, \tag{35}$$

then an improved approximation has the form

$$f'(x_0) = D_k(h) + O(h^{2k+2}) = \frac{4^k D_{k-1}(h) - D_{k-1}(2h)}{4^k - 1} + O(h^{2k+2}). \tag{36}$$

Algorithm 6.1 (Differentiation Using Limits). To approximate $f'(x)$ numerically by generating the sequence

$$f'(x) \approx D_k = \frac{f(x + h2^{-k}) - f(x - h2^{-k})}{h2^{-k+1}} \qquad \text{for } k = 0, \ldots, n$$

until $|D_{n+1} - D_n| \ge |D_n - D_{n-1}|$ or $|D_n - D_{n-1}| < \text{Tol}$, which is an attempt to find the best approximation $f'(x) \approx D_n$.

Tol := 10^{-5} {Convergence criterion}
H := 1   {Initial step size}
Max := 20   {Maximum number of terms}

INPUT X   {Abscissa for f'(x)}
D(0) := .5*[f(X+H) − f(X−H)]/H   {Compute the quotient}

FOR   N = 1 TO 2 DO   {Compute D_1 and D_2}
        H := H/2   {Reduce step size}
        D(N) := .5*[f(X+H) − f(X−H)]/H   {Compute the quotient}
        E(N) := |D(N) − D(N−1)| {Error estimate}
        R(N) := 2*E(N)/[|D(N)|+|D(N−1)|+Tol]   {Relative error}
        N := 1   {Initialize the counter}
WHILE [E(N)>E(N+1) OR R(N)≧Tol] AND N<Max   DO
        H := H/2   {Reduce the step size}
        D(N+2) := .5*[f(X+H) − f(X−H)]/H   {Compute the quotient}
        E(N+2) := |D(N+2) − D(N+1)| {Error estimate}
        R(N+2) := 2*E(N+2)/[|D(N+2)|+|D(N+1)|+Tol]
        N := N+1   {Increment the counter}

PRINT "The approximation for f'(x) is" D(N−1)   {Output}
PRINT "The accuracy is + −" E(N−1)

Algorithm 6.2   (Differentiation Using Extrapolation).   To approximate $f'(x)$ numerically by generating a table of approximations $D(J, K)$ for $K \leq J$, and using $f'(x) \approx D(N, N)$ as the final answer. The approximations $D(J, K)$ are stored in a lower-triangular matrix. The first column is

$$D(J, 0) = \frac{f(x + h2^{-J}) - f(x - h2^{-J})}{h2^{-J+1}},$$

and the elements in row J are

$$D(J, K) = D(J, K - 1) + \frac{D(J, K - 1) - D(J - 1, K - 1)}{4^K - 1} \qquad \text{for } 1 \leq K \leq J.$$

Tol := 10^{-5}, Delta := 10^{-7} {Termination criterion}
Error :=1, RelErr := 1   {Initialize the variables}
H := 1   {Initial step size}
J := 1   {Initialize the counter}

INPUT X   {Abscissa for f'(x)}
D(0,0) := .5*[f(X+H) − f(X−H)]/H   {Compute the quotient}

WHILE   RelErr>Tol AND Error>Delta AND J≤12   DO
┌
│   H := H/2   {Reduce the step size}
│   D(J,0) := .5*[f(X+H) − f(X−H)]/H   {Compute the quotient}
│   FOR K = 1 TO J DO
│   └─── D(J,K) := D(J,K−1) + [D(J,K−1)−D(J−1,K−1)]/(4^K − 1)
│   Error := |D(J,J)−D(J−1,J−1)| {Error estimate}
└───── RelErr := 2*Error/[|D(J,J)|+|D(J−1,J−1)|+Tol]
       J := J + 1, N := J

PRINT "The approximate value of f'(x) using   {Output}
       Richardson's extrapolation is" D(N−1,N−1)
PRINT "The accuracy is +−" Error

EXERCISES FOR APPROXIMATING THE DERIVATIVE

1. Let $f(x) = \sin(x)$, where x is measured in radians.
   (a) Calculate approximations to $f'(0.8)$ using formula (3) with $h = 0.1$, $h = 0.01$, $h = 0.001$. Carry eight or nine decimal places.
   (b) Compare with the value $f'(0.8) = \cos(0.8)$.
   (c) Compute bounds for the truncation error (4). Use

   $$|f^{(3)}(c)| \leq \cos(0.7) \approx 0.764842187$$

   for all the cases.

2. Let $f(x) = \exp(x)$.
   (a) Calculate approximations to $f'(2.3)$ using formula (3) with $h = 0.1$, $h = 0.01$, $h = 0.001$. Carry eight or nine decimal places.
   (b) Compare with the value $f'(2.3) = \exp(2.3)$.
   (c) Compute bounds for the truncation error (4). Use

   $$|f^{(3)}(c)| \leq \exp(2.4) \approx 11.02317638$$

   for all the cases.

3. Let $f(x) = \sin(x)$, where x is measured in radians.
   (a) Calculate approximations to $f'(0.8)$ using formula (10) with $h = 0.1$ and $h = 0.01$, and compare with $f'(0.8) = \cos(0.8)$.
   (b) Use the extrapolation formula in (29) to compute the approximations to $f'(0.8)$ in part (a).
   (c) Compute bounds for the truncation error (11). Use

   $$|f^{(5)}(c)| \leq \cos(0.6) \approx 0.825335615$$

   for both cases.

4. Let $f(x) = \exp(x)$.
   (a) Calculate approximations to $f'(2.3)$ using (10) with $h = 0.1$ and $h = 0.01$, and compare with $f'(2.3) = \exp(2.3)$.
   (b) Use the extrapolation formula in (29) to compute the approximations to $f'(2.3)$ in part (a).

(c) Compute bounds for the function error (11). Use

$$|f^{(5)}(c)| \le \exp(2.5) \approx 12.18249396$$

for both cases.

5. Compare the numerical differentiation formulas (3) and (10). Let $f(x) = x^3$ and find approximations for $f'(2)$.
   (a) Use formula (3) with $h = 0.05$.
   (b) Use formula (10) with $h = 0.05$.
   (c) Compute bounds for the truncation errors (4) and (11).

6. (a) Use Taylor's theorem to show that

$$f(x + h) = f(x) + hf'(x) + \frac{h^2 f''(c)}{2}, \qquad \text{where } |c - x| < h.$$

   (b) Use part (a) to show that the difference quotient in equation (2) has error of order $O(h) = -hf''(c)/2$.
   (c) Why is formula (3) better to use than formula (2)?

7. *Partial differentiation formulas.* The partial derivative $f_x(x, y)$ of $f(x, y)$ with respect to x is obtained holding y fixed and differentiating with respect to x. Similarly, $f_y(x, y)$ is found by holding x fixed and differentiating with respect to y. Formula (3) can be adapted to partial derivatives:

$$f_x(x, y) = \frac{f(x + h, y) - f(x - h, y)}{2h} + O(h^2), \tag{i}$$

$$f_y(x, y) = \frac{f(x, y + h) - f(x, y - h)}{2h} + O(h^2).$$

   (a) Let $f(x, y) = xy/(x + y)$. Calculate approximations to $f_x(2, 3)$ and $f_y(2, 3)$ using the formulas in (i) with $h = 0.1, 0.01,$ and 0.001. Compare with the values obtained by differentiating $f(x, y)$ partially.
   (b) Let $z = f(x, y) = \arctan(y/x)$; z is in radians. Calculate approximations to $f_x(3, 4)$ and $f_y(3, 4)$ using the formulas in (i) with $h = 0.1, 0.01,$ and 0.001. Compare with the values obtained by differentiating $f(x, y)$ partially.

8. Complete the details that show how (33) is obtained from equations (31) and (32).

9. The voltage $E = E(t)$ in an electrical circuit obeys the equation $E(t) = L(dI/dt) + RI(t)$, where R is resistance and L is inductance. Use $L = 0.05$ and $R = 2$ and values for $I(t)$ in the table below.

t	$I(t)$
1.0	8.2277
1.1	7.2428
1.2	5.9908
1.3	4.5260
1.4	2.9122

    (a) Find $I'(1.2)$ by numerical differentiation, and use it to compute $E(1.2)$.

    (b) Compare your answer with $I(t) = 10 \exp(-t/10) \sin(2t)$.

10. The distance $D = D(t)$ traveled by an object is given in the table below.

t	$D(t)$
8.0	17.453
9.0	21.460
10.0	25.752
11.0	30.301
12.0	35.084

    (a) Find the velocity $V(10)$ by numerical differentiation.

    (b) Compare your answer with $D(t) = -70 + 7t + 70 \exp(-t/10)$.

11. Let $f(x)$ be given by the table below. The inherent round-off error has the bound $|e_k| \le 5 \times 10^{-6}$. Use the rounded values in your calculations.

x	$f(x) = \cos(x)$
1.100	0.45360
1.190	0.37166
1.199	0.36329
1.200	0.36236
1.201	0.36143
1.210	0.35302
1.300	0.26750

    (a) Find approximations for $f'(1.2)$ using formula (17) with $h = 0.1$, $h = 0.01$, and $h = 0.001$.

    (b) Compare with $f'(1.2) = -\sin(1.2) \approx -0.93204$.

    (c) Find the total error bound (19) for the three cases in part (a).

12. Let $f(x)$ be given by the table below. The inherent round-off error has the bound $|e_k| \le 5 \times 10^{-6}$. Use the rounded values in your calculations.

x	$f(x) = \ln(x)$
2.900	1.06471
2.990	1.09527
2.999	1.09828
3.000	1.09861
3.001	1.09895
3.010	1.10194
3.100	1.13140

(a) Find approximations for $f'(3.0)$ using formula (17) with $h = 0.1$, $h = 0.01$, and $h = 0.0001$.

(b) Compare with $f'(3.0) = \frac{1}{3} \approx 0.33333$.

(c) Find the total error bound (19) for the three cases in part (a).

13. Let $f(x)$ be given by the table below. The inherent round-off error has the bound $|e_k| \leq 5 \times 10^{-6}$. Use the rounded values in your calculations.

x	$f(x) = x^{1/2}$
0.400	0.63246
0.490	0.70000
0.499	0.70640
0.500	0.70711
0.501	0.70781
0.510	0.71414
0.600	0.77460

(a) Find approximations for $f'(0.5)$ using formula (17) with $h = 0.1$, $h = 0.01$, and $h = 0.001$.

(b) Compare with $f'(0.5) = 0.70711$.

(c) Find the total error bound (19) for the three cases in part (a).

14. Suppose that a table of the function $f(x_k)$ is computed where the values are rounded off to three decimal places, and the inherent round-off error is 5×10^{-4}. Also, assume that $|f^{(3)}(c)| \leq 1.5$ and $|f^{(5)}(c)| \leq 1.5$.

(a) Find the best step size h for formula (17).

(b) Find the best step size h for formula (22).

15. Let $f(x)$ be given by the table below. The inherent round-off error has the bound $|e_k| \leq 5 \times 10^{-6}$. Use the rounded values in your calculations.

x	$f(x) = \cos(x)$
1.000	0.54030
1.100	0.45360
1.198	0.36422
1.199	0.36329
1.200	0.36236
1.201	0.36143
1.202	0.36049
1.300	0.26750
1.400	0.16997

(a) Approximate $f'(1.2)$ using (22) with $h = 0.1$ and $h = 0.001$.

(b) Find the total error bound (24) for the two cases in part (a).

16. Let $f(x)$ be given by the table below. The inherent round-off error has the bound $|e_k| \leq 5 \times 10^{-6}$. Use the rounded values in your calculations.

x	$f(x) = \ln(x)$
2.800	1.02962
2.900	1.06471
2.998	1.09795
2.999	1.09828
3.000	1.09861
3.001	1.09895
3.002	1.09928
3.100	1.13140
3.200	1.16315

(a) Approximate $f'(3.0)$ using (22) with $h = 0.1$ and $h = 0.001$.

(b) Find the total error bound (24) for the two cases in part (a).

17. Let $f(x)$ be given by the table below. The inherent round-off error has the bound $|e_k| \le 5 \times 10^{-6}$. Use the rounded values in your calculations.

x	$f(x) = x^{1/2}$
0.300	0.54772
0.400	0.63246
0.498	0.70569
0.499	0.70640
0.500	0.70711
0.501	0.70781
0.502	0.70852
0.600	0.77460
0.700	0.83666

(a) Approximate $f'(0.5)$ using (22) with $h = 0.1$, $h = 0.001$.

(b) Find the total error bound (24) for the two cases in part (a).

18. Write a report on forward-difference formulas. See References [9, 29, 40, 41, 51, 76, 78, 81, 85, 90, 94, 105, 117, 128, 143, 145, 153, 181, and 184].

19. Write a report on extrapolation. See References [19, 29, 35, 40, 41, 78, 117, and 153].

6.2 NUMERICAL DIFFERENTIATION FORMULAS

More Central-Difference Formulas

The formulas for $f'(x_0)$ in the preceding section required that the function can be computed at abscissas that lie on both sides of x, and were referred to as central-difference formulas. Taylor series can be used to obtain central-difference formulas for the higher derivatives. The popular choices are those of order $O(h^2)$ and $O(h^4)$ and are given in Tables 6.3 and

6.4. In these tables we use the convention that $f_k = f(x_0 + hk)$ for $k = -3, -2, -1, 0,$
1, 2, 3.

TABLE 6.3   Central-Difference Formulas of Order $O(h^2)$

$$f'(x_0) \approx \frac{f_1 - f_{-1}}{2h}$$

$$f''(x_0) \approx \frac{f_1 - 2f_0 + f_{-1}}{h^2}$$

$$f^{(3)}(x_0) \approx \frac{f_2 - 2f_1 + 2f_{-1} - f_{-2}}{2h^3}$$

$$f^{(4)}(x_0) \approx \frac{f_2 - 4f_1 + 6f_0 - 4f_{-1} + f_{-2}}{h^4}$$

TABLE 6.4   Central-Difference Formulas of Order $O(h^4)$

$$f'(x_0) \approx \frac{-f_2 + 8f_1 - 8f_{-1} + f_{-2}}{12h}$$

$$f''(x_0) \approx \frac{-f_2 + 16f_1 - 30f_0 + 16f_{-1} - f_{-2}}{12h^2}$$

$$f^{(3)}(x_0) \approx \frac{-f_3 + 8f_2 - 13f_1 + 13f_{-1} - 8f_{-2} + f_{-3}}{8h^3}$$

$$f^{(4)}(x_0) \approx \frac{-f_3 + 12f_2 - 39f_1 + 56f_0 - 39f_{-1} + 12f_{-2} - f_{-3}}{6h^4}$$

For illustration, we will derive the formula for $f''(x)$ of order $O(h^2)$ in Table 6.3.
Start with the Taylor expansions

$$f(x + h) = f(x) + hf'(x) + \frac{h^2 f''(x)}{2} + \frac{h^3 f^{(3)}(x)}{6} + \frac{h^4 f^{(4)}(x)}{24} + \cdots \qquad (1)$$

and

$$f(x - h) = f(x) - hf'(x) + \frac{h^2 f''(x)}{2} - \frac{h^3 f^{(3)}(x)}{6} + \frac{h^4 f^{(4)}(x)}{24} - \cdots . \qquad (2)$$

Adding equations (1) and (2) will eliminate the terms involving the odd derivatives $f'(x)$,
$f^{(3)}(x)$, $f^{(5)}(x)$, . . . :

$$f(x + h) + f(x - h) = 2f(x) + \frac{2h^2 f''(x)}{2} + \frac{2h^4 f^{(4)}(x)}{24} + \cdots . \qquad (3)$$

Solving equation (3) for $f''(x)$ yields

$$f''(x) = \frac{f(x + h) - 2f(x) + f(x - h)}{h^2} - \frac{2h^2 f^{(4)}(x)}{4!}$$

$$- \frac{2h^4 f^{(6)}(x)}{6!} - \cdots - \frac{2h^{2k-2} f^{(2k)}(x)}{(2k)!} - \cdots . \qquad (4)$$

If the series in (4) is truncated at the fourth derivative, there exists a value c that lies in $[x - h, x + h]$ so that

$$f''(x_0) = \frac{f_1 - 2f_0 + f_{-1}}{h^2} - \frac{h^2 f^{(4)}(c)}{12}. \qquad (5)$$

This gives us the desired formula for approximating $f''(x)$:

$$f''(x_0) \approx \frac{f_1 - 2f_0 + f_{-1}}{h^2}. \qquad (6)$$

Example 6.4

Let $f(x) = \cos(x)$.

(a) Use formula (6) with $h = 0.1, 0.01$, and 0.001 and find approximations to $f''(0.8)$. Carry nine decimal places in all the calculations.

(b) Compare with the true value $f''(0.8) = -\cos(0.8)$.

Solution.   (a) The calculation for $h = 0.01$ is

$$f''(0.8) \approx \frac{f(0.81) - 2f(0.80) + f(0.79)}{0.0001}$$

$$\approx \frac{0.689498433 - 2(0.696706709) + 0.703845316}{0.0001} \approx -0.696690000.$$

(b) The error in this approximation is -0.000016709. The other calculations are summarized in Table 6.5. The error analysis will illuminate this example and show why $h = 0.01$ was best.

TABLE 6.5   Numerical Approximations to $f''(x)$ for Example 6.4

Step size	Approximation by formula (6)	Error using formula (6)
$h = 0.1$	-0.696126300	-0.000580409
$h = 0.01$	-0.696690000	-0.000016709
$h = 0.001$	-0.696000000	-0.000706709

Error Analysis

Let $f_k = y_k + e_k$, where e_k is the error in computing $f(x_k)$, including noise in measurement and round-off error. Then formula (6) can be written

$$f''(x_0) = \frac{y_1 - 2y_0 + y_{-1}}{h^2} + E(f, h). \tag{7}$$

The error term $E(f, h)$ for the numerical derivative (7) will have a part due to round-off and a part due to truncation:

$$E(f, h) = \frac{e_1 - 2e_0 + e_{-1}}{h^2} - \frac{h^2 f^{(4)}(c)}{12}. \tag{8}$$

If it is assumed that each error e_k is of the magnitude e with signs which accumulate errors, and the $\left| f^{(4)}(x) \right| \le M$, then we get the following error bound:

$$|E(f, h)| \le \frac{4e}{h^2} + \frac{Mh^2}{12}. \tag{9}$$

If h is small, then the contribution $4e/h^2$, due to round-off, is large. When h is large, the contribution $Mh^2/12$ is large. The optimum step size will minimize the quantity

$$g(h) = \frac{4e}{h^2} + \frac{Mh^2}{12}. \tag{10}$$

Setting $g'(h) = 0$ results in $-8e/h^3 + Mh/6 = 0$, which yields the equation $h^4 = 48e/M$, from which we obtain the optimal value:

$$h = \left(\frac{48e}{M} \right)^{1/4}. \tag{11}$$

When formula (11) is applied to Example 6.4, use the bound $\left| f^{(4)}(x) \right| \le |\cos(x)| \le 1 = M$ and the value $e = 0.5 \times 10^{-9}$. The optimal step size is $h = (24 \times 10^{-9}/1)^{1/4} = 0.01244666$, and we see that $h = 0.01$ was closest to the optimal value.

Since the portion of the error due to round-off is inversely proportional to the square of h, this term grows when h gets small. This is sometimes referred to as the **step-size dilemma**. One partial solution to this problem is to use a formula of higher order so that a larger value of h will produce the desired accuracy. The formula for $f''(x_0)$ of order $O(h^4)$ in Table 6.4 is

$$f''(x_0) = \frac{-f_2 + 16f_1 - 30f_0 + 16f_{-1} - f_{-2}}{12h^2} + E(f, h). \tag{12}$$

The error term for (12) has the form

$$E(f, h) = \frac{16e}{3h^2} + \frac{h^4 f^{(6)}(c)}{90}, \tag{13}$$

where c lies in the interval $[x - 2h, x + 2h]$. A bound for $|E(f, h)|$ is

$$|E(f, h)| \leq \frac{16e}{3h^2} + \frac{Mh^4}{90}, \tag{14}$$

where $|f^{(6)}(x)| \leq M$. The optimal value for h is given by the formula

$$h = \left(\frac{240e}{M}\right)^{1/6}. \tag{15}$$

Example 6.5

Let $f(x) = \cos(x)$.
  (a) Use formula (12) with $h = 1.0, 0.1$, and 0.01 and find approximations to $f''(0.8)$. Carry nine decimal places in all the calculations.
  (b) Compare with the true value $f''(0.8) = -\cos(0.8)$.
  (c) Determine the optimal step size.

Solution.   (a) The calculation for $h = 0.1$ is

$$f''(0.8) \approx \frac{-f(1.0) + 16f(0.9) - 30f(0.8) + 16f(0.7) - f(0.6)}{0.12}$$

$$\approx \frac{-0.540302306 + 9.945759488 - 20.90120127 + 12.23747499 - 0.825335615}{0.12}$$

$$\approx -0.696705958.$$

(b) The error in this approximation is -0.000000751. The other calculations are summarized in Table 6.6.

TABLE 6.6   Numerical Approximations to $f''(x)$ for Example 6.5

Step size	Approximation by formula (12)	Error using formula (12)
$h = 1.0$	-0.689625413	-0.007081296
$h = 0.1$	-0.696705958	-0.000000751
$h = 0.001$	-0.696690000	-0.000016709

(c) When formula (15) is applied, we can use the bound $|f^{(6)}(x)| \leq |\cos(x)| \leq 1 = M$ and the value $e = 0.5 \times 10^{-9}$. These values give the optimal step size $h = (120 \times 10^{-9}/1)^{1/6} = 0.070231219$.

  Generally speaking, if numerical differentiation is performed, only about half the accuracy of which the computer is capable is obtained. This severe loss of significant digits will almost always occur unless we are fortunate to find a step size that is optimal. Hence we must always proceed with caution when numerical differentiation is performed. The difficulties are more pronounced when working with experimental data, where the function values have been rounded to only a few digits. If a numerical derivative must be obtained from data, then we should consider curve fitting, by using least-squares techniques, and differentiate the formula for the curve.

Differentiation of the Lagrange Polynomial

If the function must be evaluated at abscissas that lie on one side of x_0, the central-difference formula cannot be used. Formulas for equally spaced abscissas that lie to the right (or left) of x_0 are called the forward (or backward)-difference formulas. These formulas can be derived by differentiation of the Lagrange interpolation polynomial. Some of the common forward- and backward-difference formulas are given in Table 6.7.

TABLE 6.7   Forward/Backward-Difference Formulas of Order $O(h^2)$

$$f'(x_0) \approx \frac{-3f_0 + 4f_1 - f_2}{2h} \qquad \left(\begin{array}{l}\text{forward}\\\text{difference}\end{array}\right)$$

$$f'(x_0) \approx \frac{3f_0 - 4f_{-1} + f_{-2}}{2h} \qquad \left(\begin{array}{l}\text{backward}\\\text{difference}\end{array}\right)$$

$$f''(x_0) \approx \frac{2f_0 - 5f_1 + 4f_2 - f_3}{h^2} \qquad \left(\begin{array}{l}\text{forward}\\\text{difference}\end{array}\right)$$

$$f''(x_0) \approx \frac{2f_0 - 5f_{-1} + 4f_{-2} - f_{-3}}{h^2} \qquad \left(\begin{array}{l}\text{backward}\\\text{difference}\end{array}\right)$$

$$f^{(3)}(x_0) \approx \frac{-5f_0 + 18f_1 - 24f_2 + 14f_3 - 3f_4}{2h^3}$$

$$f^{(3)}(x_0) \approx \frac{5f_0 - 18f_{-1} + 24f_{-2} - 14f_{-3} + 3f_{-4}}{2h^3}$$

$$f^{(4)}(x_0) \approx \frac{3f_0 - 14f_1 + 26f_2 - 24f_3 + 11f_4 - 2f_5}{h^4}$$

$$f^{(4)}(x_0) \approx \frac{3f_0 - 14f_{-1} + 26f_{-2} - 24f_{-3} + 11f_{-4} - 2f_{-5}}{h^4}$$

Example 6.6

Derive the formula

$$f''(x_0) \approx \frac{2f_0 - 5f_1 + 4f_2 - f_3}{h^2}.$$

Solution.   Start with the Lagrange interpolation polynomial for $f(t)$ based on the four points $x_0, x_1, x_2,$ and x_3.

$$f(t) \approx f_0 \frac{(t - x_1)(t - x_2)(t - x_3)}{(x_0 - x_1)(x_0 - x_2)(x_0 - x_3)} + f_1 \frac{(t - x_0)(t - x_2)(t - x_3)}{(x_1 - x_0)(x_1 - x_2)(x_1 - x_3)}$$

$$+ f_2 \frac{(t - x_0)(t - x_1)(t - x_3)}{(x_2 - x_0)(x_2 - x_1)(x_2 - x_3)} + f_3 \frac{(t - x_0)(t - x_1)(t - x_2)}{(x_3 - x_0)(x_3 - x_1)(x_3 - x_2)}.$$

Differentiate the products in the numerators twice and get

$$f''(t) \approx f_0 \frac{2[(t - x_1) + (t - x_2) + (t - x_3)]}{(x_0 - x_1)(x_0 - x_2)(x_0 - x_3)} + f_1 \frac{2[(t - x_0) + (t - x_2) + (t - x_3)]}{(x_1 - x_0)(x_1 - x_2)(x_1 - x_3)}$$

$$+ f_2 \frac{2[(t - x_0) + (t - x_1) + (t - x_3)]}{(x_2 - x_0)(x_2 - x_1)(x_2 - x_3)} + f_3 \frac{2[(t - x_0) + (t - x_1) + (t - x_2)]}{(x_3 - x_0)(x_3 - x_1)(x_3 - x_2)}.$$

Then substitution of $t = x_0$ and the fact that $x_i - x_j = (i - j)h$ produces

$$f''(x_0) \approx f_0 \frac{2[(x_0 - x_1) + (x_0 - x_2) + (x_0 - x_3)]}{(x_0 - x_1)(x_0 - x_2)(x_0 - x_3)} + f_1 \frac{2[(x_0 - x_0) + (x_0 - x_2) + (x_0 - x_3)]}{(x_1 - x_0)(x_1 - x_2)(x_1 - x_3)}$$

$$+ f_2 \frac{2[(x_0 - x_0) + (x_0 - x_1) + (x_0 - x_3)]}{(x_2 - x_0)(x_2 - x_1)(x_2 - x_3)} + f_3 \frac{2[(x_0 - x_0) + (x_0 - x_1) + (x_0 - x_2)]}{(x_3 - x_0)(x_3 - x_1)(x_3 - x_2)}$$

$$= f_0 \frac{2[(-h) + (-2h) + (-3h)]}{(-h)(-2h)(-3h)} + f_1 \frac{2[(0) + (-2h) + (-3h)]}{(h)(-h)(-2h)}$$

$$+ f_2 \frac{2[(0) + (-h) + (-3h)]}{(2h)(h)(-h)} + f_3 \frac{2[(0) + (-h) + (-2h)]}{(3h)(2h)(h)}$$

$$= f_0 \frac{-12h}{-6h^3} + f_1 \frac{-10h}{2h^3} + f_2 \frac{-8h}{-2h^3} + f_3 \frac{-6h}{6h^3} = \frac{2f_0 - 5f_1 + 4f_2 - f_3}{h^2},$$

and the formula is established.

Example 6.7

Derive the formula

$$f'''(x_0) \approx \frac{-5f_0 + 18f_1 - 24f_2 + 14f_3 - 3f_4}{2h^3}.$$

Solution. Start with the Lagrange interpolation polynomial for $f(t)$ based on the four points $x_0, x_1, x_2,$ and x_3.

$$f(t) \approx f_0 \frac{(t - x_1)(t - x_2)(t - x_3)(t - x_4)}{(x_0 - x_1)(x_0 - x_2)(x_0 - x_3)(x_0 - x_4)} + f_1 \frac{(t - x_0)(t - x_2)(t - x_3)(t - x_4)}{(x_1 - x_0)(x_1 - x_2)(x_1 - x_3)(x_1 - x_4)}$$

$$+ f_2 \frac{(t - x_0)(t - x_1)(t - x_3)(t - x_4)}{(x_2 - x_0)(x_2 - x_1)(x_2 - x_3)(x_2 - x_4)} + f_3 \frac{(t - x_0)(t - x_1)(t - x_2)(t - x_4)}{(x_3 - x_0)(x_3 - x_1)(x_3 - x_2)(x_3 - x_4)}$$

$$+ f_4 \frac{(t - x_0)(t - x_1)(t - x_2)(t - x_3)}{(x_4 - x_0)(x_4 - x_1)(x_4 - x_2)(x_4 - x_3)}.$$

Differentiate the numerators three times, then use the substitution $x_i - x_j = (i - j)h$ in the denominators and get

$$f'''(t) \approx f_0 \frac{6[(t - x_1) + (t - x_2) + (t - x_3) + (t - x_4)]}{(-h)(-2h)(-3h)(-4h)}$$

$$+ f_1 \frac{6[(t - x_0) + (t - x_2) + (t - x_3) + (t - x_4)]}{(h)(-h)(-2h)(-3h)}$$

$$+ f_2 \frac{6[(t - x_0) + (t - x_1) + (t - x_3) + (t - x_4)]}{(2h)(h)(-h)(-2h)}$$

$$+ f_3 \frac{6[(t - x_0) + (t - x_1) + (t - x_2) + (t - x_4)]}{(3h)(2h)(h)(-h)}$$

$$+ f_4 \frac{6[(t - x_0) + (t - x_1) + (t - x_2) + (t - x_3)]}{(4h)(3h)(2h)(h)}.$$

Then substitution of $t = x_0$ in the form $t - x_j = x_0 - x_j = -jh$ produces

$$f'''(x_0) \approx f_0 \frac{6[(-h) + (-2h) + (-3h) + (-4h)]}{24h^4} + f_1 \frac{6[(0) + (-2h) + (-3h) + (-4h)]}{-6h^4}$$

$$+ f_2 \frac{6[(0) + (-h) + (-3h) + (-4h)]}{4h^4} + f_3 \frac{6[(0) + (-h) + (-2h) + (-4h)]}{-6h^4}$$

$$+ f_4 \frac{6[(0) + (-h) + (-2h) + (-3h)]}{24h^4},$$

$$= f_0 \frac{-60h}{24h^4} + f_1 \frac{54h}{6h^4} + f_2 \frac{-48h}{4h^4} + f_3 \frac{42h}{6h^4} + f_4 \frac{-36h}{24h^4}$$

$$= \frac{-5f_0 + 18f_1 - 24f_2 + 14f_3 - 3f_4}{2h^3},$$

and the formula is established.

Differentiation of the Newton Polynomial

In this section we show the relationship between the three formulas of order $O(h^2)$ for approximating $f'(x_0)$, and a general algorithm is given for computing the numerical derivative. In Section 4.3 we saw that the Newton polynomial $P(t)$ of degree $N = 2$ that approximates $f(t)$ using the nodes t_0, t_1, and t_2 is

$$P(t) = a_0 + a_1(t - t_0) + a_2(t - t_0)(t - t_1), \tag{16}$$

where $a_0 = f(t_0)$, $a_1 = [f(t_1) - f(t_0)]/(t_1 - t_0)$, and

$$a_2 = \frac{\dfrac{f(t_2) - f(t_1)}{t_2 - t_1} - \dfrac{f(t_1) - f(t_0)}{t_1 - t_0}}{(t_2 - t_0)}.$$

The derivative of $P(t)$ is

$$P'(t) = a_1 + a_2[(t - t_0) + (t - t_1)], \tag{17}$$

and when it is evaluated at $t = t_0$ the result is

$$P'(t_0) = a_1 + a_2(t_0 - t_1) \approx f'(t_0) \tag{18}$$

Observe that the nodes $\{t_k\}$ do not need to be equally spaced for formulas (16)–(18) to hold. Choosing the abscissas in different orders will produce different formulas for approximating $f'(x)$.

Case (i): If $t_0 = x$, $t_1 = x + h$, and $t_2 = x + 2h$, then

$$a_1 = \frac{f(x + h) - f(x)}{h},$$

$$a_2 = \frac{f(x) - 2f(x + h) + f(x + 2h)}{2h^2}.$$

When these values are substituted into (18) we get

$$P'(x) = \frac{f(x + h) - f(x)}{h} + \frac{-f(x) + 2f(x + h) - f(x + 2h)}{2h}.$$

This is simplified to obtain

$$P'(x) = \frac{-3f(x) + 4f(x + h) - f(x + 2h)}{2h} \approx f'(x), \tag{19}$$

which is the second-order forward-difference formula for $f'(x)$.

Case (ii): If $t_0 = x$, $t_1 = x + h$, and $t_2 = x - h$, then

$$a_1 = \frac{f(x + h) - f(x)}{h},$$

$$a_2 = \frac{f(x + h) - 2f(x) + f(x - h)}{2h^2}.$$

When these values are substituted into (18) we get

$$P'(x) = \frac{f(x + h) - f(x)}{h} + \frac{-f(x + h) + 2f(x) - f(x - h)}{2h}.$$

This is simplified to obtain

$$P'(x) = \frac{f(x + h) - f(x - h)}{2h} \approx f'(x), \tag{20}$$

which is the second-order central-difference formula for $f'(x)$.

Case (iii): If $t_0 = x$, $t_1 = x - h$, and $t_2 = x - 2h$, then

$$a_1 = \frac{f(x) - f(x - h)}{h},$$

$$a_2 = \frac{f(x) - 2f(x - h) + f(x - 2h)}{2h^2}.$$

These values are substituted into (18) and simplified to get

$$P'(x) = \frac{3f(x) - 4f(x - h) + f(x - 2h)}{2h} \approx f'(x), \tag{21}$$

which is the second-order backward-difference formula for $f'(x)$.

The Newton polynomial $P(t)$ of degree N that approximates $f(t)$ using the nodes t_0, $t_1, \ldots, t_N$ is

$$
\begin{aligned}
P(t) = {}& a_0 + a_1(t - t_0) + a_2(t - t_0)(t - t_1) \\
& + a_3(t - t_0)(t - t_1)(t - t_2) + \cdots + a_N(t - t_0)(t - t_1) \cdots (t - t_{N-1}).
\end{aligned} \tag{22}
$$

The derivative of $P(t)$ is

$$
\begin{aligned}
P'(t) = {}& a_1 + a_2[(t - t_0) + (t - t_1)] \\
& + a_3[(t - t_0)(t - t_1) + (t - t_0)(t - t_2) + (t - t_1)(t - t_2)] + \cdots \\
& + a_N \sum_{k=0}^{N-1} \prod_{\substack{j=0 \\ j \neq k}}^{N-1} (t - t_j).
\end{aligned} \tag{23}
$$

When $P'(t)$ is evaluated at $t = t_0$ several of the terms in the summation are zero, and $P'(t_0)$ has the simpler form

$$
\begin{aligned}
P'(t_0) = {}& a_1 + a_2(t_0 - t_1) + a_3(t_0 - t_1)(t_0 - t_2) + \cdots \\
& + a_N(t_0 - t_1)(t_0 - t_2)(t_0 - t_3) \cdots (t_0 - t_{N-1}).
\end{aligned} \tag{24}
$$

The kth partial sum on the right side of equation (24) is the derivative of the Newton polynomial of degree k based on the first k nodes. If

$$|t_0 - t_1| \leq |t_0 - t_2| \leq \cdots \leq |t_0 - t_N|, \quad \text{and} \quad \text{if } \{(t_j, 0)\}_{j=0}^{j=N}$$

forms a set of $n + 1$ equally spaced points on the real axis, the kth partial sum is an approximation to $f'(t_0)$ of order $O(h^{k-1})$.

Suppose that $N = 5$. If the five nodes are $t_k = x + hk$, then (24) is an equivalent way to compute the forward-difference formula for $f'(x)$ of order $O(h^4)$. If the five nodes $\{t_k\}$ are chosen to be $t_0 = x$, $t_1 = x + h$, $t_2 = x - h$, $t_3 = x + 2h$, and $t_4 = x - 2h$, then (24) is the central-difference formula for $f'(x_0)$ of order $O(h^4)$. When the five nodes are $t_k = x - hk$, then (24) is the backward-difference formula for $f'(x_0)$ of order $O(h^4)$.

The following algorithm is an extension of Algorithm 4.5 and can be used to implement formula (24). It uses the space-saving modifications mentioned in Exercise 13 of Section 4.4. Note that the nodes do not need to be equally spaced. Also, it computes the derivative at only one point, $f'(x_0)$.

Algorithm 6.3   (Differentiation Based on $N + 1$ nodes).   To approximate $f'(x)$ numerically by constructing the Nth-degree Newton polynomial

$$P(x) = a_0 + a_1(x - x_0) + a_2(x - x_0)(x - x_1) + a_3(x - x_0)(x - x_1)(x - x_2)$$

$$+ \cdots + a_N(x - x_0)(x - x_1) \cdots (x - x_{N-1})$$

and using $f'(x_0) \approx P'(x_0)$ as the final answer. The method must be used at x_0. The points can be rearranged $\{x_k, x_0, \ldots, x_{k-1}, x_{k+1}, \ldots, x_N\}$ to compute $f'(x_k) \approx P'(x_k)$.

```
INPUT N                                          {Degree of P(x)}
INPUT X(0), X(1), . . . , X(N)                   {Interpolation of nodes}
INPUT Y(0), Y(1), . . . ,Y(N)                    {Ordinates of the points}
FOR  K = 0  TO  N  DO                            {Initialize the coefficient array}
      A(K) := Y(K)
FOR  J = 1  TO  N  DO                            {Compute the divided
      FOR  K = N  DOWNTO  J  DO                            differences}
            A(K) := [A(K) − A(K−1)]/[X(K) − X(K−J)]

X := X(0)                                        {Compute the derivative at x₀}
Df := A(1), Prod := 1                            {Initialize the variables}
FOR  K = 2  TO  N  DO
      Prod := Prod*[X − X(K−1)]
      Df := Df + Prod*A(K)

PRINT "The numerical derivative f′(x₀) is" Df    {Output}
```

EXERCISES FOR NUMERICAL DIFFERENTIATION FORMULAS

1. Let $f(x) = \ln(x)$ and carry eight or nine decimal places.
   (a) Use formula (6) with $h = 0.05$ to approximate $f''(5)$.
   (b) Use formula (6) with $h = 0.01$ to approximate $f''(5)$.
   (c) Use formula (12) with $h = 0.1$ to approximate $f''(5)$.
   (d) Which answer—(a), (b), or (c)—is more accurate?

2. Let $f(x) = \cos(x)$ and carry eight or nine decimal places.
   (a) Use formula (6) with $h = 0.05$ to approximate $f''(1)$.
   (b) Use formula (6) with $h = 0.01$ to approximate $f''(1)$.
   (c) Use formula (12) with $h = 0.1$ to approximate $f''(1)$.
   (d) Which answer—(a), (b), or (c)—is more accurate?

3. Consider the table for $f(x) = \ln(x)$ rounded to four decimal places.

x	$f(x) = \ln(x)$
4.90	1.5892
4.95	1.5994
5.00	1.6094
5.05	1.6194
5.10	1.6292

   (a) Use formula (6) and $h = 0.05$ to approximate $f''(5)$.
   (b) Use formula (6) and $h = 0.10$ to approximate $f''(5)$.
   (c) Use formula (12) and $h = 0.05$ to approximate $f''(5)$.
   (d) Which answer—(a), (b), or (c)—is more accurate?

4. Consider the table for $f(x) = \cos(x)$ rounded to four decimal places.

x	$f(x) = \cos(x)$
0.90	0.6216
0.95	0.5817
1.00	0.5403
1.05	0.4976
1.10	0.4536

   (a) Use formula (6) and $h = 0.05$ to approximate $f''(1)$.
   (b) Use formula (6) and $h = 0.10$ to approximate $f''(1)$.
   (c) Use formula (12) and $h = 0.05$ to approximate $f''(1)$.
   (d) Which answer—(a), (b), or (c)—is more accurate?

5. Use the numerical differentiation formula (6) and $h = 0.01$ to approximate $f''(1)$ for the functions
   (a) $f(x) = x^2$ (b) $f(x) = x^4$

6. Use the numerical differentiation formula (12) and $h = 0.1$ to approximate $f''(1)$ for the functions
   (a) $f(x) = x^4$ (b) $f(x) = x^6$

7. Use the Taylor expansions for $f(x + h)$, $f(x - h)$, $f(x + 2h)$, and $f(x - 2h)$ and derive the central-difference formula:

$$f^{(3)}(x) \approx \frac{f(x + 2h) - 2f(x + h) + 2f(x - h) - f(x - 2h)}{2h^3}.$$

8. Use the Taylor expansions for $f(x + h)$, $f(x - h)$, $f(x + 2h)$, and $f(x - 2h)$ and derive the central-difference formula:

$$f^{(4)}(x) \approx \frac{f(x + 2h) - 4f(x + h) + 6f(x) - 4f(x - h) + f(x - 2h)}{h^4}.$$

9. Find the approximations to $f'(x_k)$ of order $O(h^2)$ at each of the four points in the tables.

(a)

x	$f(x)$
0.0	0.989992
0.1	0.999135
0.2	0.998295
0.3	0.987480

(b)

x	$f(x)$
0.0	0.141120
0.1	0.041581
0.2	-0.058374
0.3	-0.157746

10. Use the approximations

$$f'\left(x + \frac{h}{2}\right) \approx \frac{f_1 - f_0}{h} \quad \text{and} \quad f'\left(x - \frac{h}{2}\right) \approx \frac{f_0 - f_{-1}}{h}$$

and derive the approximation

$$f''(x) \approx \frac{f_1 - 2f_0 + f_{-1}}{h^2}.$$

11. Use formulas (16) to (18) and derive a formula for $f'(x)$ based on the abscissas $t_0 = x$, $t_1 = x + h$, and $t_2 = x + 3h$.

12. Use formulas (16) to (18) and derive a formula for $f'(x)$ based on the abscissas $t_0 = x$, $t_1 = x - h$, and $t_2 = x + 2h$.

13. The numerical solution of a certain differential equation requires an approximation to $f''(x) + f'(x)$ of order $O(h^2)$.
    (a) Find the central-difference formula for $f''(x) + f'(x)$ by adding the formulas for $f'(x)$ and $f''(x)$ of order $O(h^2)$.
    (b) Find the forward-difference formula for $f''(x) + f'(x)$ by adding the formulas for $f'(x)$ and $f''(x)$ of order $O(h^2)$.
    (c) What would happen if a formula for $f'(x)$ of order $O(h^4)$ were added to a formula for $f''(x)$ of order $O(h^2)$?

14. Critique the following argument. Taylor's formula can be used to get the representations

$$f(x + h) = f(x) + hf'(x) + \frac{h^2 f''(x)}{2} + \frac{h^3 f'''(c)}{6}$$

and

$$f(x - h) = f(x) - hf'(x) + \frac{h^2 f''(x)}{2} - \frac{h^3 f'''(c)}{6}.$$

Adding these quantities results in

$$f(x + h) + f(x - h) = 2f(x) + h^2 f''(x),$$

which can be solved to obtain an exact formula for $f''(x)$:

$$f''(x) = \frac{f(x + h) - 2f(x) + f(x - h)}{h^2}.$$

15. Write a report on the forward-difference operator for numerical differentiation.
16. Write a report on the backward-difference operator for numerical differentiation.
17. Write a report on lozenge diagrams for numerical differentiation.
18. Modify Algorithm 6.3 so that it will calculate $P'(x_M)$ for $M = 1, 2, \ldots, N + 1$.

7

Numerical Integration

Numerical integration is a primary tool used by engineers and scientists to obtain approximate answers for definite integrals that cannot be solved analytically. In the area of statistical thermodynamics, the Debye model for calculating the heat capacity of a solid involves the following function:

$$\Phi(x) = \int_0^x \frac{t^3}{e^t - 1}\, dt.$$

Since there is no analytic expression for $\Phi(x)$, numerical integration must be used to obtain approximate values. For example, the value $\Phi(5)$ is the area under the curve $y = f(t) = t^3/(e^t - 1)$ for $0 \le t \le 5$ (see Figure 7.1). The numerical approximation for $\Phi(5)$ is

$$\Phi(5) = \int_0^5 \frac{t^3}{e^t - 1}\, dt \approx 4.8998922.$$

Each additional value of $\Phi(x)$ must be determined by another numerical integration. Table 7.1 lists several of these approximations over the interval [1, 10].

The purpose of this chapter is to develop the basic principles of numerical integration. In Chapter 9, numerical integration formulas are used to derive the predictor–corrector methods for solving differential equations.

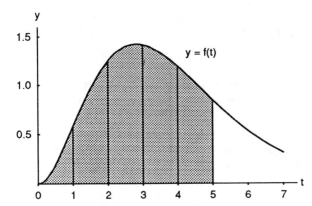

Figure 7.1   Area under the curve $y = f(t)$ for $0 \le t \le 5$.

TABLE 7.1   Values of $\Phi(x)$

x	$\Phi(x)$
1.0	0.2248052
2.0	1.1763426
3.0	2.5522185
4.0	3.8770542
5.0	4.8998922
6.0	5.5858554
7.0	6.0031690
8.0	6.2396238
9.0	6.3665739
10.0	6.4319219

7.1 INTRODUCTION TO QUADRATURE

We now approach the subject of numerical integration. The goal is to approximate the definite integral of $f(x)$ over the interval $[a, b]$ by evaluating $f(x)$ at a finite number of sample points.

Definition 7.1.   Suppose that $a = x_0 < x_1 < \cdots < x_M = b$. A formula of the form

$$Q[f] = \sum_{j=0}^{M} w_j f(x_j) = w_0 f(x_0) + w_1 f(x_1) + \cdots + w_M f(x_M) \qquad (1)$$

with the property that

$$\int_a^b f(x)\, dx = Q[f] + E[f] \qquad (2)$$

is called a numerical integration or **quadrature** formula. The term $E[f]$ is called the **truncation error** for integration. The values $\{x_j\}_{j=0}^{M}$ are called the **quadrature nodes** and $\{w_j\}_{j=0}^{M}$ are called **weights**.

Depending on the application, the nodes $\{x_k\}$ are chosen in various ways. For the trapezoidal rule, Simpson's rules, and Boole's rule, the nodes are chosen to be equally spaced. For Gauss–Legendre quadrature, the nodes are chosen to be zeros of certain Legendre polynomials. When the integration formula is used to develop a predictor formula for solving differential equations, all the nodes are chosen less than b. For all applications it is necessary to know something about the accuracy of the numerical solution.

Definition 7.2.   The **degree of precision** of a quadrature formula is the positive integer n such that $E[P_i] = 0$ for all polynomials $P_i(x)$ of degree $i \le n$, but for which $E[P_{n+1}] \ne 0$ for some polynomial $P_{n+1}(x)$ of degree $n + 1$.

The form of $E[P_i]$ can be anticipated by studying what happens when $f(x)$ is a polynomial. Consider the arbitrary polynomial $P_i(x) = a_i x^i + a_{i-1} x^{i-1} + \cdots + a_1 x + a_0$ of degree i. If $i \le n$, then $P_i^{(n+1)}(x) \equiv 0$ for all x and $P_{n+1}^{(n+1)}(x) \equiv (n + 1)!\, a_{n+1}$ for all x. Thus it is not surprising that the general form for the truncation error term is

$$E[f] = Kf^{(n+1)}(c), \tag{3}$$

where K is a suitably chosen constant and n is the degree of precision. The proof of this general result can be found in advanced books on numerical integration.

The derivation of quadrature formulas is sometimes based on polynomial interpolation. Recall that there exists a unique polynomial $P_M(x)$ of degree $\le M$ passing through the $M + 1$ equally spaced points $\{(x_j, y_j)\}_{j=0}^{M}$. When this polynomial is used to approximate $f(x)$ over $[a, b]$, and then the integral of $f(x)$ is approximated by the integral of $P_M(x)$, the resulting formula is called a **Newton–Cotes quadrature formula** (see Figure 7.2). When the sample points $x_0 = a$ and $x_M = b$ are used it is called a **closed Newton–Cotes** formula. The next result gives the formulas when approximating polynomials of degree $M = 1, 2, 3$, and 4 are used.

Theorem 7.1   (Closed Newton–Cotes Quadrature Formulas).   Assume that $x_j = x_0 + hj$ are equally spaced nodes and $f_j = f(x_j)$. The first four closed Newton–Cotes quadrature formulas are

$$\int_{x_0}^{x_1} f(x)\, dx \approx \frac{h}{2}\,(f_0 + f_1) \qquad \text{(the trapezoidal rule),} \tag{4}$$

$$\int_{x_0}^{x_2} f(x)\, dx \approx \frac{h}{3}\,(f_0 + 4f_1 + f_2) \qquad \text{(Simpson's rule),} \tag{5}$$

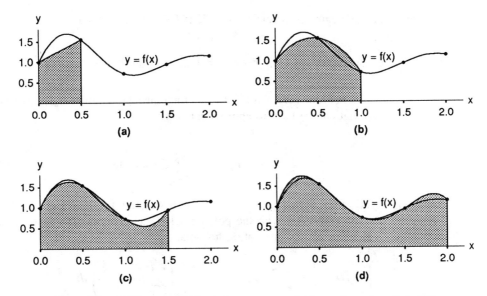

Figure 7.2 (a) The trapezoidal rule integrates $y = P_1(x)$ over $[x_0, x_1] = [0.0, 0.5]$. (b) Simpson's rule integrates $y = P_2(x)$ over $[x_0, x_2] = [0.0, 1.0]$. (c) Simpson's $\frac{3}{8}$ rule integrates $y = P_3(x)$ over $[x_0, x_3] = [0.0, 1.5]$. (d) Boole's rule integrates $y = P_4(x)$ over $[x_0, x_4] = [0.0, 2.0]$.

$$\int_{x_0}^{x_3} f(x)\, dx \approx \frac{3h}{8} (f_0 + 3f_1 + 3f_2 + f_3) \qquad \text{(Simpson's } \tfrac{3}{8} \text{ rule)}, \qquad (6)$$

$$\int_{x_0}^{x_4} f(x)\, dx \approx \frac{2h}{45} (7f_0 + 32f_1 + 12f_2 + 32f_3 + 7f_4) \qquad \text{(Boole's rule)}. \qquad (7)$$

Corollary 7.1   (Newton–Cotes Precision).   Assume that $f(x)$ is sufficiently differentiable; then $E[f]$ for Newton–Cotes quadrature involves an appropriate higher derivative. The trapezoidal rule has degree of precision $n = 1$. If $f \in C^2[a, b]$, then

$$\int_{x_0}^{x_1} f(x)\, dx = \frac{h}{2} (f_0 + f_1) - \frac{h^3}{12} f^{(2)}(c). \qquad (8)$$

Simpson's rule has degree of precision $n = 3$. If $f \in C^4[a, b]$, then

$$\int_{x_0}^{x_2} f(x)\, dx = \frac{h}{3} (f_0 + 4f_1 + f_2) - \frac{h^5}{90} f^{(4)}(c). \qquad (9)$$

Simpson's $\frac{3}{8}$ rule has degree of precision $n = 3$. If $f \in C^4[a, b]$, then

$$\int_{x_0}^{x_3} f(x)\, dx = \frac{3h}{8} (f_0 + 3f_1 + 3f_2 + f_3) - \frac{3h^5}{80} f^{(4)}(c). \qquad (10)$$

Boole's rule has degree of precision $n = 5$. If $f \in C^6[a, b]$, then

$$\int_{x_0}^{x_4} f(x) \, dx = \frac{2h}{45} (7f_0 + 32f_1 + 12f_2 + 32f_3 + 7f_4) - \frac{8h^7}{945} f^{(6)}(c). \quad (11)$$

Proof of Theorem 7.1. Start with the Lagrange polynomial $P_M(x)$ based on x_0, $x_1, \ldots, x_M$ that can be used to approximate $f(x)$:

$$f(x) \approx P_M(x) = \sum_{k=0}^{M} f_k L_{M,k}(x), \quad (12)$$

where $f_k = f(x_k)$ for $k = 0, 1, \ldots, M$. An approximation for the integral is obtained by replacing the integrand $f(x)$ with the polynomial $P_M(x)$. This is the general method for obtaining a Newton–Cotes integration formula:

$$\int_{x_0}^{x_M} f(x) \, dx \approx \int_{x_0}^{x_M} P_M(x) \, dx = \int_{x_0}^{x_M} \left(\sum_{k=0}^{M} f_k L_{M,k}(x) \right) dx$$

$$= \sum_{k=0}^{M} \left(\int_{x_0}^{x_M} f_k L_{M,k}(x) \, dx \right) = \sum_{k=0}^{M} \left(\int_{x_0}^{x_M} L_{M,k}(x) \, dx \right) f_k \quad (13)$$

$$= \sum_{k=0}^{M} w_k f_k.$$

The details for the general proof of (13) are tedious. We shall give a sample proof of Simpson's rule, which is the case $M = 2$. This case involves the approximating polynomial

$$P_2(x) = f_0 \frac{(x - x_1)(x - x_2)}{(x_0 - x_1)(x_0 - x_2)} + f_1 \frac{(x - x_0)(x - x_2)}{(x_1 - x_0)(x_1 - x_2)} + f_2 \frac{(x - x_0)(x - x_1)}{(x_2 - x_0)(x_2 - x_1)}. \quad (14)$$

Since f_0, f_1, and f_2 are constants with respect to integration, the relations in (13) lead to

$$\int_{x_0}^{x_2} f(x) \, dx \approx f_0 \int_{x_0}^{x_2} \frac{(x - x_1)(x - x_2)}{(x_0 - x_1)(x_0 - x_2)} \, dx + f_1 \int_{x_0}^{x_2} \frac{(x - x_0)(x - x_2)}{(x_1 - x_0)(x_1 - x_2)} \, dx$$

$$+ f_2 \int_{x_0}^{x_2} \frac{(x - x_0)(x - x_1)}{(x_2 - x_0)(x_2 - x_1)} \, dx. \quad (15)$$

We introduce the change of variable $x = x_0 + ht$ with $dx = h \, dt$ to assist with the evaluation of the integrals in (15). The new limits of integration are from $t = 0$ to $t = 2$. The equal spacing of the nodes $x_j - x_0 + hj$ leads to $x_j - x_k = h(j - k)$ and $x - x_j = h(t - j)$, which are used to simplify (15) and get

$$\int_{x_0}^{x_2} f(x)\, dx \approx f_0 \int_0^2 \frac{h(t-1)h(t-2)}{(-h)(-2h)} h\, dt + f_1 \int_0^2 \frac{h(t-0)h(t-2)}{(h)(-h)} h\, dt$$

$$+ f_2 \int_0^2 \frac{h(t-0)h(t-1)}{(2h)(h)} h\, dt \tag{16}$$

$$= f_0 \frac{h}{2} \int_0^2 (t^2 - 3t + 2)\, dt - f_1 h \int_0^2 (t^2 - 2t)\, dt + f_2 \frac{h}{2} \int_0^2 (t^2 - t)\, dt$$

$$= f_0 \frac{h}{2} \left(\frac{t^3}{3} - \frac{3t^2}{2} + 2t \right) \Big|_{t=0}^{t=2} - f_1 h \left(\frac{t^3}{3} - t^2 \right) \Big|_{t=0}^{t=2} + f_2 \frac{h}{2} \left(\frac{t^3}{3} - \frac{t^2}{2} \right) \Big|_{t=0}^{t=2}$$

$$= f_0 \frac{h}{2} \left(\frac{2}{3} \right) - f_1 h \left(\frac{-4}{3} \right) + f_2 \frac{h}{2} \left(\frac{2}{3} \right)$$

$$= \frac{h}{3} (f_0 + 4f_1 + f_2),$$

and the proof is complete. We postpone a sample proof of Corollary 7.1 until Section 7.3.

Example 7.1

Consider the function $f(x) = 1 + e^{-x} \sin(4x)$ and the equally spaced quadrature nodes $x_0 = 0.0$, $x_1 = 0.5$, $x_2 = 1.0$, $x_3 = 1.5$, $x_4 = 2.0$ and the corresponding function values $f_0 = 1.00000$, $f_1 = 1.55152$, $f_2 = 0.72159$, $f_3 = 0.93765$, $f_4 = 1.13390$. Apply the various quadrature formulas (4) to (7).

Solution.   The step size is $h = 0.5$, and the computations are

$$\int_0^{0.5} f(x)\, dx \approx \frac{0.5}{2} (1.00000 + 1.55152) = 0.63788$$

$$\int_0^{1.0} f(x)\, dx \approx \frac{0.5}{3} (1.00000 + 4 \times 1.55152 + 0.72159) = 1.32128$$

$$\int_0^{1.5} f(x)\, dx \approx \frac{3 \times 0.5}{8} (1.00000 + 3 \times 1.55152 + 3 \times 0.72159 + 0.93765)$$

$$= 1.64193$$

$$\int_0^{2.0} f(x)\, dx \approx \frac{2 \times 0.5}{45} (7 \times 1.00000 + 32 \times 1.55152 + 12 \times 0.72159$$

$$+ 32 \times 0.93765 + 7 \times 1.13390)$$

$$= 2.29444.$$

It is important to realize that the quadrature formulas (4) to (7) applied in the above illustration give approximations for definite integrals over different intervals. The graph of the curve $y = f(x)$ and the areas under the Lagrange polynomials $y = P_1(x)$, $y = P_2(x)$, $y = P_3(x)$, and $y = P_4(x)$ are shown in Figure 7.2(a) to (d), respectively.

In Example 7.1 we applied the quadrature rules with $h = 0.5$. If the endpoints of the interval $[a, b]$ are held fixed, the step size must be adjusted for each rule. The step sizes are $h = b - a$, $h = (b - a)/2$, $h = (b - a)/3$, and $h = (b - a)/4$ for the trapezoidal rule, Simpson's rule, Simpson's $\frac{3}{8}$ rule, and Boole's rule, respectively. The next example illustrates this point.

Example 7.2

Consider integration of the function $f(x) = 1 + e^{-x} \sin(4x)$ over the fixed interval $[a, b] = [0, 1]$. Apply the various quadrature formulas (4) to (7).

Solution.   For the trapezoidal rule, $h = 1$, and

$$\int_0^1 f(x)\, dx \approx \frac{1}{2}\,(1.00000 + 0.72159) = 0.86079.$$

For Simpson's rule, $h = \frac{1}{2}$, and we get

$$\int_0^1 f(x)\, dx \approx \frac{1}{2}\frac{1}{3}\left(f(0) + 4f\left(\tfrac{1}{2}\right) + f(1)\right)$$

$$= \frac{1}{6}\,(1.00000 + 4 \times 1.55152 + 0.72159) = 1.32128.$$

For Simpson's $\frac{3}{8}$ rule, $h = \frac{1}{3}$, and we obtain

$$\int_0^1 f(x)\, dx \approx \frac{1}{3}\frac{3}{8}\left(f(0) + 3f\left(\tfrac{1}{3}\right) + 3f\left(\tfrac{2}{3}\right) + f(1)\right)$$

$$= \frac{1}{8}\,(1.00000 + 3 \times 1.69642 + 3 \times 1.23447 + 0.72159) = 1.31440.$$

For Boole's rule, $h = \frac{1}{4}$, and the result is

$$\int_0^1 f(x)\, dx \approx \frac{1}{4}\frac{2}{45}\left(7f(0) + 32f\left(\tfrac{1}{4}\right) + 12f\left(\tfrac{1}{2}\right) + 32f\left(\tfrac{3}{4}\right) + 7f(1)\right)$$

$$= \frac{1}{90}\,(7 \times 1.00000 + 32 \times 1.65534 + 12 \times 1.55152 + 32 \times 1.06666$$

$$+ 7 \times 0.72159)$$

$$= 1.30859.$$

The true value of the definite integral is

$$\frac{21e - 4\cos(4) - \sin(4)}{17e} = 1.3082506046426.\ .\ .\ ,$$

and the approximation 1.30859 from Boole's rule is best. The area under the Lagrange polynomial $P_1(x)$, $P_2(x)$, $P_3(x)$, and $P_4(x)$ is shown in Figure 7.3(a) to (d), respectively.

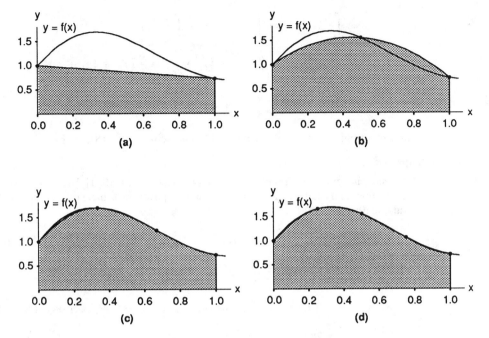

Figure 7.3   (a) The trapezoidal rule used over $[0, 1]$ yields the approximation 0.86079. (b) Simpson's rule used over $[0, 1]$ yields the approximation 1.32128. (c) Simpson's $\frac{3}{8}$ rule used over $[0, 1]$ yields the approximation 1.31440. (d) Boole's rule used over $[0, 1]$ yields the approximation 1.30859.

To make a fair comparison of quadrature methods we must use the same number of function evaluations in each method. Our final example is concerned with comparing integration over a fixed interval $[a, b]$ using exactly five function evaluations $f_k = f(x_k)$ for $k = 0, 1, \ldots, 4$ for each method. When the trapezoidal rule is applied on the four subintervals $[x_0, x_1]$, $[x_1, x_2]$, $[x_2, x_3]$, and $[x_3, x_4]$, it is called a **composite trapezoidal rule**:

$$\int_{x_0}^{x_4} f(x)\, dx = \int_{x_0}^{x_1} f(x)\, dx + \int_{x_1}^{x_2} f(x)\, dx + \int_{x_2}^{x_3} f(x)\, dx + \int_{x_3}^{x_4} f(x)\, dx$$

$$\approx \frac{h}{2}\,(f_0 + f_1) + \frac{h}{2}\,(f_1 + f_2) + \frac{h}{2}\,(f_2 + f_3) + \frac{h}{2}\,(f_3 + f_4) \qquad (17)$$

$$= \frac{h}{2}\,(f_0 + 2f_1 + 2f_2 + 2f_3 + f_4).$$

Simpson's rule can also be used in this manner. When Simpson's rule is applied on the two subintervals $[x_0, x_2]$ and $[x_2, x_4]$, it is called a **composite Simpson rule**:

$$\int_{x_0}^{x_4} f(x)\, dx = \int_{x_0}^{x_2} f(x)\, dx + \int_{x_2}^{x_4} f(x)\, dx$$

$$\approx \frac{h}{3}(f_0 + 4f_1 + f_2) + \frac{h}{3}(f_2 + 4f_3 + f_4) \tag{18}$$

$$= \frac{h}{3}(f_0 + 4f_1 + 2f_2 + 4f_3 + f_4).$$

The next example compares the values obtained with (17), (18), and (7).

Example 7.3

Consider integration of $f(x) = 1 + e^{-x}\sin(4x)$ over $[a, b] = [0, 1]$. Use exactly five function evaluations and compare the results from the composite trapezoidal rule, composite Simpson rule, and Boole's rule.

Solution.   The uniform step size is $h = \frac{1}{4}$. The composite trapezoidal rule (17) produces

$$\int_{x_0}^{x_4} f(x)\, dx \approx \frac{1}{4}\frac{1}{2}\left(f(0) + 2f(\tfrac{1}{4}) + 2f(\tfrac{1}{2}) + 2f(\tfrac{3}{4}) + f(1)\right)$$

$$= \frac{1}{8}(1.00000 + 2 \times 1.65534 + 2 \times 1.55152 + 2 \times 1.06666 + 0.72159)$$

$$= 1.28358.$$

Using the composite Simpson rule (18), we get

$$\int_{x_0}^{x_4} f(x)\, dx \approx \frac{1}{4}\frac{1}{3}\left(f(0) + 4f(\tfrac{1}{4}) + 2f(\tfrac{1}{2}) + 4f(\tfrac{3}{4}) + f(1)\right)$$

$$= \frac{1}{12}(1.00000 + 4 \times 1.65534 + 2 \times 1.55152 + 4 \times 1.06666 + 0.72159)$$

$$= 1.30938.$$

We have already seen the result of Boole's rule in Example 7.2:

$$\int_0^1 f(x)\, dx \approx \frac{1}{4}\frac{2}{45}\left(7f(0) + 32f(\tfrac{1}{4}) + 12f(\tfrac{1}{2}) + 32f(\tfrac{3}{4}) + 7f(1)\right) = 1.30859.$$

The true value of the definite integral is

$$\frac{21e - 4\cos(4) - \sin(4)}{17e} = 1.3082506046426.\ldots,$$

and the approximation 1.30938 from Simpson's rule is much better than the value 1.06943 obtained from the trapezoidal rule. Again, the approximation 1.30859 from Boole's rule is closest. Graphs for the areas under the trapezoids and parabolas are shown in Figure 7.4(a) and (b), respectively.

Example 7.4

Determine the degree of precision of Simpson's $\frac{3}{8}$ rule.

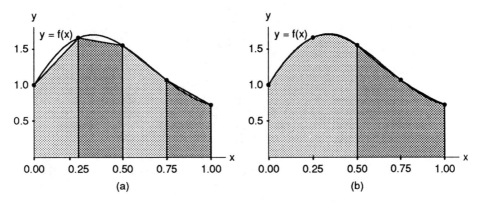

Figure 7.4 (a) The composite trapezoidal rule yields the approximation 1.06943.
(b) The composite Simpson rule yields the approximation 1.30938.

Solution.   It will suffice to apply Simpson's $\frac{3}{8}$ rule over the interval [0, 3] with the five test functions $f(x) = 1$, x, x^2, x^3, and x^4. For the first functions, Simpson's $\frac{3}{8}$ rule is exact.

$$\int_0^3 1 \, dx = 3 = \tfrac{3}{8}(1 + 3 \times 1 + 3 \times 1 + 1)$$

$$\int_0^3 x \, dx = \tfrac{3}{2} = \tfrac{3}{8}(0 + 3 \times 1 + 3 \times 2 + 3)$$

$$\int_0^3 x^2 \, dx = 9 = \tfrac{3}{8}(0 + 3 \times 1 + 3 \times 4 + 9)$$

$$\int_0^3 x^3 \, dx = \tfrac{81}{4} = \tfrac{3}{8}(0 + 3 \times 1 + 3 \times 8 + 27)$$

The function $f(x) = x^4$ is the lowest power of x for which the rule is not exact.

$$\int_0^3 x^4 \, dx = \tfrac{243}{5} \approx \tfrac{99}{2} = \tfrac{3}{8}(0 + 3 \times 1 + 3 \times 16 + 81)$$

Therefore, the degree of precision of Simpson's $\frac{3}{8}$ rule is $n = 3$.

EXERCISES FOR INTRODUCTION TO QUADRATURE

1. Consider integration of $f(x)$ over the fixed interval $[a, b] = [0, 1]$. Apply the various quadrature formulas (4) to (7). The step sizes are $h = 1$, $h = \frac{1}{2}$, $h = \frac{1}{3}$, and $h = \frac{1}{4}$ for the trapezoidal rule, Simpson's rule, Simpson's $\frac{3}{8}$ rule, and Boole's rule, respectively.
  (a) $f(x) = \sin(\pi x)$ **(b)** $f(x) = 1 + e^{-x} \cos(4x)$ **(c)** $f(x) = \sin(\sqrt{x})$

  Remark. The true values of the definite integrals are (a) $2/\pi \approx 0.636619772367.\ .\ .\ ,$ (b) $[18e - \cos(4) + 4\sin(4)]/(17e) \approx 1.007459631397.\ .\ .\ ,$ and (c) $2\sin(1) - 2\cos(1) \approx 0.602337357879.\ .\ .\ .$ Graphs of the functions are shown in Figures 7.5(a) to (c), respectively.

2. Consider integration of $f(x)$ over the fixed interval $[a, b] = [0, 1]$. Apply the various quadrature

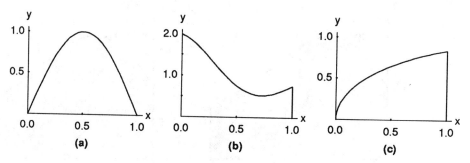

Figure 7.5   (a) $y = \sin(\pi x)$, (b) $y = 1 + e^{-x} \cos(4x)$, (c) $y = \sin(\sqrt{x})$.

formulas; the composite trapezoidal rule (17), the composite Simpson rule (18), and Boole's rule (7). Use five function evaluations at equally spaced nodes. The uniform step size is $h = \frac{1}{4}$.

(a) $f(x) = \sin(\pi x)$ **(b)** $f(x) = 1 + e^{-x} \cos(4x)$ **(c)** $f(x) = \sin(\sqrt{x})$

Remark. The true values of the definite integrals are (a) $2/\pi \approx 0.636619772367. \ldots$, (b) $[18e - \cos(4) + 4 \sin(4)]/(17e) \approx 1.007459631397. \ldots$, and (c) $2 \sin(1) - 2 \cos(1) \approx 0.602337357879. \ldots$. Graphs of the functions are shown in Figure 7.5(a) to (c), respectively.

3. Consider a general interval $[a, b]$. Show that Simpson's rule produces exact results for the functions $f(x) = x^2$ and $f(x) = x^3$; that is;

   (a) $\displaystyle\int_a^b x^2 \, dx = \frac{b^2}{3} - \frac{a^2}{3}$ **(b)** $\displaystyle\int_a^b x^3 \, dx = \frac{b^4}{4} - \frac{a^4}{4}$

4. Integrate the Lagrange interpolation polynomial

$$P_1(x) = f_0 \frac{x - x_1}{x_0 - x_1} + f_1 \frac{x - x_0}{x_1 - x_0}$$

over the interval $[x_0, x_1]$ and establish the trapezoidal rule.

5. Determine the degree of precision of the trapezoidal rule. It will suffice to apply the trapezoidal rule over $[0, 1]$ with the three test functions $f(x) = 1$, x, and x^2.

6. Determine the degree of precision of Simpson's rule. It will suffice to apply Simpson's rule over $[0, 2]$ with the five test functions $f(x) = 1, x, x^2, x^3$, and x^4. Contrast your result with the degree of precision of Simpson's $\frac{3}{8}$ rule.

7. Determine the degree of precision of Boole's rule. It will suffice to apply Boole's rule over $[0, 4]$ with the seven test functions $f(x) = 1, x, x^2, x^3, x^4, x^5$, and x^6.

8. Derive Simpson's $\frac{3}{8}$ rule using Lagrange polynomial interpolation. *Hint.* After changing the variable, integrals similar to those in (16) are obtained:

$$\int_{x_0}^{x_3} f(x) \, dx \approx -f_0 \frac{h}{6} \int_0^3 (t - 1)(t - 2)(t - 3) \, dt + f_1 \frac{h}{2} \int_0^3 (t - 0)(t - 2)(t - 3) \, dt$$

$$- f_2 \frac{h}{2} \int_0^3 (t - 0)(t - 1)(t - 3) \, dt + f_3 \frac{h}{6} \int_0^3 (t - 0)(t - 1)(t - 2) \, dt$$

$$= f_0 \frac{h}{6} \left(\frac{-t^4}{4} + 2t^3 - \frac{11t^2}{2} + 6t \right) \Bigg|_{t=0}^{t=3} + f_1 \frac{h}{2} \left(\frac{t^4}{4} - \frac{5t^3}{3} + 3t^2 \right) \Bigg|_{t=0}^{t=3}$$

$$+ f_2 \frac{h}{2} \left(\frac{-t^4}{4} + \frac{4t^3}{3} - \frac{3t^2}{2} \right) \Bigg|_{t=0}^{t=3} + f_3 \frac{h}{6} \left(\frac{t^4}{4} - t^3 + t^2 \right) \Bigg|_{t=0}^{t=3}.$$

9. Write a report on Newton–Cotes formulas. See References [9, 29, 62, 76, 78, 81, 90, 94, 97, 105, 117, 126, 128, 152, 153, 154, 160, 175, 193, and 208].

10. Write a report on Monte Carlo methods. See References [35, 41, 57, 76, 83, 87, 98, 112, 115, 135, 152, and 154].

7.2 COMPOSITE TRAPEZOIDAL AND SIMPSON'S RULE

An intuitive method of finding the area under the curve $y = f(x)$ over $[a, b]$ is by approximating that area with a series of trapezoids that lie above the intervals $\{[x_k, x_{k+1}]\}$.

Theorem 7.2   (Composite Trapezoidal Rule).   Suppose that the interval $[a, b]$ is subdivided into M subintervals $[x_k, x_{k+1}]$ of width $h = (b - a)/M$ by using the equally spaced nodes $x_k = a + kh$ for $k = 0, 1, \ldots, M$. The **composite trapezoidal rule for M subintervals** can be expressed in any of three equivalent ways:

$$T(f, h) = \frac{h}{2} \sum_{k=1}^{M} [f(x_{k-1}) + f(x_k)] \tag{1a}$$

or

$$T(f, h) = \frac{h}{2} [f_0 + 2f_1 + 2f_2 + 2f_3 + \cdots + 2f_{M-2} + 2f_{M-1} + f_M] \tag{1b}$$

or

$$T(f, h) = \frac{h}{2} [f(a) + f(b)] + h \sum_{k=1}^{M-1} f(x_k). \tag{1c}$$

This is an approximation to the integral of $f(x)$ over $[a, b]$ and we write

$$\int_a^b f(x)\, dx \approx T(f, h). \tag{2}$$

Proof. Apply the trapezoidal rule over each subinterval $[x_{k-1}, x_k]$ (see Figure 7.6). Use the additive property of the integral for subintervals:

$$\int_a^b f(x)\, dx = \sum_{k=1}^{M} \int_{x_{k-1}}^{x_k} f(x)\, dx \approx \sum_{k=1}^{M} \frac{h}{2} [f(x_{k-1}) + f(x_k)]. \tag{3}$$

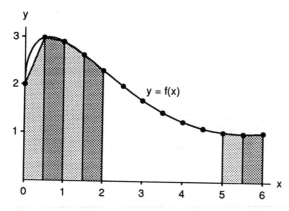

Figure 7.6 Approximating the area under the curve $y = 2 + \sin(2\sqrt{x})$ with the composite trapezoidal rule.

Since $h/2$ is a constant, the distributive law of addition can be applied to obtain (1a). Formula (1b) is the expanded version of (1a). Formula (1c) shows how to group all the intermediate terms in (1b) which are multiplied by 2.

   Approximating $f(x) = 2 + \sin(2\sqrt{x})$ with piecewise linear polynomials results in places where the approximation is close and places where it is not. To achieve accuracy the composite trapezoidal rule must be applied with many subintervals. In the next example we have chosen to numerically integrate this function over the interval [1, 6]. Investigation of the integral over [0, 1] is left as an exercise.

Example 7.5

   Consider $f(x) = 2 + \sin(2\sqrt{x})$. Use the composite trapezoidal rule with 11 sample points to compute an approximation to the integral of $f(x)$ taken over [1, 6].

Solution.   To generate 11 sample points we use $M = 10$ and $h = (6 - 1)/10 = \frac{1}{2}$. Using formula (1c), the computation is

$$T\left(f, \tfrac{1}{2}\right) = \frac{\frac{1}{2}}{2}[f(1) + f(6)]$$

$$+ \frac{1}{2}\left[f\left(\tfrac{3}{2}\right) + f(2) + f\left(\tfrac{5}{2}\right) + f(3) + f\left(\tfrac{7}{2}\right) + f(4) + f\left(\tfrac{9}{2}\right) + f(5) + f\left(\tfrac{11}{2}\right) \right]$$

$$= \tfrac{1}{4}[2.90929743 + 1.01735756]$$

$$+ \tfrac{1}{2}[2.63815764 + 2.30807174 + 1.97931647 + 1.68305284 + 1.43530410$$

$$+ 1.24319750 + 1.10831775 + 1.02872220 + 1.00024140]$$

$$= \tfrac{1}{4}[3.92665499] + \tfrac{1}{2}[14.42438165]$$

$$= 0.98166375 + 7.21219083 = 8.19385457.$$

Theorem 7.3   (Composite Simpson Rule).   Suppose that $[a, b]$ is subdivided into $2M$ subintervals $[x_k, x_{k+1}]$ of equal width $h = (b - a)/(2M)$ by using $x_k = x_0 + kh$

for $k = 0, 1, \ldots, 2M$. The **composite Simpson rule for 2M subintervals** can be expressed in any of three equivalent ways:

$$S(f, h) = \frac{h}{3} \sum_{k=1}^{M} [f(x_{2k-2}) + 4f(x_{2k-1}) + f(x_{2k})] \tag{4a}$$

or

$$S(f, h) = \frac{h}{3} [f_0 + 4f_1 + 2f_2 + 4f_3 + \cdots + 2f_{2M-2} + 4f_{2M-1} + f_{2M}] \tag{4b}$$

or

$$S(f, h) = \frac{h}{3} [f(a) + f(b)] + \frac{2h}{3} \sum_{k=1}^{M-1} f(x_{2k}) + \frac{4h}{3} \sum_{k=1}^{M} f(x_{2k-1}). \tag{4c}$$

This is an approximation to the integral of $f(x)$ over $[a, b]$ and we write

$$\int_a^b f(x)\,dx \approx S(f, h). \tag{5}$$

Proof. Apply Simpson's rule over each subinterval $[x_{2k-2}, x_{2k}]$ (see Figure 7.7). Use the additive property of the integral for subintervals:

$$\int_a^b f(x)\,dx = \sum_{k=1}^{M} \int_{x_{2k-2}}^{x_{2k}} f(x)\,dx \approx \sum_{k=1}^{M} \frac{h}{3} [f(x_{2k-2}) + 4f(x_{2k-1}) + f(x_{2k})]. \tag{6}$$

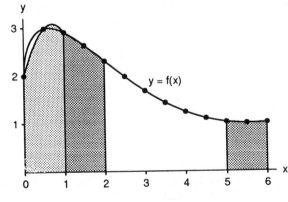

Figure 7.7 Approximating the area under the curve $y = 2 + \sin(2\sqrt{x})$ with the composite Simpson rule.

Since $h/3$ is a constant, the distributive law of addition can be applied to obtain (4a). Formula (4b) is the expanded version of (4a). Formula (4c) groups all the intermediate terms in (4b) that are multiplied by 2 and those that are multiplied by 4.

Approximating $f(x) = 2 + \sin(2\sqrt{x})$ with piecewise quadratic polynomials produces places where the approximation is close and places where it is not. To achieve accuracy the composite Simpson rule must be applied with several subintervals. In the next example we have chosen to numerically integrate this function over [1, 6] and leave investigation of the integral over [0, 1] as an exercise.

Example 7.6

Consider $f(x) = 2 + \sin(2\sqrt{x})$. Use the composite Simpson rule with 11 sample points to compute an approximation to the integral of $f(x)$ taken over [1, 6].

Solution. To generate 11 sample points we must use $M = 5$ and $h = (6 - 1)/(2 \times 5) = \frac{1}{2}$. Using formula (4c), the computation is

$$S\left(f, \tfrac{1}{2}\right) = \tfrac{1}{6}[f(1) + f(6)] + \tfrac{1}{3}[f(2) + f(3) + f(4) + f(5)]$$

$$+ \tfrac{2}{3}\left[f\left(\tfrac{3}{2}\right) + f\left(\tfrac{5}{2}\right) + f\left(\tfrac{7}{2}\right) + f\left(\tfrac{9}{2}\right) + f\left(\tfrac{11}{2}\right)\right]$$

$$= \tfrac{1}{6}[2.90929743 + 1.01735756]$$

$$+ \tfrac{1}{3}[2.30807174 + 1.68305284 + 1.24319750 + 1.02872220]$$

$$+ \tfrac{2}{3}[2.63815764 + 1.97931647 + 1.43530410 + 1.10831775 + 1.00024140]$$

$$= \tfrac{1}{6}(3.92665499) + \tfrac{1}{3}(6.26304429) + \tfrac{2}{3}(8.16133735)$$

$$= 0.65444250 + 2.08768143 + 5.44089157 = 8.18301550.$$

Error Analysis

The significance of the next two results is to understand that the error terms $E_T(f, h)$ and $E_S(f, h)$ for the composite trapezoidal rule and composite Simpson rule are of the order $O(h^2)$ and $O(h^4)$, respectively. This shows that the error for Simpson's rule converges to zero faster than the error for the trapezoidal rule as the step size h decreases to zero. In cases where the derivatives of $f(x)$ are known, the formulas

$$E_T(f, h) = \frac{-(b - a)\, f^{(2)}(c)h^2}{12} \quad \text{and} \quad E_S(f, h) = \frac{-(b - a)\, f^{(4)}(c)h^4}{180}$$

can be used to estimate the number of subintervals required to achieve a specified accuracy.

Corollary 7.2 (Trapezoidal Rule: Error Analysis). Suppose that [a, b] is subdivided into M subintervals $[x_k, x_{k+1}]$ of width $h = (b - a)/M$. The composite trapezoidal rule

$$T(f, h) = \frac{h}{2}[f(a) + f(b)] + h \sum_{k=1}^{M-1} f(x_k) \tag{7}$$

is an approximation to the integral

$$\int_a^b f(x) \, dx = T(f, h) + E_T(f, h). \tag{8}$$

Furthermore, if $f \in C^2[a, b]$, there exists a value c with $a < c < b$ so that the error term $E_T(f, h)$ has the form

$$E_T(f, h) = \frac{-(b - a) f^{(2)}(c) h^2}{12} = O(h^2). \tag{9}$$

Proof. We first determine the error term when the rule is applied over $[x_0, x_1]$. Integrating the Lagrange polynomial $P_1(x)$ and its remainder yields

$$\int_{x_0}^{x_1} f(x) \, dx = \int_{x_0}^{x_1} P_1(x) \, dx + \int_{x_0}^{x_1} \frac{(x - x_0)(x - x_1) f^{(2)}(c(x))}{2!} \, dx. \tag{10}$$

The term $(x - x_0)(x - x_1)$ does not change sign on $[x_0, x_1]$ and $f^{(2)}(c(x))$ is continuous. Hence the second mean value theorem for integrals implies that there exists a value c_1 so that

$$\int_{x_0}^{x_1} f(x) \, dx = \frac{h}{2} (f_0 + f_1) + f^{(2)}(c_1) \int_{x_0}^{x_1} \frac{(x - x_0)(x - x_1)}{2!} \, dx. \tag{11}$$

Use the change of variable $x = x_0 + ht$ in the integral on the right side of (11):

$$\int_{x_0}^{x_1} f(x) \, dx = \frac{h}{2} (f_0 + f_1) + \frac{f^{(2)}(c_1)}{2} \int_0^1 h(t - 0)h(t - 1)h \, dt$$

$$= \frac{h}{2} (f_0 + f_1) + \frac{f^{(2)}(c_1)h^3}{2} \int_0^1 (t^2 - t) \, dt \tag{12}$$

$$= \frac{h}{2} (f_0 + f_1) - \frac{f^{(2)}(c_1)h^3}{12}.$$

Now we are ready to add up the error terms for all of the intervals $[x_k, x_{k+1}]$:

$$\int_a^b f(x) \, dx = \sum_{k=1}^M \int_{x_{k-1}}^{x_k} f(x) \, dx$$

$$= \sum_{k=1}^M \frac{h}{2} [f(x_{k-1}) + f(x_k)] - \frac{h^3}{12} \sum_{k=1}^M f^{(2)}(c_k). \tag{13}$$

The first sum is the composite trapezoidal rule $T(f, h)$. In the second sum, one factor of h is replaced with its equivalent $h = (b - a)/M$, and the result is

$$\int_a^b f(x) \, dx = T(f, h) - \frac{(b - a)h^2}{12} \left[\frac{1}{M} \sum_{k=1}^M f^{(2)}(c_k) \right].$$

The term in brackets can be recognized as an average of values for the second derivative and hence is replaced by $f^{(2)}(c)$. Therefore, we have established that

$$\int_a^b f(x)\,dx = T(f,\,h) - \frac{(b-a)\,f^{(2)}(c)h^2}{12},$$

and the proof of Corollary 7.2 is complete.

Corollary 7.3 (Simpson's Rule: Error Analysis). Suppose that $[a,\,b]$ is subdivided into $2M$ subintervals $[x_k,\,x_{k+1}]$ of equal width $h = (b-a)/(2M)$. The composite Simpson rule

$$S(f,\,h) = \frac{h}{3}\,[f(a)+f(b)] + \frac{2h}{3}\sum_{k=1}^{M-1} f(x_{2k}) + \frac{4h}{3}\sum_{k=1}^{M} f(x_{2k-1}) \qquad (14)$$

is an approximation to the integral

$$\int_a^b f(x)\,dx = S(f,\,h) + E_S(f,\,h). \qquad (15)$$

Furthermore, if $f \in C^4[a,\,b]$, there exists a value c with $a < c < b$ so that the error term $E_S(f,\,h)$ has the form

$$E_S(f,\,h) = \frac{-(b-a)\,f^{(4)}(c)h^4}{180} = O(h^4). \qquad (16)$$

Example 7.7

Consider $f(x) = 2 + \sin(2\sqrt{x})$. Investigate the error when the composite trapezoidal rule is used over $[1,\,6]$ and the number of subintervals is 10, 20, 40, 80, and 160.

Solution. Table 7.2 shows the approximations $T(f,\,h)$. The antiderivative of $f(x)$ is

$$F(x) = 2x - \sqrt{x}\,\cos(2\sqrt{x}) + \frac{\sin(2\sqrt{x})}{2},$$

and the true value of the definite integral is

$$F(6) - F(1) = 10 + \cos(2) - \sqrt{6}\,\cos(2\sqrt{6}) - \frac{\sin(2)}{2} + \frac{\sin(2\sqrt{6})}{2} = 8.1834792077.$$

This value was used to compute the values $E_T(f,\,h) = 8.1834792077 - T(f,\,h)$ in Table 7.2. It is important to observe that when h is reduced by a factor of $\frac{1}{2}$ the successive errors $E_T(f,\,h)$ are diminished by approximately $\frac{1}{4}$. This confirms that the order is $O(h^2)$.

TABLE 7.2 The Composite Trapezoidal Rule for $f(x) = 2 + \sin(2\sqrt{x})$ over $[1,\,6]$

M	h	$T(f,\,h)$	$E_T(f,\,h) = O(h^2)$
10	0.5	8.19385457	-0.01037540
20	0.25	8.18604926	-0.00257006
40	0.125	8.18412019	-0.00064098
80	0.0625	8.18363936	-0.00016015
160	0.03125	8.18351924	-0.00004003

Example 7.8

Consider $f(x) = 2 + \sin(2\sqrt{x})$. Investigate the error when the composite Simpson rule is used over $[1, 6]$ and the number of subintervals is 10, 20, 40, 80, and 160.

Solution. Table 7.3 shows the approximations $S(f, h)$. The true value of the integral is 8.1834792077, which was used to compute the values $E_S(f, h) = 8.1834792077 - S(f, h)$ in Table 7.3. It is important to observe that when h is reduced by a factor of $\frac{1}{2}$ the successive errors $E_S(f, h)$ are diminished by approximately $\frac{1}{16}$. This confirms that the order is $O(h^4)$.

TABLE 7.3 The Composite Simpson Rule for
$f(x) = 2 + \sin(2\sqrt{x})$ over $[1, 6]$

M	h	$S(f, h)$	$E_S(f, h) = O(h^4)$
5	0.5	8.18301549	0.00046371
10	0.25	8.18344750	0.00003171
20	0.125	8.18347717	0.00000204
40	0.0625	8.18347908	0.00000013
80	0.03125	8.18347920	0.00000001

Example 7.9

Find the number M and the step size h so that the error $E_T(f, h)$ for the composite trapezoidal rule is less than 5×10^{-9} for the approximation $\int_2^7 dx/x \approx T(f, h)$.

Solution.   The integrand is $f(x) = 1/x$ and its derivatives are $f'(x) = -1/x^2$ and $f^{(2)}(x) = 2/x^3$. The maximum value of $|f^{(2)}(x)|$ taken over $[2, 7]$ occurs at the endpoint $x = 2$, and thus we have the bound $|f^{(2)}(c)| \leq |f^{(2)}(2)| = \frac{1}{4}$ for $2 \leq c \leq 7$. This is used with formula (9) to obtain

$$|E_T(f, h)| = \frac{|-(b - a) f^{(2)}(c) h^2|}{12} \leq \frac{(7 - 2) \frac{1}{4} h^2}{12} = \frac{5h^2}{48}. \qquad (17)$$

The step size h and number M satisfy the relation $h = 5/M$ and this is used in (17) to get the relation

$$|E_T(f, h)| \leq \frac{125}{48M^2} \leq 5 \times 10^{-9}. \qquad (18)$$

Now rewrite (18) so that is easier to solve for M:

$$\frac{25}{48} \times 10^9 \leq M^2. \qquad (19)$$

Solving (19) we find that $22821.77 \leq M$. Since M must be an integer, we choose $M = 22,822$ and the corresponding step size is $h = 5/22,822 = 0.000219086846$. When the composite trapezoidal rule is implemented with this many function evaluations there is a possibility that the rounded-off function evaluations will produce a significant amount of error. When the computation was performed, the result was

$$T\left(f, \frac{5}{22,822}\right) = 1.252762969,$$

which compares favorably with the true value $\ln(7) - \ln(2) = 1.252762968$. The error is smaller than predicted because the bound $\frac{1}{4}$ for $|f^{(2)}(c)|$ was used. Experimentation shows that it takes about 10,001 function evaluations to achieve the desired accuracy of 5×10^{-9}, and when the calculation is performed with $M = 10,000$, the result is

$$T\left(f, \frac{5}{10,000}\right) = 1.252762973.$$

The composite trapezoidal rule usually requires a large number of function evaluations to achieve an accurate answer. This is contrasted in the next example with Simpson's rule, which will require significantly fewer evaluations.

Example 7.10

Find the number M and the step size h so that the error $E_S(f, h)$ for the composite Simpson rule is less than 5×10^{-9} for the approximation $\int_2^7 dx/x \approx S(f, h)$.

Solution.   The integrand is $f(x) = 1/x$ and $f^{(4)}(x) = 24/x^5$. The maximum value of $|f^{(4)}(x)|$ taken over $[2, 7]$ occurs at the endpoint $x = 2$, and thus we have the bound $|f^{(4)}(c)| \leq |f^{(4)}(2)| = \frac{3}{4}$ for $2 \leq c \leq 7$. This is used with formula (16) to obtain

$$|E_S(f, h)| = \frac{|-(b - a) f^{(4)}(c)h^4|}{180} \leq \frac{(7 - 2)\frac{3}{4}h^4}{180} = \frac{h^4}{48}. \tag{20}$$

The step size h and number M satisfy the relation $h = 5/(2M)$ and this is used in (20) to get the relation

$$|E_S(f, h)| \leq \frac{625}{768M^4} \leq 5 \times 10^{-9}. \tag{21}$$

Now rewrite (21) so that is easier to solve for M:

$$\frac{125}{768} \times 10^9 \leq M^4. \tag{22}$$

Solving (22), we find that $112.95 \leq M$. Since M must be an integer, we chose $M = 113$ and the corresponding step size is $h = 5/226 = 0.02212389381$. When the composite Simpson rule was performed, the result was

$$S\left(f, \frac{5}{226}\right) = 1.252762969,$$

which agrees with $\ln(7) - \ln(2) = 1.252762968$. Experimentation shows that it takes about 129 function evaluations to achieve the desired accuracy of 5×10^{-9}, and when the calculation is performed with $M = 64$, the result is

$$S\left(f, \frac{5}{128}\right) = 1.252762973.$$

So we see that the composite Simpson rule using 229 evaluations of $f(x)$ and the composite trapezoidal rule using 22,823 evaluations of $f(x)$ achieve the same accuracy. In Example 7.10, Simpson's rule requires about $\frac{1}{100}$ the number of functions evaluations.

Algorithm 7.1   (Composite Trapezoidal Rule).   To approximate the integral

$$\int_A^B f(x)\,dx \approx \frac{h}{2}\,[f(A) + f(B)] + h \sum_{k=1}^{M-1} f(x_k)$$

by sampling $f(x)$ at the $M + 1$ equally spaced points $x_k = A + hk$ for $k = 0, 1,$ $2, \ldots, M$. Notice that $x_0 = A$ and $x_M = B$.

```
USES the function F(X)
INPUT   A, B, M
H := [B−A]/M
SUM := 0

FOR   K = 1   TO   M−1   DO
      X := A + H*K
      SUM := SUM + F(X)
      SUM := H*[F(A) + F(B) + 2*SUM]/2
```

PRINT 'The approximate value of the integral of f(X)' 'on the interval' A, B
'using' M 'subintervals' 'computed using the trapezoidal rule is' SUM

Algorithm 7.2   (Composite Simpson Rule).   To approximate the integral

$$\int_A^B f(x)\,dx \approx \frac{h}{3}\,[f(A) + f(B)] + \frac{2h}{3} \sum_{k=1}^{M-1} f(x_{2k}) + \frac{4h}{3} \sum_{k=1}^{M} f(x_{2k-1})$$

by sampling $f(x)$ at the $2M + 1$ equally spaced points $x_k = A + hk$ for $k = 0, 1,$ $2, \ldots, 2M$. Notice that $x_0 = A$ and $x_{2M} = B$.

```
USES the function F(X)
INPUT   A, B, M
H := [B−A]/[2*M]
SumEven := 0
FOR   K = 1   TO   M−1   DO              {Loop for even subscripts}
      X := A + H*2*K
      SumEven := SumEven + F(X)
      SumOdd := 0
```

```
FOR  K = 1  TO  M  DO                              {Loop for odd subscripts}
 |    X := A + H*(2*K-1)
 |___ SumOdd := SumOdd + F(X)
      SUM := H*[F(A) + F(B) + 2*SumEven + 4*SumOdd]/3
```

PRINT 'The approximate value of the integral of f(x)' 'on the interval' A, B
      'using' 2*M 'subintervals' 'computed using the Simpson's rule is' SUM

EXERCISES FOR COMPOSITE TRAPEZOIDAL AND SIMPSON'S RULE

1. (i) Approximate the integral using the composite trapezoidal rule with $M = 10$.
   (ii) Approximate the integral using the composite Simpson rule with $M = 5$.
   (iii) Use a computer and find the trapezoidal approximations for $M = 20, 40, 80$, and 160.
   (iv) Use a computer and find the Simpson approximations for $M = 10, 20, 40$, and 80.
   Graphs of the integrands are shown in Figure 7.8(a) to (f).

(a) $\int_{-1}^{1} \frac{1}{1 + x^2}\, dx$ (b) $\int_{0}^{1} (2 + \sin(2\sqrt{x}))\, dx$ (c) $\int_{0.25}^{4} \frac{1}{\sqrt{x}}\, dx$

(d) $\int_{0}^{4} x^2 e^{-x}\, dx$ (e) $\int_{0}^{2} 2x \cos(x)\, dx$ (f) $\int_{0}^{\pi} \sin(2x)\, e^{-x}\, dx$

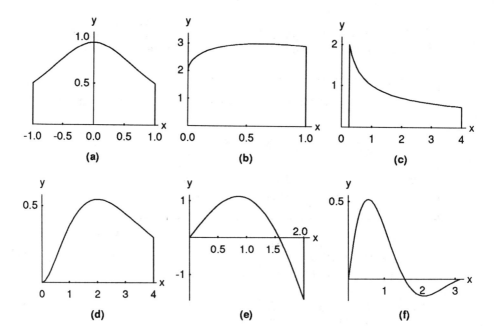

Figure 7.8 (a) $y = 1/(1 + x^2)$. (b) $y = 2 + \sin(2\sqrt{x})$. (c) $y = 1/\sqrt{x}$. (d) $y = x^2 e^{-x}$. (e) $y = 2x \cos(x)$. (f) $y = \sin(2x)e^{-x}$.

2. *Length of a curve.* The arc length of the curve $y = f(x)$ over the interval $a \leq x \leq b$ is

$$\text{length} = \int_a^b \sqrt{1 + [f'(x)]^2} \, dx.$$

Use the functions

(a) $f(x) = x^3$ for $0 \leq x \leq 1$ **(b)** $f(x) = \sin(x)$ for $0 \leq x \leq \dfrac{\pi}{4}$

(c) $f(x) = e^{-x}$ for $0 \leq x \leq 1$

  (i) Approximate the arc length using the composite trapezoidal rule with $M = 10$.

  (ii) Approximate the arc length using the composite Simpson rule with $M = 5$.

  (iii) Implement the composite trapezoidal rule on a computer and find an approximation for the arc length with eight accurate digits.

  (iv) Implement the composite Simpson rule on a computer and find an approximation for the arc length with eight accurate digits.

3. The composite trapezoidal rule can be adapted to integrate a function known only at a set of points. Adapt the trapezoidal rule to approximate the integral of a function over $[0, 14]$ that passes through the seven points: $(0, 19)$, $(1, 13)$, $(2, 10)$, $(4, 7)$, $(6, 5)$, $(10, 2)$, $(14, 1)$.

4. Sometimes the composite Simpson rule can be adapted to integrate a function known only at a set of points. Adapt Simpson's rule to approximate the integral of a function over $[0, 14]$ that passes through the seven points: $(0, 19)$, $(1, 13)$, $(2, 10)$, $(4, 7)$, $(6, 5)$, $(10, 2)$, $(14, 1)$.

5. *Surface area.* The solid of revolution obtained by rotating the region under the curve $y = f(x)$, $a \leq x \leq b$, about the x-axis has surface area given by

$$\text{area} = 2\pi \int_a^b f(x) \sqrt{1 + [f'(x)]^2} \, dx.$$

Use the functions

(a) $f(x) = x^3$ for $0 \leq x \leq 1$ **(b)** $f(x) = \sin(x)$ for $0 \leq x \leq \dfrac{\pi}{4}$

(c) $f(x) = e^{-x}$ for $0 \leq x \leq 1$

  (i) Approximate the area using the composite trapezoidal rule with $M = 10$.

  (ii) Approximate the area using the composite Simpson rule with $M = 5$.

  (iii) Implement the composite trapezoidal rule on a computer and find an approximation for the area with eight accurate digits.

  (iv) Implement the composite Simpson rule on a computer and find an approximation for the area with eight accurate digits.

6. (a) Verify that the trapezoidal rule ($M = 1$, $h = 1$) is exact for polynomials of degree ≤ 1 of the form $f(x) = c_1 x + c_0$, over $[0, 1]$.

  (b) Use the integrand $f(x) = c_2 x^2$ and verify that the error term for the trapezoidal rule ($M = 1$, $h = 1$) over the interval $[0, 1]$ is

$$E_T(f, h) = \frac{-(b - a) f^{(2)}(c) h^2}{12}.$$

7. (a) Verify that Simpson's rule ($M = 1$, $h = 1$) is exact for polynomials of degree ≤ 3 of the form $f(x) = c_3 x^3 + c_2 x^2 + c_1 x + c_0$, over $[0, 2]$.

(b) Use the integrand $f(x) = c_4 x^4$ and verify that the error term for Simpson's rule ($M = 1$, $h = 1$) over the interval $[0, 2]$ is

$$E_S(f, h) = \frac{-(b - a) f^{(4)}(c) h^4}{180}.$$

8. Derive the trapezoidal rule ($M = 1$, $h = 1$) by using the method of undetermined coefficients:

   (a) Find the constants w_0 and w_1 so that $\int_0^1 g(t) \, dt = w_0 g(0) + w_1 g(1)$ is exact for the two functions $g(t) = 1$ and $g(t) = t$.

   (b) Use the relation $f(x_0 + ht) = g(t)$ and the change of variable $x = x_0 + ht$ and $dx = h \, dt$ to translate the trapezoidal rule over $[0, 1]$ to the interval $[x_0, x_1]$.

   Hint for part (a). You will get a linear system involving the two unknowns w_0 and w_1.

9. Derive Simpson's rule ($M = 1$, $h = 1$) by using the method of undetermined coefficients:

   (a) Find the constants w_0, w_1 and w_2 so that $\int_0^2 g(t) \, dt = w_0 g(0) + w_1 g(1) + w_2 g(2)$ is exact for the three functions $g(t) = 1$, $g(t) = t$, and $g(t) = t^2$.

   (b) Use the relation $f(x_0 + ht) = g(t)$ and the change of variable $x = x_0 + ht$ and $dx = h \, dt$ to translate Simpson's rule over $[0, 2]$ to the interval $[x_0, x_2]$.

   Hint for part (a). You will get a linear system involving the unknowns w_0, w_1, and w_2.

10. Determine the number M and the interval width h so that the composite trapezoidal rule for M subintervals can be used to compute the given integral with an accuracy of 5×10^{-9}.

   (a) $\displaystyle\int_{-\pi/6}^{\pi/6} \cos(x) \, dx$ **(b)** $\displaystyle\int_2^3 \frac{1}{5 - x} \, dx$ **(c)** $\displaystyle\int_0^2 x e^{-x} \, dx$

   Hint for part (c). $f^{(2)}(x) = (x - 2)e^{-x}$.

11. Determine the number M and the interval width h so that the composite Simpson rule for $2M$ subintervals can be used to compute the given integral with an accuracy of 5×10^{-9}.

   (a) $\displaystyle\int_{-\pi/6}^{\pi/6} \cos(x) \, dx$ **(b)** $\displaystyle\int_2^3 \frac{1}{5 - x} \, dx$ **(c)** $\displaystyle\int_0^2 x e^{-x} \, dx$

   Hint for part (c). $f^{(4)}(x) = (x - 4)e^{-x}$.

12. Consider the definite integral $\displaystyle\int_{-0.1}^{0.1} \cos(x) \, dx = 2 \sin(0.1) = 0.1996668333$. The following table gives approximations using the composite trapezoidal rule. Calculate $E_T(f, h) = 0.1996668 - T(f, h)$ and confirm that the order is $O(h^2)$.

M	h	$T(f, h)$	$E_T(f, h) = O(h^2)$
1	0.2	0.1990008	
2	0.1	0.1995004	
4	0.05	0.1996252	
8	0.025	0.1996564	
16	0.0125	0.1996642	

13. Consider the definite integral $\int_{-0.75}^{0.75} \cos(x)\,dx = 2\sin(0.75) = 1.363277520$. The following table gives approximations using the composite Simpson rule. Calculate $E_S(f,\ h) = 1.3632775 - S(f,\ h)$ and confirm that the order is $O(h^4)$.

M	h	$S(f,\ h)$	$E_S(f,\ h) = O(h^4)$
1	0.75	1.3658444	
2	0.375	1.3634298	
4	0.1875	1.3632869	
8	0.09375	1.3632781	

14. *Midpoint rule*. The midpoint rule on $[x_0, x_1]$

$$\int_{x_0}^{x_1} f(x)\,dx = hf\left(x_0 + \frac{h}{2}\right) + \frac{h^3}{24} f^{(2)}(c_1), \qquad \text{where} \quad h = \frac{x_1 - x_0}{2}.$$

(a) Expand $F(x)$, the antiderivative of $f(x)$, in a Taylor series about $x_0 + h/2$ and establish the midpoint rule on $[x_0, x_1]$.

(b) Use part (a) and show that the composite midpoint rule for approximating the integral of $f(x)$ over $[a, b]$ is

$$M(f, h) = h \sum_{k=1}^{N} f\left(a + \left(k - \tfrac{1}{2}\right)h\right), \qquad \text{where} \quad h = \frac{b - a}{N}.$$

This is an approximation to the integral of $f(x)$ over $[a, b]$ and we write

$$\int_a^b f(x)\,dx \approx M(f, h).$$

(c) Show that the error term $E_M(f, h)$ for part (b) is

$$E_M(f, h) = \frac{h^3}{24} \sum_{k=1}^{N} f^{(2)}(c_k) = \frac{(b - a) f^{(2)}(c) h^2}{24} = O(h^2).$$

15. Use the midpoint rule with $M = 10$ to approximate the integrals in Exercise 1.

16. Modify Algorithm 7.1 so that it uses the composite midpoint rule to approximate the integral of $f(x)$ over $[a, b]$.

17. Prove Corollary 7.3.

18. Write a report on the corrected trapezoidal rule and corrected Simpson rule.

7.3 RECURSIVE RULES AND ROMBERG INTEGRATION

In this section we show how to compute Simpson approximations with a special linear combination of trapezoidal rules. The approximation will have greater accuracy if one uses a larger number of subintervals. How many should we choose? The sequential process helps answer this question by trying two subintervals, four subintervals, and so on, until the desired accuracy is obtained. First, a sequence $\{T(J)\}$ of trapezoidal rule approximations must be generated. As the number of subintervals is doubled, the number of function values is roughly doubled, because the function must be evaluated at all the previous points and at the midpoints of the previous subintervals (see Figure 7.9). Theorem 7.4 explains how to eliminate redundant function evaluations and additions.

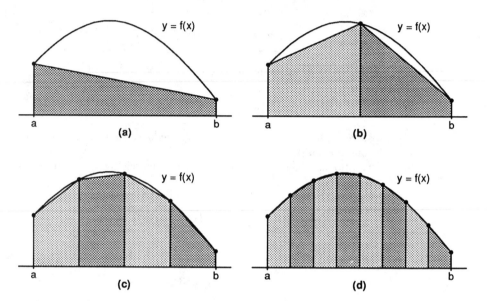

Figure 7.9   (a) $T(0)$ is the area under $2^0 = 1$ trapezoid. (b) $T(1)$ is the area under $2^1 = 2$ trapezoids. (c) $T(2)$ is the area under $2^2 = 4$ trapezoids. (d) $T(3)$ is the area under $2^3 = 8$ trapezoids.

Theorem 7.4   (Successive Trapezoidal Rules).   Suppose that $J \geq 1$ and the points $\{x_i = a + ih\}$ subdivide $[a, b]$ into $2^J = 2M$ subintervals of equal width $h = (b - a)/2^J$. The trapezoidal rules $T(f, h)$ and $T(f, 2h)$ obey the relationship

$$T(f, h) = \frac{T(f, 2h)}{2} + h \sum_{k=1}^{M} f(x_{2k-1}). \qquad (1)$$

Definition 7.3   (Sequence of Trapezoidal Rules).   Define $T(0) = (h/2)[f(a) + f(b)]$, which is the trapezoidal rule with step size $h = b - a$. Then for each $J \geq 1$ define $T(J) = T(f, h)$, where $T(f, h)$ is the trapezoidal rule with step size $h = (b - a)/2^J$.

Corollary 7.4   (Recursive Trapezoidal Rule).   Start with $T(0) = (h/2)[f(a) + f(b)]$. Then a sequence of trapezoidal rules $\{T(J)\}$ is generated by the recursive formula

$$T(J) = \frac{T(J-1)}{2} + h \sum_{k=1}^{M} f(x_{2k-1}) \qquad \text{for } J = 1, 2, \ldots, \tag{2}$$

where $h = (b-a)/2^J$ and $\{x_i = a + ih\}$.

Proof. For the even nodes $x_0 < x_2 < \cdots < x_{2M-2} < x_{2M}$ we use the trapezoidal rule with step size $2h$:

$$T(J-1) = \frac{2h}{2}(f_0 + 2f_2 + 2f_4 + \cdots + 2f_{2M-4} + 2f_{2M-2} + f_{2M}). \tag{3}$$

For all of the nodes $x_0 < x_1 < x_2 < \cdots x_{2M-2} < x_{2M-1} < x_{2M}$ we use the trapezoidal rule with step size h:

$$T(J) = \frac{h}{2}(f_0 + 2f_1 + 2f_2 + \cdots + 2f_{2M-2} + 2f_{2M-1} + f_{2M}). \tag{4}$$

Collecting the even and odd subscripts in (4) yields

$$T(J) = \frac{h}{2}(f_0 + 2f_2 + \cdots + 2f_{2M-2} + f_{2M}) + h \sum_{k=1}^{M} f_{2k-1}. \tag{5}$$

Substituting (3) in (5) results in $T(J) = T(J-1)/2 + h \sum_{k=1}^{M} f_{2k-1}$ and the proof of the theorem is complete.

Example 7.11

Use the sequential trapezoidal rule to compute the approximations $T(0)$, $T(1)$, $T(2)$, and $T(3)$ for the integral $\int_1^5 dx/x = \ln(5) = 1.609437912$.

Solution.   Table 7.4 shows the nine values required to compute $T(3)$ and the midpoints required to compute $T(1)$, $T(2)$, and $T(3)$. Details for obtaining the results are:

When $h = 4$: $T(0) = \dfrac{4}{2}(1.000000 + 0.200000) = 2.400000$.

When $h = 2$: $T(1) = \dfrac{T(0)}{2} + 2(0.333333) = 1.200000 + 0.666666 = 1.866666$.

When $h = 1$: $T(2) = \dfrac{T(1)}{2} + 1(0.500000 + 0.250000)$

$$= 0.933333 + 0.750000 = 1.683333.$$

When $h = \dfrac{1}{2}$: $T(3) = \dfrac{T(2)}{2} + \dfrac{1}{2}(0.666667 + 0.400000 + 0.285714 + 0.222222)$

$$= 0.841667 + 0.787302 = 1.628968.$$

TABLE 7.4 The Nine Points Used to Compute $T(3)$ and the Midpoints Required to Compute $T(1)$, $T(2)$, and $T(3)$

x	$f(x) = \dfrac{1}{x}$	Endpoints for computing $T(0)$	Midpoints for computing $T(1)$	Midpoints for computing $T(2)$	Midpoints for computing $T(3)$
1.0	1.000000	1.000000			
1.5	0.666667				0.666667
2.0	0.500000			0.500000	
2.5	0.400000				0.400000
3.0	0.333333		0.333333		
3.5	0.285714				0.285714
4.0	0.250000			0.250000	
4.5	0.222222				0.222222
5.0	0.200000	0.200000			

Our next result shows an important relationship between the trapezoidal rule and Simpson's rule. When the trapezoidal rule is computed using step sizes $2h$ and h the result is $T(f, 2h)$ and $T(f, h)$. These values are combined to obtain Simpson's rule:

$$S(f, h) = \frac{4T(f, h) - T(f, 2h)}{3}. \tag{6}$$

Theorem 7.5 (Recursive Simpson Rules).   Suppose that $\{T(J)\}$ is the sequence of trapezoidal rules generated by Corollary 7.4. If $J \geq 1$ and $S(J)$ is Simpson's rule for 2^J subintervals of $[a, b]$, then $S(J)$ and the trapezoidal rules $T(J - 1)$ and $T(J)$ obey the relationship

$$S(J) = \frac{4T(J) - T(J - 1)}{3} \qquad \text{for } J = 1, 2, \ldots . \tag{7}$$

Proof. The trapezoidal rule $T(J)$ with step size h yields the approximation

$$\int_a^b f(x)\, dx \approx \frac{h}{2}\left(f_0 + 2f_1 + 2f_2 + \cdots + 2f_{2M-2} + 2f_{2M-1} + f_{2M}\right) = T(J). \tag{8}$$

The trapezoidal rule $T(J - 1)$ with step size $2h$ produces

$$\int_a^b f(x)\, dx \approx h(f_0 + 2f_2 + \cdots + 2f_{2M-2} + f_{2M}) = T(J - 1). \tag{9}$$

Multiplying relation (8) by 4 yields

$$4\int_a^b f(x)\, dx \approx h(2f_0 + 4f_1 + 4f_2 + \cdots + 4f_{2M-2} + 4f_{2M-1} + 2f_{2M}) = 4T(J) \tag{10}$$

Now subtract (9) from (10) and the result is

$$3 \int_a^b f(x)\, dx \approx h(f_0 + 4f_1 + 2f_2 + \cdots + 2f_{2M-2} + 4f_{2M-1} + f_{2M}) \tag{11}$$

$$= 4T(J) - T(J - 1).$$

This can be rearranged to obtain

$$\int_a^b f(x)\, dx \approx \frac{h}{3}(f_0 + 4f_1 + 2f_2 + \cdots + 2f_{2M-2} + 4f_{2M-1} + f_{2M}) \tag{12}$$

$$= \frac{4T(J) - T(J - 1)}{3}.$$

The middle term in (12) is Simpson's rule $S(J) \equiv S(f, h)$ and the theorem is proven.

Example 7.12

Use the sequential Simpson rule to compute the approximations $S(1)$, $S(2)$, and $S(3)$ for the integral of Example 7.11.

Solution.   Using the results of Example 7.11 and formula (7) with $J = 1, 2, 3$, we compute

$$S(1) = \frac{4T(1) - T(0)}{3} = \frac{4 \times 1.866666 - 2.400000}{3} = 1.688888,$$

$$S(2) = \frac{4T(2) - T(1)}{3} = \frac{4 \times 1.683333 - 1.866666}{3} = 1.622222,$$

$$S(3) = \frac{4T(3) - T(2)}{3} = \frac{4 \times 1.628968 - 1.683333}{3} = 1.610846.$$

In Section 7.1 the formula for Boole's rule was given in Theorem 7.1. It was obtained by integrating the Lagrange polynomial of degree 4 based on the nodes x_0, x_1, x_2, x_3, and x_4. An alternative method for establishing Boole's rule is mentioned in Exercise 6. When it is applied M times over $4M$ equally spaced subintervals of $[a, b]$ of step size $h = (b - a)/(4M)$, we call it the **composite Boole rule**:

$$B(f, h) = \frac{2h}{45} \sum_{k=1}^{M} [7f_{4k-4} + 32f_{4k-3} + 12f_{4k-2} + 32f_{4k-1} + 7f_{4k}]. \tag{13}$$

The next result gives the relationship between the sequential Boole and Simpson rules.

Theorem 7.6   (Recursive Boole Rule).   Suppose that $\{S(J)\}$ is the sequence of Simpson's rules generated by Theorem 7.5. If $J \geq 2$ and $B(J)$ is Boole's rule for 2^J subintervals of $[a, b]$, then $B(J)$ and Simpson's rules $S(J - 1)$ and $S(J)$ obey the relationship

$$B(J) = \frac{16S(J) - S(J - 1)}{15} \qquad \text{for } J = 2, 3, \ldots. \tag{14}$$

Proof.   The proof is left for the reader (see Exercise 7).

Example 7.13

Use the sequential Boole rule to compute the approximations $B(2)$ and $B(3)$ for the integral of Example 7.11.

Solution. Using the results of Example 7.12 and formula (14) with $J = 2, 3$ we compute

$$B(2) = \frac{16S(2) - S(1)}{15} = \frac{16 \times 1.622222 - 1.688888}{15} = 1.617778,$$

$$B(3) = \frac{16S(3) - S(2)}{15} = \frac{16 \times 1.610846 - 1.622222}{15} = 1.610088.$$

The reader may wonder what we are leading up to. We will now show that formulas (7) and (14) are special cases of the process of Romberg integration. Let us announce that the next level of approximation for the integral of Example 7.11 is

$$\frac{64B(3) - B(2)}{63} = \frac{64 \times 1.610088 - 1.617778}{63} = 1.609490,$$

and this answer gives an accuracy of five decimal places.

Romberg Integration

In Section 7.2 we saw that the error terms $E_T(f, h)$ and $E_S(f, h)$ for the composite trapezoidal rule and composite Simpson rule are of the order $O(h^2)$ and $O(h^4)$, respectively. It is not too difficult to show that the error term $E_B(f, h)$ for the composite Boole rule is of the order $O(h^6)$. Thus we have the pattern

$$\int_a^b f(x)\, dx = T(f, h) + O(h^2), \tag{15}$$

$$\int_a^b f(x)\, dx = S(f, h) + O(h^4), \tag{16}$$

$$\int_a^b f(x)\, dx = B(f, h) + O(h^6). \tag{17}$$

The pattern for the remainders in (15) to (17) is extended in the following sense. Suppose that an approximation rule is used with step sizes h and $2h$; then an algebraic manipulation of the two answers is used to produce an improved answer. Each successive level of improvement increases the order of the error term from $O(h^{2N})$ to $O(h^{2N+2})$. This process, called **Romberg integration**, has its strengths and weaknesses.

The Newton–Cotes rules are seldom used past Boole's rule. This is because the nine-point Newton–Cotes quadrature rule involves negative weights, and all the rules past the 10-point rule involve negative weights. This could introduce loss of significance error due to round-off. The Romberg method has the advantage that all of the weights are positive and the equally spaced abscissas are easy to compute.

A computational weakness of Romberg integration is that twice as many function evaluations are needed to decrease the error from $O(h^{2N})$ to $O(h^{2N+2})$. The use of the sequential rules will help keep the number of computations down.

The development of Romberg integration relies on the theoretical assumption that if $f \in C^N[a, b]$ for all N, then the error term for the trapezoidal rule can be represented in a series involving only even powers of h; that is,

$$\int_a^b f(x) \, dx = T(f, h) + E_T(f, h),$$ (18)

where

$$E_T(f, h) = a_1 h^2 + a_2 h^4 + a_3 h^6 + \cdots .$$ (19)

A derivation of formula (19) can be found in Reference [153].

Since only even powers of h occur in (19), the Richardson improvement process is used successively first to eliminate a_1, next to eliminate a_2, then to eliminate a_3, and so on. This process generates quadrature formulas whose error terms have the even orders $O(h^4)$, $O(h^6)$, $O(h^8)$, and so on. We shall show that the first improvement is Simpson's rule for $2M$ intervals. Start with $T(f, 2h)$ and $T(f, h)$ and the equations

$$\int_a^b f(x) \, dx = T(f, 2h) + a_1 4h^2 + a_2 16h^4 + a_3 64h^6 + \cdots$$ (20)

and

$$\int_a^b f(x) \, dx = T(f, h) + a_1 h^2 + a_2 h^4 + a_3 h^6 + \cdots .$$ (21)

Multiply equation (21) by 4 and obtain

$$4 \int_a^b f(x) \, dx = 4T(f, h) + a_1 4h^2 + a_2 4h^4 + a_3 4h^6 + \cdots .$$ (22)

Eliminate a_1 by subtracting (20) from (22). The result is

$$3 \int_a^b f(x) \, dx = 4T(f, h) - T(f, 2h) - a_2 12h^4 - a_3 60h^6 - \cdots .$$ (23)

Now divide equation (23) by 3 and rename the coefficients in the series

$$\int_a^b f(x) \, dx = \frac{4T(f, h) - T(f, 2h)}{3} + b_1 h^4 + b_2 h^6 + \cdots .$$ (24)

As noted in (6), the first quantity on the right side of (24) is Simpson's rule $S(f, h)$. This shows that $E_S(f, h)$ involves only even powers of h:

$$\int_a^b f(x) \, dx = S(f, h) + b_1 h^4 + b_2 h^6 + b_3 h^8 + \cdots .$$ (25)

To show that the second improvement is Boole's rule, start with (25) and write down the formula involving $S(f, 2h)$:

$$\int_a^b f(x)\, dx = S(f, 2h) + b_1 16h^4 + b_2 64h^6 + b_3 256h^6 + \cdots . \qquad (26)$$

When b_1 is eliminated from (25) and (26), the result involves Boole's rule:

$$\int_a^b f(x)\, dx = \frac{16S(f, h) - S(f, 2h)}{15} - \frac{b_2 48h^4}{15} - \frac{b_3 240h^6}{15} - \cdots \qquad (27)$$

$$= B(f, h) - \frac{b_2 48h^4}{15} - \frac{b_3 240h^6}{15} - \cdots .$$

The general pattern for Romberg integration relies on Lemma 7.1.

Lemma 7.1   (Richardson's Improvement, for Romberg Integration).   Given two approximations $R(2h, K - 1)$ and $R(h, K - 1)$ for the quantity Q that satisfy

$$Q = R(h, K - 1) + c_1 h^{2K} + c_2 h^{2K+2} + \cdots \qquad (28)$$

and

$$Q = R(2h, K - 1) + c_1 4^K h^{2K} + c_2 4^{K+1} h^{2K+2} + \cdots , \qquad (29)$$

an improved approximation has the form

$$Q = \frac{4^K R(h, K - 1) - R(2h, K - 1)}{4^K - 1} + O(h^{2K+2}). \qquad (30)$$

The proof is straightforward and is left for the reader.

Definition 7.4.   Define the sequence $\{R(J, K): J \ge K\}_{J=0}^{\infty}$ of quadrature formulas for $f(x)$ over $[a, b]$ as follows:

$$R(J, 0) = T(J) \qquad \text{for } J \ge 0 \quad \text{is the sequential trapezoidal rule.}$$

$$R(J, 1) = S(J) \qquad \text{for } J \ge 1 \quad \text{is the sequential Simpson rules.} \qquad (31)$$

$$R(J, 2) = B(J) \qquad \text{for } J \ge 2 \quad \text{is the sequential Boole rule.}$$

The starting rules, $\{R(J, 0)\}$, are used to generate the first improvement, $\{R(J, 1)\}$, which in turn is used to generate the second improvement, $\{R(J, 2)\}$. We have already seen the patterns

$$R(J, 1) = \frac{4^1 R(J, 0) - R(J - 1, 0)}{4^1 - 1} \qquad \text{for } J \ge 1$$

$$\qquad (32)$$

$$R(J, 2) = \frac{4^2 R(J, 1) - R(J - 1, 1)}{4^2 - 1} \qquad \text{for } J \ge 2,$$

which are the rules in (24) and (27) stated using the notation in (31). The general rule for constructing improvements is

$$R(J, K) = \frac{4^K R(J, K - 1) - R(J - 1, K - 1)}{4^K - 1} \qquad \text{for } J \geq K. \qquad (33)$$

For computational purposes, the values $R(J, K)$ are arranged in the Romberg integration tableau given in Table 7.5.

TABLE 7.5   The Romberg Integration Tableau

J	$R(J, 0)$ Trapezoidal rule	$R(J, 1)$ Simpson's rule	$R(J, 2)$ Boole's rule	$R(J, 3)$ Third improvement	$R(J, 4)$ Fourth improvement
0	$R(0, 0)$				
1	$R(1, 0)$	$R(1, 1)$			
2	$R(2, 0)$	$R(2, 1)$	$R(2, 2)$		
3	$R(3, 0)$	$R(3, 1)$	$R(3, 2)$	$R(3, 3)$	
4	$R(4, 0)$	$R(4, 1)$	$R(4, 2)$	$R(4, 3)$	$R(4, 4)$

Example 7.14

Use Romberg integration to find approximations for the definite integral

$$\int_0^{\pi/2} (x^2 + x + 1) \cos(x) \, dx = -2 + \frac{\pi}{2} + \frac{\pi^2}{4} = 2.038197427067. \ldots$$

Solution.   The computations are given in Table 7.6. In each column the numbers are converging to the value 2.038197427067. . . .The values in Simpson's rule column converge faster that the values in the trapezoidal rule column. For this example, convergence in columns to the right are faster than the adjacent column to the left.

TABLE 7.6   The Romberg Integration Tableau for Example 7.14

J	$R(J, 0)$ Trapezoidal rule	$R(J, 1)$ Simpson's rule	$R(J, 2)$ Boole's rule	$R(J, 3)$ Third improvement
0	0.785398163397			
1	1.726812656758	2.040617487878		
2	1.960534166564	2.038441336499	2.038296259740	
3	2.018793948078	2.038213875249	2.038198711166	2.038197162776
4	2.033347341805	2.038198473047	2.038197446234	2.038197426156
5	2.036984954990	2.038197492719	2.038197427363	2.038197427064

Convergence of the the Romberg values in Table 7.6 is easier to see if we look at the error terms; $E(J, K) = -2 + \pi/2 + \pi^2/4 - R(J, K)$. Suppose that the interval width is $H = b - a$, and that the higher derivatives of $f(x)$ are of the same magnitude. The error in column K of the Romberg table diminishes by about a factor of $1/2^{2K+2} = 1/4^{K+1}$ as one progresses down its rows. The errors $E(J, 0)$ diminish by a factor of $\frac{1}{4}$, the errors $E(J, 1)$

diminish by a factor of $\frac{1}{16}$, and so on. This can be observed by inspecting the entries $\{E(J, K)\}$ in Table 7.7.

TABLE 7.7   The Romberg Error Tableau for Example 7.14

J	H	$E(J, 0) = O(H^2)$	$E(J, 1) = O(H^4)$	$E(J, 2) = O(H^6)$	$E(J, 3) = O(H^8)$
0	$b - a$	-1.252799263670			
1	$\dfrac{b - a}{2}$	-0.311384770309	0.002420060811		
2	$\dfrac{b - a}{4}$	-0.077663260503	0.000243909432	0.000098832673	
3	$\dfrac{b - a}{8}$	-0.019403478989	0.000016448182	0.000001284099	-0.000000264291
4	$\dfrac{b - a}{16}$	-0.004850085262	0.000001045980	0.000000019167	-0.000000000912
5	$\dfrac{b - a}{32}$	-0.001212472077	0.000000065651	0.000000000296	-0.000000000003

Theorem 7.7 (Precision of Romberg Integration).   Assume that $f \in C^{2K+2}[a, b]$. Then the truncation error term for the Romberg approximation is given in the formula

$$\int_a^b f(x)\, dx = R(J, K) + b_K h^{2K+2} f^{(2K+2)}(c_{J,K}) = R(J, K) + O(h^{2K+2}), \quad (34)$$

where $h = (b - a)/2^J$, b_K is a constant that depends on K, and $c_{J,K} \in [a, b]$, see Reference [153], page 126.

Example 7.15

Apply Theorem 7.7 and show that

$$\int_0^2 10x^9\, dx = 1024 \equiv R(4, 4).$$

Solution.   The integrand is $f(x) = 10x^9$ and $f^{(10)}(x) \equiv 0$. Thus the value $K = 4$ will make the error term identically zero. A numerical computation will produce $R(4, 4) = 1024$.

Algorithm 7.3   (Recursive Trapezoidal Rule).   To approximate the integral

$$\int_A^B f(x)\, dx \approx \frac{h}{2} \sum_{k=1}^{2^J} [f(x_{k-1}) + f(x_k)]$$

by using the trapezoidal rule and successively and increasing number of subintervals of $[A, B]$. The Jth iteration samples $f(x)$ at $2^J + 1$ equally spaced points.

```
       INPUT   A, B                                    {Endpoints of the interval}
       INPUT   N                         {Number of times subintervals are bisected}
       VARiable declaration REAL   T[0 . . N]
       M := 1
       H := B − A
       T(0) := H*[F(A) + F(B)]/2                             {Compute one trapezoid}

PRINT 'The approximate value of the integral using' 1
      'subinterval and the trapezoidal rule is' T(0)

FOR   J = 1  TO  N  DO                                   {Do recursive calculations}
       M := 2*M                                    {Use twice as many subintervals}
       H := H/2                                    {Reduce the step size one-half}
       SUM := 0
       FOR  K = 1  TO  M/2  DO                          {Loop for odd subscripts}
              SUM := SUM + F(A + H*(2*K−1))
       T(J) := T(J−1)/2 + H*SUM
       PRINT 'The approximate value of the integral using'  M
             'subintervals and the trapezoidal rule is'   T(J)
```

Algorithm 7.4   (Romberg Integration).   To approximate the integral

$$\int_A^B f(x)\ dx \approx R(J, J)$$

by generating a table of approximations $R(I, K)$ for $K \le I$, and using $R(J, J)$ as the final answer. The approximations $R(J, K)$ are stored in a special lower-triangular matrix. The elements $R(J, 0)$ of column 0 are computed using the sequential trapezoidal rule based on 2^J subintervals of $[A, B]$; then $R(J, K)$ is computed using Romberg's rule.

The elements of row J are

$$R(J, K) = R(J, K - 1) + \frac{R(J, K - 1) - R(J - 1, K - 1)}{4^K - 1} \qquad \text{for } 1 \le K \le J.$$

The algorithm is terminated in the Jth row when $|R(J, J) - R(J - 1, J - 1)| < \text{Tol}.$

```
INPUT A, B                                              {Endpoints of the interval}
INPUT N                                    {Maximum number of rows, e.g., 14}
INPUT Tol                                 {Termination criterion, e.g., 10⁻⁷}

VARiable declaration REAL   R[0 . . N,0 . . N]          {Array passed to the subroutine}
```

```
SUBROUTINE  TrRule (A,H,J,M,R)                    {Sequential trapezoidal rule}
    H := H/2                                 {Reduce step size for Jth refinement}
    SUM := 0
    FOR  P = 1  TO  M  DO                             {Loop for odd subscripts}
        SUM := SUM + F(A + H*(2*P−1))
    R(J,0) := R(J−1,0)/2 + H*SUM
    M := 2*M                                 {Update number of subintervals}
    END                                          {End of the procedure TrRule}
```

{The main program starts here.}

```
    M := 1                                  {Initialize the number of subintervals}
    H := B − A                                       {Initialize the step size}
    Close := 1                                       {Initialize the variable}
    J := 0                                           {Initialize the counter}

    R(0,0) := H*[F(A) + F(B)]/2                          {Compute one trapezoid}

WHILE  [Close>Tol and J<N] or [J<4]  DO                         {Do sequential
    J := J + 1                                                  calculations}
    CALL Procedure  TrRule (A,H,J,M,R)                 {Compute new R(J,0)}

    FOR  K = 1  TO  J  DO                         {Richardson's improvements}
        R(J,K) := R(J,K−1) + [R(J,K−1)−R(J−1,K−1)]/[4^K − 1]

    PRINT 'The entries in row' J 'for Romberg integration are'
            R(J,0), R(J,1), R(J,2), . . . , R(J,J)
    Close := |R(J−1,J−1) − R(J,J)|

    PRINT 'The best approximation for the value of the integral
            using Romberg integration is'; R(J,J)
```

EXERCISES FOR RECURSIVE RULES AND ROMBERG INTEGRATION

1.   (i) Compute $T(J) = R(J, 0)$ for $J = 0, 1, 2, 3$ by using the sequential trapezoidal rule.
(ii) Use values $\{R(J, 0)\}$ from the sequential trapezoidal rule in part (i) and compute the values for the sequential Simpson rule $\{R(J, 1)\}$ sequential Boole rule $\{R(J, 2)\}$, and the third improvement $\{R(J, 3)\}$.
(iii) Use a computer to find the approximations in part (i) and extend the table to include nine rows. Use the following functions:

(a) $\int_0^3 \dfrac{\sin(2x)}{1 + x^5} \, dx = 0.6711575864b \ldots$

J	$R(J, 0)$ Trapezoidal rule	$R(J, 1)$ Simpson's rule	$R(J, 2)$ Boole's rule	$R(J, 3)$ Third improvement
0	−0.00171772			
1	0.02377300	————		
2	0.60402717	————	————	
3	0.64844713	————	————	————
4	0.66591329	————	————	————

(b) $\int_0^3 \sin(4x)e^{-2x}\, dx = 0.1997146621\ldots$

J	$R(J, 0)$ Trapezoidal rule	$R(J, 1)$ Simpson's rule	$R(J, 2)$ Boole's rule	$R(J, 3)$ Third improvement
0	−0.00199505			
1	−0.02186444	————		
2	0.01611754	————	————	
3	0.15265719	————	————	————
4	0.18800280	————	————	————

(c) $\int_{0.04}^1 \dfrac{1}{\sqrt{x}}\, dx = 1.6$

J	$R(J, 0)$ Trapezoidal rule	$R(J, 1)$ Simpson's rule	$R(J, 2)$ Boole's rule	$R(J, 3)$ Third improvement
0	2.88			
1	2.10564024	————		
2	1.78167637	————	————	
3	1.65849527	————	————	————
4	1.61691082	————	————	————

(d) $\int_0^2 \dfrac{1}{x^2 + \frac{1}{10}}\, dx = 4.4713993943\ldots$

J	$R(J, 0)$ Trapezoidal rule	$R(J, 1)$ Simpson's rule	$R(J, 2)$ Boole's rule	$R(J, 3)$ Third improvement
0	10.24390244			
1	6.03104213	————		
2	4.65685845	————	————	
3	4.47367658	————	————	————
4	4.47109102	————	————	————

(e) $\displaystyle\int_{1/(2\pi)}^{2} \sin\!\left(\frac{1}{x}\right)\,dx = 1.1140744942\ldots$

J	$R(J, 0)$ Trapezoidal rule	$R(J, 1)$ Simpson's rule	$R(J, 2)$ Boole's rule	$R(J, 3)$ Third improvement
0	0.44127407			
1	0.95641862	_____		
2	1.21628836	_____	_____	
3	1.22765253	_____	_____	_____
4	1.15845291	_____	_____	_____

(f) $\displaystyle\int_{0}^{2} \sqrt{4x - x^2}\,dx = \pi = 3.1415926535\ldots$

J	$R(J, 0)$ Trapezoidal rule	$R(J, 1)$ Simpson's rule	$R(J, 2)$ Boole's rule	$R(J, 3)$ Third improvement
0	2.			
1	2.73205081	_____		
2	2.99570907	_____	_____	
3	3.08981914	_____	_____	_____
4	3.12325304	_____	_____	_____

2. Consider the integral

$$\int_{0}^{1} \frac{4}{1 + x^2}\,dx = \pi = 3.1415926535897932385\ldots.$$

Use Romberg integration and compute the approximations $R(K, K)$ for $K = 0, 1, 2, \ldots, 8$.

3. The normal probability density function is $f(t) = (1/\sqrt{2\pi})\,e^{-t^2/2}$, and the cumulative distribution a function defined by the integral $\Phi(x) = \frac{1}{2} + (1/\sqrt{2\pi})\displaystyle\int_{0}^{x} e^{-t^2/2}\,dt$. Compute values for $\Phi(0.5)$, $\Phi(1.0)$, $\Phi(1.5)$, $\Phi(2.0)$, $\Phi(2.5)$, $\Phi(3.0)$, $\Phi(3.5)$, and $\Phi(4.0)$ that have eight digits of accuracy.

4. Assume that the sequential trapezoidal rule converges to L [i.e., $\lim_{J\to\infty} T(J) = L$].
   (a) Show that the sequential Simpson rule converges to L [i.e., $\lim_{J\to\infty} S(J) = L$].
   (b) Show that the sequential Boole rule converges to L [i.e., $\lim_{J\to\infty} B(J) = L$].

5. (a) Verify that Boole's rule ($M = 1$, $h = 1$) is exact for polynomials of degree ≤ 5 of the form $f(x) = c_5 x^5 + c_4 x^4 + \cdots + c_1 x + c_0$, over $[0, 4]$.
   (b) Use the integrand $f(x) = c_6 x^6$ and verify that the error term for Boole's rule ($M = 1$, $h = 1$) over the interval $[0, 4]$ is $E_B(f, h) = [-2(b - a)\,f^{(6)}(c)h^6]/945$.

6. Derive Boole's rule ($M = 1$, $h = 1$) by using the method of undetermined coefficients. Find the constants w_0, w_1, w_2, w_3, and w_4 so that

$$\int_0^4 g(t) \, dt = w_0 g(0) + w_1 g(1) + w_2 g(2) + w_3 g(3) + w_4 g(4)$$

is exact for the five functions $g(t) = 1, t, t^2, t^3,$ and t^4.
Hint: You will get the linear system:

$$w_0 + w_1 + \quad w_2 + \quad w_3 + \quad w_4 = 4$$
$$w_1 + 2w_2 + 3w_3 + \quad 4w_4 = 8$$
$$w_1 + 4w_2 + 9w_3 + 16w_4 = \frac{64}{3}$$
$$w_1 + 8w_2 + 27w_3 + 64w_4 = 64$$
$$w_1 + 16w_2 + 81w_3 + 256w_4 = \frac{1024}{5}.$$

7. Establish the relation $B(J) = [16S(J) - S(J - 1)]/16$ for the case $J = 2$. Use the following information:

$$S(1) = \frac{2h}{3}(f_0 + 4f_2 + f_4) \quad \text{and} \quad S(2) = \frac{h}{3}(f_0 + 4f_1 + 2f_2 + 4f_3 + f_4).$$

8. *Simpson's $\frac{3}{8}$ rule.* Consider the trapezoidal rules over $[x_0, x_3]$: $T(f, 3h) = (3h/2)(f_0 + f_3)$ with step size $3h$, and $T(f, h) = (h/2)(f_0 + 2f_1 + 2f_2 + f_3)$ with step size h. Show that the linear combination $[9T(f, h) - T(f, 3h)]/8$ produces Simpson's $\frac{3}{8}$ rule.

9. Modify Algorithm 7.3 so that it will stop when consecutive values $T(K - 1)$ and $T(K)$ for the sequential trapezoidal rule differ by less than 5×10^{-6}.

10. Extend Algorithm 7.3 so that it will compute values for the sequential Simpson and Boole rules.

11. Use equations (25) and (26) to establish equation (27).

12. Use equations (28) and (29) to establish equation (30).

13. Determine the smallest integer K for which

(a) $\int_0^2 8x^7 \, dx = 256 \equiv R(K, K)$ (b) $\int_0^2 11x^{10} \, dx = 2048 \equiv R(K, K)$

14. Modify Algorithm 7.4 so that it uses the relative error stopping criterion instead of the absolute error criterion.

15. Consider the functions $f(x) = \sqrt{x}$ and $g(t) = 2t^2$ and the integrals (a) $\int_0^1 \sqrt{x} \, dx$ and

(b) $\int_0^1 2t^2 \, dt$. Romberg integration was used to approximate these integrals and the results are given in the table below.

Approximations for (a)	Approximations for (b)
$R(0, 0) = 0.5000000$	$R(0, 0) = 1.0000000$
$R(1, 1) = 0.6380712$	$R(1, 1) = 0.6666667$
$R(2, 2) = 0.6577566$	$R(2, 2) = 0.6666667$
$R(3, 3) = 0.6636076$	$R(3, 3) = 0.6666667$
$R(4, 4) = 0.6655929$	$R(4, 4) = 0.6666667$

(i) Use the change of variable $x = t^2$ and $dx = 2t \, dt$ and show that the two integrals have the same numerical value.

(ii) Discuss why convergence of the Romberg sequence is slower for integral (a) and faster for integral (b).

16. Verify that the following modification of Algorithm 7.4 will sequentially compute the rows of the Romberg integration tableau in the one-dimensional array R[0..N]; hence it saves space. The algorithm will stop when the relative error in the consecutive approximations is less than the preassigned value of Tol (e.g., 10^{-7}).

```
SUBROUTINE  TrRule(A,H,M,R)                     {Sequential trapezoidal rule}
   H := H/2                               {Reduce the step size for refinement}
   SUM := 0
   FOR  P = 1  TO  M  DO                           {Loop for odd subscripts}
        SUM := SUM + F(A + H*(2P−1))
   M := 2M                                 {Update the number of subintervals}
   R(0) := R(0)/2 + H*SUM                         {End of the procedure TrRule}
```

{The main program starts here.}

```
   M := 1                                {Initialize the number of subintervals}
   H := B − A                                       {Initialize the step size}
   RelErr := 1                                       {Initialize the variable}
   J := 0                               {Initialize the loop control counter}
   R(0) := H*[F(A) + F(B)]/2                         {Compute one trapezoid}

WHILE [RelErr > Tol and J < N] or [J < 4] DO
   OldR := R(J)                              {Old Romberg approximation}
   J := J + 1
   Temp := R(0)
   CALL Procedure  TrRule(A,H,M,R)                     {Compute new R(0)}
   FOR  K = 1  TO  J  DO                       {Richardson's improvements}
        Last := R(K)
        R(K) := R(K−1) + [R(K−1) − Temp]/[4^K − 1]
        Temp := Last
   PRINT 'The entries in row' J 'for Romberg integration are'
        R(0), R(1), R(2), . . . , R(J)
   RelErr := 2*|OldR − R(J)|/[|OldR| + |R(J)|]
PRINT 'The Romberg approximation for the integral is" R(J)
```

17. *Romberg integration based on the midpoint rule.* The composite midpoint rule is competitive with the composite trapezoidal rule with respect to efficiency and the speed of convergence.

Use the following facts about the midpoint rule: $\int_a^b f(x) \, dx = M(f, h) + E_M(f, h)$. The rule $M(f, h)$ and error term $E_M(f, h)$ are given by

$$M(f, h) = h \sum_{k=1}^{N} f\left(a + \left(k - \tfrac{1}{2}\right)h\right), \qquad \text{where} \quad h = \frac{b - a}{N}$$

and

$$E_M(f, h) = a_1 h^2 + a_2 h^4 + a_3 h^6 + \cdots .$$

(a) Start with

$$M(0) = \frac{b - a}{2} f\left(\frac{a + b}{2}\right).$$

Develop the sequential midpoint rule for computing

$$M(J) = M(f, h_J) = h_J \sum_{k=1}^{2^J} f\left(a + \left(k - \tfrac{1}{2}\right)h_J\right), \qquad \text{where} \quad h_J = \frac{b - a}{2^J}.$$

(b) Show how the sequential midpoint rule can be used in place of the sequential trapezoidal rule in Romberg integration.

18. Modify Algorithm 7.4 so that it uses the sequential midpoint rule to perform Romberg integration (use the results of Exercise 9).

19. Write a computer program based on the sequential midpoint rule and use it to evaluate the definite integrals

(a) $\displaystyle \int_0^1 \frac{\sin(x)}{x}\,dx$ **(b)** $\displaystyle \int_{-1}^1 \sqrt{1 - x^2}\,dx$

20. Write a report on the numerical solution of multiple integrals. See References [29, 62, 67, 85, 96, 112, 117, 152, and 153].

7.4 ADAPTIVE QUADRATURE

The composite quadrature rules necessitate the use of equally spaced points. This does not take into account that some portions of the curve have large functional variation and hence require more attention and other portions do not. Thus a small step size was used uniformly across the entire interval to ensure the overall accuracy. It is useful to introduce a method that adjusts the step size to be smaller over portions of the curve where a larger functional variation occurs. This technique is called **adaptive quadrature**. The method is based on Simpson's rule.

Simpson's rule uses two subintervals over $[a_k, b_k]$:

$$S(a_k, b_k) = \frac{h}{3} [f(a_k) + 4f(c_k) + f(b_k)], \tag{1}$$

where $c_k = \tfrac{1}{2}(a_k + b_k)$ is the center of $[a_k, b_k]$ and $h = (b_k - a_k)/2$. Furthermore, if $f \in C^4[a_k, b_k]$ then there exists a value $d_1 \in [a_k, b_k]$ so that

$$\int_{a_k}^{b_k} f(x)\,dx = S(a_k, b_k) - h^5 \frac{f^{(4)}(d_1)}{90}. \tag{2}$$

Refinement

A composite Simpson rule using four subintervals of $[a_k, b_k]$ can be performed by bisecting this interval into two equal subintervals $[a_{k1}, b_{k1}]$ and $[a_{k2}, b_{k2}]$ and applying formula (1) recursively over each piece. Only two additional evaluations of $f(x)$ are needed, and the result is

$$S(a_{k1}, b_{k1}) + S(a_{k2}, b_{k2})$$

$$= \frac{h}{6} [f(a_{k1}) + 4f(c_{k1}) + f(b_{k1})] + \frac{h}{6} [f(a_{k2}) + 4f(c_{k2}) + f(b_{k2})], \qquad (3)$$

where $a_{k1} = a_k$, $b_{k1} = a_{k2} = c_k$, $b_{k2} = b_k$, c_{k1} is the midpoint of $[a_{k1}, b_{k1}]$, and c_{k2} is the midpoint of $[a_{k2}, b_{k2}]$. In formula (3) the step size is $h/2$, which accounts for the factors $h/6$ on the right side of the equation. Furthermore, if $f \in C^4[a, b]$, there exists a value $d_2 \in [a_k, b_k]$ so that

$$\int_{a_k}^{b_k} f(x)\, dx = S(a_{k1}, b_{k1}) + S(a_{k2}, b_{k2}) - \frac{h^5}{16} \frac{f^{(4)}(d_2)}{90}. \qquad (4)$$

Assume that $f^{(4)}(d_1) \approx f^{(4)}(d_2)$; then the right sides of equations (2) and (4) are used to obtain the relation

$$S(a_k, b_k) - h^5 \frac{f^{(4)}(d_2)}{90} \approx S(a_{k1}, b_{k1}) + S(a_{k2}, b_{k2}) - \frac{h^5}{16} \frac{f^{(4)}(d_2)}{90}, \qquad (5)$$

which can be written as

$$-h^5 \frac{f^{(4)}(d_2)}{90} \approx \tfrac{16}{15} [S(a_{k1}, b_{k1}) + S(a_{k2}, b_{k2}) - S(f, a_k, b_k)]. \qquad (6)$$

Then (6) is substituted in (4) to obtain the error estimate:

$$\left| \int_{a_k}^{b_k} f(x)\, dx - S(a_{k1}, b_{k1}) - S(a_{k2}, b_{k2}) \right|$$

$$\approx \tfrac{1}{15} |S(a_{k1}, b_{k1}) + S(a_{k2}, b_{k2}) - S(a_k, b_k)|. \qquad (7)$$

Because of the assumption $f^{(4)}(d_1) \approx f^{(4)}(d_2)$, the fraction $\tfrac{1}{15}$ is replaced with $\tfrac{1}{10}$ on the right side of (7) when implementing the method. This justifies the following test.

Accuracy Test

Assume that the tolerance $\epsilon_k > 0$ is specified for the interval $[a_k, b_k]$. If

$$\tfrac{1}{10} |S(a_{k1}, b_{k1}) + S(a_{k2}, b_{k2}) - S(a_k, b_k)| < \epsilon_k, \qquad (8)$$

we infer that

$$\left| \int_{a_k}^{b_k} f(x)\, dx - S(a_{k1}, b_{k1}) - S(a_{k2}, b_{k2}) \right| < \epsilon_k. \tag{9}$$

Thus the composite Simpson rule (3) is used to approximate the integral

$$\int_{a_k}^{b_k} f(x)\, dx \approx S(a_{k1}, b_{k1}) + S(a_{k2}, b_{k2}), \tag{10}$$

and the error bound for this approximation over $[a_k, b_k]$ is ϵ_k.

Adaptive quadrature is implemented by applying Simpson's rules (1) and (3). Start with $\{[a_0, b_0], \epsilon_0\}$, where ϵ_0 is the tolerance for numerical quadrature over $[a_0, b_0]$. The interval is refined into subintervals labeled $[a_{01}, b_{01}]$ and $[a_{02}, b_{02}]$. If the accuracy test (8) is passed, quadrature formula (3) is applied to $[a_0, b_0]$ and we are done. If the test in (8) fails, the subintervals are labeled $[a_1, b_1]$ and $[a_2, b_2]$, over which we use the tolerances $\epsilon_1 = \frac{1}{2}\epsilon_0$ and $\epsilon_2 = \frac{1}{2}\epsilon_0$, respectively. Thus we have two intervals with their associated tolerances to consider for further refinement and testing: $\{\{[a_1, b_1], \epsilon_1\},$ $\{[a_2, b_2], \epsilon_2\}\}$, where $\epsilon_1 + \epsilon_2 = \epsilon_0$. If adaptive quadrature must be continued, the smaller intervals must be refined and tested, each with its own associated tolerance.

In the second step we first consider $\{[a_1, b_1], \epsilon_1\}$ and refine the interval $[a_1, b_1]$ into $[a_{11}, b_{11}]$ and $[a_{12}, b_{12}]$. If they pass the accuracy test (8) with the tolerance ϵ_1, quadrature formula (3) is applied to $[a_1, b_1]$ and accuracy has been achieved over this interval. If they fail the test in (8) with the tolerance ϵ_1, each subinterval $[a_{11}, b_{11}]$ and $[a_{12}, b_{12}]$ must be refined and tested in the third step with the reduced tolerance $\frac{1}{2}\epsilon_1$. Moreover, the second step involves looking at $\{[a_2, b_2], \epsilon_2\}$ and refining $[a_2, b_2]$ into $[a_{21}, b_{21}]$ and $[a_{22}, b_{22}]$. If they pass the accuracy test (8) with the tolerance ϵ_2, quadrature formula (3) is applied to $[a_2, b_2]$ and accuracy is achieved over this interval. If they fail the test in (8) with tolerance ϵ_2, each subinterval $[a_{21}, b_{21}]$ and $[a_{22}, b_{22}]$ must be refined and tested in the third step with the reduced tolerance $\frac{1}{2}\epsilon_2$. Therefore, the second step produces either three or four intervals, which we relabel consecutively. The three intervals would be relabeled to produce $\{\{[a_1, b_1], \epsilon_1\}, \{[a_2, b_2], \epsilon_2\}, \{[a_3, b_3], \epsilon_3\}\}$, where $\epsilon_1 + \epsilon_2 + \epsilon_3 = \epsilon_0$. In the case of four intervals, we would obtain $\{\{[a_1, b_1], \epsilon_1\}, \{[a_2, b_2], \epsilon_2\}, \{[a_3, b_3], \epsilon_3\},$ $\{[a_4, b_4], \epsilon_4\}\}$, where $\epsilon_1 + \epsilon_2 + \epsilon_3 + \epsilon_4 = \epsilon_0$.

If adaptive quadrature must be continued, the smaller intervals must be tested, each with its own associated tolerance. The error term in (4) shows that each time a refinement is made over a small subinterval there is a reduction of error by about a factor of $\frac{1}{16}$. Thus the process will terminate after a finite number of steps. The bookkeeping for implementing the method includes a sentinel variable which indicates if a particular subinterval has passed its test. To avoid unnecessary additional evaluations of $f(x)$, the function values can be included in a data list corresponding to each subinterval. The details are shown in Algorithm 7.5.

Example 7.16

Use adaptive quadrature to numerically approximate the value of the definite integral $\int_0^4 13(x - x^2)e^{-3x/2}\, dx$ with the starting tolerance $\epsilon_0 = 0.00001$.

Solution.   Implementation of the method revealed that 20 subintervals are needed. Table 7.8 lists each interval $[a_k, b_k]$, composite Simpson rule $S(a_{k1}, b_{k1}) + S(a_{k2}, b_{k2})$, the error bound for this approximation, and the associated tolerance ϵ_k. The approximate value of the integral is obtained by summing the Simpson rule approximations to get

$$\int_0^4 13(x - x^2)e^{-3x/2}\, dx \approx -1.54878823413. \tag{11}$$

TABLE 7.8   Adaptive Quadrature Computations for $f(x) = 13(x - x^2)e^{-3x/2}$

a_k	b_k	$S(a_{k1}, b_{k1}) + S(a_{k2}, b_{k2})$	Error bound on the left side of (8)	Tolerance ϵ_k for $[a_k, b_k]$
0.0	0.0625	0.02287184840	0.00000001522	0.00000015625
0.0625	0.125	0.05948686456	0.00000001316	0.00000015625
0.125	0.1875	0.08434213630	0.00000001137	0.00000015625
0.1875	0.25	0.09969871532	0.0000000981	0.00000015625
0.25	0.375	0.21672136781	0.00000025055	0.0000003125
0.375	0.5	0.20646391592	0.00000018402	0.0000003125
0.5	0.625	0.17150617231	0.00000013381	0.0000003125
0.625	0.75	0.12433363793	0.00000009611	0.0000003125
0.75	0.875	0.07324515141	0.00000006799	0.0000003125
0.875	1.0	0.02352883215	0.00000004718	0.0000003125
1.0	1.125	-0.02166038952	0.00000003192	0.0000003125
1.125	1.25	-0.06065079384	0.00000002084	0.0000003125
1.25	1.5	-0.21080823822	0.00000031714	0.000000625
1.5	2.0	-0.60550965007	0.00000003195	0.00000125
2.0	2.25	-0.31985720175	0.00000008106	0.000000625
2.25	2.5	-0.30061749228	0.00000008301	0.000000625
2.5	2.75	-0.27009962412	0.00000007071	0.000000625
2.75	3.0	-0.23474721177	0.00000005447	0.000000625
3.0	3.5	-0.36389799695	0.00000103699	0.00000125
3.5	4.0	-0.24313827772	0.00000041708	0.00000125
	Totals	-1.54878823413	0.00000296809	0.00001

The true value of the integral is

$$\int_0^4 13(x - x^2)e^{-3x/2}\, dx = \frac{4108e^{-6} - 52}{27} = -1.5487883725279481333. \tag{12}$$

Therefore, the error for adaptive quadrature is

$$-1.54878837253 - (-1.54878823413) = -0.00000013840, \tag{13}$$

which is smaller than the specified tolerance $\epsilon_0 = 0.00001$. The adaptive method involves 20 subintervals of $[0, 4]$ and 81 function evaluations were used. Figure 7.10 shows the graph of $y = f(x)$ and these 20 subintervals. The intervals are smaller where a larger functional variation occurs near the origin.

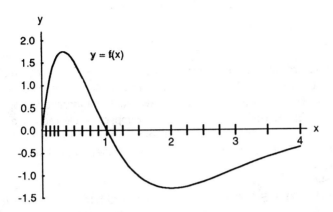

Figure 7.10 The subintervals of [0, 4] used in adaptive quadrature.

In the refinement and testing process in the adaptive method, the first four intervals were bisected into eight subintervals of width 0.03125. If this uniform spacing is continued throughout the interval [0, 4], $m = 128$ subintervals are required for the composite Simpson rule, which yields the approximation -1.54878844029, which is in error by the amount 0.00000006776. Although the composite Simpson contains half the error of the adaptive quadrature, 176 more functions evaluations are required. This gain of accuracy is negligible; hence there is a considerable saving of computing effort with the adaptive method.

Algorithm 7.5   (Adaptive Quadrature Using Simpson's Rule).   To approximate the integral

$$\int_A^B f(x)\,dx \approx \sum_{k=1}^{M} \frac{h_k}{3}\,[f(x_{4k-4}) + 4f(x_{4k-3}) + 2f(x_{4k-2}) + 4f(x_{4k-1}) + f(x_{4k})].$$

The composite Simpson rule is applied to the $4M$ subintervals $[x_{4k-4}, x_{4k}]$, where $[A, B] = [x_0, x_{4M}]$ and $x_{4k-4+j} = x_{4k-4} + jh_k$ for each $k = 1, \ldots, M$ and $j = 1, \ldots, 4$.

VARiable declaration   Real SR[1..100,1..11]

```
SUBROUTINE SIMPSONRULE(A0,B0,Tol0)          {This is a vector function}
        A := A0,  B := B0,  H := (B − A)/2,  C := A + H
        Fa := F(A),  Fc := F(C),  Fb := F(B)         {Construct the Simpson's
        S := H*(Fa + 4Fc + Fb)/3,  S2 := S,             rule approximation, and
        Tol := Tol0,  Err := Tol0,  Check := 0          record the information}
        RETURN({A,C,B,Fa,Fc,Fb,S,S2,Err,Tol,Check})
```

```
SUBROUTINE REFINE(P)
        State := Done
        Sr0 := SR(P)                                              {Vector
        {A,C,B,Fa,Fc,Fb,S,S2,Err,Tol,Check} := Sr0           replacements}
        IF   Check = 1   THEN   RETURN        {Return if refinement is not needed}
        Sr1 := SIMPSONRULE(A,C,Tol/2)                  {Bisect the interval and apply
        Sr2 := SIMPSONRULE(C,B,Tol/2)                    Simpson's rule recursively}
        Err := |Sr0(7)−Sr1(7)−Sr2(7)|/10            {Determine the error bound and
        IF   Err < Tol   THEN   Sr0(11)   := 1          if the interval meets tolerance}
        IF   Err < Tol   THEN
                SR(P) := Sr0                                 {Vector replacement}
                SR(P,8) := Sr1(7) + Sr2(7)           {Record Simpson composite
                SR(P,9) := Err                          rule and the error bound}
        ELSE
                FOR J=M+1 DOWNTO P DO   SR(J):=SR(J−1)      {Shift the list of
                                                              records for the
                M := M+1                                 subinterals, and quadrature
                SR(P) := Sr1                             information and insert
                SR(P+1) := Sr2                           the new vector records
                state := Iterating                       for the bisected interval}
        ENDIF
```

{The main program starts here.}
```
        INPUT A,B                                  {Endpoints of the interval}
        Tol := 10^{-6}                                    {Initial tolerance}
        SR(1) := SIMPSONRULE(A0,B0,Tol0)               {Vector replacement}
        M := 1
        State := Iterating
        N := M
        WHILE   State = Iterating   Do             {Recursively refine the
                FOR J = N DOWNTO 1 DO   CALL REFINE(J)      the subintervals}
        SUM := 0                                   {Add up the Simpson
        FOR J=1 TO M DO   SUM := SUM + SR(J,8)          rule approximations}
        PRINT "The approximate value of the integral is"; SUM      {Print result}
```

EXERCISES FOR ADAPTIVE QUADRATURE

1. Use Algorithm 7.5 (adaptive quadrature using Simpson's rule) to approximate the value of the definite integral. Use the starting tolerance $\epsilon_0 = 0.00001$. Graphs of the integrands are shown in Figure 7.11.

   (a) $\int_0^3 \dfrac{\sin(2x)}{1 + x^5}\, dx$ **(b)** $\int_0^3 \sin(4x)e^{-2x}\, dx$ **(c)** $\int_{0.04}^1 \dfrac{1}{\sqrt{x}}\, dx$

(d) $\displaystyle\int_0^2 \frac{1}{x^2 + \frac{1}{10}}\, dx$ **(e)** $\displaystyle\int_{1/(2\pi)}^2 \sin\left(\frac{1}{x}\right) dx$ **(f)** $\displaystyle\int_0^2 \sqrt{4x - x^2}\, dx$

2. Modify Algorithm 7.5 so that Boole's rule is used in each subinterval $[a_k, b_k]$.
3. Use the modified algorithm in Exercise 2 to compute approximations for the definite integrals in Exercise 1.

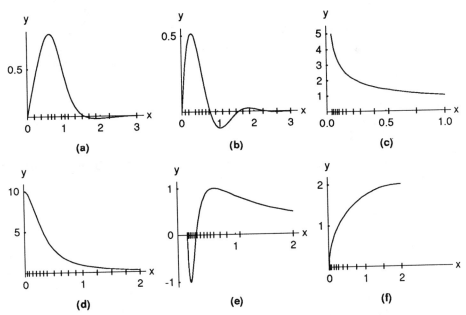

Figure 7.11   (a) $y = \sin(2x)/(1 + x^5)$. (b) $y = \sin(4x)e^{-2x}$. (c) $y = 1/\sqrt{x}$. (d) $y = 1/\left(x^2 + \frac{1}{10}\right)$. (e) $y = \sin(1/x)$. (f) $y = \sqrt{4x - x^2}$.

7.5 GAUSS–LEGENDRE INTEGRATION (OPTIONAL)

We wish to find the area under the curve

$$y = f(x), \qquad -1 \le x \le 1.$$

What method gives the best answer if only two function evaluations are to be made? We have already seen that the trapezoidal rule is a method for finding the area under the curve and that it uses two function evaluations at the endpoints $(-1, f(-1))$, $(1, f(1))$. But if the graph $y = f(x)$ is concave down, the error in approximation is the entire region that lies between the curve and the line segment joining the points [see Figure 7.12(a)].

    If we can use nodes x_1, x_2 that lie inside the interval, the line through the two points $(x_1, f(x_1))$, $(x_2, f(x_2))$ crosses the curve and the area under the line more closely

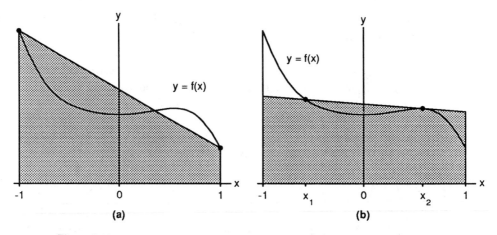

Figure 7.12 (a) Trapezoidal approximation using the abscissas -1 and 1. (b) Trapezoidal approximation using the abscissas x_1 and x_2.

approximates the area under the curve [see Figure 7.12(b)]. The equation of this line is

$$y = f(x_1) + \frac{(x - x_1)[f(x_2) - f(x_1)]}{x_2 - x_1} \tag{1}$$

and the area of the trapezoid under this line is

$$\text{area} = \frac{2x_2}{x_2 - x_1} f(x_1) - \frac{2x_1}{x_2 - x_1} f(x_2). \tag{2}$$

Notice that the trapezoidal rule is a special case of (2). When we choose $x_1 = -1$, $x_2 = 1$, and $h = 2$, then

$$T(f, h) = \frac{2}{2} f(x_1) - \frac{-2}{2} f(x_2) = f(x_1) + f(x_2).$$

We shall use the method of undetermined coefficients to find the absicssas x_1, x_2 and weights w_1, w_2 so that the formula

$$\int_{-1}^{1} f(x)\, dx \approx w_1 f(x_1) + w_2 f(x_2) \tag{3}$$

is exact for cubic polynomials [i.e., $f(x) = a_3 x^3 + a_2 x^2 + a_1 x + a_0$]. Since four coefficients w_1, w_2, x_1, and x_2 need to be determined in equation (3), we can select four conditions to be satisfied. Using the fact that integration is additive, it will suffice to require that (3) is exact for the four functions $f(x) = 1, x, x^2, x^3$. The four integral conditions are:

$$f(x) = 1; \qquad \int_{-1}^{1} 1\, dx = 2 = w_1 + w_2.$$

$$f(x) = x: \qquad \int_{-1}^{1} x \, dx = 0 = w_1 x_1 + w_2 x_2.$$

(4)

$$f(x) = x^2: \qquad \int_{-1}^{1} x^2 \, dx = \tfrac{2}{3} = w_1 x_1^2 + w_2 x_2^2.$$

$$f(x) = x^3: \qquad \int_{-1}^{1} x^3 \, dx = 0 = w_1 x_1^3 + w_2 x_2^3.$$

Now solve the system of nonlinear equations

$$w_1 \quad + \quad w_2 \quad = \ 2 \qquad\qquad (5)$$

$$w_1 x_1 \quad = \ -w_2 x_2 \qquad\qquad (6)$$

$$w_1 x_1^2 \ + \ w_2 x_2^2 \ = \ \tfrac{2}{3} \qquad\qquad (7)$$

$$w_1 x_1^3 \quad = \ - w_2 x_2^3. \qquad\qquad (8)$$

We can divide (8) by (6), and the result is

$$x_1^2 = x_2^2 \quad \text{or} \quad x_1 = -x_2. \qquad\qquad (9)$$

Use (9) and divide (6) by x_1 on the left and $-x_2$ on the right to get

$$w_1 = w_2. \qquad\qquad (10)$$

Substituting (10) into (5) results in $w_1 + w_1 = 2$. Hence

$$w_1 = w_2 = 1. \qquad\qquad (11)$$

Now using (11) and (9) in (7), we write

$$w_1 x_1^2 + w_2 x_2^2 = x_2^2 + x_2^2 = \tfrac{2}{3} \quad \text{or} \quad x_2^2 = \tfrac{1}{3}. \qquad\qquad (12)$$

Finally, from (12) and (9), we see that the modes are

$$-x_1 = x_2 = 1/3^{1/2} \approx 0.5773502692.$$

We have found the nodes and the weights that make up the two-point Gauss–Legendre rule. Since the formula is exact for cubic equations, the error term will involve the fourth derivative. A discussion of the error term can be found in Reference [41].

Theorem 7.8   (Gauss–Legendre Two-Point Rule).   If f is continuous on $[-1, 1]$, then

$$\int_{-1}^{1} f(x) \, dx \approx G_2(f) = f\!\left(\frac{-1}{\sqrt{3}}\right) + f\!\left(\frac{1}{\sqrt{3}}\right). \qquad\qquad (13)$$

The Gauss–Legendre rule $G_2(f)$ has degree of precision $n = 3$. If $f \in C^4[-1, 1]$, then

$$\int_{-1}^{1} f(x) \, dx = f\!\left(\frac{-1}{\sqrt{3}}\right) + f\!\left(\frac{1}{\sqrt{3}}\right) + E_2(f), \qquad\qquad (14)$$

where

$$E_2(f) = \frac{f^{(4)}(c)}{135}. \tag{15}$$

Example 7.17

Use the two-point Gauss–Legendre rule to approximate

$$\int_{-1}^{1} \frac{dx}{x + 2} = \ln(3) \approx 1.09861$$

and compare the result with the trapezoidal rule $T(f, h)$ with $h = 2$, and Simpson's rule $S(f, h)$ with $h = 1$.

Solution.   Let $G_2(f)$ denote the two-point Gauss–Legendre rule; then

$$G_2(f) = f(-0.57735) + f(0.57735) = 0.70291 + 0.38800 = 1.09091,$$

$$T(f, 2) = f(-1.0000) + f(1.0000) = 1.00000 + 0.33333 = 1.33333,$$

$$S(f, 1) = \frac{f(-1) + 4f(0) + f(1)}{3} = \frac{1 + 2 + \frac{1}{3}}{3} = 1.11111.$$

The errors are 0.00770, −0.23472, and −0.01250, respectively, so that the Gauss–Legendre rule is seen to be best. Notice that the Gauss–Legendre rule required only two function evaluations and Simpson's rule required three. In this example the size of the error for $G_2(f)$ is about 61% of the size of the error for $S(f, 1)$.

The general N-point Gauss–Legendre rule is exact for polynomial functions of degree $\leq 2N - 1$ and the numerical integration formula is

$$G_N(f) = w_{N,1}f(x_{N,1}) + w_{N,2}f(x_{N,2}) + \cdots + w_{N,N}f(x_{N,N}). \tag{16}$$

The abscissas $x_{N,k}$ and weights $w_{N,k}$ to be used have been tabulated and are easily available; Table 7.9 gives the values up to eight points. Also included in the table is the form of the error term $E_N(f)$ that corresponds to $G_N(f)$, and it can be used to determine the accuracy of the Gauss–Legendre integration formula.

The values in Table 7.9 in general have no easy representation. This fact makes the method less attractive for human beings to use when hand calculations are required. But once the values are stored in a computer it is easy to call them up when needed. The nodes are actually roots of the Legendre polynomials and the corresponding weights must be obtained by solving a system of equations. For the three-point Gauss–Legendre rule the nodes are $-(0.6)^{1/2}$, 0, and $(0.6)^{1/2}$ and the corresponding weights are 5/9, 8/9, and 5/9.

Theorem 7.9   (Gauss–Legendre Three-Point Rule).   If f is continuous on $[-1, 1]$, then

$$\int_{-1}^{1} f(x)\, dx \sim G_3(f) = \frac{5f(-\sqrt{3/5}) + 8f(0) + 5f(\sqrt{3/5})}{9}. \tag{17}$$

The Gauss–Legendre rule $G_3(f)$ has degree of precision $n = 5$. If $f \in C^6[-1, 1]$, then

$$\int_{-1}^{1} f(x) \, dx = \frac{5f(-\sqrt{3/5}) + 8f(0) + 5f(\sqrt{3/5})}{9} + E_3(f), \qquad (18)$$

where

$$E_3(f) = \frac{f^{(6)}(c)}{15{,}750}. \qquad (19)$$

TABLE 7.9   Gauss–Legendre Abscissas and Weights

$$\int_{-1}^{1} f(x) \, dx = \sum_{k=1}^{N} w_{N,k} f(x_{N,k}) + E_N(f)$$

N	Abscissas, $x_{N,k}$	Weights, $w_{N,k}$	Truncation error, $E(f, N)$
2	−0.5773502692 0.5773502692	1.0000000000 1.0000000000	$\dfrac{f^{(4)}(c)}{135}$
3	±0.7745966692 0.0000000000	0.5555555556 0.8888888888	$\dfrac{f^{(6)}(c)}{15{,}750}$
4	±0.8611363116 ±0.3399810436	0.3478548451 0.6521451549	$\dfrac{f^{(8)}(c)}{3{,}472{,}875}$
5	±0.9061798459 ±0.5384693101 0.0000000000	0.2369268851 0.4786286705 0.5688888888	$\dfrac{f^{(10)}(c)}{1{,}237{,}732{,}650}$
6	±0.9324695142 ±0.6612093865 ±0.2386191861	0.1713244924 0.3607615730 0.4679139346	$\dfrac{f^{(12)}(c)2^{13}[6!]^4}{[12!]^3 13!}$
7	±0.9491079123 ±0.7415311856 ±0.4058451514 0.0000000000	0.1294849662 0.2797053915 0.3818300505 0.4179591837	$\dfrac{f^{(14)}(c)2^{15}[7!]^4}{[14!]^3 15!}$
8	±0.9602898565 ±0.7966664774 ±0.5255324099 ±0.1834346425	0.1012285363 0.2223810345 0.3137066459 0.3626837834	$\dfrac{f^{(16)}(c)2^{17}[8!]^4}{[16!]^3 17!}$

Example 7.18

Show that the three-point Gauss–Legendre rule is exact for

$$\int_{-1}^{1} 5x^4 \, dx = 2 = G_3(f)$$

Solution.   Since the integrand is $f(x) = 5x^4$ and $f^{(6)}(x) = 0$, we can use (19) to see that $E_3(f) = 0$. But it is instructive to use (17) and do the calculation in this case.

$$G_3(f) = \frac{5(5)(0.6)^2 + 0 + 5(5)(0.6)^2}{9} = \frac{18}{9} = 2.$$

The next result shows how to change the variable of integration so that the Gauss–Legendre rules can be used on the interval $[a, b]$.

Theorem 7.10   (The Gauss–Legendre Translation).   Suppose that the abscissas $\{x_{N,k}\}_{k=1}^N$ and weights $\{w_{N,k}\}_{k=1}^N$ are given for the N-point Gauss–Legendre rule over $[-1, 1]$. To apply the rule over the interval $[a, b]$, use the change of variable

$$t = \frac{a+b}{2} + \frac{b-a}{2}x \quad \text{and} \quad dt = \frac{b-a}{2}dx. \tag{20}$$

Then the relationship

$$\int_a^b f(t)\, dt = \int_{-1}^1 f\left(\frac{a+b}{2} + \frac{b-a}{2}x\right)\frac{b-a}{2}\, dx \tag{21}$$

is used to obtain the quadrature formula

$$\int_a^b f(t)\, dt = \frac{b-a}{2}\sum_{k=1}^N w_{N,k} f\left(\frac{a+b}{2} + \frac{b-a}{2}x_{N,k}\right). \tag{22}$$

Example 7.19

Use the three-point Gauss–Legendre rule to approximate

$$\int_1^5 \frac{dt}{t} = \ln(5) \approx 1.609438$$

and compare the result with Boole's rule $B(2)$ with $h = 1$.

Solution.   Here $a = 1$ and $b = 5$, so that the rule in (22) yields

$$G_3(f) = (2)\frac{5f(3 - 2(0.6)^{1/2}) + 8f(3 + 0) + 5f(3 + 2(0.6)^{1/2})}{9}$$

$$= (2)\frac{3.446359 + 2.666667 + 1.099096}{9} = 1.602694.$$

In Example 7.13 we saw that Boole's rule gave $B(2) = 1.617778$. The errors are 0.006744 and -0.008340, respectively, so that the Gauss–Legendre rule is slightly better in this case. Notice that the Gauss–Legendre rule requires three function evaluations and Boole's rule requires five. In this example the size of the two errors is about the same.

Gauss–Legendre integration formulas are extremely accurate; and they should be considered seriously when many integrals of a similar nature are to be evaluated. In this case proceed as follows. Pick a few representative integrals, including some with the worst behavior that is likely to occur. Determine the number of sample points N that are needed to obtain accuracy. Then fix the value N, and use the Gauss–Legendre rule with N sample points for all the integrals.

Algorithm 7.6   (Gauss–Legendre Quadrature).   To approximate the integral

$$\int_A^B f(x)\,dx \approx \frac{B-A}{2}\,[w_{N,1}f(t_{N,1}) + w_{N,2}f(t_{N,2}) + \cdots + w_{N,N}f(t_{N,N})]$$

by sampling $f(x)$ at the N unequally spaced points $t_{N,1}, t_{N,2}, \ldots, t_{N,N}$. The change of variable

$$t = \frac{a+b}{2} + \frac{b-a}{2}x \quad \text{and} \quad dt = \frac{b-a}{2}dx$$

are used. The abscissas $x_{N,1}, x_{N,2}, \ldots, x_{N,N}$ and the corresponding weights $w_{N,1}, w_{N,2}, \ldots, w_{N,N}$ must be obtained from a table of known values.

```
        USES the function   F(X)
        USES the Table of Abscissa   X(N,K)
        USES the Table of Weights W(N,K)
        INPUT  A,  B,  N
        SUM := 0


FOR   K = 1  TO  N  DO
   |    T := [A+B]/2 + X(N,K)*[B−A]/2
   |___ SUM := SUM + W(N,K)*F(T)

        SUM := SUM*[B−A]/2


PRINT 'The approximate value of the integral of f(t)'
        'on the interval' A,B, 'using' N 'sample points'
        'computed using the' N 'point Gauss-Legendre rule is'   SUM
```

EXERCISES FOR GAUSS–LEGENDRE INTEGRATION

In Exercises 1–4:
(a) Show that the two integrals are equivalent.
(b) Find $G_2(f)$. (c) Find $G_3(f)$. (d) Find $G_4(f)$.
(e) Use a computer to find the values in parts (b) to (d).

1. $\int_0^2 6t^5\,dt = \int_{-1}^1 6(x+1)^5\,dx = 64$

2. $\int_0^2 \sin(t)\, dt = \int_{-1}^1 \sin(x+1)\, dx = -\cos(2) + \cos(0) \approx 1.416147$

3. $\int_0^1 \dfrac{\sin(t)}{t}\, dt = \int_{-1}^1 \dfrac{\sin((x+1)/2)}{x+1}\, dx \approx 0.9460831$

4. $(2\pi)^{-1/2} \int_0^1 \exp\!\left(\dfrac{-t^2}{2}\right) dt = (2\pi)^{-1/2} \int_{-1}^1 \dfrac{\exp(-(x+1)^2/8)}{2}\, dx$

5. $\pi^{-1} \int_0^\pi \cos(0.6 \sin(t))\, dt = 0.5 \int_{-1}^1 \cos\!\left(0.6 \sin\!\left((x+1)\dfrac{\pi}{2}\right)\right) dx$

(a) Show that the two integrals are equivalent.
(b) Find $G_4(f)$. (c) Find $G_6(f)$.
(d) Use a computer to find the values in parts (b) and (c).

6. Use $E_N(f)$ in Table 7.11 and the change of variable $dt = dx$ in Theorem 7.9 to find the smallest integer N so that $E_N(f) = 0$ for

(a) $\int_0^2 8x^7\, dx = 256 = G_N(f)$ (b) $\int_0^2 11x^{10}\, dx = 2048 = G_N(f)$

7. Find the roots of the following Legendre polynomials and compare them with the abscissa in Table 7.11.
(a) $P_2(x) = (3x^2 - 1)/2$
(b) $P_3(x) = (5x^3 - 3x)/2$
(c) $P_4(x) = (35x^4 - 30x^2 + 3)/8$

8. The truncation error term for two-point Gauss–Legendre rule on the interval $[-1, 1]$ is $f^{(4)}(c_1)/135$. The truncation error term for Simpson's rule on $[a, b]$ is $-h^5 f^{(4)}(c_2)/90$. Compare the truncation error terms when $[a, b] = [-1, 1]$. Which method do you think is best? Why?

9. The three-point Gauss–Legendre rule is

$$\int_{-1}^1 f(x)\, dx \approx \dfrac{5f(-(0.6)^{1/2}) + 8f(0) + 5f((0.6)^{1/2})}{9}.$$

Show that the formula is exact for $f(x) = 1, x, x^2, x^3, x^4, x^5$. *Hint.* If f is an odd function [i.e, $f(-x) = -f(x)$], the integral of f over $[-1, 1]$ is zero.

10. The truncation error term for the three-point Gauss–Legendre rule on the interval $[-1, 1]$ is $f^{(6)}(c_1)/15{,}750$. The truncation error term for Boole's rule on $[a, b]$ is $-8h^7 f^{(6)}(c_2)/945$. Compare the error terms when $[a, b] = [-1, 1]$. Which method do you think is better? Why?

11. Derive the three-point Gauss–Legendre rule using the following steps. Use the fact that the abscissas are the roots of the Legendre polynomial of degree 3.

$$x_1 = -(0.6)^{1/2} \qquad x_2 = 0 \qquad x_3 = (0.6)^{1/2}$$

Find the weights w_1, w_2, w_3 so that the relation

$$\int_{-1}^1 f(x)\, dx \approx w_1 f(-(0.6)^{1/2}) + w_2 f(0) + w_3 f((0.6)^{1/2})$$

is exact for the functions $f(x) = 1, x, x^2$. *Hint.* First obtain, then solve, the linear system of equations

$$w_1 + w_2 + \qquad w_3 = ?$$

$$-(0.6)^{1/2}w_1 \qquad + \quad (0.6)^{1/2}w_3 \ = \ 0$$

$$0.6w_1 \qquad + \qquad 0.6w_3 \ = \ \tfrac{2}{3}.$$

12. Modify Algorithm 7.5 so that it will compute $G_1(f)$, $G_2(f)$, . . . , and stop when the relative error in the approximations $G_{N-1}(f)$ and $G_N(f)$ is less than the preassigned value Tol, that is,

$$\frac{2|G_{N-1}(f) - G_N(f)|}{|G_{N-1}(f) + G_N(f)|} < \text{Tol}.$$

13. In practice, if many integrals of a similar type are evaluated, a preliminary analysis is made to determine the number of function evaluations required to obtain the desired accuracy. Suppose that 17 function evaluations are to be made. Compare the Romberg answer $R(4, 4)$ with the Gauss–Legendre answer $G_{17}(f)$.

8

Numerical Optimization

The two-dimensional wave equation is used in mechanical engineering to model vibrations in rectangular plates. If the plates have all four edges clamped, the sinusoidal vibrations are described with a double Fourier series. Suppose that at a certain instant of time the height $z = f(x, y)$ over the point (x, y) is given by the function

$$z = f(x, y) = 0.02 \sin(x) \sin(y) - 0.03 \sin(2x) \sin(y)$$
$$+ 0.04 \sin(x) \sin(2y) + 0.08 \sin(2x) \sin(2y).$$

Where are the points of maximum deflection located? Looking at the three-dimensional graph and the companion contour plot in Figure 8.1(a) and (b), respectively, we see that there are two local minima and two local maxima over the square $0 \le x \le \pi$, $0 \le y \le \pi$. Numerical methods can be used to determine their approximate location:

$f(0.8278, 2.3322) = -0.1200$ and $f(2.5351, 0.6298) = -0.0264$ are the local minima

and

$f(0.9241, 0.7640) = 0.0998$ and $f(2.3979, 2.2287) = 0.0853$ are the local maxima.

In this chapter we give a brief introduction to some of the basic methods for locating extrema of functions of one or several variables.

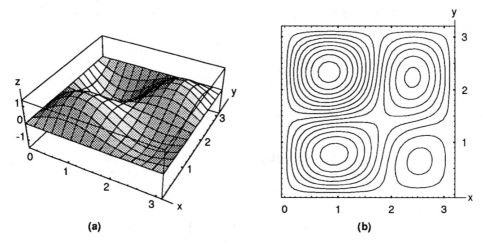

Figure 8.1   (a) The displacement $z = f(x, y)$ of a vibrating plate. (b) The contour plot $f(x, y) = C$ for a vibrating plate.

8.1 MINIMIZATION OF A FUNCTION

An important topic in calculus is finding the local extrema of a function. For differentiable functions a primary method is to locate points where the derivative is zero. Numerical methods for locating roots can be adapted for this situation. To start, we review some definitions and theorems.

Functions of One Variable

Definition 8.1   (Local Extremum).   The function f is said to have a **local minimum value** at $x = p$ if there exists an open interval I containing p so that $f(p) \leq f(x)$ for all $x \in I$. Similarly, f is said to have a **local maximum value** at $x = p$ if $f(x) \leq f(p)$ for all $x \in I$. If f has either a local minimum or maximum value at $x = p$, it is said to have a **local extremum** at $x = p$.

Definition 8.2   (Increasing and Decreasing).   Assume that $f(x)$ defined on the interval I.

(i)   If $x_1 < x_2$ implies that $f(x_1) < f(x_2)$ for all $x_1, x_2 \in I$, then f is said to be **increasing** on I.

(ii)   If $x_1 < x_2$ implies that $f(x_1) > f(x_2)$ for all $x_1, x_2 \in I$, then f is said to be **decreasing** on I.

Theorem 8.1.   Suppose that $f(x)$ is continuous on $I = [a, b]$ and is differentiable on (a, b).

(i) If $f'(x) > 0$ for all $x \in (a, b)$, then $f(x)$ is increasing on I.
(ii) If $f'(x) < 0$ for all $x \in (a, b)$, then $f(x)$ is decreasing on I.

Theorem 8.2.   Assume that $f(x)$ is defined on $I = [a, b]$ and has a local extremum at an interior point $p \in (a, b)$. If $f(x)$ is differentiable at $x = p$, then $f'(p) = 0$.

Theorem 8.3   (First Derivative Test).   Assume that $f(x)$ is continuous on $I = [a, b]$. Furthermore, suppose that $f'(x)$ is defined for all $x \in (a, b)$, except possibly at $x = p$.

(i) If $f'(x) < 0$ on (a, p) and $f'(x) > 0$ on (p, b), then $f(p)$ is a local minimum.
(ii) If $f'(x) > 0$ on (a, p) and $f'(x) < 0$ on (p, b), then $f(p)$ is a local maximum.

Theorem 8.4   (Second Derivative Test).   Assume that f is continuous on $[a, b]$ and f' and f'' are defined on (a, b). Also, suppose that $p \in (a, b)$ is a critical point where $f'(p) = 0$.

(i) If $f''(p) > 0$, then $f(p)$ is a local minimum of f.
(ii) If $f''(p) < 0$, then $f(p)$ is a local maximum of f.
(iii) If $f''(p) = 0$, then this test is inconclusive.

Example 8.1

Use the second derivative test to classify the local extrema of $f(x) = x^3 + x^2 - x + 1$ on the interval $[-2, 2]$.

Solution.   The derivative is $f'(x) = 3x^2 + 2x - 1 = (3x - 1)(x + 1)$, and the second derivative is $f''(x) = 6x + 2$. There are two points where $f'(x) = 0$ (i.e., $x = \frac{1}{3}, -1$).
  Case (i): At $x = \frac{1}{3}$ we find that $f'\left(\frac{1}{3}\right) = 0$ and $f''\left(\frac{1}{3}\right) = 4 > 0$, so that $f(x)$ has a local minimum at $x = \frac{1}{3}$.
  Case (ii): At $x = -1$ we find that $f'(-1) = 0$ and $f''(-1) = -4 < 0$, so that $f(x)$ has a local maximum at $x = -1$.

Search Method

Another method for finding the minimum of $f(x)$ is to evaluate the function many times and search for a local minimum. To reduce the number of function evaluations, it is important to have a good strategy for determining where $f(x)$ is evaluated. One of the most efficient methods is called the **golden ratio search**, which is named for the ratio's involvement in selecting the points.

The Golden Ratio

Let the initial interval be [0, 1]. If $0.5 < r < 1$, then $0 < 1 - r < 0.5$ and the interval is divided into three subintervals $[0, 1 - r]$, $[1 - r, r]$, and $[r, 1]$. A decision process is used to either squeeze from the right and get the new interval $[0, r]$ or squeeze from the left and get $[1 - r, 1]$. Then this new subinterval is divided into three subintervals in the same ratio as was $[0, 1]$.

We want to choose r so that one of the old points will be in the correct position with respect to the new interval as shown in Figure 8.2. This implies that the ratio $(1 - r):r$ be the same as $r:1$. Hence r satisfies the equation $1 - r = r^2$, which can be expressed as a quadratic equation $r^2 + r - 1 = 0$. The solution r satisfying $0.5 < r < 1$ is found to be $r = (5^{1/2} - 1)/2$.

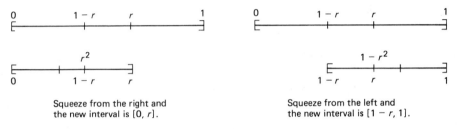

Squeeze from the right and   Squeeze from the left and
the new interval is $[0, r]$.   the new interval is $[1 - r, 1]$.

Figure 8.2   The intervals involved in the golden ratio search.

To use the golden search for finding the minimum of $f(x)$, a special condition must be met to ensure that there is a proper minimum in the interval.

Definition 8.3   (Unimodal Function).   The function $f(x)$ is **unimodal** on $I = [a, b]$ if there exists a unique number $p \in I$ such that

$$f(x) \quad \text{is decreasing on } [a, p] \tag{1}$$

$$f(x) \quad \text{is increasing on } [p, b]. \tag{2}$$

If $f(x)$ is known to be unimodal on $[a, b]$, it is possible to replace the interval with a subinterval on which $f(x)$ takes on its minimum value. The golden search requires that two interior points $c = a + (1 - r)(b - a)$ and $d = a + r(b - a)$ be used, where r is the golden ratio mentioned above. This results in $a < c < d < b$. The condition that $f(x)$ is unimodal guarantees that the function values $f(c)$ and $f(d)$ are less than max $\{f(a), f(b)\}$. We have two cases to consider (see Figure 8.3).

If $f(c) \le f(d)$, the minimum must occur in the subinterval $[a, d]$ and we replace b with d and continue the search in the new subinterval. If $f(d) < f(c)$, the minimum must occur in $[c, b]$ and we replace a with c and continue the search. The next example compares the root-finding method with the golden search method.

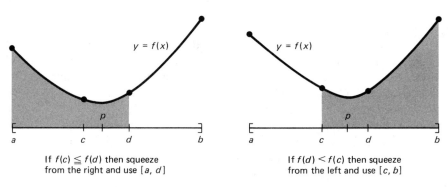

If $f(c) \leq f(d)$ then squeeze
from the right and use $[a, d]$

If $f(d) < f(c)$ then squeeze
from the left and use $[c, b]$

Figure 8.3   The decision process for the golden ratio search.

Example 8.2

Find the minimum of the unimodal function $f(x) = x^2 - \sin(x)$ in $[0, 1]$.

Solution by solving $f'(x) = 0$.   A root-finding method can be used to determine where the derivative $f'(x) = 2x - \cos(x)$ is zero. Since $f'(0) = -1$ and $f'(1) = 1.4596977$, a root of $f'(x)$ lies in the interval $[0, 1]$. Starting with $p_0 = 0$ and $p_1 = 1$, Table 8.1 shows the iterations.

TABLE 8.1   Secant Method for
Solving $f'(x) = 2x - \cos(x) = 0$

k	p_k	$2p_k - \cos(p_k)$
0	0.0000000	-1.00000000
1	1.0000000	1.45969769
2	0.4065540	-0.10538092
3	0.4465123	-0.00893398
4	0.4502137	0.00007329
5	0.4501836	-0.00000005

The conclusion from applying the secant method is that $f'(0.4501836) = 0$. The second derivative is $f''(x) = x + \sin(x)$ and we compute $f''(0.4501863) = 0.8853196 > 0$. Hence the minimum value is $f(0.4501863) = -0.2324656$.

Solution using the golden search.   At each step, the function values $f(c)$ and $f(d)$ are compared and a decision is made as to whether to continue the search in $[a, b]$ or $[c, b]$. Some of the computations are shown in Table 8.2.

At the twenty-third iteration the interval has been narrowed down to $[a_{23}, b_{23}] = [0.4501827, 0.4501983]$. This interval has width 0.0000156. However, the computed function values at the endpoints agree to eight decimal places [i.e., $f(a_{23}) \approx -0.23246558 \approx f(b_{23})$], hence the algorithm is terminated. A problem in using search methods is that the function may be flat near the minimum and this limits the accuracy that can be obtained. The secant method was able to find the more accurate answer $p_5 = 0.4501836$.

Although the golden search is slower in this example, it has the desirable feature that it can apply in cases where $f(x)$ is not differentiable.

TABLE 8.2   Golden Search for the Minimum of $f(x) = x^2 - \sin(x)$

k	a_k	c_k	d_k	b_k	$f(c_k)$	$f(d_k)$
0	0.0000000	0.3819660	0.6180340	1	−0.22684748	−0.19746793
1	0.0000000	0.2360680	0.3819660	0.6180340	−0.17815339	−0.22684748
2	0.2360680	0.3819660	0.4721360	0.6180340	−0.22684748	−0.23187724
3	0.3819660	0.4721360	0.5278640	0.6180340	−0.23187724	−0.22504882
4	0.3819660	0.4376941	0.4721360	0.5278640	−0.23227594	−0.23187724
5	0.3819660	0.4164079	0.4376941	0.4721360	−0.23108238	−0.23227594
6	0.4164079	0.4376941	0.4508497	0.4721360	−0.23227594	−0.23246503
.	.	.	.	.	.	.
.	.	.	.	.	.	.
.	.	.	.	.	.	.
21	0.4501574	0.4501730	0.4501827	0.4501983	−0.23246558	−0.23246558
22	0.4501730	0.4501827	0.4501886	0.4501983	−0.23246558	−0.23246558
23	0.4501827	0.4501886	0.4501923	0.4501983	−0.23246558	−0.23246558

Finding Extreme Values of f(x, y)

Definition 8.1 is easily extended to functions of several variables. Suppose that $f(x, y)$ is defined in the region

$$R = \{(x, y): (x - p)^2 + (y - q)^2 < r^2\}. \tag{3}$$

The function $f(x, y)$ has a local minimum at (p, q) provided that

$$f(p, q) \le f(x, y) \text{ for each point } (x, y) \in R. \tag{4}$$

The function $f(x, y)$ has a local maximum at (p, q) provided that

$$f(x, y) \le f(p, q) \text{ for each point } (x, y) \in R. \tag{5}$$

The second derivative test for an extreme value is an extension of Theorem 8.4.

Theorem 8.5   (Second Derivative Test).   Assume that $f(x, y)$ and its first- and second-order partial derivatives are continuous on a region R. Suppose that $(p, q) \in R$ is a critical point where both $f_x(p, q) = 0$ and $f_y(p, q) = 0$. The higher-order partial derivatives are used to determine the nature of the critical point.

(i) If $f_{xx}(p, q)f_{yy}(p, q) - f_{xy}^2(p, q) > 0$ and $f_{xx}(p, q) > 0$, then $f(p, q)$ is a local minimum of f.

(ii) If $f_{xx}(p, q)f_{yy}(p, q) - f_{xy}^2(p, q) > 0$ and $f_{xx}(p, q) < 0$, then $f(p, q)$ is a local maximum of f.

(iii) If $f_{xx}(p, q)f_{yy}(p, q) - f_{xy}^2(p, q) < 0$, then $f(x, y)$ does not have a local extremum at (p, q).

(iv) If $f_{xx}(p, q)f_{yy}(p, q) - f_{xy}^2(p, q) = 0$, this test is inconclusive.

Example 8.3

Find the minimum of $f(x, y) = x^2 - 4x + y^2 - y - xy$.

Solution. The first-order partial derivatives are

$$f_x(x, y) = 2x - 4 - y \quad \text{and} \quad f_y(x, y) = 2y - 1 - x. \tag{6}$$

Setting these derivatives equal to zero yields the linear system

$$2x - y = 4 \quad -x + 2y = 1. \tag{7}$$

The solution point to (7) is found to be $x = 3$, $y = 2$. The second-order partial derviatives of $f(x, y)$ are

$$f_{xx}(x, y) = 2 \quad f_{yy}(x, y) = 2 \quad f_{xy}(x, y) = -1.$$

It is easy to see that we have case (i) of Theorem 8.5, that is,

$$f_{xx}(3, 2)f_{yy}(3, 2) - f_{xy}^2(3, 2) = 3 > 0 \quad \text{and} \quad f_{xx}(3, 2) = 2 > 0.$$

Hence $f(x, y)$ has a local minimum $f(3, 2) = -7$ at the point $(3, 2)$.

The Nelder–Mead Method

A simplex method for finding a local minimum of a function of several variables has been devised by Nelder and Mead. For two variables, a simplex is a triangle, and the method is a pattern search that compares function values at the three vertices of a triangle. The worst vertex, where $f(x, y)$ is largest, is rejected and replaced with a new vertex. A new triangle is formed and the search is continued. The process generates a sequence of triangles (which might have different shapes), for which the function values at the vertices get smaller and smaller. The size of the triangles is reduced and the coordinates of the minimum point are found.

The algorithm is stated using the term simplex (a generalized triangle in N dimensions) and will find the minimum of a function of N variables. It is effective and computationally compact.

The Initial Triangle **BGW**

Let $f(x, y)$ be the function that is to be minimized. To start we are given three vertices of a triangle; $V_k = (x_k, y_k)$, $k = 1, 2, 3$. The function $f(x, y)$ is then evaluated at each of the three points $z_k = f(x_k, y_k)$ for $k = 1, 2, 3$. The subscripts are then reordered so that $z_1 \leq z_2 \leq z_3$. We use the notation

$$\mathbf{B} = (x_1, y_1) \quad \mathbf{G} = (x_2, y_2) \quad \mathbf{W} = (x_3, y_3) \tag{8}$$

to help remember that **B** is the best vertex, **G** is good (next to best), and **W** is the worst vertex.

Midpoint of the Good Side

The construction process uses the midpoint of the line segment joining **B** and **G**. It is found by averaging the coordinates;

$$\mathbf{M} = \frac{\mathbf{B} + \mathbf{G}}{2} = \left(\frac{x_1 + x_2}{2}, \frac{y_1 + y_2}{2}\right). \tag{9}$$

Reflection Using the Point **R**

The function decreases as we move along the side of the triangle from **W** to **B**, and it decreases as we move along the side from **W** to **G**. Hence it is feasible that $f(x, y)$ takes on smaller values at points that lie away from **W** on the opposite side of the line between **B** and **G**. We choose a test point **R** that is obtained by ''reflecting'' the triangle through the side $\widehat{BG}$. To determine **R**, we first find the midpoint **M** of the side $\widehat{BG}$. Then draw the line segment from **W** to **M** and call its length d. This last segment is extended a distance d through **M** to locate the point **R** (see Figure 8.4). The vector formula for **R** is

$$\mathbf{R} = \mathbf{M} + (\mathbf{M} - \mathbf{W}) = 2\mathbf{M} - \mathbf{W}. \tag{10}$$

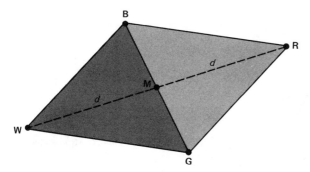

Figure 8.4   The triangle $\widehat{BGW}$ and midpoints **M** and reflected point **R** for the Nelder–Mead method.

Expansion Using the Point **E**

If the function value at **R** is smaller than the function value at **W** then we have moved in the correct direction toward the minimum. Perhaps the minimum is just a bit farther than the point **R**. So we extend the line segment through **M** and **R** to the point **E**. This forms an expanded triangle BGE. The point **E** is found by moving an additional distance d along the line joining **M** and **R** (see Figure 8.5). If the function value at **E** is

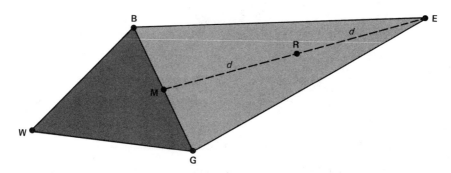

Figure 8.5   The triangle $\widehat{BGW}$ and point **R** and extended point **E**.

less than the function value at $\mathbf{R}$, then we have found a better vertex than $\mathbf{R}$. The vector formula for $\mathbf{E}$ is

$$\mathbf{E} = \mathbf{R} + (\mathbf{R} - \mathbf{M}) = 2\mathbf{R} - \mathbf{M}. \tag{11}$$

Contraction Using the Point C

If the function values at $\mathbf{R}$ and $\mathbf{W}$ are the same, another point must be tested. Perhaps the function is smaller at $\mathbf{M}$, but we cannot replace $\mathbf{W}$ with $\mathbf{M}$ because we must have a triangle. Consider the two midpoints $\mathbf{C}_1$ and $\mathbf{C}_2$ of the line segments $\widehat{WM}$ and $\widehat{MR}$, respectively (see Figure 8.6). The point with the smaller function value is called $\mathbf{C}$, and the new triangle is *BGC*. *Note:* The choice between $\mathbf{C}_1$ and $\mathbf{C}_2$ might seem inappropriate for the two-dimensional case, but it is important in higher dimensions.

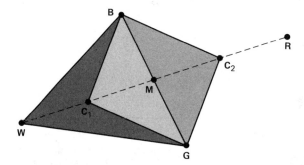

Figure 8.6   The contraction point $\mathbf{C}_1$ or $\mathbf{C}_2$ for the Nelder–Mead method.

Shrink toward B

If the function value at $\mathbf{C}$ is not less than the value at $\mathbf{W}$, the points $\mathbf{G}$ and $\mathbf{W}$ must be shrunk toward $\mathbf{B}$ (see Figure 8.7). The point $\mathbf{G}$ is replaced with $\mathbf{M}$, and $\mathbf{W}$ is replaced with $\mathbf{S}$, which is the midpoint of the line segment joining $\mathbf{B}$ with $\mathbf{W}$.

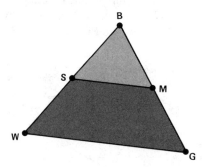

Figure 8.7   Shrinking the triangle toward $\mathbf{B}$.

Logical Decisions for Each Step

A computationally efficient algorithm should perform function evaluations only if needed. In each step, a new vertex is found which replaces $\mathbf{W}$. As soon as it is found, further investigation is not needed, and the iteration step is completed. In the two-dimensional cases, the logical details are explained in Table 8.3.

TABLE 8.3   Logical Decisions for the Nelder–Mead Algorithm

IF   f(R) < f(G),   THEN   Perform Case (i) {either reflect or extend}
                   ELSE   Perform Case (ii) {either contract or shrink}

BEGIN {Case (i).}	BEGIN {Case (ii).}
IF f(B) < f(R) THEN	IF f(R) < f(W) THEN
replace W with R	replace W with R
ELSE	Compute C = [W+M]/2
	or C = [M+R]/2 and f(C)
Compute E and f(E)	IF f(C) < f(W) THEN
IF f(E) < f(B) THEN	replace W with C
replace W with E	ELSE
ELSE	Compute S and f(S)
replace W with R	replace W with S
ENDIF	replace G with M
ENDIF	ENDIF
END {Case (i).}	END {Case (ii).}

Example 8.4

Use the Nelder–Mead algorithm to find the minimum of $f(x, y) = x^2 - 4x + y^2 - y - xy$. Start with the three vertices

$$\mathbf{V}_1 = (0, 0) \qquad \mathbf{V}_2 = (1.2, 0.0) \qquad \mathbf{V}_3 = (0.0, 0.8).$$

Solution.   The function $f(x, y)$ takes on the values

$$f(0, 0) = 0.0 \qquad f(1.2, 0.0) = -3.36 \qquad f(0.0, 0.8) = -0.16.$$

The function values must be compared to determine **B**, **G**, and **W**;

$$\mathbf{B} = (1.2, 0.0) \qquad \mathbf{G} = (0.0, 0.8) \qquad \mathbf{W} = (0, 0).$$

The vertex $\mathbf{W} = (0, 0)$ will be replaced. The points **M** and **R** are

$$\mathbf{M} = \frac{\mathbf{B} + \mathbf{G}}{2} = (0.6, 0.4) \quad \text{and} \quad \mathbf{R} = 2\mathbf{M} - \mathbf{W} = (1.2, 0.8).$$

The function value $f(\mathbf{R}) = f(1.2, 0.8) = -4.48$ is less than $f(\mathbf{G})$, so the situation is case (i). Since $f(\mathbf{R}) \leq f(\mathbf{B})$ we have moved in the right direction and the vertex **E** must be constructed;

$$\mathbf{E} = 2\mathbf{R} - \mathbf{M} = 2(1.2, 0.8) - (0.6, 0.4) = (1.8, 1.2).$$

The function value $f(\mathbf{E}) = f(1.8, 1.2) = -5.88$ is less than $f(\mathbf{B})$, and the new triangle has vertices

$$\mathbf{V}_1 = (1.8, 1.2) \qquad \mathbf{V}_2 = (1.2, 0.0) \qquad \mathbf{V}_3 = (0.0, 0.8).$$

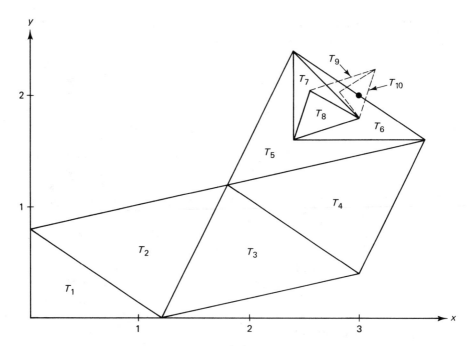

Figure 8.8 The sequence of triangles $\{T_k\}$ converging to the point (3, 2) for the Nelder–Mead method.

The process continues and generates a sequence of triangles that converge down on solution point (3, 2) (see Figure 8.8). Table 8.4 gives the function values at vertices of the triangle for several steps in the iteration. A computer implementation of the algorithm continued until the thirty-third step, where the best vertex was $\mathbf{B}$ = (2.99996456, 1.99983839) and $f(\mathbf{B})$ = −6.99999998. These values are approximations to $f(3, 2)$ = −7 found in Example 8.3. The reason that the iteration quit before (3, 2) was obtained is that the

TABLE 8.4 Function Values at Various Triangles for Example 8.4

k	Best point	Good point	Worst point
1	$f(1.2, 0.0)$ = −3.36	$f(0.0, 0.8)$ = −0.16	$f(0.0, 0.0)$ = 0.00
2	$f(1.8, 1.2)$ = −5.88	$f(1.2, 0.0)$ = −3.36	$f(0.0, 0.8)$ = −0.16
3	$f(1.8, 1.2)$ = −5.88	$f(3.0, 0.4)$ = −4.44	$f(1.2, 0.0)$ = −3.36
4	$f(3.6, 1.6)$ = −6.24	$f(1.8, 1.2)$ = −5.88	$f(3.0, 0.4)$ = −4.44
5	$f(3.6, 1.6)$ = −6.24	$f(2.4, 2.4)$ = −6.24	$f(1.8, 1.2)$ = −5.88
6	$f(2.4, 1.6)$ = −6.72	$f(3.6, 1.6)$ = −6.24	$f(2.4, 2.4)$ = −6.24
7	$f(3.0, 1.8)$ = −6.96	$f(2.4, 1.6)$ = −6.72	$f(2.4, 2.4)$ = −6.24
8	$f(3.0, 1.8)$ = −6.96	$f(2.55, 2.05)$ = −6.7725	$f(2.4, 1.6)$ = −6.72
9	$f(3.0, 1.8)$ = −6.96	$f(3.15, 2.25)$ = −6.9525	$f(2.55, 2.05)$ = −6.7725
10	$f(3.0, 1.8)$ = −6.96	$f(2.8125, 2.0375)$ = −6.95640625	$f(3.15, 2.25)$ = −6.9525

function is flat near the minimum. The function values $f(\mathbf{B})$, $f(\mathbf{G})$, and $f(\mathbf{W})$ were checked and found to be the same (this is an example of round-off error), and the algorithm was terminated.

Minimization Using Derivatives

Suppose that $f(x)$ is unimodal over $[a, b]$ and has a unique minimum at $x = p$. Also, assume that $f'(x)$ is defined at all points in (a, b). Let the starting value p_0 lie in (a, b). If $f'(p_0) < 0$, the minimum point p lies to the right of p_0. If $f'(p_0) > 0$, p lies to the left of p_0 (see Figure 8.9).

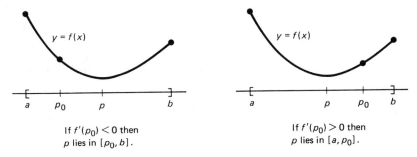

Figure 8.9   Using $f'(x)$ to find the minimum value of the unimodal function $f(x)$ on the interval $[a, b]$.

Bracketing the Minimum

Our first task is to obtain three test values:

$$p_0, \quad p_1 = p_0 + h, \quad \text{and} \quad p_2 = p_0 + 2h \tag{12}$$

so that

$$f(p_0) > f(p_1) \quad \text{and} \quad f(p_1) < f(p_2). \tag{13}$$

Suppose that $f'(p_0) < 0$; then $p_0 < p$ and the step size h should be chosen positive. It is an easy task to find a value for h so that the three points in (12) satisfy (13). Start with $h = 1$ in formula (12) (provided that $a + 1 < b$).

Case (i):   If (13) is satisfied, we are done.

Case (ii):   If $f(p_0) > f(p_1)$ and $f(p_1) > f(p_2)$, then $p_2 < p$. We need to check points that lie farther to the right. Double the step size and repeat the process.

Case (iii):   If $f(p_0) \le f(p_1)$, we have jumped over p and h is too large. We need to check values closer to p_0. Reduce the step size by a factor of $\frac{1}{2}$ and repeat the process.

When $f'(p_0) > 0$ the step size h should be chosen negative and then cases similar to (i) to (iii) can be used.

Quadratic Approximation to Find p

Finally, we have three points (12) that satisfy (13). We will use quadratic interpolation to find p_{min}, which is an approximation to p. The Lagrange polynomial based on the nodes in (12) is

$$Q(x) = \frac{y_0(x - p_1)(x - p_2)}{2h^2} - \frac{y_1(x - p_0)(x - p_2)}{h^2} + \frac{y_2(x - p_0)(x - p_1)}{2h^2}. \quad (14)$$

The derivative of $Q(x)$ is

$$Q'(x) = \frac{y_0(2x - p_1 - p_2)}{2h^2} - \frac{y_1(2x - p_0 - p_2)}{h^2} + \frac{y_2(2x - p_0 - p_1)}{2h^2}. \quad (15)$$

Solving $Q'(x) = 0$ in the form $Q'(p_0 + h_{min}) = 0$ yields

$$0 = \frac{y_0[2(p_0 + h_{min}) - p_1 - p_2]}{2h^2} - \frac{y_1[4(p_0 + h_{min}) - 2p_0 - 2p_2]}{2h^2}$$

$$+ \frac{y_2[2(p_0 + h_{min}) - p_0 - p_1]}{2h^2}. \quad (16)$$

Multiply each term in (16) by $2h^2$ and collect terms involving h_{min}:

$$-h_{min}(2y_0 - 4y_1 + 2y_2) = y_0(2p_0 - p_1 - p_2) - y_1(4p_0 - 2p_0 - 2p_2)$$

$$+ y_2(2p_0 - p_0 - p_1)$$

$$= y_0(-3h) - y_1(-4h) + y_2(-h).$$

This last quantity is easily solved for h_{min}:

$$h_{min} = \frac{h(4y_1 - 3y_0 - y_2)}{4y_1 - 2y_0 - 2y_2}. \quad (17)$$

The value $p_{min} = p_0 + h_{min}$ is a better approximation to p than p_0. Hence we can replace p_0 with p_{min} and repeat the two processes outlined above to determine a new h and a new h_{min}. Continue the iteration until the desired accuracy is achieved. The details are outlined in Algorithm 8.3.

Steepest Descent or Gradient Method

Now let us turn to the minimization of a function $f(\mathbf{X})$ of N variables, where $\mathbf{X} = (x_1, x_2, \ldots, x_N)$. The gradient of $f(\mathbf{X})$ is a vector function defined as follows:

$$\text{grad } f(\mathbf{X}) = (f_1, f_2, \ldots, f_N), \quad (18)$$

where the partial derivatives $f_k = \partial f / \partial x_k$ are evaluated at $\mathbf{X}$.

Recall that the gradient vector (18) points locally in the direction of the greatest rate of increase of $f(\mathbf{X})$. Hence $-\text{grad } f(\mathbf{X})$ points locally in the direction of the greatest decrease. Start at the point $\mathbf{P}_0$ and search along the line through $\mathbf{P}_0$ in the direction $\mathbf{S}_0 = -\mathbf{G}/\|\mathbf{G}\|$, where $\mathbf{G} = \text{grad}(\mathbf{P}_0)$. You will arrive at a point $\mathbf{P}_1$ which is a local minimum when the point $\mathbf{X}$ is constrained to lie on the line $\mathbf{X} = \mathbf{P}_0 + t\mathbf{S}_0$.

Next, we can compute $\mathbf{G} = \text{grad}(\mathbf{P}_1)$ and move in the search direction $\mathbf{S}_1 = -\mathbf{G}/\|\mathbf{G}\|$. You will come to $\mathbf{P}_2$, which is a local minimum when $\mathbf{X}$ is constrained to lie on the line $\mathbf{X} = \mathbf{P}_1 + t\mathbf{S}_1$. Iteration will produce a sequence $\{\mathbf{P}_k\}$ of points with the property $f(\mathbf{P}_0) > f(\mathbf{P}_1) > \cdots > f(\mathbf{P}_k) > \cdots$. If $\lim_{k \to \infty} \mathbf{P}_k = \mathbf{P}$, then $f(\mathbf{P})$ will be a local minimum for $f(\mathbf{X})$.

Outline of the Gradient Method

Suppose that $\mathbf{P}_k$ has been obtained.

Step 1. Evaluate the gradient vector $\mathbf{G} = \text{grad } f(\mathbf{P}_k)$.

Step 2. Compute the search direction $\mathbf{S} = -\mathbf{G}/\|\mathbf{G}\|$.

Step 3. Perform a single parameter minimization of $\phi(t) = f(\mathbf{P}_k + t\mathbf{S})$ on the interval $[0, b]$, where b is large. This will produce a value $t = h_{\min}$ that is a local minimum for $\phi(t)$. The relation $\phi(h_{\min}) = f(\mathbf{P}_k + h_{\min}\mathbf{S})$ shows that this is a minimum for $f(\mathbf{X})$ along the search line $\mathbf{X} = \mathbf{P}_k + t\mathbf{S}$.

Step 4. Construct the next point $\mathbf{P}_{k+1} = \mathbf{P}_k + h_{\min}\mathbf{S}$.

Step 5. Perform the termination test for minimization; that is, are the function values $f(\mathbf{P}_k)$ and $f(\mathbf{P}_{k+1})$ sufficiently close and the distance $\|\mathbf{P}_{k+1} - \mathbf{P}_k\|$ small enough?

Repeat the process.

Algorithm 8.1   (Golden Search for a Minimum).   To numerically approximate the minimum of $f(x)$ on the interval $[a, b]$ by using a golden search. Proceed with the method only if $f(x)$ is a unimodal function on the interval $[a, b]$.

Delta := 10^{-5}	{Tolerance for interval width}
Epsilon := 10^{-7}	{Tolerance for $|f(b) - f(a)|$}
Rone := $[5^{1/2} - 1]/2$	{Determine constants
Rtwo := Rone*Rone	for golden search}
INPUT A, B	{Input endpoints of interval}
YA := F(A), YB := F(B), H := B−A	{Compute function values}
C := A+Rtwo*H, YC := F(C)	{Compute two
D := A+Rone*H, YD := F(D)	interior points}

```
WHILE  |YB−YA|>Epsilon OR H>Delta  DO
         IF  YC < YD  THEN
              B := D, YB := YD                           {Squeeze from the right}
              D := C, YD := YC, H := B−A
              C := A + Rtwo*H, YC := F(C)
         ELSE
              A := C, YA := YC                            {Squeeze from the left}
              C := D, YC := YD, H := B−A
              D := A + Rone*H, YD := F(D)
         ENDIF


         P := A, YP := YA
         IF  YB < YA  THEN  P := B, YP := YB

         PRINT 'The minimum of f(x) occurs at' P               {Output}
         PRINT 'The accuracy is +−' H
         PRINT 'The minimum of the function f(P) is' YP
```

Algorithm 8.2 (Nelder–Mead's Minimization Method).   To approximate a local minimum of $f(x_1, x_2, \ldots, x_N)$, where f is a continuous function of N real variables, and given the $N + 1$ initial starting points $\mathbf{V}_k = (v_{k,1}, \ldots, v_{k,N})$ for $k = 0, 1, \ldots, N$.

```
SUBROUTINE  Size(V,Lo,Norm)                              {Size of simplex}
        Norm := 0
FOR  J = 0  TO  N  DO
        S := 0
        FOR  K = 1  TO  N  DO  S := S+[V(Lo,K)−V(J,K)]²
        IF     S > Norm  THEN  Norm := S
        Norm := [Norm]¹ᐟ²                                {End of procedure Size}

SUBROUTINE  Order(Y,Lo,Li,Ho,Hi)                         {Find indices:
        Lo := 0, Hi := 0                                  Lo, best vertex;
FOR  J = 1  TO  N  DO                                     Hi, worst vertex}
        IF  Y(J)<Y(Lo)  THEN  Lo := J
        IF  Y(J)>Y(Hi)  THEN  Hi := J
        Li := Hi, Ho := Lo                               {Li, next to best;
FOR  J = 0  TO  N  DO                                     Ho, next to worst}
        IF  J≠Lo  AND  Y(J)<Y(Li)  THEN  Li := J
        IF  J≠Hi  AND  Y(J)>Y(Ho)  THEN  Ho := J         {End of procedure Order}
```

```
SUBROUTINE  Newpoints(V,Hi,M,R,YR)                          {Compute M and R}
FOR  K = 1  TO  N  DO
     │   S := 0
     │   FOR  J=0  TO  N  DO   S := S+V(J,K)                 {Construct median M}
     └── M(K) := [S − V(Hi,K)]/N
         FOR  K = 1  TO  N  DO  R(K) := 2*M(K)−V(Hi,K)       {Construct R}
         YR := F(R(1), . . . , R(N))                         {End of procedure Newpoints}

SUBROUTINE  Shrink(V,Y,Lo)                                  {Shrink simplex}
FOR  J = 0  TO  N  DO
     │   IF  J≠Lo  THEN
     │       │   FOR  K=1  TO  N  DO  V(J,K) := [V(J,K)+V(Lo,K)]/2
     └────── └── Y(J) := F(V(J,1), . . . , V(J,N))           {End of procedure Shrink}

SUBROUTINE  Replace(V,R,YR,Hi)                              {Replace W with R}
         FOR  K=1  TO  N  DO  V(Hi,K) := R(K)
         Y(Hi) := YR                                        {End of procedure Replace}

SUBROUTINE  Improve(V,Y,M,R,YR,Lo,Li,Ho,Hi)                {Improve worst vertex}

IF  YR < Y(Ho)  THEN
    │   IF  Y(Li) < YR  THEN                                {N−M: use Y(Lo), not Y(Li)}
    │       │   CALL Procedure Replace(V,R,YR,Hi)           {Replace W with R}
    │       ELSE
    │           FOR  K=1  TO  N  DO  E(K) := 2*R(K)−M(K)    {Construct E}
    │           YE := F(E(1), . . . , E(N))
    │           IF  YE < Y(Li)  THEN                        {N−M: use Y(Lo), not Y(Li)}
    │               │   FOR  K=1  TO  N  DO  V(Hi,K) := E(K) {Replace W with E}
    │               │   Y(Hi) := YE
    │               ELSE
    │                   CALL Procedure Replace(V,R,YR,Hi)   {Replace W with R}
    │               ENDIF
    │   ENDIF
    ELSE
    │   IF  YR < Y(Hi)  THEN
    │   └── CALL Procedure Replace(V,R,YR,Hi)               {Replace W with R}
    │           FOR  K=1  TO  N  DO  C(K) := [V(Hi,K)+M(K)]/2 {Construct C}
    │           YC := F(C(1), . . . , C(N))
    │   IF  YC < Y(Hi)  THEN
    │       │   FOR  K=1  TO  N  DO  V(Hi,K) := C(K)        {Replace W with C}
    │       │   Y(Hi) := YC
    │       ELSE
    │       │   CALL Procedure Shrink(V,Y,Lo)               {Shrink simplex}
    │   ENDIF
ENDIF                                                       {End of the procedure improve}
```

{The main program starts here.}

Tol := 10^{-6}	{Convergence criterion}
Min := 10	{Minimum number of iterations}
Max := 200	{Maximum number of iterations}
M := 0	{Initialize the counter}
INPUT N	{Number of variables}

FOR J = 0 TO N DO   {Get N+1 initial points}
└──── GET V(J,1), . . . , V(J,N)   {Get N coordinates of V_J}

FOR J = 0 TO N DO   {Compute function}
└──── Y(J) := F(V(J,1), . . . , V(J,N))   value at V_J}

CALL Procedure Order(Y,Lo,Li,Ho,Hi)

WHILE [Y(Hi)>Y(Lo)+Tol AND M<Max] OR [M<Min] DO
│   CALL Procedure Newpoints(V,Hi,M,R,YR)
│   CALL Procedure Improve(V,Y,M,R,YR,Lo,Li,Ho,Hi)
│   CALL Procedure Order(Y,Lo,Li,Ho,Hi)
└──── M := M+1   {Increment the counter}

CALL Procedure Size(V,Lo,Norm)

PRINT "Coordinates of local minimum are"   {Output}
          V(Lo,1), V(Lo,2), . . . , V(Lo,N)
PRINT "The function value at this point is" Y(Lo)
PRINT "The size of the final simplex is" Norm

Algorithm 8.3   (Local Minimum Search Using Quadratic Interpolation).   To find a local minimum of the function $f(x)$ over the interval $[a, b]$, by starting with one initial approximation p_0 and then searching the intervals $[a, p_0]$ and $[p_0, b]$.

Delta := 10^{-5}	{Convergence tolerance for the abscissas}
Epsilon := 10^{-7}	{Convergence tolerance for the ordinates}
Jmax := 20, Kmax := 50	{Maximum number of iterations}

INPUT P0   {Get the starting value}

Err := 1	{Initialize the variable}
H := 1	{Initialize the step size}
K := 0	{Initialize the counter}

```
WHILE  (K<Kmax) AND (Err>Epsilon)  DO                    {Loop to find Ymin}
    IF  f'(P0)>0  THEN  H := −|H|                        {Slope points downhill}
    P1 := P0+H, P2 := P0+2*H                             {Two more test points}
    Y0 := F(P0), Y1 := F(P1), Y2 := F(P2)                {Compute the function values}
    Cond := 0, J := 0                                    {Initialize the variables}
    WHILE  (J<Jmax) AND (H>Delta) AND (Cond=0)  DO       {Determine H so
        IF  Y0 ≤ Y1  THEN                                that Y1<Y0 & Y1<Y2}
            P2 := P1, Y2 := Y1, H := H/2                  {Make H smaller}
            P1 := P0+H, Y1 := F(P1)
        ELSE
            IF  Y2 < Y1  THEN
                P1 := P2, Y1 := Y2, H := 2*H             {Make H larger}
                P2 := P0+2*H, Y2 := F(P2)
            ELSE
                Cond := −1                               {Loop termination}
        ENDIF
        J := J+1                                         {Increment the counter}
    ENDWHILE                                             {End loop to determine proper H}
    D := 4*Y1−2*Y0−2*Y2                                  {Use quadratic
    IF  D<0  THEN                                        interpolation to
        Hmin := H*(4*Y1−3*Y0−Y2)/D                       compute the abscissa
    ELSE                                                 at the minimum}
        Hmin := H/3                                      {Check division by zero}
        Cond := 4
    Pmin := P0+Hmin                                      {Coordinates of
    Ymin := F(Pmin)                                      the minimum}
    H := |H|                                             {How determine the
    IF  |Hmin|<H  THEN  H := |Hmin|                      magnitude of
    IF  |Hmin−H|<H  THEN  H := |Hmin−H|                  the next H}
    IF  |Hmin−2*H|<H  THEN  H := |Hmin−2*H|
    IF  H < Delta  THEN  Cond := 1                       {Convergence of abscissa}
    Err := |Y0−Ymin|
    IF  |Y1−Ymin|<Err  THEN  Err := |Y1−Ymin|
    IF  |Y2−Ymin|<Err  THEN  Err := |Y2−Ymin|
    IF  Err<Epsilon  THEN  Cond := 2                     {Convergence of ordinate}
    K := K+1
    P0 := Pmin                                           {Update the value P0}
    IF  (Cond=2) AND (H<Delta)  THEN  Cond := 3
ENDWHILE                                                 {End loop to find Ymin}
    PRINT 'An approximation for the minimum is', Pmin    {Output}
    PRINT 'An estimate for its accuracy is', |H|
    PRINT 'The function value f(Pmin) is', Ymin
```

```
IF   Cond = 0   THEN
     PRINT 'Convergence is doubtful because the maximum'
           'number of iterations was exceeded.'
IF   Cond = 1   THEN
     PRINT 'Convergence of the abscissas has been achieved.'
IF   Cond = 2   THEN
     PRINT 'Convergence of the ordinates has been achieved.'
IF   Cond = 3   THEN
     PRINT 'Convergence of both coordinates has been achieved.'
IF   Cond = 4   THEN
     PRINT 'Convergence is doubtful because'
           'division by zero was encountered.'
```

Algorithm 8.4 (Steepest Descent or Gradient Method).   To numerically approximate a local minimum of $f(\mathbf{X})$, where f is a continuous function of N real variables and $\mathbf{X} = (x_1, x_2, \ldots, x_N)$, by starting with one point $\mathbf{P}_0$ and using the gradient method.

Delta := 10^{-5} {Convergence tolerance for points}
Epsilon := 10^{-7} {Tolerance for the function values}
Jmax := 20, Max := 50   {Maximum number of iterations}

PROCEDURE GRADIENT $\{\vec{S} = (s_1, s_2, \ldots, s_N)$ is a
    $\vec{G} := \text{grad } f(\vec{P})$ unit vector pointing
    $\vec{S} := -\vec{G}/\|\vec{G}\|$ toward the minimum$\}$

PROCEDURE QUADMIN

Cond := 0, J := 0

$\vec{P}_1 := \vec{P}_0 + H*\vec{S}$, $\vec{P}_2 := \vec{P}_0 + 2*H*\vec{S}$

Y1 := $f(\vec{P}_1)$, Y2 := $f(\vec{P}_2)$

WHILE (J<Jmax) AND (Cond=0) DO   {Determine H so that

   IF Y0 ≤ Y1 THEN   Y1<Y0 and Y1<Y2}

     $\vec{P}_2 := \vec{P}_1$, Y2 := Y1, H := H/2   {Make H smaller}

     $\vec{P}_1 := \vec{P}_0 + H*\vec{S}$, Y1 := $f(\vec{P}_1)$

   ELSE

     IF Y2 < Y1 THEN

       $\vec{P}_1 := \vec{P}_2$, Y1 := Y2, H := 2*H   {Make H larger}

       $\vec{P}_2 := \vec{P}_0 + 2*H*\vec{S}$, Y2 := $f(\vec{P}_2)$

     ELSE

       Cond := −1

IF H<Delta THEN Cond := 1

D := 4*Y1−2*Y0−2*Y2

IF D<0 THEN   {Quadratic interpolation

   Hmin := H*(4*Y1−3*Y0−Y2)/D   to find Hmin)}

   ELSE

     Cond := 4, Hmin := H/3   {Check division by zero}

$\vec{P}_{min} := \vec{P}_0 + Hmin*\vec{S}$, Ymin := $f(\vec{P}_{min})$

H0 := |Hmin|, H1 := |Hmin−H|, H2 := |Hmin−2*H| {Convergence

IF H0<H THEN H := H0

IF H1<H THEN H := H1   test for

IF H2<H THEN H := H2

IF H<Delta THEN Cond := 1   the points}

E0 := |Y0−Ymin|, E1 := |Y1−Ymin|, E2 := |Y2−Ymin| {Convergence

IF E0 < Err THEN Err := E0

IF E1 < Err THEN Err := E1   test for the

IF E2 < Err THEN Err := E2

IF (E0=0) AND (E1=0) AND (E2=0) THEN Err := 0   function values}

IF Err<Epsilon THEN Cond := 2

IF (Cond=2) AND (H<Delta) THEN Cond := 3

J := J+1

END   {End of Procedure Qmin}

{The main Gradient Search program starts here.}
INPUT $\vec{P}_0$ {Get the starting point}
Count := 0, H := 1, Err := 1   {Initialize variables}
WHILE (Count<Max) AND ((H>Delta) OR (Err>Epsilon)) DO
    | CALL PROCEDURE GRADIENT
    | CALL PROCEDURE QUADMIN
    └ $\vec{P}_0 = \vec{P}_{min}$, Y0 := Ymin, Count := Count+1

PRINT 'The local minimum is', $\vec{P}_{min}$ {Output}
PRINT 'The minimum value of the function is', Ymin

    See Algorithm 8.3 for the output remarks when Cond $= 0, 1, 2, 3, 4$.

EXERCISES FOR MINIMIZATION OF A FUNCTION

1. Use Definition 8.2 to show that the following functions are increasing on the interval $[0, 4]$.
   (a) $f(x) = x^2$ (b) $f(x) = x^{1/2}$

2. Use Theorem 8.1 to show that the following functions are decreasing on the interval $[1, 3]$.
   (a) $f(x) = x^{-1}$ (b) $f(x) = \cos(x)$

3. Use Definition 8.3 to show that the following functions are unimodal on the interval $[0, 4]$.
   (a) $f(x) = x^2 - 2x + 1$ (b) $f(x) = \cos(x)$

In Exercises 4–11:
(a) Use Theorem 8.4 to find the local minimum.
(b) Use Algorithm 8.1 to find the local minimum.
(c) Use Algorithm 8.3 to find the local minimum, starting with the midpoint of the given interval.

4. $f(x) = 3x^2 - 2x + 5$ on the interval $[0, 1]$.
5. $f(x) = 2x^3 - 3x^2 - 12x + 1$ on the interval $[0, 3]$.
6. $f(x) = 4x^3 - 8x^2 - 11x + 5$ on the interval $[0, 2]$.
7. $f(x) = x + 3/x^2$ on the interval $[0.5, 3]$.
8. $f(x) = (x + 2.5)/(4 - x^2)$ on the interval $[-1.9, 1.9]$.
9. $f(x) = \exp(x)/x^2$ on the interval $[0.5, 3]$.
10. $f(x) = -\sin(x) - \sin(3x)/3$ on the interval $[0, 2]$.
11. $f(x) = -2\sin(x) + \sin(2x) - 2\sin(3x)/3$ on $[1, 3]$.
12. Find the point on the parabola $y = x^2$ that is closest to the point $(3, 1)$.
13. Find the point on the curve $y = \sin(x)$ that is closest to the point $(2, 1)$.

In Exercises 14–21:
(a) Use Theorem 8.5 to find the local minimum.
(b) Use Algorithm 8.2 to find the local minimum.
(c) Use Algorithm 8.4 to find the local minimum, starting with the first vertex given.

14. (a) $f(x, y) = x^3 + y^3 - 3x - 3y + 5$
    (b) Use the starting vertices $(1, 2)$, $(2, 0)$, and $(2, 2)$.

15. (a) $f(x, y) = x^2 + y^2 + x - 2y - xy + 1$
  (b) Use the starting vertices $(0, 0)$, $(2, 0)$, and $(2, 1)$.

16. (a) $f(x, y) = x^2y + xy^2 - 3xy$
  (b) Use the starting vertices $(0, 0)$, $(2, 0)$, and $(2, 1)$.

17. (a) $f(x, y) = x^4 - 8xy + 2y^2$
  (b) i. Use the starting vertices $(2, 2)$, $(3, 2)$, and $(3, 3)$.
      ii. Use the starting vertices $(-1, -2)$, $(-1, -3)$, and $(-2, -2)$.

18. (a) $f(x, y) = (x - y)/(2 + x^2 + y^2)$
  (b) Use the starting vertices $(0, 0)$, $(0, 1)$, and $(1, 1)$.

19. (a) $f(x, y) = x^4 + y^4 - (x + y)^2$
  (b) i. Use the starting vertices $(0, 0)$, $(0, 1)$, and $(2, 0)$.
      ii. Use the starting vertices $(0, 0)$, $(-2, -2)$, and $(-2, 0)$.

20. (a) $f(x, y) = (x - y)^4 + (x + y - 2)^2$
  (b) Use the starting vertices $(0, 0)$, $(0, 2)$, and $(1, 0)$.

21. [Rosenbrock's parabolic valley, circa 1960]
  (a) $f(x, y) = 100(y - x^2)^2 + (1 - x)^2$
  (b) Use the starting vertices $(0, 0)$, $(1, 0)$, and $(0, 2)$.

22. Let $\mathbf{B} = (2, -3)$, $\mathbf{G} = (1, 1)$, and $\mathbf{W} = (5, 2)$. Find the points $\mathbf{M}$, $\mathbf{R}$, and $\mathbf{E}$ and sketch the triangles that are involved.

23. Let $\mathbf{B} = (-1, 2)$, $\mathbf{G} = (-2, -5)$, and $\mathbf{W} = (3, 1)$. Find the points $\mathbf{M}$, $\mathbf{R}$, and $\mathbf{E}$ and sketch the triangles that are involved.

24. Give a vector proof that $\mathbf{M} = (\mathbf{B} + \mathbf{G})/2$ is the midpoint of the line segment joining the points $\mathbf{B}$ and $\mathbf{G}$.

25. Give a vector proof of equation (10).

26. Give a vector proof of equation (11).

27. Give a vector proof that the medians of any triangle intersect at a point which is two-thirds of the distance from each vertex to the midpoint of the opposite side.

28. Let $\mathbf{B} = (0, 0, 0)$, $\mathbf{G} = (1, 1, 0)$, $\mathbf{P} = (0, 0, 1)$, and $\mathbf{W} = (1, 0, 0)$.
  (a) Sketch the tetrahedron $BGPW$.
  (b) Find $\mathbf{M} = (\mathbf{B} + \mathbf{G} + \mathbf{P})/3$.
  (c) Find $\mathbf{R} = 2\mathbf{M} - \mathbf{W}$ and sketch the tetrahedron $BGPR$.
  (d) Find $\mathbf{E} = 2\mathbf{R} - \mathbf{M}$ and sketch the tetrahedron $BGPE$.

29. Let $\mathbf{B} = (0, 0, 0)$, $\mathbf{G} = (0, 2, 0)$, $\mathbf{P} = (0, 1, 1)$, and $\mathbf{W} = (2, 1, 0)$. Follow the instructions in Exercise 28.

30. Write a report on the Fibonacci search algorithm.

31. Write a report on the method of steepest descent.

In Exercises 32–37:
(a) Set all partial derivatives equal to zero and solve for the local minimum.
(b) Use Algorithm 8.2 to find the local minimum.
(c) Use Algorithm 8.4 to find the local minimum, starting with the first vertex given.

32. (a) $f(x, y, z) = 2x^2 + 2y^2 + z^2 - 2xy + yz - 7y - 4z$
  (b) Start with $(1, 1, 1)$, $(0, 1, 0)$, $(1, 0, 1)$, and $(0, 0, 1)$.

33. (a) $f(x, y, z) = 2x^2 + 2y^2 + z^2 + xy - xz - 2x - 5y - 5z$
   (b) Start with $(1, 1, 1)$, $(0, 1, 0)$, $(1, 0, 1)$, and $(0, 0, 1)$.

34. (a) $f(x, y, z) = x^4 + y^4 + z^4 - 4xyz$
   (b) Use $(0.5, 0.5, 0.5)$, $(1, 0.5, 0.5)$, $(0.5, 2, 0.5)$, and $(0.5, 0.5, 3)$.

35. (a) $f(x, y, z, u) = 2(x^2 + y^2 + z^2 + u^2) - x(y + z - u) + yz - 3x - 8y - 5z - 9u$
   (b) Start the search near $(1, 1, 1, 1)$.

36. (a) $f(x, y, z, u) = 2(x^2 + y^2 + z^2 + u^2) + x(y + z - u) + yz - 5x - 6y - 6z - 3u$
   (b) Start the search near $(0, 0, 0, 0)$.

37. (a) $f(x, y, z, u) = xyzu + 1/x + 1/y + 1/z + 1/u$
   (b) Use $(0.7, 0.7, 0.7, 0.7)$, $(1.2, 0.7, 0.7, 0.7)$, $(0.7, 1.2, 0.7, 0.7)$, $(0.7, 0.7, 1.2, 0.7)$, and $(0.7, 0.7, 0.7, 1.2)$.

38. Write a report on minimization of multivariable functions. Include a discussion of the quasi-Newton methods. See References [29, 96, 97, 139, 152, and 153].

9

Solution of Differential Equations

Differential equations are commonly used for mathematical modeling in science and engineering. Often there is no known analytic solution and numerical approximations are required. As an illustration, we consider population dynamics and a nonlinear system that is a modification of the Lotka–Volterra equations:

$$x' = f(t, x, y) = x - xy - \frac{x^2}{10} \quad \text{and} \quad y' = g(t, x, y) = yx - y - \frac{y^2}{20},$$

with the initial condition $x(0) = 2$ and $y(0) = 1$ for $0 \leq t \leq 30$. Although the numerical solution is a list of numbers, it is helpful to plot the polygonal path joining the approximation points $\{(x_j, y_j)\}$ and plot the trajectory shown in Figure 9.1. In this chapter

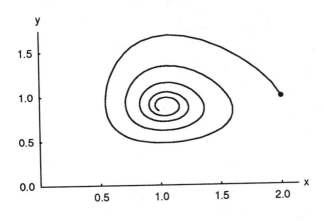

Figure 9.1 The trajectory for a nonlinear system of differential equations $x' = f(t, x, y)$ and $y' = g(t, x, y)$.

we present the standard methods for solving ordinary differential equations, systems of differential equations, and boundary value problems.

9.1 INTRODUCTION TO DIFFERENTIAL EQUATIONS

Consider the equation

$$\frac{dy}{dt} = 1 - \exp(-t). \tag{1}$$

It is a differential equation because it involves the derivative dy/dt of the "unknown function" $y = y(t)$. Only the independent variable t appears on the right side of equation (1); hence a solution is an antiderivative of $1 - \exp(-t)$. The rules for integration can be used to find $y(t)$:

$$y(t) = t + \exp(-t) + C, \tag{2}$$

where C is the constant of integration. All the functions in (2) are solutions of (1) because they satisfy the requirement that $y'(t) = 1 - \exp(-t)$. They form the family of curves in Figure 9.2.

Integration was the technique used to find the explicit formula for the functions in (2), and Figure 9.2 emphasizes that there is one degree of freedom involved in the solution, namely the constant of integration C. By varying the value of C one "moves the solution curve" up or down and a particular curve can be found that will pass through any desired point.

The secrets of the world are seldom observed as explicit formulas. Instead, one usually measures how a change in one variable affects another variable. When this is translated into a mathematical model, the result is an equation involving the rate of change of the unknown function and the independent and/or dependent variable.

Consider the temperature $y(t)$ of a cooling object. It might be conjectured that the rate of change of the temperature of the body is related to the temperature difference

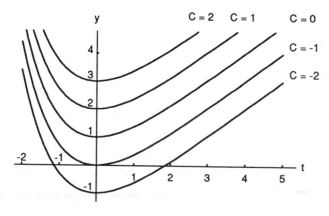

Figure 9.2 The solution curves $y(t) = t + e^{-t} + C$.

between its temperature and that of the surrounding medium. Experimental evidence verifies this conjecture. Newton's law of cooling asserts that the rate of change is directly proportional to the difference in these temperatures. If A is the temperature of the surrounding medium and $y(t)$ is the temperature of the body at time t, then

$$\frac{dy}{dt} = -k(y - A), \tag{3}$$

where k is a positive constant. The negative sign is required because dy/dt will be negative when the temperature of the body is greater than the temperature of the medium.

If the temperature of the object is known at time $t = 0$, we call this an initial condition and include this information in the statement of the problem. Usually we are asked to solve

$$\frac{dy}{dt} = -k(y - A) \quad \text{with} \quad y(0) = y_0. \tag{4}$$

The technique of separation of variables can be used to find the solution

$$y = A + (y_0 - A) \exp(-kt). \tag{5}$$

For each choice of y_0, the solution curve will be different, and there is no simple way to move one curve around to get another one. The initial value is a point where the desired solution is "nailed down." Several solution curves are shown in Figure 9.3, and it can be observed that as t gets large, the temperature of the object approaches room temperature. If $y_0 < A$, the body is warming instead of cooling.

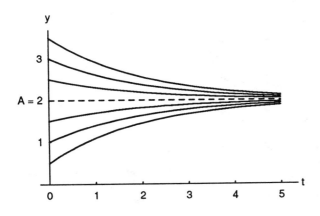

Figure 9.3   The solution curves $y = A + (y_0 - A)e^{-kt}$ for Newton's law of cooling (and warming).

The Initial Value Problem

Definition 9.1.   A **solution** to the **initial value problem (I.V.P.)**

$$y' = f(t, y) \quad \text{with} \quad y(t_0) = y_0 \tag{6}$$

on an interval $[t_0, b]$ is a differentiable function $y = y(t)$ such that

$$y(t_0) = y_0 \quad \text{and} \quad y'(t) = f(t, y(t)) \qquad \text{for all } t \in [t_0, b]. \tag{7}$$

Notice that the solution curve $y = y(t)$ must pass through the initial point (t_0, y_0).

The Geometric Interpretation

At each point (t, y) in the rectangular region R: $a \le t \le b$, $c \le y \le d$, the slope of a solution curve $y = y(t)$ can be found by using the implicit formula $f(t, y(t))$. Hence the values $m_{i,j} = f(t_i, y_j)$ can be computed throughout the rectangle, and each value $m_{i,j}$ represents the slope of the line tangent to a solution curve that passes through the point (t_i, y_j).

A slope field or direction field is a graph that indicates the slopes $\{m_{i,j}\}$ over the region. It can be used to visualize how a solution curve "fits" the slope constraint. To move along a solution curve one must start at the initial point and check the slope field to determine which direction to move. Then take a small step from t_0 to $t_0 + h$ horizontally and move the appropriate vertical distance $hf(t_0, y_0)$ so that the resulting displacement has the required slope. The next point on the solution curve is (t_1, y_1). Repeat the process to continue your journey along the curve. Since a finite number of steps will be used, the method will produce an approximation to the solution.

Example 9.1

The slope field for $y' = (t - y)/2$ over the rectangle R: $0 \le t \le 5$, $0 \le y \le 3$ is shown in Figure 9.4. The solution curves with the following initial values are also shown:
1. For $y(0) = 1$, the solution is $y(t) = 3 \exp(-t/2) - 2 + t$.
2. For $y(0) = 4$, the solution is $y(t) = 6 \exp(-t/2) - 2 + t$.

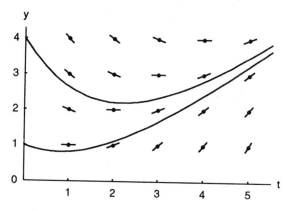

Figure 9.4 The slope field for the differential equation $y' = f(t, y) = (t - y)/2$.

Definition 9.2.   Given the rectangle $R = \{(t, y): a \le t \le b, c \le y \le d\}$, assume that $f(t, y)$ is continuous on R. The function f is said to satisfy a **Lipschitz condition** in the variable y on R provided that a constant $L > 0$ exists with the property that

$$|f(t, y_1) - f(t, y_2)| \le L|y_1 - y_2| \tag{8}$$

whenever $(t, y_1), (t, y_2) \in R$. The constant L is called a **Lipschitz constant** for f.

Theorem 9.1.   Suppose that $f(t, y)$ is defined on the region R. If there exists a constant $L > 0$ so that

$$|f_y(t, y)| \leq L \qquad \text{for all } (t, y) \in R, \tag{9}$$

then f satisfies a Lipschitz condition in the variable y with Lipschitz constant L over the rectangle R.

Proof. Fix t and use the mean value theorem to get c_1 with $y_1 < c_1 < y_2$ so that

$$|f(t, y_1) - f(t, y_2)| = |f_y(t, c_1)(y_1 - y_2)| = |f_y(t, c_1)| \, |(y_1 - y_2)| \leq L|y_1 - y_2|.$$

Theorem 9.2   (Existence and Uniqueness).   Assume that $f(t, y)$ is continuous in a region $R = \{(t, y): t_0 \leq t \leq b, c \leq y \leq d\}$. If f satisfies a Lipschitz condition on R in the variable y and $(t_0, y_0) \in R$, then the initial value problem (6) $y' = f(t, y)$ with $y(t_0) = y_0$ has a unique solution $y = y(t)$ on some subinterval $t_0 \leq t \leq t_0 + \delta$.

Proof. See an advanced text on differential equations such as Reference [38].

Let us apply Theorems 9.1 and 9.2 to the function $f(t, y) = (t - y)/2$ in Example 9.1. The partial derivative is $f_y(t, y) = -1/2$. Hence $|f_y(t, y)| \leq \frac{1}{2}$, and according to Theorem 9.1 the Lipschitz constant is $L = \frac{1}{2}$. Therefore, by Theorem 9.2 the I.V.P. has a unique solution.

EXERCISES FOR INTRODUCTION TO DIFFERENTIAL EQUATIONS

In Exercises 1–5:
(a) Show that $y(t)$ is the solution to the differential equation by substituting $y(t)$ and $y'(t)$ into the differential equation $y'(t) = f(t, y(t))$.
(b) Use Theorem 9.1 to find a Lipschitz constant L for the rectangle R: $0 \leq t \leq 3, 0 \leq y \leq 5$.

1. $y' = t^2 - y$, $y(t) = C \exp(-t) + t^2 - 2t + 2$
2. $y' = 3y + 3t$, $y(t) = C \exp(3t) - t - \frac{1}{3}$
3. $y' = -ty$, $y(t) = C \exp(-t^2/2)$
4. $y' = \exp(-2t) - 2y$, $y(t) = C \exp(-2t) + t \exp(-2t)$
5. $y' = 2ty^2$, $y(t) = 1/(C - t^2)$

In Exercises 6–9:
(a) Draw the slope field $m_{i,j} = f(t_i, y_j)$ over the rectangle $0 \leq t \leq 4, 0 \leq y \leq 4$.
(b) Sketch the indicated solution curves.

6. $y' = f(t, y) = -t/y$. Sketch the solution curves (circles):

$$y(t) = (C - t^2)^{1/2} \quad \text{for} \quad C = 1, 2, 4, 9.$$

7. $y' = f(t, y) = t/y$. Sketch the solution curves (hyperbolas):

$$y(t) = (C + t^2)^{1/2} \quad \text{for} \quad C = -4, -1, 1, 4.$$

8. $y' = f(t, y) = 1/y$. Sketch the solution curves (parabolas):

$$y(t) = (C + 2t)^{1/2} \quad \text{for} \quad C = -4, -2, 0, 2.$$

9. $y' = f(t, y) = y^2$. Sketch the solution curves (hyperbolas):

$$y(t) = \frac{1}{C - t} \quad \text{for} \quad C = 1, 2, 3, 4.$$

10. Here is an example of an initial value problem that has "two solutions": $y' = \frac{3}{2} y^{1/3}$ with $y(0) = 0$.
   (a) Verify that $y(t) = 0$ for $t \geq 0$ is a solution.
   (b) Verify that $y(t) = t^{3/2}$ for $t \geq 0$ is a solution.
   (c) Does this violate Theorem 9.2? Why?

11. Consider the initial value problem

$$y' = (1 - y^2)^{1/2} \qquad y(0) = 0.$$

   (a) Verify that $y(t) = \sin(t)$ is a solution on $[0, \pi/4]$.
   (b) Determine the largest interval over which the solution exists.

12. Show that the definite integral

$$\int_a^b f(t)\, dt$$

can be computed by solving the initial value problem

$$y' = f(t) \quad \text{for} \quad a \leq t \leq b \quad \text{with} \quad y(a) = 0.$$

In Exercises 13–15, find the solution to the I.V.P.

13. $y' = 3t^2 + \sin(t), \quad y(0) = 2$

14. $y' = 1/(1 + t^2), \quad y(0) = 0$

15. $y' = \exp(-t^2/2), \quad y(0) = 0$. *Hint.* This answer must be expressed as a certain integral.

16. Consider the first-order differential equation

$$y'(t) + p(t) y(t) = q(t).$$

Show that the general solution $y(t)$ can be found by using two special integrals. First define $F(t)$ as follows:

$$F(t) = \exp\left(\int p(t)\, dt\right).$$

Second, define $y(t)$ as follows:

$$y(t) = \frac{1}{F(t)} \left[\int F(t) q(t)\, dt + C \right].$$

Hint. Differentiate the product $F(t) y(t)$.

17. Consider the decay of a radioactive substance. If $y(t)$ is the amount of substance present at time t, then $y(t)$ decreases and experiments have verified that the rate of change of $y(t)$ is proportional to the amount of undecayed material. Hence the I.V.P. for the decay of a radioactive substance is

$$y' = -ky \quad \text{with} \quad y(0) = y_0.$$

(a) Show that the solution is $y(t) = y_0 \exp(-kt)$.

The half-life of a radioactive substance is the time required for half of an initial amount to decay. The half-life of ^{14}C is 5730 years.

(b) Find the formula $y(t)$ that gives the amount of ^{14}C present at time t. *Hint.* Find k so that $y(5730) = 0.5y_0$.

(c) A piece of wood is analyzed and the amount of ^{14}C present is 0.712 of the amount that was present when the tree was alive. How old is the sample of wood?

(d) At a certain instant 10 mg of a radioactive substance is present. After 23 sec only 1 mg is present. What is the half-life of the substance?

18. Write a report on numerical solution of differential equations. See References [7, 31, 33, 39, 42, 99, 104, 136, 138, 152, 171, and 173].

9.2 EULER'S METHOD

The reader should be convinced that not all initial value problems can be solved explicitly, and often it is impossible to find a formula for the solution $y(t)$, for example, there is no "closed-form expression" for the solution to $y' = x^3 + y^2$ with $y(0) = 0$. Hence for engineering and scientific purposes it is necessary to have methods for approximating the solution. If a solution with many significant digits is required, then more computing effort and a sophisticated algorithm must be used.

The first approach is called Euler's method and serves to illustrate the concepts involved in the advanced methods. It has limited usage because of the larger error that is accumulated as the process proceeds. However, it is important to study because the error analysis is easier to understand.

Let $[a, b]$ be the interval over which we want to find the solution to the well-posed I.V.P. $y' = f(t, y)$ with $y(a) = y_0$. In actuality, we will *not* find a differentiable function that satisfies the I.V.P. Instead, a set of points $\{(t_k, y_k)\}$ are generated which are used for an approximation [i.e., $y(t_k) \approx y_k$]. How can we proceed to construct a "set of points" that will "satisfy a differential equation approximately"? First we choose the abscissas for the points. For convenience we subdivide the interval $[a, b]$ into M equal subintervals and select the mesh points

$$t_k = a + hk \quad \text{for } k = 0, 1, \ldots, M \quad \text{where} \quad h = \frac{b - a}{M}. \tag{1}$$

The value h is called the *step size*. We now proceed to solve approximately

$$y' = f(t, y) \quad \text{over} \quad [t_0, t_M] \quad \text{with} \quad y(t_0) = y_0. \tag{2}$$

Assume that $y(t)$, $y'(t)$, and $y''(t)$ are continuous and use Taylor's theorem to expand $y(t)$ about $t = t_0$. For each value t, there exists a value c_1 that lies between t_0 and t so that

$$y(t) = y(t_0) + y'(t_0)(t - t_0) + \frac{y''(c_1)(t - t_0)^2}{2}. \tag{3}$$

When $y'(t_0) = f(t_0, y(t_0))$ and $h = t_1 - t_0$ are substituted in equation (3), the result is an expression for $y(t_1)$:

$$y(t_1) = y(t_0) + hf(t_0, y(t_0)) + y''(c_1)\frac{h^2}{2}. \tag{4}$$

If the step size h is chosen small enough, then we may neglect the second-order term (involving h^2) and get

$$y_1 = y_0 + hf(t_0, y_0), \tag{5}$$

which is **Euler's approximation**.

The process is repeated and generates a sequence of points that approximate the solution curve $y = y(t)$. The general step for Euler's method is

$$t_{k+1} = t_k + h, \quad y_{k+1} = y_k + hf(t_k, y_k) \qquad \text{for } k = 0, 1, \ldots, M - 1. \tag{6}$$

Example 9.2

Use Euler's method to solve approximately the initial value problem

$$y' = Ry \quad \text{over} \quad [0, 1] \quad \text{with} \quad y(0) = y_0 \text{ and } R \text{ constant.} \tag{7}$$

Solution. The step size must be chosen and then the second formula in (6) can be determined for computing the ordinates. This formula is sometimes called a difference equation and in this case it is

$y_{k+1} = y_k + h \cdot R y_k$

$$y_{k+1} = y_k(1 + hR) \qquad \text{for } k = 0, 1, \ldots, M - 1. \tag{8}$$

$= y_k(1 + hR)$

If we trace the solution values recursively, we see that

$$y_1 = y_0(1 + hR)$$

$$y_2 = y_1(1 + hR) = y_0(1 + hR)^2$$
$$\vdots \tag{9}$$

$$y_M = y_{M-1}(1 + hR) = y_0(1 + hR)^M.$$

For most problems there is no explicit formula for determining the solution points and each new point must be computed successively from the previous point. However, for the initial value problem (7) we are fortunate; Euler's method has the explicit solution:

$$t_k = hk \qquad y_k = y_0(1 + hR)^k \qquad \text{for } k = 0, 1, 2, \ldots, M. \tag{10}$$

الفائدة المركبة

Formula (10) can be viewed as the "compound interest" formula, and the Euler approximation gives the future value of a deposit.

Example 9.3

Suppose that $1000 is deposited and earns 10% interest compounded continuously over 5 years. What is the value at the end of 5 years?

Solution. We choose to use Euler approximations with $h = 1, \frac{1}{12}, \frac{1}{360}$ to approximate $y(5)$ for the I.V.P.:

$$y' = 0.1y \quad \text{over} \quad [0, 5] \quad \text{with} \quad y(0) = 1000.$$

Formula (10) with $R = 0.1$ produces Table 9.1.

TABLE 9.1 Compound Interest in Example 9.4

Step size, h	Number of iterations, M	Approximation to $y(5)$, y_M
1	5	$1000\left(1 + \dfrac{0.1}{1}\right)^5 = 1610.51$
$\frac{1}{12}$	60	$1000\left(1 + \dfrac{0.1}{12}\right)^{60} = 1645.31$
$\frac{1}{360}$	1800	$1000\left(1 + \dfrac{0.1}{360}\right)^{1800} = 1648.61$

Think about the different values y_5, y_{60}, y_{1800} that are used to determine the future value after 5 years. These values are obtained using different step sizes and reflect different amounts of computing effort to obtain an approximation to $y(5)$. The solution to the I.V.P. is $y(5) = 1000 \exp(0.5) = 1648.72$. If we did not use the closed-form solution (10), then it would have required 1800 iterations of Euler's method to obtain y_{1800}, and we still have only about five digits of accuracy in the answer!

If bankers had to approximate the solution to the I.V.P. (7), they would choose Euler's method because of the explicit formula in (10). The more sophisticated methods for approximating solutions do not have an explicit formula for finding y_k, but they will require less computing effort.

The Geometric Description

If you start at the point (t_0, y_0) and compute the value of the slope $m_0 = f(t_0, y_0)$ and move horizontally the amount h and vertically $hf(t_0, y_0)$, then you are moving along the tangent line to $y(t)$ and will end up at the point (t_1, y_1) (see Figure 9.5). Notice that (t_1, y_1) is *not* on the desired solution curve! But this is the approximation that we are generating. Hence

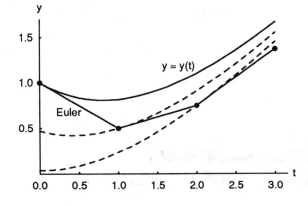

Figure 9.5 Euler's approximations $y_{k+1} = y_k + hf(t_k, y_k)$.

we must use (t_1, y_1) as though it were correct and proceed by computing the slope $m_1 = f(t_1, y_1)$ and using it to obtain the next vertical displacement $hf(t_1, y_1)$ to locate (t_2, y_2), and so on.

Step Size versus Error

The methods we introduce for approximating the solution of an initial value problem are called **difference methods** or **discrete variable methods**. The solution is approximated at a set of discrete points called a grid (or mesh) of points. A elementary single-step method has the form $y_{k+1} = y_k + h\Phi(t_k, y_k)$ for some function Φ called an **increment function**.

When using any discrete variable method to approximately solve an initial value problem, there are two sources of error: discretization and round-off.

Definition 9.3   (Discretization Error).   Assume that $\{(t_k, y_k)\}_{k=0}^{M}$ is the set of discrete approximations and that $y = y(t)$ is the unique solution to the initial value problem.

The **global discretization error** e_k is defined by

$$e_k = y(t_k) - y_k \qquad \text{for } k = 1, 2, \ldots, M. \tag{11}$$

It is the difference between the unique solution and the solution obtained by the discrete variable method.

The **local discretization error** ϵ_{k+1} is defined by

$$\epsilon_{k+1} = y(t_{k+1}) - y_k - h\Phi(t_k, y_k) \qquad \text{for } k = 0, 1, \ldots, M - 1. \tag{12}$$

It is the error committed in the single step from t_k to t_{k+1}.

When we obtained equation (6) for Euler's method, the neglected term for each step was $y^{(2)}(c_k) (h^2/2)$. If this was the only error at each step, then at the end of the interval $[a, b]$, after M steps have been made, the accumulated error would be

$$\sum_{k=1}^{M} y^{(2)}(c_k) \frac{h^2}{2} = My^{(2)}(c) \frac{h^2}{2} = \frac{hM}{2} y^{(2)}(c)h = \frac{(b - a)y^{(2)}(c)}{2} h = O(h^1).$$

There could be more error, but this estimate predominates. A detailed discussion on this topic can be found in Reference [75].

Theorem 9.3   (Precision of Euler's Method).   Assume that $y(t)$ is the solution to the I.V.P. given in (2). If $y(t) \in C^2[t_0, b]$ and $\{(t_k, y_k)\}_{k=0}^{M}$ is the sequence of approximations generated by Euler's method, then

$$|e_k| = |y(t_k) - y_k| = O(h),$$
$$|\epsilon_{k+1}| = |y(t_{k+1}) - y_k - hf(t_k, y_k)| = O(h^2). \tag{13}$$

The error at the end of the interval is called the **final global error (F.G.E.)**:

$$E(y(b), h) = |y(b) - y_M| = O(h). \tag{14}$$

Remark.   The final global error $E(y(b), h)$ is used to study the behavior of the error for various step sizes. It can be used to give us an idea of how much computing effort must be done to obtain an accurate approximation.

Examples 9.4 and 9.5 illustrate the concepts in Theorem 9.3. If approximations are computed using the step sizes h and $h/2$, we should have

$$E(y(b), h) \approx Ch \tag{15}$$

for the larger step size, and

$$E\left(y(b), \frac{h}{2}\right) \approx C\frac{h}{2} \approx \frac{1}{2}Ch \approx \frac{1}{2}E(y(b), h). \tag{16}$$

Hence the idea in Theorem 9.3 is that if the step size in Euler's method is reduced by a factor of $\frac{1}{2}$, we can expect that the overall F.G.E. will be reduced by a factor of $\frac{1}{2}$.

Example 9.4

Use Euler's method to solve the I.V.P.

$$y' = \frac{t - y}{2} \quad \text{on } [0, 3] \quad \text{with} \quad y(0) = 1.$$

Compare solutions for $h = 1, \frac{1}{2}, \frac{1}{4},$ and $\frac{1}{8}$.

Solution.   Figure 9.6 shows graphs of the four Euler solutions and the exact solution curve $y(t) = 3\exp(-t/2) - 2 + t$. Table 9.2 gives the values for the four solutions at selected abscissas. For the step size $h = 0.25$ the calculations are

$$y_1 = 1.0 + 0.25\frac{0.0 - 1.0}{2} = 0.875, \qquad y_1 = y_0 + h\left(\frac{t_0 - y}{2}\right)$$

$$y_2 = 0.875 + 0.25\frac{0.25 - 0.875}{2} = 0.796875, \quad \text{etc.} \qquad y_2 = y_1 + h\left(\frac{t_1 - y_1}{2}\right)$$

$$y_3 = y_2 + h\left(\frac{t_2 - y_2}{2}\right)$$

$$y_3 = 0.796875 + 0.25\left(\frac{0.5 - 0.796875}{2}\right) = 0.7597656$$

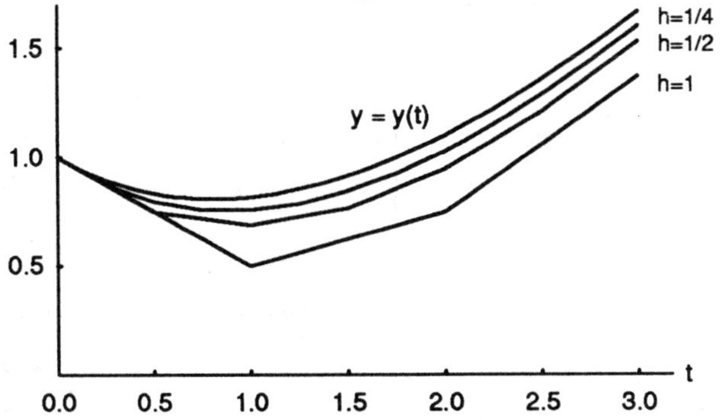

Figure 9.6   Comparison of Euler solutions with different step sizes for $y' = (t - y)/2$ over $[0, 3]$ with the initial condition $y(0) = 1$.

TABLE 9.2 Comparison of Euler Solutions with Different Step Sizes for $y' = (t - y)/2$ over [0, 3] with $y(0) = 1$

t_k	y_k				$y(t_k)$ Exact
	$h = 1$	$h = \frac{1}{2}$	$h = \frac{1}{4}$	$h = \frac{1}{8}$	
0	1.0	1.0	1.0	1.0	1.0
0.125				0.9375	0.943239
0.25			0.875	0.886719	0.897491
0.375				0.846924	0.862087
0.50		0.75	0.796875	0.817429	0.836402
0.75			0.759766	0.786802	0.811868
1.00	0.5	0.6875	0.758545	0.790158	0.819592
1.50		0.765625	0.846386	0.882855	0.917100
2.00	0.75	0.949219	1.030827	1.068222	1.103638
2.50		1.211914	1.289227	1.325176	1.359514
3.00	1.375	1.533936	1.604252	1.637429	1.669390

This iteration continues until we arrive at the last step:

$$y(3) \approx y_{12} = 1.440573 + 0.25 \frac{2.75 - 1.440573}{2} = 1.604252.$$

Example 9.5

Compare the F.G.E when Euler's method is used to solve $y' = (t - y)/2$ over [0, 3] with $y(0) = 1$ using step sizes $1, \frac{1}{2}, \ldots, \frac{1}{64}$.

Solution. Table 9.3 gives the F.G.E. for several step sizes and shows that the error in the approximation to $y(3)$ decreases by about $\frac{1}{2}$ when the step size is reduced

TABLE 9.3 Relation between Step Size and F.G.E. for Euler Solutions to $y' = (t - y)/2$ over [0, 3] with $y(0) = 1$

Step size, h	Number of steps, M	Approximation to $y(3)$, y_M	F.G.E. Error at $t = 3$, $y(3) - y_M$	$O(h) \approx Ch$ where $C = 0.256$
1	3	1.375	0.294390	0.256
$\frac{1}{2}$	6	1.533936	0.135454	0.128
$\frac{1}{4}$	12	1.604252	0.065138	0.064
$\frac{1}{8}$	24	1.637429	0.031961	0.032
$\frac{1}{16}$	48	1.653557	0.015833	0.016
$\frac{1}{32}$	96	1.661510	0.007880	0.008
$\frac{1}{64}$	192	1.665459	0.003931	0.004

by a factor of $\frac{1}{2}$. For the smaller step sizes the conclusion of Theorem 9.3 is easy to see:

$$E(y(3), h) = y(3) - y_M = O(h^1) \approx Ch, \qquad \text{where } C = 0.256.$$

Algorithm 9.1   (Euler's Method).   To approximate the solution of the initial value problem $y' = f(t, y)$ with $y(a) = y_0$ over $[a, b]$ by computing

$$y_{k+1} = y_k + hf(t_k, y_k) \qquad \text{for } k = 0, 1, \ldots, M - 1.$$

```
INPUT A, B, Y(0)                                    {Endpoints and initial value}
INPUT M                                             {Number of steps}
H := [B - A]/M                                      {Compute the step size}
T(0) := A                                           {Initialize the variable}

FOR  K = 0  TO  M−1  DO
     Y(K+1) := Y(K) + H*F(T(K),Y(K))                {Euler solution y_{k+1}}
     T(K+1) := A + H*[K+1]                           {Compute the mesh point t_{k+1}}

FOR  K = 0  TO  M  DO                               {Output}
     PRINT T(K), Y(K)
```

EXERCISES FOR EULER'S METHOD

In Exercises 1–5, use Euler's method to find approximations to the I.V.P.

(a) Compute $y_1, y_2, \ldots, y_M$ for each of the three cases: (i) $h = 0.2$, $M = 1$, (ii) $h = 0.1$, $M = 2$, and (iii) $h = 0.05$, $M = 4$.

(b) Compare the exact solution $y(0.2)$ with the three approximations in part (a).

(c) Does the F.G.E. in part (a) behave as expected when h is halved?

(d) Use a computer to carry out the computations on the larger interval $[a, b]$ that is given.

1. (a) Solve $y' = t^2 - y$ over $[0, 0.2]$ with $y(0) = 1$.
   (b) and (c) Compare with $y(t) = -\exp(-t) + t^2 - 2t + 2$.
   (d) Use $[a, b] = [0, 2]$ with $h = 0.2, 0.1, 0.05$.

2. (a) Solve $y' = 3y + 3t$ over $[0, 0.2]$ with $y(0) = 1$.
   (b) and (c) Compare with $y(t) = \frac{4}{3}\exp(3t) - t - \frac{1}{3}$.
   (d) Use $[a, b] = [0, 2]$ with $h = 0.2, 0.1,$ and 0.05.

3. (a) Solve $y' = -ty$ over $[0, 0.2]$ with $y(0) = 1$.
   (b) and (c) Compare with $y(t) = \exp(-t^2/2)$.
   (d) Use $[a, b] = [0, 2]$ with $h = 0.2, 0.1,$ and 0.05.

4. (a) Solve $y' = \exp(-2t) - 2y$ over $[0, 0.2]$ with $y(0) = \frac{1}{10}$.
   (b) and (c) Compare with $y(t) = \frac{1}{10}\exp(-2t) + t\exp(-2t)$.
   (d) Use $[a, b] = [0, 2]$ with $h = 0.2, 0.1,$ and 0.05.

5. **(a)** Solve $y' = 2ty^2$ over $[0, 0.2]$ with $y(0) = 1$.
   (b) and **(c)** Compare with $y(t) = 1/(1 - t^2)$.
   (d) Use $[a, b] = [0, 1]$ with $h = 0.1, 0.05$, and 0.025.
   Notice that Euler's method will generate an approximation to $y(1)$ even though the solution curve is not defined at $t = 1$.

6. Consider $y' = 0.12y$ over $[0, 5]$ with $y(0) = 1000$.
   (a) Apply formula (10) to find Euler's approximation to $y(5)$ using the step sizes $h = 1, \frac{1}{12}$, and $\frac{1}{360}$.
   (b) What is the limit in part (a) when h goes to zero?

7. *Exponential population growth.* The population of a certain species grows at a rate that is proportional to the current population, and obeys the I.V.P.

$$y' = 0.02y \quad \text{over } [0, 5] \quad \text{with} \quad y(0) = 5000.$$

   (a) Apply formula (10) to find Euler's approximation to $y(5)$ using the step sizes $h = 1, \frac{1}{12}$, and $\frac{1}{360}$.
   (b) What is the limit in part (a) when h goes to zero?

8. *Logistic population growth.* The population curve $P(t)$ for the United States is assumed to obey the differential equation for a logistic curve $P' = aP - bP^2$. Let t denote the year past 1900, and let the step be $h = 10$. The values $a = 0.02$ and $b = 0.00004$ produce a model for the population. Find the Euler approximations to $P(t)$ and fill in the table below. Round off each value P_k to the nearest tenth.

Year	t_k	$P(t_k)$ Actual	P_k Euler approximation
1900	0.0	76.1	76.1
1910	10.0	92.4	89.0
1920	20.0	106.5	_____
1930	30.0	123.1	_____
1940	40.0	132.6	138.2
1950	50.0	152.3	_____
1960	60.0	180.7	_____
1970	70.0	204.9	202.8
1980	80.0	226.5	_____

9. Show that when Euler's method is used to solve the I.V.P.

$$y' = f(t) \quad \text{over } [a, b] \text{ with } y(a) = y_0 = 0,$$

   the result is

$$y(b) \approx \sum_{k=0}^{M-1} f(t_k)h,$$

   which is a Riemann sum that approximates the definite integral of $f(t)$ taken over the interval $[a, b]$.

10. A skydiver jumps from a plane, and up to the moment he opens the parachute the air resistance is proportional to $v^{3/2}$. Assume that the time inteval is [0, 6] and that the differential equation for the downard direction is

$$v' = 32 - 0.032v^{3/2} \quad \text{over } [0, 6] \text{ with } v(0) = 0.$$

Use Euler's method with $h = 0.1$ and find the solution.

11. *Epidemic model*. The mathematical model for epidemics is described as follows. Assume that there is a community of L members which contains P infected individuals and Q uninfected individuals. Let $y(t)$ denote the number of infected individuals at time t. For a mild illness such as the common cold, everyone continues to be active, and the epidemic spreads from those who are infected to those uninfected. Since there are PQ possible contacts between these two groups, the rate of change of $y(t)$ is proportional to PQ. Hence the problem can be stated as the I.V.P.

$$y' = ky(L - y) \quad \text{with } y(0) = y_0.$$

Use $L = 25,000$, $k = 0.00003$, and $h = 0.2$ with the initial condition $y(0) = 250$, and compute Euler's approximate solution over [0, 12].

12. Show that Euler's method fails to approximate the solution $y(t) = t^{3/2}$ of the I.V.P.

$$y' = f(t, y) = 1.5y^{1/3} \quad \text{with } y(0) = 0.$$

Justify your answer. What difficulties were encountered?

13. Can Euler's method be used to solve the I.V.P.

$$y' = 1 + y^2 \quad \text{over } [0, 3] \text{ with } y(0) = 0?$$

Hint. The exact solution curve is $y(t) = \tan(t)$.

14. Write a report on the modified Euler method.

9.3 HEUN'S METHOD

The next approach, Heun's method, introduces a new idea for constructing an algorithm to solve the I.V.P.

$$y'(t) = f(t, y(t)) \quad \text{over} \quad [a, b] \quad \text{with} \quad y(t_0) = y_0. \tag{1}$$

To obtain the solution point (t_1, y_1) we can use the fundamental theorem of calculus, and integrate $y'(t)$ over $[t_0, t_1]$ and get

$$\int_{t_0}^{t_1} f(t, y(t)) \, dt = \int_{t_0}^{t_1} y'(t) \, dt = y(t_1) - y(t_0), \tag{2}$$

where the antiderivative of $y'(t)$ is the desired function $y(t)$. When equation (2) is solved for $y(t_1)$, the result is

$$y(t_1) = y(t_0) + \int_{t_0}^{t_1} f(t, y(t)) \, dt. \tag{3}$$

Now a numerical integration method can be used to approximate the definite integral in (3). If the trapezoidal rule is used with step size $h = t_1 - t_0$, then the result is

$$y(t_1) \approx y(t_0) + \frac{h}{2} [f(t_0, y(t_0)) + f(t_1, y(t_1))]. \tag{4}$$

Notice that the formula on the right-hand side of (4) involves the yet to be determined value $y(t_1)$. To proceed, we use an estimate for $y(t_1)$. Euler's solution will suffice for this purpose. After it is substituted into (4), the resulting formula for finding (t_1, y_1) is called **Heun's method**:

$$y_1 = y(t_0) + \frac{h}{2} [f(t_0, y_0) + f(t_1, y_0 + hf(t_0, y_0))]. \tag{5}$$

The process is repeated and generates a sequence of points that approximates the solution curve $y = y(t)$. At each step, Euler's method is used as a prediction, and then the trapezoidal rule is used to make a correction to obtain the final value. The general step for Heun's method is

$$p_{k+1} = y_k + hf(t_k, y_k), \qquad t_{k+1} = t_k + h,$$

$$y_{k+1} = y_k + \frac{h}{2} [f(t_k, y_k) + f(t_{k+1}, p_{k+1})]. \tag{6}$$

Notice the role played by differentiation and integration in Heun's method. Draw the line tangent to the solution curve $y = y(t)$ at the point (t_0, y_0), and use it to find the predicted point (t_1, p_1). Now look at the graph $z = f(t, y(t))$ and consider the points (t_0, f_0) and (t_1, f_1), where $f_0 = f(t_0, y_0)$, $f_1 = f(t_1, p_1)$. The area of the trapezoid with vertices (t_0, f_0) and (t_1, f_1) is an approximation to the integral in (3), which is used to obtain the final value in equation (5). The graphs are shown in Figure 9.7.

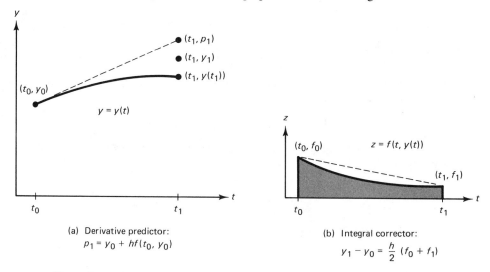

(a) Derivative predictor:
$$p_1 = y_0 + hf(t_0, y_0)$$

(b) Integral corrector:
$$y_1 - y_0 = \frac{h}{2}(f_0 + f_1)$$

Figure 9.7 The graphs $y = y(t)$ and $z = f(t, y(t))$ in the derivation of Heun's method.

Step Size versus Error

The error term for the trapezoidal rule used to approximate the integral in (3) is

$$-y^{(2)}(c_k)\,\frac{h^3}{12}. \tag{7}$$

If the only error at each step is that given in (7), after M steps the acumulated error for Heun's method would be

$$-\sum_{k=1}^{M} y^{(2)}(c_k)\,\frac{h^3}{12} \approx -\frac{b-a}{12}\,y^{(2)}(c)h^2 \approx O(h^2). \tag{8}$$

The next theorem is important, because it states the relationship between F.G.E. and step size. It is used to give us an idea of how much computing effort must be done to obtain an accurate approximation using Heun's method.

Theorem 9.4   (Precision of Heun's Method).   Assume that $y(t)$ is the solution to the I.V.P. (1). If $y(t) \in C^3[t_0, b]$ and $\{(t_k, y_k)\}_{k=0}^{M}$ is the sequence of approximations generated by Heun's method, then

$$\begin{aligned}
|e_k| &= |y(t_k) - y_k| = O(h^2), \\
|\epsilon_{k+1}| &= |y(t_{k+1}) - y_k - h\Phi(t_k, y_k)| = O(h^3),
\end{aligned} \tag{9}$$

where $\Phi(t_k, y_k) = y_k + (h/2)[f(t_k, y_k) + f(t_{k+1}, y_k + hf(t_k, y_k))]$.
    In particular, the final global error (F.G.E.) at the end of the interval will satisfy

$$E(y(b), h) = |y(b) - y_M| = O(h^2). \tag{10}$$

Examples 9.6 and 9.7 illustrate Theorem 9.4. If approximations are computed using the step sizes h and $h/2$, we should have

$$E(y(b), h) \approx Ch^2 \tag{11}$$

for the larger step size, and

$$E\left(y(b), \frac{h}{2}\right) \approx C\,\frac{h^2}{4} \approx \frac{1}{4}\,Ch^2 \approx \frac{1}{4}\,E(y(b), h). \tag{12}$$

Hence the idea in Theorem 9.4 is that if the step size in Heun's method is reduced by a factor of $\frac{1}{2}$, we can expect that the overall F.G.E. will be reduced by a factor of $\frac{1}{4}$.

Example 9.6

Use Heun's method to solve the I.V.P.

$$y' = \frac{t - y}{2} \quad \text{on } [0, 3] \text{ with } y(0) = 1.$$

Compare solutions for $h = 1, \frac{1}{2}, \frac{1}{4}$, and $\frac{1}{8}$.

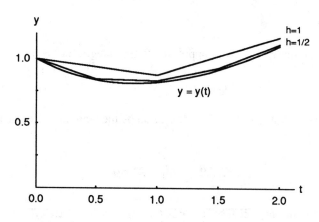

Figure 9.8 Comparison of Heun solutions with different step sizes for $y' = (t - y)/2$ over $[0, 2]$ with the initial condition $y(0) = 1$.

Solution. Figure 9.8 shows graphs of the first two Heun solutions and the exact solution curve $y(t) = 3 \exp(-t/2) - 2 + t$. Table 9.4 gives the values for the four solutions at selected abscissas. For the step size $h = 0.25$ a sample calculation is

$$f(t_0, y_0) = \frac{0 - 1}{2} = -0.5$$

$$p_1 = 1.0 + 0.25(-0.5) = 0.875,$$

$$f(t_1, p_1) = \frac{0.25 - 0.875}{2} = -0.3125,$$

$$y_1 = 1.0 + 0.125(-0.5 - 0.3125) = 0.8984375.$$

TABLE 9.4 Comparison of Heun Solutions with Different Step Sizes for $y' = (t - y)/2$ over $[0, 3]$ with $y(0) = 1$

t_k	y_k				$y(t_k)$ Exact
	$h = 1$	$h = \frac{1}{2}$	$h = \frac{1}{4}$	$h = \frac{1}{8}$	
0	1.0	1.0	1.0	1.0	1.0
0.125				0.943359	0.943239
0.25			0.898438	0.897717	0.897491
0.375				0.862406	0.862087
0.50		0.84375	0.838074	0.836801	0.836402
0.75			0.814081	0.812395	0.811868
1.00	0.875	0.831055	0.822196	0.820213	0.819592
1.50		0.930511	0.920143	0.917825	0.917100
2.00	1.171875	1.117587	1.106800	1.104392	1.103638
2.50		1.373115	1.362593	1.360248	1.359514
3.00	1.732422	1.682121	1.672269	1.670076	1.669390

This iteration continues until we arrive at the last step:

$$y(3) \approx y_{12} = 1.511508 + 0.125(0.619246 + 0.666840) = 1.672269.$$

Example 9.7

Compare the F.G.E. when Heun's method is used to solve $y' = (t - y)/2$ over $[0, 3]$ with $y(0) = 1$ using step sizes $1, \frac{1}{2}, \ldots, \frac{1}{64}$.

Solution.   Table 9.5 gives the F.G.E. and shows that the error in the approximation to $y(3)$ decreases by about $\frac{1}{4}$ when the step size is reduced by a factor of $\frac{1}{2}$:

$$E(y(3), h) = y(3) - y_M = O(h^2) \approx Ch^2, \qquad \text{where} \quad C = -0.0432.$$

TABLE 9.5   Relation between Step Size and F.G.E. for Heun Solutions to $y' = (t - y)/2$ over $[0, 3]$ with $y(0) = 1$

Step size, h	Number of steps, M	Approximation to $y(3)$, y_M	Error at $t = 3$, $y(3) - y_M$	$O(h^2) \approx Ch^2$ where $C = -0.0432$
1	3	1.732422	−0.063032	−0.043200
$\frac{1}{2}$	6	1.682121	−0.012731	−0.010800
$\frac{1}{4}$	12	1.672269	−0.002879	−0.002700
$\frac{1}{8}$	24	1.670076	−0.000686	−0.000675
$\frac{1}{16}$	48	1.669558	−0.000168	−0.000169
$\frac{1}{32}$	96	1.669432	−0.000042	−0.000042
$\frac{1}{64}$	192	1.669401	−0.000011	−0.000011

Algorithm 9.2   (Heun's Method).   To approximate the solution of the initial value problem $y' = f(t, y)$ with $y(a) = y_0$ over $[a, b]$ by computing

$$y_{k+1} = y_k + \frac{h}{2} [f(t_k, y_k) + f(t_{k+1}, y_k + hf(t_k, y_k))]$$

$$\text{for } k = 0, 1, \ldots, M - 1.$$

```
INPUT A, B, Y(0)                                    {Endpoints and initial value}
INPUT M                                             {Number of steps}
H := [B − A]/M                                      {Compute the step size}
T(0) := A                                           {Initialize the variable}
```

```
FOR  J = 0  TO  M−1  DO
    ┌  K₁ := F(T(J),Y(J))                              {Function value at tⱼ}
    │  P := Y(J) + H*K₁                                {Euler predictor for yⱼ₊₁}
    │  T(J+1) := A + H*[J+1]                           {Compute the mesh point tⱼ₊₁}
    │  K₂ := F(T(J+1),P)                               {Function value at tⱼ₊₁}
    └  Y(J+1) := Y(J) + H*[K₁ + K₂]/2                  {Trapezoidal corrector}


FOR  J = 0  TO  M  DO                                  {Output}
    └  PRINT T(J), Y(J)
```

EXERCISES FOR HEUN'S METHOD

In Exercises 1–5, use Heun's method to find approximations to the I.V.P.
(a) Compute y_1, y_2, . . . , y_M for each of the three cases: (i) $h = 0.2$, $M = 1$, (ii) $h = 0.1$, $M = 2$, and (iii) $h = 0.05$, $M = 4$.
(b) Compare with the exact solution $y(0.2)$ with the three approximations in part (a).
(c) Does the F.G.E. in part (a) behave as expected when h is halved?
(d) Use a computer to carry out the computations on the larger interval $[a, b]$ that is given.

1. (a) Solve $y' = t^2 - y$ over $[0, 0.2]$ with $y(0) = 1$.
   (b) and (c) Compare with $y(t) = -\exp(-t) + t^2 - 2t + 2$.
   (d) Use $[a, b] = [0, 2]$ with $h = 0.2$, 0.1, and 0.05.

2. (a) Solve $y' = 3y + 3t$ over $[0, 0.2]$ with $y(0) = 1$.
   (b) and (c) Compare with $y(t) = \frac{4}{3}\exp(3t) - t - \frac{1}{3}$.
   (d) Use $[a, b] = [0, 2]$ with $h = 0.2$, 0.1, and 0.05.

3. (a) Solve $y' = -ty$ over $[0, 0.2]$ with $y(0) = 1$.
   (b) and (c) Compare with $y(t) = \exp(-t^2/2)$.
   (d) Use $[a, b] = [0, 2]$ with $h = 0.2$, 0.1, and 0.05.

4. (a) Solve $y' = \exp(-2t) - 2y$ over $[0, 0.2]$ with $y(0) = \frac{1}{10}$.
   (b) and (c) Compare with $y(t) = \frac{1}{10}\exp(-2t) + t\exp(-2t)$.
   (d) Use $[a, b] = [0, 2]$ with $h = 0.2$, 0.1, and 0.05.

5. (a) Solve $y' = 2ty^2$ over $[0, 0.2]$ with $y(0) = 1$.
   (b) and (c) Compare with $y(t) = 1/(1 - t^2)$.
   (d) Use $[a, b] = [0, 1]$ with $h = 0.1$, 0.05, and 0.025.
   Notice that Heun's method will generate an approximation to $y(1)$ even though the solution curve is not defined at $t = 1$.

6. Consider a projectile that is fired straight up and falls straight down. If air resistance is proportional to the velocity, the I.V.P. for the velocity $v(t)$ is

$$v' = -32 - \frac{K}{M}v \quad \text{with} \quad v(0) = v_0,$$

where v_0 is the initial velocity, M the mass, and K the coefficient of air resistance. Suppose that $v_0 = 160$ ft/sec and $K/M = 0.1$. Use Heun's method with $h = 0.5$ to solve

$$v' = -32 - \frac{v}{10} \quad \text{over} \quad [0, 30] \quad \text{with} \quad v(0) = 160.$$

Remark. You can compare your computer solution with the exact solution $v(t) = 480 \exp(-t/10) - 320$, which is the velocity component in the vertical direction for Example 2.10. Observe that the limiting velocity is -320 ft/sec.

7. In psychology, the Weber–Fechner law for stimulus–response states that the rate of change dR/dS of the reaction R is inversely proportional to the stimulus. The threshold value is the lowest level of the stimulus that can be consistently detected. The I.V.P. for this model is

$$R' = \frac{k}{S} \quad \text{with} \quad R(s_0) = 0.$$

Suppose that $s_0 = 0.1$ and $R(0.1) = 0$. Use Heun's method with $h = 0.1$ to solve

$$R' = \frac{1}{S} \quad \text{over} \quad [0.1, 5.1] \quad \text{with} \quad R(0.1) = 0.$$

8. Show that when Heun's method is used to solve the I.V.P. $y' = f(t)$ over $[a, b]$ with $y(a) = y_0 = 0$, the result is

$$y(b) \approx \frac{h}{2} \sum_{k=0}^{M-1} [f(t_k) + f(t_{k+1})],$$

which is the trapezoidal rule approximation for the definite integral of $f(t)$ taken over the interval $[a, b]$.

9. The Richardson improvement method discussed in Lemma 7.1 (Section 7.3) can be used in conjunction with Heun's method. If Heun's method is used with step size h, then we have

$$y(b) \approx y_h + Ch^2.$$

If Heun's method is used with $2h$, we have

$$y(b) \approx y_{2h} + 4Ch^2.$$

The terms involving Ch^2 can be eliminated to obtain an improved approximation for $y(b)$ and the result is

$$y(b) \approx \frac{4y_h - y_{2h}}{3}.$$

This improvement scheme can be used with the values in Example 9.7 to obtain better approximations to $y(3)$. Find the missing entries in the table below.

h	y_h	$(4y_h - y_{2h})/3$
1	1.732422	
1/2	1.682121	1.665354
1/4	1.672269	_____
1/8	1.670076	_____
1/16	1.669558	1.669385
1/32	1.669432	_____
1/64	1.669401	_____

10. Show that Heun's method fails to approximate the solution $y(t) = t^{3/2}$ of the I.V.P.

$$y' = f(t, y) = 1.5y^{1/3} \quad \text{with} \quad y(0) = 0.$$

Justify your answer. What difficulties were encountered?

11. Write a report on the improved Heun's method.

9.4 TAYLOR SERIES METHOD

The Taylor series method is of general applicability and it is the standard to which we compare the accuracy of the various other numerical methods for solving an I.V.P. It can be devised to have any specified degree of accuracy. We start by reformulating Taylor's theorem in a form that is suitable for solving differential equations.

Theorem 9.5 (Taylor's Theorem). Assume that $y(t) \in C^{N+1}[t_0, b]$ and that $y(t)$ has a Taylor series expansion of order N about the fixed value $t = t_k \in [t_0, b]$:

$$y(t_k + h) = y(t_k) + hT_N(t_k, y(t_k)) + O(h^{N+1}), \tag{1}$$

where

$$T_N(t_k, y(t_k)) = \sum_{j=1}^{N} \frac{y^{(j)}(t_k)}{j!} h^{j-1} \tag{2}$$

and $y^{(j)}(t) = f^{(j-1)}(t, y(t))$ denotes the $(j - 1)$st total derivative of the function f with respect t. The formulas for the derivatives can be computed recursively:

$$y'(t) = f,$$

$$y''(t) = f_t + f_y y' = f_t + f_y f,$$

$$y^{(3)}(t) = f_{tt} + 2f_{ty}y' + f_y y'' + f_{yy} [y']^2$$

$$\qquad = f_{tt} + 2f_{ty}f + f_{yy}f^2 + f_y(f_t + f_y f),$$

$$y^{(4)}(t) = f_{ttt} + 3f_{tty}y' + 3f_{tyy} [y']^2 + 3f_{ty}y'' + f_y y''' + 3f_{yyy}y'y'' + f_{yyy} [y']^3 \tag{3}$$

$$\qquad = (f_{ttt} + 3f_{tty}f + 3f_{tyy}f^2 + f_{yyy}f^3) + f_t(f_{tt} + 2f_{ty}f + f_{yy}f^2)$$

$$\qquad + 3(f_t + f_y f)(f_{ty} + f_{yy}f) + f_y^2(f_t + f_y f)$$

and in general,

$$y^{(N)}(t) = P^{(N-1)}f(t, y(t)), \tag{4}$$

where P is the derivative operator

$$P = \left(\frac{\partial}{\partial t} + f \frac{\partial}{\partial y} \right).$$

The approximate numerical solution to the I.V.P. $y'(t) = f(t, y)$ over $[t_0, t_M]$ is derived by using formula (1) on each subinterval $[t_k, t_{k+1}]$. The general step for Taylor's method of order N is

$$y_{k+1} = y_k + d_1 h + \frac{d_2 h^2}{2!} + \frac{d_3 h^3}{3!} + \cdots + \frac{d_N h^N}{N!}, \tag{5}$$

where $d_j = y^{(j)}(t_k)$ for $j = 1, 2, \ldots, N$ at each step $k = 0, 1, \ldots, M - 1$.

The Taylor method of order N has the property that the final global error (F.G.E.) is of the order $O(h^{N+1})$; hence N can be chosen as large as necessary to make this error as small as desired. If the order N is fixed, it is theoretically possible to a priori determine the step size h so that the F.G.E. will be as small as desired. However, in practice one usually computes two sets of approximations using step sizes h and $h/2$ and compares the results.

Theorem 9.6   (Precision of Taylor's Method of Order N).   Assume that $y(t)$ is the solution to the I.V.P. If $y(t) \in C^{N+1}[t_0, b]$ and $\{(t_k, y_k)\}_{k=0}^{M}$ is the sequence of approximations generated by Taylor's method of order N, then

$$|e_k| = |y(t_k) - y_k| = O(h^{N+1}),$$
$$|\epsilon_{k+1}| = |y(t_{k+1}) - y_k - hT_n(t_k, y_k)| = O(h^N). \tag{6}$$

In particular, the final global error (F.G.E.) at the end of the interval will satisfy

$$E(y(b), h) = |y(b) - y_M| = O(h^N). \tag{7}$$

The reader is encouraged to read the proof, which can be found in Reference [78].

Examples 9.8 and 9.9 illustrate Theorem 9.6 for the case $N = 4$. If approximations are computed using the step sizes h and $h/2$, we should have

$$E(y(b), h) \approx Ch^4 \tag{8}$$

for the larger step size, and

$$E\left(y(b), \frac{h}{2}\right) \approx C \frac{h^4}{16} \approx \frac{1}{16} Ch^4 \approx \frac{1}{16} E(y(b), h). \tag{9}$$

Hence the idea in Theorem 9.6 is that if the step size in the Taylor method of order 4 is reduced by a factor of $\frac{1}{2}$, the overall F.G.E. will be reduced by about $\frac{1}{16}$.

Example 9.8

Use the Taylor method of order $N = 4$ to solve $y' = (t - y)/2$ on $[0, 3]$ with $y(0) = 1$. Compare solutions for $h = 1, \frac{1}{2}, \frac{1}{4}$, and $\frac{1}{8}$.

Solution.   The derivatives of $y(t)$ must first be determined. Recall that the solution $y(t)$ is a function of t, and differentiate the formula $y'(t) = f(t, y(t))$ with respect to t to get $y^{(2)}(t)$.

Then continue the process to obtain the higher derivatives:

$$y'(t) = \frac{t - y}{2},$$

$$y^{(2)}(t) = \frac{d}{dt}\frac{t - y}{2} = \frac{1 - y'}{2} = \frac{1 - (t - y)/2}{2} = \frac{2 - t + y}{4},$$

$$y^{(3)}(t) = \frac{d}{dt}\frac{2 - t + y}{4} = \frac{0 - 1 + y'}{4} = \frac{-1 + (t - y)/2}{4} = \frac{-2 + t - y}{8},$$

$$y^{(4)}(t) = \frac{d}{dt}\frac{-2 + t - y}{8} = \frac{-0 + 1 - y'}{8} = \frac{1 - (t - y)/2}{8} = \frac{2 - t + y}{16}.$$

To find y_1, the derivatives given above must be evaluated at the point $(t_0, y_0) = (0, 1)$. Calculation reveals that

$$d_1 = y'(0) = \frac{0.0 - 1.0}{2} = -0.5,$$

$$d_2 = y^{(2)}(0) = \frac{2.0 - 0.0 + 1.0}{4} = 0.75,$$

$$d_3 = y^{(3)}(0) = \frac{-2.0 + 0.0 - 1.0}{8} = -0.375,$$

$$d_4 = y^{(4)}(0) = \frac{2.0 - 0.0 + 1.0}{16} = 0.1875.$$

Next the derivatives $\{d_j\}$ are substituted into (4), with $h = 0.25$, and nested multiplication is used to compute the value y_1:

$$y_1 = 1.0 + 0.25\left\{-0.5 + 0.25\left[\frac{0.75}{2} + 0.25\left(\frac{-0.375}{6} + 0.25\frac{0.1875}{24}\right)\right]\right\}$$

$$= 0.8974915.$$

The computed solution point is $(t_1, y_1) = (0.25, 0.8974915)$.

To find y_2, the derivatives $\{d_j\}$ must now be evaluated at the point $(t_1, y_1) = (0.25, 0.8974915)$. The calculations are starting to require a considerable amount of computational effort and are tedious to do by hand. Calculation reveals that

$$d_1 = y'(0) = \frac{0.25 - 0.8974915}{2} = -0.3237458,$$

$$d_2 = y^{(2)}(0) = \frac{2.0 - 0.25 + 0.8974915}{4} = 0.6618729,$$

$$d_3 = y^{(3)}(0) = \frac{-2.0 + 0.25 - 0.8974915}{8} = -0.3309364,$$

$$d_4 = y^{(4)}(0) = \frac{2.0 - 0.25 + 0.8974915}{16} = 0.1654682.$$

Now these derivatives $\{d_j\}$ are substituted into (5), with $h = 0.25$, and nested multiplication is used to compute the value y_2:

$y_2 = 0.8974915$

$$+ 0.25 \left\{ -0.3237458 + 0.25 \left[\frac{0.6618729}{2} + 0.25 \left(\frac{-0.3309364}{6} + 0.25 \frac{0.1654682}{24} \right) \right] \right\}$$

$$= 0.8364037.$$

The solution point is $(t_1, y_1) = (0.50, 0.8364037)$. Table 9.6 gives solution values at selected abscissas, using various step sizes.

TABLE 9.6 Comparison of the Taylor Solutions of Order $N = 4$ for $y' = (t - y)/2$ over $[0, 3]$ with $y(0) = 1$

t_k	$h = 1$	$h = \frac{1}{2}$	$h = \frac{1}{4}$	$h = \frac{1}{8}$	$y(t_k)$ Exact
0	1.0	1.0	1.0	1.0	1.0
0.125				0.9432392	0.9432392
0.25			0.8974915	0.8974908	0.8974917
0.375				0.8620874	0.8620874
0.50		0.8364258	0.8364037	0.8364024	0.8364023
0.75			0.8118696	0.8118679	0.8118678
1.00	0.8203125	0.8196285	0.8195940	0.8195921	0.8195920
1.50		0.9171423	0.9171021	0.9170998	0.9170997
2.00	1.1045125	1.1036826	1.1036408	1.1036385	1.1036383
2.50		1.3595575	1.3595168	1.3595145	1.3595144
3.00	1.6701860	1.6694308	1.6693928	1.6693906	1.6693905

Example 9.9

Compare the F.G.E. for the Taylor solutions to $y' = (t - y)/2$ over $[0, 3]$ with $y(0) = 1$ given in Example 9.8.

Solution.   Table 9.7 gives the F.G.E. for these step sizes and shows that the error in the

TABLE 9.7 Relation between Step Size and F.G.E. for the Taylor Solutions to $y' = (t - y)/2$ over $[0, 3]$

Step size, h	Number of steps, N	Approximation to $y(3)$, y_N	Error at $t = 3$, $y(3) - y_N$	$O(h^2) \approx Ch^4$ where $C = -0.000614$
1	3	1.6701860	-0.0007955	-0.0006140
$\frac{1}{2}$	6	1.6694308	-0.0000403	-0.0000384
$\frac{1}{4}$	12	1.6693928	-0.0000023	-0.0000024
$\frac{1}{8}$	24	1.6693906	-0.0000001	-0.0000001

approximation to $y(3)$ decreases by about $\frac{1}{16}$ when the step size is reduced by a factor of $\frac{1}{2}$:

$$E(y(3), h) = y(3) - y_N = \boldsymbol{O}(h^4) \approx Ch^4, \qquad \text{where} \quad C = -0.000614.$$

Algorithm 9.3   Taylor's Method of Order 4).   To approximate the solution of the initial value problem $y' = f(t, y)$ with $y(a) = y_0$ over $[a, b]$ by evaluating y'', y''', and y'''' and using the Taylor polynomial at each step.

```
INPUT A, B, Y(0)                                {Endpoints and initial value}
INPUT M                                         {Number of steps}
H := [B − A]/M                                  {Compute the step size}
T(0) := A                                       {Initialize the variable}

FOR  K = 0  TO  M−1  DO
    T := T(K) and Y := Y(K)
    D₁ := F₁(T,Y)                               {Slope function f(t,y(t))}
    D₂ := F₂(T,Y)                               {Derivative of f(t,y(t))}
    D₃ := F₃(T,Y)                               {Second derivative of f(t,y(t))}
    D₄ := F₄(T,Y)                               {Third derivative of f(t,y(t))}
    Y(K+1) := Y + H*[D₁ + H*[D₂/2 + H*[D₃/6 + H*D₄/24]]]
    T(K+1) := A + H*[K+1]

FOR  K = 1  TO  M  DO                           {Output}
    PRINT T(K), Y(K)
```

EXERCISES FOR TAYLOR METHODS

In Exercises 1–5, use the Taylor method, of order $N = 4$, to find approximations to the I.V.P.
(a) Compute $y_1, y_2, \ldots, y_M$ for each of the two cases (i) $h = 0.2$, $M = 1$ and (ii) $h = 0.1$, $M = 2$.
(b) Compare with the exact solution $y(0.2)$ with the two approximations in part (a).
(c) Does the F.G.E. in part (a) behave as expected when h is halved?
(d) Use a computer to carry out the computations on the larger interval $[a, b]$ that is given.

1. (a) Solve $y' = t^2 - y$ over $[0, 0.2]$ with $y(0) = 1$.
   (b) and (c) Compare with $y(t) = -\exp(-t) + t^2 - 2t + 2$.
   (d) Use $[a, b] = [0, 2]$ with $h = 0.2, 0.1,$ and 0.05.

2. (a) Solve $y' = 3y + 3t$ over $[0, 0.2]$ with $y(0) = 1$.
   (b) and (c) Compare with $y(t) = \frac{4}{3}\exp(3t) - t - \frac{1}{3}$.
   (d) Use $[a, b] = [0, 2]$ with $h = 0.2, 0.1,$ and 0.05.

3. (a) Solve $y' = -ty$ over [0, 0.2] with $y(0) = 1$.
   (b) and (c) Compare with $y(t) = \exp(-t^2/2)$.
   (d) Use $[a, b] = [0, 2]$ with $h = 0.2, 0.1,$ and 0.05.

4. (a) Solve $y' = \exp(-2t) - 2y$ over [0, 0.2] with $y(0) = \frac{1}{10}$.
   (b) and (c) Compare with $y(t) = \frac{1}{10}\exp(-2t) + t\exp(-2t)$.
   (d) Use $[a, b] = [0, 2]$ with $h = 0.2, 0.1,$ and 0.05.

5. (a) Solve $y' = 2ty^2$ over [0, 0.2] with $y(0) = 1$.
   (b) and (c) Compare with $y(t) = 1/(1 - t^2)$.
   (d) Use $[a, b] = [0, 1]$ with $h = 0.1, 0.05,$ and 0.025.
   Notice that Taylor's method will generate an approximation to $y(1)$ even though the solution curve is not defined at $t = 1$.

6. The Richardson improvement method discussed in Lemma 7.1 (Section 7.3) can be used in conjunction with Taylor's method. If Taylor's method of order $N = 4$ is used with step size h, then $y(b) \approx y_h + Ch^4$. If Taylor's method of order $N = 4$ is used with step size $2h$, then $y(b) \approx y_{2h} + 16Ch^4$. The terms involving Ch^4 can be eliminated to obtain an improved approximation for $y(b)$:

$$y(b) \approx \frac{16y_h - y_{2h}}{15}.$$

This improvement scheme can be used with the values in Example 9.9 to obtain better approximations to $y(3)$. Find the missing entries in the table below.

h	y_h	$(16y_h - y_{2h})/15$
1.0	1.6701860	
0.5	1.6694308	_____
0.25	1.6693928	_____
0.125	1.6693906	_____

7. Show that when Taylor's method of order N is used with step sizes h and $h/2$, then the overall F.G.E. will be reduced by a factor of about 2^{-N} for the smaller step size.

8. Show that Taylor's method fails to approximate the solution $y(t) = t^{3/2}$ of the I.V.P. $y' = f(t, y) = 1.5y^{1/3}$ with $y(0) = 0$. Justify your answer. What difficulties were encountered?

9. (a) Verify that the solution to the I.V.P. $y' = y^2$, $y(0) = 1$ over the interval [0, 1) is $y(t) = 1/(1 - t)$.
   (b) Verify that the solution to the I.V.P. $y' = 1 + y^2$, $y(0) = 1$ over the interval [0, $\pi/4$) is $y = \tan(t + \pi/4)$.
   (c) Use the result of parts (a) and (b) to argue that the solution to the I.V.P. $y' = t^2 + y^2$, $y(0) = 1$ has a vertical asymptote between $\pi/4$ and 1. (Its location is near $t = 0.96981$.)

10. Consider the I.V.P. $y' = 1 + y^2$, $y(0) = 1$.
    (a) Find an expression for $y^{(2)}(t)$, $y^{(3)}(t)$, and $y^{(4)}(t)$.
    (b) Evaluate these derivatives at $t = 0$, and use them to find the first five terms in the Maclaurin expansion for $\tan(t)$.

11. The following table of values is the Taylor solution of order $N = 4$ for the I.V.P. $y' = t^2 + y^2$, $y(0) = 1$ over [0, 0.8]. The true solution at $t = 0.8$ is known to be $y(0.8) = 5.8486168$.

Step size, h	Number of steps, M	Approximation to $y(0.8)$, y_M	Error at $t = 0.8$, $y(0.8) - y_M$
0.05	16	5.8410780	0.0075388
0.025	32	5.8479637	0.0006531
0.0125	64	5.8485686	0.0000482
0.00625	128	5.8486136	0.0000032
0.003125	256	5.8486166	0.0000002

(a) Does the F.G.E. behave as expected when h is halved?

(b) Use the Richardson improvement scheme to find improvements for the approximations.

12. The following table of values is the Taylor solution of order $N = 3$ for the I.V.P. $y' = t^2 + y^2$, $y(0) = 1$ over $[0, 0.8]$.

Step size, h	Number of steps, M	Approximatioin to $y(0.8)$, y_M	Error at $t = 0.8$, $y(0.8) - y_M$
0.05	16	5.8050504	0.0435664
0.025	32	5.8416192	0.0069976
0.0125	64	5.8476250	0.0009918
0.00625	128	5.8484848	0.0001320
0.003125	256	5.8485998	0.0000170

(a) Does the F.G.E. behave as expected when h is halved?

(b) Develop the Richardson improvement scheme for a Taylor method of order 3 method and find improvements for these approximations.

9.5 RUNGE–KUTTA METHODS

The Taylor methods in the preceding section have the desirable feature that the F.G.E. is of order $O(h^N)$, and N can be chosen large so that this error is small. However, the shortcoming of the Taylor methods is the a priori determination of N and the computation of the higher derivatives, which can be very complicated. Each Runge–Kutta method is derived from an appropriate Taylor method in such a way that the F.G.E. is of order $O(h^N)$. A trade-off is made to perform several function evaluations at each step and eliminate the necessity to compute the higher derivatives. These methods can be constructed for any order N. The Runge–Kutta method of order $N = 4$ is most popular. It is a good choice for common purposes because it is quite accurate, stable, and easy to program. Most authorities proclaim that it is not necessary to go to a higher-order method because the increased accuracy is offset by additional computational effort. If more accuracy is required, then either a smaller step size or an adaptive method should be used.

The fourth-order Runge–Kutta method (RK4) simulates the accuracy of the Taylor series method of order $N = 4$. The method is based on computing y_{k+1} as follows:

$$y_{k+1} = y_k + w_1 k_1 + w_2 k_2 + w_3 k_3 + w_4 k_4, \tag{1}$$

where k_1, k_2, k_3, and k_4 have the form

$$
\begin{aligned}
k_1 &= hf(t_k, y_k), \\
k_2 &= hf(t_k + a_1 h, y_k + b_1 k_1), \\
k_3 &= hf(t_k + a_2 h, y_k + b_2 k_1 + b_3 k_2), \\
k_4 &= hf(t_k + a_3 h, y_k + b_4 k_1 + b_5 k_2 + b_6 k_3).
\end{aligned}
\tag{2}
$$

By matching coefficients with those of the Taylor series method of order $N = 4$ so that the local truncation error is of order $O(h^5)$, Runge and Kutta were able to obtain the following system of equations:

$$
\begin{aligned}
b_1 &= a_1, \\
b_2 + b_3 &= a_2, \\
b_4 + b_5 + b_6 &= a_3, \\
w_1 + w_2 + w_3 + w_4 &= 1, \\
w_2 a_1 + w_3 a_2 + w_4 a_3 &= \tfrac{1}{2}, \\
w_2 a_1^2 + w_3 a_2^2 + w_4 a_3^2 &= \tfrac{1}{3}, \\
w_2 a_1^3 + w_3 a_2^3 + w_4 a_3^3 &= \tfrac{1}{4}, \\
w_3 a_1 b_3 + w_4 (a_1 b_5 + a_2 b_6) &= \tfrac{1}{6}, \\
w_3 a_1 a_2 b_3 + w_4 a_3 (a_1 b_5 + a_2 b_6) &= \tfrac{1}{8}, \\
w_3 a_1^2 b_3 + w_4 (a_1^2 b_5 + a_2^2 b_6) &= \tfrac{1}{12}, \\
w_4 a_1 b_3 b_6 &= \tfrac{1}{24}.
\end{aligned}
\tag{3}
$$

The system involves 11 equations in 13 unknowns. Two additional conditions must be supplied to solve the system. The most useful choice is

$$a_1 = \tfrac{1}{2} \quad \text{and} \quad b_2 = 0. \tag{4}$$

Then the solution for the remaining variables are

$$a_2 = \tfrac{1}{2}, \quad a_3 = 1, \quad b_1 = \tfrac{1}{2}, \quad b_3 = \tfrac{1}{2}, \quad b_4 = 0, \quad b_5 = 0, \quad b_6 = 1,$$

$$w_1 = \tfrac{1}{6}, \quad w_2 = \tfrac{1}{3}, \quad w_3 = \tfrac{1}{3}, \quad \text{and } w_4 = \tfrac{1}{6}. \tag{5}$$

The values in (4) and (5) are substituted into (2) and (1) to obtain the formula for the standard Runge–Kutta method of order $N = 4$, which is stated as follows. Start with the initial point (t_0, y_0) and generate the sequence of approximations using

$$y_{k+1} = y_k + \frac{h(f_1 + 2f_2 + 2f_3 + f_4)}{6}. \tag{6}$$

where

$$f_1 = f(t_k, y_k),$$

$$f_2 = f\left(t_k + \frac{h}{2}, y_k + \frac{h}{2}f_1\right),$$

$$f_3 = f\left(t_k + \frac{h}{2}, y_k + \frac{h}{2}f_2\right), \tag{7}$$

$$f_4 = f(t_k + h, y_k + hf_3).$$

Discussion about the Method

The complete development of the equations in (7) is beyond the scope of this book and can be found in advanced texts, but we can get some insights. Consider the graph of the solution curve $y = y(t)$ over the first subinterval $[t_0, t_1]$. The function values in (7) are approximations for slopes to this curve. Here f_1 is the slope at the left, f_2 and f_3 are two estimates for the slope in the middle, and f_4 is the slope at the right [see Figure 9.9(a)]. The next point (t_1, y_1) is obtained by integrating the slope function

$$y(t_1) - y(t_0) = \int_{t_0}^{t_1} f(t, y(t)) \, dt. \tag{8}$$

If Simpson's rule is applied with step size $h/2$, the approximation to the integral in (8) is

$$\int_{t_0}^{t_1} f(t, y(t)) \, dt \approx \frac{h}{6} \left[f(t_0, y(t_0)) + 4f(t_{1/2}, y(t_{1/2})) + f(t_1, y(y_1)) \right], \tag{9}$$

where $t_{1/2}$ is the midpoint of the interval. Three function values are needed; hence we make the obvious choices $f(t_0, y(t_0)) = f_1$ and $f(t_1, y(t_1)) \approx f_4$. For the value in the middle we chose the average of f_2 and f_3:

$$f(t_{1/2}, y(t_{1/2})) \approx \frac{f_2 + f_3}{2}.$$

These values are substituted into (9), which is used in equation (8) to get y_1.

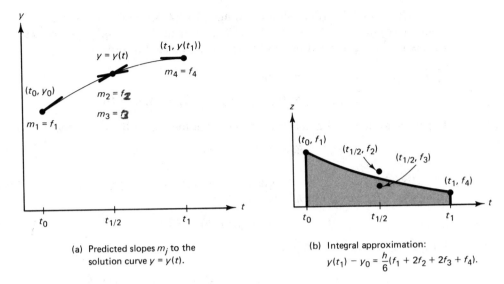

(a) Predicted slopes m_j to the solution curve $y = y(t)$.

(b) Integral approximation:
$$y(t_1) - y_0 = \frac{h}{6}(f_1 + 2f_2 + 2f_3 + f_4).$$

Figure 9.9   The graphs $y = y(t)$ and $z = f(t, y(t))$ in the discussion of the Runge–Kutta method of order $N = 4$.

$$y_1 = y_0 + \frac{h}{6}\left[f_1 + \frac{4(f_2 + f_3)}{2} + f_4 \right]. \tag{10}$$

When this formula is simplified, it is seen to be the first equation in (7) with $k = 0$. The graph for the integral in (9) is shown in Figure 9.9(b).

Step Size versus Error

The error term for Simpson's rule with step size $h/2$ is

$$-y^{(4)}(c_1)\frac{h^5}{2880}. \tag{11}$$

If the only error at each step is that given in (11), after M steps the accumulated error for the RK4 method would be

$$-\sum_{k=1}^{M} y^{(4)}(c_k)\frac{h^5}{2880} \approx -\frac{b-a}{5760} y^{(4)}(c)h^4 \approx O(h^4). \tag{12}$$

The next theorem states the relationship between F.G.E. and step size. It is used to give us an idea of how much computing effort must be done when using the RK4 method.

Theorem 9.7   (Precision of Runge–Kutta Method).   Assume that $y(t)$ is the solution to the I.V.P. If $y(t) \in C^5[t_0, b]$ and $\{(t_k, y_k)\}_{k=0}^M$ is the sequence of approximations generated by the Runge–Kutta method of order 4, then

$$|e_k| = |y(t_k) - y_k| = O(h^4),$$
$$|\epsilon_{k+1}| = |y(t_{k+1}) - y_k - hT_n(t_k, y_k)| = O(h^5). \tag{13}$$

In particular, the final global error (F.G.E.) at the end of the interval will satisfy

$$E(y(b), h) = |y(b) - y_M| = O(h^4). \tag{14}$$

Examples 9.10 and 9.11 illustrate Theorem 9.7. If approximations are computed using the step sizes h and $h/2$, we should have

$$E(y(b), h) \approx Ch^4 \tag{15}$$

for the larger step size, and

$$E\left(y(b), \frac{h}{2}\right) \approx C\frac{h^4}{16} \approx \frac{1}{16}Ch^4 \approx \frac{1}{16}E(y(b), h). \tag{16}$$

Hence the idea in Theorem 9.7 is that if the step size in the RK4 method is reduced by a factor of $\frac{1}{2}$, we can expect that the overall F.G.E. will be reduced by a factor of $\frac{1}{16}$.

Example 9.10

Use the RK4 method to solve the I.V.P. $y' = (t - y)/2$ on $[0, 3]$ with $y(0) = 1$. Compare solutions for $h = 1, \frac{1}{2}, \frac{1}{4},$ and $\frac{1}{8}$.

Solution.   Table 9.8 gives the solution values at selected abscissas. For the step size $h = 0.25$ a sample calculation is

$$f_1 = \frac{0.0 - 1.0}{2} = -0.5,$$

$$f_2 = \frac{0.125 - [1 + 0.25(0.5)(-0.5)]}{2} = -0.40625,$$

$$f_3 = \frac{0.125 - [1 + 0.25(0.5)(-0.40625)]}{2} = -0.4121094,$$

$$f_4 = \frac{0.25 - [1 + 0.25(-0.4121094)]}{2} = -0.3234863,$$

$$y_1 = 1.0 + 0.25 \frac{-0.5 + 2(-0.40625) + 2(-0.4121094) - 0.3234863}{6}$$

$$= 0.8974915.$$

TABLE 9.8 Comparison of the RK4 Solutions with Different Step Sizes for $y' = (t - y)/2$ over [0, 3] with $y(0) = 1$

t_k	y_k				$y(t_k)$ Exact
	$h = 1$	$h = \frac{1}{2}$	$h = \frac{1}{4}$	$h = \frac{1}{8}$	
0	1.0	1.0	1.0	1.0	1.0
0.125				0.9432392	0.9432392
0.25			0.8974915	0.8974908	0.8974917
0.375				0.8620874	0.8620874
0.50		0.8364258	0.8364037	0.8364024	0.8364023
0.75			0.8118696	0.8118679	0.8118678
1.00	0.8203125	0.8196285	0.8195940	0.8195921	0.8195920
1.50		0.9171423	0.9171021	0.9170998	0.9170997
2.00	1.1045125	1.1036826	1.1036408	1.1036385	1.1036383
2.50		1.3595575	1.3595168	1.3595145	1.3595144
3.00	1.6701860	1.6694308	1.6693928	1.6693906	1.6693905

Example 9.11

Compare the F.G.E. when the RK4 method is used to solve $y' = (t - y)/2$ over [0, 3] with $y(0) = 1$ using sizes $1, \frac{1}{2}, \frac{1}{4},$ and $\frac{1}{8}$.

Solution. Table 9.9 gives the F.G.E. for the various step sizes and shows that the error in the approximation to $y(3)$ decreases by about $\frac{1}{16}$ when the step size is reduced by a factor of $\frac{1}{2}$:

$$E(y(3), h) = y(3) - y_M = \boldsymbol{O}(h^4) \approx Ch^4 \quad \text{where } C = -0.000614.$$

TABLE 9.9 Relation between Step Size and F.G.E. for the RK4 Solutions to $y' = (t - y)/2$ over [0, 3] with $y(0) = 1$

Step size, h	Number of steps, M	Approximation to $y(3)$, y_M	Error at $t = 3$, $y(3) - y_M$	$\boldsymbol{O}(h^4) \approx Ch^4$ where $C = -0.000614$
1	3	1.6701860	−0.0007955	−0.0006140
$\frac{1}{2}$	6	1.6694308	−0.0000403	−0.0000384
$\frac{1}{4}$	12	1.6693928	−0.0000023	−0.0000024
$\frac{1}{8}$	24	1.6693906	−0.0000001	−0.0000001

A comparison of Examples 9.10 and 9.11 and Examples 9.8 and 9.9 shows what is meant by the statement ''The RK4 method simulates the Taylor series method of order $N = 4$.'' For these examples, the two methods generate identical solution sets $\{(t_k, y_k)\}$

over the given interval. The advantage of the RK4 method is obvious; no formulas for the higher derivatives need to be computed nor do they have to be in a program.

It is not easy to determine the accuracy to which a Runge–Kutta solution has been computed. We could estimate the size of $y^{(4)}(c)$ and use formula (12). Another way is to repeat the algorithm using a smaller step size and compare results. A third way is to adaptively determine the step size, which is done in Algorithm 9.5. In Section 9.6 we will see how to change the step size for a multistep method.

Runge–Kutta Methods of Order N = 2

The second-order Runge–Kutta method (denoted RK2) simulates the accuracy of the Taylor series method of order 2. Although this method is not as good to use as the RK4 method, its proof is easier to understand and illustrates the principles involved. To start, we write down the Taylor series formula for $y(t + h)$:

$$y(t + h) = y(t) + hy'(t) + \frac{1}{2} h^2 y''(t) + C_T h^3 + \cdots, \tag{17}$$

where C_T is a constant involving the third derivative of $y(t)$ and the other terms in the series involve powers of h^j for $j > 3$.

The derivatives $y'(t)$ and $y''(t)$ in equation (17) must be expressed in terms of $f(t, y)$ and its partial derivatives. Recall that

$$y'(t) = f(t, y). \tag{18}$$

The chain rule for differentiating a function of two variables can be used to differentiate (18) with respect to t, and the result is

$$y''(t) = f_t(t, y) + f_y(t, y) \, y'(t).$$

Using (18), this can be written

$$y''(t) = f_t(t, y) + f_y(t, y) f(t, y). \tag{19}$$

The derivatives (18) and (19) are substituted in (17) to give the Taylor expression for $y(t + h)$:

$$y(t + h) = y(t) + hf(t, y) + \frac{1}{2} h^2 f_t(t, y) + \frac{1}{2} h^2 f_y(t, y) f(t, y) + C_T h^3 + \cdots. \tag{20}$$

Now consider the Runge–Kutta method of order $N = 2$ which uses a linear combination of two function values to express $y(t + h)$:

$$y(t + h) = y(t) + Ahf_0 + Bhf_1, \tag{21}$$

where

$$\begin{aligned} f_0 &= f(t, y), \\ f_1 &= f(t + Ph, y + Qhf_0). \end{aligned} \tag{22}$$

Next the Taylor polynomial approximation for a function of two independent variables is used to expand $f(t, y)$ (see Exercises 17 and 18 of Section 4.1). This gives the following representation for f_1:

$$f_1 = f(t, y) + Phf_t(t, y) + Qhf_y(t, y) f(t, y) + C_P h^2 + \cdots , \tag{23}$$

where C_P involves the second-order partial derivatives of $f(t, y)$. Then (23) is used in (21) to get the RK2 expression for $y(t + h)$:

$$y(t + h) = y(t) + (A + B)hf(t, y) + BPh^2 f_t(t, y)$$
$$+ BQh^2 f_y(t, y) f(t, y) + BC_P h^3 + \cdots . \tag{24}$$

A comparison of similar terms in equations (20) and (24) will produce the following conclusions:

$$hf(t, y) = (A + B)hf(t, y) \quad \text{implies that } 1 = A + B,$$

$$\tfrac{1}{2} h^2 f_t(t, y) = BPh^2 f_t(t, y) \quad \text{implies that } \tfrac{1}{2} = BP,$$

$$\tfrac{1}{2} h^2 f_y(t, y) f(t, y) = BQh^2 f_y(t, y) f(t, y) \quad \text{implies that } \tfrac{1}{2} = BQ.$$

Hence if we require that A, B, P, and Q satisfy the relations

$$A + B = 1 \qquad BP = \tfrac{1}{2} \qquad BQ = \tfrac{1}{2}, \tag{25}$$

then the RK2 method in (24) will have the same order of accuracy as the Taylor's method in (20).

Since there are only three equations in four unknowns, the system of equations (25) is underdetermined, and we are permitted to choose one of the coefficients. There are several special choices that have been studied in the literature; we mention two of them.

Case (i): Choose $A = \tfrac{1}{2}$. This choice leads to $B = \tfrac{1}{2}$, $P = 1$, and $Q = 1$. If equation (21) is written with these parameters, the formula is

$$y(t + h) \approx y(t) + \frac{h}{2} [f(t, y) + f(t + h, y + hf(t, y))]. \tag{26}$$

When this scheme is used to generate $\{(t_k, y_k)\}$, the result is Heun's method.

Case (ii): Choose $A = 0$. This choice leads to $B = 1$, $P = \tfrac{1}{2}$, and $Q = \tfrac{1}{2}$. If equation (21) is written with these paramters, the formula is

$$y(t + h) = y(t) + hf\left(t + \frac{h}{2}, y + \frac{h}{2} f(t, y)\right). \tag{27}$$

When this scheme is used to generate $\{(t_k, y_k)\}$, it is called the **modified Euler–Cauchy method**.

The Runge–Kutta–Fehlberg Method (RKF45)

One way to guarantee accuracy in the solution of an I.V.P. is to solve the problem twice using step sizes h asnd $h/2$ and compare answers at the mesh points corresponding to the larger step size. But this requires a significant amount of computation for the smaller step size and must be repeated if it is determined that the agreement is not good enough.

The Runge–Kutta–Fehlberg method (denoted RKF45), is one way to try to resolve this problem. It has a procedure to determine if the proper step size h is being used. At each step two different approximations for the solution are made and compared. If the two answers are in close agreement, the approximation is accepted. If the answers do not agree to a specified accuracy, the step size is reduced. If the answers agree to more significant digits than required, the step size is increased.

Each step requires the use of the following six values:

$$
\begin{aligned}
k_1 &:= hf(t_k, y_k), \\[6pt]
k_2 &:= hf\left(t_k + \frac{1}{4}h,\ y_k + \frac{1}{4}k_1\right), \\[6pt]
k_3 &:= hf\left(t_k + \frac{3}{8}h,\ y_k + \frac{3}{32}k_1 + \frac{9}{32}k_2\right), \\[6pt]
k_4 &:= hf\left(t_k + \frac{12}{13}h,\ y_k + \frac{1932}{2197}k_1 - \frac{7200}{2197}k_2 + \frac{7296}{2197}k_3\right), \\[6pt]
k_5 &:= hf\left(t_k + h,\ y_k + \frac{439}{216}k_1 - 8k_2 + \frac{3680}{513}k_3 - \frac{845}{4104}k_4\right), \\[6pt]
k_6 &:= hf\left(t_k + \frac{1}{2}h,\ y_k - \frac{8}{27}k_1 + 2k_2 - \frac{3544}{2565}k_3 + \frac{1859}{4104}k_4 - \frac{11}{40}k_5\right).
\end{aligned}
\tag{28}
$$

Then an approximation to the solution of the I.V.P. is made using a Runge–Kutta method of order 4:

$$
y_{k+1} = y_k + \frac{25}{216}k_1 + \frac{1408}{2565}k_3 + \frac{2197}{4104}k_4 - \frac{1}{5}k_5.
\tag{29}
$$

where the four function values f_1, f_3, f_4, and f_5 are used. Notice that f_2 is not used in formula (29). A better value for the solution is determined using a Runge–Kutta method of order 5:

$$
z_{k+1} = y_k + \frac{16}{135}k_1 + \frac{6656}{12,825}k_3 + \frac{28,561}{56,430}k_4 - \frac{9}{50}k_5 + \frac{2}{55}k_6.
\tag{30}
$$

The optimal step size sh can be determined by multiplying the scalar s times the current step size h. The scalar s is

$$s = \left(\frac{\text{Tol } h}{2|z_{k+1} - y_{k+1}|}\right)^{1/4}$$

$$\approx 0.84\left(\frac{\text{Tol } h}{|z_{k+1} - y_{k+1}|}\right)^{1/4}$$

(31)

where Tol is the specified error control tolerance.

The derivation of formula (31) can be found in advanced books on numerical analysis. It is important to learn that a fixed step size is not the best strategy even though it would give a nicer-appearing table of values. If values are needed that are not in the table, polynomial interpolation should be used.

Example 9.12

Compare RKF45 and RK4 solutions to the I.V.P.

$$y' = 1 + y^2 \qquad y(0) = 0 \quad \text{on} \quad [0, 1.4].$$

Solution. An RKF45 program was used with the value Tol $= 2 \times 10^{-5}$ for the error control tolerance. It automatically changed the step size and generated the 10 approximations to the solution in Table 9.10. An RK4 program was used with the a priori step size of $h = 0.1$, which required the computer to generate 14 approximations at the equally spaced points in Table 9.11. The approximations at the endpoint are

$$y(1.4) \approx y_{10} = 5.7985045 \quad \text{and} \quad y(1.4) \approx y_{14} = 5.7919748$$

and the errors are

$$E_{10} = -0.0006208 \quad \text{and} \quad E_{14} = 0.0059089$$

for the RKF45 and RK4 methods, respectively. The RKF45 method has the smaller error.

TABLE 9.10   RKF45 Solution to $y' = 1 + y^2$, $y(0) = 0$

k	t_k	RKF45 approximation y_k	True solution, $y(t_k) = \tan(t_k)$	Error, $y(t_k) - y_k$
0	0.0	0.0000000	0.0000000	0.0000000
1	0.2	0.2027100	0.2027100	0.0000000
2	0.4	0.4227933	0.4227931	−0.0000002
3	0.6	0.6841376	0.6841368	−0.0000008
4	0.8	1.0296434	1.0296386	−0.0000048
5	1.0	1.5574398	1.5774077	−0.0000321
6	1.1	1.9648085	1.9647597	−0.0000488
7	1.2	2.5722408	2.5721516	−0.0000892
8	1.3	3.6023295	3.6021024	−0.0002271
9	1.35	4.4555714	4.4552218	−0.0003496
10	1.4	5.7985045	5.7978837	−0.0006208

TABLE 9.11 RK4 Solution to $y' = 1 + y^2$, $y(0) = 0$

k	t_k	RK4 approximation y_k	True solution, $y(t_k) = \tan(t_k)$	Error, $y(t_k) - y_k$
0	0.0	0.0000000	0.0000000	0.0000000
1	0.1	0.1003346	0.1003347	0.0000001
2	0.2	0.2027099	0.2027100	0.0000001
3	0.3	0.3093360	0.3093362	0.0000002
4	0.4	0.4227930	0.4227932	0.0000002
5	0.5	0.5463023	0.5463025	0.0000002
6	0.6	0.6841368	0.6841368	0.0000000
7	0.7	0.8422886	0.8422884	−0.0000002
8	0.8	1.0296391	1.0296386	−0.0000005
9	0.9	1.2601588	1.2601582	−0.0000006
10	1.0	1.5574064	1.5574077	0.0000013
11	1.1	1.9647466	1.9647597	0.0000131
12	1.2	2.5720718	2.5721516	0.0000798
13	1.3	3.6015634	3.6021024	0.0005390
14	1.4	5.7919748	5.7978837	0.0059089

Algorithm 9.4 (Runge–Kutta Method of Order 4).   To approximate the solution of the initial value problem $y' = f(t, y)$ with $y(a) = y_0$ over $[a, b]$ by using the formula

$$y_{k+1} = y_k + \frac{h}{6}[K_1 + 2K_2 + 2K_3 + K_4].$$

```
INPUT A, B, Y(0)                                {Endpoints and initial value}
INPUT M                                         {Number of steps}
H := [B − A]/M                                  {Compute the step size}
T(0) := A                                       {Initialize the value}

FOR  J = 0  TO  M−1  DO
     T := T(J) and Y := Y(J)                    {Local variables}
     K₁ := H*F(T,Y)                             {Function value at tⱼ}
     K₂ := H*F(T+H/2, Y + .5*K₁)                {Function value at t_{j+1/2}}
     K₃ := H*F(T+H/2, Y + .5*K₂)                {Function value at t_{j+1/2}}
     K₄ := H*F(T+H, Y + K₃)                     {Function value at t_{j+1}}
     Y(J+1) := Y + [K₁+2*K₂+2*K₃+K₄]/6          {Integrate f(t,y)}
     T(J+1) := A + H*[J+1]                       {Generate the mesh point}

FOR  J = 0  TO  M  DO                           {Output}
     PRINT T(J), Y(J)
```

Algorithm 9.5 [Runge–Kutta–Fehlberg Method (RKF45)]. To approximate the solution of the initial value problem $y' = f(t, y)$ with $y(a) = y_0$ over $[a, b]$ with an error control and variable step-size method.

Tol := $2 \ast 10^{-5}$ {Error control tolerance}

INPUT A, B, Y(0)   {Endpoints and initial value}

INPUT N   {Tentative number of steps}

H := [B − A]/N   {Initial the step size}

Hmin := H/64 and Hmax := h$\ast$64   {Minimum and maximum step sizes }

T(0) := A and J := 0   {Initialize}

WHILE T(J) < B   DO

    IF T(J)+H > B   THEN   H := B−T(J)   {The last step}

    T := T(J) and Y := Y(J)

    K_1 := H$\ast$F(T,Y)   {Compute

    K_2 := H$\ast$F(T + $\frac{1}{4}$H, Y + $\frac{1}{4}$$K_1$)   the function

    K_3 := H$\ast$F(T + $\frac{3}{8}$H, Y + $\frac{3}{32}$$K_1$ + $\frac{9}{32}$$K_2$)   values}

    K_4 := H$\ast$F(T + $\frac{12}{13}$H, Y + $\frac{1932}{2197}$$K_1$ − $\frac{7200}{2197}$$K_2$ + $\frac{7296}{2197}$$K_3$)

    K_5 := H$\ast$F(T + H, Y + $\frac{439}{216}$$K_1$ − 8K_2 + $\frac{3680}{513}$$K_3$ − $\frac{845}{4104}$$K_4$)

    K_6 := H$\ast$F(T + $\frac{1}{2}$H, Y − $\frac{8}{27}$$K_1$ + 2K_2 − $\frac{3544}{2565}$$K_3$ + $\frac{1859}{4104}$$K_4$ − $\frac{11}{40}$$K_5$)

    Err := $\left| \frac{1}{360}K_1 - \frac{128}{4275}K_3 - \frac{2197}{75240}K_4 + \frac{1}{50}K_5 + \frac{2}{55}K_6 \right|$ {$|z_{k+1} - y_{k+1}|$}

    IF   Err<Tol OR H $\leqq$2$\ast$Hmin   THEN   {Accept

        Y(J+1) := Y + $\frac{25}{216}$$K_1$ + $\frac{1408}{2565}$$K_3$ + $\frac{2197}{4104}$$K_4$ − $\frac{1}{5}$$K_5$ the

        T(J+1) := T+H, J := J+1   approximation}

    IF   Err=0   THEN

        S := 0   {Trap division by 0}

        ELSE

        S := .84$\ast$[Tol$\ast$H/Err]$^{1/4}$ {Step size scalar}

    ENDIF

    IF   S<.75 AND H>2$\ast$Hmin   THEN   H := H/2   {Reduce step}

    IF   S>1.5 AND 2$\ast$H<Hmax   THEN   H := H$\ast$2   {Increase step}

END

FOR I = 0 TO J DO   {Output}

    PRINT T(I), Y(I)

EXERCISES FOR RUNGE–KUTTA METHODS

In Exercises 1–5, use the Runge–Kutta method, of order $N = 4$, to find approximations to the I.V.P.

(a) Compute $y_1, y_2, \ldots, y_M$ for each of the two cases (i) $h = 0.2$, $M = 1$ and (ii) $h = 0.1$, $M = 2$.

(b) Compare with the exact solution $y(0.2)$ with the two approximations in part (a).

(c) Does the F.G.E. in part (a) behave as expected when h is halved?

(d) Use a computer to carry out the computations on the larger interval $[a, b]$ that is given.

1. (a) Solve $y' = t^2 - y$ over $[0, 0.2]$ with $y(0) = 1$.
   (b) and (c) Compare with $y(t) = -\exp(-t) + t^2 - 2t + 2$.
   (d) Use $[a, b] = [0, 2]$ with $h = 0.2, 0.1$, and 0.05.

2. (a) Solve $y' = 3y + 3t$ over $[0, 0.2]$ with $y(0) = 1$.
   (b) and (c) Compare with $y(t) = \frac{4}{3} \exp(3t) - t - \frac{1}{3}$.
   (d) Use $[a, b] = [0, 2]$ with $h = 0.2, 0.1$, and 0.05.

3. (a) Solve $y' = -ty$ over $[0, 0.2]$ with $y(0) = 1$.
   (b) and (c) Compare with $y(t) = \exp(-t^2/2)$.
   (d) Use $[a, b] = [0, 2]$ with $h = 0.2, 0.1$, and 0.05.

4. (a) Solve $y' = \exp(-2t) - 2y$ over $[0, 0.2]$ with $y(0) = \frac{1}{10}$.
   (b) and (c) Compare with $y(t) = \frac{1}{10} \exp(-2t) + t \exp(-2t)$.
   (d) Use $[a, b] = [0, 2]$ with $h = 0.2, 0.1$, and 0.05.

5. (a) Solve $y' = 2ty^2$ over $[0, 0.2]$ with $y(0) = 1$.
   (b) and (c) Compare with $y(t) = 1/(1 - t^2)$.
   (d) Use $[a, b] = [0, 1]$ with $h = 0.1, 0.5$, and 0.025.
   Notice that the Runge–Kutta method will generate an approximation to $y(1)$ even though the solution curve is not defined at $t = 1$.

6. In a chemical reaction, one molecule of A combines with one molecule of B to form one molecule of the chemical product C. It is found that the concentration $y(t)$ of C, at time t, is the solution to the I.V.P.

$$y' = k(a - y)(b - y) \quad \text{with} \quad y(0) = 0,$$

where k is a positive constant and a and b are the initial concentrations of A and B, respectively. Suppose that $k = 0.01$, $a = 70$ millimoles/liter, and $b = 50$ millimoles/liter. Use the Runge–Kutta method of order $N = 4$ with $h = 0.5$ to find the solution over $[0, 20]$.

   Remark. You can compare your computer solution with the exact solution $y(t) = 350[1 - \exp(-0.2t)]/[7 - 5 \exp(-0.2t)]$. Observe that the limiting value is 50 as $t \to +\infty$.

7. By solving an appropriate initial value problem, make a table of values of the function $f(x)$ given by the following integral:

$$f(x) = \frac{1}{2} + (2\pi)^{-1/2} \int_0^x \exp\left(\frac{-t^2}{2}\right) dt \quad \text{for } 0 \le x \le 3.$$

Use the Runge–Kutta method of order $N = 4$ with $h = 0.1$ for your computations. Your solution should agree with the values in the table below.

   Remark. This is a good way to generate the table of areas for a standard normal distribution (see Exercise 8).

x	$f(x)$
0.0	0.5
0.5	0.6914625
1.0	0.8413448
1.5	0.9331928
2.0	0.9772499
2.5	0.9937903
3.0	0.9986501

8. Show that when the Runge–Kutta method of order $N = 4$ is used to solve the I.V.P. $y' = f(t)$ over $[a, b]$ with $y(a) = 0$, the result is

$$y(b) \approx \frac{h}{6} \sum_{k=0}^{M-1} [f(t_k) + 4f(t_{k+1/2}) + f(t_{k+1})],$$

where $h = (b - a)/M$ and $t_k = a + kh$, and $t_{k+1/2} = a + \left(k + \frac{1}{2}\right)h$, which is Simpson's approximation (with step size $h/2$) for the definite integral of $f(t)$ taken over the interval $[a, b]$.

9. The Richardson improvement method discussed in Lemma 7.1 (Section 7.3) can be used in conjunction with the Runge–Kutta method. If the Runge–Kutta method of order $N = 4$ is used with step size h, we have

$$y(b) \approx y_h + Ch^4.$$

If the Runge–Kutta method of order $N = 4$ is used with $2h$, we have

$$y(b) \approx y_{2h} + 16Ch^4.$$

The terms involving Ch^4 can be eliminated to obtain an improved approximation for $y(b)$ and the result is

$$y(b) \approx \frac{16y_h - y_{2h}}{15}.$$

This improvement scheme can be used with the values in Example 9.11 to obtain better approximations to $y(3)$. Find the missing entries in the table below.

h	y_h	$\dfrac{16y_h - y_{2h}}{15}$
1	1.6701860	
$\frac{1}{2}$	1.6694308	——————
$\frac{1}{4}$	1.6693928	——————
$\frac{1}{8}$	1.6693906	——————

10. Write a report about the proof of the Runge–Kutta method of order $N = 4$. Be sure to mention the parameters and the system of eight equations involving them.

11. Write a report on the Runge–Kutta–Fehlberg method.

9.6 PREDICTOR–CORRECTOR METHODS

The methods of Euler, Heun, Taylor, and Runge–Kutta are called **single-step methods** because they use only the information from one previous point to compute the successive point, that is, only the initial point (t_0, y_0) is used to compute (t_1, y_1) and in general y_k is needed to compute y_{k+1}. After several points have been found it is feasible to use several prior points in the calculation. For illustration, we develop the Adams–Bashforth four-step method, which requires y_{k-3}, y_{k-2}, y_{k-1}, and y_k in the calculation of y_{k+1}. This method is not self-starting; four initial points (t_0, y_0), (t_1, y_1), (t_2, y_2), and (t_3, y_3) must be given in advance in order to generate the points $\{(t_k, y_k): \text{for } k \geq 4\}$.

A desirable feature of a multistep method is that the local truncation error (L.T.E) can be determined and a correction term can be included, which improves the accuracy of the answer at each step. Also, it is possible to determine if the step size is small enough to obtain an accurate value for y_{k+1}, yet large enough so that unnecessary and time-consuming calculations are eliminated. Using the combination of a predictor and corrector requires only two function evaluations of $f(t, y)$ per step.

The Adams–Bashforth–Moulton Method

The Adams–Bashforth–Moulton predictor–corrector method is a multistep method based on the fundamental theorem of calculus:

$$y(t_{k+1}) = y(t_k) + \int_{t_k}^{t_{k+1}} f(t, y(t))\, dt. \tag{1}$$

The predictor uses the Lagrange polynomial approximation for $f(t, y(t))$ based on the points (t_{k-3}, f_{k-3}), (t_{k-2}, f_{k-2}), (t_{k-1}, f_{k-1}), and (t_k, f_k). It is integrated over the interval $[t_k, t_{k+1}]$ in (1). This produces the Adams–Bashforth predictor:

$$p_{k+1} = y_k + \frac{h}{24}(-9f_{k-3} + 37f_{k-2} - 59f_{k-1} + 55f_k). \tag{2}$$

The corrector is developed similarly. The value p_{k+1} can now be used. A second Lagrange polynomial for $f(t, y(t))$ is constructed which is based on the points (t_{k-2}, f_{k-2}), (t_{k-1}, f_{k-1}), (t_k, f_k), and the new point $(t_{k+1}, f_{k+1}) = (t_{k+1}, f(t_{k+1}, p_{k+1}))$. It is integrated over $[t_k, t_{k+1}]$. This produces the Adams–Moulton corrector:

$$y_{k+1} = y_k + \frac{h}{24}(f_{k-2} - 5f_{k-1} + 19f_k + 9f_{k+1}). \tag{3}$$

Figure 9.10 shows the nodes for the Lagrange polynomials that are used in developing formulas (2) and (3), respectively.

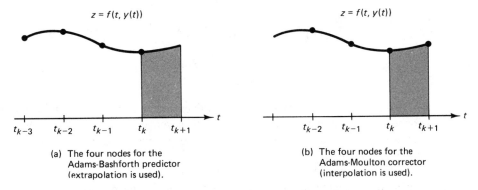

Figure 9.10   Integration over $[t_k, t_{k+1}]$ in the Adams–Bashforth method.

Error Estimation and Correction

The error terms for the numerical integration formulas used to obtain both the predictor and corrector are of the order $O(h^5)$. The L.T.E. for formulas (2) and (3) are

$$y(t_{k+1}) - p_{k+1} = \tfrac{251}{720} y^{(5)}(c_{k+1})h^5 \qquad \text{(L.T.E. for the predictor)}, \qquad (4)$$

$$y(t_{k+1}) - y_{k+1} = \tfrac{-19}{720} y^{(5)}(d_{k+1})h^5 \qquad \text{(L.T.E. for the corrector)}. \qquad (5)$$

Suppose that h is small and $y^{(5)}(t)$ is nearly constant over the interval; then the terms involving the fifth derivative in (4)–(5) can be eliminated and the result is

$$y(t_{k+1}) - y_{k+1} \approx \tfrac{-19}{270} (y_{k+1} - p_{k+1}). \qquad (6)$$

The importance of the predictor–corrector method should now be evident. Formula (6) gives an approximate error estimate based on the two computed values p_{k+1}, y_{k+1} and does not use $y^{(5)}(t)$.

Practical Considerations

The corrector (3) used the approximation $f_{k+1} \approx f(t_{k+1}, p_{k+1})$ in the calculation of y_{k+1}. Since y_{k+1} is also an estimate for $y(t_{k+1})$, it could be used in the corrector (3) to generate a new approximation for f_{k+1}, which in turn will generate a new value for y_{k+1}. However, when this iteration on the corrector is continued, it will converge to a fixed point of (3) rather than the differential equation. It is more efficient to reduce the step size if more accuracy is needed.

Formula (6) can be used to determine when to change step size. Although elaborate methods are available, we show how to reduce the step size to $h/2$ or increase it to $2h$. Let RelErr $= 5 \times 10^{-6}$ be our relative error criterion, and let Small $= 10^{-5}$.

$$\text{IF } \frac{19}{270} \frac{|y_{k+1} - p_{k+1}|}{|y_{k+1}| + \text{Small}} > \text{RelErr} \quad \text{THEN} \quad \text{Set } h = \frac{h}{2}. \tag{7}$$

$$\text{IF } \frac{19}{270} \frac{|y_{k+1} - p_{k+1}|}{|y_{k+1}| + \text{Small}} < \frac{\text{RelErr}}{100} \quad \text{THEN} \quad \text{Set } h = 2h. \tag{8}$$

When the predicted and corrected values do not agree to five significant digits, then (7) reduces the step size. If they agree to seven or more significant digits, then (8) increases the step size. Fine tuning of these parameters should be made to suit your particular computer.

Reducing the step size requires four new starting values. Interpolation of $f(t, y(t))$ with a fourth-degree polynomial is used to supply the missing values that bisect the intervals $[t_{k-2}, t_{k-1}]$ and $[t_{k-1}, t_k]$. The four mesh points $t_{k-3/2}, t_{k-1}, t_{k-1/2},$ and t_k used in the successive calculations are shown in Figure 9.11.

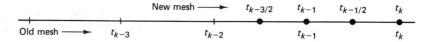

Figure 9.11   Reduction of the step size to $h/2$ in an adaptive method.

The interpolation formulas needed to obtain the new starting values for the step size $h/2$ are

$$\begin{aligned}
f_{k-1/2} &= \frac{-5f_{k-4} + 28f_{k-3} - 70f_{k-2} + 140f_{k-1} + 35f_k}{128}, \\
f_{k-3/2} &= \frac{3f_{k-4} - 20f_{k-3} + 90f_{k-2} + 60f_{k-1} - 5f_k}{128}.
\end{aligned} \tag{9}$$

Increasing the step size is an easier task. Seven prior points are needed to double the step size. The four new points are obtained by omitting every second one, as shown in Figure 9.12.

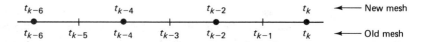

Figure 9.12   Increasing the step size to $2h$ in an adaptive method.

The Milne–Simpson Method

Another popular predictor–corrector scheme is known as the Milne–Simpson method. Its predictor is based on integration of $f(t, y(t))$ over the interval $[t_{k-3}, t_{k+1}]$:

$$y(t_{k+1}) = y(t_{k-3}) + \int_{t_{k-3}}^{t_{k+1}} f(t, y(t)) \, dt. \tag{10}$$

The predictor uses the Lagrange polynomial approximation for $f(t, y(t))$ based on the points (t_{k-3}, f_{k-3}), (t_{k-2}, f_{k-2}), (t_{k-1}, f_{k-1}), and (t_k, f_k). It is integrated over the interval $[t_{k-3}, t_{k+1}]$. This produces the Milne predictor:

$$p_{k+1} = y_{k-3} + \frac{4h}{3} (2f_{k-2} - f_{k-1} + 2f_k). \tag{11}$$

The corrector is developed similarly. The value p_{k+1} can now be used. A second Lagrange polynomial for $f(t, y(t))$ is constructed which is based on the points (t_{k-2}, f_{k-2}), (t_{k-1}, f_{k-1}), and the new point $(t_{k+1}, f_{k+1}) = (t_{k+1}, f(t_{k+1}, p_{k+1}))$. The polynomial is integrated over $[t_{k-1}, t_{k+1}]$ and the result is the familiar Simpson's rule:

$$y_{k+1} = y_{k-1} + \frac{h}{3} (f_{k-1} + 4f_k + f_{k+1}). \tag{12}$$

Error Estimation and Correction

The error terms for the numerical integration formulas used to obtain both the predictor and corrector are of the order $O(h^5)$. The L.T.E. for the formulas in (11) and (12) are

$$y(t_{k+1}) - p_{k+1} = \frac{28}{90} y^{(5)}(c_{k+1})h^5 \qquad \text{(L.T.E. for the predictor)}, \tag{13}$$

$$y(t_{k+1}) - y_{k+1} = \frac{-1}{90} y^{(5)}(d_{k+1})h^5 \qquad \text{(L.T.E. for the corrector)}. \tag{14}$$

Suppose that h is small enough so that $y^{(5)}(t)$ is nearly constant over the interval $[t_{k-3}, t_{k+1}]$. Then the terms involving the fifth derivative can be eliminated in (13) and (14) and the result is

$$y(t_{k+1}) - p_{k+1} \approx \frac{28}{29} (y_{k+1} - p_{k+1}). \tag{15}$$

Formula (15) gives an error estimate for the predictor that is based on the two computed values p_{k+1}, y_{k+1} and does not use $y^{(5)}(t)$. It can be used to improve the predicted value. Under the assumption that the difference between the predicted and corrected values at each step changes slowly, we can substitute p_k and y_k for p_{k+1} and y_{k+1} in (15) and get the following modifier:

$$m_{k+1} = p_{k+1} + 28 \frac{y_k - p_k}{29}. \tag{16}$$

This modified value is used in place of p_{k+1} in the correction step, and equation (12) becomes

$$y_{k+1} = y_{k-1} + \frac{h}{3} [f_{k-1} + 4f_k + f(t_{k+1}, m_{k+1})]. \tag{17}$$

Therefore, the improved (modified) Milne–Simpson method is

$$p_{k+1} = y_{k-3} + \frac{4h}{3}(2f_{k-2} - f_{k-1} + 2f_k) \qquad \text{(predictor)}$$

$$m_{k+1} = p_{k+1} + 28\frac{y_k - p_k}{29} \qquad \text{(modifier)}$$

$$\tag{18}$$

$$f_{k+1} = f(t_{k+1}, m_{k+1})$$

$$y_{k+1} = y_{k-1} + \frac{h}{3}(f_{k-1} + 4f_k + f_{k+1}) \qquad \text{(corrector)}.$$

Hamming's method is another important method. We shall omit its derivation but present it in algorithmic form. As a final precaution we mention that all the predictor–corrector methods have stability problems. Stability is an advanced topic and the serious reader should research this subject.

Example 9.13

Use the Adams–Bashforth–Moulton, Milne–Simpson, and Hamming methods with $h = \frac{1}{8}$ and compute approximations for the solution of the I.V.P.

$$y' = \frac{t - y}{2} \quad \text{with} \quad y(0) = 1 \quad \text{over} \quad [0, 3].$$

Solution. A Runge–Kutta method was used to obtain the starting values

$$y_1 = 0.94323919, \quad y_2 = 0.89749071, \quad \text{and} \quad y_3 = 0.86208736.$$

Then a computer implementation of Algorithms 9.6 to 9.8 produced the values in Table 9.12. The error for each entry in the table is given as a multiple of 10^{-8}. In all entries there are at least six digits of accuracy. In this example, the best answers were produced by Hamming's method.

TABLE 9.12 Comparison of the Adams–Bashforth–Moulton, Milne–Simpson, and Hamming Methods for Solving $y' = (t - y)/2$, $y(0) = 1$

k	Adams–Bashforth–Moulton	Error	Milne–Simpson	Error	Hamming's method	Error
0.0	1.00000000	0E-8	1.00000000	0E-8	1.00000000	0E-8
0.5	0.83640227	8E-8	0.83640231	4E-8	0.83640234	1E-8
0.625	0.81984673	16E-8	0.81984687	2E-8	0.81984688	1E-8
0.75	0.81186762	22E-8	0.81186778	6E-8	0.81186783	1E-8
0.875	0.81194530	28E-8	0.81194555	3E-8	0.81194558	0E-8
1.0	0.81959166	32E-8	0.81959190	8E-8	0.81959198	0E-8
1.5	0.91709920	46E-8	0.91709957	9E-8	0.91709967	−1E-8
2.0	1.10363781	51E-8	1.10363822	10E-8	1.10363834	−2E-8
2.5	1.35951387	52E-8	1.35951429	10E-8	1.35951441	−2E-8
2.625	1.43243853	52E-8	1.43243899	6E-8	1.43243907	−2E-8
2.75	1.50851827	52E-8	1.50851869	10E-8	1.50851881	−2E-8
2.875	1.58756195	51E-8	1.58756240	6E-8	1.58756248	−2E-8
3.0	1.66938998	50E-8	1.66939038	10E-8	1.66939050	−2E-8

The Right Step

Our selection of methods has a purpose: first, their development is easy enough for a first course; second, more advanced methods have a similar development; third, most undergraduate problems can be solved by one of the methods. However, when a predictor–corrector method is used to solve the I.V.P. $y' = f(t, y)$, $y(t_0) = y_0$ over a large interval, difficulties sometimes occur.

If $f_y(t, y) < 0$ and the step size is too large, a predictor–corrector method might be unstable. As a rule of thumb, stability exists when a small error is propagated as a decreasing error and instability exists when a small error is propagated as an increasing error. When too large a step size is used over a large interval, instability will result and is sometimes manifest by oscillations in the computed solution. They can be attenuated by changing to a smaller step size. Formulas (7) to (9) suggest how to modify the algorithm(s). When step-size control is included, the following error estimate(s) should be used:

$$y(t_k) - y_k \approx 19 \frac{p_k - y_k}{270} \qquad \text{(Adams–Bashforth–Moulton)}, \tag{19}$$

$$y(t_k) - y_k \approx \frac{p_k - y_k}{29} \qquad \text{(Milne–Simpson)}, \tag{20}$$

$$y(t_k) - y_k \approx 9 \frac{p_k - y_k}{121} \qquad \text{(Hamming)}. \tag{21}$$

In all methods, the corrector step is a type of fixed-point iteration. It can be proven that the step size h for the methods must satisfy the following conditions:

$$h \ll \frac{2.66667}{|f_y(t, y)|} \qquad \text{(Adams–Bashforth–Moulton)}, \tag{22}$$

$$h \ll \frac{3.00000}{|f_y(t, y)|} \qquad \text{(Milne–Simpson)}, \tag{23}$$

$$h \ll \frac{2.66667}{|f_y(t, y)|} \qquad \text{(Hamming)}. \tag{24}$$

The notation $\ll$ in (22)–(24) means "much smaller than." The next example shows that more stringent inequalities should be used:

$$h < \frac{0.75}{|f_y(t, y)|} \qquad \text{(Adams–Bashforth–Moulton)}, \tag{25}$$

$$h < \frac{0.45}{|f_y(t, y)|} \qquad \text{(Milne–Simpson)}, \tag{26}$$

$$h < \frac{0.69}{|f_y(t, y)|} \qquad \text{(Hamming)}. \tag{27}$$

Inequality (27) is found in advanced books on numerical analysis. The other two inequalities seem appropriate for the example.

Example 9.14

Use the Adams–Bashforth–Moulton, Milne–Simpson, and Hamming methods and compute approximations for the solution of

$$y' = 30 - 5y, \qquad y(0) = 1 \qquad \text{over the interval } [0, 10].$$

Solution.   All three methods are of the order $O(h^4)$. When $N = 120$ steps was used for all three methods, the maximum error for each method occurred at a different place:

$$y(0.41666667) - y_5 \approx -0.00277037 \qquad \text{(Adams–Bashforth–Moulton)},$$

$$y(0.33333333) - y_4 \approx -0.00139255 \qquad \text{(Milne–Simpson)},$$

$$y(0.33333333) - y_4 \approx -0.00104982 \qquad \text{(Hamming)}.$$

At the right endpoint $t = 10$, the error was

$$y(10) - y_{120} \approx 0.00000000 \qquad \text{(Adams–Bashforth–Moulton)},$$

$$y(10) - y_{120} \approx 0.00001015 \qquad \text{(Milne–Simpson)},$$

$$y(10) - y_{120} \approx 0.00000000 \qquad \text{(Hamming)}.$$

Both the Adams–Bashforth–Moulton and Hamming methods gave approximate solutions with eight digits of accuracy at the right endpoint.

It is instructive to see that if the step size is too large, the computed solution oscillates about the true solution. Figure 9.13 illustrates this phenomenon. The small number of steps were determined experimentally so that the oscillations were about the same magnitude. The large number of steps required to attenuate the oscillations were determined with equations (25) to (27).

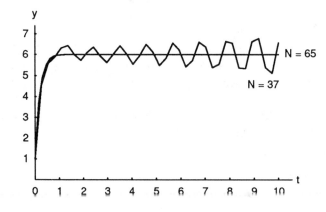

Figure 9.13   (a) The Adams–Bashforth–Moulton solution to $y' = 30 - 5y$ with $N = 37$ steps produces oscillation. It is stabilized when $N = 65$ because $h = 10/65 = 0.1538 \approx 0.15 = 0.75/5 = 0.75/|f_y(t, y)|$.

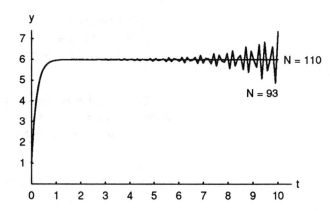

Figure 9.13   (b) The Milne–Simpson solution to $y' = 30 - 5y$ with $N = 93$ steps produces oscillation. It is stabilized when $N = 110$ because $h = 10/110 = 0.0909 \approx 0.09 = 0.45/5 = 0.45/|f_y(t, y)|$.

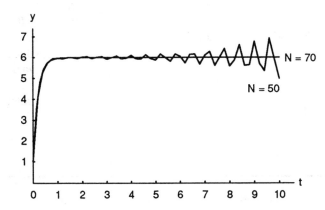

Figure 9.13   (c) Hamming's solution to $y' = 30 - 5y$ with $N = 50$ steps produces oscillation. It is stabilized when $N = 70$ because $h = 10/70 = 0.1428 \approx 0.138 = 0.69/5 = 0.69/|f_y(t, y)|$.

Algorithm 9.6   (Adams–Bashforth–Moulton Method).   To approximate the solution of the initial value problem $y' = f(t, y)$ with $y(a) = y_0$ over $[a, b]$ by using the predictor

$$p_{k+1} = y_k + \frac{h}{24}\left[-9f_{k-3} + 37f_{k-2} - 59f_{k-1} + 55f_k\right]$$

and the corrector

$$y_{k+1} = y_k + \frac{h}{24}\left[f_{k-2} - 5f_{k-1} + 19f_k + 9f_{k+1}\right].$$

```
          INPUT A, B, Y(0)                           {Endpoint and initial value}
          INPUT N                                    {Number of steps, N > 3}
          H := [B − A]/N                             {Compute the step size}
          T(0) := A, F₀ := F(T(0),Y(0))

FOR   K = 1  TO  3  DO                               {Either input three additional
      | T(K) := A + K*H                               starting values or compute them
      |___ Get Y(K)                                   using the Runge–Kutta method}

          F₁ := F(T(1),Y(1)), F₂ := F(T(2),Y(2)), F₃ := F(T(3),Y(3))
          H2 := H/24                                 {Saves wasted computations}

FOR   K = 3  TO  N−1  DO
      | P := Y(K) + H2*[−9*F₀+37*F₁−59*F₂+55*F₃]     {Predictor}
      | T(K+1) := A + H*[K+1]                        {Next abscissa}
      | F₄ := F(T(K+1),P)                            {Evaluate f(t,y)}
      | Y(K+1) := Y(K) + H2*[F₁−5*F₂+19*F₃+9*F₄]     {Corrector}
      | F₀ := F₁, F₁ := F₂, F₂ := F₃                 {Update
      |___ F₃ := F(T(K+1),Y(K+1))                     the values}

FOR   K = 0  TO  N  DO                               {Output}
      |___ PRINT T(K), Y(K)
```

Algorithm 9.7   (Milne–Simpson Method).   To approximate the solution of the initial value problem $y' = f(t, y)$ with $y(a) = y_0$ over $[a, b]$ by using the predictor

$$p_{k+1} = y_{k-3} + \frac{4h}{3} [2f_{k-2} - f_{k-1} + 2f_k]$$

and the corrector

$$y_{k+1} = y_{k-1} + \frac{h}{3} [f_{k-1} + 4f_k + f_{k+1}].$$

```
          INPUT A, B, Y(0)                           {Endpoints and initial value}
          INPUT N                                    {Number of steps, n > 3}
          H := [B − A]/N                             {Compute the step size}
          T(0) := A                                  {Initialize}
```

```
FOR  K = 1  TO  3  DO                           {Either input three additional
   │   T(K) := A + K*H                            starting values or compute them
   └── Get Y(K)                                   using the Runge–Kutta method}


       F₁ := F(T(1),Y(1)), F₂ := F(T(2),Y(2)), F₃ := F(T(3),Y(3))
       Pold := 0, Yold := 0                                                 {Initialize}

FOR  K = 3  TO  N−1  DO
   │   Pnew := Y(K−3) + 4*H*[2*F₁−F₂+2*F₃]/3                          {Milne predictor}
   │   Pmod := Pnew + 28*[Yold − Pold]/29                                  {Modifier}
   │   T(K+1) := A + H*[K+1]                                        {New mesh point}
   │   F₄ := F(T(K+1),Pmod)                                         {Function value}
   │   Y(K+1) := Y(K−1) + H*[F₂+4*F₃+F₄]/3                         {Simpson corrector}
   │   Pold := Pnew, Yold := Y(K+1)                                         {Update
   │   F₁ := F₂, F₂ := F₃, F₃ := F(T(K+1),Y(K+1))                            values}
END

FOR  K = 0  TO  N  DO                                                      {Output}
   └── PRINT T(K), Y(K)
```

Algorithm 9.8   (The Hamming Method).   To approximate the solution of the
initial value problem $y' = f(t, y)$ with $y(a) = y_0$ over $[a, b]$ by using the predictor

$$p_{k+1} = y_{k-3} + \frac{4h}{3}\,[2f_{k-2} - f_{k-1} + 2f_k]$$

and the corrector

$$y_{k+1} = \frac{-y_{k-2} + 9y_k}{8} + \frac{3h}{8}\,[-f_{k-1} + 2f_k + f_{k+1}].$$

```
INPUT A, B, Y(0)                                {Endpoints and initial value}
INPUT N                                         {Number of steps, N > 3}
H := [B − A]/N                                  {Compute the step size}
T(0) := A                                       {Initialize}

FOR  K = 1  TO  3  DO                           {Either input three additional
   │   T(K) := A + K*H                            starting values or compute them
   └── Get Y(K)                                   using the Runge–Kutta method}


       F₁ := F(T(1),Y(1)), F₂ := F(T(2),Y(2)), F₃ := F(T(3),Y(3))
       Pold := 0, Cold := 0                                                 {Initialize}
```

```
FOR  K = 3  TO  N−1  DO
     Pnew := Y(K−3) + 4*H*[2*F₁−F₂+2*F₃]/3              {Milne predictor}
     Pmod := Pnew + 112*[Cold − Pold]/121                     {Modifier}
     T(K+1) := A + H*[K+1]                              {New mesh point}
     F₄ := F(T(K+1),Pmod)                             {Function value}
     Cnew := [9*Y(K)−Y(K−2) + 3*H*[−F₂+2*F₃+F₄]]/8    {Hamming corrector}
     Y(K+1) := Cnew + 9*[Pnew − Cnew]/121           {New value yₖ₊₁}
     Pold := Pnew, Cold := Cnew                              {Update
     F₁ := F₂, F₂ := F₃, F₃ := F(T(K+1),Y(K+1))            values}
END

FOR  K = 0  TO  N  DO                                        {Output}
     PRINT T(K), Y(K)
```

EXERCISES FOR PREDICTOR–CORRECTOR METHODS

In Exercises 1–10:

(a) Use any one of the three predictor–corrector methods and the three starting values y_1, y_2, and y_3 and $h = 0.05$ to calculate the next value y_4 for the I.V.P.

(b) Use a computer implementation of any predictor–corrector method and solve the I.V.P. in part (a) over $[a, b]$. Use the three starting values y_1, y_2, and y_3 that are given or generate them with a Runge–Kutta method.

(c) Compare your answer with the true solution $y(t)$.

1. (a) $y' = t^2 - y$ with $y(0) = 1$
   (b) $[a, b] = [0, 5]$

   $$y(0.05) = 0.95127058, \quad y(0.10) = 0.90516258, \quad y(0.15) = 0.86179202$$

   (c) $y(t) = -\exp(-t) + t^2 - 2t + 2$

2. (a) $y' = y + 3t - t^2$ with $y(0) = 1$
   (b) $[a, b] = [0, 5]$

   $$y(0.05) = 1.0550422, \quad y(0.10) = 1.1203418, \quad y(0.15) = 1.1961685$$

   (c) $y(t) = 2\exp(t) + t^2 - t - 1$

3. (a) $y' = -t/y$ with $y(1) = 1$
   (b) $[a, b] = [1, 1.4]$

   $$y(1.05) = 0.94736477, \quad y(1.10) = 0.88881944, \quad y(1.15) = 0.82310388$$

   (c) $y(t) = (2 - t^2)^{1/2}$

4. (a) $y' = \exp(-t) - y$ with $y(0) = 1$
   (b) $[a, b] = [0, 5]$

   $$y(0.05) = 0.99879090, \quad y(0.10) = 0.99532116, \quad y(0.15) = 0.98981417$$

   (c) $y(t) = t\exp(-t) + \exp(-t)$

5. (a) $y' = 2ty^2$ with $y(0) = 1$
   (b) $[a, b] = [0, 0.95]$

$$y(0.05) = 1.0025063, \quad y(0.10) = 1.0101010, \quad y(0.15) = 1.0230179$$

  (c) $y(t) = 1/(1 - t^2)$

6. (a) $y' = 1 + y^2$ with $y(0) = 1$
   (b) $[a, b] = [0, 0.75]$

$$y(0.05) = 1.1053556, \quad y(0.10) = 1.2230489, \quad y(0.15) = 1.3560879$$

  (c) $y(t) = \tan(t + \pi/4)$

7. (a) $y' = 2y - y^2$ with $y(0) = 1$
   (b) $[a, b] = [0, 5]$

$$y(0.05) = 1.0499584, \quad y(0.10) = 1.0996680, \quad y(0.15) = 1.1488850$$

  (c) $y(t) = 1 + \tanh(t)$

8. (a) $y' = [1 - y^2]^{1/2}$ with $y(0) = 0$
   (b) $[a, b] = [0, 1.55]$

$$y(0.05) = 0.049979169, \quad y(0.10) = 0.099833417, \quad y(0.15) = 0.14943813$$

  (c) $y(t) = \sin(t)$

9. (a) $y' = y^2 \sin(t)$ with $y(0) = 1$
   (b) $[a, b] = [0, 1.55]$

$$y(0.05) = 1.0012513, \quad y(0.10) = 1.0050209, \quad y(0.15) = 1.0113564$$

  (c) $y(t) = \sec(t)$

10. (a) $y' = 1 - y^2$ with $y(0) = 0$
    (b) $[a, b] = [0, 5]$

$$y(0.05) = 0.049958375, \quad y(0.10) = 0.099667995, \quad y(0.15) = 0.14888503$$

  (c) $y(t) = [1 - \exp(-2t)]/[1 + \exp(-2t)]$

11. Write a report on step-size control for the numerical solution of differential equations. See References [29, 40, 60, 75, 101, 117, and 160].

12. Write a report on the stability of numerical solutions for differential equations. See References [3, 8, 9, 29, 40, 60, 76, 78, 79, 96, 101, 128, 146, 152, 153, and 160].

13. Write a report on the numerical solution of stiff differential equations. See References [9, 29, 40, 57, 60, 98, 117, 152, 153, 160, and 173].

9.7 SYSTEMS OF DIFFERENTIAL EQUATIONS

This section is an introduction to systems of differential equations. To illustrate the concepts, we consider the initial value problem

$$\frac{dx}{dt} = f(t, x, y) \qquad \text{with} \quad \begin{cases} x(t_0) = x_0, \\ \\ y(t_0) = y_0. \end{cases} \tag{1}$$

$$\frac{dy}{dt} = g(t, x, y)$$

A solution to (1) is a pair of differentiable functions $x(t)$ and $y(t)$ with the property that when t, $x(t)$, and $y(t)$ are substituted in $f(t, x, y)$ and $g(t, x, y)$, the result is equal to the derivative $x'(t)$ and $y'(t)$, respectively, that is,

$$
\begin{aligned}
x'(t) &= f(t, x(t), y(t)) \\
y'(t) &= g(t, x(t), y(t))
\end{aligned}
\quad \text{with} \quad
\begin{cases}
x(t_0) = x_0, \\
y(t_0) = y_0.
\end{cases}
\tag{2}
$$

For example, consider the system of differential equations

$$
\begin{aligned}
\frac{dx}{dt} &= x + 2y \\
\frac{dy}{dt} &= 3x + 2y
\end{aligned}
\quad \text{with} \quad
\begin{cases}
x(0) = 6, \\
y(0) = 4.
\end{cases}
\tag{3}
$$

The solution to the I.V.P. (3) is

$$
\begin{aligned}
x(t) &= 4e^{4t} + 2e^{-t}, \\
y(t) &= 6e^{4t} - 2e^{-t}.
\end{aligned}
\tag{4}
$$

This is verified by directly substituting $x(t)$ and $y(t)$ into the right-hand side of (3) and computing the derivatives of (4) and substituting them in the left-hand side of (4) to get

$$
16e^{4t} - 2e^{-t} = (4e^{4t} + 2e^{-t}) + 2(6e^{4t} - 2e^{-t}),
$$

$$
24e^{4t} + 2e^{-t} = 3(4e^{4t} + 2e^{-t}) + 2(6e^{4t} - 2e^{-t}).
$$

Numerical Solutions

A numerical solution to (1) over the interval $a \le t \le b$ is found by considering the differentials

$$
dx = f(t, x, y)\, dt \quad \text{and} \quad dy = g(t, x, y)\, dt.
\tag{5}
$$

Euler's method for solving the system is easy to formulate. The differentials $dt = t_{k+1} - t_k$, $dx = x_{k+1} - x_k$, and $dy = y_{k+1} - y_k$ are substituted into (5) to get

$$
\begin{aligned}
x_{k+1} - x_k &\approx f(t_k, x_k, y_k)(t_{k+1} - t_k), \\
y_{k+1} - y_k &\approx g(t_k, x_k, y_k)(t_{k+1} - t_k).
\end{aligned}
\tag{6}
$$

The interval is divided into M subintervals of width $h = (b - a)/M$, and the mesh points are $t_{k+1} = t_k + h$. This is used in (6) to get the recursive formulas for Euler's method:

$$
\begin{aligned}
t_{k+1} &= t_k + h, \\
x_{k+1} &= x_k + hf(t_k, x_k, y_k), \\
y_{k+1} &= y_k + hg(t_k, x_k, y_k) \quad \text{for } k = 0, 1, \ldots, M - 1.
\end{aligned}
\tag{7}
$$

A higher-order method should be used to achieve a reasonable amount of accuracy. For example, the Runge–Kutta formulas of order 4 are

$$x_{k+1} = x_k + \frac{h}{6}(f_1 + 2f_2 + 2f_3 + f_4)$$

$$y_{k+1} = y_k + \frac{h}{6}(g_1 + 2g_2 + 2g_3 + g_4),$$

(8)

where

$$f_1 = f(t_k, x_k, y_k), \qquad\qquad g_1 = g(t_k, x_k, y_k),$$

$$f_2 = f\left(t_k + \frac{h}{2}, x_k + \frac{h}{2}f_1, y_k + \frac{h}{2}g_1\right), \quad g_2 = g\left(t_k + \frac{h}{2}, x_k + \frac{h}{2}f_1, y_k + \frac{h}{2}g_1\right),$$

$$f_3 = f\left(t_k + \frac{h}{2}, x_k + \frac{h}{2}f_2, y_k + \frac{h}{2}g_2\right), \quad g_3 = g\left(t_k + \frac{h}{2}, x_k + \frac{h}{2}f_2, y_k + \frac{h}{2}g_2\right),$$

$$f_4 = f(t_k + h, x_k + hf_3, y_k + hg_3), \quad g_4 = g(t_k + h, x_k + hf_3, y_k + hg_3).$$

Example 9.15

Use the Runge–Kutta method given in (8) and compute the numerical solution to (3) over the interval [0.0, 0.2] using 10 subintervals and the step size $h = 0.02$.

Solution.   For the first point we have $t_1 = 0.02$ and the intermediate calculations required to compute x_1 and y_1 are

$f(t,x,y) = x + 2y$
$= 6 + 2\times4 = 14$

$g(t,x,y) = 3x + 2y$
$= 3\times6 + 2\times4 = 26$

$$f_1 = f(0.00, 6.0, 4.0) = 14.0 \qquad g_1 = g(0.00, 6.0, 4.0) = 26.0$$

$6 + \frac{0.02}{2}\times14 = 6.14$

$t_0 + \frac{h}{2}$

$$x_0 + \frac{h}{2}f_1 = 6.14 \qquad y_0 + \frac{h}{2}g_1 = 4.26, \qquad 4 + \frac{0.02}{2}\times26 = 4.26$$

$$f_2 = f(0.01, 6.14, 4.26) = 14.66 \qquad g_2 = g(0.01, 6.14, 4.26) = 26.94$$

$$x_0 + \frac{h}{2}f_2 = 6.1466 \qquad y_0 + \frac{h}{2}g_2 = 4.2694,$$

$$f_3 = f(0.01, 6.1466, 4.2694) = 14.6854,$$

$$g_3 = g(0.01, 6.1466, 4.2694) = 26.9786,$$

$$x_0 + hf_3 = 6.293708 \qquad y_0 + hg_3 = 4.539572,$$

$$f_4 = f(0.02, 6.293708, 4.539572) = 15.372852,$$

$$g_4 = g(0.02, 6.293708, 4.539572) = 27.960268.$$

These values are used in the final computation:

$$x_1 = 6 + \frac{0.02}{6}(14.0 + 2 \times 14.66 + 2 \times 14.6854 + 15.372852) = 6.29354551,$$

$$y_1 = 4 + \frac{0.02}{6}(26.0 + 2 \times 26.94 + 2 \times 26.9786 + 27.960268) = 4.53932490.$$

The calculations are summarized in Table 9.13.

TABLE 9.13   Runge–Kutta Solution to $x'(t) = x + 2y$, $y'(t) = 3x + 2y$ with the Initial Values $x(0) = 6$ and $y(0) = 4$

k	t_k	x_k	y_k
0	0.00	6.00000000	4.00000000
1	0.02	6.29354551	4.53932490
2	0.04	6.61562213	5.11948599
3	0.06	6.96852528	5.74396525
4	0.08	7.35474319	6.41653305
5	0.10	7.77697287	7.14127221
6	0.12	8.23813750	7.92260406
7	0.14	8.74140523	8.76531667
8	0.16	9.29020955	9.67459538
9	0.18	9.88827138	10.6560560
10	0.20	10.5396230	11.7157807

The numerical solutions contain a certain amount of error at each step. For the example above, the error grows and at the right endpoint $t = 0.2$ it reaches its maximum:

$$x(0.2) - x_{10} = 10.5396252 - 10.5396230 = 0.0000022,$$

$$y(0.2) - y_{10} = 11.7157841 - 11.7157807 = 0.0000034.$$

Higher-Order Differential Equations

Higher-order differential equations involve the higher derivatives $x''(t)$, $x'''(t)$, and so on. They arise in mathematical models for problems in physics and engineering. For example,

$$mx''(t) + cx'(t) + kx(t) = f(t),$$

represents a mechanical system in which a spring with spring constant k restores a displaced mass m. Damping is assumed to be proportional to the velocity and the function $f(t)$ is an external force. It is often the case that the position $x(t_0)$ and velocity $x'(t_0)$ are known at a certain time t_0.

By solving for the second derivative, we can write a second-order initial value problem in the form

$$x''(t) = f(t, x(t), x'(t)) \quad \text{with} \quad x(t_0) = x_0 \quad \text{and} \quad x'(t_0) = y_0. \tag{9}$$

The second-order differential equation can be reformulated as a system of two first-order equations if we use the substitution

$$x'(t) = y(t) \tag{10}$$

then $x''(t) = y'(t)$ and the differential equation in (9) becomes a system

$$\frac{dx}{dt} = y \qquad \text{with} \quad \begin{cases} x(t_0) = x_0, \\[2mm] y(t_0) = y_0. \end{cases} \tag{11}$$
$$\frac{dy}{dt} = f(t, x, y)$$

A numerical procedure such as the Runge–Kutta method can be used to solve (11) and will generate two sequences, $\{x_k\}$ and $\{y_k\}$. The first sequence is the numerical solution to (9). The next example can be interpreted as damped harmonic motion.

Example 9.16

Consider the second-order initial value problem

$$x''(t) + 4x'(t) + 5x(t) = 0 \quad \text{with} \quad x(0) = 3 \quad \text{and} \quad x'(0) = -5.$$

(a) Write down the equivalent system of two first-order equations.
(b) Use the Runge–Kutta method to solve the reformulated problem over $[0, 5]$ using $M = 50$ subintervals of width $h = 0.1$.
(c) Compare the numerical solution with the true solution:

$$x(t) = 3e^{-2t}\cos(t) + e^{-2t}\sin(t).$$

Solution.   The differential equation has the form

$$x''(t) = f(t, x(t), x'(t)) = -4x'(t) - 5x(t).$$

Using the substitutions in (10), we get the reformulated problem:

$$\frac{dx}{dt} = y \qquad \text{with} \begin{cases} x(0) = 3, \\ y(0) = -5. \end{cases}$$
$$\frac{dy}{dt} = -5x - 4y$$

Samples of the numerical computations are given in Table 9.14. The values $\{y_k\}$ are extraneous and are not included. Instead, the true solution values $\{x(t_k)\}$ are included for comparison.

TABLE 9.14   Runge–Kutta Solution to $x''(t) + 4x'(t) + 5x(t) = 0$ with the Initial Conditions $x(0) = 3$ and $x'(0) = -5$

k	t_k	x_k	$x(t_k)$
0	0.0	3.00000000	3.00000000
1	0.1	2.52564583	2.52565822
2	0.2	2.10402783	2.10404686
3	0.3	1.73506269	1.73508427
4	0.4	1.41653369	1.41655509
5	0.5	1.14488509	1.14490455
10	1.0	0.33324302	0.33324661
20	2.0	-0.00620684	-0.00621162
30	3.0	-0.00701079	-0.00701204
40	4.0	-0.00091163	-0.00091170
48	4.8	-0.00004972	-0.00004969
49	4.9	-0.00002348	-0.00002345
50	5.0	-0.00000493	-0.00000490

EXERCISES FOR SYSTEMS OF DIFFERENTIAL EQUATIONS

Instructions for Exercises 1–4:

(a) Use $h = 0.05$ and Euler's method to find $\{x_1, y_1\}$ and $\{x_2, y_2\}$.

(b) Use $h = 0.05$ and the Runge–Kutta method to find $\{x_1, y_1\}$.

(c) Use a computer implementation of the Runge–Kutta method to solve the system over the interval that is indicated.

1. Solve the system $x' = 2x + 3y$, $y' = 2x + y$ with the initial condition $x(0) = -2.7$ and $y(0) = 2.8$ over the interval $0 \le t \le 1.0$ using the step size $h = 0.05$. The polygonal path formed by the solution set is given in Figure 9.14 and can be compared with the analytic solution:

$$x(t) = -\frac{69}{25} e^{-t} + \frac{3}{50} e^{4t} \quad \text{and} \quad y(t) = \frac{69}{25} e^{-t} + \frac{1}{25} e^{4t}.$$

2. Solve the system $x' = 3x - y$, $y' = 4x - y$ with the initial condition $x(0) = 0.2$ and $y(0) = 0.5$ over the interval $0 \le t \le 2$ using the step size $h = 0.05$. The polygonal path formed by the solution set is given in Figure 9.15 and can be compared with the analytic solution:

$$x(t) = \frac{1}{5} e^t - \frac{1}{10} te^t \quad \text{and} \quad y(t) = \frac{1}{2} e^t - \frac{1}{5} te^t.$$

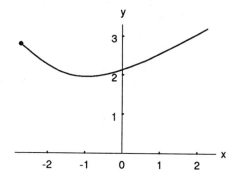

Figure 9.14   The solution to the system $x' = 2x + 3y$ and $y' = 2x + y$ over [0.0, 1.0].

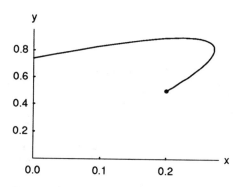

Figure 9.15   The solution to the system $x' = 3x - y$ and $y' = 4x - y$ over [0.0, 2.0].

3. Solve the system $x' = x - 4y$, $y' = x + y$ with the initial condition $x(0) = 2$ and $y(0) = 3$ over the interval $0 \le t \le 2$ using the step size $h = 0.05$. The polygonal path formed by the solution set is given in Figure 9.16 and can be compared with the analytic solution:

$$x(t) = -2e^t + 4e^t \cos^2(t) - 12e^t \cos(t) \sin(t)$$

and

$$y(t) = -3e^t + 6e^t \cos^2(t) + 2e^t \cos(t) \sin(t).$$

4. Solve the system $x' = y - 4x$, $y' = x + y$ with the initial condition $x(0) = 1$ and $y(0) = 1$ over the interval $0 \leq t \leq 1.2$ using the step size $h = 0.05$. The polygonal path formed by the solution set is given in Figure 9.17 and can be compared with the analytic solution:

$$x(t) = \frac{3 \exp(-\sqrt{29}\ t/2) - 3 \exp(\sqrt{29}\ t/2)}{2\sqrt{29}\ \exp(3t/2)} + \frac{\exp(-\sqrt{29}\ t/2 + \exp(\sqrt{29}\ t/2)}{2 \exp(3t/2)}$$

and

$$y(t) = \frac{-7 \exp(-\sqrt{29}\ t/2) + 7 \exp(\sqrt{29}\ t/2)}{2\sqrt{29}\ \exp(3t/2)} + \frac{\exp(-\sqrt{29}\ t/2) + \exp(\sqrt{29}\ t/2)}{2 \exp(3t/2)}.$$

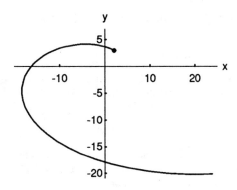

Figure 9.16   The solution to the system $x' = x - 4y$ and $y' = x + y$ over [0.0, 2.0].

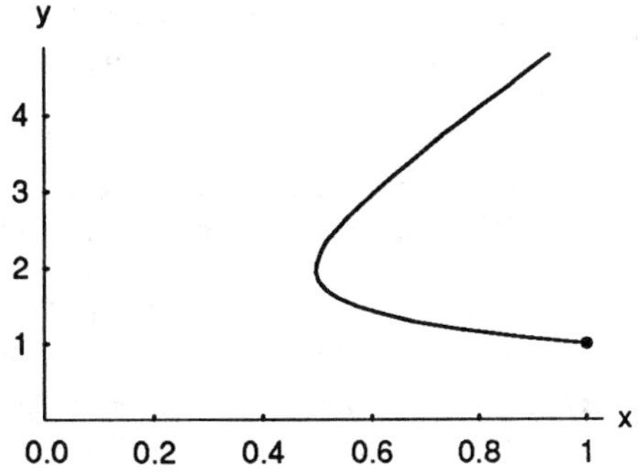

Figure 9.17   The solution to the system $x' = y - 4x$ and $y' = x + y$ over [0.0, 1.2].

In Exercises 5–8:
(a) Verify that the function $x(t)$ is the solution.
(b) Reformulate the second-order differential equation as a system of two first-order equations.
(c) Use $h = 0.1$ and Euler's method to find x_1 and x_2.
(d) Use $h = 0.05$ and the Runge–Kutta method to find x_1.
(e) Use the Runge–Kutta method to solve the differential equation over the interval [0, 2] using $M = 40$ steps and $h = 0.05$.

5. $2x''(t) - 5x'(t) - 3x(t) = 45e^{2t}$ with $x(0) = 2$ and $x'(0) = 1$.

$x(t) = 4e^{-t/2} + 7e^{3t} - 9e^{2t}$

6. $x''(t) + 6x'(t) + 9x(t) = 0$ with $x(0) = 4$ and $x'(0) = -4$.

$x(t) = 4e^{-3t} + 8te^{-3t}$

7. $x''(t) + x(t) = 6 \cos(t)$ with $x(0) = 2$ and $x'(0) = 3$.

$x(t) = 2 \cos(t) + 3 \sin(t) + 3t \sin(t)$

8. $x''(t) + 3x'(t) = 12$ with $x(0) = 5$ and $x'(0) = 1$.

$x(t) = 4 + 4t + e^{-3t}$

9. A certain resonant spring system with a periodic forcing function is modeled by

$$x''(t) + 25x(t) = 8 \sin(5t) \quad \text{with} \quad x(0) = 0 \quad \text{and} \quad x'(0) = 0.$$

Use the Runge–Kutta method to solve the differential equation over the interval $[0, 2]$ using $M = 40$ steps and $h = 0.05$.

10. The mathematical model of a certain *RLC* electrical circuit is

$$Q''(t) + 20Q'(t) + 125Q(t) = 9 \sin(5t) \quad \text{with} \quad Q(0) = 0 \quad \text{and} \quad Q'(0) = 0.$$

Use the Runge–Kutta method to solve the differential equation over the interval $[0, 2]$ using $M = 40$ steps and $h = 0.05$.

Remark. $I(t) = Q'(t)$ is the current at time t.

11. At time t, a pendulum makes an angle $x(t)$ with the vertical axis. Assuming that there is no friction, the equation of motion is

$$mlx''(t) = -mg \sin(x(t)),$$

where m is the mass and l is the length of the string. Use the Runge–Kutta method to solve the differential equation over the interval $[0, 2]$ using $M = 40$ steps and $h = 0.05$ if $g = 32$ ft/sec^2 and

(a) $l = 3.2$ ft and $x(0) = 0.3$ and $x'(0) = 0$
(b) $l = 0.8$ ft and $x(0) = 0.3$ and $x'(0) = 0$

12. *Predator–prey model.* An example of a system of nonlinear differential equations is the predator–prey problem. Let $x(t)$ and $y(t)$ denote the population of rabbits and foxes, respectively, at time t. The predator–prey model asserts that $x(t)$ and $y(t)$ satisfy

$$x'(t) = Ax(t) - Bx(t)y(t),$$

$$y'(t) = Cx(t)y(t) - Dy(t).$$

A typical computer simulation might use the coefficients

$$A = 2, \quad B = 0.02, \quad C = 0.0002, \quad D = 0.8.$$

Use the Runge–Kutta method to solve the differential equation over the interval $[0, 5]$ using $M = 50$ steps and $h = 0.1$ if
(a) $x(0) = 3000$ rabbits and $y(0) = 120$ foxes
(b) $x(0) = 5000$ rabbits and $y(0) = 100$ foxes

Instructions for Exercises 13–18: Use a computer implementation of the Runge–Kutta method to solve the system with the given initial condition over the interval that is indicated.

13. Solve $x' = x - xy$, $y' = -y + xy$ with $x(0) = 4$ and $y(0) = 1$ over $[0, 8]$ using $h = 0.1$. The trajectories of this system form closed paths. The polygonal path formed by the solution set is one of the curves shown in Figure 9.18.

14. Solve $x' = -3x - 2y - 2xy^2$, $y' = 2x - y + 2y^3$ with $x(0) = 0.8$ and $y(0) = 0.6$ over $[0, 4]$ using $h = 0.1$. For this system, the origin is classified as a spiral point that is asymptotically stable. The polygonal path formed by the solution set is one of the curves shown in Figure 9.19.

15. Solve $x' = y^2 - x^2$, $y' = 2xy$ with $x(0) = 2.0$ and $y(0) = 0.1$ over $[0.0, 1.5]$ using $h = 0.05$. For this linear system, there is an unstable saddle point at the origin. The polygonal path formed by the solution set is one of the curves shown in Figure 9.20.

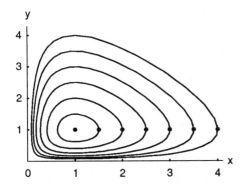

Figure 9.18   Solutions to the system $x' = x - xy$ and $y' = -y + xy$.

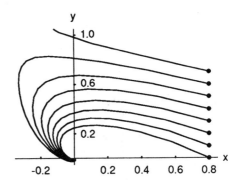

Figure 9.19   Solutions to the system $x' = -3x - 2y - 2xy^2$ and $y' = 2x - y + 2y^3$.

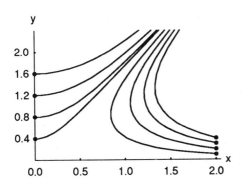

Figure 9.20   Solutions to the system $x' = y^2 - x^2$ and $y' = 2xy$.

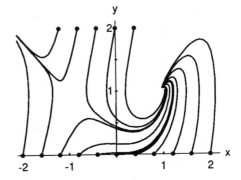

Figure 9.21   Solutions to the system $x' = 1 - y$ and $y' = x^2 - y^2$.

16. Solve $x' = 1 - y$, $y' = x^2 - y^2$ with $x(0) = -1.2$ and $y(0) = 0.0$ over $[0, 5]$ using $h = 0.1$. The point $(1, 1)$ is a spiral point that is asymptotically stable and the point $(-1, 1)$ is an unstable saddle point. The polygonal path formed by the solution set is one of the curves shown in Figure 9.21.

17. Solve $x' = x^3 - 2xy^2$, $y' = 2x^2y - y^3$ with $x(0) = 1.0$ and $y(0) = 0.2$ over $[0, 2]$ using $h = 0.025$. This linear system has an unstable critical point at the origin. The polygonal path formed by the solution set is one of the curves shown in Figure 9.22.

18. Solve $x' = x^2 - y^2$, $y' = 2xy$ with $x(0) = 2.0$ and $y(0) = 0.6$ over $[0.0, 1.6]$ using $h = 0.02$. The origin is an unstable critical point. The polygonal path formed by the solution set is one of the curves shown in Figure 9.23.

19. Write a report on dynamical systems. See References [2, 17, 48, and 164].

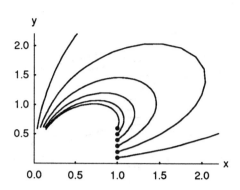

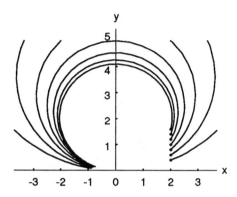

Figure 9.22   Solutions to the system $x' = x^3 - 2xy^2$ and $y' = 2x^2y - y^3$.

Figure 9.23   Solutions to the system $x' = x^2 - y^2$ and $y' = 2xy$.

9.8 BOUNDARY VALUE PROBLEMS

Another type of differential equation has the form

$$x'' = f(t, x, x') \qquad \text{for } a \leq t \leq b, \tag{1}$$

with the boundary conditions

$$x(a) = \alpha \quad \text{and} \quad x(b) = \beta. \tag{2}$$

This is called a **boundary value problem**.

The conditions which guarantee that a solution to (1) exists should be checked before any numerical scheme is applied; otherwise, a list of meaningless output may be generated. The general conditions are stated in the following theorem.

Theorem 9.8   (Boundary Value Problem).   Assume that $f(t, x, y)$ is continuous on the region $R = \{(t, x, y): a \leq t \leq b, -\infty < x < \infty, -\infty < y < \infty\}$ and that $\partial f/\partial x = f_x(t, x, y)$ and $\partial f/\partial y = f_y(t, x, y)$ are continuous on R. If there exists a constant $M > 0$ for which f_x and f_y satisfy

$$f_x(t, x, y) > 0 \qquad \text{for all } (t, x, y) \in R \tag{3}$$

and

$$|f_y(t, x, y)| \leq M \qquad \text{for all } (t, x, y) \in R, \tag{4}$$

then the boundary value problem

$$x'' = f(t, x, x') \qquad \text{with} \quad x(a) = \alpha \quad \text{and} \quad x(b) = \beta \tag{5}$$

has a unique solution $x = x(t)$ for $a \leq t \leq b$.

The notation $y = x'(t)$ has been used to distinguish the third variable of the function $f(t, x, x')$. Finally, the special case of linear differential equations is worthy of mention.

Corollary 9.1   (Linear Boundary Value Problem).   Assume that f in Theorem 9.8 has the form $f(t, x, y) = p(t)y + q(t)x + r(t)$ and that f and its partial derivatives $\partial f/\partial x = q(t)$ and $\partial f/\partial y = p(t)$ are continuous on R. If there exists a constant $M > 0$ for which $p(t)$ and $q(t)$ satisfy

$$q(t) > 0 \qquad \text{for all } t \in [a, b], \tag{6}$$

and

$$|p(t)| \le M = \max_{a \le t \le b} \{|p(t)|\}, \tag{7}$$

then the **linear boundary value problem**

$$x'' = p(t)x'(t) + q(t)x(t) + r(t) \qquad \text{with} \quad x(a) = \alpha \quad \text{and} \quad x(b) = \beta \tag{8}$$

has a unique solution $x = x(t)$ over $a \le t \le b$.

Reduction to Two I.V.P.'s: The Linear Shooting Method

Finding the solution of a linear boundary problem is assisted by the linear structure of the equation and the use of two special initial value problems. Suppose that u is the unique solution to the I.V.P.

$$u'' = p(t)u'(t) + q(t)u(t) + r(t) \qquad \text{with} \quad u(a) = \alpha \quad \text{and} \quad u'(a) = 0. \tag{9}$$

Furthermore, suppose that v is the unique solution to the I.V.P.

$$v'' = p(t)v'(t) + q(t)v(t) \qquad \text{with} \quad v(a) = 0 \quad \text{and} \quad v'(a) = 1. \tag{10}$$

Then the linear combination

$$x(t) = u(t) + Cv(t) \tag{11}$$

is a solution to $x'' = p(t)x'(t) + q(t)x(t) + r(t)$, as seen by the computation

$$x'' = u'' + Cv'' = p(t)u'(t) + q(t)u(t) + r(t) + p(t)Cv'(t) + q(t)Cv(t)$$
$$= p(t)(u'(t) + Cv'(t)) + q(t)(u(t) + Cv(t)) + r(t)$$
$$= p(t)x'(t) + q(t)x(t) + r(t).$$

The solution x in equation (11) takes on the boundary values

$$x(a) = u(a) + Cv(a) = \alpha + 0 = \alpha,$$
$$x(b) = u(b) + Cv(b). \tag{12}$$

Imposing the boundary condition $x(b) = \beta$ in (12) produces $C = [\beta - u(b)]/v(b)$. Therefore, if $v(b) \ne 0$, the unique solution to (8) is

$$x(t) = u(t) + \frac{\beta - u(b)}{v(b)} v(t). \tag{13}$$

Remark.   If q fulfills the hypothesis of Corollary 9.1, this rules out the troublesome solution $v(t) \equiv 0$, so that (13) is the form of the required solution. The details are left for the reader to investigate in Exercise 9.

Example 9.17

Solve the boundary value problem

$$x''(t) = \frac{2t}{1 + t^2} x'(t) - \frac{2}{1 + t^2} x(t) + 1$$

with $x(0) = 1.25$ and $x(4) = -0.95$ over the interval $[0, 4]$.

Solution.   The functions p, q, and r are $p(t) = \dfrac{2t}{1 + t^2}$, $q(t) = \dfrac{-2}{1 + t^2}$, and $r(t) = 1$, respectively. The Runge–Kutta method of order 4 with step size $h = 0.2$ is used to construct numerical solutions $\{u_j\}$ and $\{v_j\}$ to equations (9) and (10), respectively. The approximations $\{u_j\}$ for $u(t)$ are given in the first column of Table 9.15. Then $u(4) \approx u_{20} = -2.893535$ and $v(4) \approx v_{20} = 4$ are used with (13) to construct

$$w_j = \frac{b - u(4)}{v(4)} v_j = 0.485884 v_j.$$

TABLE 9.15   The Approximate Solutions $\{x_j\} = \{u_j + w_j\}$ to the Equation $x''(t) = \dfrac{2t}{1 + t^2} x'(t) - \dfrac{2}{1 + t^2} x(t) + 1$

t_j	u_j	w_j	$x_j = u_j + w_j$
0.0	1.250000	0.000000	1.250000
0.2	1.220131	0.097177	1.317308
0.4	1.132073	0.194353	1.326426
0.6	0.990122	0.291530	1.281652
0.8	0.800569	0.388707	1.189276
1.0	0.570844	0.485884	1.056728
1.2	0.308850	0.583061	0.891911
1.4	0.022522	0.680237	0.702759
1.6	−0.280424	0.777413	0.496989
1.8	−0.592609	0.874591	0.281982
2.0	−0.907039	0.971767	0.064728
2.2	−1.217121	1.068944	−0.148177
2.4	−1.516639	1.166121	−0.350518
2.6	−1.799740	1.263297	−0.536443
2.8	−2.060904	1.360474	−0.700430
3.0	−2.294916	1.457651	−0.837265
3.2	−2.496842	1.554828	−0.942014
3.4	−2.662004	1.652004	−1.010000
3.6	−2.785960	1.749181	−1.036779
3.8	−2.864481	1.846358	−1.018123
4.0	−2.893535	1.943535	−0.950000

Then the required approximate solution is $\{x_j\} = \{u_j + w_j\}$. Sample computations are given in Table 9.15, and Figure 9.24 shows their graphs. The reader can verify that $v(t) = t$ is the analytic solution for boundary value problem (10): that is,

$$v''(t) = \frac{2t}{1 + t^2}\, v'(t) - \frac{2}{1 + t^2}\, v(t)$$

with the initial conditions $v(0) = 0$ and $v'(0) = 1$.

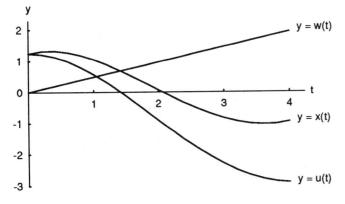

Figure 9.24 Numerical approximations $u(t)$ and $w(t)$ used to form $x(t) = u(t) + w(t)$ which is the solution to

$$x''(t) = \frac{2t}{1 + t^2}x'(t) - \frac{2}{1 + t^2}x(t) + 1.$$

The approximations in Table 9.16 compare numerical solutions obtained with the linear shooting method with the step sizes $h = 0.2$ and $h = 0.1$ and the analytic solution

TABLE 9.16 Numerical Approximations for $x''(t) = \dfrac{2t}{1 + t^2}\, x'(t) - \dfrac{2}{1 + t^2}\, x(t) + 1$

t_j	x_j $h = 0.2$	$x(t_j)$ exact	$x(t_j) - x_j$ error	t_j	x_j $h = 0.1$	$x(t_j)$ exact	$x(t_j) - x_j$ error
0.0	1.250000	1.250000	0.000000	0.0	1.250000	1.250000	0.000000
				0.1	1.291116	1.291117	0.000001
0.2	1.317308	1.317350	0.000042	0.2	1.317348	1.317350	0.000002
				0.3	1.328986	1.328990	0.000004
0.4	1.326426	1.326505	0.000079	0.4	1.326500	1.326505	0.000005
				0.5	1.310508	1.310514	0.000006
0.6	1.281652	1.281762	0.000110	0.6	1.281756	1.281762	0.000006
0.8	1.189276	1.189412	0.000136	0.8	1.189404	1.189412	0.000008
1.0	1.056728	1.056886	0.000158	1.0	1.056876	1.056886	0.000010
1.2	0.891911	0.892086	0.000175	1.2	0.892076	0.892086	0.000010
1.6	0.496989	0.497187	0.000198	1.6	0.497175	0.497187	0.000012
2.0	0.064728	0.064931	0.000203	2.0	0.064919	0.064931	0.000012
2.4	−0.350518	−0.350325	0.000193	2.4	−0.350337	−0.350325	0.000012
2.8	−0.700430	−0.700262	0.000168	2.8	−0.700273	−0.700262	0.000011
3.2	−0.942014	−0.941888	0.000126	3.2	−0.941895	−0.941888	0.000007
3.6	−1.036779	−1.036708	0.000071	3.6	−1.036713	−1.036708	0.000005
4.0	−0.950000	−0.950000	0.000000	4.0	−0.950000	−0.950000	0.000000

$$x(t) = 1.25 + 0.4860896526t - 2.25t^2 + 2t \arctan(t) - \frac{1}{2}\ln(1 + t^2) + \frac{1}{2}t^2\ln(1 + t^2).$$

A graph of the approximate solution when $h = 0.2$ is given in Figure 9.25. Included in the table are columns for the error. Since the Runge–Kutta solutions have error of order $O(h^4)$, the error in the solution with the smaller step size $h = 0.1$ is about $\frac{1}{16}$ the error of the solution with the large step size $h = 0.2$.

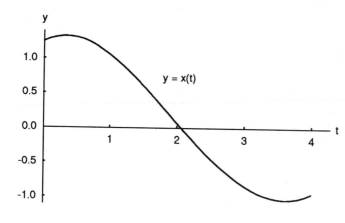

Figure 9.25 The graph of the numerical approximation for

$$x''(t) = \frac{2t}{1 + t^2}x'(t) - \frac{2}{1 + t^2}x(t) + 1$$

(using $h = 0.2$).

Algorithm 9.9 (Linear Shooting Method). To approximate the solution of the boundary value problem $x'' = p(t)x'(t) + q(t)x(t) + r(t)$ with $x(a) = \alpha$ and $x(b) = \beta$ over the interval $[a, b]$ by using the Runge–Kutta method of order 4.

```
INPUT A, B                                    {Endpoints of interval}
INPUT Alpha, Beta                             {Boundary values}
INPUT M                                       {Number of steps}

F1(t,x,y) := P(t) y + Q(t) x + R(t)
F2(t,x,y) := P(t) y + Q(t) x

Subroutine RKbdd4(F,A,B,Alpha,Beta,T,X,M)     {Runge–Kutta order 4
      H :=   (B−A)/M                              subroutine}
      T(0)   := A
      X(0)   := Alpha
      Y(0)   := Beta
```

```
FOR  J=0  TO  M−1  DO
     Tj  := T(J)
     Xj  := X(J)
     Yj  := Y(J)
     K1 := H*Yj
     R1 := H*F(Tj,Xj,Yj)
     K2 := H*(Yj + R1/2)
     R2 := H*F(Tj+H/2,Xj+K1/2,Yj+R1/2)
     K3 := H*(Yj + R2/2)
     R3 := H*F(Tj+H/2,Xj+K2/2,Yj+R2/2)
     K4 := H*(Yj + R3 )
     R4 := H*F(Tj+ H ,Xj+ K3 ,Yj+ R3 )
     X(J+1) := Xj + (K1 + 2 K2 + 2 K3 + K4)/6
     Y(J+1) := Yj + (R1 + 2 R2 + 2 R3 + R4)/6
     T(J+1) := A + H*(J+1)
```

{The main program starts here.}
CALL   RKbdd4(F1,A,B,Alpha,0,T,X1,M);
CALL   RKbdd4(F2,A,B,0,1,T,X2,M);

```
FOR  J=0  TO  M  DO
     X(J) := X1(J) + (Beta−X1(M))*X2(J)/X2(M)
```

OUTPUT   {T(J),X(J): J = 0, . . . M}.

EXERCISES FOR BOUNDARY VALUE PROBLEMS

Instructions for Exercises 1–9:
(a) Use the linear shooting method to solve the given boundary value problem using the step size $h = 0.2$ (and $h = 0.1$ and $h = 0.05$).
(b) Show that $f(t)$ is the general solution to the D.E. (without boundary conditions).
(c) Compare the solutions in part (a) with the particular solution $g(t)$.

1. Solve $x'' = (-2/t)x'(t) + (2/t^2)x(t) + [10 \cos(\ln(t))]/t^2$ over $[1, 3]$ with $x(1) = 1$ and $x(3) = -1$.

$$f(t) = \frac{-3t^2 \cos(\ln(t)) + t^2 \sin(\ln(t)) + C_2 t^3 + C_1}{t^2}$$

$$g(t) = \frac{4.335950689 - 0.3359506908\ t^3 - 3t^2 \cos(\ln(t)) + t^2 \sin(\ln(t))}{t^2}$$

2. Solve $x'' = -5x'(t) - 6x(t) + te^{-2t} + 3.9 \cos(3t)$ over $[0, 3]$ with $x(0) = 0.95$ and $x(3) = 0.15$.

$$f(t) = C_1 e^{-3t} + e^{-2t} + C_2 e^{-2t} - te^{-2t} + \frac{1}{2}t^2 e^{-2t}$$

$$+ \frac{3}{20}\cos(t) - \frac{1}{5}\cos^3(t) - \frac{1}{4}\sin(t) + \cos^2(t)\sin(t)$$

$$g(t) = 2.030708977e^{-3t} - 1.030708977e^{-2t} - te^{-2t} + \frac{1}{2}t^2 e^{-2t}$$

$$+ \frac{3}{20}\cos(t) - \frac{1}{5}\cos^3(t) - \frac{1}{4}\sin(t) + \cos^2(t)\sin(t)$$

3. Solve $x'' = -2x'(t) - 2x(t) + e^{-t} + \sin(2t)$ over $[0, 4]$ with $x(0) = 0.6$ and $x(4) = -0.1$.

$$f(t) = \frac{1}{5} + e^{-t} + C_1 e^{-t}\cos(t) - \frac{2}{5}\cos^2(t) + C_2 e^{-t}\sin(t) - \frac{1}{5}\cos(t)\sin(t)$$

$$g(t) = \frac{1}{5} + e^{-t} - \frac{1}{5}e^{-t}\cos(t) - \frac{2}{5}\cos^2(t) + 3.670227413e^{-t}\sin(t) - \frac{1}{5}\cos(t)\sin(t)$$

4. Solve $x'' = -4x'(t) - 4x(t) + 5\cos(4t) + \sin(2t)$ over $[0, 2]$ with $x(0) = 0.75$ and $x(2) = 0.25$.

$$f(t) = -\frac{1}{40} + C_2 e^{-2t} + C_1 te^{-2t} + \frac{19}{20}\cos^2(t) - \frac{6}{5}\cos^4(t) - \frac{4}{5}\cos(t)\sin(t) + \frac{8}{5}\cos^3(t)\sin(t)$$

$$g(t) = -\frac{1}{40} + 1.025e^{-2t} - 1.915729975te^{-2t} + \frac{19}{20}\cos^2(t)$$

$$- \frac{6}{5}\cos^4(t) - \frac{4}{5}\cos(t)\sin(t) + \frac{8}{5}\cos^3(t)\sin(t)$$

5. Solve $x'' + (1/t)x' + [1 - 1/(4t^2)]x = 0$ over $[1, 6]$ with $x(1) = 1$ and $x(6) - 0$.

$$f(t) = \frac{C_1\cos(t) - C_2\sin(t)}{\sqrt{t}}, \qquad g(t) = \frac{0.2913843206\cos(t) + 1.001299385\sin(t)}{\sqrt{t}}$$

6. Solve $x'' + (2/t)x' - (2/t^2)x = \dfrac{\sin(t)}{t^2}$ over $[1, 6]$ with $x(1) = -0.02$ and $x(6) = 0.02$.

$$f(t) = \frac{-t^2\sin(t) + t^3 \operatorname{Ci}(t) - \operatorname{Sin}(t) + t\operatorname{Cos}(t) + C_2 t^3 + C_1}{3t^2}$$

$$g(t) = \frac{-t^2\sin(t) + t^3 \operatorname{Ci}(t) - \operatorname{Sin}(t) + t\operatorname{Cos}(t) + 0.0000731808t^3 + 0.7451625603}{3t^2}$$

7. Solve $x'' + (1/t)x' + [1 - 1/(4t^2)]x = \sqrt{t}\cos(t)$ over $[1, 6]$ with $x(1) = 1.0$ and $x(6) = -0.5$.

$$f(t) = \frac{t\cos(t) + t^2\sin(t) - 4C_2\sin(t) + 4C_1\cos(t)}{4\sqrt{t}}$$

$$g(t) = \frac{t\cos(t) + t^2\sin(t) + 2.960284381\sin(t) + 0.2354853843\cos(t)}{4\sqrt{t}}$$

8. Solve $x'' - (1/t)x' + (1/t^2)x = 1$ over $[0.5, 4.5]$ with $x(0.5) = 1$ and $x(4.5) = 2$.

$$f(t) = t^2 + C_2 t + C_1 t \ln(t), \qquad g(t) = t^2 - 0.2525826491t - 2.528442297t \ln(t)$$

9. If q fulfills the hypothesis of Corollary 9.1, show that $v(t) \equiv 0$ is the unique solution to the boundary value problem

$$v'' = p(t)v'(t) + q(t)v(t) \quad \text{with} \quad v(a) = 0 \quad \text{and} \quad v(b) = 0.$$

10. Does the problem in Exercise 8 satisfy the hypothesis of Corollary 9.1? Why?

11. Construct an algorithm similar to Algorithm 9.9 based on Heun's method.

12. Consider the following problem:

$$x'' + x = 0 \quad \text{over} \quad [0, 2\pi] \quad \text{with} \quad x(0) = 1 \quad \text{and} \quad x(2\pi) = 2.$$

Discuss why there cannot be a solution to this problem.

13. Derive a "shooting method" for solving the nonlinear boundary value problem

$$x'' = f(t, x, x') \quad \text{over} \quad [a, b] \quad \text{with} \quad x(a) = \alpha \quad \text{and} \quad x(b) = \beta.$$

14. Use the method you developed in Exercise 13 to solve the following problems.
   (a) $x(t)x''(t) + [x'(t)]^2 + 1 = 0$ over $[0, 3]$ with $x(0) = 1.8$ and $x(3) = 0.0$. Use the step size $h = 0.1$.
   (b) $x''(t) = x'(t)x(t)$ over $[0, 2]$ with $x(0) = -4.4$ and $x(2) = 5.5$. Use the step size $h = 0.1$.
   (c) $x''(t) = 3\sqrt{1 + [x'(t)]^2}$ over $[0, 2]$ with $x(0) = 3$ and $x(2) = 3$. Use the step size $h = 0.1$.

9.9 FINITE-DIFFERENCE METHOD

Methods involving difference quotient approximations for derivatives can be used for solving certain second-order boundary value problems. Consider the linear equation

$$x'' = p(t)x'(t) + q(t)x(t) + r(t) \quad \text{over} \quad [a, b] \quad \text{with} \quad x(a) = \alpha \quad \text{and} \quad x(b) = \beta. \quad (1)$$

Form a partition of $[a, b]$ using the points $a = t_0 < t_1 < \cdots t_N = b$, where $h = (b - a)/N$ and $t_j = a + hj$ for $j = 0, 1, \ldots, N$. The central-difference formulas discussed in Chapter 6 are used to approximate the derivatives

$$x'(t_j) = \frac{x(t_{j+1}) - x(t_{j-1})}{2h} + O(h^2), \quad (2)$$

and

$$x''(t_j) = \frac{x(t_{j+1}) - 2x(t_j) + x(t_{j-1})}{h^2} + O(h^2). \quad (3)$$

To start the derivation we replace each term $x(t_j)$ on the right side of (2) and (3) with x_j and the resulting equations are substituted into (1) to obtain the relation

$$\frac{x_{j+1} - 2x_j + x_{j-1}}{h^2} + O(h^2) = p(t_j)\left(\frac{x_{j+1} - x_{j-1}}{2h} + O(h^2)\right) + q(t_j)x_j + r(t_j). \quad (4)$$

Next, we drop the two terms $O(h^2)$ in (4) and introduce the notation $p_j = p(t_j)$, $q_j = q(t_j)$, and $r_j = r(t_j)$; this produces the difference equation

$$\frac{x_{j+1} - 2x_j + x_{j-1}}{h^2} = p_j \frac{x_{j+1} - x_{j-1}}{2h} + q_j x_j + r_j, \tag{5}$$

which is used to compute numerical approximations to the differential equation (1). This is carried out by multiplying each side of (5) by h^2 and then collecting terms involving x_{j-1}, x_j, and x_{j+1} and arranging them in a system of linear equations:

$$\left(\frac{-h}{2} p_j - 1\right) x_{j-1} + (2 + h^2 q_j) x_j + \left(\frac{h}{2} p_j - 1\right) x_{j+1} = -h^2 r_j, \tag{6}$$

for $j = 1, 2, \ldots, N - 1$ where $x_0 = \alpha$ and $x_N = \beta$. The system in (6) has the familiar tridiagonal form, which is more visible when displayed with matrix notation:

$$
\begin{pmatrix}
2 + h^2 q_1 & \frac{h}{2} p_1 - 1 & & & & \\
\frac{-h}{2} p_2 - 1 & 2 + h^2 q_2 & \frac{h}{2} p_2 - 1 & & \mathbf{0} & \\
& \frac{-h}{2} p_j - 1 & 2 + h^2 q_j & \frac{h}{2} p_j - 1 & & \\
\mathbf{0} & & \frac{-h}{2} p_{N-2} - 1 & 2 + h^2 q_{N-2} & \frac{h}{2} p_{N-2} - 1 & \\
& & & \frac{-h}{2} p_{N-1} - 1 & 2 + h^2 q_{N-1}
\end{pmatrix}
\begin{pmatrix}
x_1 \\
x_2 \\
x_j \\
x_{N-2} \\
x_{N-1}
\end{pmatrix}
$$

$$
= \begin{pmatrix}
-h^2 r_1 + e_0 \\
-h^2 r_2 \\
-h^2 r_j \\
-h^2 r_{N-2} \\
-h^2 r_{N-1} + e_N
\end{pmatrix},
$$

where

$$e_0 = \left(\frac{h}{2} p_1 + 1\right)\alpha \quad \text{and} \quad e_N = \left(\frac{-h}{2} p_{N-1} + 1\right)\beta.$$

When computations with step size h are used, the numerical approximation to the solution is a set of discrete points $\{(t_j, x_j)\}$; if the analytic solution $x(t_j)$ is known, we can compare x_j and $x(t_j)$.

Example 9.18

Solve the boundary value problem

$$x''(t) = \frac{2t}{1 + t^2} x'(t) - \frac{2}{1 + t^2} x(t) + 1$$

with $x(0) = 1.25$ and $x(4) = -0.95$ over the interval $[0, 4]$.

Solution.   The functions p, q, and r are $p(t) = 2t/(1 + t^2)$, $q(t) = -2/(1 + t^2)$, and $r(t) = 1$, respectively. The finite-difference method is used to construct numerical solutions $\{x_j\}$ using the system of equation (6). Sample values of the approximations $\{x_{j,1}\}$, $\{x_{j,2}\}$, $\{x_{j,3}\}$, and $\{x_{j,4}\}$ corresponding to the step sizes $h_1 = 0.2$, $h_2 = 0.1$, $h_3 = 0.05$, and $h_4 = 0.025$ are given in Table 9.17. Figure 9.26 shows the graph of polygonal path formed from $\{(t_j, x_{j,1})\}$ for the case $h_1 = 0.2$. There are 41 terms in the sequence generated with $h = 0.1$, and the sequence $\{x_{j,2}\}$ only includes every other term from these computations; they correspond to the 21 values of $\{t_j\}$ given in Table 9.17. Similarly, the sequences $\{x_{j,3}\}$ and $\{x_{j,4}\}$ are a portion of the values generated with step sizes $h_3 = 0.05$ and $h_4 = 0.025$, respectively, and they correspond to the 21 values of $\{t_j\}$ in Table 9.17.

Next we compare numerical solutions in Table 9.17 with the analytic solution: $x(t) = 1.25 + 0.4860896526t - 2.25t^2 + 2t \arctan(t) - \frac{1}{2} \ln(1 + t^2) + \frac{1}{2} t^2 \ln(1 + t^2)$. The numerical solutions can be shown to have error of order $O(h^2)$. Hence reducing the step size by a factor of $\frac{1}{2}$ results in the error being reduced by about $\frac{1}{4}$. A careful scrutiny of Table 9.18 will reveal that this is happening. For instance, at $t_j = 1.0$ the errors incurred with step sizes

TABLE 9.17   Numerical Approximations for $x'' = \dfrac{2t}{1 + t^2} x' - \dfrac{2}{1 + t^2} x + 1$

t_j	$x_{j,1}$ $h_1 = 0.2$	$x_{j,2}$ $h_2 = 0.1$	$x_{j,3}$ $h_3 = 0.05$	$x_{j,4}$ $h_4 = 0.025$	$x(t_j)$ Exact
0.0	1.250000	1.250000	1.250000	1.250000	1.250000
0.2	1.314503	1.316646	1.317174	1.317306	1.317350
0.4	1.320607	1.325045	1.326141	1.326414	1.326505
0.6	1.272755	1.279533	1.281206	1.281623	1.281762
0.8	1.177399	1.186438	1.188670	1.189227	1.189412
1.0	1.042106	1.053226	1.055973	1.056658	1.056886
1.2	0.874878	0.887823	0.891023	0.891821	0.892086
1.4	0.683712	0.698181	0.701758	0.702650	0.702947
1.6	0.476372	0.492027	0.495900	0.496865	0.497187
1.8	0.260264	0.276749	0.280828	0.281846	0.282184
2.0	0.042399	0.059343	0.063537	0.064583	0.064931
2.2	-0.170616	-0.153592	-0.149378	-0.148327	-0.147977
2.4	-0.372557	-0.355841	-0.351702	-0.350669	-0.350325
2.6	-0.557565	-0.541546	-0.537580	-0.536590	-0.536261
2.8	-0.720114	-0.705188	-0.701492	-0.700570	-0.700262
3.0	-0.854988	-0.841551	-0.838223	-0.837393	-0.837116
3.2	-0.957250	-0.945700	-0.942839	-0.942125	-0.941888
3.4	-1.022221	-1.012958	-1.010662	-1.010090	-1.009899
3.6	-1.045457	-1.038880	-1.037250	-1.036844	-1.036709
3.8	-1.022727	-1.019238	-1.018373	-1.018158	-1.018086
4.0	-0.950000	-0.950000	-0.950000	-0.950000	-0.950000

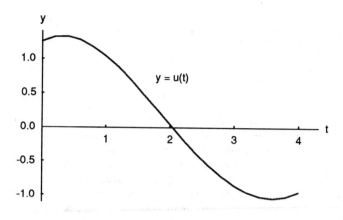

Figure 9.26 The graph of the numerical approximation for

$$x''(t) = \frac{2t}{1 + t^2}x'(t) - \frac{2}{1 + t^2}x(t) + 1$$

(using $h = 0.2$).

h_1, h_2, h_3, and h_4 are $e_{j,1} = 0.014780$, $e_{j,2} = 0.003660$, $e_{j,3} = 0.000913$, and $e_{j,4} = 0.000228$, respectively. Their successive ratios $e_{j,2}/e_{j,1} = 0.003660/0.014780 = 0.2476$, $e_{j,3}/e_{j,2} = 0.000913/0.003660 = 0.2495$, and $e_{j,4}/e_{j,3} = 0.000228/0.000913 = 0.2497$ are approaching $\frac{1}{4}$.

Finally, we show how Richardson's improvement scheme can be used to extrapolate the seemingly inaccurate sequences $\{x_{j,1}\}$, $\{x_{j,2}\}$, $\{x_{j,3}\}$, and $\{x_{j,4}\}$ and obtain six digits of precision. Eliminate the error terms $O(h^2)$ and $O((h/2)^2)$ in the approximations $\{x_{j,1}\}$

TABLE 9.18   Errors in Numerical Approximations Using the Finite-Difference Method

t_j	$x(t_j) - x_{j,1}$ $= e_{j,1}$	$x(t_j) - x_{j,2}$ $= e_{j,2}$	$x(t_j) - x_{j,3}$ $= e_{j,3}$	$x(t_j) - x_{j,4}$ $= e_{j,4}$
	$h_1 = 0.2$	$h_2 = 0.1$	$h_3 = 0.05$	$h_4 = 0.025$
0.0	0.000000	0.000000	0.000000	0.000000
0.2	0.002847	0.000704	0.000176	0.000044
0.4	0.005898	0.001460	0.000364	0.000091
0.6	0.009007	0.002229	0.000556	0.000139
0.8	0.012013	0.002974	0.000742	0.000185
1.0	0.014780	0.003660	0.000913	0.000228
1.2	0.017208	0.004263	0.001063	0.000265
1.4	0.019235	0.004766	0.001189	0.000297
1.6	0.020815	0.005160	0.001287	0.000322
1.8	0.021920	0.005435	0.001356	0.000338
2.0	0.022533	0.005588	0.001394	0.000348
2.2	0.022639	0.005615	0.001401	0.000350
2.4	0.022232	0.005516	0.001377	0.000344
2.6	0.021304	0.005285	0.001319	0.000329
2.8	0.019852	0.004926	0.001230	0.000308
3.0	0.017872	0.004435	0.001107	0.000277
3.2	0.015362	0.003812	0.000951	0.000237
3.4	0.012322	0.003059	0.000763	0.000191
3.6	0.008749	0.002171	0.000541	0.000135
3.8	0.004641	0.001152	0.000287	0.000072
4.0	0.000000	0.000000	0.000000	0.000000

and $\{x_{j,2}\}$ by generating the extrapolated sequence $\{z_{j,1}\} = \{(4x_{j,2} - x_{j,1})/3\}$. Similarly, the error terms $O((h/2)^2)$ and $O((h/4)^2)$ for $\{x_{j,2}\}$ and $\{x_{j,3}\}$ are eliminated by generating $\{z_{j,2}\} = \{(4x_{j,3} - x_{j,2})/3\}$. It has been shown that the second level of Richardson improvement scheme applies to the sequences $\{z_{j,1}\}$ and $\{z_{j,2}\}$, so that the third improvement is $\{(16z_{j,2} - z_{j,1})/15\}$ (see Reference [41]). Let us illustrate the situation by finding the extrapolated values which correspond to $t_j = 1.0$. The first extrapolated value is

$$\frac{4x_{j,2} - x_{j,1}}{3} = \frac{4 \times 1.053226 - 1.042106}{3} = 1.056932 = z_{j,1}.$$

The second extrapolated value is

$$\frac{4x_{j,3} - x_{j,2}}{3} = \frac{4 \times 1.055973 - 1.053226}{3} = 1.056889 = z_{j,2}.$$

Finally, the third extrapolation involves the terms $z_{j,1}$ and $z_{j,2}$:

$$\frac{16z_{j,2} - z_{j,1}}{15} = \frac{16 \times 1.056889 - 1.056932}{15} = 1.056886.$$

This last computation contains six decimal places of accuracy. The values at the other points are given in Table 9.19.

TABLE 9.19 Extrapolation of the Numerical Approximations $\{x_{j,1}\}$, $\{x_{j,2}\}$, and $\{x_{j,3}\}$ Obtained with the Finite-Difference Method

t_j	$\dfrac{4x_{j,2} - x_{j,1}}{3}$ $= z_{j,1}$	$\dfrac{4x_{j,3} - x_{j,2}}{3}$ $= z_{j,2}$	$\dfrac{16z_{j,2} - z_{j,1}}{15}$	$x(t_j)$ Exact solution
0.0	1.250000	1.250000	1.250000	1.250000
0.2	1.317360	1.317351	1.317350	1.317350
0.4	1.326524	1.326506	1.326504	1.326505
0.6	1.281792	1.281764	1.281762	1.281762
0.8	1.189451	1.189414	1.189412	1.189412
1.0	1.056932	1.056889	1.056886	1.056886
1.2	0.892138	0.892090	0.892086	0.892086
1.4	0.703003	0.702951	0.702947	0.702948
1.6	0.497246	0.497191	0.497187	0.497187
1.8	0.282244	0.282188	0.282184	0.282184
2.0	0.064991	0.064935	0.064931	0.064931
2.2	−0.147918	−0.147973	−0.147977	−0.147977
2.4	−0.350268	−0.350322	−0.350325	−0.350325
2.6	−0.536207	−0.536258	−0.536261	−0.536261
2.8	−0.700213	−0.700259	−0.700263	−0.700262
3.0	−0.837072	−0.837113	−0.837116	−0.837116
3.2	−0.941850	−0.941885	−0.941888	−0.941888
3.4	−1.009870	−1.009898	−1.009899	−1.009899
3.6	−1.036688	−1.036707	−1.036708	−1.036708
3.8	−1.018075	−1.018085	−1.018086	−1.018086
4.0	−0.950000	−0.950000	−0.950000	−0.950000

Algorithm 9.10   (Finite-Difference Method).   To approximate the solution of the boundary value problem $x'' = p(t)x'(t) + q(t)x(t) + r(t)$ with $x(a) = \alpha$ and $x(b) = \beta$ over the interval $[a, b]$ by using the finite-difference method of order $O(h^2)$.

   Remark. The mesh is $a = t_1 < \cdots < t_{N+1} = b$ and the solution points are $\{(t_j, x_j)\}_{j=1}^{N+1}$.

INPUT A, B   {Endpoints of interval}
INPUT Alpha, Beta   {Boundary values}
INPUT N   {Number of steps}

Subroutine   CoeffMat(A,B,Alpha,Beta,Vt,Va,Vb,Vc,Vd,N)
       H := [B−A]/N
       FOR   J=1   TO   N−1   DO
       └──── Vt(J) := A + H*J
       FOR   J=1   TO   N−1   DO
       └──── Vb(J) := − H^2*R(Vt(J))
       Vb(1) := Vb(1) + [1 + H/2*P(Vt(1))]*Alpha
       Vb(N−1) := Vb(N−1) + [1 − H/2*P(Vt(N−1))]*Beta
       FOR   J=1   TO   N−1   DO
       └──── Vd(J) := 2 + H^2*Q(Vt(J))
       FOR   J=1   TO   N−2   DO
       └──── Va(J) := − 1 − H*P(Vt(J+1))/2
       FOR   J=1   TO   N−2   DO
       └──── Vc(J) := − 1 + H*P(Vt(J))/2

Subroutine TriMat(Va,Vb,Vc,Vd,X0,N)
       FOR   K=2   TO   N−1   DO
       │   Temp := Va(K−1)/Vd(K−1)
       │   Vd(K) := Vd(K) − Temp*Vc(K−1)
       └──── Vb(K) := Vb(K) − Temp*Vb(K−1)
             X0(N−1) := Vb(N−1)/Vd(N−1)
       FOR   K=N−2   DOWNTO   1   DO
       └──── X0(K) := (Vb(K) − Vc(K)*X0(K+1))/Vd(K)

{The main program starts here.}
CALL   CoeffMat (A,B,Alpha,Beta,Vt,Va,Vb,Vo,Vd,N)
CALL   TriMatTriMat(Va,Vb,Vc,Vd,X0,N)

```
FOR  i=1  TO  N−1  DO
     T(i+1) := Vt(i)
     X(i+1) := X0(i)
T(1) := A,   T(N+1)  := B
X(1) := Alpha,  X(N+1) := Beta
```

OUTPUT {T(J), X(J): for J=1,2,...,N+1}

EXERCISES FOR THE FINITE-DIFFERENCE METHOD

Instructions for Exercises 1−9:
(a) Use the finite-difference method to solve the given boundary value problem using the step size $h = 0.2$, $h = 0.1$, and $h = 0.05$.
(b) Use extrapolation of the values in part (a) to obtain a better answer (i.e., construct a table of improvements similar to Table 9.19).

1. Solve $x'' = (-2/t)x'(t) + (2/t^2)x(t) + [10 \cos(\ln(t))]/t^2$ over $[1, 3]$ with $x(1) = 1$ and $x(3) = -1$.

2. Solve $x'' = -5x'(t) - 6x(t) + te^{-2t} + 3.9 \cos(3t)$ over $[0, 3]$ with $x(0) = 0.95$ and $x(3) = 0.15$.

3. Solve $x'' = -2x'(t) - 2x(t) + e^{-t} + \sin(2t)$ over $[0, 4]$ with $x(0) = 0.6$ and $x(4) = -0.1$.

4. Solve $x'' = -4x'(t) - 4x(t) + 5 \cos(4t) + \sin(2t)$ over $[0, 2]$ with $x(0) = 0.75$ and $x(2) = 0.25$.

5. Solve $x'' + (1/t)x' + [1 - 1/(4t^2)]x = 0$ over $[1, 6]$ with $x(1) = 1$ and $x(6) = 0$.

6. Solve $x'' + (2/t)x' - (2/t^2)x = \sin(t)/t^2$ over $[1, 6]$ with $x(1) = -0.02$ and $x(6) = 0.02$.

7. Solve $x'' + (1/t)x' + [1 - 1/(4t^2)]x = \sqrt{t} \cos(t)$ over $[1, 6]$ with $x(1) = 1.0$ and $x(6) = -0.5$.

8. Solve $x'' - (1/t)x' + (1/t^2)x = 1$ over $[0.5, 4.5]$ with $x(0.5) = 1$ and $x(4.5) = 2$.

9. Solve $x'' + (1/t)x' + (16/t^2)x = 1/t^2$ over $[1, 7]$ with $x(1) = 0.75$ and $x(7) = 0.30$.

10. Assume that p, q, and r are continuous over the interval $[a, b]$ and that $q(t) \geq 0$ for $a \leq t \leq b$. If h satisfies $0 < h < 2/M$, where $M = \max_{a \leq t \leq b} \{|p(t)|\}$, prove that the coefficient matrix of (6) is diagonally dominant and that there is a unique solution.

11. Assume that $p(t) \equiv C_1 > 0$ and $q(t) \equiv C_2 > 0$. (a) Write out the tridiagonal linear system for this situation. (b) Prove that the tridiagonal system is diagonally dominant and hence has a unique solution provided that $C_1/C_2 < h$.

10

Solution of Partial Differential Equations

Many problems in applied science, physics, and engineering are modeled mathematically with partial differential equations. A differential equation involving more than one independent variable is called a **partial differential equation** (PDE). It is not necessary to have taken a specialized course in PDEs to understand the rudimentary principles involved in obtaining computer solutions. In this chapter we will study finite-difference methods which are based on formulas for approximating the first and second derivatives of a function. We start by classifying the three types of equations under investigation and introduce a physical problem for each case. A partial differential equation of the form

$$A\phi_{xx} + B\phi_{xy} + C\phi_{yy} = F(x, y, \phi, \phi_x, \phi_y), \tag{1}$$

where A, B, and C are constants, is called **quasilinear**. There are three types of quasilinear equations:

If $B^2 - 4AC < 0$, the equation is called **elliptic**. $\tag{2}$

If $B^2 - 4AC = 0$, the equation is called **parabolic**. $\tag{3}$

If $B^2 - 4AC > 0$, the equation is called **hyperbolic**. $\tag{4}$

As an example of a hyperbolic equation, we consider the one-dimensional model for a vibrating string. The displacement $u(x, t)$ is governed by the wave equation

$$\rho u_{tt}(x, t) = T u_{xx}(x, t) \qquad \text{for } 0 < x < L \quad \text{and} \quad 0 < t < \infty, \tag{5}$$

498

with the given initial position and velocity functions

$$u(x, 0) = f(x) \qquad \text{for } t = 0 \quad \text{and} \quad 0 \le x \le L,$$
$$u_t(x, 0) = g(x) \qquad \text{for } t = 0 \quad \text{and} \quad 0 < x < L,$$

(6)

and the boundary values

$$u(0, t) = 0 \qquad \text{for } x = 0 \quad \text{and} \quad 0 \le t < \infty,$$
$$u(L, t) = 0 \qquad \text{for } x = L \quad \text{and} \quad 0 \le t < \infty.$$

(7)

The constant ρ is the mass of the string per unit length and T is the tension in the string. A diagram of a string with fixed ends at the locations $(0, 0)$ and $(L, 0)$ is shown in Figure 10.1.

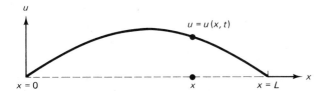

Figure 10.1 The wave equation models a vibrating string.

As an example of a parabolic equation, we consider the one-dimensional model for heat flow in an insulated rod of length L (see Figure 10.2). The heat equation, which involves the temperature $u(x, t)$ in the rod at the position x and time t, is

$$\kappa u_{xx}(x, t) = \sigma \rho u_t(x, t) \qquad \text{for } 0 < x < L \quad \text{and} \quad 0 < t < \infty,$$

(8)

the initial temperature distribution at $t = 0$ is

$$u(x, 0) = f(x) \qquad \text{for } t = 0 \quad \text{and} \quad 0 \le x \le L,$$

(9)

and the boundary values at the ends of the rod are

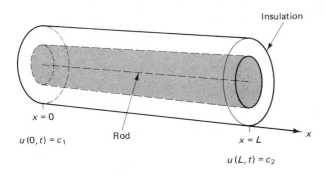

Figure 10.2 The heat equation models the temperature in an insulated rod.

$$u(0, t) = c_1 \qquad \text{for } x = 0 \quad \text{and} \quad 0 \le t < \infty,$$
$$u(L, t) = c_2 \qquad \text{for } x = L \quad \text{and} \quad 0 \le t < \infty. \tag{10}$$

The constant κ is the coefficient of thermal conductivity, σ is the specific heat, and ρ is the density of the material in the rod.

As an example of a elliptic equation, consider the potential function $u(x, y)$, which might represent a steady-state electrostatic potential or a steady-state temperature distribution in a rectangular region in the plane. These situations are modeled with Laplace's equation in a rectangle:

$$u_{xx}(x, y) + u_{yy}(x, y) = 0 \qquad \text{for } 0 < x < 1 \text{ and } 0 < y < 1, \tag{11}$$

with boundary conditions specified:

$$u(x, 0) = f_1(x) \qquad \text{for } y = 0 \quad \text{and} \quad 0 \le x \le 1 \quad \text{(on the bottom)},$$
$$u(x, 1) = f_2(x) \qquad \text{for } y = 1 \quad \text{and} \quad 0 \le x \le 1 \quad \text{(on the top)},$$
$$u(0, y) = f_3(y) \qquad \text{for } x = 0 \quad \text{and} \quad 0 \le y \le 1 \quad \text{(on the left)},$$
$$u(1, y) = f_4(y) \qquad \text{for } x = 1 \quad \text{and} \quad 0 \le y \le 1 \quad \text{(on the right)}.$$

A contour plot for $u(x, y)$ with boundary functions $f_1(x) = 0$, $f_2(x) = \sin(\pi x)$, $f_3(y) = 0$, and $f_4(y) = 0$ over the square $R = \{(x, y): 0 \le x \le 1, 0 \le y \le 1\}$ is shown in Figure 10.3.

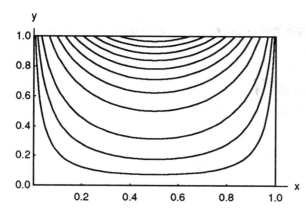

Figure 10.3   Solution curves $u(x, y) = C$ to Laplace's equation.

10.1 HYPERBOLIC EQUATIONS

The Wave Equation

As an example of a hyperbolic partial differential equation, we consider the wave equation

$$u_{tt}(x, t) = c^2 u_{xx}(x, t) \qquad \text{for } 0 < x < a \quad \text{and} \quad 0 < t < b, \tag{1}$$

with the boundary conditions

$$u(0,\ t) = 0 \quad \text{and} \quad u(a,\ t) = 0 \qquad \text{for } 0 \le t \le b,$$

$$u(x,\ 0) = f(x) \qquad\qquad\qquad \text{for } 0 \le x \le a, \qquad (2)$$

$$u_t(x,\ 0) = g(x) \qquad\qquad\qquad \text{for } 0 < x < a.$$

The wave equation models the displacement u of a vibrating elastic string with fixed ends at $x = 0$ and $x = a$. Although analytic solutions to the wave equation can be obtained with Fourier series, we use the problem as a prototype of a hyperbolic equation.

Derivation of the Difference Equation

Partition the rectangle $R = \{(x,\ t)\colon 0 \le x \le a,\ 0 \le t \le b\}$ into a grid consisting of $n - 1$ by $m - 1$ rectangles with sides $\Delta x = h$ and $\Delta t = k$, as shown in Figure 10.4. Start at the bottom row, where $t = t_1 = 0$ and the solution is known to be $u(x_i,\ t_1) = f(x_i)$. We shall use a difference-equation method to compute approximations

$$\{u_{i,j}\colon i = 1,\ 2,\ \ldots,\ n\} \qquad \text{in successive rows for } j = 2,\ 3,\ \ldots,\ m.$$

The true solution value at the grid points is $u(x_i,\ t_j)$.

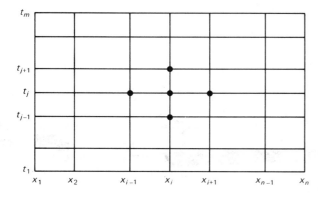

Figure 10.4   The grid for solving $u_{tt}(x,\ t) = c^2 u_{xx}(x,\ t)$ over R.

The central-difference formulas for approximating $u_{tt}(x,\ t)$ and $u_{xx}(x,\ t)$ are

$$u_{tt}(x,\ t) = \frac{u(x,\ t + k) - 2u(x,\ t) + u(x,\ t - k)}{k^2} + O(k^2) \qquad (3)$$

and

$$u_{xx}(x,\ t) = \frac{u(x + h,\ t) - 2u(x,\ t) + u(x - h,\ t)}{h^2} + O(h^2). \qquad (4)$$

The grid spacing is uniform in every row: $x_{i+1} = x_i + h$ (and $x_{i-1} = x_i - h$) and it is uniform in every column: $t_{j+1} = t_j + k$ (and $t_{j-1} = t_j - k$). Next, we drop the terms $O(k^2)$

and $O(h^2)$ and use the approximation $u_{i,j}$ for $u(x_i, t_j)$ in equations (3) and (4), which in turn are substituted into (1); this produces the difference equation

$$\frac{u_{i,j+1} - 2u_{i,j} + u_{i,j-1}}{k^2} = c^2 \frac{u_{i+1,j} - 2u_{i,j} + u_{i-1,j}}{h^2}, \tag{5}$$

which approximates the solution to (1). For convenience, the substitution $r = ck/h$ is introduced in (5) and we obtain the relation

$$u_{i,j+1} - 2u_{i,j} + u_{i,j-1} = r^2(u_{i+1,j} - 2u_{i,j} + u_{i-1,j}). \tag{6}$$

Equation (6) is employed to find row $j + 1$ across the grid, assuming that approximations in both rows j and $j - 1$ are known:

$$u_{i,j+1} = (2 - 2r^2)u_{i,j} + r^2(u_{i+1,j} + u_{i-1,j}) - u_{i,j-1} \qquad \text{for } i = 2, 3, \ldots, n - 1. \tag{7}$$

The four known values on the right side of equation (7), which are used to create the approximation $u_{i,j+1}$, are shown in Figure 10.5.

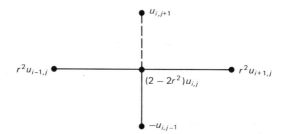

Figure 10.5 The wave equation stencil.

Caution must be taken when using formula (7). If the error made at one stage of the calculations is eventually dampened out, the method is called stable. To guarantee stability in formula (7) it is necessary that $r = ck/h \le 1$. There are other schemes, called implicit methods, which are more complicated to implement, but do not have stability restrictions for r (see Reference [90]).

Starting Values

Two starting rows of values corresponding to $j = 1$ and $j = 2$ must be supplied in order to use formula (7) to compute the third row. Since the second row is not usually given, the boundary function $g(x)$ is used to help produce starting approximations in the second row. Fix $x = x_i$ at the boundary and apply Taylor's formula of order 1 for expanding $u(x, t)$ about $(x_i, 0)$. The value $u(x_i, k)$ satisfies

$$u(x_i, k) = u(x_i, 0) + u_t(x_i, 0) k + O(k^2). \tag{8}$$

Then use $u(x_i, 0) = f(x_i) = f_i$ and $u_t(x_i, 0) = g(x_i) = g_i$ in (8) to produce the formula for computing the numerical approximations in the second row:

$$u_{i,2} = f_i + kg_i \quad \text{for } i = 2, 3, \ldots, n - 1. \tag{9}$$

Usually, $u(x_i, t_2) \neq u_{i,2}$, and such errors introduced by formula (9) will propagate throughout the grid and will not be dampened out when the scheme in (7) is implemented. Hence it is prudent to use a very small step size for k so that the values for $u_{i,2}$ given in (9) do not contain a large amount of truncation error.

Often, the boundary function $f(x)$ has a second derivative $f''(x)$ over the interval. In this case we have $u_{xx}(x, 0) = f''(x)$ and it is beneficial to use the Taylor formula of order $n = 2$ to help construct the second row. To do this, we go back to the wave equation and use the relationship between the second-order partial derivatives and obtain

$$u_{tt}(x_i, 0) = c^2 u_{xx}(x_i, 0) = c^2 f_{xx}(x_i) = c^2 \frac{f_{i+1} - 2f_i + f_{i-1}}{h^2} + O(h^2). \tag{10}$$

Recall that Taylor's formula of order 2 is

$$u(x, k) = u(x, 0) + u_t(x, 0) \, k + \frac{u_{tt}(x, 0) \, k^2}{2} + O(k^3). \tag{11}$$

Applying formula (11) at $x = x_i$, together with (9) and (10), we get

$$u(x_i, k) = f_i + kg_i + \frac{c^2 k^2}{2h^2} (f_{i+1} - 2f_i + f_{i-1}) + O(h^2) \, O(k^2) + O(k^3). \tag{12}$$

Using $r = ck/h$, formula (12) can be simplified to obtain a difference formula for the improved numerical approximations in the second row:

$$u_{i,2} = (1 - r^2) f_i + kg_i + \frac{r^2}{2} (f_{i+1} + f_{i-1}) \quad \text{for } i = 2, 3, \ldots, n - 1. \tag{13}$$

D'Alembert's Solution

The French mathematician Jean Le Rond d'Alembert (1717–1783) discovered that

$$u(x, t) = F(x + ct) + G(x - ct) \tag{14}$$

is a solution to the wave equation (1), over the interval $0 \leq x \leq a$, provided that F', F'', G', and G'' all exist and F and G have periodic $2a$ and obey the relationships $F(-z) = -F(z)$, $F(z + 2a) = F(z)$, $G(-z) = -G(z)$, and $G(z + 2a) = G(z)$ for all z. We can check this out by direct substitution. The second-order partial derivatives of the solution (14) are

$$u_{tt}(x, t) = c^2 F''(x + ct) + c^2 G''(x - ct), \tag{15}$$

$$u_{xx}(x, t) = F''(x + ct) + G''(x - ct). \tag{16}$$

Substitution of these quantities into (1) produces the desired relationship:

$$u_{tt}(x, t) = c^2 F''(x + ct) + c^2 G''(x - ct)$$

$$= c^2 [F''(x + ct) + G''(x - ct)]$$

$$= c^2 u_{xx}(x, t).$$

The particular solution which has the boundary values $u(x, 0) = f(x)$ and $u_t(x, 0) \equiv 0$ requires that $F(x) = G(x) = f(x)/2$ and is left for the reader to verify.

Two Exact Rows Given

The accuracy of the numerical approximations produced by the equations in (7) depends on the truncation errors in the formulas used to convert the partial differential equation into a difference equation. Although it is unlikely to know values of the exact solution for the second row of the grid, if such knowledge were available, using the increment $k = ch$ along the t-axis will generate an exact solution at all the other points through the grid.

Theorem 10.1.   Assume that the two rows of values $u_{i,1} = u(x_i, 0)$ and $u_{i,2} = u(x_i, k)$ for $i = 1, 2, \ldots, n$ are the exact solutions to the wave equation (1). If the step size $k = h/c$ is chosen along the t-axis, then $r = 1$ and formula (7) becomes

$$u_{i,j+1} = u_{i+1,j} + u_{i-1,j} - u_{i,j-1}. \tag{17}$$

Furthermore, the finite-difference solutions produced by (17) throughout the grid are exact solution values to the differential equation (neglecting computer round-off error).

Proof. Use d'Alembert's solution and the relation $ck = h$. The calculation $x_i - ct_j = (i - 1)h - c(j - 1)k = (i - 1)h - (j - 1)h = (i - j)h$ and a similar one producing $x_i + ct_j = (i + j - 2)h$ are used in equation (14) to produce the following special form for $u_{i,j}$:

$$u_{i,j} = F((i - j)h) + G((i + j - 2)h) \qquad \text{for } i = 1, 2, \ldots, n \quad \text{and}$$
$$j = 1, 2, \ldots, m. \tag{18}$$

Applying this formula to the terms $u_{i+1,j}$, $u_{i-1,j}$, and $u_{i,j-1}$ on the right side of (17) yields

$$u_{i+1,j} + u_{i-1,j} - u_{i,j-1} = F((i + 1 - j)h) + F((i - 1 - j)h)$$
$$- F((i - (j - 1))h) + G((i + 1 + j - 2)h)$$
$$+ G((i - 1 + j - 2)h) - G((i + j - 1 - 2)h)$$
$$= F((i - (j + 1))h) + G((i + j + 1 - 2)h) = u_{i,j+1}$$

for $i = 1, 2, \ldots, n$ and $j = 1, 2, \ldots, m$.

Warning.   Theorem 10.1 does not guarantee that the numerical solutions are exact when numerical calculations based on (9) or (13) are used to construct approximations $u_{i,2}$ in the second row. Indeed, truncation error will be introduced if $u_{i,2} \neq u(x_i, k)$ for some i where $1 \leq i \leq n$. This is why we endeavor to obtain the best possible values for the second row by using the second-order Taylor approximations in equation (13).

Example 10.1

Use the finite-difference method to solve the wave equation for a vibrating string

$$u_{tt}(x, t) = 4u_{xx}(x, t) \qquad \text{for } 0 < x < 1 \quad \text{and} \quad 0 < t < 0.5 \tag{19}$$

with the boundary conditions

$$u(0, t) = 0 \quad \text{and} \quad u(1, t) = 0 \qquad \text{for } 0 \le t \le 0.5,$$

$$u(x, 0) = f(x) = \sin(\pi x) + \sin(2\pi x) \qquad \text{for } 0 \le x \le 1, \qquad (20)$$

$$u_t(x, 0) = g(x) = 0 \qquad \text{for } 0 \le x \le 1.$$

Solution.   For convenience we choose $h = 0.1$ and $k = 0.05$. Since $c = 2$ this yields $r = ck/h = 2(0.05)/0.1 = 1$. Since $g(x) = 0$ and $r = 1$, formula (13) for creating the second row is

$$u_{i,2} = \frac{f_{i-1} + f_{i+1}}{2} \qquad \text{for } i = 2, 3, \ldots, 9. \qquad (21)$$

Substituting $r = 1$ into equation (7) gives the simplified difference equation

$$u_{i,j+1} = u_{i+1,j} + u_{i-1,j} - u_{i,j-1}. \qquad (22)$$

Applying formulas (21) and (22) successively to generate rows will produce the approximations to $u(x, t)$ given in Table 10.1 for $0 < x_i < 1$ and $0 \le t_j \le 0.50$.

TABLE 10.1   Solution of the Wave Equation (19) with Boundary Conditions (20)

t_j	x_2	x_3	x_4	x_5	x_6	x_7	x_8	x_9	x_{10}
0.00	0.896802	1.538842	1.760074	1.538842	1.000000	0.363271	−0.142040	−0.363271	−0.278768
0.05	0.769421	1.328438	1.538842	1.380037	0.951056	0.428980	0.000000	−0.210404	−0.181636
0.10	0.431636	0.769421	0.948401	0.951056	0.809017	0.587785	0.360616	0.181636	0.068364
0.15	0.000000	0.051599	0.181636	0.377381	0.587785	0.740653	0.769421	0.639384	0.363271
0.20	−0.380037	−0.587785	−0.519421	−0.181636	0.309017	0.769421	1.019421	0.951056	0.571020
0.25	−0.587785	−0.951056	−0.951056	−0.587785	0.000000	0.587785	0.951056	0.951056	0.587785
0.30	−0.571020	−0.951056	−1.019421	−0.769421	−0.309017	0.181636	0.519421	0.587785	0.380037
0.35	−0.363271	−0.639384	−0.769421	−0.740653	−0.587785	−0.377381	−0.181636	−0.051599	0.000000
0.40	−0.068364	−0.181636	−0.360616	−0.587785	−0.809017	−0.951056	−0.948401	−0.769421	−0.431636
0.45	0.181636	0.210404	0.000000	−0.428980	−0.951056	−1.380037	−1.538842	−1.328438	−0.769421
0.50	0.278768	0.363271	0.142040	−0.363271	−1.000000	−1.538842	−1.760074	−1.538842	−0.896802

The numerical values in Table 10.1 agree to more than six decimal places of accuracy with those obtained with the analytic solution $u(x, t) = \sin(\pi x) \cos(2\pi t) + \sin(2\pi x) \cos(4\pi t)$. A three-dimensional presentation of the data in Table 10.1 is given in Figure 10.6.

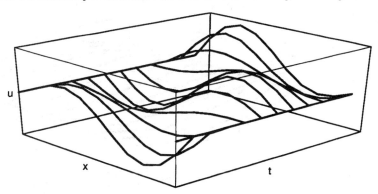

Figure 10.6   The vibrating string for equations (19) and (20).

Example 10.2

Use the finite-difference method to solve the wave equation for a vibrating string:

$$u_{tt}(x, t) = 4u_{xx}(x, t) \qquad \text{for } 0 < x < 1 \quad \text{and} \quad 0 < t < 0.5 \tag{23}$$

with the boundary conditions

$$u(0, t) = 0 \quad \text{and} \quad u(1, t) = 0 \qquad \text{for } 0 \le t \le 1,$$

$$u(x, 0) = f(x) = \begin{cases} \dfrac{5x}{3} & \text{for } 0 \le x \le \dfrac{3}{5}, \\[2mm] \dfrac{5 - 5x}{2} & \text{for } \dfrac{3}{5} \le x \le 1, \end{cases} \tag{24}$$

$$u_t(x, 0) = g(x) = 0 \qquad \text{for } 0 < x < 1.$$

Solution. For convenience we choose $h = 0.1$ and $k = 0.05$. Since $c = 2$ this again yields $r = 1$. Applying formulas (21) and (22) successively to generate rows will produce the approximations to $u(x, t)$ given in Table 10.2 for $0 \le x_i \le 1$ and $0 \le t_j \le 0.50$.

TABLE 10.2 Solution of the Wave Equation (23) with Boundary Conditions (24)

t_j	x_2	x_3	x_4	x_5	x_6	x_7	x_8	x_9	x_{10}
0.00	0.100	0.200	0.300	0.400	0.500	0.600	0.450	0.300	0.150
0.05	0.100	0.200	0.300	0.400	0.500	0.475	0.450	0.300	0.150
0.10	0.100	0.200	0.300	0.400	0.375	0.350	0.325	0.300	0.150
0.15	0.100	0.200	0.300	0.275	0.250	0.225	0.200	0.175	0.150
0.20	0.100	0.200	0.175	0.150	0.125	0.100	0.075	0.050	0.025
0.25	0.100	0.075	0.050	0.025	0.000	−0.025	−0.050	−0.075	−0.100
0.30	−0.025	−0.050	−0.075	−0.100	−0.125	−0.150	−0.175	−0.200	−0.100
0.35	−0.150	−0.175	−0.200	−0.225	−0.250	−0.275	−0.300	−0.200	−0.100
0.40	−0.150	−0.300	−0.325	−0.350	−0.375	−0.400	−0.300	−0.200	−0.100
0.45	−0.150	−0.300	−0.450	−0.475	−0.500	−0.400	−0.300	−0.200	−0.100
0.50	−0.150	−0.300	−0.450	−0.600	−0.500	−0.400	−0.300	−0.200	−0.100

A three-dimensional presentation of the data in Table 10.2 is given in Figure 10.7.

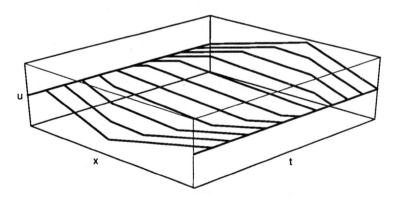

Figure 10.7 The vibrating string for equations (23) and (24).

Algorithm 10.1   (Finite-Difference Solution for the Wave Equation).   To approximate the solution of $u_{tt}(x, t) = c^2 u_{xx}(x, t)$ over $R = \{(x, t): 0 \le x \le a, 0 \le t \le b\}$ with $u(0, t) = 0$, $u(a, t) = 0$ for $0 \le t \le b$ and $u(x, 0) = f(x)$, $u_t(x, 0) = g(x)$ for $0 \le x \le a$.

```
INPUT A, B                                    {Width and height of R}
INPUT C                                       {Wave equation constant}
INPUT N, M                                    {Dimensions of the grid}
H := A/(N−1)                                   {Compute step sizes}
K := B/(M−1)
R := C*K/H                                          {Compute ratio}
R2 := R*R,   R22 := R*R/2                        {Compute constants}
S1 := 1 − R*R,   S2 := 2 − 2*R*R

Fi(I) := F(H*(I−1))                             {Boundary functions
Gi(I) := G(H*(I−1))                              at the grid points}

FOR  J = 1  TO  M  DO                         {Boundary conditions}
 └─────U(1,J) := 0,   U(N,J) := 0

FOR  I = 2  TO  N−1  DO                        {First and second rows}
 │      U(I,1) := Fi(I)
 └─────U(I,2) := S1*Fi(I) + K*Gi(I) + R22*(Fi(I+1) + Fi(I−1))

FOR  J= 3  TO  M  DO                            {Generate new waves}
 │      FOR  I = 2  TO  N−1  DO
 └──────└────U(I,J) := S2*U(I,J−1) + R2*(U(I−1,J−1) + U(I+1,J−1)) − U(I,J−2)

FOR  J = 1  TO  M  DO                          {Output the solution}
 └─────PRINT  K*(J−1),  U(1,J),  U(2,J), . . . , U(N,J)
```

EXERCISES FOR HYPERBOLIC EQUATIONS

1. (a) Verify by direct substitution that $u(x, t) = \sin(n\pi x)\cos(2n\pi t)$ is a solution to the wave equation $u_{tt}(x, t) = 4u_{xx}(x, t)$ for each positive integer $n = 1, 2, \ldots$.

   (b) Verify by direct substitution that $u(x, t) = \sin(n\pi x)\cos(cn\pi t)$ is a solution to the wave equation $u_{tt}(x, t) = c^2 u_{xx}(x, t)$ for each positive integer $n = 1, 2, \ldots$.

2. Assume that the initial position and velocity are $u(x, 0) = f(x)$ and $u_t(x, 0) \equiv 0$, respectively. Show that the d'Alembert solution for this case is

$$u(x, t) = \frac{f(x + ct) + f(x - ct)}{2}.$$

3. Obtain a simplified form of the difference equation (7) in the case $h = 2ck$.

4. Use the finite-difference method to solve the wave equation $u_{tt}(x, t) = 4u_{xx}(x, t)$ for $0 \le x \le 1$ and $0 \le t \le 0.5$ with the boundary conditions

$$u(0, t) = 0 \quad \text{and} \quad u(1, t) = 0 \qquad \text{for } 0 \le t \le 0.5,$$

$$u(x, 0) = f(x) = \sin(\pi x) \qquad \text{for } 0 \le x \le 1,$$

$$u_t(x, 0) = g(x) = 0 \qquad \text{for } 0 \le x \le 1.$$

Let $h = 0.2$, $k = 0.1$, and $r = 1$. Use a calculator and apply formulas (21) and (22) to compute approximations to $u(x, t)$ for $0 \le t \le 0.5$.

Hint. The starting row is

$$0.000000, \ 0.587785, \ 0.951057, \ 0.951057, \ 0.587785, \ 0.000000$$

5. Use the finite-difference method to solve the wave equation $u_{tt}(x, t) = 4u_{xx}(x, t)$ for $0 \le x \le 1$ and $0 \le t \le 0.5$ with the boundary conditions

$$u(0, t) = 0 \quad \text{and} \quad u(1, t) = 0 \qquad \text{for } 0 \le t \le 0.5,$$

$$u(x, 0) = f(x) = \begin{cases} \dfrac{5x}{2} & \text{for } 0 \le x \le \dfrac{3}{5}, \\[2mm] \dfrac{15 - 15x}{4} & \text{for } \dfrac{3}{5} \le x \le 1, \end{cases}$$

$$u_t(x, 0) = g(x) = 0 \qquad \text{for } 0 \le x \le 1.$$

Let $h = 0.2$, $k = 0.1$, and $r = 1$. Use a calculator and apply formulas (21) and (22) to compute approximations to $u(x, t)$ for $0 \le t \le 0.5$.

Hint. The starting row is

$$0.000, \ 0.500, \ 1.000, \ 1.500, \ 0.750, \ 0.000$$

6. Assume that the initial position and velocity are $u(x, 0) = f(x)$ and $u_t(x, 0) \equiv g(x)$, respectively. Show that the d'Alembert solution for this case is

$$u(x, t) = \frac{f(x + ct) + f(x - ct)}{2} + \frac{1}{2c} \int_{x-ct}^{x+ct} g(s) \, ds.$$

7. For the equation $u_{tt}(x, t) = 9u_{xx}(x, t)$, what relationship between h and k must occur in order to produce the difference equation $u_{i,j+1} = u_{i+1,j} + u_{i-1,j} - u_{i,j-1}$?

8. What difficulty might occur when trying to use the finite-difference method to solve $u_{tt}(x, t) = 4u_{xx}(x, t)$ with the choice $k = 0.02$ and $h = 0.03$?

9. Use a computer implementation of the finite-difference method and solve $u_{tt}(x, t) = u_{xx}(x, t)$ for $0 \le x \le 1$ and $0 \le t \le 1$ with the boundary conditions

$$u(0, t) = 0 \quad \text{and} \quad u(1, t) = 0 \qquad \text{for } 0 \le t \le 1,$$

$$u(x, 0) = f(x) = \sin(\pi x) \qquad \text{for } 0 \le x \le 1,$$

$$u_t(x, 0) = g(x) = 0 \qquad\qquad \text{for } 0 \leq x \leq 1.$$

For convenience choose $h = 0.1$ and $k = 0.1$.

10. Repeat Exercise 9, but use the boundary conditions

$$u(0, t) = 0 \quad \text{and} \quad u(1, t) = 0 \qquad \text{for } 0 \leq t \leq 1,$$
$$u(x, 0) = f(x) = x - x^2 \qquad \text{for } 0 \leq x \leq 1,$$
$$u_t(x, 0) = g(x) = 0 \qquad \text{for } 0 \leq x \leq 1.$$

11. Repeat Exercise 9, but use the boundary conditions

$$u(0, t) = 0 \quad \text{and} \quad u(1, t) = 0 \qquad \text{for } 0 \leq t \leq 1,$$

$$u(x, 0) = f(x) = \begin{cases} 2x & \text{for } 0 \leq x \leq \frac{1}{2}, \\ 2 - 2x & \text{for } \frac{1}{2} \leq x \leq 1, \end{cases}$$

$$u_t(x, 0) = g(x) = 0 \qquad \text{for } 0 \leq x \leq 1.$$

12. Repeat Exercise 9 but with the equation $u_{tt}(x, t) = 4u_{xx}(x, t)$, use $k = 0.05$.
13. Repeat Exercise 10 but with the equation $u_{tt}(x, t) = 4u_{xx}(x, t)$, use $k = 0.05$.
14. Repeat Exercise 11 but with the equation $u_{tt}(x, t) = 4u_{xx}(x, t)$, use $k = 0.05$.
15. Repeat Exercise 9 but with the boundary function $f(x) = \sin(2\pi x) + \sin(4\pi x)$.
16. Repeat Exercise 9 but with the equation $u_{tt}(x, t) = 4u_{xx}(x, t)$ and the boundary function $f(x) = \sin(2\pi x) + \sin(4\pi x)$, use $k = 0.05$.

10.2 PARABOLIC EQUATIONS

The Heat Equation

As an example of parabolic differential equations, we consider the one-dimensional heat equation

$$u_t(x, t) = c^2 u_{xx}(x, t) \qquad \text{for } 0 < x < a \quad \text{and} \quad 0 < t < b \tag{1}$$

with the initial condition

$$u(x, 0) = f(x) \qquad \text{for } t = 0 \quad \text{and} \quad 0 \leq x \leq a \tag{2}$$

and the boundary conditions

$$u(0, t) = g_1(t) \equiv c_1 \qquad \text{for } x = 0 \quad \text{and} \quad 0 \leq t \leq b,$$
$$u(a, t) = g_2(t) \equiv c_2 \qquad \text{for } x = a \quad \text{and} \quad 0 \leq t \leq b. \tag{3}$$

The heat equation models the temperature in an insulated rod with ends held at constant temperatures c_1 and c_2 and the initial temperature distribution along the rod being $f(x)$. Although analytic solutions to the heat equation can be obtained with Fourier series, we use the problem as a prototype of a parabolic equation for numerical solution.

Derivation of the Difference Equation

Assume that the rectangle $R = \{(x, t): 0 \le x \le a, 0 \le t \le b\}$ is subdivided into $n - 1$ by $m - 1$ rectangles with sides $\Delta x = h$ and $\Delta t = k$, as shown in Figure 10.8. Start at the bottom row, where $t = t_1 = 0$ and the solution is $u(x_i, t_1) = f(x_i)$. A method for computing the approximations to $u(x, t)$ at grid points in successive rows $\{u(x_i, t_j): i = 1, 2, \ldots, n\}$ for $j = 2, 3, \ldots, m$ will be developed.

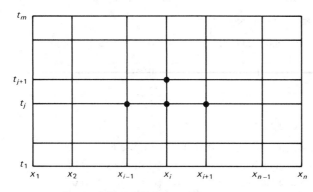

Figure 10.8 The grid for solving $u_t(x, t) = c^2 u_{xx}(x, t)$ over R.

The difference formulas used for $u_t(x, t)$ and $u_{xx}(x, t)$ are

$$u_t(x, t) = \frac{u(x, t + k) - u(x, t)}{k} + O(k) \tag{4}$$

and

$$u_{xx}(x, t) = \frac{u(x - h, t) - 2u(x, t) + u(x + h, t)}{h^2} + O(h^2). \tag{5}$$

The grid spacing is uniform in every row: $x_{i+1} = x_i + h$ (and $x_{i-1} = x_i - h$), and it is uniform in every column: $t_{j+1} = t_j + k$. Next, we drop the terms $O(k)$ and $O(h^2)$ and use the approximation $u_{i,j}$ for $u(x_i, t_j)$ in equations (4) and (5), which are in turn substituted into equation (1) to obtain

$$\frac{u_{i,j+1} - u_{i,j}}{k} = c^2 \frac{u_{i-1,j} - 2u_{i,j} + u_{i+1,j}}{h^2}, \tag{6}$$

which approximates the solution to (1). For convenience, the substitution $r = c^2 k/h^2$ is introduced in (6) and the result is the explicit forward difference equation

$$u_{i,j+1} = (1 - 2r)u_{i,j} + r(u_{i-1,j} + u_{i+1,j}). \tag{7}$$

Equation (7) is employed to create the $(j + 1)$th row across the grid, assuming that approximations in the jth row are known. Notice that this formula explicitly gives the value $u_{i,j+1}$ in terms of $u_{i-1,j}$, $u_{i,j}$, and $u_{i+1,j}$. The computational stencil representing the situation in formula (7) is given in Figure 10.9.

The simplicity of formula (7) makes it appealing to use. However, it is important to use numerical techniques that are stable. If any error made at one stage of the

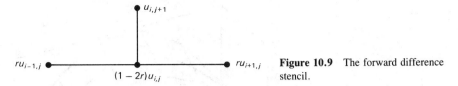

Figure 10.9   The forward difference stencil.

calculations is eventually dampened out, the method is called stable. The explicit forward-difference equation (7) is stable if and only if r is restricted to the interval $0 \le r \le \frac{1}{2}$. This means that the step size k must satisfy $k \le h^2/(2c^2)$. If this condition is not fulfilled, errors committed at one row $\{u_{i,j}\}$ might be magnified in subsequent rows $\{u_{i,p}\}$ for some $p > j$. The next example illustrates this point.

Example 10.3

Use the forward-difference method to solve the heat equation

$$u_t(x, t) = u_{xx}(x, t) \qquad \text{for } 0 < x < 1 \quad \text{and} \quad 0 < t < 0.20 \qquad (8)$$

with the initial condition

$$u(x, 0) = f(x) = 4x - 4x^2 \qquad \text{for } t = 0 \quad \text{and} \quad 0 \le x \le 1 \qquad (9)$$

and the boundary conditions

$$
\begin{aligned}
u(0, t) &= g_1(t) \equiv 0 && \text{for } x = 0 \quad \text{and} \quad 0 \le t \le 0.20, \\
u(1, t) &= g_2(t) \equiv 0 && \text{for } x = 1 \quad \text{and} \quad 0 \le t \le 0.20.
\end{aligned}
\qquad (10)
$$

Solution.   For the first illustration, we use the step sizes $\Delta x = h = 0.2$ and $\Delta t = k = 0.02$ and $c = 1$ so that the ratio is $r = 0.5$. The grid will be $n = 6$ columns wide by $m = 11$ rows high. In this case, formula (7) becomes

$$u_{i,j+1} = \frac{u_{i-1,j} + u_{i+1,j}}{2}. \qquad (11)$$

Formula (11) is stable for $r = 0.5$ and can be used successfully to generate reasonably accurate approximations to $u(x, t)$. Successive rows in the grid are given in Table 10.3. A three-dimensional presentation of the data in Table 10.3 is given in Figure 10.10.

TABLE 10.3   Using the Forward-Difference Method with $r = 0.5$

	$x_1 = 0.00$	$x_2 = 0.20$	$x_3 = 0.40$	$x_4 = 0.60$	$x_5 = 0.80$	$x_6 = 1.00$
$t_1 = 0.00$	0.000000	0.640000	0.960000	0.960000	0.640000	0.000000
$t_2 = 0.02$	0.000000	0.480000	0.800000	0.800000	0.480000	0.000000
$t_3 = 0.04$	0.000000	0.400000	0.640000	0.640000	0.400000	0.000000
$t_4 = 0.06$	0.000000	0.320000	0.520000	0.520000	0.320000	0.000000
$t_5 = 0.08$	0.000000	0.260000	0.420000	0.420000	0.260000	0.000000
$t_6 = 0.10$	0.000000	0.210000	0.340000	0.340000	0.210000	0.000000
$t_7 = 0.12$	0.000000	0.170000	0.275000	0.275000	0.170000	0.000000
$t_8 = 0.14$	0.000000	0.137500	0.222500	0.222500	0.137500	0.000000
$t_9 = 0.16$	0.000000	0.111250	0.180000	0.180000	0.111250	0.000000
$t_{10} = 0.18$	0.000000	0.090000	0.145625	0.145625	0.090000	0.000000
$t_{11} = 0.20$	0.000000	0.072812	0.117813	0.117813	0.072812	0.000000

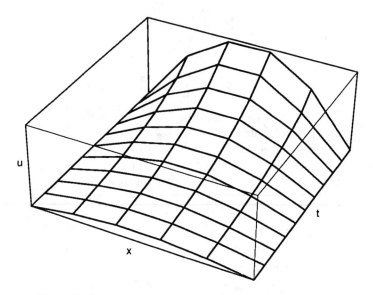

Figure 10.10   Using the forward difference method with $r = 0.5$.

For our second illustration, we use the step sizes $\Delta x = h = 0.2$ and $\Delta t = k = \frac{1}{30} = 0.033333$, so that the ratio is $r = 0.833333$. In this case, formula (7) becomes

$$u_{i,j+1} = -0.666665u_{i,j} + 0.833333(u_{i-1,j} + u_{i+1,j}). \qquad (12)$$

Formula (12) is unstable stable in this case, because $r > \frac{1}{2}$, and errors committed at one row will be magnified in successive rows. Numerical values which turn out to be imprecise approximations to $u(x, t)$ for $0 \le t \le 0.333333$ are given in Table 10.4. A three-dimensional presentation of the data in Table 10.4 is given in Figure 10.11.

TABLE 10.4   Using the Forward-Difference Method with $r = 0.833333$

	$x_1 = 0.00$	$x_2 = 0.20$	$x_3 = 0.40$	$x_4 = 0.60$	$x_5 = 0.80$	$x_6 = 1.00$
$t_1 = 0.000000$	0.000000	0.640000	0.960000	0.960000	0.640000	0.000000
$t_2 = 0.033333$	0.000000	0.373333	0.693333	0.693333	0.373333	0.000000
$t_3 = 0.066667$	0.000000	0.328889	0.426667	0.426667	0.328889	0.000000
$t_4 = 0.100000$	0.000000	0.136296	0.345185	0.345185	0.136296	0.000000
$t_5 = 0.133333$	0.000000	0.196790	0.171111	0.171111	0.196790	0.000000
$t_6 = 0.166667$	0.000000	0.011399	0.192510	0.192510	0.011399	0.000000
$t_7 = 0.200000$	0.000000	0.152826	0.041584	0.041584	0.152826	0.000000
$t_8 = 0.233333$	0.000000	−0.067230	0.134286	0.134286	−0.067230	0.000000
$t_9 = 0.266667$	0.000000	0.156725	−0.033644	−0.033644	0.156725	0.000000
$t_{10} = 0.300000$	0.000000	−0.132520	0.124997	0.124997	−0.132520	0.000000
$t_{11} = 0.333333$	0.000000	0.192511	−0.089601	−0.089601	0.192511	0.000000

The difference equation (7) has accuracy of the order $O(k) + O(h^2)$. Because the term $O(k)$ decreases linearly as k tends to zero, it is not surprising that it must be made

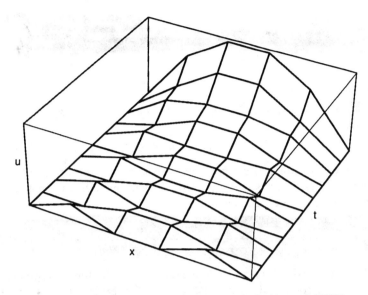

Figure 10.11   Using the forward difference method with $r = 0.833333$.

small to produce good approximations. However, the stability requirement introduces further considerations. Suppose that the solutions over the grid are not sufficiently accurate and that both the increments $\Delta x = h_0$ and $\Delta t = k_0$ must be reduced. For simplicity, suppose that the new x-increment is $\Delta x = h_1 = h_0/2$. If the same ratio r is used, k_1 must satisfy

$$k_1 = \frac{r(h_1)^2}{c^2} = \frac{r(h_0)^2}{4c^2} = \frac{k_0}{4}.$$

This results in a doubling and quadrupling of the number of grid points along the x-axis and t-axis, respectively. Consequently, there must be an eightfold increase in the total computational effort when reducing the grid size in this manner. This extra effort is usually prohibitive and demands that we explore a more efficient method which does not have stability restrictions. The method proposed will be implicit rather than explicit. The apparent rise in the level of complexity will have the immediate payoff of being unconditionally stable.

The Crank–Nicholson Method

An implicit scheme, invented by John Crank and Phyllis Nicholson (see Reference [29]), is based on numerical approximations for solutions of equation (1) at the point $(x, t + k/2)$ which lies between the rows in the grid. Specifically, the approximation used for $u_t(x, t + k/2)$ is obtained from the central-difference formula,

$$u_t\!\left(x, t + \frac{k}{2}\right) = \frac{u(x, t + k) - u(x, t)}{k} + O(k^2). \tag{13}$$

The approximation used for $u_{xx}(x, t + k/2)$ is the average of the approximations $u_{xx}(x, t)$ and $u_{xx}(x, t + k)$, which has an accuracy of the order $O(h^2)$:

$$u_{xx}\left(x, t + \frac{k}{2}\right)$$

$$= \frac{1}{2h^2} [u(x - h, t + k) - 2u(x, t + k) + u(x + h, t + k) \tag{14}$$

$$+ u(x - h, t) - 2u(x, t) + u(x + h, t)] + O(h^2).$$

In a fashion similar to the previous derivation, we substitute (13) and (14) into (1) and neglect the error terms $O(h^2)$ and $O(k^2)$. Then employing the notation $u_{i,j} = u(x_i, t_j)$ will produce the difference equations

$$\frac{u_{i,j+1} - u_{i,j}}{k} = c^2 \frac{u_{i-1,j+1} - 2u_{i,j+1} + u_{i+1,j+1} + u_{i-1,j} - 2u_{i,j} + u_{i+1,j}}{2h^2}. \tag{15}$$

Also, the substitution $r = c^2 k/h^2$ is used in (15). But this time we must solve for the three "yet to be computed" values $u_{i-1,j+1}$, $u_{i,j+1}$, and $u_{i+1,j+1}$. This is accomplished by placing them all on the left side of the equation. Then rearrangement of the terms in equation (15) results in the implicit difference formula

Unknowns   *Knowns*

$$-ru_{i-1,j+1} + (2 + 2r)u_{i,j+1} - ru_{i+1,j+1} = (2 - 2r)u_{i,j} + r(u_{i-1,j} + u_{i+1,j}) \tag{16}$$

for $i = 2, 3, \ldots, n - 1$. The terms on the right-hand side of equation (16) are all known. Hence the equations in (16) form a tridiagonal linear system $AX = B$. The six points used in the Crank–Nicholson formula (16), together with the intermediate grid point where the numerical approximations are based, are shown in Figure 10.12.

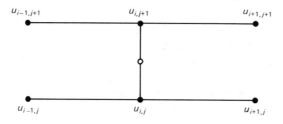

$u_{i-1,j+1}$ $u_{i,j+1}$ $u_{i+1,j+1}$

$u_{i-1,j}$ $u_{i,j}$ $u_{i+1,j}$

Figure 10.12   The Crank–Nicholson stencil.

Implementation of formula (16) is sometimes done by using the ratio $r = 1$. In this case the increment along the t-axis is $\Delta t = k = h^2/c^2$, and the equations in (16) simplify and become

$$-u_{i-1,j+1} + 4u_{i,j+1} - u_{i+1,j+1} = u_{i-1,j} + u_{i+1,j}, \tag{17}$$

for $i = 2, 3, \ldots, n - 1$. The boundary conditions are used in the first and last equations (i.e., $u_{1,j} = u_{1,j+1} = c_1$ and $u_{n,j} = u_{n,j+1} = c_2$, respectively). Equations (17) are especially pleasing to view in their tridiagonal matrix form $AX = B$:

$$
\begin{pmatrix}
4 & -1 & & & & & \\
-1 & 4 & -1 & & & \mathbf{O} & \\
 & & \cdot & & & & \\
 & & \cdot & & & & \\
 & & -1 & 4 & -1 & & \\
 & & & \cdot & & & \\
 & & & \cdot & & & \\
\mathbf{O} & & & -1 & 4 & -1 \\
 & & & & -1 & 4
\end{pmatrix}
\begin{pmatrix}
u_{2,j+1} \\
u_{3,j+1} \\
\cdot \\
\cdot \\
u_{p,j+1} \\
\cdot \\
\cdot \\
u_{n-2,j+1} \\
u_{n-1,j+1}
\end{pmatrix}
=
\begin{pmatrix}
2c_1 + u_{3,j} \\
u_{2,j} + u_{4,j} \\
\cdot \\
\cdot \\
u_{p-1,j} + u_{p+1,j} \\
\cdot \\
\cdot \\
u_{n-3,j} + u_{n-1,j} \\
u_{n-2,j} + 2c_2
\end{pmatrix}
$$

When the Crank–Nicholson method is implemented with a computer, the linear system $A\mathbf{X} = \mathbf{B}$ can be solved by either direct means or by iteration.

Example 10.4

Use the Crank–Nicholson method to solve the heat equation

$$u_t(x, t) = u_{xx}(x, t) \qquad \text{for } 0 < x < 1 \quad \text{and} \quad 0 < t < 0.1 \qquad (18)$$

with the initial condition

$$u(x, 0) = f(x) = \sin(\pi x) + \sin(3\pi x) \qquad \text{for } t = 0 \quad \text{and} \quad 0 \le x \le 1 \qquad (19)$$

and the boundary conditions

$$
\begin{aligned}
u(0, t) &= g_1(t) \equiv 0 \qquad \text{for } x = 0 \quad \text{and} \quad 0 \le t \le 0.1, \\
u(1, t) &= g_2(t) \equiv 0 \qquad \text{for } x = 1 \quad \text{and} \quad 0 \le t \le 0.1.
\end{aligned}
\qquad (20)
$$

Solution. For simplicity, we use the step sizes $\Delta x = h = 0.1$ and $\Delta t = k = 0.01$ so that the ratio is $r = 1$. The grid will be $n = 11$ columns wide by $m = 11$ rows high. Applying the algorithm generates the values in Table 10.5 for $0 < x_i < 1$ and $0 \le t_j \le 0.1$.

TABLE 10.5 The Values $u(x_i, t_j)$ Using the Crank–Nicholson Method with $t_j = (j - 1)/100$

	$x_2 = 0.1$	$x_3 = 0.2$	$x_4 = 0.3$	$x_5 = 0.4$	$x_6 = 0.5$	$x_7 = 0.6$	$x_8 = 0.7$	$x_9 = 0.8$	$x_{10} = 0.9$
t_1	1.118034	1.538842	1.118034	0.363271	0.000000	0.363271	1.118034	1.538842	1.118034
t_2	0.616905	0.928778	0.862137	0.617659	0.490465	0.617659	0.862137	0.928778	0.616905
t_3	0.394184	0.647957	0.718601	0.680009	0.648834	0.680009	0.718601	0.647957	0.394184
t_4	0.288660	0.506682	0.625285	0.666493	0.673251	0.666493	0.625285	0.506682	0.288660
t_5	0.233112	0.425766	0.556006	0.625082	0.645788	0.625082	0.556006	0.425766	0.233112
t_6	0.199450	0.372035	0.499571	0.575402	0.600242	0.575402	0.499571	0.372035	0.199450
t_7	0.175881	0.331490	0.451058	0.525306	0.550354	0.525306	0.451058	0.331490	0.175881
t_8	0.157405	0.298131	0.408178	0.477784	0.501545	0.477784	0.408178	0.298131	0.157405
t_9	0.141858	0.269300	0.369759	0.433821	0.455802	0.433821	0.369759	0.269300	0.141858
t_{10}	0.128262	0.243749	0.335117	0.393597	0.413709	0.393597	0.335117	0.243749	0.128262
t_{11}	0.116144	0.220827	0.303787	0.356974	0.375286	0.356974	0.303787	0.220827	0.116144

The values obtained with the Crank–Nicholson method compare favorably with the analytic solution $u(x, t) = \sin(\pi x) \exp(-t\pi^2) + \sin(3\pi x) \exp(-t9\pi^2)$, the true values for the final row being

t_{11}	0.115285	0.219204	0.301570	0.354385	0.372569	0.354385	0.301570	0.219204	0.115285

A three-dimensional presentation of the data in Table 10.5 is given in Figure 10.13.

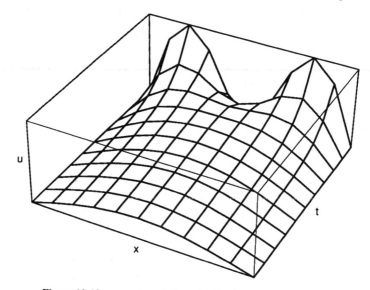

Figure 10.13 $u = u(x_i, t_j)$ from the Crank–Nicholson method.

Algorithm 10.2 (Forward-Difference Method for the Heat Equation). To approximate the solution of $u_t(x, t) = c^2 u_{xx}(x, t)$ over $R = \{(x, t): 0 \le x \le a, 0 \le t \le b\}$ with $u(x, 0) = f(x)$ for $0 \le x \le a$ and $u(0, t) = c_1$, $u(a, t) = c_2$ for $0 \le t \le b$.

```
INPUT A, B,                      {Width and height of R}
INPUT C, C1, C2                  {Supply constants}
INPUT N, M                       {Dimensions of the grid}
H := A/(N−1)                     {Compute step sizes}
K := B/(M−1)
R := C^2*K/H^2                   {Compute ratio}
S := 1 − 2*R                     {Compute constant}
Fi(I) := F(H*(I−1))              {Boundary function
                                  at the grid points}
```

```
FOR  J = 1  TO  M  DO                              {Boundary conditions}
└───── U(1,J) := C1,  U(N,J) := C2


FOR  I = 2  TO  N−1  DO                                          {First row}
└───── U(I,1) := Fi(I)


FOR  J = 2  TO  M  DO                                   {Generate new waves}
│      FOR  I = 2  TO  N−1  DO
└──────└───── U(I,J) := S*U(I,J−1) + R*(U(I−1,J−1) + U(I+1,J−1))


FOR  J = 1  TO  M  DO                                    {Output the solution}
└───── PRINT  K*(J−1),  U(1,J),  U(2,J), . . . ,  U(N,J)
```

Algorithm 10.3   (Crank−Nicholson Method for the Heat Equation).   To approximate the solution of $u_t(x,\ t) = c^2 u_{xx}(x,\ t)$ over $R = \{(x,\ t) \colon 0 \le x \le a,\ 0 \le t \le b\}$ with $u(x,\ 0) = f(x)$ for $0 \le x \le a$ and $u(0,\ t) = c_1$, $u(a,\ t) = c_2$ for $0 \le t \le b$.

```
INPUT A, B                                       {Width and height of R}
INPUT C, C1, C2                                       {Supply constants}
INPUT N, M                                        {Dimensions of the grid}
H := A/(N−1)                                         {Compute step sizes}
K := B/(M−1)
R := C^2*K/H^2                                            {Compute ratio}
S1 := 2 + 2/R                                        {Compute constants}
S2 := 2/R − 2


Fi(I) := F(H*(I−1))                                   {Boundary function
                                                     at the grid points}


FOR  J = 1  TO  M  DO                              {Boundary conditions}
└───── U(1,J) := C1,  U(N,J) := C2


FOR  I = 2  TO  N−1  DO                                          {First row}
└───── U(I,1) := Fi(I)


FOR  I = 1  TO  N  DO                                  {Form the diagonal
└───── Vd(I)  := S1                                     elements of the
       Vd(1)  := 1                                          matrix A}
       Vd(N)  := 1
```

```
FOR  I = 1  TO  N−1  DO                              {Form the off-
     Va(I) := −1, Vc(I) := −1,                       diagonal elements
     Va(N−1) := 0                                    of the matrix A}
     Vc(1) := 0


SUBROUTINE  TriSystem(Va,Vd,Vc,Vb,N)
     A0 := Va, B0 := Vb, C0 := Vc, D0 := Vd          {Vector replacements}
     FOR  K = 2  TO  N  DO
          T := A0(K−1)/D0(K−1)
          D0(K) := D0(K) − T*C0(K−1)
          B0(K) := B0(K) − T*B0(K−1)
     X0(N) := B0(N)/D0(N)
     FOR  K = N−1  DOWNTO  1  DO
          X0(K) := (B0(K) − C0(K)*X0(K+1))/D0(K)
     Return X0                                       {Return solution vector}
```

{The main program starts here.}

```
          Vb(1) := C1,   Vb(N) := C2


FOR  J = 2  TO  M  DO                                {Construct successive
     FOR  I = 2  TO  N−1  DO                          rows in the grid}
          Vb(I) := U(I−1,J−1) + U(I+1,J−1) + S2*U(I,J−1)
     X := TriSystem(Va,Vd,Vc,Vb,N)                   {Solve tridiagonal
                                                     system, pass vectors}
                                                     {Next row in grid}
     FOR  I = 1  TO  N  DO
          U(I,J) := X(I)


FOR  J = 1  TO  M  DO                                {Output the solution}
     PRINT  K*(J−1),  U(1,J),  U(2,J), . . . , U(N,J)
```

EXERCISES FOR PARABOLIC EQUATIONS

1. (a) Verify by direct substitution that $u(x, t) = \sin(n\pi x)e^{-4tn^2\pi^2}$ is a solution to the heat equation $u_t(x, t) = 4u_{xx}(x, t)$ for each positive integer $n = 1, 2, \ldots$.
   (b) Verify by direct substitution that $u(x, t) = \sin(n\pi x)e^{-tc^2n^2\pi^2}$ is a solution to the heat equation $u_t(x, t) = c^2 u_{xx}(x, t)$ for each positive integer $n = 1, 2, \ldots$.

2. What difficulty might occur if $\Delta t = k = h^2/c^2$ is used with formula (7)?

3. Use the forward-difference method to solve the heat equation $u_t(x, t) = u_{xx}(x, t)$ for $0 < x < 1$ and $0 \le t \le 0.1$ with the initial condition

$$u(x, 0) = f(x) = \sin(\pi x) \qquad \text{for } t = 0 \quad \text{and} \quad 0 \le x \le 1$$

and the boundary conditions

$$u(0, t) = g_1(t) \equiv 0 \qquad \text{for } x = 0 \quad \text{and} \quad 0 \le t \le 0.1,$$

$$u(1, t) = g_2(t) \equiv 0 \qquad \text{for } x = 1 \quad \text{and} \quad 0 \le t \le 0.1.$$

Let $h = 0.2$, $k = 0.02$, and $r = 0.5$. Use a calculator and apply formula (11) to compute the approximations to $u(x, t)$ for $0 \le t \le 0.1$. *Hint.* The starting row is

$$0.000000, \ 0.587785, \ 0.951057, \ 0.951057, \ 0.587785, \ 0.000000$$

4. Use the forward-difference method to solve the heat equation $u_t(x, \ t) = u_{xx}(x, \ t)$ for $0 < x < 1$ and $0 \le t \le 0.1$, with the initial condition

$$u(x, 0) = f(x) = 1 - |2x - 1| \qquad \text{for } t = 0 \quad \text{and} \quad 0 \le x \le 1,$$

and the boundary conditions

$$u(0, t) = g_1(t) \equiv 0 \qquad \text{for } x = 0 \quad \text{and} \quad 0 \le t \le 0.1,$$

$$u(1, t) = g_2(t) \equiv 0 \qquad \text{for } x = 1 \quad \text{and} \quad 0 \le t \le 0.1.$$

Let $h = 0.2$, $k = 0.02$, and $r = 0.5$. Use a calculator and apply formula (11) to compute the approximations to $u(x, t)$ for $0 \le t \le 0.1$. *Hint.* The starting row is

$$0.00, \ 0.40, \ 0.80, \ 0.80, \ 0.40, \ 0.00$$

5. Use a computer to implement the Crank–Nicholson method for solving the heat equation $u_t(x, \ t) = u_{xx}(x, \ t)$ for $0 < x < 1$ and $0 < t < 0.1$, with the initial condition

$$u(x, 0) = f(x) = \sin(\pi x) + \sin(2\pi x) \qquad \text{for } t = 0 \quad \text{and} \quad 0 \le x \le 1$$

and boundary conditions

$$u(0, t) = g_1(t) \equiv 0 \qquad \text{for } x = 0 \quad \text{and} \quad 0 \le t \le 0.1,$$

$$u(1, t) = g_2(t) \equiv 0 \qquad \text{for } x = 1 \quad \text{and} \quad 0 \le t \le 0.1.$$

Use the values $h = 0.1$, $k = 0.01$, and $r = 1$.

6. Use a computer to implement the Crank–Nicholson method for solving the heat equation $u_t(x, \ t) = u_{xx}(x, \ t)$ for $0 < x < 1$ and $0 < t < 0.1$, with the initial condition

$$u(x, 0) = f(x) = 3 - |3x - 1| - |3x - 2| \qquad \text{for } t = 0 \quad \text{and} \quad 0 \le x \le 1$$

and boundary conditions

$$u(0, t) = g_1(t) \equiv 0 \qquad \text{for } x = 0 \quad \text{and} \quad 0 \le t \le 0.1,$$

$$u(1, t) = g_2(t) \equiv 0 \qquad \text{for } x = 1 \quad \text{and} \quad 0 \le t \le 0.1.$$

Use the values $h = 0.1$, $k = 0.01$, and $r = 1$.

7. Suppose that $\Delta t = k = h^2/(2c^2)$.
   (a) Use this in formula (16) and simplify.
   (b) Express the equations in part (a) in matrix form $A\mathbf{X} = \mathbf{B}$.
   (c) Is the matrix in part (b) diagonally dominant? Why?

8. Show that $u(x,\ t) = \displaystyle\sum_{j=1}^{N} a_j e^{-tj^2\pi^2} \sin(j\pi x)$ is a solution to $u_t = u_{xx}$ for $0 \le x \le 1$ and $0 < t$ and has the boundary values $u(0,\ t) = 0$, $u(1,\ t) = 0$, and

$$u(x,\ 0) = \sum_{j=1}^{N} a_j \sin(j\pi x).$$

9. Consider the analytic solution $u(x,\ t) = \sin(\pi x)\exp(-t\pi^2) + \sin(3\pi x)\exp(-t9\pi^2)$ that was discussed in Example 4.
   (a) Hold x fixed and determine $\lim_{t \to \infty} u(x,\ t)$.
   (b) What does this mean physically?

10. Suppose that we wish to solve the parabolic equation $u_t(x,\ t) - u_{xx}(x,\ t) = h(x)$.
   (a) Derive the explicit forward-difference equation for this situation.
   (b) Derive the implicit difference formula for this situation.

11. Suppose that $g_1(t) \ne 0$ and $g_2(t) \ne 0$.
   (a) Discuss how to incorporate this situation into a forward-difference scheme.
   (b) Discuss how to incorporate this situation into an implicit difference scheme.

12. Suppose that equation (11) is used and that $f(x) \ge 0$, $g_1(x) = 0$, and $g_2(x) = 0$.
   (a) Show that the maximum value of $u(x_i,\ t_{j+1})$ in row $j + 1$ is less than or equal to the maximum of $u(x_i,\ t_j)$ in row j.
   (b) Make a conjecture concerning the maximum of $u(x_i,\ t_n)$ in row n as n tends to infinity.

10.3 ELLIPTIC EQUATIONS

As examples of elliptic partial differential equations, we consider the Laplace, Poisson, and Helmholtz equations. Recall that the Laplacian of the function $u(x,\ y)$ is

$$\nabla^2 u \equiv u_{xx} + u_{yy}. \tag{1}$$

With this notation, we can write the Laplace, Poisson, and Helmholtz equations in the following forms:

$$\nabla^2 u = 0 \qquad \text{Laplace's equation,} \tag{2}$$

$$\nabla^2 u = g(x,\ y) \qquad \text{Poisson's equation,} \tag{3}$$

$$\nabla^2 u + f(x,\ y)u = g(x,\ y) \qquad \text{Helmholtz's equation.} \tag{4}$$

It is often the case that the boundary values for the functions g and f are known at all points on the sides of a rectangular region R in the plane. In this case, each of these equations can be solved by the numerical technique known as the finite-difference method.

The Laplacian Difference Equation

The Laplacian operator must be expressed in a discrete form suitable for numerical computations. The formula for approximating $f''(x)$ is obtained from

$$f''(x) = \frac{f(x + h) - 2f(x) + f(x - h)}{h^2} + O(h^2). \tag{5}$$

When this is applied to the function $u(x, y)$ to approximate $u_{xx}(x, y)$ and $u_{yy}(x, y)$ and the results are added, we obtain

$$\nabla^2 u = \frac{u(x + h, y) + u(x - h, y) + u(x, y + h) + u(x, y - h) - 4u(x, y)}{h^2} + O(h^2). \tag{6}$$

Assume that the rectangle $R = \{(x, y): 0 \le x \le a, 0 \le y \le b$ where $b/a = m/n\}$ is subdivided into $n - 1 \times m - 1$ squares with side h (i.e., $a = nh$ and $b = mh$), as shown in Figure 10.14.

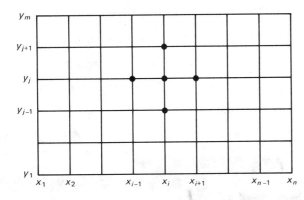

Figure 10.14   The grid used with Laplace's difference equation.

To solve Laplace's equation, we impose the approximation

$$\frac{u(x + h, y) + u(x - h, y) + u(x, y + h) + u(x, y - h) - 4u(x, y)}{h^2} = 0, \tag{7}$$

which has order of accuracy $O(h^2)$ at all interior grid points $(x, y) = (x_i, y_j)$ for $i = 2$, ..., $n - 1$ and $j = 2, \ldots , m - 1$. The grid points are uniformly spaced $x_{i+1} = x_i + h$, $x_{i-1} = x_i - h$, $y_{i+1} = y_i + h$, and $y_{i-1} = y_i - h$. Using the approximation $u_{i,j}$ for $u(x_i, y_j)$, equation (7) can be written in the form

$$\nabla^2 u_{i,j} \approx \frac{u_{i+1,j} + u_{i-1,j} + u_{i,j+1} + u_{i,j-1} - 4u_{i,j}}{h^2} = 0, \tag{8}$$

which is known as the **five-point difference formula** for Laplace's equation. This formula relates the function value $u_{i,j}$ to its four neighboring values $u_{i+1,j}$, $u_{i-1,j}$, $u_{i,j+1}$, and $u_{i,j-1}$, as shown in Figure 10.15. The term h^2 can be eliminated in (8) to obtain the Laplacian computational formula

$$u_{i+1,j} + u_{i-1,j} + u_{i,j+1} + u_{i,j-1} - 4u_{i,j} = 0. \tag{9}$$

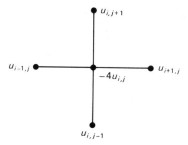

Figure 10.15   The Laplace stencil.

Setting Up the Linear System

Assume that the values $u(x, y)$ are known at the following boundary grid points:

$$u(x_1, y_j) = u_{1,j} \quad \text{for } 2 \le j \le m - 1 \quad \text{(on the left)},$$

$$u(x_i, y_1) = u_{i,1} \quad \text{for } 2 \le i \le n - 1 \quad \text{(on the bottom)},$$

$$u(x_n, y_j) = u_{n,j} \quad \text{for } 2 \le j \le m - 1 \quad \text{(on the right)},$$

$$u(x_i, y_m) = u_{i,m} \quad \text{for } 2 \le i \le n - 1 \quad \text{(on the top)}.$$

Then applying the Laplacian computational formula (9) at each of the interior points of R will create a linear system of $(n - 2)$ equations in $(n - 2)$ unknowns, which is solved to obtain approximations to $u(x, y)$ at the interior points of R. For example, suppose that the region is a square and that $n = m = 5$ and that the unknown values of $u(x_i, y_j)$ at the nine interior grid points are labeled $p_1, p_2, \ldots, p_9$ and positioned in the grid as shown in Figure 10.16.

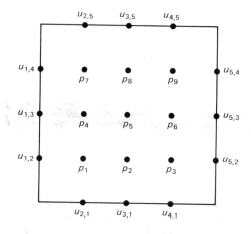

Figure 10.16   A 5×5 grid for boundary values only.

The Laplacian computational formula (9) is applied at each of the interior grid points and the result is the system $A\mathbf{P} = \mathbf{B}$ of nine linear equations:

$$
\begin{aligned}
-4p_1 + p_2 \quad\quad + p_4 \quad\quad\quad\quad\quad\quad\quad\quad\quad\quad &= -u_{2,1} - u_{1,2} \\
p_1 - 4p_2 + p_3 \quad\quad + p_5 \quad\quad\quad\quad\quad\quad\quad &= -u_{3,1} \\
p_2 - 4p_3 \quad\quad\quad\quad + p_6 \quad\quad\quad\quad\quad &= -u_{4,1} - u_{5,2} \\
p_1 \quad\quad\quad\quad -4p_4 + p_5 \quad\quad + p_7 \quad\quad\quad &= -u_{1,3} \\
p_2 \quad\quad + p_4 - 4p_5 + p_6 \quad\quad + p_8 \quad &= 0 \\
p_3 \quad\quad + p_5 - 4p_6 \quad\quad\quad\quad + p_9 &= -u_{5,3} \\
p_4 \quad\quad\quad\quad -4p_7 + p_8 \quad\quad\quad &= -u_{2,5} - u_{1,4} \\
p_5 \quad\quad\quad\quad + p_7 - 4p_8 + p_9 &= -u_{3,5} \\
p_6 \quad\quad\quad\quad + p_8 - 4p_9 &= -u_{4,5} - u_{5,4}.
\end{aligned}
$$

Example 10.5

Find an approximate solution to Laplace's equation $\nabla^2 u = 0$ in the rectangle $R = \{(x, y):$ $0 \le x \le 4,\ 0 \le y \le 4\}$ where $u(x, y)$ denotes the temperature at the point (x, y) and the boundary values are:

$$u(x, 0) = 20 \quad \text{and} \quad u(x, 4) = 180 \quad\quad \text{for } 0 < x < 4,$$

and

$$u(0, y) = 80 \quad \text{and} \quad u(4, y) = 0 \quad\quad \text{for } 0 < y < 4.$$

See Figure 10.17 for the grid to be used.

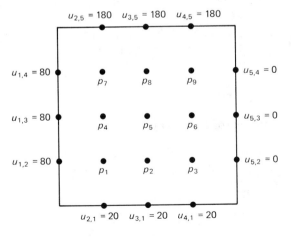

Figure 10.17   The 5 × 5 grid in Example 10.5.

Solution. Applying formula (9) in this case, the linear $A\mathbf{P} = \mathbf{B}$ is

$$
\begin{aligned}
-4p_1 + p_2 \qquad\quad + p_4 \qquad\qquad\qquad\qquad\qquad &= -100 \\
p_1 - 4p_2 + p_3 \qquad\quad + p_5 \qquad\qquad\qquad &= -20 \\
p_2 - 4p_3 \qquad\qquad\quad + p_6 \qquad\qquad &= -20 \\
p_1 \qquad\qquad - 4p_4 + p_5 \qquad\quad + p_7 \qquad &= -80 \\
p_2 \qquad\quad + p_4 - 4p_5 + p_6 \qquad\quad + p_8 \quad &= 0 \\
p_3 \qquad\quad + p_5 - 4p_6 \qquad\qquad\quad + p_9 &= 0 \\
p_4 \qquad\qquad\qquad - 4p_7 + p_8 \qquad\quad &= -260 \\
p_5 \qquad\qquad\quad + p_7 - 4p_8 + p_9 &= -180 \\
p_6 \qquad\qquad\qquad + p_8 - 4p_9 &= -180.
\end{aligned}
$$

The solution vector $\mathbf{P}$ can be obtained by Gaussian elimination (or more efficient schemes can be devised, such as the extension of the tridiagonal algorithm to pentadiagonal systems). The temperatures at the interior grid points are given expressed in vector form:

$$
\mathbf{P} = (p_1, p_2, p_3, p_4, p_5, p_6, p_7, p_8, p_9)^T
$$

$$
= (55.7143, 43.2143, 27.1429, 79.6429, 70.0000, 45.3571, 112.857, 111.786, 84.2857)^T.
$$

Derivative Boundary Conditions

The Neumann boundary conditions specify the directional derivative of $u(x, y)$ normal to an edge. For our illustration we will use the zero normal derivative condition,

$$
\frac{\partial}{\partial N} u(x, y) = 0. \tag{10}
$$

For applications in the area of heat flow, this means that the edge is thermally insulated and the heat flux throughout the edge is zero.

Suppose that the $x = x_n$ is held fixed and that we are considering the right edge $x = a$ of the rectangle $R = \{(x, y): 0 \le x \le a, 0 \le y \le b\}$. The normal boundary condition to be used along this edge is

$$
\frac{\partial}{\partial x} u(x_n, y_j) = u_x(x_n, y_j) = 0. \tag{11}
$$

Then the Laplace difference equation for the point (x_n, y_j) is

$$
u_{n+1,j} + u_{n-1,j} + u_{n,j+1} + u_{n,j-1} - 4u_{n,j} = 0. \tag{12}
$$

The value $u_{n+1,j}$ is unknown, because it lies outside the region R. However, we can use the numerical differentiation formula

$$
\frac{u_{n+1,j} - u_{n-1,j}}{2h} \approx u_x(x_n, y_j) = 0 \tag{13}
$$

and obtain the approximation $u_{n+1,j} \approx u_{n-1,j}$, which has order of accuracy $O(h^2)$. When this approximation is used in (12) the result is

$$2u_{n-1,j} + u_{n,j+1} + u_{n,j-1} - 4u_{n,j} = 0.$$

This formula relates the function value $u_{n,j}$ to its three neighboring values $u_{n-1,j}$, $u_{n,j+1}$, and $u_{n,j-1}$.

The computational stencils for the other edges can be derived similarly (see Figure 10.18). The four cases for the Neumann computational stencils are summarized below:

$$2u_{i,2} \quad + u_{i-1,1} + u_{i+1,1} - 4u_{i,1} = 0 \qquad \text{(bottom edge)}, \qquad (14)$$

$$2u_{i,m-1} + u_{i-1,m} + u_{i+1,m} - 4u_{i,m} = 0 \qquad \text{(top edge)}, \qquad (15)$$

$$2u_{2,j} \quad + u_{1,j-1} + u_{1,j+1} - 4u_{1,j} = 0 \qquad \text{(left edge)}, \qquad (16)$$

$$2u_{n-1,j} + u_{n,j-1} + u_{n,j+1} - 4u_{n,j} = 0 \qquad \text{(right edge)}. \qquad (17)$$

Suppose that the derivative condition $\dfrac{\partial}{\partial N} u(x, y) = 0$ is used along part of the boundary of R, and that known boundary values of $u(x, y)$ are used on the other portions of the boundary; then we have a mixed problem. The equations for determining approximations for $u(x_i, y_j)$ at boundary points will involve appropriate Neumann computational stencils (14) to (17). The Laplacian computational formula (9) is still used to determine approximations for $u(x_i, y_j)$ at the interior points of R.

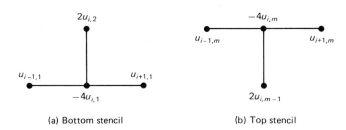

(a) Bottom stencil   (b) Top stencil

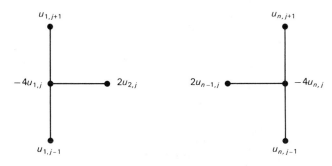

(c) Left stencil   (d) Right stencil   **Figure 10.18** The Neumann stencils.

Example 10.6

Find an approximate solution to Laplace's equation $\nabla^2 u = 0$ in the rectangle $R = \{(x, y): 0 \le x \le 4, 0 \le y \le 4\}$, where $u(x, y)$ denotes the temperature at the point (x, y) and the boundary values are shown in Figure 10.19:

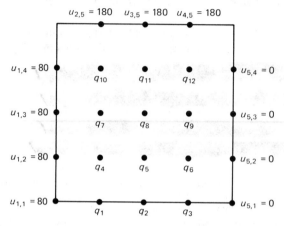

$u_{2,5} = 180 \quad u_{3,5} = 180 \quad u_{4,5} = 180$

$u_{1,4} = 80 \qquad \bullet \quad \bullet \quad \bullet \qquad u_{5,4} = 0$
$\qquad\qquad q_{10} \quad q_{11} \quad q_{12}$

$u_{1,3} = 80 \qquad \bullet \quad \bullet \quad \bullet \qquad u_{5,3} = 0$
$\qquad\qquad q_7 \quad q_8 \quad q_9$

$u_{1,2} = 80 \qquad \bullet \quad \bullet \quad \bullet \qquad u_{5,2} = 0$
$\qquad\qquad q_4 \quad q_5 \quad q_6$

$u_{1,1} = 80 \qquad \bullet \quad \bullet \quad \bullet \qquad u_{5,1} = 0$
$\qquad\qquad q_1 \quad q_2 \quad q_3$

Figure 10.19 The 5×5 grid in Example 10.6.

$$u(x, 4) = 180 \qquad \text{for } 0 < x < 4, \text{ and}$$
$$u_y(x, 0) = 0 \qquad \text{for } 0 < x < 4, \text{ and}$$
$$u(0, y) = 80 \qquad \text{for } 0 \le y < 4, \text{ and}$$
$$u(4, y) = 0 \qquad \text{for } 0 \le y < 4.$$

Solution. The Neumann computational formula (14) is applied at the boundary points q_1, q_2, and q_3 and the Laplace computational stencil (9) is applied at the other points, q_4, q_5, ..., q_{12}. The result is a linear system $A\mathbf{Q} = \mathbf{B}$ involving 12 equations in 12 unknowns:

$$
\begin{aligned}
-4q_1 + q_2 \qquad\qquad + 2q_4 \qquad\qquad\qquad\qquad\qquad\qquad\qquad &= -80 \\
q_1 - 4q_2 + q_3 \qquad\qquad + 2q_5 \qquad\qquad\qquad\qquad\qquad\qquad &= 0 \\
q_2 - 4q_3 \qquad\qquad\qquad + 2q_6 \qquad\qquad\qquad\qquad\qquad &= 0 \\
q_1 \qquad\qquad - 4q_4 + q_5 \qquad + q_7 \qquad\qquad\qquad\qquad &= -80 \\
q_2 \qquad\qquad + q_4 - 4q_5 + q_6 \qquad + q_8 \qquad\qquad\qquad &= 0 \\
q_3 \qquad\qquad + q_5 - 4q_6 \qquad\qquad + q_9 \qquad\qquad &= 0 \\
q_4 \qquad\qquad - 4q_7 + q_8 \qquad + q_{10} \qquad\qquad &= -80 \\
q_5 \qquad\qquad + q_7 - 4q_8 + q_9 \qquad + q_{11} \qquad &= 0 \\
q_6 \qquad\qquad + q_8 - 4q_9 \qquad\qquad + q_{12} &= 0 \\
q_7 \qquad\qquad\qquad\qquad - 4q_{10} + q_{11} \qquad &= -260 \\
q_8 \qquad\qquad\qquad\qquad + q_{10} - 4q_{11} + q_{12} &= -180 \\
q_9 \qquad\qquad\qquad\qquad + q_{11} - 4q_{12} &= -180.
\end{aligned}
$$

The solution vector $\mathbf{Q}$ can be obtained by Gaussian elimination (or more efficient schemes can be devised, such as the extension of the tridiagonal algorithm to pentadiagonal systems). The temperatures at the interior grid points and along the lower edge are expressed in vector form:

$$\mathbf{Q} = (q_1, q_2, q_3, q_4, q_5, q_6, q_7, q_8, q_9, q_{10}, q_{11}, q_{12})^T$$
$$= (71.8218, 56.8543, 32.2342, 75.2165, 61.6806, 36.0412,$$
$$87.3636, 78.6103, 50.2502, 115.628, 115.147, 86.3492)^T.$$

Iterative Methods

The preceding method showed how to solve Laplace's difference equation by constructing a certain system of linear equations and solving it. The shortcoming of this method is storage; each interior grid point introduces an equation to be solved. Since better approximations require a finer mesh grid, many equations might be needed. For example, the solution of Laplace's equation with the Dirichlet boundary conditions requires solving a system of $(n - 2)(m - 2)$ equations. If R is divided into a modest number of squares, say 10 by 10, there would be 91 equations involving 91 unknowns. Hence it is sensible to develop techniques that will reduce the amount of storage. An iterative method would require only the storage of the 100 numerical approximations $\{u_{i,j}\}$ through the grid.

Let us start with Laplace's difference equation:

$$u_{i+1,j} + u_{i-1,j} + u_{i,j+1} + u_{i,j-1} - 4u_{i,j} = 0 \tag{18}$$

and suppose that the boundary values $u(x, y)$ are known at the following grid points:

$$u(x_1, y_j) = u_{1,j} \qquad \text{for } 2 \le j \le m - 1 \quad \text{(on the left)},$$
$$u(x_i, y_1) = u_{i,1} \qquad \text{for } 2 \le i \le n - 1 \quad \text{(on the bottom)},$$
$$u(x_n, y_j) = u_{n,j} \qquad \text{for } 2 \le j \le m - 1 \quad \text{(on the right)}, \tag{19}$$
$$u(x_i, y_m) = u_{i,m} \qquad \text{for } 2 \le i \le n - 1 \quad \text{(on the top)}.$$

Equation (18) is rewritten in the following form that is suitable for iteration:

$$u_{i,j} = u_{i,j} + r_{i,j}, \tag{20}$$

where

$$r_{i,j} = \frac{u_{i+1,j} + u_{i-1,j} + u_{i,j+1} + u_{i,j-1} - 4u_{i,j}}{4} \tag{21}$$

for $2 \le i \le n - 1$ and $2 \le j \le m - 1$.

Starting values for all interior grid points must be supplied. The constant K, which is the average of the $2n + 2m - 4$ boundary values given in (19), can be used for this purpose. One iteration consists of sweeping formula (20) throughout all of the interior points of the grid. Successive iterations sweep the interior of the grid with the Laplace iterative operator (20) until the residual term $r_{i,j}$ on the right side of equation (20) is "reduced to zero" (i.e., $|r_{i,j}| < \epsilon$ holds for each $2 \le i \le n - 1$ and $2 \le j \le m - 1$). The speed of convergence for reducing all the residuals $\{r_{i,j}\}$ to zero is increased by using the method called successive over-relaxation (SOR). The SOR method uses the iteration formula

$$u_{i,j} = u_{i,j} + \omega \left[\frac{u_{i+1,j} + u_{i-1,j} + u_{i,j+1} + u_{i,j-1} - 4u_{i,j}}{4} \right] = u_{i,j} + \omega r_{i,j}, \quad (22)$$

where the parameter ω lies in the range $1 \le \omega < 2$. In the SOR method, formula (22) is swept across the grid until $|r_{i,j}| < \epsilon$. The optimal choice for ω is based on the study of eigenvalues of iteration matrices for linear systems and is given in this case by the formula

$$\omega = \frac{4}{2 + \sqrt{4 - \left[\cos\left(\frac{\pi}{n-1}\right) + \cos\left(\frac{\pi}{m-1}\right) \right]^2}} \quad (23)$$

If the Neumann boundary condition is specified on some portion of the boundary, we must rewrite equations (14) to (17) in a form that is suitable for iteration: The four cases are summarized below and include the relaxation parameter ω:

$$u_{i,1} = u_{i,1} + \omega \left[\frac{2u_{i,2} + u_{i-1,1} + u_{i+1,1} - 4u_{i,1}}{4} \right] \qquad \text{(bottom edge)}, \quad (24)$$

$$u_{i,m} = u_{i,m} + \omega \left[\frac{2u_{i,m-1} + u_{i-1,m} + u_{i+1,m} - 4u_{i,m}}{4} \right] \qquad \text{(top edge)}, \quad (25)$$

$$u_{1,j} = u_{1,j} + \omega \left[\frac{2u_{2,j} + u_{1,j-1} + u_{1,j+1} - 4u_{1,j}}{4} \right] \qquad \text{(left edge)}, \quad (26)$$

$$u_{n,j} = u_{n,j} + \omega \left[\frac{2u_{n-1,j} + u_{n,j-1} + u_{n,j+1} - 4u_{n,j}}{4} \right] \qquad \text{(right edge)}. \quad (27)$$

Example 10.7

Use an iterative method to compute an approximate solution to Laplace's equation $\nabla^2 u = 0$ in $R = \{(x, y): 0 \le x \le 4, 0 \le y \le 4\}$ where the boundary values are

$$u(x, 0) = 20 \quad \text{and} \quad u(x, 4) = 180 \qquad \text{for } 0 < x < 4$$

and

$$u(0, y) = 80 \quad \text{and} \quad u(4, y) = 0 \qquad \text{for } 0 < y < 4.$$

Solution. For illustration, the square is divided into 64 squares with sides $\Delta x = h = 0.5$ and $\Delta y = h = 0.5$. The initial value at the interior grid points was set at $u_{i,j} = 70$ for each $i = 2, \ldots , 8$ and $j = 2, \ldots , 8$. Then the SOR method was used with the parameter $\omega = 1.44646$ [substitute $n = 9$ and $m = 9$ in formula (23)]. After 19 iterations, the residual was uniformly reduced (i.e., $|r_{i,j}| \le 0.000606 < 0.001$). The resulting approximations are given in Table 10.6. Because of the discontinuity of the boundary function at the corners, the boundary values $u_{1,1} = 50$, $u_{9,1} = 10$, $u_{1,9} = 130$, and $u_{9,9} = 90$ have been introduced in

Table 10.6 and Figure 10.20; they were not used in the computations at the interior grid points. A three-dimensional presentation of the data in Table 10.6 is given in Figure 10.20.

TABLE 10.6 Approximate Solution to Laplace's Equation with Dirichlet Conditions

	x_1	x_2	x_3	x_4	x_5	x_6	x_7	x_8	x_9
y_9	130.000	180.000	180.000	180.000	180.000	180.000	180.000	180.000	90.0000
y_8	80.000	124.821	141.172	145.414	144.005	137.478	122.642	88.6070	0.0000
y_7	80.000	102.112	113.453	116.479	113.126	103.266	84.4844	51.7856	0.0000
y_6	80.000	89.1736	94.0499	93.9210	88.7553	77.9737	60.2439	34.0510	0.0000
y_5	80.000	80.5319	79.6515	76.3999	70.0003	59.6301	44.4667	24.1744	0.0000
y_4	80.000	73.3023	67.6241	62.0267	55.2159	46.0796	33.8184	18.1798	0.0000
y_3	80.000	65.0528	55.5159	48.8671	42.7568	35.6543	26.5473	14.7266	0.0000
y_2	80.000	51.3931	40.5195	35.1691	31.2899	27.2335	21.9900	14.1791	0.0000
y_1	50.000	20.0000	20.0000	20.0000	20.0000	20.0000	20.0000	20.0000	10.0000

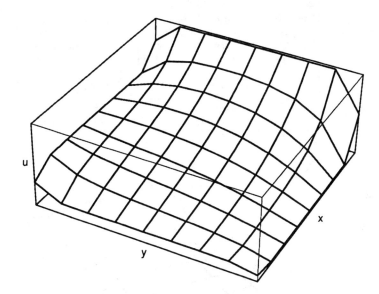

Figure 10.20 $u = u(x, y)$ with Dirichlet boundary values.

Example 10.8

Use an iterative method to compute an approximate solution to Laplace's equation $\nabla^2 u = 0$ in $R = \{(x, y): 0 \le x \le 4, 0 \le y \le 4\}$, where the boundary values are

$$u(x, 4) = 180 \qquad \text{for } y = 4 \quad \text{and} \quad 0 < x < 4, \text{ and}$$

$$u_y(x, 0) = 0 \qquad \text{for } y = 0 \quad \text{and} \quad 0 < x < 4, \text{ and}$$

$$u(0, y) = 80 \qquad \text{for } x = 0 \quad \text{and} \quad 0 \le y < 4, \text{ and}$$

$$u(4, y) = 0 \qquad \text{for } x = 4 \quad \text{and} \quad 0 \le y < 4.$$

Solution. For illustration, the square is divided into 64 squares with sides $\Delta x = h = 0.5$ and $\Delta y = h = 0.5$. Starting values using linear interpolation were used along the edge where $y = y_1 = 0$. The initial value at the interior grid points was set at $u_{i,j} = 70$ for each $i = 2$, . . . , 8 and $j = 2$, . . . , 8. Then the SOR method was employed with the parameter $\omega = 1.44646$ (as in Example 10.7). After 29 iterations, the residual was uniformly reduced; (i.e., $|r_{i,j}| \leq 0.000998 < 0.001$). The resulting approximations are given in Table 10.7. Because of the discontinuity of the boundary function at the corners, the boundary values $u_{1,9} = 130$ and $u_{9,9} = 90$ have been introduced in Table 10.7 and Figure 10.21; they were not used in the computations at the interior grid points. A three-dimensional presentation of the data in Table 10.7 is given in Figure 10.21.

TABLE 10.7 Approximate Solution to Laplace's Equation with Mixed Boundary Conditions

	x_1	x_2	x_3	x_4	x_5	x_6	x_7	x_8	x_9
y_9	130.000	180.000	180.000	180.000	180.000	180.000	180.000	180.000	90.0000
y_8	80.000	126.457	142.311	146.837	145.468	138.762	123.583	89.1008	0.0000
y_7	80.000	103.518	115.951	119.568	116.270	105.999	86.4683	52.8201	0.0000
y_6	80.000	91.6621	98.4053	99.2137	94.0461	82.4936	63.4715	35.7113	0.0000
y_5	80.000	84.7247	86.7936	84.8347	78.2063	66.4578	49.2124	26.5538	0.0000
y_4	80.000	80.4424	79.2089	75.1245	67.4860	55.9185	40.3665	21.2915	0.0000
y_3	80.000	77.8354	74.4742	68.9677	60.6944	49.3635	35.0435	18.2459	0.0000
y_2	80.000	76.4244	71.8842	65.5772	56.9600	45.7972	32.1981	16.6485	0.0000
y_1	80.000	75.9774	71.0605	64.4964	55.7707	44.6670	31.3032	16.1500	0.0000

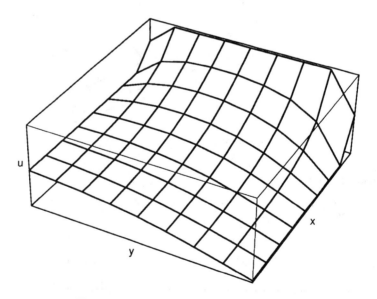

Figure 10.21 $u = u(x, y)$ for a mixed problem.

Poisson's and Helmholtz's Equations

Consider Poisson's equation

$$\nabla^2 u = g(x, y). \tag{28}$$

Using the notation $g_{i,j} = g(x_i, y_j)$, the generalization of formula (20) for solving (28) over the rectangular grid is

$$u_{i,j} = u_{i,j} + \frac{u_{i+1,j} + u_{i-1,j} + u_{i,j+1} + u_{i,j-1} - 4u_{i,j} - h^2 g_{i,j}}{4}. \tag{29}$$

Consider Helmholtz's equation

$$\nabla^2 u + f(x, y)u = g(x, y). \tag{30}$$

Using the notation $f_{i,j} = f(x_i, y_j)$, the generalization of formula (20) for solving (30) over the rectangular grid is

$$u_{i,j} = u_{i,j} + \frac{u_{i+1,j} + u_{i-1,j} + u_{i,j+1} + u_{i,j-1} - (4 - h^2 f_{i,j})u_{i,j} - h^2 g_{i,j}}{4 - h^2 f_{i,j}}. \tag{31}$$

These formulas are explored in greater detail in the exercises.

Improvements

A modification of (8) that can be employed is the **nine-point difference formula** for Laplace's equation:

$$\nabla^2 u_{i,j} \approx \frac{1}{6h^2} [u_{i+1,j-1} + u_{i-1,j-1} + u_{i+1,j+1} + u_{i-1,j-1}$$

$$+ 4u_{i+1,j} + 4u_{i-1,j} + 4u_{i,j+1} + 4u_{i,j-1} - 20u_{i,j}]$$

$$= 0.$$

The truncation error for the nine-point difference formula is of the order $O(h^4)$ when it is used to solve the Poisson or Helmholtz equation; thus there is no improvement if the nine-point difference formula is used instead of the five-point difference formula. However, when the nine-point formula is used to solve Laplace's equation $\nabla^2 u = 0$, the truncation error is of the order $O(h^6)$ and there is an advantage to using the nine-point difference formula.

Algorithm 10.4   (Dirichlet Method for Laplace's Equation).   To approximate the solution of $u_{xx}(x, y) + u_{yy}(x, y) = 0$ over $R = \{(x, y): 0 \leq x \leq a, 0 \leq y \leq b\}$ with $u(x, 0) = f_1(x)$, $u(x, b) = f_2(x)$ for $0 \leq x \leq a$ and $u(0, y) = f_3(y)$, $u(a, y) = f_4(y)$ for $0 \leq y \leq b$. It is assumed that $\Delta x = \Delta y = h$ and that integers n and m exist so that $a = nh$ and $b = mh$.

```
INPUT A, B                                          {Width and height of R}
INPUT H                                                    {Grid spacing}
INPUT N, M                                          {Dimensions of the grid}
INPUT Ave                                          {Initial approximation}
Pi := 3.1415926535                                   {Approximation for π}
F1i(I) := F1(H*(I−1)), F2i(I) := F2(H*(I−1))        {Boundary functions
F3i(I) := F3(H*(I−1)), F4i(I) := F4(H*(I−1))          at the grid points}

FOR  I = 2  TO  N−1  DO                             {Initialize starting
    FOR  J = 2  TO  M−1  DO                              values at the
         U(I,J) := Ave                                 interior points}
FOR  J = 1  TO  M  DO                              {Store boundary values
    U(1,J) := F3i(J) , U(N,J) := F4i(J)              in the solution matrix}
FOR  I = 1  TO  N  DO
    U(I,1) := F1i(I), U(I,M) := F2i(I)
w := 4/(2+SQRT(4−(cos(Pi/(N−1))+cos(Pi/(M−1)))^2))  {The SOR parameter}
Tol := 1, Count := 0                               {Initialize loop
                                                    control parameters}
WHILE  Tol > 0.001  and  Count ≤ 25  DO            {Refine the
    Tol := 0.0                                       approximations}
    FOR  J = 2  TO  M−1  DO                         {Sweep the operator
        FOR  I = 2  TO  N−1  DO                      throughout the grid}
            Relax := w*(  u(I,J+1) + u(I,J−1) + u(I+1,J) +
                          u(I−1,J) − 4.0*u(I,J)  )/4.0
            u(I,J) := u(I,J) + Relax
            Tol := MAX(Tol,ABS(Relax))
    Count := Count+1

PRINT  The table of solutions {U(I,J): I=1, . . . ,N and J=1, . . . ,M}
```

EXERCISES FOR ELLIPTIC EQUATIONS

1. (a) Determine the system of four equations in the four unknowns p_1, p_2, p_3, and p_4 for computing approximations for the harmonic function $u(x, y)$ in the rectangle $R = \{(x, y): 0 \leq x \leq 3, 0 \leq y \leq 3\}$, (see Figure 10.22). The boundary values are

$$u(x, 0) = 10 \quad \text{and} \quad u(x, 3) = 90 \quad \text{for } 0 < x < 3,$$

$$u(0, y) = 70 \quad \text{and} \quad u(3, y) = 0 \quad \text{for } 0 < y < 3.$$

   (b) Solve the equations in part (a) for p_1, p_2, p_3, and p_4.

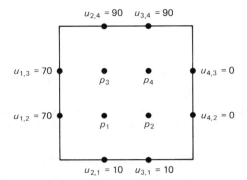

Figure 10.22   The grid for Exercise 1.

2. **(a)** Determine the system of six equations in the six unknowns $q_1, q_2, \ldots, q_6$ for computing approximations for the harmonic function $u(x, y)$ in the rectangle $R = \{(x, y): 0 \le x \le 3, 0 \le y \le 3\}$ (see Figure 10.23). The boundary values are

$$u(x, 3) = 90 \quad \text{for } 0 < x < 3, \qquad u_y(x, 0) = 0 \quad \text{for } 0 < x < 3,$$

$$u(0, y) = 70 \quad \text{for } 0 \le y < 3, \qquad u(3, y) = 0 \quad \text{for } 0 \le y < 3.$$

(b) Solve the equations in part (a) for $q_1, q_2, \ldots, q_6$.

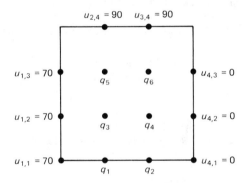

Figure 10.23   The grid for Exercise 2.

3. **(a)** Compute approximations for the harmonic function $u(x, y)$ in the rectangle $R = \{(x, y): 0 \le x \le 1.5, 0 \le y \le 1.5\}$, use $h = 0.5$. The boundary values are

$$u(x, 0) = x^4 \quad \text{and} \quad u(x, 1.5) = x^4 - 13.5x^2 + 5.0625 \qquad \text{for } 0 \le x \le 1.5$$

$$u(0, y) = y^4 \quad \text{and} \quad u(1.5, y) = 5.0625 - 13.5y^2 + y^4 \qquad \text{for } 0 \le y \le 1.5.$$

(b) Compare with the exact solution $u(x, y) = x^4 - 6x^2y^2 + y^4$.

4. **(a)** Show that $u(x, y) = a_1 \sin(x) \sinh(y) + b_1 \sinh(x) \sin(y)$ is a solution of Laplace's equation.

(b) Show that $u(x, y) = a_n \sin(nx) \sinh(ny) + b_n \sinh(nx) \sin(ny)$ is a solution of Laplace's equation for each positive integer $n = 1, 2, \ldots$.

5. **(a)** Use a 5 × 5 grid similar to that in Example 10.5 and determine the system of nine equations in the nine unknowns $p_1, p_2, p_3, \ldots, p_9$ for computing approximations for the

harmonic function $u(x, y)$ in the rectangle $R = \{(x, y): 0 \le x \le 4, 0 \le y \le 4\}$. The boundary values are

$$u(x, 0) = 10 \quad \text{and} \quad u(x, 4) = 120 \qquad \text{for } 0 < x < 4$$

$$u(0, y) = 90 \quad \text{and} \quad u(4, y) = 40 \qquad \text{for } 0 < y < 4.$$

(b) Use a computer and a direct method to solve for $p_1, p_2, p_3, \ldots, p_9$.
(c) Use a computer and an iterative method to solve for the approximations.
(d) Use a 9×9 grid similar to that in Example 10.7 and an iterative method to solve for the approximations.

6. (a) Use a 5×5 grid similar to Example 10.6 and determine the system of 12 equations in the 12 unknowns $q_1, q_2, q_3, \ldots, q_{12}$ for computing approximations for the harmonic function $u(x, y)$ in the rectangle $R = \{(x, y): 0 \le x \le 4, 0 \le y \le 4\}$. The boundary values are

$$u(x, 4) = 120 \qquad \text{for } 0 < x < 4,$$

$$u_y(x, 0) = 0 \qquad \text{for } 0 < x < 4,$$

$$u(0, y) = 90 \qquad \text{for } 0 \le y < 4,$$

$$u(4, y) = 40 \qquad \text{for } 0 \le y < 4.$$

(b) Use a computer and a direct method to solve for $q_1, q_2, q_3, \ldots, q_{12}$.
(c) Use a computer and an iterative method to solve for the approximations.
(d) Use a 9×9 grid similar to that in Example 10.8 and an iterative method to solve for the approximations.

7. Let $u(x,y) = x^2 - y^2$. Determine the quantities $u(x + h, y)$, $u(x - h, y)$, $u(x, y + h)$, and $u(x, y - h)$ and substitute them into equation (7) and simplify.

8. Suppose that u has the form $u(x, y) = ax^2 + bxy + cy^2 + dx + ey + f$. Find a relationship among the coefficients which guarantees that $u_{xx} + u_{yy} = 0$.

9. (a) Using a 5×5 grid, derive the nine equations involving $p_1, p_2, \ldots, p_9$ for computing approximations for the solution $u(x, y)$ to Poisson's equation with $g(x, y) = 2$ in the rectangle $R = \{(x, y): 0 \le x \le 1, 0 \le y \le 1\}$. The boundary values are:

$$u(x, 0) = x^2 \quad \text{and} \quad u(x, 1) = (x - 1)^2 \qquad \text{for } 0 \le x \le 1$$

$$u(0, y) = y^2 \quad \text{and} \quad u(1, y) = (y - 1)^2 \qquad \text{for } 0 \le y \le 1.$$

(b) Use a computer and a direct method to solve for $p_1, p_2, p_3, \ldots, p_9$.
(c) Use a computer and an iterative method to solve for the approximations.
(d) Use a 9×9 grid and an iterative method to solve for the approximations.

10. (a) Using a 5×5 grid, derive the nine equations involving $p_1, p_2, \ldots, p_9$ for computing approximations for the solution $u(x, y)$ to Poisson's equation with $g(x, y) = y$ in the rectangle $R = \{(x, y): 0 \le x \le 1, 0 \le y \le 1\}$. The boundary values are

$$u(x, 0) = x^3 \quad \text{and} \quad u(x, 1) = x^3 \qquad \text{for } 0 \le x \le 1$$

$$u(0, y) = 0 \quad \text{and} \quad u(1, y) = 1 \qquad \text{for } 0 \le y \le 1.$$

(b) Use a computer and a direct method to solve for $p_1, p_2, p_3, \ldots, p_9$.

(c) Use a computer and an iterative method to solve for the approximations.

(d) Use a 9×9 grid and an iterative method to solve for the approximations.

11. Solve $u_{xx} + u_{yy} = -4u$ over $R = \{(x, y): 0 \le x \le 1, 0 \le y \le 1\}$ with the boundary values

$$u(x, y) = \cos(2x) + \sin(2y).$$

12. Determine the system of four equations in the four unknowns p_1, p_2, p_3, and p_4 for implementing the Laplace nine-point difference equation on the 4×4 grid shown in Figure 10.24.

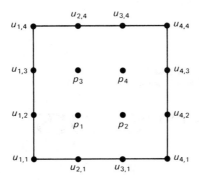

Figure 10.24   The grid for Exercise 12.

11

Eigenvalues and Eigenvectors

The design of certain engineering systems involves the **maximum stress theory of failure**. This theory is based on the assumption that the maximum principal stress acting on a body determines its failure. The related mathematical result is the principal axes theorem for a linear transformation $\mathbf{Y} = A\mathbf{X}$. In two dimensions there exists basis vectors $\mathbf{U}_1$ and $\mathbf{U}_2$ so that the effect of this transformation is to stretch space in the directions parallel to $\mathbf{U}_1$ and $\mathbf{U}_2$ by the amount λ_1 and λ_2, respectively. Consider the symmetric matrix

$$A = \begin{pmatrix} 3.8 & 0.6 \\ 0.6 & 2.2 \end{pmatrix};$$

the principal directions are $\mathbf{U}_1 = (3, 1)^T$ and $\mathbf{U}_2 = (-1, 3)^T$ with corresponding eigenvalues $\lambda_1 = 4$ and $\lambda_2 = 2$, respectively. Images of these vectors are $\mathbf{V}_1 = A\mathbf{U}_1 = (12, 4)^T = 4(3, 1)^T$ and $\mathbf{V}_2 = A\mathbf{U}_2 = (-2, 6)^T = 2(-1, 3)^T$. This transformation stretches the quarter circle shown in Figure 11.1(a) into the quarter ellipse shown in Figure 11.1(b).

11.1 HOMOGENEOUS SYSTEMS: THE EIGENVALUE PROBLEM

Background

We will now review some ideas from linear algebra. Proofs of the theorems are either left as exercises or can be found in any standard text on linear algebra (see Reference [132]).

536

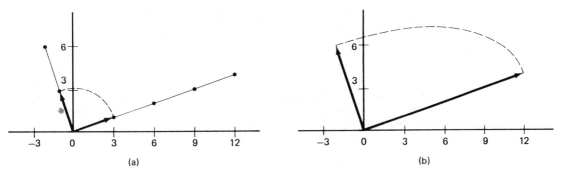

Figure 11.1 (a) Preimages $\mathbf{U}_1 = (3, 1)^T$ and $\mathbf{U}_2 = (-1, 3)^T$ for the transformation $\mathbf{Y} = A\mathbf{X}$. (b) The image vectors $\mathbf{V}_1 = A\mathbf{U}_1 = (12, 4)^T$ and $\mathbf{V}_2 = A\mathbf{U}_2 = (-2, 6)^T$.

In Chapter 3 we saw how to solve n linear equations in n unknowns. It was assumed that the determinant of the matrix was nonzero, and hence that the solution was unique. In the case of a homogeneous system $A\mathbf{X} = \mathbf{0}$, if det $(A) \neq 0$, the unique solution is the trivial solution $\mathbf{X} = \mathbf{0}$. If det $(A) = 0$, there exist nontrivial solutions to $A\mathbf{X} = \mathbf{0}$.

Suppose that det $(A) = 0$, and consider solutions to the homogeneous linear system

$$
\begin{aligned}
a_{11}x_1 + a_{12}x_2 + \cdots + a_{1n}x_n &= 0 \\
a_{21}x_1 + a_{22}x_2 + \cdots + a_{2n}x_n &= 0 \\
\vdots \qquad \vdots \qquad\quad \vdots \qquad \vdots \\
a_{n1}x_1 + a_{n2}x_2 + \cdots + a_{nn}x_n &= 0.
\end{aligned}
\tag{1}
$$

The system of equations (1) always has the trivial solution $x_1 = 0$, $x_2 = 0$, . . . , $x_n = 0$. Gaussian elimination can be used to obtain a solution by forming a set of relationships between the variables.

Example 11.1

Find the nontrivial solution to the homogeneous linear system

$$
\begin{aligned}
x_1 + 2x_2 - x_3 &= 0 \\
2x_1 + x_2 + x_3 &= 0 \\
5x_1 + 4x_2 + x_3 &= 0.
\end{aligned}
$$

Solution.   Use Gaussian elimination to eliminate x_1 and the result is

$$
\begin{aligned}
x_1 + 2x_2 - x_3 &= 0 \\
-3x_2 + 3x_3 &= 0 \\
-6x_2 + 6x_3 &= 0.
\end{aligned}
$$

Since the third equation is a multiple of the second equation, this system reduces to two equations in three unknowns:

$$x_1 + x_2 = 0$$

$$-x_2 + x_3 = 0.$$

We can select one unknown and use it as a parameter. For instance, let $x_3 = t$; then the second equation implies that $x_2 = t$ and the first equation is used to compute $x_1 = -t$. Therefore, the solution can be expressed as the set of relations

$$\begin{matrix} x_1 = & -t \\ x_2 = & t \\ x_3 = & t \end{matrix} \quad \text{or} \quad X = \begin{pmatrix} -t \\ t \\ t \end{pmatrix} = t \begin{pmatrix} -1 \\ 1 \\ 1 \end{pmatrix},$$

where t is any real number.

Definition 11.1 (Linear Independence). The vector $U_1, U_2, \ldots, U_n$ are said to be **linearly independent** if the equation

$$c_1 U_1 + c_2 U_2 + \cdots + c_n U_n = 0 \tag{2}$$

implies that $c_1 = 0, c_2 = 0, \ldots, c_n = 0$. If the vectors are not linearly independent, they are said to be **linearly dependent**. In other words, the vectors are linearly dependent if there exists a set of numbers $\{c_1, c_2, \ldots, c_n\}$ not all zero, such that (2) holds.

Two vectors in $\mathfrak{R}^2$ are linearly independent if and only if they are not parallel. Three vectors in $\mathfrak{R}^3$ are linearly independent if and only if they do not lie in the same plane.

Theorem 11.1. The vectors $U_1, U_2, \ldots, U_n$ are linearly dependent if and only if at least one of them is a linear combination of the others.

A desirable feature for a vector space is the ability to express each vector as a linear combination of vectors chosen from a small subset of vectors. This motivates the next definition.

Definition 11.2 (Basis). Suppose that $S = \{U_1, U_2, \ldots, U_m\}$ is a set of m vectors in $\mathfrak{R}^n$. The set S is called a basis for $\mathfrak{R}^n$ if for every vector X in $\mathfrak{R}^n$, there exists a unique set of scalars $\{c_1, c_2 \ldots, c_m\}$ so that X can be expressed as the linear combination

$$X = c_1 U_1 + c_2 U_2 + \cdots + c_m U_m. \tag{3}$$

Theorem 11.2. In $\mathfrak{R}^n$, any set of n linearly independent vectors forms a basis of $\mathfrak{R}^n$. Each vector X in $\mathfrak{R}^n$ is uniquely expressed as a linear combination of the basis vectors as shown in equation (3).

Theorem 11.3. Let $K_1, K_2, \ldots, K_m$ be vectors in $\mathfrak{R}^n$.

If $m > n$, then the vectors are linearly dependent,   (4)

If $m = n$, the vectors are linearly dependent if and only if det $(K) = 0$, where $K = [\mathbf{K}_1, \mathbf{K}_2, \ldots, \mathbf{K}_n]$.   (5)

Eigenvalues

Applications of mathematics sometimes encounter the following equations: What are the singularities of $A - \lambda I$, where λ is a parameter? What is the behavior of the sequence of vectors $\{A^j \mathbf{X}_0\}$? What are the geometric features of a linear transformation? Solutions for problems in many different disciplines, such as economics, engineering, and physics, can involve ideas related to these equations. The theory of eigenvalues and eigenvectors is powerful enough to help solve these otherwise intractable problems.

Let A be a square matrix of dimension $n \times n$ and let $\mathbf{X}$ be a vector of dimension n. The product $\mathbf{Y} = A\mathbf{X}$ can be viewed as a linear transformation from n-dimensional space into itself. We want to find scalars λ for which there exists a nonzero vector $\mathbf{X}$ such that

$$A\mathbf{X} = \lambda\mathbf{X};   (6)$$

that is, the linear transformation $T(\mathbf{X}) = A\mathbf{X}$ maps $\mathbf{X}$ onto the multiple $\lambda\mathbf{X}$. When this occurs, we call $\mathbf{X}$ an eigenvector that corresponds to the eigenvalue λ and together they form the eigenpair λ, $\mathbf{X}$ for A. In general, the scalar λ and vector $\mathbf{X}$ can involve complex numbers. For simplicity, most of our illustrations will involve real calculations. However, the techniques are easily extended to the complex case. The identity matrix I can be used to express equation (6) as $A\mathbf{X} = \lambda I\mathbf{X}$, which is then rewritten in the standard form for a linear system

$$(A - \lambda I)\mathbf{X} = 0.   (7)$$

The significance of equation (7) is that the product of the matrix $(A - \lambda I)$ and the nonzero vector $\mathbf{X}$ is the zero vector! According to Theorem 3.5, this linear system has nontrivial solutions if and only if the matrix $A - \lambda I$ is singular, that is,

$$\det (A - \lambda I) = 0.   (8)$$

This determinant can be written in the form

$$\begin{vmatrix} a_{11} - \lambda & a_{12} & \cdots & a_{1n} \\ a_{21} & a_{22} - \lambda & \cdots & a_{2n} \\ \vdots & \vdots & \cdots & \vdots \\ a_{n1} & a_{n2} & \cdots & a_{nn} - \lambda \end{vmatrix} = 0.   (9)$$

When the determinant in (9) is expanded, it becomes a polynomial of degree n, which is called the characteristic polynomial

$$\begin{aligned} p(\lambda) &= \det(A - \lambda I) \\ &= (-1)^n(\lambda^n + c_1\lambda^{n-1} + c_2\lambda^{n-2} + \cdots + c_{n-1}\lambda + c_n). \end{aligned}   (10)$$

There exist exactly n roots (not necessarily distinct) of a polynomial of degree n. Each root λ can be substituted into equation (7) to obtain an underdetermined system of equations that has a corresponding nontrivial solution vector $\mathbf{X}$. If λ is real, a real eigenvector $\mathbf{X}$ can be constructed. For emphasis, we state the following definitions.

Definition 11.3   (Eigenvalue).   If A is an $n \times n$ real matrix, then its n eigenvalues $\lambda_1, \lambda_2, \ldots, \lambda_n$ are the real and complex roots of the characteristic polynomial

$$p(\lambda) = \det(A - \lambda I). \tag{11}$$

Definition 11.4   (Eigenvector).   If λ is an eigenvalue of A and the nonzero vector $\mathbf{V}$ has the property that

$$A\mathbf{V} = \lambda\mathbf{V}, \tag{12}$$

then $\mathbf{V}$ is called an eigenvector of A corresponding to the eigenvalue λ.

The characteristic polynomial (11) can be factored in the form

$$p(\lambda) = (-1)^n(\lambda - \lambda_1)^{m_1}(\lambda - \lambda_2)^{m_2} \cdots (\lambda - \lambda_k)^{m_k}, \tag{13}$$

where m_j is called the multiplicity of the eigenvalue λ_j. The sum of the multiplicities of all eigenvalues is n; that is

$$n = m_1 + m_2 + \cdots + m_k.$$

The next three results concern the existence of eigenvectors.

Theorem 11.4.   (a) For each distinct eigenvalue λ there exists at least one eigenvector $\mathbf{V}$ corresponding to λ.
     (b) If λ has multiplicity r, then there exist at most r linearly independent eigenvectors $\mathbf{V}_1, \mathbf{V}_2, \ldots, \mathbf{V}_r$ which correspond to λ.

Theorem 11.5.   Suppose that A is a square matrix and $\lambda_1, \lambda_2, \ldots, \lambda_k$ are distinct eigenvalues of A with associated eigenvectors $\mathbf{V}_1, \mathbf{V}_2, \ldots, \mathbf{V}_k$, respectively; then $\{\mathbf{V}_1, \mathbf{V}_2, \ldots, \mathbf{V}_k\}$ is a set of linearly independent vectors.

Theorem 11.6.   If the eigenvalues of the $n \times n$ matrix A are all distinct, then there exists n eigenvectors $\mathbf{V}_j$ for $j = 1, 2, \ldots, n$.

Theorem 11.4 is usually applied for hand computations in the following manner. The eigenvalue λ of multiplicity $r \geq 1$ is substituted into the equation

$$(A - \lambda I)\mathbf{V} = 0. \tag{14}$$

Then Gaussian elimination can be performed to obtain the Gauss reduced form which will involve $n - k$ equations in n unknowns, where $1 \leq k \leq r$. Hence there are k free variables to choose. The free variables can be selected in a judicious manner to produce k linearly independent solution vectors $\mathbf{V}_1, \mathbf{V}_2, \ldots, \mathbf{V}_k$ which correspond to λ.

Example 11.2

Find the eigenpairs λ_j, $\mathbf{V}_j$ for the matrix

$$A = \begin{pmatrix} 3 & -1 & 0 \\ -1 & 2 & -1 \\ 0 & -1 & 3 \end{pmatrix}.$$

Also, show that the eigenvectors are linearly independent.

Solution.   The characteristic equation $\det(A - \lambda I) = 0$ is

$$\begin{vmatrix} 3 - \lambda & -1 & 0 \\ -1 & 2 - \lambda & -1 \\ 0 & -1 & 3 - \lambda \end{vmatrix} = -\lambda^3 + 8\lambda^2 - 19\lambda + 12 = 0, \tag{15}$$

which can be written $-(\lambda - 1)(\lambda - 3)(\lambda - 4) = 0$. Therefore, the three eigenvalues are $\lambda_1 = 1$, $\lambda_2 = 3$, and $\lambda_3 = 4$.

Case (i): Substitute $\lambda_1 = 1$ to equation (15) and obtain

$$2x_1 - x_2 = 0$$
$$-x_1 + x_2 - x_3 = 0$$
$$-x_2 + 2x_3 = 0.$$

Since the sum of the first equation plus two times the second equation plus the third equation is identically zero, the system can be reduced to two equations in three unknowns:

$$2x_1 - x_2 = 0$$
$$-x_2 + 2x_3 = 0.$$

Choose $x_2 = 2a$, where a is an arbitrary constant; then the first and second equations are used to compute $x_1 = a$ and $x_3 = a$, respectively. Thus the first eigenpair is $\lambda_1 = 1$, $\mathbf{V}_1 = (a, 2a, a)^T = a(1, 2, 1)^T$.

Case (ii): Substitute $\lambda_2 = 3$ into equation (15) and obtain

$$-x_2 = 0$$
$$-x_1 - x_2 - x_3 = 0$$
$$-x_2 = 0.$$

This is equivalent to the system of two equations

$$x_1 + x_3 = 0$$
$$x_2 = 0.$$

Choose $x_1 = b$, where b is an arbitrary constant, and compute $x_3 = -b$. Hence the second eigenpair is $\lambda_2 = 3$; $\mathbf{V}_2 = (b, 0, -b)^T = b(1, 0, -1)^T$.

Case (iii): Substitute $\lambda_3 = 4$ into (15); the result is

$$-x_1 - x_2 = 0$$

$$-x_1 - 2x_2 - x_3 = 0$$

$$-x_2 - x_3 = 0.$$

This is equivalent to the two equations

$$x_1 + x_2 = 0$$

$$x_2 + x_3 = 0.$$

Choose $x_3 = c$, where c is a constant, then use the second equation to compute $x_2 = -c$. Then use the first equation to get $x_1 = c$. Thus the third eigenpair is $\lambda_3 = 4$, $\mathbf{V}_3 = (c, -c, c)^T = c(1, -1, 1)^T$.

To prove that the vectors are linearly independent, it suffices to apply Theorem 11.5. However, it is beneficial to review techniques from linear algebra and use Theorem 11.3. Form the determinant

$$\det[\mathbf{V}_1, \mathbf{V}_2, \mathbf{V}_3] = \begin{vmatrix} a & b & c \\ 2a & 0 & -c \\ a & -b & c \end{vmatrix} = -6abc.$$

Since $\det[\mathbf{V}_1, \mathbf{V}_2, \mathbf{V}_3] \neq 0$, Theorem 11.3 implies that the vectors $\mathbf{V}_1$, $\mathbf{V}_2$, and $\mathbf{V}_3$ are linearly independent.

Example 11.2 shows how hand computations are used to find eigenvalues when the dimension n is small. First, find the coefficients of the characteristic polynomial; second, find its roots; and third, find the nonzero solutions of the homogeneous linear system $(A - \lambda I)\mathbf{V} = 0$. We will take the prevalent approach of studying the power and Jacobi methods and the QR and QL algorithms. The QR and QL algorithms and their improvements are used in professional software packages such as EISPACK and Matlab (see Reference [178]).

Since $\mathbf{V}$ in (12) is multiplied on the right side of the matrix A, it is called a **right eigenvector** corresponding to λ. There also exists a left eigenvector $\mathbf{Y}$ such that

$$\mathbf{Y}^T A = \lambda \mathbf{Y}^T. \tag{16}$$

In general, the left eigenvector $\mathbf{Y}$ is not equal to the right eigenvector $\mathbf{V}$. However, if A is real and symmetric ($A^T = A$), then

$$(A\mathbf{V})^T = \mathbf{V}^T A^T = \mathbf{V}^T A,$$

$$(\lambda\mathbf{V})^T = \lambda \mathbf{V}^T. \tag{17}$$

Therefore, the right eigenvector $\mathbf{V}$ is a left eigenvector when A is symmetric. In the remainder of the book we consider only right eigenvectors.

An eigenvector $\mathbf{V}$ is unique only up to a constant multiple. Suppose that c is a scalar; then the following calculation shows that $c\mathbf{V}$ is an eigenvector:

$$A(c\mathbf{V}) = c(A\mathbf{V}) = c\lambda\mathbf{V} = \lambda(c\mathbf{V}). \tag{18}$$

To regain the some semblance of uniqueness, we normalize the eigenvector in one of the following ways. Use one of the vector norms

$$\| \mathbf{X} \|_{\infty} = \max_{1 \leq k \leq n} \{|x_k|\} \tag{19}$$

or

$$\| \mathbf{X} \|_{2} = \left(\sum_{k=1}^{n} |x_k|^2 \right)^{1/2} \tag{20}$$

and require that either $\| X \|_{\infty} = 1$ or $\| X \|_{2} = 1$.

Diagonalizability

The eigenvalue situation is easiest to understand for a diagonal matrix D which has the form

$$D = \text{diag}(\lambda_1, \lambda_2, \ldots, \lambda_n) = \begin{pmatrix} \lambda_1 & 0 & \cdots & 0 \\ 0 & \lambda_2 & \cdots & 0 \\ \cdot & \cdot & \cdots & \cdot \\ \cdot & \cdot & & \cdot \\ \cdot & \cdot & & \cdot \\ 0 & 0 & \cdots & \lambda_n \end{pmatrix}. \tag{21}$$

Let $\mathbf{E}_j = (0, \ldots, 1_j, \ldots, 0)^T$ be the standard base vector where the jth component is 1 and all other components are 0. Then

$$D\mathbf{E}_j = (0, \ldots, \lambda_j, \ldots, 0)^T = \lambda_j \mathbf{E}_j, \tag{22}$$

which implies that the eigenpairs of D are λ_j, $\mathbf{E}_j$ for $j = 1, 2, \ldots, n$. It is desirable to invent a simple way of transforming the matrix A into diagonal form so that the eigenvalues are left invariant. This is the motivation for the following definition.

Definition 11.5.   Two $n \times n$ matrices A and B are said to be similar if there exists a nonsingular matrix K so that

$$B = K^{-1}AK. \tag{23}$$

Theorem 11.7.   Suppose that A and B are similar matrices and that λ is an eigenvalue of A with corresponding eigenvector $\mathbf{V}$. Then λ is also an eigenvalue of B. If $K^{-1}AK = B$, then $\mathbf{Y} = K^{-1}\mathbf{V}$ is an eigenvector of $\mathbf{B}$ associated with the eigenvalue λ.

An $n \times n$ matrix A is called **diagonalizable** if it is similar to a diagonal matrix. The next theorem illuminates the intimate role of eigenvectors in this process.

Theorem 11.8 (Diagonalization).   The matrix A is similar to a diagonal matrix D if and only if it has n linearly independent eigenvectors. If A is similar to D, then

$$V^{-1}AV = D = \text{diag}(\lambda_1, \lambda_2, \ldots, \lambda_n)$$
$$V = [\mathbf{V}_1, \mathbf{V}_2, \ldots, \mathbf{V}_n], \tag{24}$$

where the n eigenpairs are λ_j, $\mathbf{V}_j$ for $j = 1, 2, \ldots, n$.

Theorem 11.8 implies that every matrix A which has n distinct eigenvalues is diagonalizable.

Example 11.3

Show that the following matrix is diagonalizable.

$$A = \begin{pmatrix} 3 & -1 & 0 \\ -1 & 2 & -1 \\ 0 & -1 & 3 \end{pmatrix}.$$

Solution.   In Example 11.2 we found the eigenvalues $\lambda_1 = 1$, $\lambda_2 = 3$, and $\lambda_3 = 4$, and the matrix of eigenvectors

$$V = [\mathbf{V}_1, \mathbf{V}_2, \mathbf{V}_3] = \begin{pmatrix} 1 & 1 & 1 \\ 2 & 0 & -1 \\ 1 & -1 & 1 \end{pmatrix}.$$

The inverse matrix V^{-1} is

$$V^{-1} = \begin{pmatrix} \frac{1}{6} & \frac{1}{3} & \frac{1}{6} \\ \frac{1}{2} & 0 & -\frac{1}{2} \\ \frac{1}{3} & -\frac{1}{3} & \frac{1}{3} \end{pmatrix}.$$

It is left for the reader to check the details in computing the product in (22):

$$\begin{pmatrix} \frac{1}{6} & \frac{1}{3} & \frac{1}{6} \\ \frac{1}{2} & 0 & -\frac{1}{2} \\ \frac{1}{3} & -\frac{1}{3} & \frac{1}{3} \end{pmatrix} \begin{pmatrix} 3 & -1 & 0 \\ -1 & 2 & -1 \\ 0 & -1 & 3 \end{pmatrix} \begin{pmatrix} 1 & 1 & 1 \\ 2 & 0 & -1 \\ 1 & -1 & 1 \end{pmatrix} = \begin{pmatrix} 1 & 0 & 0 \\ 0 & 3 & 0 \\ 0 & 0 & 4 \end{pmatrix}.$$

Hence we have verified that A can be diagonalized; that is, $V^{-1}AV = D = \text{diag}(1, 3, 4)$.

A more general result relating the structure of a matrix to its eigenvalues is the following theorem.

Theorem 11.9   (Schur).   Suppose that A is an arbitrary $n \times n$ matrix. A non-singular matrix P exists with the property that $T = P^{-1}AP$, where T is an upper-triangular matrix whose diagonal entries consist of the eigenvalues of A.

Certain types of structural analysis in engineering require that a basis of $\mathfrak{R}^n$ be selected which consists of the eigenvectors of A. This choice makes it easier to visualize how space is transformed by the mapping $\mathbf{Y} = T(\mathbf{X}) = A\mathbf{X}$. Recall that the eigenpair λ_j, $\mathbf{V}_j$ has the property that T maps $\mathbf{V}_j$ onto the multiple of $\lambda\mathbf{V}_j$. This characteristic is exploited in the following theorem.

Theorem 11.10.   Suppose that A is an $n \times n$ matrix that possesses n linearly independent eigenpairs λ_j, $\mathbf{V}_j$ for $j = 1, 2, \ldots, n$; then any vector $\mathbf{X}$ in $\mathfrak{R}^n$ has a unique representation as a linear combination of the eigenvectors

$$\mathbf{X} = c_1\mathbf{V}_1 + c_2\mathbf{V}_2 + \cdots + c_n\mathbf{V}_n. \tag{25}$$

The linear transformation $T(\mathbf{X}) = A\mathbf{X}$ maps $\mathbf{X}$ onto the vector

$$\mathbf{Y} = T(\mathbf{X}) = c_1\lambda_1\mathbf{V}_1 + c_2\lambda_2\mathbf{V}_2 + \cdots + c_n\lambda_n\mathbf{V}_n. \tag{26}$$

Example 11.4

Suppose that the 3×3 matrix A has eigenvalues $\lambda_1 = 2$, $\lambda_2 = -1$, and $\lambda_3 = 4$ which correspond to the eigenvectors $\mathbf{V}_1 = (1, 2, -2)^T$, $\mathbf{V}_2 = (-2, 1, 1)^T$, and $\mathbf{V}_3 = (1, 3, -4)^T$, respectively. If $\mathbf{X} = (-1, 2, 1)^T$, find the image of $\mathbf{X}$ under the mapping $T(\mathbf{X}) = A\mathbf{X}$.

Solution.   We must first express $\mathbf{X}$ as a linear combination of the eigenvectors. This is accomplished by solving the equation

$$(-1, 2, 1)^T = c_1(1, 2, -2)^T + c_2(-2, 1, 1)^T + c_3(1, 3, -4)^T,$$

for c_1, c_2, and c_3. Observe that this is equivalent to solving the linear system

$$c_1 - 2c_2 + c_3 = -1$$
$$2c_1 + c_2 + 3c_3 = 2$$
$$-2c_1 + c_2 - 4c_3 = 1.$$

The solution is $c_1 = 2$, $c_2 = 1$, and $c_3 = -1$. Using Definition 11.4, for eigenvectors, $T(\mathbf{X})$ is found by the computation

$$T(\mathbf{X}) = A(2\mathbf{V}_1 + \mathbf{V}_2 - \mathbf{V}_3)$$
$$= 2A\mathbf{V}_1 + A\mathbf{V}_2 - A\mathbf{V}_3$$
$$= 2(2\mathbf{V}_1) - \mathbf{V}_2 - 4\mathbf{V}_3$$
$$= (2, -5, 7)^T.$$

Virtues of Symmetry

There is no easy way to determine how many linearly independent eigenvectors a matrix possesses without resorting to using the most effective algorithms in a professional software package such as EISPACK or MATLAB. However, it is known that a real symmetric matrix has n real eigenvectors and that for each eigenvalue of multiplicity m_j there corresponds m_j linearly independent eigenvectors. Hence every real symmetric matrix is diagonalizable.

Definition 11.6   (Orthogonal).   A set of vectors $\{\mathbf{V}_1, \mathbf{V}_2, \ldots, \mathbf{V}_n\}$ is said to be orthogonal provided that

$$\mathbf{V}_j^T \mathbf{V}_k = 0 \qquad \text{whenever} \quad j \neq k. \tag{27}$$

Definition 11.7   (Orthonormal).   Suppose that $\{\mathbf{V}_1, \mathbf{V}_2, \ldots, \mathbf{V}_n\}$ is a set of orthogonal vectors; then we say that they are orthonormal if they are all of unit norm, that is,

$$\begin{aligned} \mathbf{V}_j^T \mathbf{V}_k &= 0 \qquad \text{whenever } j \neq k. \\ \mathbf{V}_j^T \mathbf{V}_j &= 1 \qquad \text{for all } j = 1, 2, \ldots, n. \end{aligned} \tag{28}$$

Theorem 11.11.   An orthonormal set of vectors is linearly independent.

Remark.   The zero vector cannot belong to an orthonormal set of vectors.

Definition 11.8   (Orthogonal Matrix).   An $n \times n$ matrix A is said to be orthogonal provided that A^T is the inverse of A, that is,

$$A^T A = I, \tag{29}$$

which is equivalent to

$$A^{-1} = A^T. \tag{30}$$

Also, A is orthogonal if and only if the columns (and rows) of A form a set of orthonormal vectors.

Theorem 11.12.   If A is a real symmetric matrix, there exists an orthogonal matrix K such that

$$K^T A K = K^{-1} A K = D, \tag{31}$$

where D is a diagonal matrix consisting of the eigenvalues of A.

Corollary 11.1.   If A is an $n \times n$ real symmetric matrix, there exists n linearly independent eigenvectors for A, and they form an orthonormal set.

Corollary 11.2.   The eigenvalues of a real symmetric matrix are all real numbers.

Theorem 11.13.   Eigenvectors corresponding to distinct eigenvalues of a symmetric matrix are orthogonal.

Theorem 11.14.   A symmetric matrix A is positive definite if and only if all the eigenvalues of A are positive.

Estimates for the Size of Eigenvalues

It is useful to find a bound for the magnitude of the eigenvalues of A. The following results will give some insights.

Definition 11.9   (Matrix Norm).   Let $\| \mathbf{X} \|$ be a vector norm. Then a corresponding natural matrix norm is

$$\| A \| = \max_{\| \mathbf{X} \| = 1} \left\{ \frac{\| A\mathbf{X} \|}{\| \mathbf{X} \|} \right\}. \tag{32}$$

For the norm $\| A \|_\infty$ the following formula holds:

$$\| A \|_\infty = \max_{1 \le i \le n} \left\{ \sum_{j=1}^n |a_{ij}| \right\}, \tag{33}$$

Theorem 11.15.   If λ is any eigenvalue of A, then

$$|\lambda| \le \| A \|, \tag{34}$$

for any natural matrix norm $\| A \|$.

Theorem 11.16   (Gerschgorin's Circle Theorem).   Assume that A is an $n \times n$ matrix and let C_j denote the disk in the complex plane with center a_{jj} and radius

$$r_j = \sum_{\substack{k=1 \\ k \ne j}}^n |a_{jk}| \qquad \text{for each } j = 1, 2, \ldots, n; \tag{35}$$

that is, C_j consists of all complex numbers $z = x + iy$ such that

$$C_j = \{z\colon |z - a_{jj}| \le r_j\}. \tag{36}$$

If $S = \bigcup_{i=1}^n C_i$, then all of the eigenvalues of A lie in the set S. Moreover, the union of any k of these disks that do not intersect the remaining $n - k$ must contain precisely k (counting multiplicities) of the eigenvalues.

Theorem 11.17   (Spectral Radius Theorem).   Let A be a symmetric matrix. The spectral radius of A is $\| A \|_2$ and obeys the relationship

$$\| A \|_2 = \max\{|\lambda_1|, |\lambda_2|, \ldots, |\lambda_n|\}. \tag{37}$$

An Overview of Methods

For problems involving moderate-sized symmetric matrices it is safe to use Jacobi's method. For problems involving large symmetric matrices (for n up to several hundred) it is best to use Householder's method to produce a tridiagonal form, followed by the QL algorithm. Unlike real symmetric matrices, real unsymmetric matrices can have complex eigenvalues and eigenvectors.

For matrices that possess a dominant eigenvalue, the power method can be used to find the dominant eigenvector. Deflation techniques can be used thereafter to find the first few subdominant eigenvectors. For real unsymmetric matrices, Householder's method is used to produce a Hessenberg matrix, followed by the LR or QR algorithm.

EXERCISES FOR HOMOGENEOUS SYSTEMS: THE EIGENVALUE PROBLEM

1. Suppose that V is an eigenvector of A that corresponds to the eigenvalue $\lambda = 3$. Prove that $\lambda = 9$ is an eigenvalue of the matrix A^2 corresponding to V.

2. Suppose that V is an eigenvector of A that corresponds to the eigenvalue $\lambda = 2$. Prove that $\lambda = \frac{1}{2}$ is an eigenvalue of the matrix A^{-1} corresponding to V.

3. Suppose that V is an eigenvector of A that corresponds to the eigenvalue $\lambda = 5$. Prove that $\lambda = 4$ is an eigenvalue of the matrix $A - I$ corresponding to V.

4. For any fixed θ, show that

$$R = \begin{pmatrix} \cos \theta & \sin \theta \\ -\sin \theta & \cos \theta \end{pmatrix}$$

is an orthogonal matrix.

Remark. The matrix R is called a rotation matrix.

5. Assume that λ, V form an eigenpair of the matrix A. If k is a positive integer, prove that λ^k, V are an eigenpair of the matrix A^k.

6. Find the characteristic polynomial $p(\lambda)$ and eigenpairs λ_1, V_1 and λ_2, V_2 for the following matrices.

(a) $A = \begin{pmatrix} 1 & 2 \\ 3 & 2 \end{pmatrix}$ (b) $A = \begin{pmatrix} -1 & -3 \\ 4 & 6 \end{pmatrix}$ (c) $A = \begin{pmatrix} 1 & 3 \\ 3 & 1 \end{pmatrix}$ (d) $A = \begin{pmatrix} -2 & 3 \\ 3 & -2 \end{pmatrix}$

7. Let $A = \begin{pmatrix} a + 3 & 2 \\ 2 & a \end{pmatrix}$.

(a) Show that the characteristic polynomial is $p(\lambda) = \lambda^2 - (3 + 2a)\lambda + a^2 - 3a - 4$.
(b) Show that the eigenvalues of A are $\lambda_1 = a + 4$ and $\lambda_2 = a - 1$.
(c) Show that the eigenvectors of A are $V_1 = (2, 1)^T$ and $V_2 = (-1, 2)^T$.

8. What is the spectral radius of the matrix in Example 11.2?

9. Let A be an $n \times n$ square matrix with characteristic polynomial $p(\lambda)$ given by

$$p(\lambda) = \det(A - \lambda I) = (-1)^n(\lambda^n + c_1\lambda^{n-1} + c_2\lambda^{n-2} + \cdots + c_{n-1}\lambda + c_n).$$

(a) Show that the constant term of $p(\lambda)$ is $c_n = (-1)^n \det(A)$.

(b) Show that the coefficient of λ^{n-1} is $c_1 = -(a_{11} + a_{22} + \cdots + a_{nn})$.

Remark. This quantity is called the **trace** of the matrix A and it is the sum of the diagonal elements.

10. Find the characteristic polynomial $p(\lambda)$ and eigenpairs for:

(a) $A = \begin{pmatrix} -2 & 1 & 1 \\ -6 & 1 & 3 \\ -12 & -2 & 8 \end{pmatrix}$
(b) $A = \begin{pmatrix} -3 & -7 & -2 \\ 12 & 20 & 6 \\ -20 & -31 & -9 \end{pmatrix}$

11. Assume that A is similar to a diagonal matrix; that is,

$$\mathbf{V}^{-1}A\mathbf{V} = D = \mathrm{diag}(\lambda_1, \lambda_2, \ldots, \lambda_n).$$

If k is a positive integer, prove that $A^k = \mathbf{V}\mathrm{diag}(\lambda_1^k, \lambda_2^k, \ldots, \lambda_n^k)\mathbf{V}^{-1}$.

11.2 THE POWER METHOD

We now describe the power method for computing the dominant eigenpair. Its extension to the inverse power method is practical for finding any eigenvalue provided that a good initial approximation is known. Some schemes for finding eigenvalues use other methods which converge fast but have limited precision. The inverse power method is then invoked to refine the numerical values and gain full precision. To discuss the situation, we will need the following definitions.

Definition 11.10.   If λ_1 is an eigenvalue of A that is larger in absolute value than any other eigenvalue, it is called the **dominant eigenvalue**. An eigenvector $\mathbf{V}_1$ corresponding to λ_1 is called a **dominant eigenvector**.

Definition 11.11.   An eigenvector $\mathbf{V}$ is said to be normalized if the coordinate of largest magnitude is equal to unity (i.e., the largest coordinate in $\mathbf{V}$ is the number 1).

It is easy to normalize an eigenvector $(v_1, v_2, \ldots, v_n)^T$ by forming the new vector $\mathbf{V} = (1/c)(v_1, v_2, \ldots, v_n)^T$, where $c = v_j$ and $|v_j| = \max_{1 \le i \le n} \{|v_i|\}$.

Suppose that the matrix A has a dominant eigenvalue λ and that there is a unique normalized eigenvector $\mathbf{V}$ that corresponds to λ. This eigenpair λ, $\mathbf{V}$ can be found by the following iterative procedure called the **power method**. Start with the vector

$$\mathbf{X}_0 = (1, 1, \ldots, 1)^T. \tag{1}$$

Generate the sequence $\{\mathbf{X}_k\}$ recursively, using

$$\mathbf{Y}_k = A\mathbf{X}_k,$$

$$\mathbf{X}_{k+1} = \frac{1}{c_{k+1}} \mathbf{Y}_k, \tag{2}$$

where c_{k+1} is the coordinate of $\mathbf{Y}_k$ of largest magnitude. (In the case of a tie, choose the coordinate that comes first.) The sequences $\{\mathbf{X}_k\}$ and $\{c_k\}$ will converge to $\mathbf{V}$ and λ, respectively:

$$\lim_{k \to \infty} \mathbf{X}_k = \mathbf{V} \quad \text{and} \quad \lim_{k \to \infty} c_k = \lambda. \tag{3}$$

Remark. If $\mathbf{X}_0$ is an eigenvector and $\mathbf{X}_0 \neq \mathbf{V}$, then some other starting vector must be chosen.

Example 11.5

Use the power method to find the dominant eigenvalue and eigenvector for the matrix

$$A = \begin{pmatrix} 0 & 11 & -5 \\ -2 & 17 & -7 \\ -4 & 26 & -10 \end{pmatrix}.$$

Solution. Start with $\mathbf{X}_0 = (1, 1, 1)^T$ and use the formulas in (2) to generate the sequence of vectors $\{\mathbf{X}_k\}$ and constants $\{c_k\}$. The first iteration produces

$$\begin{pmatrix} 0 & 11 & -5 \\ -2 & 17 & -7 \\ -4 & 26 & -10 \end{pmatrix} \begin{pmatrix} 1 \\ 1 \\ 1 \end{pmatrix} = \begin{pmatrix} 6 \\ 8 \\ 12 \end{pmatrix} = 12 \begin{pmatrix} \frac{1}{2} \\ \frac{2}{3} \\ 1 \end{pmatrix} = c_1 \mathbf{X}_1.$$

The second iteration produces

$$\begin{pmatrix} 0 & 11 & -5 \\ -2 & 17 & -7 \\ -4 & 26 & -10 \end{pmatrix} \begin{pmatrix} \frac{1}{2} \\ \frac{2}{3} \\ 1 \end{pmatrix} = \begin{pmatrix} \frac{7}{3} \\ \frac{10}{3} \\ \frac{16}{3} \end{pmatrix} = \frac{16}{3} \begin{pmatrix} \frac{7}{16} \\ \frac{5}{8} \\ 1 \end{pmatrix} = c_2 \mathbf{X}_2.$$

Iteration generates the sequence $\{\mathbf{X}_k\}$ (where $\mathbf{X}_k$ is a normalized vector)

$$12 \begin{pmatrix} \frac{1}{2} \\ \frac{2}{3} \\ 1 \end{pmatrix}, \frac{16}{3} \begin{pmatrix} \frac{7}{16} \\ \frac{5}{8} \\ 1 \end{pmatrix}, \frac{9}{2} \begin{pmatrix} \frac{5}{12} \\ \frac{11}{18} \\ 1 \end{pmatrix}, \frac{38}{9} \begin{pmatrix} \frac{31}{76} \\ \frac{23}{38} \\ 1 \end{pmatrix}, \frac{78}{19} \begin{pmatrix} \frac{21}{52} \\ \frac{47}{78} \\ 1 \end{pmatrix}, \frac{158}{39} \begin{pmatrix} \frac{127}{316} \\ \frac{95}{158} \\ 1 \end{pmatrix}, \frac{318}{79} \begin{pmatrix} \frac{85}{212} \\ \frac{191}{318} \\ 1 \end{pmatrix}, \dots$$

The sequence of vectors converges to $\mathbf{V} = \left(\frac{2}{3}, \frac{3}{5}, 1\right)^T$, and the sequence of constants converges to $\lambda = 4$ (see Table 11.1). It can be proven that the rate of convergence is linear.

TABLE 11.1   The Power Method Used in Example 11.5 to Find the Normalized Dominant Eigenvector $\mathbf{V} = \left(\frac{2}{5}, \frac{3}{5}, 1\right)^T$ and Corresponding Eigenvalue $\lambda = 4$

$A\mathbf{X}_k =$	$\mathbf{Y}_k$	$=$	$c_{k+1}\mathbf{X}_{k+1}$
$A\mathbf{X}_0 = (6.000000, 8.000000, 12.00000)^T =$	12.00000	$(0.500000, 0.666667, 1)^T$	$= c_1\mathbf{X}_1$
$A\mathbf{X}_1 = (2.333333, 3.333333, 5.333333)^T =$	5.333333	$(0.437500, 0.625000, 1)^T$	$= c_2\mathbf{X}_2$
$A\mathbf{X}_2 = (1.875000, 2.750000, 4.500000)^T =$	4.500000	$(0.416667, 0.611111, 1)^T$	$= c_3\mathbf{X}_3$
$A\mathbf{X}_3 = (1.722222, 2.555556, 4.222222)^T =$	4.222222	$(0.407895, 0.605263, 1)^T$	$= c_4\mathbf{X}_4$
$A\mathbf{X}_4 = (1.657895, 2.473684, 4.105263)^T =$	4.105263	$(0.403846, 0.602564, 1)^T$	$= c_5\mathbf{X}_5$
$A\mathbf{X}_5 = (1.628205, 2.435897, 4.051282)^T =$	4.051282	$(0.401899, 0.601266, 1)^T$	$= c_6\mathbf{X}_6$
$A\mathbf{X}_6 = (1.613924, 2.417722, 4.025316)^T =$	4.025316	$(0.400943, 0.600629, 1)^T$	$= c_7\mathbf{X}_7$
$A\mathbf{X}_7 = (1.606918, 2.408805, 4.012579)^T =$	4.012579	$(0.400470, 0.600313, 1)^T$	$= c_8\mathbf{X}_8$
$A\mathbf{X}_8 = (1.603448, 2.404389, 4.006270)^T =$	4.006270	$(0.400235, 0.600156, 1)^T$	$= c_9\mathbf{X}_9$
$A\mathbf{X}_9 = (1.601721, 2.402191, 4.003130)^T =$	4.003130	$(0.400117, 0.600078, 1)^T$	$= c_{10}\mathbf{X}_{10}$
$A\mathbf{X}_{10} = (1.600860, 2.401095, 4.001564)^T =$	4.001564	$(0.400059, 0.600039, 1)^T$	$= c_{11}\mathbf{X}_{11}$

Theorem 11.18   (Power Method).   Assume that the $n \times n$ matrix A has n distinct eigenvalues $\lambda_1, \lambda_2, \ldots, \lambda_n$ and that they are ordered in decreasing magnitude, that is,

$$|\lambda_1| > |\lambda_2| \geq |\lambda_3| \geq \cdots \geq |\lambda_n|. \tag{4}$$

If $\mathbf{X}_0$ is chosen appropriately, then the sequence $\{\mathbf{X}_k = (x_1^{(k)}, x_2^{(k)}, \ldots, x_n^{(k)})^T\}$ and $\{c_k\}$ generated recursively by

$$\mathbf{Y}_k = A\mathbf{X}_k \tag{5}$$

and

$$\mathbf{X}_{k+1} = \frac{1}{c_{k+1}} \mathbf{Y}_k, \tag{6}$$

where

$$c_{k+1} = x_j^{(k)} \quad \text{and} \quad x_j^{(k)} = \max_{1 \leq i \leq n} \{|x_i^{(k)}|\} \tag{7}$$

will converge to the dominant eigenvector $\mathbf{V}_1$ and eigenvalue λ_1, respectively. That is,

$$\lim_{k \to \infty} \mathbf{X}_k = \mathbf{V}_1 \quad \text{and} \quad \lim_{k \to \infty} c_k = \lambda_1. \tag{8}$$

Proof. Since A has n eigenvalues, there are n corresponding eigenvectors $\{\mathbf{V}_j\}_{j=1}^n$ which are linearly independent, normalized, and form a basis for n-dimensional space. Hence the starting vector $\mathbf{X}_0$ can be expressed as the linear combination

$$\mathbf{X}_0 = b_1\mathbf{V}_1 + b_2\mathbf{V}_2 + \cdots + b_n\mathbf{V}_n. \tag{9}$$

Assume that $\mathbf{X}_0 = (x_1, x_2, \ldots, x_n)^T$ was chosen in such a manner that $b_1 \neq 0$. Also, assume that the coordinates of $\mathbf{X}_0$ are scaled so that $\max\limits_{1 \leq j \leq n} \{|x_j|\} = 1$. Because $\mathbf{V}_j$ are eigenvectors of A, the multiplication $A\mathbf{X}_0$, followed by normalization, produce

$$\begin{aligned}
\mathbf{Y}_0 = A\mathbf{X}_0 &= A(b_1\mathbf{V}_1 + b_2\mathbf{V}_2 + \cdots + b_n\mathbf{V}_n) \\
&= b_1 A\mathbf{V}_1 + b_2 A\mathbf{V}_2 + \cdots + b_n A\mathbf{V}_n \\
&= b_1\lambda_1\mathbf{V}_1 + b_2\lambda_2\mathbf{V}_2 + \cdots + b_n\lambda_n\mathbf{V}_n \\
&= \lambda_1\left(b_1\mathbf{V}_1 + b_2\left(\frac{\lambda_2}{\lambda_1}\right)\mathbf{V}_2 + \cdots + b_n\left(\frac{\lambda_n}{\lambda_1}\right)\mathbf{V}_n\right),
\end{aligned} \tag{10}$$

and

$$\mathbf{X}_1 = \frac{\lambda_1}{c_1}\left(b_1\mathbf{V}_1 + b_2\left(\frac{\lambda_2}{\lambda_1}\right)\mathbf{V}_2 + \cdots + b_n\left(\frac{\lambda_n}{\lambda_1}\right)\mathbf{V}_n\right).$$

After k iterations we arrive at

$$\begin{aligned}
\mathbf{Y}_{k-1} = A\mathbf{X}_{k-1} &= A\,\frac{\lambda_1^{k-1}}{c_1 c_2 \cdots c_{k-1}}\left(b_1\mathbf{V}_1 + b_2\left(\frac{\lambda_2}{\lambda_1}\right)^{k-1}\mathbf{V}_2 + \cdots + b_n\left(\frac{\lambda_n}{\lambda_1}\right)^{k-1}\mathbf{V}_n\right) \\
&= \frac{\lambda_1^{k-1}}{c_1 c_2 \cdots c_{k-1}}\left(b_1 A\mathbf{V}_1 + b_2\left(\frac{\lambda_2}{\lambda_1}\right)^{k-1}A\mathbf{V}_2 + \cdots + b_n\left(\frac{\lambda_n}{\lambda_1}\right)^{k-1}A\mathbf{V}_n\right) \\
&= \frac{\lambda_1^{k-1}}{c_1 c_2 \cdots c_{k-1}}\left(b_1\lambda_1\mathbf{V}_1 + b_2\left(\frac{\lambda_2}{\lambda_1}\right)^{k-1}\lambda_2\mathbf{V}_2 + \cdots + b_n\left(\frac{\lambda_n}{\lambda_1}\right)^{k-1}\lambda_n\mathbf{V}_n\right) \\
&= \frac{\lambda_1^{k}}{c_1 c_2 \cdots c_{k-1}}\left(b_1\mathbf{V}_1 + b_2\left(\frac{\lambda_2}{\lambda_1}\right)^{k}\mathbf{V}_2 + \cdots + b_n\left(\frac{\lambda_n}{\lambda_1}\right)^{k}\mathbf{V}_n\right),
\end{aligned} \tag{11}$$

and

$$\mathbf{X}_k = \frac{\lambda_1^{k}}{c_1 c_2 \cdots c_k}\left(b_1\mathbf{V}_1 + b_2\left(\frac{\lambda_2}{\lambda_1}\right)^{k}\mathbf{V}_2 + \cdots + b_n\left(\frac{\lambda_n}{\lambda_1}\right)^{k}\mathbf{V}_n\right).$$

Since we assumed that $|\lambda_j|/|\lambda_1| < 1$ for each $j = 2, 3, \ldots, n$, we have

$$\lim_{k \to \infty} b_j\left(\frac{\lambda_j}{\lambda_1}\right)^{k}\mathbf{V}_j = \mathbf{0} \qquad \text{for each } j = 2, 3, \ldots, n. \tag{12}$$

Hence it follows that

$$\lim_{k \to \infty} \mathbf{X}_k = \lim_{k \to \infty} \frac{b_1 \lambda_1^k}{c_1 c_2 \cdots c_k} \mathbf{V}_1. \tag{13}$$

We have required that both $\mathbf{X}_k$ and $\mathbf{V}_1$ are normalized and their largest component is 1. Hence the limiting vector on the left side of (13) will be normalized, with its largest component being 1. Consequently, the limit of the scalar multiple of $\mathbf{V}_1$ on the right side of (13) exists and its value must be 1; that is,

$$\lim_{k \to \infty} \frac{b_1 \lambda_1^k}{c_1 c_2 \cdots c_k} = 1. \tag{14}$$

Therefore, the sequence of vectors $\{\mathbf{X}_k\}$ converges to the dominant eigenvector:

$$\lim_{k \to \infty} \mathbf{X}_k = \mathbf{V}_1. \tag{15}$$

Replacing k with $k - 1$ in the terms of the sequence in (14) yields

$$\lim_{k \to \infty} b_1 \lambda_1^{k-1} / (c_1 c_2 \cdots c_{k-1}) = 1,$$

and dividing both sides of this result into (14) yields

$$\lim_{k \to \infty} \frac{\lambda_1}{c_k} = \lim_{k \to \infty} \frac{b_1 \lambda_1^k / (c_1 c_2 \cdots c_k)}{b_1 \lambda_1^{k-1} / (c_1 c_2 \cdots c_{k-1})} = \frac{1}{1} = 1.$$

Therefore, the sequence of constants $\{c_k\}$ converges to the dominant eigenvalue:

$$\lim_{k \to \infty} c_k = \lambda_1, \tag{16}$$

and the proof of the theorem is complete.

Speed of Convergence

In light of equation (12) we see that the coefficient of $\mathbf{V}_j$ in $\mathbf{X}_k$ goes to zero in proportion to $(\lambda_j / \lambda_1)^k$ and that the speed of convergence of $\{\mathbf{X}_k\}$ to $\mathbf{V}_1$ is governed by the terms $(\lambda_2 / \lambda_1)^k$. Consequently, the rate of convergence is linear. Similarly, the convergence of the sequence of constants $\{c_k\}$ to λ_1 is linear. The Aitken Δ^2 method can be used for any linearly convergent sequence $\{p_k\}$ to form a new sequence,

$$\left\{ \hat{p}_k = \frac{(p_{k+1} - p_k)^2}{p_{k+2} - 2p_{k+1} + p_k} \right\},$$

which converges faster. In Example 11.4, this Aitken Δ^2 method can be applied to speed up convergence of the sequence of constants $\{c_k\}$ as well as the first two components of the sequence of vectors $\{\mathbf{X}_k\}$. A comparison of the results obtained with this technique and the original sequences is shown in Table 11.2.

TABLE 11.2 Comparison of the Rate of Convergence of the Power Method and Acceleration of the Power Method Using Aitken's Δ^2 Technique

$c_k \mathbf{X}_k$	$\hat{c}_k \hat{\mathbf{X}}_k$
$c_1 \mathbf{X}_1$ = 12.000000 $(0.5000000, 0.6666667, 1)^T$; 4.3809524 $(0.4062500, 0.6041667, 1)^T = \hat{c}_1 \hat{\mathbf{X}}_1$	
$c_2 \mathbf{X}_2$ = 5.3333333 $(0.4375000, 0.6250000, 1)^T$; 4.0833333 $(0.4015152, 0.6010101, 1)^T = \hat{c}_2 \hat{\mathbf{X}}_2$	
$c_3 \mathbf{X}_3$ = 4.5000000 $(0.4166667, 0.6111111, 1)^T$; 4.0202020 $(0.4003759, 0.6002506, 1)^T = \hat{c}_3 \hat{\mathbf{X}}_3$	
$c_4 \mathbf{X}_4$ = 4.2222222 $(0.4078947, 0.6052632, 1)^T$; 4.0050125 $(0.4000938, 0.6000625, 1)^T = \hat{c}_4 \hat{\mathbf{X}}_4$	
$c_5 \mathbf{X}_5$ = 4.1052632 $(0.4038462, 0.6025641, 1)^T$; 4.0012508 $(0.4000234, 0.6000156, 1)^T = \hat{c}_5 \hat{\mathbf{X}}_5$	
$c_6 \mathbf{X}_6$ = 4.0512821 $(0.4018987, 0.6012658, 1)^T$; 4.0003125 $(0.4000059, 0.6000039, 1)^T = \hat{c}_6 \hat{\mathbf{X}}_6$	
$c_7 \mathbf{X}_7$ = 4.0253165 $(0.4009434, 0.6006289, 1)^T$; 4.0000781 $(0.4000015, 0.6000010, 1)^T = \hat{c}_7 \hat{\mathbf{X}}_7$	
$c_8 \mathbf{X}_8$ = 4.0125786 $(0.4004702, 0.6003135, 1)^T$; 4.0000195 $(0.4000004, 0.6000002, 1)^T = \hat{c}_8 \hat{\mathbf{X}}_8$	
$c_9 \mathbf{X}_9$ = 4.0062696 $(0.4002347, 0.6001565, 1)^T$; 4.0000049 $(0.4000001, 0.6000001, 1)^T = \hat{c}_9 \hat{\mathbf{X}}_9$	
$c_{10} \mathbf{X}_{10}$ = 4.0031299 $(0.4001173, 0.6000782, 1)^T$; 4.0000012 $(0.4000000, 0.6000000, 1)^T = \hat{c}_{10} \hat{\mathbf{X}}_{10}$	

Shifted Inverse Power Method

We will now discuss the shifted inverse power method. It requires a good starting approximation for an eigenvalue and then iteration is used to obtain a precise solution. Other procedures such as the QL and Given's method are used first to obtain the starting approximations. Cases involving complex eigenvalues, multiple eigenvalues, or the presence of two eigenvalues with the same magnitude or approximately the same magnitude will cause computational difficulties and require more advanced methods. Our illustrations will focus on the case where the eigenvalues are distinct. The shifted inverse power method is based on the following three results. (The proofs are left as exercises.)

Theorem 11.19 (Shifting Eigenvalues). Suppose that λ, $\mathbf{V}$ is an eigenpair of A. If α is any constant, then $\lambda - \alpha$, $\mathbf{V}$ is an eigenpair of the matrix $A - \alpha I$.

Theorem 11.20 (Inverse Eigenvalues). Suppose that λ, $\mathbf{V}$ is an eigenpair of A. If $\lambda \neq 0$, then $1/\lambda$, $\mathbf{V}$ is an eigenpair of the matrix A^{-1}.

Theorem 11.21. Suppose that λ, $\mathbf{V}$ is an eigenpair of A. If $\alpha \neq \lambda$, then $1/(\lambda - \alpha)$, $\mathbf{V}$ is an eigenpair of the matrix $(A - \alpha I)^{-1}$.

Theorem 11.22 (Shifted-Inverse Power Method). Assume that the $n \times n$ matrix A has distinct eigenvalues $\lambda_1, \lambda_2, \ldots, \lambda_n$ and consider the eigenvalue λ_j. Then a constant α can be chosen so that $\mu_1 = 1/(\lambda_j - \alpha)$ is the dominant eigenvalue of $(A - \alpha I)^{-1}$. Furthermore, if $\mathbf{X}_0$ is chosen appropriately, then the sequences $\{\mathbf{X}_k = (x_1^{(k)}, x_2^{(k)}, \ldots, x_n^{(k)})^T\}$ and $\{c_k\}$ are generated recursively by

$$\mathbf{Y}_k = (A - \alpha I)^{-1} \mathbf{X}_k \tag{17}$$

and

$$\mathbf{X}_{k+1} = \frac{1}{c_{k+1}} \mathbf{Y}_k, \tag{18}$$

where

$$c_{k+1} = x_j^{(k)} \quad \text{and} \quad x_j^{(k)} = \max_{1 \le i \le n} \{|x_i^{(k)}|\} \tag{19}$$

will converge to the dominant eigenpair μ_1, $\mathbf{V}_j$ of matrix $(A - \alpha I)^{-1}$. Finally, the corresponding eigenvalue for the matrix A is given by the calculation

$$\lambda_j = \frac{1}{\mu_1} + \alpha. \tag{20}$$

Remark.   For practical implementations of Theorem 11.22, a linear system solver is used to compute $\mathbf{Y}_k$ in each step by solving the linear system $(A - \alpha I)\mathbf{Y}_k = \mathbf{X}_k$.

Proof.   Without loss of generality, we may assume that $\lambda_1 < \lambda_2 < \cdots < \lambda_n$. Select a number α ($\alpha \ne \lambda$) that is closer to λ_j than any of the other eigenvalues (see Figure 11.2), that is,

$$|\lambda_j - \alpha| < |\lambda_i - \alpha| \quad \text{for each } i = 1, 2, \ldots, j - 1, j + 1, \ldots, n. \tag{21}$$

Figure 11.2   The location of α for the shifted inverse power method.

According to Theorem 11.21, $1/(\lambda_j - \alpha)$, $\mathbf{V}_j$ is an eigenpair of the matrix $(A - \alpha I)^{-1}$. Relation (21) implies that $1/|\lambda_i - \alpha| < 1/|\lambda_j - \alpha|$ for each $i \ne j$ so that $\mu_1 = 1/(\lambda_j - \alpha)$ is the dominant eigenvalue of the matrix $(A - \alpha I)^{-1}$. The shifted inverse power method uses a modification of power method to determine the eigenpair μ_1, $\mathbf{V}_j$. Then the calculation $\lambda_j = 1/\mu_1 + \alpha$ produces the desired eigenvalue of the matrix A.

Example 11.6

Employ the shifted inverse power method to find the eigenpairs of the matrix

$$A = \begin{pmatrix} 0 & 11 & -5 \\ -2 & 17 & -7 \\ -4 & 26 & -10 \end{pmatrix}.$$

Use the fact that the eigenvalues of A are $\lambda_1 = 4$, $\lambda_2 = 2$, and $\lambda_3 = 1$ and select an appropriate α and starting vector for each case.

Solution.   *Case (i)*: For the eigenvalue $\lambda_1 = 4$, we select $\alpha = 4.2$ and the starting vector $X_0 = (1, 1, 1)^T$. First, form the matrix $A - 4.2I$, compute the solution to

$$\begin{pmatrix} -4.2 & 11 & -5 \\ -2 & 12.8 & -7 \\ -4 & 26 & -14.2 \end{pmatrix} Y_0 = X_0 = \begin{pmatrix} 1 \\ 1 \\ 1 \end{pmatrix},$$

and get the vector $Y_0 = (-9.545454545, -14.09090909, -23.18181818)^T$. Then compute $c_1 = -23.18181818$ and $X_1 = (0.4117647059, 0.6078431373, 1)^T$. Iteration generates the values given in Table 11.3. The sequence $\{c_k\}$ converges to $\mu_1 = -5$, which is the dominant eigenvalue of $(A - 4.2I)^{-1}$, and $\{X_k\}$ converges to $V_1 = \left(\frac{2}{5}, \frac{3}{5}, 1\right)^T$. The eigenvalue λ_1 of A is given by the computation $\lambda_1 = 1/\mu_1 + \alpha = 1/(-5) + 4.2 = -0.2 + 4.2 = 4$.

TABLE 11.3   Shifted Inverse Power Method for the Matrix $(A - 4.2I)^{-1}$ in Example 11.6: Convergence to the Eigenvector $V = \left(\frac{2}{5}, \frac{3}{5}, 1\right)^T$ and $\mu_1 = -5$

$(A - \alpha I)^{-1}X_k =$	$c_{k+1}X_{k+1}$
$(A - \alpha I)^{-1}X_0 = -23.18181818\ (0.4117647059, 0.6078431373, 1)^T = c_1X_1$	
$(A - \alpha I)^{-1}X_1 = -5.356506239\ (0.4009983361, 0.6006655574, 1)^T = c_2X_2$	
$(A - \alpha I)^{-1}X_2 = -5.030252609\ (0.4000902120, 0.6000601413, 1)^T = c_3X_3$	
$(A - \alpha I)^{-1}X_3 = -5.002733697\ (0.4000081966, 0.6000054644, 1)^T = c_4X_4$	
$(A - \alpha I)^{-1}X_4 = -5.000248382\ (0.4000007451, 0.6000004967, 1)^T = c_5X_5$	
$(A - \alpha I)^{-1}X_5 = -5.000022579\ (0.4000000677, 0.6000000452, 1)^T = c_6X_6$	
$(A - \alpha I)^{-1}X_6 = -5.000002053\ (0.4000000062, 0.6000000041, 1)^T = c_7X_7$	
$(A - \alpha I)^{-1}X_7 = -5.000000187\ (0.4000000006, 0.6000000004, 1)^T = c_8X_8$	
$(A - \alpha I)^{-1}X_8 = -5.000000017\ (0.4000000001, 0.6000000000, 1)^T = c_9X_9$	

Case (ii): For the eigenvalue $\lambda_2 = 2$, we select $\alpha = 2.1$ and the starting vector $X_0 = (1, 1, 1)^T$. Form the matrix $A - 2.1I$, compute the solution to

$$\begin{pmatrix} -2.1 & 11 & -5 \\ -2 & 14.9 & -7 \\ -4 & 26 & -12.1 \end{pmatrix} Y_0 = X_0 = \begin{pmatrix} 1 \\ 1 \\ 1 \end{pmatrix},$$

and get the vector $Y_0 = (11.05263158, 21.57894737, 42.63157895)^T$. Then $c_1 = 42.63157895$ and vector $X_1 = (0.2592592593, 0.5061728395, 1)^T$. Iteration produces the values given in Table 11.4. The dominant eigenvalue of $(A - 2.1I)^{-1}$ is $\mu_1 = -10$ and the eigenpair of the matrix A are $\lambda_2 = \dfrac{1}{-10} + 2.1 = -0.1 + 2.1 = 2$ and $V_2 = \left(\frac{1}{4}, \frac{1}{2}, 1\right)^T$.

TABLE 11.4 Shifted Inverse Power Method for the Matrix $(A - 2.1I)^{-1}$ in Example 11.6: Convergence to the Dominant Eigenvector $\mathbf{V} = \left(\frac{1}{4}, \frac{1}{2}, 1\right)^T$ and $\mu_1 = -10$

$(A - \alpha I)^{-1}\mathbf{X}_k =$	$c_{k+1}\mathbf{X}_{k+1}$
$(A - \alpha I)^{-1}\mathbf{X}_0 =$	$42.63157895 \ (0.2592592593, 0.5061728395, 1)^T = c_1\mathbf{X}_1$
$(A - \alpha I)^{-1}\mathbf{X}_1 =$	$-9.350227420 \ (0.2494788047, 0.4996525365, 1)^T = c_2\mathbf{X}_2$
$(A - \alpha I)^{-1}\mathbf{X}_2 =$	$-10.03657511 \ (0.2500273314, 0.5000182209, 1)^T = c_3\mathbf{X}_3$
$(A - \alpha I)^{-1}\mathbf{X}_3 =$	$-9.998082009 \ (0.2499985612, 0.4999990408, 1)^T = c_4\mathbf{X}_4$
$(A - \alpha I)^{-1}\mathbf{X}_4 =$	$-10.00010097 \ (0.2500000757, 0.5000000505, 1)^T = c_5\mathbf{X}_5$
$(A - \alpha I)^{-1}\mathbf{X}_5 =$	$-9.999994686 \ (0.2499999960, 0.4999999973, 1)^T = c_6\mathbf{X}_6$
$(A - \alpha I)^{-1}\mathbf{X}_6 =$	$-10.00000028 \ (0.2500000002, 0.5000000001, 1)^T = c_7\mathbf{X}_7$

Case (iii): For the eigenvalue $\lambda_3 = 1$, we select $\alpha = 0.875$ and the starting vector $\mathbf{X}_0 = (0, 1, 1)^T$. Iteration produces the values given in Table 11.5. The dominant eigenvalue of $(A + 0.875I)^{-1}$ is $\mu_1 = -10$ and the eigenpair of the matrix A is $\lambda_3 = \frac{1}{8} + 0.875 = -0.125 + 0.875 = 1$ and $\mathbf{V}_3 = \left(\frac{1}{2}, \frac{1}{2}, 1\right)^T$. The sequence $\{\mathbf{X}_k\}$ of vectors with the starting vector $(0, 1, 1)^T$ converged in seven iterations. [Computational difficulties were encountered when $\mathbf{X}_0 = (1, 1, 1)^T$ was used, and convergence took significantly longer.]

TABLE 11.5 Shifted Inverse Power Method for the Matrix $(A + 0.875I)^{-1}$ in Example 11.6: Convergence to the Dominant Eigenvector $\mathbf{V} = \left(\frac{1}{2}, \frac{1}{2}, 1\right)^T$ and $\mu_1 = 8$

$(A - \alpha I)^{-1}\mathbf{X}_k =$	$c_{k+1}\mathbf{X}_{k+1}$
$(A - \alpha I)^{-1}\mathbf{X}_0 =$	$-30.40000000 \ (0.5052631579, 0.4947368421, 1)^T = c_1\mathbf{X}_1$
$(A - \alpha I)^{-1}\mathbf{X}_1 =$	$8.404210526 \ (0.5002004008, 0.4997995992, 1)^T = c_2\mathbf{X}_2$
$(A - \alpha I)^{-1}\mathbf{X}_2 =$	$8.015390782 \ (0.5000080006, 0.4999919994, 1)^T = c_3\mathbf{X}_3$
$(A - \alpha I)^{-1}\mathbf{X}_3 =$	$8.000614449 \ (0.5000003200, 0.4999996800, 1)^T = c_4\mathbf{X}_4$
$(A - \alpha I)^{-1}\mathbf{X}_4 =$	$8.000024576 \ (0.5000000128, 0.4999999872, 1)^T = c_5\mathbf{X}_5$
$(A - \alpha I)^{-1}\mathbf{X}_5 =$	$8.000000983 \ (0.5000000005, 0.4999999995, 1)^T = c_6\mathbf{X}_6$
$(A - \alpha I)^{-1}\mathbf{X}_6 =$	$8.000000039 \ (0.5000000000, 0.5000000000, 1)^T = c_7\mathbf{X}_7$

Algorithm 11.1 (Power Method). To compute the dominant value λ_1 and its associated eigenvector $\mathbf{V}_1$ for the $n \times n$ matrix A. It is assumed that the n eigenvalues have the dominance property $|\lambda_1| > |\lambda_2| \geq |\lambda_3| \geq \cdots \geq |\lambda_n| > 0$.

```
Epsilon := 1E−7                                              {Tolerance}
Max := 50                                    {Maximum number of iterations}

FUNCTION    MaxElement(X,N)
    MaxElement := 0
    FOR  J = 1   TO   N   DO
        │   IF   ABS(X[J]) > ABS(MaxElement)   THEN
        └────────── MaxElement := X[J]

FUNCTION    DIST(X,Y,N)
    Sum := 0
    FOR   J = 1   TO   N   DO
        └────Sum := Sum + (Y[J]−X[J])^2
    DIST := SQRT(Sum)

{The main program starts here.}

    FOR   J = 1   TO   N   DO                       {Initialize the matrix V}
        └────X[J] := 1

Lambda := 0, Count := 0, Err := 1

WHILE   Count ≤ Max   AND   State = Iterating
    │   Y := A X                            {Perform matrix multiplication}
    │   C1 := MaxElement(Y,N)                    {Find largest element of Y}
    │   DC := ABS(Lambda−C1)
    │   Y := (1/C1)*Y                       {Perform scalar multiplication}
    │   DV := DIST(X,Y,N)
    │   Err := MAXIMUM(DC,DV)
    │   X := Y                                            {Update vector X
    │   Lambda := C1                                   and scalar Lambda}
    │   State := Done
    │   IF   Err > Epsilon   THEN                    {Check for convergence}
    │       └────State := Iterating
    └────────Count := Count + 1

OUTPUT   vector   X   and   scalar   Lambda
```

Algorithm 11.2 (Shifted Inverse Power Method). To compute the dominant value λ_j and its associated eigenvector $\mathbf{V}_j$ for the $n \times n$ matrix A. It is assumed that the n eigenvalues are $\lambda_1 < \lambda_2 < \cdots < \lambda_n$ and that α is a real number such that $|\lambda_j - \alpha| < |\lambda_i - \alpha|$ for each $i = 1, 2, \ldots, j-1, j+1, \ldots, n$.

Epsilon   := 1E−7   {Tolerance}
Max   := 50   {Maximum number of iterations}
INPUT   Alpha   {Input the shifting parameter}

FUNCTION   MaxElement(X,N)
    MaxElement := 0
    FOR   J = 1   TO   N   DO
    │   IF   ABS(X[J]) > ABS(MaxElement)   THEN
    └────────MaxElement := X[J]

FUNCTION   DIST(X,Y,N)
    Sum   := 0
    FOR   J = 1   TO   N   DO
    └────────Sum := Sum + (Y[J]−X[J])^2
    DIST   := SQRT(Sum)

{The main program starts here.}

FOR   J = 1   TO   N   DO   {Initialize the matrix V}
└────────X[J] := 1

A1   := A − Alpha*I   {Form the matrix A−αI}

Lambda := 0, Count := 0, Err := 1

WHILE   Count ≤ Max   AND   State = Iterating
    │   Y := Solve(A1 Y = X)   {Solve a linear system}
    │   C1 := MaxElement(Y,N)   {Find largest element of Y}
    │   DC := ABS(Lambda−C1)
    │   Y := (1/C1)*Y   {Perform scalar multiplication}
    │   DV := DIST(X,Y,N)
    │   Err := MAXIMUM(DC,DV)
    │   X := Y   {Update vector X
    │   Lambda := C1   and scalar Lambda}
    │   State := Done
    │   IF   Err > Epsilon   THEN   {Check for convergence}
    │   └────State := Iterating
    └────────Count := Count + 1

Lambda := Alpha + 1/C1

OUTPUT   vector   X   and   scalar   Lambda

EXERCISES FOR THE POWER METHOD

1. Let λ, $\mathbf{V}$ be an eigenpair of A. If α is any constant, show that $\lambda - \alpha$, $\mathbf{V}$ is an eigenpair of the matrix $A - \alpha I$.

2. Let λ, $\mathbf{V}$ be an eigenpair of A. If $\lambda \neq 0$ show that $1/\lambda$, $\mathbf{V}$ is an eigenpair of the matrix A^{-1}.

3. Let λ, $\mathbf{V}$ be an eigenpair of A. If $\alpha \neq \lambda$ show that $1/(\lambda - \alpha)$, $\mathbf{V}$ is an eigenpair of the matrix $(A - \alpha I)^{-1}$.

Instructions for Exercises 4–9:

(a) Use the power method to find the dominant eigenpair of the given matrices.

(b) Use the shifted inverse power method to find the other eigenpairs.

4. (a) $A = \begin{pmatrix} 7 & 6 & -3 \\ -12 & -20 & 24 \\ -6 & -12 & 16 \end{pmatrix}$.   **(b)** Use the shifts $\alpha = 0, -3$.

5. (a) $A = \begin{pmatrix} -14 & -30 & 42 \\ 24 & 49 & -66 \\ 12 & 24 & -32 \end{pmatrix}$.   **(b)** Use the shifts $\alpha = 0, -3$.

6. (a) $A = \begin{pmatrix} 16 & 30 & -42 \\ -24 & -47 & 66 \\ -12 & -24 & 34 \end{pmatrix}$.   **(b)** Use the shifts $\alpha = 0, -3$.

7. (a) $A = \begin{pmatrix} -12 & -72 & -59 \\ 5 & 29 & 23 \\ -2 & -12 & -9 \end{pmatrix}$.   **(b)** Use the shifts $\alpha = 0, 3$.

8. (a) $A = \begin{pmatrix} 2.5 & -2.5 & 3.0 & 0.5 \\ 0.0 & 5.0 & -2.0 & 2.0 \\ -0.5 & -0.5 & 4.0 & 2.5 \\ -2.5 & -2.5 & 5.0 & 3.5 \end{pmatrix}$.   **(b)** Use the shifts $\alpha = 0, 2.5, 4.5$.

9. (a) $A = \begin{pmatrix} 2.5 & -2.0 & 2.5 & 0.5 \\ 0.5 & 5.0 & -2.5 & -0.5 \\ -1.5 & 1.0 & 3.5 & -2.5 \\ 2.0 & 3.0 & -5.0 & 3.0 \end{pmatrix}$.   **(b)** Use the shifts $\alpha = 0, 2.5, 4.5$.

10. *Markov processes and eigenvalues.* A Markov process can be described by a square matrix whose entries are all positive and the column sums all equal 1. For illustration, let $\mathbf{P}_0 = (x^{(0)}, y^{(0)}, z^{(0)})^T$ record the number of people in a certain city who use brand X, Y, and Z, respectively. Each month people decide to keep using the same brand or switch brands. The probability that a user of brand X will switch to brand Y or Z is 0.3 and 0.3, respectively. The probability that a user of brand Y will switch to brand X or Z is 0.3 and 0.2, respectively.

The probability that a user of brand Z will switch to brand X or Y is 0.1 and 0.3, respectively. The transition matrix for this process is

$$\mathbf{P}_{k+1} = A\mathbf{P}_k = \begin{pmatrix} 0.4 & 0.3 & 0.1 \\ 0.3 & 0.5 & 0.3 \\ 0.3 & 0.2 & 0.6 \end{pmatrix} \begin{pmatrix} x^{(k)} \\ y^{(k)} \\ z^{(k)} \end{pmatrix}$$

Given that $\mathbf{P}_0 = (2000, 6000, 4000)^T$, find $\mathbf{P}_1$ and $\mathbf{P}_2$. Show that the limit of the sequence $\{\mathbf{P}_k\}$ is $\mathbf{V}_1 = (3000, 4500, 4500)^T$, which represents the steady-state distribution.

Remark. For this matrix A the eigenvalues are 1, 0.3, and 0.2.

11. Suppose that the following changes are made in Exercise 10. The probability that a user of brand X will switch to brand Y or Z is 0.3 and 0.1, respectively. The probability that a user of brand Y will switch to brand X or Z is 0.2 and 0.3, respectively. The probability that a user of brand Z will switch to brand X or Y is 0.1 and 0.1, respectively. The transition matrix for this process is

$$\mathbf{P}_{k+1} = A\mathbf{P}_k = \begin{pmatrix} 0.6 & 0.2 & 0.1 \\ 0.3 & 0.5 & 0.1 \\ 0.1 & 0.3 & 0.8 \end{pmatrix} \begin{pmatrix} x^{(k)} \\ y^{(k)} \\ z^{(k)} \end{pmatrix}$$

Given that $\mathbf{P}_0 = (2000, 4000, 4000)^T$, find $\mathbf{P}_1$ and $\mathbf{P}_2$ and the limit of the sequence $\{\mathbf{P}_k\}$.

Remark. For this matrix A the eigenvalues are 1, 0.6, and 0.3.

12. Suppose that the following changes are made in Exercise 10. The probability that a user of brand X will switch to brand Y or Z is 0.4 and 0.2, respectively. The probability that a user of brand Y will switch to brand X or Z is 0.2 and 0.2, respectively. The probability that a user of brand Z will switch to brand X or Y is 0.1 and 0.1, respectively. The transition matrix for this process is

$$\mathbf{P}_{k+1} = A\mathbf{P}_k = \begin{pmatrix} 0.4 & 0.2 & 0.1 \\ 0.4 & 0.6 & 0.1 \\ 0.2 & 0.2 & 0.8 \end{pmatrix} \begin{pmatrix} x^{(k)} \\ y^{(k)} \\ z^{(k)} \end{pmatrix}$$

Given that $\mathbf{P}_0 = (6000, 6000, 8000)^T$, find $\mathbf{P}_1$ and $\mathbf{P}_2$ and the limit of the sequence $\{\mathbf{P}_k\}$.

Remark. For this matrix A the eigenvalues are 1, 0.6, and 0.2.

13. Write a report on the Given's method.

14. *Deflation techniques.* Suppose that $\lambda_1, \lambda_2, \lambda_3, \ldots, \lambda_n$ are the eigenvalues of A with associated eigenvectors $\mathbf{V}^{(1)}, \mathbf{V}^{(2)}, \mathbf{V}^{(3)}, \ldots, \mathbf{V}^{(n)}$, and that λ_1 has multiplicity 1. If $\mathbf{X}$ is any vector with the property that $\mathbf{X}^T \mathbf{V}^{(1)} = 1$, prove that the matrix

$$B = A - \lambda_1 \mathbf{V}^{(1)} \mathbf{X}^T$$

has eigenvalues $0, \lambda_2, \lambda_3, \ldots, \lambda_n$ with associated eigenvectors $\mathbf{V}^{(1)}, \mathbf{W}^{(2)}, \mathbf{W}^{(3)}, \ldots, \mathbf{W}^{(n)}$, where $\mathbf{V}^{(j)}$ and $\mathbf{W}^{(j)}$ are related by the equation

$$\mathbf{V}^{(j)} = (\lambda - \lambda_1)\mathbf{W}^{(j)} + \lambda_1(\mathbf{X}^T \mathbf{W}^{(j)})\mathbf{V}^{(1)} \qquad \text{for each} \quad j = 2, 3, \ldots, n.$$

15. Write a report on Wielandt's deflation method.

16. Write a report on Hotelling's deflation method for symmetric matrices.

11.3 JACOBI'S METHOD

Jacobi's method is an easily understood algorithm for finding all eigenpairs for a symmetric matrix. It is a reliable method that produces uniformly accurate answers for the results. For matrices of order up to 10, the algorithm is competitive with more sophisticated ones. If speed is not a major consideration, it is quite acceptable for matrices up to order 20.

A solution is guaranteed for all real symmetric matrices when Jacobi's method is used. This limitation is not severe since many practical problems of applied mathematics and engineering involve symmetric matrices. From a theoretical viewpoint, the method embodies techniques that are found in more sophisticated algorithms. For instructive purposes, it is worthwhile to investigate the details of Jacobi's method.

Plane Rotations

We start with some geometrical background about coordinate transformations. Let $\mathbf{X}$ denote a vector in n-dimensional space and consider the linear transformation $\mathbf{Y} = R\mathbf{X}$ where R is an $n \times n$ matrix:

$$
R = \begin{pmatrix}
1 & \cdots & 0 & \cdots & 0 & \cdots & 0 \\
\vdots & & \vdots & & \vdots & & \vdots \\
0 & \cdots & \cos \phi & \cdots & \sin \phi & \cdots & 0 \\
\vdots & & \vdots & & \vdots & & \vdots \\
0 & \cdots & -\sin \phi & \cdots & \cos \phi & \cdots & 0 \\
\vdots & & \vdots & & \vdots & & \vdots \\
0 & \cdots & 0 & \cdots & 0 & \cdots & 1
\end{pmatrix}
\begin{array}{l}
\\ \\ \leftarrow \text{row } p \\ \\ \leftarrow \text{row } q \\ \\ \\
\end{array}
$$

$$\underset{\text{col } p}{\uparrow} \qquad \underset{\text{col } q}{\uparrow}$$

Here all off-diagonal elements of R are zero except for the values $\pm\sin\phi$, and all diagonal elements are 1 except for $\cos\phi$. The effect of the transformation $\mathbf{Y} = R\mathbf{X}$ is easy to grasp:

$$y_j = x_j \quad \text{when } j \neq p \text{ and } j \neq q,$$

$$y_p = x_p \cos\phi + x_q \sin\phi,$$

$$y_q = -x_p \sin\phi + x_q \cos\phi.$$

The transformation is seen to be a rotation of n-dimensional space in the $x_p x_q$-plane through the angle ϕ. By selecting an appropriate angle ϕ, we could make either $y_p = 0$ or $y_q = 0$ in the image. The inverse transformation $\mathbf{X} = R^{-1}\mathbf{Y}$ rotates space in the same $x_p x_q$-plane through the angle $-\phi$. Observe that R is an orthogonal matrix, that is,

$$R^{-1} = R^T \quad \text{or} \quad R^T R = I.$$

Similarity and Orthogonal Transformations

Consider the eigenproblem

$$A\mathbf{X} = \lambda \mathbf{X}. \tag{1}$$

Suppose that K is a nonsingular matrix and that B is defined by

$$B = K^{-1}AK. \tag{2}$$

Multiply both members of (2) on the right by the quantity $K^{-1}\mathbf{X}$. This produces

$$BK^{-1}\mathbf{X} = K^{-1}AKK^{-1}\mathbf{X} = K^{-1}A\mathbf{X} = K^{-1}\lambda\mathbf{X} = \lambda K^{-1}\mathbf{X}. \tag{3}$$

We define the change of variable

$$\mathbf{Y} = K^{-1}\mathbf{X} \quad \text{or} \quad \mathbf{X} = K\mathbf{Y}. \tag{4}$$

When (4) is used in (3) the new eigenproblem is

$$B\mathbf{Y} = \lambda\mathbf{Y}. \tag{5}$$

Comparing (1) and (5), we see that the similarity transformation (2) preserved the eigenvalue λ and that the eigenvectors are different, but are related by the change of variable in (4).

Suppose that the matrix R is an orthogonal matrix (i.e., $R^{-1} = R^T$) and that D is defined by

$$D = R^TAR. \tag{6}$$

Multiply both terms in (6) on the right by $R^T\mathbf{X}$ to obtain

$$DR^T\mathbf{X} = R^TARR^T\mathbf{X} = R^TA\mathbf{X} = R^T\lambda\mathbf{X} = \lambda R^T\mathbf{X}. \tag{7}$$

We define the change of variable

$$\mathbf{Y} = R^T\mathbf{X} \quad \text{or} \quad \mathbf{X} = R\mathbf{Y}. \tag{8}$$

Now use (8) in (7) to obtain a new eigenproblem,

$$D\mathbf{Y} = \lambda\mathbf{Y}. \tag{9}$$

As before, the eigenvalues of (1) and (9) are the same. However, for equation (9) the change of variable (8) makes it easier to convert $\mathbf{X}$ to $\mathbf{Y}$ and $\mathbf{Y}$ back into $\mathbf{X}$ because $R^{-1} = R^T$.

In addition, suppose that A is a symmetric matrix (i.e., $A = A^T$). Then we find that

$$D^T = (R^TAR)^T = R^TA(R^T)^T = R^TAR = D. \tag{10}$$

Hence D is a symmetric matrix. Therefore, we conclude that if A is a symmetric matrix and R is an orthogonal matrix, the transformation of A into D given by (6) preserves symmetry as well as eigenvalues. The relationship between their eigenvectors is given by the change of variables (8).

The Jacobi Series of Transformations

Start with the real symmetric matrix A. Then construct the sequence of orthogonal matrices $R_1, R_2, \ldots, R_n$ as follows:

$$
\begin{aligned}
D_0 &= A, \\
D_j &= R_j^T D_{j-1} R_j \qquad \text{for } j = 1, 2, \ldots.
\end{aligned}
\tag{11}
$$

We will show how to construct the sequence $\{R_j\}$ so that

$$
\lim_{j \to \infty} D_j = D = \text{diag}(\lambda_1, \lambda_2, \ldots, \lambda_n).
\tag{12}
$$

In practice we will stop when the off-diagonal elements are close to zero. Then we will have

$$
D_n \approx D.
\tag{13}
$$

The construction produces

$$
D_n = R_n^T R_{n-1}^T \cdots R_1^T A R_1 R_2 \cdots R_{n-1} R_n.
\tag{14}
$$

If we define

$$
R = R_1 R_2 \cdots R_{n-1} R_n,
\tag{15}
$$

then $R^{-1}AR = D$, which implies that

$$
AR = RD = R \, \text{diag}(\lambda_1, \lambda_2, \ldots, \lambda_n).
\tag{16}
$$

Let the columns of R be denoted by the vectors $\mathbf{X}_1, \mathbf{X}_2, \ldots, \mathbf{X}_n$. Then R can be expressed as a row vector of column vectors,

$$
R = [\mathbf{X}_1, \mathbf{X}_2, \ldots, \mathbf{X}_n].
\tag{17}
$$

The columns of the products in (16) now take on the form

$$
[A\mathbf{X}_1, A\mathbf{X}_2, \ldots, A\mathbf{X}_n] = [\lambda_1 \mathbf{X}_1, \lambda_2 \mathbf{X}_2, \ldots, \lambda_n \mathbf{X}_n].
\tag{18}
$$

From (17) and (18), we see that the vector $\mathbf{X}_j$, which is the jth column of R, is an eigenvector that corresponds to the eigenvector λ_j.

The General Step

Each step in the Jacobi iteration will accomplish the limited objective of reduction of the two off-diagonal elements a_{pq} and a_{qp} to zero. Let R_1 denote the first orthogonal matrix used. Suppose that

$$
D_1 = R_1^T A R_1
\tag{19}
$$

reduces the elements a_{pq} and a_{qp} to zero, where R_1 has the form

$$R_1 = \begin{pmatrix} 1 & \cdots & 0 & \cdots & 0 & \cdots & 0 \\ \vdots & & & & & & \vdots \\ 0 & \cdots & c & \cdots & s & \cdots & 0 \\ \vdots & & & & & & \vdots \\ 0 & \cdots & -s & \cdots & c & \cdots & 0 \\ \vdots & & & & & & \vdots \\ 0 & \cdots & 0 & \cdots & 0 & \cdots & 1 \end{pmatrix} \begin{matrix} \\ \\ \leftarrow \text{row } p \\ \\ \leftarrow \text{row } q \\ \\ \\ \end{matrix} \tag{20}$$

$$\begin{matrix} \uparrow & & \uparrow \\ \text{col } p & & \text{col } q \end{matrix}$$

Here all off-diagonal elements of R_1 are zero except for the element s located in row p and column q, and the element $-s$ located in row q and column p. Also note that all diagonal elements are 1 except for the element c, which appears at two locations, in row p column p and in row q column q. The matrix is a plane rotation where we have used the notation $c = \cos\phi$ and $s = \sin\phi$.

We must verify that the transformation (19) will produce a change only to rows p and q and columns p and q. Consider postmultiplication of A by R_1 and the product $B = AR_1$:

$$B = \begin{pmatrix} a_{11} & \cdots & a_{1p} & \cdots & a_{1q} & \cdots & a_{1n} \\ a_{p1} & \cdots & a_{pp} & \cdots & a_{pq} & \cdots & a_{pn} \\ a_{q1} & \cdots & a_{qp} & \cdots & a_{qq} & \cdots & a_{qn} \\ a_{n1} & \cdots & a_{np} & \cdots & a_{nq} & \cdots & a_{nn} \end{pmatrix} \begin{pmatrix} 1 & \cdots & 0 & \cdots & 0 & \cdots & 0 \\ 0 & \cdots & c & \cdots & s & \cdots & 0 \\ 0 & \cdots & -s & \cdots & c & \cdots & 0 \\ 0 & \cdots & 0 & \cdots & 0 & \cdots & 1 \end{pmatrix} \tag{21}$$

The row by column rule for multiplication applies and we observe that there is no change to columns 1 to $p - 1$ and $p + 1$ to $q - 1$ and $q + 1$ to n. Hence only columns p and q are altered.

$$\begin{aligned} b_{jk} &= a_{jk} & \text{when } k \neq p \text{ and } k \neq q, \\ b_{jp} &= ca_{jp} - sa_{jq} & \text{for } j = 1, 2, \ldots, n, \\ b_{jq} &= sa_{jp} + ca_{jq} & \text{for } j = 1, 2, \ldots, n. \end{aligned} \tag{22}$$

A similar argument shows that premultiplication of A by R_1^T will only alter rows p and q. Therefore, the transformation

$$D_1 = R_1^T A R_1 \tag{23}$$

will alter only columns p and q and rows p and q of A. The elements d_{jk} of D_1 are computed with the formulas

$$
\begin{aligned}
d_{jp} &= ca_{jp} - sa_{jq} &&\text{when } j \neq p \text{ and } j \neq q, \\
d_{jq} &= sa_{jp} + ca_{jq} &&\text{when } j \neq p \text{ and } j \neq q, \\
d_{pp} &= c^2 a_{pp} + s^2 a_{qq} - 2csa_{pq}, \\
d_{qq} &= s^2 a_{pp} + c^2 a_{qq} + 2csa_{pq}, \\
d_{pq} &= (c^2 - s^2)a_{pq} + cs(a_{pp} - a_{qq}),
\end{aligned}
\tag{24}
$$

and the other elements of D_1 are found by symmetry.

Zeroing Out d_{pq} and d_{qp}

The goal for each step of Jacobi's iteration is to make the two off-diagonal elements d_{pq} and d_{qp} zero. The obvious strategy would be to observe the fact that

$$
c = \cos \phi \quad \text{and} \quad s = \sin \phi, \tag{25}
$$

where ϕ is the angle of rotation that produces the desired effect. However, some ingenious maneuvers with trigonometric identities are now required. The identity for $\cot \phi$ is used with (25) to define

$$
\theta = \cot 2\phi = \frac{c^2 - s^2}{2cs}. \tag{26}
$$

Suppose that $a_{pq} \neq 0$ and we want to produce $d_{pq} = 0$. Then using the last equation in (24), we obtain

$$
0 = (c^2 - s^2)a_{pq} + cs(a_{pp} - a_{qq}). \tag{27}
$$

This can be rearranged to yield $(c^2 - s^2)/(cs) = (a_{qq} - a_{pp})/a_{pq}$, which is used in (26) to solve for θ:

$$
\theta = \frac{a_{qq} - a_{pp}}{2a_{pq}}. \tag{28}
$$

Although we can use (28) with formulas (25) and (26) to compute c and s, less round-off error is propagated if we compute $\tan \phi$ and use it in later computations. So we define

$$
t = \tan \phi = \frac{s}{c}. \tag{29}
$$

Now divide the numerator and denominator in (26) by c^2 to obtain

$$
\theta = \frac{1 - s^2/c^2}{2s/c} = \frac{1 - t^2}{2t},
$$

which yields the equation

$$t^2 + 2t\theta - 1 = 0. \tag{30}$$

Since $t = \tan \phi$, the smaller root of (30) corresponds to the smaller angle of rotation with $|\phi| \leq \pi/4$. The special form of the quadratic formula for finding this root is

$$t = -\theta \pm (\theta^2 + 1)^{1/2} = \frac{\text{sign}(\theta)}{|\theta| + (\theta^2 + 1)^{1/2}}, \tag{31}$$

where $\text{sign}(\theta) = 1$ when $\theta \geq 0$ and $\text{sign}(\theta) = -1$ when $\theta < 0$. Then c and s are computed with the formulas

$$c = \frac{1}{(t^2 + 1)^{1/2}} \tag{32}$$

$$s = ct.$$

Summary of the General Step

We can now outline the calculations required to zero-out the element d_{pq}. First, select row p and column q for which $a_{pq} \neq 0$. Second, form the preliminary quantities

$$\theta = \frac{a_{qq} - a_{pp}}{2a_{pq}},$$

$$t = \frac{\text{sign}(\theta)}{(|\theta| + \theta^2 + 1)^{1/2}}, \tag{33}$$

$$c = \frac{1}{(t^2 + 1)^{1/2}},$$

$$s = ct.$$

Third, to construct $D = D_1$, use

$$d_{pq} = 0,$$
$$d_{qp} = 0,$$
$$d_{pp} = c^2 a_{pp} + s^2 a_{qq} - 2cs a_{pq}, \tag{34}$$
$$d_{qq} = s^2 a_{pp} + c^2 a_{qq} + 2cs a_{pq}.$$

```
FOR  j = 1  TO  N  DO
        IF  (j ≠ p) AND (j ≠ q) THEN
                d_jp = c a_jp − s a_jq
                d_pj = d_jp
                d_jq = c a_jq + s a_jp
                d_qj = d_jq
```

Updating the Matrix of Eigenvectors

We need to keep track of the matrix product $R_1 R_2 \cdots R_n$. When we stop at the nth iteration, we will have computed

$$V_n = R_1 R_2 \cdots R_n, \tag{35}$$

where V_n is an orthogonal matrix. We need only keep track of the current matrix V_j for $j = 1, 2, \ldots, n$. Start by initializing $V = I$. Use the vector variables **XP** and **XQ** to store columns p and q of A, respectively. Then for each step perform the calculation

```
FOR  j = 1  TO  N  DO
     XP_j = v_jp
     XQ_j = v_jq
```

$$\tag{36}$$

```
FOR  j = 1  TO  N  DO
     v_jp = c XP_j − s XQ_j,
     v_jq = s XP_j + c XQ_j.
```

Strategy for Eliminating a_{pq}

The speed of convergence of Jacobi's method is seen by considering the sum of the squares of the off-diagonal elements

$$S_1 = \sum_{\substack{j,k=1 \\ k \neq j}}^{n} |a_{jk}|^2 \tag{37}$$

$$S_2 = \sum_{\substack{j,k=1 \\ k \neq j}}^{n} |d_{jk}|^2, \qquad \text{where } D_1 = R^T A R. \tag{38}$$

The reader can verify that the equations given in (34) can be used to prove that

$$S_2 = S_1 - 2|a_{pq}|^2. \tag{39}$$

At each step we let S_j denote the sum of the squares of the off-diagonal elements of D_j. Then the sequence $\{S_j\}$ decreases monotonically and is bounded below by zero. Jacobi's original algorithm of 1846 selected, at each step, the off-diagonal element a_{pq} of largest magnitude to zero-out, and involved a search to compute the value

$$\max \{A\} = \max_{p<q} \{|a_{pq}|\}. \tag{40}$$

This choice will guarantee that $\{S_j\}$ converges to zero. As a consequence, this proves that

$\{D_j\}$ converges to D and $\{V_j\}$ converges to the matrix V of eigenvectors (for more details, see Reference [68]).

Jacobi's search can become time consuming since it requires an order of $(n^2 - n)/2$ comparisons in a loop. It is prohibitive for larger values of n. A better strategy is the cyclic Jacobi method, where one annihilates elements in a strict order across the rows. A tolerance value ϵ is selected, then a sweep is made throughout the matrix and if an element a_{pq} is found to be larger than ϵ, it is zeroed-out. For one sweep through the matrix the elements are checked in row 1; $a_{12}, a_{13}, \ldots, a_{1n}$; then row 2; $a_{23}, a_{24}, \ldots, a_{2n}$; and so on. It has been proven that the convergence rate is quadratic for both the original and cyclic Jacobi methods. An implementation of the cyclic Jacobi method starts by observing that the sum of the squares of the diagonal elements increases with each iteration; that is, if

$$T_0 = \sum_{j=1}^{n} |a_{jj}|^2 \tag{41}$$

and

$$T_1 = \sum_{j=1}^{n} |d_{jj}|^2$$

then

$$T_1 = T_0 + 2|a_{pq}|^2.$$

Consequently, the sequence $\{D_j\}$ converges to the diagonal matrix D. Notice that the average size of a diagonal element can be computed with the formula $(T_0/n)^{1/2}$. The magnitudes of the off-diagonal elements are compared to $\epsilon(T_0/n)^{1/2}$, where ϵ is the preassigned tolerance. Therefore, the element a_{pq} is zeroed-out if

$$|a_{pq}| > \epsilon \left(\frac{T_0}{n}\right)^{1/2} \tag{42}$$

Another variation of the method, called the threshold Jacobi method, is left for the reader to investigate (see Reference [178]).

Example 11.7

Use Jacobi iteration to transform the following symmetric matrix into diagonal form.

$$\begin{pmatrix} 8 & -1 & 3 & -1 \\ -1 & 6 & 2 & 0 \\ 3 & 2 & 9 & 1 \\ -1 & 0 & 1 & 7 \end{pmatrix}$$

Solution.   The computational details are left for the reader. The first rotation matrix that will zero-out $a_{13} = 3$ is

$$R_1 = \begin{pmatrix} 0.763020 & 0.000000 & 0.646375 & 0.000000 \\ 0.000000 & 0.000000 & 0.000000 & 0.000000 \\ -0.646375 & 0.000000 & 0.763020 & 0.000000 \\ 0.000000 & 0.000000 & 0.000000 & 0.000000 \end{pmatrix}.$$

Calculation reveals that $A_2 = R_1 A_1 R_1$ is

$$A_2 = \begin{pmatrix} 5.458619 & -2.055770 & 0.000000 & -1.409395 \\ -2.055770 & 6.000000 & 0.879665 & 0.000000 \\ 0.000000 & 0.879665 & 11.541381 & 0.116645 \\ -1.409395 & 0.000000 & 0.116645 & 7.000000 \end{pmatrix}.$$

Next, the element $a_{12} = -2.055770$ is zeroed-out and we get

$$A_3 = \begin{pmatrix} 3.655795 & 0.000000 & 0.579997 & -1.059649 \\ 0.000000 & 7.802824 & 0.661373 & 0.929268 \\ 0.579997 & 0.661373 & 11.541381 & 0.116645 \\ -1.059649 & 0.929268 & 0.116645 & 7.000000 \end{pmatrix}.$$

After 10 iterations we arrive at

$$A_{10} = \begin{pmatrix} 3.295870 & 0.002521 & 0.037859 & 0.000000 \\ 0.002521 & 8.405210 & -0.004957 & 0.066758 \\ 0.037859 & -0.004957 & 11.704123 & -0.001430 \\ 0.000000 & 0.066758 & -0.001430 & 6.594797 \end{pmatrix}.$$

It will take six more iterations for the diagonal elements to get close to the diagonal matrix

$$D = \text{diag}(3.295699, 8.407662, 11.704301, 6.592338).$$

However, the off-diagonal elements are not small enough and it will take three more iterations for them to be less than 10^{-6} in magnitude. Then the eigenvectors are the columns of the matrix $V = R_1 R_2 \cdots R_{18}$, which is

$$V = \begin{pmatrix} 0.528779 & -0.573042 & 0.582298 & 0.230097 \\ 0.591967 & 0.472301 & 0.175776 & -0.628975 \\ -0.536039 & 0.282050 & 0.792487 & -0.071235 \\ 0.287454 & 0.607455 & 0.044680 & 0.739169 \end{pmatrix}.$$

Algorithm 11.3   (Jacobi Iteration for Eigenvalues and Eigenvectors).   To compute the full set of eigenpairs $\{\lambda_j, \mathbf{V}_j\}_{j=1}^n$ of the $n \times n$ real symmetric matrix A. Jacobi iteration is used to find all eigenpairs.

```
Epsilon   := 1E−7                              {Off-diagonal element tolerance}
MaxSweep  := 50                                {Maximum number of sweeps}

PROCEDURE  INITIALIZE(V,N)                     {Initialize the matrix V
    FOR  J = 1  TO  N  DO                       as the identity matrix}
        FOR  K = 1  TO  N  DO
            IF  J ≠ K  THEN  V[J,K] := 0
                       ELSE  V[J,K] := 1

FUNCTION  RMS(A,N)
    Sum := 0
        FOR  J:=1  TO  N  DO
            Sum := Sum + A[J,J]*A[J,J]
    RMS := SQRT(Sum/N)

PROCEDURE  ZEROOUT(A,V,P,Q)
    Theta := (A[Q,Q] − A[P,P])/(2*A[P,Q])
    T := SIGN(Theta)/(ABS(Theta) + SQRT(Theta*Theta + 1))
    C := 1/SQRT(T*T + 1)
    S := C*T

    FOR    J = 1  TO  N  DO
        XP[J] := A[J,P], XQ[J] := A[J,Q]
    A[P,Q] := 0, A[Q,P] := 0
    A[P,P] := C*C*XP[P] + S*S*XQ[Q] − 2*C*S*XQ[P]
    A[Q,Q] := S*S*XP[P] + C*C*XQ[Q] + 2*C*S*XQ[P]

    FOR    J = 1  TO  N  DO
        IF  (J<>P)  AND  (J<>Q)  THEN
            A[J,P] := C*XP[J] − S*XQ[J], A[P,J] := A[J,P]
            A[J,Q] := C*XQ[J] + S*XP[J], A[Q,J] := A[J,Q]

    FOR    J = 1  TO  N  DO
        YP[J] := V[J,P], YQ[J] := V[J,Q]

    FOR    J = 1  TO  N  DO
        V[J,P] := C*YP[J] − S*YQ[J], V[J,Q] := S*YP[J] + C*YQ[J]
```

{The main program for cyclic Jacobi iteration starts here.}
INITIALIZE(V,N), CountS :=0
REPEAT

      T:=RMS(A,N), CountS :=CountS+1, State := Done
      FOR P:= 1 TO N−1 DO
                  FOR Q := P+1 TO N DO
                          IF ABS(A[P,Q])/T > Epsilon   THEN
                          ZEROPQ(A,V,N,P,Q)
                          State := Iterating
UNTIL   (State = Done) OR (MaxS < CountS)

EXERCISES FOR JACOBI'S METHOD

Instructions for Exercises 1–9: Use Jacobi's method to find the eigenpairs of the given matrices.

1. $A = \begin{pmatrix} 3 & 2 & 1 \\ 2 & 3 & 2 \\ 1 & 2 & 3 \end{pmatrix}$

2. $A = \begin{pmatrix} 5 & -4 & 6 \\ -4 & 9 & 2 \\ 6 & 2 & 4 \end{pmatrix}$

3. $A = \begin{pmatrix} -2 & -2 & 6 \\ -2 & 3 & 4 \\ 6 & 4 & -1 \end{pmatrix}$

4. $A = \begin{pmatrix} 3 & 2 & -1 \\ 2 & 1 & \frac{1}{2} \\ -1 & \frac{1}{2} & \frac{-1}{3} \end{pmatrix}$

5. $A = \begin{pmatrix} 4 & 3 & 2 & 1 \\ 3 & 4 & 3 & 2 \\ 2 & 3 & 4 & 3 \\ 1 & 2 & 3 & 4 \end{pmatrix}$

6. $A = \begin{pmatrix} 2.75 & -0.25 & -0.75 & 1.25 \\ -0.25 & 2.75 & 1.25 & -0.75 \\ -0.75 & 1.25 & 2.75 & -0.25 \\ 1.25 & -0.75 & -0.25 & 2.75 \end{pmatrix}$

7. $A = \begin{pmatrix} 2.25 & -0.25 & -1.25 & 2.75 \\ -0.25 & 2.25 & 2.75 & 1.25 \\ 1.25 & 2.75 & 2.25 & -0.25 \\ 2.75 & 1.25 & -0.25 & 2.25 \end{pmatrix}$

8. $A = \begin{pmatrix} 5 & 4 & 3 & 2 & 1 \\ 4 & 5 & 4 & 3 & 2 \\ 3 & 4 & 5 & 4 & 3 \\ 2 & 3 & 4 & 5 & 4 \\ 1 & 2 & 3 & 4 & 5 \end{pmatrix}$

9. $A = \begin{pmatrix} 3.6 & 4.4 & 0.8 & -1.6 & -2.8 \\ 4.4 & 2.6 & 1.2 & -0.4 & 0.8 \\ 0.8 & 1.2 & 0.8 & -4.0 & -2.8 \\ -1.6 & -0.4 & -4.0 & 1.2 & 2.0 \\ -2.8 & 0.8 & -2.8 & 2.0 & 1.8 \end{pmatrix}$

10. *Mass–spring systems.* Consider the undamped mass–spring system shown in Figure 11.3. The mathematical model describing the displacements from static equilibrium is

$$\begin{pmatrix} k_1 + k_2 & -k_2 & 0 \\ -k_2 & k_2 + k_3 & -k_3 \\ 0 & -k_3 & k_3 \end{pmatrix} \begin{pmatrix} x_1(t) \\ x_2(t) \\ x_3(t) \end{pmatrix} + \begin{pmatrix} m_1 & 0 & 0 \\ 0 & m_2 & 0 \\ 0 & 0 & m_3 \end{pmatrix} \begin{pmatrix} x_1''(t) \\ x_2''(t) \\ x_3''(t) \end{pmatrix} = \begin{pmatrix} 0 \\ 0 \\ 0 \end{pmatrix}$$

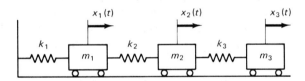

Figure 11.3   An undamped mass–spring system.

(a) Use the substitutions $x_j(t) = v_j \sin(\omega t + \theta)$ for $j = 1, 2, 3$, where θ is a constant, and show that the solution to the mathematical model can be reformulated as follows:

$$\begin{pmatrix} \dfrac{k_1 + k_2}{m_1} & \dfrac{-k_2}{m_1} & 0 \\ \dfrac{-k_2}{m_2} & \dfrac{k_2 + k_3}{m_2} & \dfrac{-k_3}{m_2} \\ 0 & \dfrac{-k_3}{m_3} & \dfrac{k_3}{m_3} \end{pmatrix} \begin{pmatrix} v_1 \\ v_2 \\ v_3 \end{pmatrix} = \omega^2 \begin{pmatrix} v_1 \\ v_2 \\ v_3 \end{pmatrix}.$$

(b) Set $\lambda = \omega^2$; then the three solutions to part (a) are the eigenpairs: λ_j, $\mathbf{V}_j = (v_1^{(j)}, v_2^{(j)}, v_3^{(j)})^T$ for $j = 1, 2, 3$. Show that they are used to form the three fundamental solutions:

$$\mathbf{X}_j(t) = \begin{pmatrix} v_1^{(j)} \sin(\omega_j t + \theta) \\ v_2^{(j)} \sin(\omega_j t + \theta) \\ v_3^{(j)} \sin(\omega_j t + \theta) \end{pmatrix} = \begin{pmatrix} v_1^{(j)} \\ v_2^{(j)} \\ v_3^{(j)} \end{pmatrix} \sin(\omega_j t + \theta),$$

where $\omega_j = \sqrt{\lambda_j}$ for $j = 1, 2, 3$.

Remark. These three solutions are referred to as the **three principal modes of vibration**.

11. Use the technique outlined in Exercise 10, together with Jacobi iteration, to find the eigenpairs and the three principal modes of vibration for the undamped mass–spring systems with the following coefficients:

(a) $k_1 = 3, k_2 = 2, k_3 = 1, m_1 = 1, m_2 = 1, m_3 = 1$
(b) $k_1 = \frac{1}{2}, k_2 = \frac{1}{4}, k_3 = \frac{1}{4}, m_1 = 4, m_2 = 4, m_3 = 4$
(c) $k_1 = 0.2, k_2 = 0.4, k_3 = 0.3, m_1 = 2.5, m_2 = 2.5, m_3 = 2.5$

11.4 EIGENVALUES OF SYMMETRIC MATRICES

Householder's Method

Each transformation in Jacobi's method produced two zero off-diagonal elements, but subsequent iterations might make them nonzero. Hence many iterations are required to make the off-diagonal entries sufficiently close to zero. We now develop a method that produces several zero off-diagonal elements in each iteration, and they remain zero in subsequent iterations. We start by developing an important step in the process.

Theorem 11.23   (Householder Reflection).   If $\mathbf{X}$ and $\mathbf{Y}$ are vectors with the same norm, there exists an orthogonal symmetric matrix P such that

$$\mathbf{Y} = P\mathbf{X}, \tag{1}$$

where

$$P = I - 2\mathbf{W}\mathbf{W}^T \tag{2}$$

and

$$\mathbf{W} = \frac{\mathbf{X} - \mathbf{Y}}{\|\mathbf{X} - \mathbf{Y}\|_2}. \tag{3}$$

Since P is both orthogonal and symmetric, it follows that

$$P^{-1} = P. \tag{4}$$

Proof. Equation (3) is used and defines $\mathbf{W}$ to be the unit vector in the direction $\mathbf{X} - \mathbf{Y}$; hence

$$\mathbf{W}^T\mathbf{W} = 1 \tag{5}$$

and

$$\mathbf{Y} = \mathbf{X} + c\mathbf{W}, \tag{6}$$

where $c = -\parallel \mathbf{X} - \mathbf{Y} \parallel_2$. Since $\mathbf{X}$ and $\mathbf{Y}$ have the same norm, the parallelogram rule for vector addition can be used to see that $\mathbf{Z} = (\mathbf{X} + \mathbf{Y})/2 = \mathbf{X} + (c/2)\mathbf{W}$ is orthogonal to vector $\mathbf{W}$ (see Figure 11.4). This implies that

$$\mathbf{W}^T\left(\mathbf{X} + \frac{c}{2}\,\mathbf{W}\right) = 0.$$

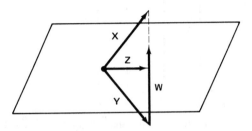

Figure 11.4 The vectors $\mathbf{W}$, $\mathbf{X}$, $\mathbf{Y}$, and $\mathbf{Z}$ involved in the Householder reflection.

Now we can use (5) to expand the preceding equation and get

$$\mathbf{W}^T\mathbf{X} + \frac{c}{2}\,\mathbf{W}^T\mathbf{W} = \mathbf{W}^T\mathbf{X} + \frac{c}{2} = 0. \tag{7}$$

The crucial step is to use (7) and express c in the form

$$c = -2(\mathbf{W}^T\mathbf{X}). \tag{8}$$

Now (8) can be used in (6) to see that

$$\mathbf{Y} = \mathbf{X} + c\mathbf{W} = \mathbf{X} - 2\mathbf{W}^T\mathbf{X}\mathbf{W}.$$

Since the quantity $\mathbf{W}^T\mathbf{X}$ is a scalar, the last equation can be written

$$\mathbf{Y} = \mathbf{X} - 2\mathbf{W}\mathbf{W}^T\mathbf{X} = (I - 2\mathbf{W}\mathbf{W}^T)\mathbf{X}. \tag{9}$$

Looking at (9), we see that $P = I - 2\mathbf{W}\mathbf{W}^T$. The matrix P is symmetric because

$$P^T = (I - 2\mathbf{W}\mathbf{W}^T)^T = I - 2(\mathbf{W}\mathbf{W}^T)^T$$

$$= I - 2\mathbf{W}\mathbf{W}^T = P.$$

The following calculation shows that P is orthogonal:

$$P^TP = (I - 2\mathbf{W}\mathbf{W}^T)(I - 2\mathbf{W}\mathbf{W}^T)$$

$$= I - 4\mathbf{W}\mathbf{W}^T + 4\mathbf{W}\mathbf{W}^T\mathbf{W}\mathbf{W}^T$$

$$= I - 4\mathbf{W}\mathbf{W}^T + 4\mathbf{W}\mathbf{W}^T = I,$$

and the proof is complete.

It should be observed that the effect of the mapping $\mathbf{Y} = P\mathbf{X}$ is to reflect $\mathbf{X}$ through the line whose direction is $\mathbf{Z}$, hence the name **Householder reflection**.

Corollary 11.3   (kth Householder Matrix).   Let A be an $n \times n$ matrix, and $\mathbf{X}$ any vector. If k is an integer with $1 \le k \le n - 2$, we can construct a vector $\mathbf{W}_k$ and matrix $P_k = I - 2\mathbf{W}_k\mathbf{W}_k^T$ so that

$$P_k\mathbf{X} = P_k \begin{pmatrix} x_1 \\ \cdot \\ \cdot \\ \cdot \\ x_k \\ x_{k+1} \\ x_{k+2} \\ \cdot \\ \cdot \\ \cdot \\ x_n \end{pmatrix} = \begin{pmatrix} x_1 \\ \cdot \\ \cdot \\ \cdot \\ x_k \\ -S \\ 0 \\ \cdot \\ \cdot \\ \cdot \\ 0 \end{pmatrix} = \mathbf{Y}. \tag{10}$$

Proof. The key is to define the value S so that $\|\mathbf{X}\|_2 = \|\mathbf{Y}\|_2$, and then invoke Theorem 11.23. The proper value for S must satisfy

$$S^2 = x_{k+1}^2 + x_{k+2}^2 + \cdots + x_n^2, \tag{11}$$

which is readily verified by computing the norms of $\mathbf{X}$ and $\mathbf{Y}$,

$$\begin{aligned} \|\mathbf{X}\|_2 &= x_1^2 + x_2^2 + \cdots + x_n^2 \\ &= x_1^2 + x_2^2 + \cdots + x_k^2 + S^2 \\ &= \|\mathbf{Y}\|_2. \end{aligned} \tag{12}$$

The vector $\mathbf{W}$ is found by using equation (3) of Theorem 11.23:

$$\begin{aligned} \mathbf{W} &= \frac{1}{R}(\mathbf{X} - \mathbf{Y}) \\ &= \frac{1}{R}(0, \ldots, 0, x_{k+1} + S, x_{k+2}, \ldots, x_n)^T. \end{aligned} \tag{13}$$

Less round-off error is propagated when the sign of S is chosen to be the same as the sign of x_{k+1}; hence we compute

$$S = \text{sign}(x_{k+1})(x_{k+1}^2 + x_{k+2}^2 + \cdots + x_n^2)^{1/2}. \tag{14}$$

The number R in (13) is chosen so that $\|\mathbf{W}\|_2 = 1$, and must satisfy

$$\begin{aligned} R^2 &= (x_{k+1} + S)^2 + x_{k+2}^2 + \cdots + x_n^2 \\ &= 2x_{k+1}S + S^2 + x_{k+1}^2 + x_{k+2}^2 + \cdots + x_n^2 \\ &= 2x_{k+1}S + 2S^2. \end{aligned} \tag{15}$$

Therefore, the matrix P_k is given by the formula

$$P_k = I - 2\mathbf{W}\mathbf{W}^T, \tag{16}$$

and the proof is complete.

Householder Transformations

Suppose that A is a symmetric $n \times n$ matrix. Then a sequence of $n - 2$ transformations of the form PAP will reduce A to a symmetric tridiagonal matrix. Let us visualize the process when $n = 5$. The first transformation is defined to be $P_1 A P_1$, where P_1 is constructed by applying Corollary 11.3, with the vector $\mathbf{X}$ being the first column of the matrix A. The general form of P_1 is

$$P_1 = \begin{pmatrix} 1 & 0 & 0 & 0 & 0 \\ 0 & p & p & p & p \\ 0 & p & p & p & p \\ 0 & p & p & p & p \\ 0 & p & p & p & p \end{pmatrix}, \tag{17}$$

where the letter p stands for some element in P_1. As a result, the transformation $P_1 A P_1$ does not effect the element a_{11} of A,

$$P_1 A P_1 = \begin{pmatrix} a_{11} & v_1 & 0 & 0 & 0 \\ u_1 & w_1 & w & w & w \\ 0 & w & w & w & w \\ 0 & w & w & w & w \\ 0 & w & w & w & w \end{pmatrix} = A_1. \tag{18}$$

The element denoted u_1 is changed because of premultiplication by P_1, and v_1 is changed because of postmultiplication by P_1; since A_1 is symmetric, we have $u_1 = v_1$. The changes to the elements denoted w have been affected by both premultiplication and postmultiplication. Also, since $\mathbf{X}$ is the first column of A, equation (10) implies that $u_1 = -S$.

The second Householder transformation is applied to the matrix A_1 defined in (18), and is denoted $P_2 A_1 P_2$, where P_2 is constructed by applying Corollary 11.3, with the vector $\mathbf{X}$ being the second column of the matrix A_1. The form of P_2 is

$$P_2 = \begin{pmatrix} 1 & 0 & 0 & 0 & 0 \\ 0 & 1 & 0 & 0 & 0 \\ 0 & 0 & p & p & p \\ 0 & 0 & p & p & p \\ 0 & 0 & p & p & p \end{pmatrix}, \tag{19}$$

where p stands for some element in P_2. The 2×2 identity block in the upper left corner ensures that the partial tridiagonalization achieved in the first step will not be altered by the second transformation $P_2A_1P_2$. The outcome of this transformation is

$$P_2A_1P_2 = \begin{pmatrix} a_{11} & v_1 & 0 & 0 & 0 \\ u_1 & w_1 & v_2 & 0 & 0 \\ 0 & u_2 & w_2 & w & w \\ 0 & 0 & w & w & w \\ 0 & 0 & w & w & w \end{pmatrix} = A_2. \tag{20}$$

The elements u_2 and v_2 were affected by premultiplication and postmultiplication by P_2. Additional changes have been introduced to the other elements w by the transformation.

The third Householder transformation, $P_3A_2P_3$, is applied to the matrix A_2 defined in (20), where the corollary is used with **X** being the third column of A_2. The form of P_3 is

$$P_3 = \begin{pmatrix} 1 & 0 & 0 & 0 & 0 \\ 0 & 1 & 0 & 0 & 0 \\ 0 & 0 & 1 & 0 & 0 \\ 0 & 0 & 0 & p & p \\ 0 & 0 & 0 & p & p \end{pmatrix}. \tag{21}$$

Again, the 3×3 identity block ensures that $P_3A_2P_3$ does not affect the elements of A_2 which lie in the upper left 3×3 corner, and we obtain

$$P_3A_2P_3 = \begin{pmatrix} a_{11} & v_1 & 0 & 0 & 0 \\ u_1 & w_1 & v_2 & 0 & 0 \\ 0 & u_2 & w_2 & v_3 & 0 \\ 0 & 0 & u_3 & w & w \\ 0 & 0 & 0 & w & w \end{pmatrix} = A_3. \tag{22}$$

Thus it has taken three transformations to reduce A to tridiagonal form.

For efficiency, the transformation PAP is not performed in matrix form. The next result shows that it is more efficiently carried out via some clever vector manipulations.

Theorem 11.24   (Computation of One Householder Transformation).   If P is a Householder matrix, the transformation PAP is accomplished as follows. Let

$$\mathbf{V} = A\mathbf{W} \tag{23}$$

and compute

$$c = \mathbf{W}^T\mathbf{V} \tag{24}$$

and

$$Q = \mathbf{V} - c\mathbf{W}. \tag{25}$$

Then

$$PAP = A - 2\mathbf{W}Q^T - 2Q\mathbf{W}^T. \tag{26}$$

Proof. First, form the product

$$AP = A(I - 2\mathbf{W}\mathbf{W}^T) = A - 2A\mathbf{W}\mathbf{W}^T.$$

Using equation (23), this is written

$$AP = A - 2\mathbf{V}\mathbf{W}^T. \tag{27}$$

Now use (27) and write

$$PAP = (I - 2\mathbf{W}\mathbf{W}^T)(A - 2\mathbf{V}\mathbf{W}^T). \tag{28}$$

When this quantity is expanded, the term $2(2\mathbf{W}\mathbf{W}^T\mathbf{V}\mathbf{W}^T)$ is divided into two portions and (28) can be rewritten as

$$PAP = A - 2\mathbf{W}(\mathbf{W}^T A) + 2\mathbf{W}(\mathbf{W}^T\mathbf{V}\mathbf{W}^T) - 2\mathbf{V}\mathbf{W}^T + 2\mathbf{W}(\mathbf{W}^T\mathbf{V})\mathbf{W}^T. \tag{29}$$

Under the assumption that A is symmetric, we can use the identity $(\mathbf{W}^T A) = (\mathbf{W}^T A^T) = \mathbf{V}^T$. The tricky part is to observe that $(\mathbf{W}^T\mathbf{V})$ is a scalar quantity; hence it can commute freely about in any term. Another scalar identity $\mathbf{W}^T\mathbf{V} = (\mathbf{W}^T\mathbf{V})^T$ is used to obtain the relation $\mathbf{W}^T\mathbf{V}\mathbf{W}^T = (\mathbf{W}^T\mathbf{V})\mathbf{W}^T = \mathbf{W}^T(\mathbf{W}^T\mathbf{V}) = \mathbf{W}^T(\mathbf{W}^T\mathbf{V})^T = [(\mathbf{W}^T\mathbf{V})\mathbf{W}]^T = [\mathbf{W}^T\mathbf{V}\mathbf{W}]^T$. These results are used in the terms of (29) in parentheses to get

$$PAP = A - 2\mathbf{W}\mathbf{V}^T + 2\mathbf{W}(\mathbf{W}^T\mathbf{V}\mathbf{W})^T - 2\mathbf{V}\mathbf{W}^T + 2\mathbf{W}^T\mathbf{V}\mathbf{W}\mathbf{W}^T. \tag{30}$$

Now the distributive law is used in (30) and we obtain

$$PAP = A - 2\mathbf{W}[\mathbf{V}^T - (\mathbf{W}^T\mathbf{V}\mathbf{W})^T] - 2[\mathbf{V} - \mathbf{W}^T\mathbf{V}\mathbf{W}]\mathbf{W}^T. \tag{31}$$

Finally, the definition for $\mathbf{Q}$ given in (25) is used in (31) and the outcome is equation (26), and the proof is complete.

Reduction to Tridiagonal Form

Suppose that A is a symmetric $n \times n$ matrix. Start with

$$A_0 = A. \tag{32}$$

Construct the sequence $P_1, P_2, \ldots, P_{n-1}$ of Householder matrices, so that

$$A_k = P_k A_{k-1} P_k \quad \text{for } i = 1, 2, \ldots, n - 2, \tag{33}$$

where A_k has zeros below the subdiagonal in columns $1, 2, \ldots, k$. Then A_{n-2} is a symmetric tridiagonal matrix that is similar to A. This process is called **Householder's method**.

Example 11.8

Use Householder's method to reduce the following matrix to symmetric tridiagonal form:

$$A_0 = \begin{pmatrix} 4 & 2 & 2 & 1 \\ 2 & -3 & 1 & 1 \\ 2 & 1 & 3 & 1 \\ 1 & 1 & 1 & 2 \end{pmatrix}$$

Solution. The details are left for the reader. The constants $S = 3$ and $R = 30^{1/2} = 5.477226$ are used to construct the vector

$$\mathbf{W}^T = \frac{1}{\sqrt{30}} (0, 5, 2, 1) = (0.000000, 0.912871, 0.365148, 0.182574).$$

Then matrix multiplication $\mathbf{V} = A\mathbf{W}$ is used to form

$$\mathbf{V}^T = \frac{1}{\sqrt{30}} (0, -12, 12, 9) = (0.000000, -2.190890, 2.190890, 1.643168).$$

The constant $c = \mathbf{W}^T\mathbf{V}$ is then found to be

$$c = -0.9.$$

Then the vector $\mathbf{Q} = \mathbf{V} - c\mathbf{W} = \mathbf{V} + 0.9\mathbf{W}$ is formed:

$$\mathbf{Q}^T = \frac{1}{\sqrt{30}} (0.000000, -7.500000, 13.80000, 9.900000)$$

$$= (0.000000, -1.369306, 2.519524, 1.807484).$$

The computation $A_1 = A_0 - 2\mathbf{W}\mathbf{Q}^T - 2\mathbf{Q}\mathbf{W}^T$ produces

$$A_1 = \begin{pmatrix} 4.0 & -3.0 & 0.0 & 0.0 \\ -3.0 & 2.0 & -2.6 & -1.8 \\ 0.0 & -2.6 & -0.68 & -1.24 \\ 0.0 & -1.8 & -1.24 & 0.68 \end{pmatrix}$$

The final step uses the constants $S = -3.1622777$, $R = 6.0368737$, $C = -1.2649111$ and the vectors $\mathbf{W}^T = (0.000000, 0.000000, -0.954514, -0.298168)$, $\mathbf{V}^T = (0.000000, 0.000000, 1.018797, 0.980843)$, $\mathbf{Q}^T = (0.000000, 0.000000, -0.188578, 0.603687)$. The tridiagonal matrix $A_2 = A_1 - 2\mathbf{W}\mathbf{Q}^T - 2\mathbf{Q}\mathbf{W}^T$ is

$$A_2 = \begin{pmatrix} 4.0 & -3.0 & 0.0 & 0.0 \\ -3.0 & 2.0 & 3.162278 & 0.0 \\ 0.0 & 3.162278 & -1.4 & -0.2 \\ 0.0 & 0.0 & -0.2 & 1.4 \end{pmatrix}$$

> **Algorithm 11.4   (Reduction to Tridiagonal Form).**   To reduce the $n \times n$ symmetric matrix A to tridiagonal form by using $n - 2$ Householder transformations.

```
SUBROUTINE  FormW                              {Construct the vector W}
     Sum := 0
     FOR  J = K+1  TO  N  DO
     |____ Sum := Sum + A[J,K]*A[J,K]
     S := SQRT(Sum)
     IF    A[K+1,K] < 0        THEN
     |____ S := − S
     R := SQRT(2*A[K+1,K] + 2*S*S)
     FOR  J = 1  TO  K  DO  W[J] := 0          {First k elements are 0}
     W[K+1] := (A[K+1,K] + S)/R
     FOR  J = K+2  TO  N  DO
     |____ W[J] := A[J,K]/R                    {End of procedure FormW}


SUBROUTINE  FormV                              {Construct the vector V}
     FOR  I = 1  TO  K  DO  V[I] := 0          {first k elements are 0}
     FOR  I = K+1  TO  N  DO
     |    Sum := 0
     |    FOR  J = K+1  TO  N  DO
     |    |____ Sum := Sum + A[I,J]*W[J]
     |_____ V[I] := Sum                  {End of procedure FormV}


SUBROUTINE  FormQ                              {Construct the vector Q}
     C := 0
     FOR  J = K+1  TO  N  DO
     |____    C := C + W[J]*V[J]
     FOR  J = 1  TO  K  DO  Q[J] := 0          {First k elements are 0}
     FOR  J = K+1  TO  N  DO
     |____ Q[J] := V[J] − C*W[J]               {End of procedure FormQ}


SUBROUTINE  FormA                              {Construct the matrix A}
     FOR  J = K+2  TO  N  DO
     |____ A[J,K] := 0  and  A[K,J] := 0
     A[K+1,K] := −S  and  A[K,K+1] := −S
     FOR  J = K  TO  N  DO
     |____ A[J,J] := A[J,J] − 4*Q[J]*W[J]
```

```
FOR  I = K+1  TO  N  DO
   FOR  J = I+1  TO  N  DO
      A[I,J] := A[I,J] − 2*W[I]*Q[J] − 2*Q[I]*W[J]
      A[J,I] := A[I,J]                                {End of procedure FormA}

{The main program starts here.}
FOR  K = 1  TO  N−2  DO
      CALL Procedure FormW(W,A,S,K,N)
      CALL Procedure FormV(V,A,W,K,N)
      CALL Procedure FormQ(Q,V,W,K,N)
      CALL Procedure FormA(A,Q,W,S,K,N)
```

The QL Method

Suppose that A is a real symmetric matrix. In the preceding section we saw how Householder's method is used to construct a similar tridiagonal matrix. The QL method is used to find all eigenvalues of a tridiagonal matrix. Plane rotations similar to those that were introduced in Jacobi's method are used to construct an orthogonal matrix $Q_1 = Q$ and lower-triangular matrix $L_1 = L$ so that $A_1 = A$ has the factorization

$$A_1 = Q_1L_1. \tag{34}$$

Then form the product

$$A_2 = L_1Q_1. \tag{35}$$

Since Q_1 is orthogonal, we can use (34) to see that

$$Q_1^TA_1 = Q_1^TQ_1L_1 = L_1. \tag{36}$$

Therefore, A_2 can be computed with the formula

$$A_2 = Q_1^TA_1Q_1. \tag{37}$$

Since $Q_1^T = Q_1^{-1}$ it follows that A_2 is similar to A_1 and has the same eigenvalues. In general, construct the orthogonal matrix Q_k and lower-triangular matrix L_k so that

$$A_k = Q_kL_k. \tag{38}$$

Then define

$$A_{k+1} = L_kQ_k = Q_k^TA_kQ_k. \tag{39}$$

Again, we have $Q_k^T = Q_k^{-1}$, which implies that A_{k+1} and A_k are similar. An important consequence is that A_k is similar to A and hence has the same structure. Specifically, we can conclude that if A is tridiagonal, A_k is also tridiagonal for all k. Now suppose that A is written as

$$A = \begin{pmatrix} d_1 & e_1 \\ e_1 & d_2 & e_2 \\ & e_2 & d_3 & \cdots \\ & & \vdots & d_{n-2} & e_{n-2} \\ & & & e_{n-2} & d_{n-1} & e_{n-1} \\ & & & & e_{n-1} & d_n \end{pmatrix}. \tag{40}$$

We can find a plane rotation P_{n-1} that reduces to zero the element of A in location $(n-1, n)$, that is,

$$P_{n-1}A = \begin{pmatrix} d_1 & e_1 \\ e_1 & d_2 & e_2 \\ & e_2 & d_3 & \cdots \\ & & \vdots & d_{n-2} & e_{n-2} \\ & & & q_{n-2} & p_{n-1} & 0 \\ & & & r_{n-2} & q_{n-1} & p_n \end{pmatrix}. \tag{41}$$

Continuing in a similar fashion, we can construct a plane rotation P_{n-2} that will reduce to zero the element of $P_{n-1}A$ located in position $(n-2, n-1)$. After $n-1$ steps we arrive at

$$P_1 \cdots P_{n-1}A \begin{pmatrix} p_1 & 0 & 0 \\ q_1 & p_2 & 0 \\ r_1 & q_2 & p_3 & \cdots \\ \vdots \\ & & r_{n-4} & q_{n-3} & p_{n-2} & 0 & 0 \\ & & & r_{n-3} & q_{n-2} & p_{n-1} & 0 \\ & & & & r_{n-2} & q_{n-1} & p_n \end{pmatrix} = L. \tag{42}$$

Since each plane rotation is represented by an orthogonal matrix, equation (42) implies that

$$Q = P_{n-1}^T P_{n-2}^T \cdots P_1^T. \tag{43}$$

Direct multiplication of L by Q will produce all zero elements above the upper second diagonal. The tridiagonal form of A_2 implies that it also has zeros below the lower second diagonal. Investigation will reveal that the terms r_j are used only to compute these zero elements. Consequently, the numbers $\{r_j\}$ do not need to be stored or used in the computer.

For each plane rotation P_j it is assumed that we store the coefficients c_j and s_j which define it. Then we do not need to compute and store Q explicitly; instead, we can use the sequences $\{c_j\}$ and $\{s_j\}$ together with the correct formulas to unravel the product

$$A_2 = LQ = LP_{n-1}^T P_{n-2}^T \cdots P_1^T. \tag{44}$$

The Acceleration Shifts

As outlined above, the QL method will work, but convergence is slow even for matrices of small dimension. We can add a shifting technique which speeds up the rate of convergence. Recall that if λ_j is an eigenvalue of A, then $\lambda_j - s_i$ is an eigenvalue of the matrix $B = A - s_i I$. This idea is incorporated in the modified step

$$A_i - s_i I = Q_i L_i; \tag{45}$$

then form

$$A_{i+1} = L_i Q_i \qquad \text{for } i = 1, 2, \ldots, k_j, \tag{46}$$

where $\{s_i\}$ is a sequence whose sum is λ_j; that is, $\lambda_j = s_1 + s_2 + \cdots s_{k_j}$.

At each stage, the correct amount of shift is found by using the four elements in the upper left corner of the matrix. Start by finding λ_1 and compute the eigenvalues of the 2×2 matrix

$$\begin{pmatrix} d_1 & e_1 \\ e_1 & d_2 \end{pmatrix}. \tag{47}$$

They are x_1 and x_2 and are the roots of the quadratic equation

$$x^2 - (d_1 + d_2)x + d_1 d_2 - e_1 e_1 = 0. \tag{48}$$

The value s_i in equation (45) is chosen to be the root of (48) that is closest to d_1.

Then QL iterating with shifting is repeated until we have $e_1 \approx 0$. This will produce the first eigenvalue, $\lambda_1 = s_1 + s_2 + \cdots + s_{k_1}$. A similar process is repeated with the lower $n - 1$ rows to obtain $e_2 \approx 0$, and the next eigenvalue is λ_2. Successive iteration is applied to smaller submatrices until we obtain $e_{n-2} \approx 0$ and the eigenvalue λ_{n-2}. Finally, the quadratic formula is used to find the last two eigenvalues. The details can be gleaned from the algorithm.

Example 11.9

Find the eigenvalues of the matrix

$$M = \begin{pmatrix} 4 & 2 & 2 & 1 \\ 2 & -3 & 1 & 1 \\ 2 & 1 & 3 & 1 \\ 1 & 1 & 1 & 2 \end{pmatrix}$$

Solution.   In Example 11.8 a tridiagonal matrix A_1 was constructed which is similar to M. We start our diagonalization process with this matrix:

$$A_1 = \begin{pmatrix} 4 & -3 & 0 & 0 \\ -3 & 2 & 3.1622777 & 0 \\ 0 & 3.1622777 & -1.4 & -0.2 \\ 0 & 0 & -0.2 & 1.4 \end{pmatrix}$$

The four elements in the upper left corner are $d_1 = 4$, $d_2 = 2$, and $e_1 = -3$ and are used to form the quadratic equation

$$x^2 - (4 + 2)x + (4)(2) - (-3)(-3) = x^2 - 6x - 1 = 0.$$

Calculation produces the roots $x_1 = 6.162277$ and $x_2 = -0.162277$. The root closest to d_1 is chosen as the first shift $s_1 = 6.162277$, and the first shifted matrix is

$$A_1 - s_1 I = \begin{pmatrix} -2.162277 & -3 & 0 & 0 \\ -3 & -4.162277 & 3.162277 & 0 \\ 0 & 3.162277 & -7.562277 & -0.2 \\ 0 & 0 & -0.2 & -4.762277 \end{pmatrix}$$

Next, the factorization $A_1 - s_1 I = Q_1 L_1$ is computed:

$$Q_1 L_1 = \begin{pmatrix} -0.657500 & -0.753455 & 0 & 0 \\ 0.694920 & -0.606419 & 0.386447 & 0 \\ 0.290914 & -0.253865 & -0.921499 & -0.041960 \\ -0.012217 & 0.010662 & 0.038799 & -0.999119 \end{pmatrix} \times$$

$$\begin{pmatrix} -0.663063 & 0 & 0 & 0 \\ 3.448437 & 3.981659 & 0 & 0 \\ -1.159342 & -4.522537 & 8.182947 & 0 \\ 0 & -0.132688 & 0.517135 & 4.766475 \end{pmatrix}$$

Then the matrix product is computed in the reverse order to obtain

$$A_2 = L_1 Q_1 = \begin{pmatrix} 0.435964 & 0.499588 & 0 & 0 \\ 0.499588 & -5.012797 & 1.538701 & 0 \\ 0 & 1.538701 & -9.288301 & -0.343354 \\ 0 & 0 & -0.343354 & -4.783977 \end{pmatrix}$$

The second shift is $s_2 = 0.48139$, the second shifted matrix is $A_2 - s_2 I = Q_2 L_2$, and

$$A_3 = L_2 Q_2 = \begin{pmatrix} 0.002082 & -0.000201 & 0 & 0 \\ -0.000201 & -5.170593 & 0.814194 & 0 \\ 0 & 0.814194 & -10.077135 & -0.640800 \\ 0 & 0 & -0.640800 & -5.329032 \end{pmatrix}.$$

The third shift is $s_3 = 0.002082$, the third shifted matrix is $A_3 - s_3 I = Q_3 L_3$, and

$$A_4 = L_3 Q_3 = \begin{pmatrix} 0 & 0 & 0 & 0 \\ 0 & -5.072302 & 0.415865 & 0 \\ 0 & 0.415865 & -9.960112 & -1.189118 \\ 0 & 0 & -1.189118 & -5.550592 \end{pmatrix}.$$

The computer implementation carried 11 digits; thus one more shift, $s_4 = 0.0000000001009$, was used to reduce the diagonal to zero. The first seven decimal places of the first eigenvalue is given in the calculation

$$\lambda_1 = s_1 + s_2 + s_3 + s_4 = 6.162277 + 0.48139 + 0.002082 + 0.0000000 = 6.645751.$$

Next, λ_1 is replaced as the first diagonal element, and the process is repeated, but changes are made only in the lower right 3×3 corner of the matrix.

$$A_5 = \begin{pmatrix} 6.645751 & 0 & 0 & 0 \\ 0 & -5.044795 & -0.221069 & 0 \\ 0 & -0.221069 & -9.306984 & -1.989874 \\ 0 & 0 & -1.989874 & -6.231227 \end{pmatrix}.$$

In a similar manner, three more shifts are computed:

$$s_5 = -5.033359, \qquad A_5 - s_5 I = Q_5 L_5, \; A_6 = L_5 Q_5,$$
$$s_6 = \;\;\; 0.033271, \qquad A_6 - s_6 I = Q_6 L_6, \; A_7 = L_6 Q_6,$$
$$s_7 = \;\;\; 0.000089, \qquad A_7 - s_7 I = Q_7 L_7, \; A_8 = L_7 Q_7.$$

An eighth shift, s_8 (which was rounded to zero in six decimal places), was used to reduce the element in the second position on the diagonal to zero. Hence the second eigenvalue is

$$\lambda_2 = \lambda_1 + s_5 + s_6 + s_7 + s_8$$
$$= 6.645751 - 5.033359 + 0.033271 + 0.000089 + 0.000000 = 1.645751.$$

The final step, $A_8 - s_8 I = Q_8 L_8$, $A_9 = L_8 Q_8$, produced the matrix

$$A_9 = \begin{pmatrix} 6.645751 & 0 & 0 & 0 \\ 0 & 1.645751 & 0 & 0 \\ 0 & 0 & -0.291503 & -0.000079 \\ 0 & 0 & -0.000079 & -5.291503 \end{pmatrix}.$$

The final computation requires finding the eigenvalues of the 2×2 matrix in the right-hand corner. The characteristic equation is

$$x^2 - (-0.291503 - 5.291503)x + (-0.291503)(-5.291503) - (-0.000079)(-0.000079),$$

which reduces to

$$x^2 - 5.583006x - 1.542489 = 0.$$

The roots are $x_1 = -5.291502$ and $x_2 = -0.291502$, and the last two eigenvalues are computed with the calculations

$$\lambda_3 = \lambda_2 + x_1 = 1.645751 - 5.291503 = -3.645751$$

and

$$\lambda_4 = \lambda_2 + x_2 = 1.645751 - 0.291503 = 1.354249.$$

Algorithm 11.5   (The *QL* Method with Shifts).   Given the symmetric tridiagonal matrix A whose diagonal elements are stored in D[1], D[2], . . . , D[N] and whose subdiagonal elements are stored in E[1], E[2], . . . , E[N−1], to construct the eigenvalues of A and store them in D[1], D[2], . . . , D[N].

```
SUBROUTINE  FindShift                          {Find the shift parameter}
    B1 := - ( D[M+1] + D[M] )
    C1 := D[M+1]*D[M] - E[M]*E[M]
    D1 := SQRT( ABS( B1*B1 - 4*C1) )
    IF  B1 > 0 THEN
      │ R1 := (- B1 - D1)/2
      │ R2 := 2*C1/(- B1 - D1)
      │ ELSE
      │ R1 := (- B1 + D1)/2
      │ R2 := 2*C1/(- B1 + D1)
    ENDIF
    SH := R1
    IF ABS(D[M]-R2) < ABS(D[M]-R1) THEN SH := R2
```

```
SUBROUTINE  FormL                                   {Construct L and Q implicitly}
    FOR J=1  TO  N  DO  P[J]  := D[J]                   {Make copy of vector D}
    FOR J=1  TO  N-1  DO   Q[J] := E[J]                 {Make copy of vector E}
    PJ1 := P[N]                                     {PJ1 is old value of P[J+1] and}
    QJ := Q[N-1]                                       {QJ is the old value of Q[J]}
    FOR J = N-1  DOWNTO  M  DO                                        {in the loop}
    ┌   P[J+1] := SQRT( PJ1*PJ1 + Q[J]*Q[J] )
    │   C[J] := PJ1/P[J+1]
    │   S[J] := Q[J]/P[J+1]
    │   Q[J] := C[J]*QJ + S[J]*D[J]
    │   PJ1 := C[J]*D[J] − S[J]*QJ
    │      IF  J > M THEN
    │      └──── QJ := C[J]*Q[J−1]
    └────────────────── { R[J−1] := S[J]*Q[J−1]              would be dead code}
    P[M] := PJ1

SUBROUTINE  FormA                                     {Construct the product LQ}
    D[N] := S[N−1]*Q[N−1] + C[N−1]*P[N]
    E[N−1] := S[N−1]*P[N−1]
    FOR     J = N−2  DOWNTO  M  DO
    ┌   D[J+1] := S[J]*Q[J] + C[J]*C[J+1]*P[J+1]
    └──── E[J] := S[J]*P[J]
    D[M] := C[M]*P[M]                                   {End of procedure FormA}
```

{The main program starts here.}

```
DELTA := 0.00000001                                              {Tolerance}
SHIFT := 0
M := 1
```

```
WHILE  M <= N−2 DO
    K := 1
    Cond := 0
    WHILE   (K < 50) AND (Cond = 0)  DO
            CALL   Procedure FindShift
            IF     ABS(E[M]) > DELTA THEN
                   SHIFT := SHIFT + SH
                   FOR  J = M  TO  N  DO
                        D[J] := D[J] − SH
            ELSE
                   D[M] := D[M] + SHIFT
                   M := M + 1
                   Cond := 1
            ENDIF
            CALL   Procedure FormL
            CALL   Procedure FormA
            K := K+1
    ENDWHILE
ENDWHILE
CALL   Procedure Findshift
D[N−1] := R1 + SHIFT
D[N] := R2 + SHIFT
FOR   J = 1   TO   N−1   DO
       E[J] := 0
OUTPUT   the eigenvalues   D[1], D[2], . . . , D[N].
```

EXERCISES FOR SYMMETRIC MATRICES AND THE QL METHOD

1. In the proof of Theorem 11.23, carefully explain why $\mathbf{Z}$ is perpendicular to $\mathbf{W}$.

2. If $\mathbf{X}$ is any vector and $P = I - 2\mathbf{X}\mathbf{X}^T$, show that P is a symmetric matrix.

3. Let $\mathbf{X}$ be any vector and set $P = I - 2\mathbf{X}\mathbf{X}^T$.
   (a) Find the quantity $P^T P$.
   (b) What additional condition is necessary in order that P is an orthogonal matrix?

4. Use Householder's method to reduce the following symmetric matrices to tridiagonal form:

(a) $\begin{pmatrix} 3 & 2 & 1 \\ 2 & 3 & 2 \\ 1 & 2 & 3 \end{pmatrix}$
(b) $\begin{pmatrix} 4 & 3 & 2 & 1 \\ 3 & 4 & 3 & 2 \\ 2 & 3 & 4 & 3 \\ 1 & 2 & 3 & 4 \end{pmatrix}$
(c) $\begin{pmatrix} 5 & 4 & 3 & 2 & 1 \\ 4 & 5 & 4 & 3 & 2 \\ 3 & 4 & 5 & 4 & 3 \\ 2 & 3 & 4 & 5 & 4 \\ 1 & 2 & 3 & 4 & 5 \end{pmatrix}$

5. Use Householder's method to reduce the following symmetric matrices to tridiagonal form:

(a) $\begin{pmatrix} 2.75 & -0.25 & -0.75 & 1.25 \\ -0.25 & 2.75 & 1.25 & -0.75 \\ -0.75 & 1.25 & 2.75 & -0.25 \\ 1.25 & -0.75 & -0.25 & 2.75 \end{pmatrix}$

(b) $\begin{pmatrix} 2.25 & -0.25 & -1.25 & 2.75 \\ -0.25 & 2.25 & 2.75 & 1.25 \\ 1.25 & 2.75 & 2.25 & -0.25 \\ 2.75 & 1.25 & -0.25 & 2.25 \end{pmatrix}$

(c) $\begin{pmatrix} 3.6 & 4.4 & 0.8 & -1.6 & -2.8 \\ 4.4 & 2.6 & 1.2 & -0.4 & 0.8 \\ 0.8 & 1.2 & 0.8 & -4.0 & -2.8 \\ -1.6 & -0.4 & -4.0 & 1.2 & 2.0 \\ -2.8 & 0.8 & -2.8 & 2.0 & 1.8 \end{pmatrix}$

6. Use the QL method to find the eigenvalues of the following tridiagonal matrices:

(a) $\begin{pmatrix} 2 & 1 & 0 \\ 1 & 2 & 1 \\ 0 & 1 & 2 \end{pmatrix}$

(b) $\begin{pmatrix} 4 & 1 & 0 \\ 1 & 4 & 1 \\ 0 & 1 & 4 \end{pmatrix}$

7. Use the QL method to find the eigenvalues of the following tridiagonal matrices:

(a) $\begin{pmatrix} 2 & 1 & 0 & 0 \\ 1 & 2 & 1 & 0 \\ 0 & 1 & 2 & 1 \\ 0 & 0 & 1 & 2 \end{pmatrix}$

(b) $\begin{pmatrix} 2 & -1 & 0 & 0 \\ -1 & 2 & -1 & 0 \\ 0 & -1 & 2 & -1 \\ 0 & 0 & -1 & 2 \end{pmatrix}$

8. Use the QL method to find the eigenvalues of the following tridiagonal matrices:

(a) $\begin{pmatrix} 4 & 1 & 0 & 0 \\ 1 & 4 & 1 & 0 \\ 0 & 1 & 4 & 1 \\ 0 & 0 & 1 & 4 \end{pmatrix}$

(b) $\begin{pmatrix} 4 & -1 & 0 & 0 \\ -1 & 4 & -1 & 0 \\ 0 & -1 & 4 & -1 \\ 0 & 0 & -1 & 4 \end{pmatrix}$

9. Use the QL method to find the eigenvalues of the following tridiagonal matrices:

(a) $\begin{pmatrix} 2 & 1 & 0 & 0 & 0 \\ 1 & 2 & 1 & 0 & 0 \\ 0 & 1 & 2 & 1 & 0 \\ 0 & 0 & 1 & 2 & 1 \\ 0 & 0 & 0 & 1 & 2 \end{pmatrix}$

(b) $\begin{pmatrix} 2 & -1 & 0 & 0 & 0 \\ -1 & 2 & -1 & 0 & 0 \\ 0 & -1 & 2 & -1 & 0 \\ 0 & 0 & -1 & 2 & -1 \\ 0 & 0 & 0 & -1 & 2 \end{pmatrix}$

10. Use the *QL* method to find the eigenvalues of the following tridiagonal matrices:

(a) $\begin{pmatrix} 4 & 1 & 0 & 0 & 0 \\ 1 & 4 & 1 & 0 & 0 \\ 0 & 1 & 4 & 1 & 0 \\ 0 & 0 & 1 & 4 & 1 \\ 0 & 0 & 0 & 1 & 4 \end{pmatrix}$ **(b)** $\begin{pmatrix} 4 & -1 & 0 & 0 & 0 \\ -1 & 4 & -1 & 0 & 0 \\ 0 & -1 & 4 & -1 & 0 \\ 0 & 0 & -1 & 4 & -1 \\ 0 & 0 & 0 & -1 & 4 \end{pmatrix}$

11. Write a report on the *QR* algorithm. See References [3, 9, 10, 19, 29, 40, 41, 74, 85, 92, 97, 104, 128, 152, 153, 169, 175, 192, and 203].

Some Suggested
References
for Reports

Approximation of Functions [34, 44, 114, 149, 157, 161, 182]

Band Systems of Equations [29, 35, 41, 128, 160, 192]

Basic Splines (B-Splines) [35, 96, 101, 149, 160]

Calculus and Computers [13, 18, 36, 55, 110, 111, 120, 122, 134, 162, 176, 179]

Choleski's Factorization [9, 29, 40, 41, 51, 90, 97, 152, 153, 160]

Condition Number of a Matrix [9, 19, 29, 40, 41, 57, 62, 74, 94, 96, 98, 101, 117, 128, 145, 152, 153, 160, 192]

Differential Equation [7, 31, 33, 39, 42, 99, 104, 136, 138, 152, 171, 173]

Dynamical Systems [2, 17, 48, 164]

Economization of Power Series [3, 9, 29, 41, 51, 62, 76, 85, 88, 117, 153, 184]

Engineering Usage of Numerical Methods [6, 17, 20, 31, 33, 39, 54, 59, 71, 88, 93, 104, 131, 136, 141, 163, 174, 183, 190, 195]

Error Propagation [4, 9, 40, 41, 49, 51, 78, 79, 81, 133, 142, 145, 153, 204]

Extrapolation [19, 29, 35, 40, 41, 78, 117, 153]

Fast Fourier Transform [25, 29, 33, 40, 51, 62, 79, 96, 98, 112, 136, 141, 145, 149, 150, 152, 153, 155, 169, 210]

Floating-Point Arithmetic [8, 9, 35, 40, 41, 51, 57, 62, 90, 101, 103, 128, 129, 142, 153, 181, 184, 208]

Forward-Difference Formulas [9, 29, 40, 41, 51, 76, 78, 81, 85, 90, 94, 105, 117, 128, 143, 145, 153, 181, 184]

Gauss–Jordan Method [29, 44, 51, 62, 79, 85, 90, 117, 152]

Hermite Interpolation [9, 29, 40, 41, 79, 81, 90, 92, 128, 153, 191, 193, 208]

Hexadecimal Numbers [8, 35, 51, 101, 142]

Ill-Conditioned Matrices [9, 19, 29, 40, 41, 47, 49, 62, 94, 101, 128, 145, 153, 192, 197]

Inverse Interpolation [9, 19, 29, 35, 41, 62, 81, 128, 153, 166, 181, 191]

Iterated Interpolation [29, 78, 81, 90, 126, 128, 129, 181, 184, 208]

Iterative Improvement (Residual Correction) [8, 9, 19, 29, 40, 41, 49, 51, 58, 72, 90, 94, 96, 97, 117, 137, 152, 153, 160]

Least Squares [39, 92, 109, 112, 152]

Legendre Polynomials [9, 29, 40, 41, 75, 152, 153]

Linear Programming (Simplex Method) [19, 27, 35, 37, 41, 44, 50, 53, 79, 83, 94, 104, 115, 135, 152, 153, 154, 165, 169]

Linear Systems [61, 66, 74, 82, 152, 159]

Loss of Significance (Cancellation) [3, 8, 35, 40, 79, 142]

Mathematical Modeling [15, 17, 22, 23, 32, 39, 42, 64, 72, 83, 95, 98, 102, 104, 107, 113, 115, 116, 131, 135, 136, 190]

Monte Carlo Methods [35, 41, 57, 76, 83, 87, 98, 112, 115, 135, 152, 154]

Multiple Integrals [29, 62, 67, 85, 96, 112, 117, 152, 153]

Newton–Cotes Formulas [9, 29, 62, 76, 78, 81, 90, 94, 97, 105, 117, 126, 128, 152, 153, 154, 160, 175, 193, 208]

Norms of Vectors and Matrices [9, 19, 29, 40, 49, 62, 90, 94, 96, 101, 117, 128, 145, 153, 192]

Orthogonal Polynomials [9, 19, 29, 34, 40, 41, 44, 76, 81, 90, 96, 126, 128, 143, 145, 149, 152, 153, 169]

Pivoting Strategies [9, 29, 35, 40, 41, 58, 79, 96, 101, 117, 128, 145, 146, 152, 153, 160]

Programming [12, 103, 119, 150, 151, 152]

QR Algorithm [3, 9, 10, 19, 29, 40, 41, 74, 85, 92, 97, 104, 128, 152, 153, 169, 175, 192, 203]

Quasi–Newton Methods [29, 96, 97, 139, 152, 153]

Quotient Difference Algorithm [3, 29, 62, 78, 79, 86, 112, 152, 200]

Relaxation Methods [19, 29, 40, 41, 62, 90, 139, 152, 199, 207]

Remes Algorithm [9, 19, 56, 88, 128, 149, 152, 153]

Round-Off Errors [4, 9, 29, 35, 41, 51, 76, 79, 81, 90, 94, 101, 117, 128, 146, 153, 160, 181, 184, 186, 204]

Secant Method (Convergence) [9, 35, 40, 41, 153, 160]

Scientific Computing [5, 71, 98, 103, 150, 151, 152, 158, 159, 160]

Software for Numerical Analysis [32, 52, 82, 84, 95, 97, 98, 124, 125, 150, 151, 152, 158, 159, 160, 178]

SOR Method [10, 29, 40, 41, 49, 137, 139, 152, 160, 175, 199, 207]

Stability of Differential Equations [3, 8, 9, 29, 40, 60, 76, 78, 79, 96, 101, 128, 146, 152, 153, 160]

Step-Size Control for Differential Equations [29, 40, 60, 75, 101, 117, 160]

Stiff Differential Equations [9, 29, 40, 57, 60, 98, 117, 152, 153, 160, 173]

Bibliography
and References

1. Aberth, Oliver (1988). *Precise Numerical Analysis,* Wm. C. Brown, Dubuque, Ia.
2. Aburdene, Maurice F. (1988). *Computer Simulation of Dynamic Systems,* Wm. C. Brown, Dubuque, Ia.
3. Acton, Forman S. (1970). *Numerical Methods that Work,* Harper & Row, New York.
4. Adby, P. R., and M. A. H. Dempster (1974). *Introduction to Optimization Methods,* Halsted Press, New York.
5. Aho, Alfred V., John E. Hopcroft, and Jeffrey D. Ullman (1974). *The Design and Analysis of Computer Algorithms,* Addison-Wesley, Reading, Mass.
6. Al-Khafaji, Amir Wadi, and John R. Tooley (1986). *Numerical Methods in Engineering Practice,* Holt, Rinehart and Winston, New York.
7. Ascher, Uri M., Robert M. M. Mattheij, and Robert D. Russell (1988). *Numerical Solution of Boundary Value Problems for Ordinary Differential Equations,* Prentice Hall, Englewood Cliffs, N.J.
8. Atkinson, Kendall E. (1985). *Elementary Numerical Analysis,* John Wiley, New York.
9. Atkinson, Kendall E. (1988). *An Introduction to Numerical Analysis,* 2nd ed., John Wiley, New York.
10. Atkinson, Laurence V., and P. J. Harley (1983). *An Introduction to Numerical Methods with Pascal,* Addison-Wesley, Reading, Mass.
11. Bailey, Paul B., Lawrence F. Shampine, and Paul E. Waltman (1968). *Nonlinear Two Point Boundary Value Problems,* Academic Press, New York.
12. Barnard, David T., and Robert G. Crawford (1982). *Pascal Programming Problems and Applications,* Reston, Reston, Va.

13. Beckman, Charlene E., and Ted Sundstrom (1990). *Graphing Calculator Laboratory Manual for Calculus,* Addison-Wesley, New York.

14. Bender, Carl M., and Steven A. Orszag (1978). *Advanced Mathematical Methods for Scientists and Engineers,* McGraw-Hill, New York.

15. Bender, Edward A. (1978). *An Introduction to Mathematical Modeling,* John Wiley, New York.

16. Bennett, William Ralph (1976). *Introduction to Computer Applications for Nonscience Students,* Prentice Hall, Englewood Cliffs, N.J.

17. Bennett, William Ralph (1976). *Scientific and Engineering Problem-Solving with the Computer,* Prentice Hall, Englewood Cliffs, N.J.

18. Bitter, Gary G. (1983). *Microcomputer Applications for Calculus,* Prindle, Weber & Schmidt, Boston, Mass.

19. Blum, E. K. (1972). *Numerical Analysis and Computation: Theory and Practice,* Addison-Wesley, Reading, Mass.

20. Borse, G. J. (1985). *FORTRAN 77 and Numerical Methods for Engineers,* Prindle, Weber & Schmidt, Boston, Mass.

21. Brainerd, Walter S., and Lawrence H. Landweber (1974). *Theory of Computation,* John Wiley, New York.

22. Brams, Steven J., William F. Lucas, and Philip D. Straffin, eds. (1983). *Political and Related Models,* Springer-Verlag, New York.

23. Braun, Martin, Courtney S. Coleman, and Donald A. Drew, eds. (1983). *Differential Equation Models,* Springer-Verlag, New York.

24. Brent, Richard P. (1973). *Algorithms for Minimization without Derivatives,* Prentice Hall, Englewood Cliffs, N.J.

25. Brigham, E. Oran (1988). *The Fast Fourier Transform and Its Applications,* Prentice Hall, Englewood Cliffs, N.J.

26. Buck, R. Creighton (1978). *Advanced Calculus,* 3rd ed., McGraw-Hill, New York.

27. Bunday, Brian D. (1984). *Basic Linear Programming,* Edward Arnold, Baltimore Md.

28. Bunday, Brian D. (1984). *Basic Optimisation Methods,* Edward Arnold, Baltimore, Md.

29. Burden, Richard L., and J. Douglas Faires (1985). *Numerical Analysis,* 3rd ed., Prindle, Weber & Schmidt, Boston, Mass.

30. Burnett, David S. (1987). *Finite Element Analysis: From Concepts to Applications,* Addison-Wesley, Reading, Mass.

31. Carnahan, Brice, H. A. Luther, and James O. Wilkes (1969). *Applied Numerical Methods,* John Wiley, New York.

32. Carroll, John M. (1987). *Simulation Using Personal Computers,* Prentice Hall, Englewood Cliffs, N.J.

33. Chapra, Steven C. (1985). *Numerical Methods for Engineers: With Personal Computer Applications,* McGraw-Hill, New York.

34. Cheney, Ward (1966). *Introduction to Approximation Theory,* McGraw-Hill, New York.

35. Cheney, Ward, and David Kincaid (1985). *Numerical Mathematics and Computing,* 2nd ed. Brooks/Cole, Monterey, Calif.

36. Christensen, Mark J. (1981). *Computing for Calculus,* Academic Press, New York.

37. Chvatal, Vasek (1980). *Linear Programming,* W. H. Freeman & Company Publishers, New York.

38. Coddington, Earl A., and Norman Levinson (1955). *Theory of Ordinary Differential Equations,* McGraw-Hill, New York.

39. Constantinides, Alkis (1987). *Applied Numerical Methods with Personal Computers,* McGraw-Hill, New York.

40. Conte, S. D., and Carl de Boor (1980). *Elementary Numerical Analysis: An Algorithmic Approach,* McGraw-Hill, New York.

41. Dahlquist, Germund, and Ake Bjorck (1974). *Numerical Methods,* Prentice Hall, Englewood Cliffs, N.J.

42. Danby, J. M. A. (1985). *Computing Applications to Differential Equations: Modelling in the Physical and Social Sciences,* Reston, Reston, Va.

43. Daniels, Richard W. (1978). *An Introduction to Numerical Methods and Optimization Techniques,* North-Holland, New York.

44. Davis, Philip J. (1963). *Interpolation and Approximation,* Blaisdell, New York.

45. Davis, Philip J., and Philip Rabinowitz (1984). *Methods of Numerical Integration,* 2nd ed., Academic Press, New York.

46. deBoor, Carl (1978). *A Practical Guide to Splines,* Springer-Verlag, New York.

47. Deif, Assem S., (1986). *Sensitivity Analysis in Linear Systems,* Springer-Verlag, New York.

48. Devaney, Robert L. (1990). *Chaos, Fractals, and Dynamics: Computer Experiments in Mathematics,* Addison-Wesley, Menlo Park, Calif.

49. Dew, P. M., and K. R. James (1983). *Introduction to Numerical Computation in Pascal,* Springer-Verlag, New York.

50. Dixon, L. C. W., (1972). *Nonlinear Optimisation,* Crane, Russak & Co., New York.

51. Dodes, Irving Allen (1978). *Numerical Analysis for Computer Science,* North-Holland, New York.

52. Dongarra, J. J. (1979). *LINPACK: Users' Guide,* SIAM, Philadelphia, Pa.

53. Dorn, William S., and Daniel D. McCracken (1976). *Introductory Finite Mathematics with Computing,* John Wiley, New York.

54. Dorn, William S., and Daniel D. McCracken (1972). *Numerical Methods with Fortran IV Case Studies,* John Wiley, New York.

55. Edwards, C. H. (1986). *Calculus and the Personal Computer,* Prentice Hall, Englewood Cliffs, N.J.

56. Fike, C. T. (1968). *Computer Evaluation of Mathematical Functions,* Prentice Hall, Englewood Cliffs, N.J.

57. Forsythe, George E., Michael A. Malcolm, and Cleve B. Moler (1977). *Computer Methods for Mathematical Computations,* Prentice Hall, Englewood Cliffs, N.J.

58. Forsythe, George E., and Cleve B. Moler (1967). *Computer Solution of Linear Algebraic Systems,* Prentice Hall, Englewood Cliffs, N.J.

59. Fox, L., and D. F. Mayers (1968). *Computing Methods for Scientists and Engineers,* Clarendon Press, Oxford.

60. Gear, C. William (1971). *Numerical Initial Value Problems in Ordinary Differential Equations,* Prentice Hall, Englewood Cliffs, N.J.

61. George, Alan, and Joseph W. H. Liu (1981). *Computer Solution of Large Sparse Positive Definite Systems,* Prentice Hall, Englewood Cliffs, N.J.

62. Gerald, Curtis F., and Patrick O. Wheatley (1984). *Applied Numerical Analysis,* 3rd ed., Addison-Wesley, Reading, Mass.

63. Gill, Philip E., Walter Murray, and Margaret H. Wright (1981). *Practical Optimization,* Academic Press, New York.

64. Giordano, Frank R., and Maurice D. Weir (1985). *A First Course in Mathematical Modeling,* Brooks/Cole, Monterey, Calif.

65. Goldstine, Herman H. (1977). *A History of Numerical Analysis from the 16th through the 19th Century,* Springer-Verlag, New York.

66. Golub, Gene H., and Charles F. VanLoan (1989). *Matrix Computations,* The John Hopkins University Press, Baltimore, Md.

67. Gordon, Sheldon P. (1986). Simpson's Rule for Double Integrals, *The UMAP Journal,* Vol. 7, No. 4, pp. 319–328.

68. Gourlay, A. R., and G. A. Watson (1973). *Computational Methods for Matrix Eigenproblems,* John Wiley, New York.

69. Greenspan, Donald, and Vincenzo Casulli (1988). *Numerical Analysis for Applied Mathematics, Science and Engineering,* Addison-Wesley, New York.

70. Grove, Wendell E. (1966). *Brief Numerical Methods,* Prentice Hall, Englewood Cliffs, N.J.

71. Guggenheimer, H. (1987). *BASIC Mathematical Programs for Engineers and Scientists,* Petrocelli Books, West Hempstead, N.Y.

72. Haberman, Richard (1977). *Mathematical Models: Mechanical Vibrations, Population Dynamics, and Traffic Flow,* Prentice Hall, Englewood Cliffs, N.J.

73. Hageman, Louis A., and David M. Young (1981). *Applied Iterative Methods,* Academic Press, New York.

74. Hager, William W. (1988). *Applied Numerical Linear Algebra,* Prentice Hall, Englewood Cliffs, N.J.

75. Hamming, Richard W. (1971). *Introduction to Applied Numerical Analysis,* McGraw-Hill, New York.

76. Hamming, Richard W. (1973). *Numerical Methods for Scientists and Engineers,* 2nd ed., McGraw-Hill, New York.

77. Henrici, Peter (1974). *Applied and Computational Complex Analysis,* Vol. 1, John Wiley, New York.

78. Henrici, Peter (1964). *Elements of Numerical Analysis,* John Wiley, New York.

79. Henrici, Peter (1982). *Essentials of Numerical Analysis with Pocket Calculator Demonstrations,* John Wiley, New York.

80. Hildebrand, Francis B. (1976). *Advanced Calculus for Applications,* 2nd ed., Prentice Hall, Englewood Cliffs, N.J.

81. Hildebrand, Francis B. (1974). *Introduction to Numerical Analysis,* 2nd ed., McGraw-Hill, New York.

82. Hill, David R., and Cleve B. Moler (1989). *Experiments in Computational Matrix Algebra,* Random House, New York.

83. Hillier, Frederick S., and Gerald J. Lieberman (1974). *Operations Research,* 2nd ed., Holden-Day, San Francisco, Calif.

84. Hopkins, Tim, and Chris Philips (1988). *Numerical Methods in Practice Using the NAG Library,* Addison-Wesley, New York.

85. Hornbeck, Robert W. (1975). *Numerical Methods,* Quantum, New York.

86. Householder, Alston S. (1970). *The Numerical Treatment of a Single Nonlinear Equation,* McGraw-Hill, New York.

87. Householder, Alston S. (1953). *Principles of Numerical Analysis,* McGraw-Hill, New York.

88. Hultquist, Paul F. (1988). *Numerical Methods for Engineers and Computer Scientists,* Benjamin/Cummings, Menlo Park, Calif.

89. Hundhausen, Joan R., and Robert A. Walsh (1985). Unconstrained Optimization, *The UMAP Journal,* Vol. 6, No. 4, pp. 57–90.

90. Isaacson, Eugene, and Herbert Bishop Keller (1966). *Analysis of Numerical Methods,* John Wiley, New York.

91. Jacobs, D., ed. (1977). *The State of the Art in Numerical Analysis,* Academic Press, New York.

92. Jacques, Ian, and Colin Judd (1987). *Numerical Analysis,* Chapman and Hall, New York.

93. James, M. L., G. M. Smith, and J. C. Wolford, (1985). *Applied Numerical Methods for Digital Computation,* 3rd ed., Harper & Row, New York.

94. Jensen, Jens A., and John H. Rowland (1975). *Methods of Computation: The Linear Space Approach to Numerical Analysis,* Scott, Foresman, Glenview, Ill.

95. Jepsen, Charles H., and Eugene Herman (1988). *The Matrix Algebra Calculator: Linear Algebra Problems for Computer Solution,* Brooks/Cole, Pacific Grove, Calif.

96. Johnson, Lee W., and R. Dean Riess (1982). *Numerical Analysis,* 2nd ed., Addison-Wesley, Reading, Mass.

97. Johnston, R. L. (1982). *Numerical Methods: a Software Approach,* John Wiley, New York.

98. Kahaner, David, Cleve Moler, and Stephen Nash (1989). *Numerical Methods and Software,* Prentice Hall, Englewood Cliffs, N.J.

99. Keller, Herbert Bishop (1976). *Numerical Solution of Two Point Boundary Value Problems,* SIAM, Philadelphia, Pa.

100. Kincaid, David, and Ward Cheney (1991). *Numerical Analysis Mathematics of Scientific Computing,* Brooks/Cole, Pacific Grove, Calif.

101. King, J. Thomas (1984). *Introduction to Numerical Computation,* McGraw-Hill, New York.

102. Klamkin, Murray S. (1987). *Mathematical Modeling: Classroom Notes in Applied Mathematics,* SIAM, Philadelphia, Pa.

103. Knuth, Donald E. (1981). *The Art of Computer Programming,* Vol. 2, Seminumerical Algorithms, 2nd ed., Addison-Wesley, Reading, Mass.

104. Kreyszig, Erwin (1983). *Advanced Engineering Mathematics,* 5th ed., John Wiley, New York.

105. Kunz, Kaiser S. (1957). *Numerical Analysis,* McGraw-Hill, New York.

106. Lambert, J. D. (1973). *Computational Methods in Ordinary Differential Equations,* John Wiley, New York.

107. Lancaster, Peter (1976). *Mathematics: Models of the Real World,* Prentice Hall, Englewood Cliffs, N.J.

108. Lapidus, L., and J. H. Seinfeld (1971). *Numerical Solution of Ordinary Differential Equations*, Academic Press, New York.

109. Lawson, C. L., and R. J. Hanson (1974). *Solving Least-Squares Problems*, Prentice Hall, Englewood Cliffs, N.J.

110. Lax, Peter, Samuel Burstein, and Anneli Lax (1976). *Calculus with Applications and Computing*, Springer-Verlag, New York.

111. Leinbach, L. Carl (1991). *Calculus Laboratories Using DERIVE*, Wadsworth, Belmont, Calif.

112. Lindfield, G. R., and J. E. T. Penny (1989). *Microcomputers in Numerical Analysis*, Halsted Press, New York.

113. Lucas, William F., Fred S. Roberts, and Robert M. Thrall, eds. (1983). *Discrete and System Models*, Springer-Verlag, New York.

114. Luke, Yudell L. (1975). *Mathematical Functions and Their Applications*, Academic Press, New York.

115. Maki, Daniel P., and Maynard Thompson (1973). *Mathematical Models and Applications*, Prentice Hall, Englewood Cliffs, N.J.

116. Marcus-Roberts, Helen, and Maynard Thompson, eds. (1983). *Life Science Models*, Springer-Verlag, New York.

117. Maron, Melvin J., and Robert J. Lopez (1991). *Numerical Analysis: a Practical Approach*, 3rd ed., Wadsworth, Belmont, Calif.

118. Mathews, John H. (1988). *Complex Variables for Mathematics and Engineering*, Wm. C. Brown, Dubuque, Ia.

119. McCalla, Thomas Richard (1967). *Introduction to Numerical Methods and FORTRAN Programming*, John Wiley, New York.

120. McCarty, George (1975). *Calculator Calculus*, Page-Ficklin Publications, Palo Alto, Calif.

121. McCormick, John M., and Mario G. Salvadori (1964). *Numerical Methods in FORTRAN*, Prentice Hall, Englewood Cliffs, N.J.

122. McNeary, Samuel S. (1973). *Introduction to Computational Methods for Students of Calculus*, Prentice Hall, Englewood Cliffs, N.J.

123. Miel, George J. (1981). Calculator Demonstrations of Numerical Stability, *The UMAP Journal*, Vol. 2, No. 2, pp. 3–7.

124. Miller, Webb (1984). *The Engineering of Numerical Software*, Prentice Hall, Englewood Cliffs, N.J.

125. Miller, Webb (1987). *A Software Tools Sampler*, Prentice Hall, Englewood Cliffs, N.J.

126. Milne, William Edmund (1949). *Numerical Calculus*, Princeton University Press, Princeton, N.J.

127. Moore, Ramon E. (1966). *Interval Analysis*, Prentice Hall, Englewood Cliffs, N.J.

128. Morris, John Ll. (1983). *Computational Methods in Elementary Numerical Analysis*, John Wiley, New York.

129. Moursund, David G., and Charles S. Duris (1967). *Elementary Theory and Application of Numerical Analysis*, McGraw-Hill, New York.

130. Murphy, J., D. Ridout, and Brigid McShane (1988). *Numerical Analysis, Algorithms and Computation*, Halsted Press, New York.

131. Noble, Ben (1967). *Applications of Undergraduate Mathematics in Engineering,* Macmillan, New York.

132. Noble, Ben, and James W. Daniel (1977). *Applied Linear Algebra,* 2nd ed., Prentice Hall, Englewood Cliffs, N.J.

133. Nonweiler, T. R. F. (1984). *Computational Mathematics: an Introduction to Numerical Approximation,* Halsted Press, New York.

134. Oldknow, Adrian, and Derek Smith (1983). *Learning Mathematics with Micros,* Halsted Press, New York.

135. Olinick, Michael (1978). *An Introduction to Mathematical Models in the Social and Life Sciences,* Addison-Wesley, Reading, Mass.

136. O'Neil, Peter V. (1991). *Advanced Engineering Mathematics,* 3rd ed., Wadsworth, Belmont, Calif.

137. Ortega, James M. (1972). *Numerical Analysis,* Academic Press, New York.

138. Ortega, James M., and William G. Poole (1981). *An Introduction to Numerical Methods for Differential Equations,* Pitman, Marshfield, Mass.

139. Ortega, James M., and W. C. Rheinboldt (1970). *Iterative Solution of Nonlinear Equations in Several Variables,* Academic Press, New York.

140. Parlett, Beresford N. (1980). *The Symmetric Eigenvalue Problem,* Prentice Hall, Englewood Cliffs, N.J.

141. Pearson, Carl E. (1986). *Numerical Methods in Engineering and Science,* Van Nostrand Reinhold, New York.

142. Pennington, Ralph H. (1970). *Introductory Computer Methods and Numerical Analysis,* 2nd ed., Macmillan, New York.

143. Pettofrezzo, Anthony J. (1984). *Introductory Numerical Analysis,* Orange Publishers, Winter Park, Fla.

144. Phillips, G. M., and P. J. Taylor (1974). *Theory and Applications of Numerical Analysis,* Academic Press, New York.

145. Pizer, Stephen M. (1975). *Numerical Computing and Mathematical Analysis,* Science Research Associates, Chicago, Ill.

146. Pizer, Stephen M., with Victor L. Wallace (1983). *To Compute Numerically: Concepts and Strategies,* Little, Brown, Boston, Mass.

147. Pokorny, Cornel K., and Curtis F. Gerald (1989). *Computer Graphics: the Principles Behind the Art and Science,* Franklin, Beedle & Associates, Irvine, Calif.

148. Potts, J. Frank, and J. Walter Oler (1989). *Finite Element Applications with Microcomputers,* Prentice Hall, Englewood Cliffs, N.J.

149. Powell, Michael James David (1981). *Approximation Theory and Methods,* Cambridge University Press, Cambridge, Mass.

150. Press, William H., Brian P. Flannery, Saul A. Teukolsky, and William T. Vetterling (1988). *Numerical Recipes in C: the Art of Scientific Computing,* Cambridge University Press, New York.

151. Press, William H., Brian P. Flannery, Saul A. Teukolsky, and William T. Vetterling (1989). *Numerical Recipes in Pascal: the Art of Scientific Computing,* Cambridge University Press, New York.

152. Press, William H., Brian P. Flannery, Saul A. Teukolsky, and William T. Vetterling (1986). *Numerical Recipes: the Art of Scientific Computing,* Cambridge University Press, New York.

153. Ralston, Anthony, and Philip Rabinowitz (1978). *A First Course in Numerical Analysis,* 2nd ed., McGraw-Hill, New York.

154. Ralston, Anthony, and Herbert S. Wilf (1960). *Mathematical Methods for Digital Computers,* John Wiley, New York.

155. Ramirez, Robert W. (1985). *The FFT, Fundamentals and Concepts,* Prentice Hall, Englewood Cliffs, N.J.

156. Rheinboldt, Werner C. (1981). Algorithms for Finding Zeros of Functions, *The UMAP Journal,* Vol. 2, No. 1, pp. 43–72.

157. Rice, John R. (1969). *The Approximation of Functions,* Addison-Wesley, Reading, Mass.

158. Rice, John R. (1980). *Mathematical Aspects of Scientific Software,* Springer-Verlag, New York.

159. Rice, John R. (1981). *Matrix Computations and Mathematical Software,* McGraw-Hill, New York.

160. Rice, John Rischard (1983). *Numerical Methods, Software and Analysis: IMSL Reference Edition,* McGraw-Hill, New York.

161. Rivlin, Theodore J. (1969). *An Introduction to the Approximation of Functions,* Blaisdell, Waltham, Mass.

162. Rosser, J. Barkley, and Carl de Boor (1979). *Pocket Calculator Supplement for Calculus,* Addison-Wesley, Reading, Mass.

163. Salvadori, Mario G., and Melvin L. Baron (1961). *Numerical Methods in Engineering,* 2nd ed., Prentice Hall, Englewood Cliffs, N.J.

164. Sandefur, James T. (1990). *Discrete Dynamical Systems: Theory and Applications,* Oxford University Press, New York.

165. Scalzo, Frank, and Rowland Hughes (1977). *A Computer Approach to Introductory College Mathematics,* Mason/Charter, New York.

166. Scarborough, James B. (1966). *Numerical Mathematical Analysis,* The Johns Hopkins University Press, Baltimore, Md.

167. Scheid, Francis (1968). *Theory and Problems of Numerical Analysis,* McGraw-Hill, New York.

168. Schultz, M. H. (1966). *Spline Analysis,* Prentice Hall, Englewood Cliffs, N.J.

169. Schwarz, Hans Rudolf, and J. Waldvogel (1989). *Numerical Analysis: a Comprehensive Introduction,* John Wiley, New York.

170. Scraton, R. E. (1984). *Basic Numerical Methods: An Introduction to Numerical Mathematics on a Microcomputer,* Edward Arnold, Baltimore, Md.

171. Sewell, Granville (1988). *The Numerical Solution to Ordinary and Partial Differential Equations,* Harcourt Brace Jovanovich, San Diego, Calif.

172. Shampine, Lawrence F., and Richard C. Allen (1973). *Numerical Computing: an Introduction,* Saunders, Philadelphia, Pa.

173. Shampine, Lawrence F., and M. K. Gordon (1975). *Computer Solution of Ordinary Differential Equations: the Initial Value Problem,* W. H. Freeman & Company Publishers, San Francisco, Calif.

174. Shoup, Terry E. (1979). *A Practical Guide to Computer Methods for Engineers,* Prentice Hall, Englewood Cliffs, N.J.

175. Shoup, Terry E. (1983). *Numerical Methods for the Personal Computer,* Prentice Hall, Englewood Cliffs, N.J.

176. Sicks, Jon L. (1985). *Investigating Secondary Mathematics with Computers,* Prentice Hall, Englewood Cliffs, N.J.

177. Simmons, George F. (1972). *Differential Equations: With Applications and Historical Notes,* McGraw-Hill, New York.

178. Smith, B. T., J. M. Boyle, J. Dongarra, B. Garbow, Y. Ikebe, V. C. Klema, and C. B. Moler (1976). *Matrix Eigensystem Routines: EISPACK Guide,* 2nd ed., Vol. 6 of *Lecture Notes in Computer Science,* Springer-Verlag, New York.

179. Smith, David A. (1976). *INTERFACE: Calculus and the Computer,* Houghton Mifflin, Boston, Mass.

180. Smith, G. D. (1978). *The Numerical Solution of Partial Differential Equations,* 2nd ed., Oxford University Press, Oxford.

181. Smith, W. Allen (1986). *Elementary Numerical Analysis,* Prentice Hall, Englewood Cliffs, N.J.

182. Snyder, Martin Avery (1966). *Chebyshev Methods in Numerical Approximation,* Prentice Hall, Englewood Cliffs, N.J.

183. Stanton, Ralph G. (1961). *Numerical Methods for Science and Engineering,* Prentice Hall, Englewood Cliffs, N.J.

184. Stark, Peter A. (1970). *Introduction to Numerical Methods,* Macmillan, Toronto, Ontario.

185. Strang, G., and G. Fix (1973). *An Analysis of the Finite Element Method,* Prentice Hall, Englewood Cliffs, N.J.

186. Strecker, George E. (1982). Round Numbers: An Introduction to Numerical Expression, *The UMAP Journal,* Vol. 3, No. 4, pp. 425–454.

187. Stroud, A. H. (1971). *Approximate Calculation of Multiple Integrals,* Prentice Hall, Englewood Cliffs, N.J.

188. Stroud, A. H., and Don Secrest (1966). *Gaussian Quadrature Formulas,* Prentice Hall, Englewood Cliffs, N.J.

189. Szidarovszky, Ferenec, and Sidney Yakowitz (1978). *Principles and Procedures of Numerical Analysis,* Plenum Press, New York.

190. Thompson, William J. (1984). *Computing in Applied Science,* John Wiley, New York.

191. Todd, John (1979). *Basic Numerical Mathematics,* Vol. 1: Numerical Analysis, Academic Press, New York.

192. Todd, John (1977). *Basic Numerical Mathematics,* Vol. 2: Numerical Algebra, Academic Press, New York.

193. Tompkins, Charles B., and Walter L. Wilson (1969). *Elementary Numerical Analysis,* Prentice Hall, Englewood Cliffs, N.J.

194. Traub, J. F. (1964). *Iterative Methods for the Solution of Equations,* Prentice Hall, Englewood Cliffs, N.J.

195. Tuma, Jan J. (1989). *Handbook of Numerical Calculations in Engineering,* McGraw-Hill, New York.

196. Turner, Peter R. (1989). *Guide to Numerical Analysis,* CRC Press, Boca Raton, Fla.

197. Vandergraft, James S. (1983). *Introduction to Numerical Computations,* Academic Press, New York.

198. VanIwaarden, John L. (1985). *Ordinary Differential Equations with Numerical Techniques,* Harcourt Brace Jovanovich, San Diego, Calif.

199. Varga, Richard S. (1962). *Matrix Iterative Analysis,* Prentice Hall, Englewood Cliffs, N.J.

200. Wachspress, Eugene L. (1966). *Iterative Solution of Elliptic Systems,* Prentice Hall, Englewood Cliffs, N.J.

201. Wendroff, Burton (1966). *Theoretical Numerical Analysis,* Academic Press, New York.

202. Wilkes, Maurice Vincent (1966). *A Short Introduction to Numerical Analysis,* Cambridge University Press, New York.

203. Wilkinson, J. H. (1965). *The Algebraic Eigenvalue Problem,* Clarendon Press, Oxford.

204. Wilkinson, J. H. (1963). *Rounding Errors in Algebraic Processes,* Prentice Hall, Englewood Cliffs, N.J.

205. Wilkinson, J. H., and C. Reinsch (1971). *Handbook for Automatic Computation,* Vols. 1 and 2, Springer-Verlag, New York.

206. Yakowitz, Sidney, and Ferenec Szidarovszky (1989). *An Introduction to Numerical Computations,* 2nd ed., Macmillan, New York.

207. Young, David M. (1971). *Iterative Solution of Large Linear Systems,* Academic Press, New York.

208. Young, David M., and Robert Todd Gregory (1972). *A Survey of Numerical Mathematics,* Vols. 1 and 2, Addison-Wesley, Reading, Mass.

209. Zienkiewicz, O. C., and R. L. Taylor (1989). *The Finite Element Method,* 4th ed., McGraw-Hill, New York.

210. Zohar, Shalhav (1979). *Faster Fourier Transformation: the Algorithm of S. Winograd,* Jet Propulsion Laboratory, Pasadena, Calif.

Answers to Selected Exercises

1. (a) $L = 2$ (b) $L = \frac{1}{2}$

2. (a) $\sin(2)$ (b) $\ln(4)$

3. (a) $c = -0.414214$ (b) $c = 2.414214$

4. $\min\{f(-1), f(1), f(2)\} = \min\{3, -1, 3\} = -1$
$\max\{f(-1), f(1), f(2)\} = \max\{3, -1, 3\} = 3$

5. Solving $f'(c) = 0$ yields $c = 1/\sqrt{3} = 0.577350$.

6. Solution $m = [f(3) - f(1)]/(3 - 1) = 10$. Solving $f'(c) = 10$ yields $c = \sqrt{13/3} = 2.081666$.

7. Compute $f''(x) = -8 + 6x$. The solution to $f''(c) = 0$ is $c = \frac{4}{3}$.

8. $F(x) = xe^x - e^x$ and $F(2) - F(0) = 1 + e^2 = 8.389056$

9. $\dfrac{d}{dx} \displaystyle\int_0^x t^2 \cos(t)\, dt = x^2 \cos(x)$

10. $\dfrac{1}{\pi/2 - 0} \displaystyle\int_0^{\pi/2} \sin(x)\, dx = \dfrac{2}{\pi} = f(c)$ implies that $c = \arcsin(2/\pi) = 0.690107$.

11. $S = \cos(0.5) + \cos(1.5) + \cos(2.5) + \cos(3.5) + \cos(4.5) = -1.000076$.
Compare this with $-\sin(5) + \sin(0) = -0.958924$. The error is $-0.958924 + 1.000076 = 0.041152$.

12. $\displaystyle\sum_{n=1}^{\infty} \left(\frac{2}{3}\right)^n = 2$

13. $P(x) = 1 - x^2/2! + x^4/4! - x^6/6! + x^8/8!$

15. (a) $(x_1, x_2) = (14/5, -27/5)$ **(c)** $(x_1, x_2) = \left(\frac{1}{5}, \frac{8}{5}\right)$

16. (a) $Q_0(x) = -4 - x + 4x^2 + x^3$ **(b)** $Q_0(x) = -5 - x + 5x^2 + x^3$

SECTION 1.2 BINARY NUMBERS: page 25

1. (a) The computer's answer is not 0 because 0.1 is not an exact binary fraction   **(b)** 0 (exactly)

2. (a) 21   **(c)** 109

3. (a) 0.84375   **(c)** 0.6640625

4. (a) 1.4140625

5. (a) $2^{1/2} - 1.4140625 = 0.000151062.\ .\ .$

6. (a) 10111_{two} **(c)** 101101000_{two}

7. (a) 0.0111_{two} **(c)** 0.10111_{two}

8. (a)
$$
\begin{aligned}
2R &= 0.2, & d_1 &= 0 = \text{int}(0.2), & F_1 &= 0.2 = \text{frac}(0.2)\\
2F_1 &= 0.4, & d_2 &= 0 = \text{int}(0.4), & F_2 &= 0.4 = \text{frac}(0.4)\\
2F_2 &= 0.8, & d_3 &= 0 = \text{int}(0.8), & F_3 &= 0.8 = \text{frac}(0.8)\\
2F_3 &= 1.6, & d_4 &= 1 = \text{int}(1.6), & F_2 &= 0.6 = \text{frac}(1.6)\\
2F_4 &= 1.2, & d_5 &= 1 = \text{int}(1.2), & F_5 &= 0.2 = \text{frac}(1.2)
\end{aligned}
$$

$$\tfrac{1}{10} = 0.d_1d_2d_3d_4d_5 \ .\ .\ .\ _{two} = 0.000110\overline{0011}\ .\ .\ .\ _{two}$$

9. (a) $\tfrac{1}{10} - 0.0001100_{two} = 0.006250000\ .\ .\ .$

11. Use $c = \tfrac{3}{16}$ and $r = \tfrac{1}{16}$ to get $S = \dfrac{\frac{3}{16}}{1 - \frac{1}{16}} = \tfrac{1}{5}$

14. (a) $\tfrac{1}{3} \approx 0.1011_{two} \times 2^{-1} = 0.1011_{two} \times 2^{-1}$

$$\frac{\frac{1}{5}}{\frac{8}{15}} \approx 0.1101_{two} \times 2^{-2} = \frac{0.01101_{two} \times 2^{-1}}{0.100011_{two} \times 2^{0}}$$

$$\frac{8}{15} \approx 0.1001_{two} \times 2^{0} = 0.1001_{two} \times 2^{0}$$

$$\frac{\frac{1}{6}}{\frac{7}{10}} \approx 0.1011_{two} \times 2^{-2} = \frac{0.001011_{two} \times 2^{0}}{0.101111_{two} \times 2^{0}} \approx \boxed{0.1100_{two}}$$

15. (b) $10 = 101_{three}$ **(d)** $49 = 1211_{three}$

16. (c) $\tfrac{1}{2} = 0.11111111\overline{1}\ .\ .\ .\ _{three}$

17. (b) $10 = 20_{five}$ **(d)** $49 = 144_{five}$

18. (c) $\tfrac{1}{2} = 0.2222222\overline{2}\ .\ .\ .\ _{five}$

24. (a) $213_{16} = 531_{ten}$ **(c)** $1ABE_{16} = 6846_{ten}$
 (e) $0.2_{16} = 0.125_{ten}$ **(g)** $0.A4B_{16} = 0.643310546875_{ten}$

25. (a) $512_{ten} = 200_{16}$ **(c)** $51264_{ten} = C840_{16}$

26. (a) $\tfrac{1}{3} = 0.55555\overline{5}\ .\ .\ .\ _{16}$ **(c)** $\tfrac{1}{10} = 0.19999\overline{9}\ .\ .\ .\ _{16}$

SECTION 1.3 ERROR ANALYSIS: page 39

1. (a) $x = 2.71828182$, $\bar{x} = 2.7182$, $(x - \bar{x}) = 0.00008182$, $(x - \bar{x})/x = 0.00003010$, four significant digits

2. $\dfrac{1}{4} + \dfrac{1}{4^3 3} + \dfrac{1}{4^5 5(2!)} + \dfrac{1}{4^7 7(3!)} = \dfrac{292{,}807}{1{,}146{,}880} = 0.2553074428 = \bar{p}$,

$p - \bar{p} = 0.0000000178$, $(p - \bar{p})/p = 0.0000000699$

3. (a) $p_1 + p_2 = 1.414 + 0.09125 = 1.505$, $p_1 p_2 = (1.414)(0.09125) = 0.1290$

4. The error involves loss of significance.

   (a) $\dfrac{0.70711385222 - 0.70710678119}{0.00001} = \dfrac{0.00000707103}{0.000001} = 0.707103$

5. (a) $\ln((x + 1)/x)$ or $\ln(1 + 1/x)$ **(c)** $\cos(2x)$

6. (a) $P(2.72) = (((2.72)^3 - 3(2.72)^2) + 3(2.72)) - 1$

$= ((20.12 - 3(7.398)) + 8.16) - 1 = ((20.12 - 22.19) + 8.16) - 1$

$= (-2.07 + 8.16) - 1 = 6.09 - 1 = 5.09$

$Q(2.72) = ((2.72 - 3)2.72 + 3)2.72 - 1$

$= ((-0.28)2.72 + 3)2.72 - 1 = (-0.7616 + 3)2.72 - 1$

$= (2.238)2.72 - 1 = 6.087 - 1 = 5.087$

$R(2.72) = (2.72 - 1)^3 = (1.72)^3 = 5.088$

7. (a) 0.498 **(b)** 0.499

9. (a) $\dfrac{1}{1 - h} + \cos(h) = 2 + h + \dfrac{h^2}{2} + h^3 + O(h^4)$

   (b) $\dfrac{1}{1 - h}\cos(h) = 1 + h + \dfrac{h^2}{2} + \dfrac{h^3}{2} + O(h^4)$

SECTION 2.1 ITERATION FOR SOLVING $x = g(x)$: page 52

1. (a) $g(2) = -4 + 8 - 2 = 2$, $g(4) = -4 + 16 - 8 = 4$

   (b) $p_0 = 1.9$, $E_0 = 0.1$, $R_0 = 0.05$

   $p_1 = 1.795$, $E_1 = 0.205$, $R_1 = 0.1025$

   $p_2 = 1.5689875$, $E_2 = 0.4310125$, $R_2 = 0.21550625$

   $p_3 = 1.04508911$, $E_3 = 0.95491089$, $R_3 = 0.477455444$

   (e) The sequence in part (b) does not converge to $P = 2$. The sequence in part (c) converges to $P = 4$.

4. (a) $p_0 = 1.9$, $E_0 = 0.1$, $R_0 = 0.5$

   $p_1 = 1.939$, $E_1 = 0.061$, $R_1 = 0.0305$

   The sequence converges to $P = 2$ because $|g'(2)| = 0.6 < 1$.

   (b) $p_0 = -1.9$, $E_0 = -0.1$, $R_0 = 0.05$

   $p_1 = -1.861$, $E_1 = -0.139$, $R_1 = 0.0695$

   The sequence does not converge because $|g'(-2)| = 1.4 > 1$.

5. $P = 2$, $g'(2) = 5$, iteration will not converge to $P = 2$.

   $P = -2$, $g'(-2) = 5$, iteration will not converge to $P = -2$.

10. (d) $p_1 = 0.99990000$, $p_2 = 0.99980002$

   $p_3 = 0.99970006$, $p_4 = 0.99960012$

   $p_5 = 0.99950020$, $\{p_k\}$ converges very slowly to zero.

13. (b) $p_0 = 4.4$

   $p_1 = 4.39886349$

   $p_2 = 4.39770070$

   (c) $p_0 = 4.6$

   $p_1 = 4.39891292$

   $p_2 = 4.59784909$

$p_3 = 4.39651069$
$p_4 = 4.39529251$
The sequence does not
converge to $P = 4.5$

$p_3 = 4.59680778$
$p_4 = 4.59578829$
The sequence converges
slowly to $P = 4.5$

15. (a) $r_1 = -3$, $r_2 = 3$
   (c) One real root, $r_1 = 3$; ignore the complex roots.
16. (a) $g'(x) = (1 - 9x^{-2})/2$, $g'(3) = 0$
   (c) $g'(x) = 3x^2$, $g'(3) = 27$
17. (a) $p_1 = 3.00161290$, $p_2 = 3.00000043$
   (c) $p_1 = 5.79100000$, $p_2 = 170.205129$

SECTION 2.2 BRACKETING METHODS FOR LOCATING A ROOT: page 63

1. $I_0 = (0.11 + 0.12)/2 = 0.115$, $A(0.115) = 254{,}403$
   $I_1 = (0.11 + 0.115)/2 = 0.1125$, $A(0.1125) = 246{,}072$
   $I_2 = (0.1125 + 0.125)/2 = 0.11375$, $A(0.11375) = 250{,}198$
4. There are many choices for intervals $[a, b]$ on which $f(a)$ and $f(b)$ have opposite sign. The following
   answers are one such choice.
   (a) $f(1) < 0$ and $f(2) > 0$, so there is a root in $[1, 2]$; also
        $f(-1) < 0$ and $f(-2) > 0$, so there is a root in $[-2, -1]$.
   (c) $f(3) < 0$ and $f(4) > 0$, so there is a root in $[3, 4]$.
5. (a) $[1.0, 1.8]$, $[1.0, 1.4]$, $[1.0, 1.2]$, $[1.1, 1.2]$, $[1.10, 1.15]$
7. $[3.2, 4.0]$, $[3.6, 4.0]$, $[3.6, 3.8]$, $[3.6, 3.7]$, $[3.65, 3.70]$
8. (a) $[3.2, 4.0]$, $[3.2, 3.6]$, $[3.4, 3.6]$, $[3.5, 3.6]$, $[3.55, 3.60]$
   (b) $[6.0, 6.8]$, $[6.4, 6.8]$, $[6.4, 6.6]$, $[6.4, 6.5]$, $[6.40, 6.45]$
9. (a) $r_1 = 1.14619322$, $r_2 = -1.84140566$ (c) $r = 3.69344136$
10. $c_0 = -1.8300782$, $c_1 = -1.8409252$, $c_2 = -1.8413854$, $c_3 = -1.8414048$
12. $c_0 = 3.6979549$, $c_1 = 3.6935108$, $c_2 = 3.6934424$, $c_3 = 3.6934414$
14. (a) $r_1 = 1.14619322$, $r_2 = -1.84140566$ (c) $r = 3.69344136$

SECTION 2.3 INITIAL APPROXIMATIONS AND CONVERGENCE CRITERIA: page 70

1. $g(x) = x^2$, $h(x) = \exp(x)$ for $-2 \le x \le 2$
   Approximate root location -0.7. Computed root -0.7034674225.
3. $g(x) = \sin(x)$, $h(x) = 2\cos(2x)$ for $-2 \le x \le 2$
   Approximate root locations -1.0, 0.6.
   Computed roots -1.002966954, 0.6348668712.
5. $g(x) = \ln(x)$, $h(x) = (x - 2)^2$ for $0.5 \le x \le 4.5$
   Approximate root locations 1.4, 3.0.
   Computed roots 1.412391172, 3.057103550.
7. (a) Ymin := Y(0)
         DO FOR K = 1 TO N
               IF Y(K) < Ymin THEN   Ymin := Y(K)

SECTION 2.4 NEWTON–RAPHSON AND SECANT METHODS: page 86

1. (a) $p_{k+1} = (p_k + 8/p_k)/2$
      $p_0 = 3$
      $p_1 = 2.833333334$

   (c) $p_{k+1} = (p_k + 91/p_k)/2$
      $p_0 = 10$
      $p_1 = 9.55$

$$p_2 = 2.828431373$$
$$p_3 = 2.828427125$$

3. (a) $p_{k+1} = (2p_k + 7/p_k^2)/3$
$$p_0 = 2$$
$$p_1 = 1.916666667$$
$$p_2 = 1.912938458$$
$$p_3 = 1.912931183$$

7. (a) $g(p_{k-1}) = (p_{k-1}^2 + 1)/(2p_{k-1} - 2)$
(b) $p_0 = 2.5$
$$p_1 = 2.41666667$$
$$p_2 = 2.41421569$$
$$p_3 = 2.41421356$$

10. (a) $f'(t) = 320 \exp(-t/5) - 160$
$p_1 = 7.96819027$, $p_2 = 7.96812129$, $p_3 = 7.96812130$
(b) Range $= r(7.96812130) = 637.4497040$

12. (a) $g(p_{k-1}) = p_{k-1} - 0.25(p_{k-1} - 2) = 0.5 + 0.75p_{k-1}$
(b) $p_0 = 2.1$
$$p_1 = 2.075$$
$$p_2 = 2.05625$$
$$p_3 = 2.0421875$$
$$p_4 = 2.031640625$$

15. (c) $g(p_{k-1}) = (4p_{k-1}^3 + 3)/(6p_{k-1}^2 - 1)$
$$p_0 = 1.0$$
$$p_1 = 1.4$$
$$p_2 = 1.29888476$$
$$p_3 = 1.28969699$$
$$p_4 = 1.28962390$$

The point on the parabola is
$(p, p^2) = (1.28962390, 1.66312980)$.

18. No, because $f'(x)$ is not continuous at the root $p = 0$. You could also try computing terms with $g(p_{k-1}) = -2p_{k-1}$ and see that the sequence diverges.

20. There are two solutions, $x = 0.839018884$ and $x = 3.401748648$.

22. (a) $g(p_{k-1}) = p_{k-1} - p_{k-1} \exp(-p_{k-1})/[(1 - p_{k-1}) \exp(-p_{k-1})]$
$g(p_{k-1}) = p_{k-1}^2/(p_{k-1} - 1)$
(b) $p_0 = \quad 0.20$
$$p_1 = -0.05$$
$$p_2 = -0.002380952$$
$$p_3 = -0.000005655$$
$$p_4 = -0.000000000$$
$$\lim_{k \to \infty} p_k = 0.0$$

(d) The value of the function in part (c) is $f(p_4) = 7.515 \times 10^{-10}$

24. $g(p_{k-1}) = [2 + \sin(p_{k-1}) - p_{k-1} \cos(p_{k-1})]/[2 - \cos(p_{k-1})]$
$$p_0 = 1.5$$
$$p_1 = 1.49870157$$
$$p_2 = 1.49870113$$

31. $p_0 = 2.6$
$$p_1 = 2.5$$
$$p_2 = 2.41935484$$
$$p_3 = 2.41436464$$

(c) $p_{k+1} = (2p_k + 200/p_k^2)/3$
$$p_0 = 6$$
$$p_1 = 5.851851853$$
$$p_2 = 5.848037967$$
$$p_3 = 5.848035477$$

(c) $p_0 = -0.5$
$$p_1 = -0.41666667$$
$$p_2 = -0.41421569$$
$$p_3 = -0.41421356$$

(c) Convergence is linear. The error is reduced by a factor of $\frac{3}{4}$ with each iteration.

(c) $p_0 = 20.0$
$$p_1 = 21.05263158$$
$$p_2 = 22.10250034$$
$$p_3 = 23.14988809$$
$$p_4 = 24.19503505$$
$$\lim_{k \to \infty} p_k = \infty$$

34. (b) $g(x) = x - \tan(x^3)/x^2$
$$p_0 = \quad 1.0$$
$$p_1 = \quad 0.557407725$$
$$p_2 = \quad 0.005640693$$

$p_4 = 2.41421384$ $\qquad\qquad\qquad\qquad\qquad p_3 = 0.000000000$

$p_5 = 2.41421356$

36. (a) $g(x) = x - \dfrac{x^2 - a}{2x}\left[1 - \dfrac{(x^2 - a)2}{2[2x]^2}\right]^{-1} = \dfrac{x(x^2 + 3a)}{3x^2 + a}$

$\qquad g(x) = \dfrac{15x + x^3}{5 + 3x^2}$

$\qquad\quad p_1 = 2.2352941176,\ p_2 = 2.2360679775,\ p_3 = 2.2360679775$

(b) $g(x) = \dfrac{2 + 4x + 2x^2 + x^3}{3 + 4x + 2x^2}$

$\qquad\quad p_1 = -2.0130081301,\ p_2 = -2.0000007211,\ p_3 = -2.0000000000$

SECTION 2.5 AITKEN'S PROCESS AND STEFFENSEN'S AND MULLER'S METHODS: page 98

4. $p_n = 1/(4^n + 4^{-n})$

n	p_n	q_n Aitken's
0	0.5	-0.26437542
1	0.23529412	-0.00158492
2	0.06225681	-0.00002390
3	0.01562119	-0.00000037
4	0.00390619	
5	0.00097656	

5. $g(x) = (6 + x)^{1/2}$

n	p_n	q_n Aitken's
0	2.5	3.00024351
1	2.91547595	3.00000667
2	2.98587943	3.00000018
3	2.99764565	3.00000001
4	2.99960758	
5	2.99993460	

7. Solution of $\cos(x) - 1 = 0$.

n	p_n Steffensen's
0	0.5
1	0.24465808
2	0.12171517
3	0.00755300
4	0.00377648
5	0.00188824
6	0.00000003

9. Solution of $\sin(x^3) = 0$. The accuracy of the table values depends on the correct evaluation of $\sin(x)$ and $\cos(x)$!

n	p_n Steffensen's
0	0.5
1	0.33245982
2	0.22158997
3	0.00468419
4	0.00312280
5	0.00208187
6	0.00000000

11. The sum of the infinite series is $S = 99$.

n	S_n	T_n
1	0.99	98.9999988
2	1.9701	99.0000017
3	2.940399	98.9999988
4	3.90099501	98.9999992
5	4.85198506	
6	5.79346521	

13. The sum of the infinite series is $S = 4$.

15. Muller's method for $f(x) = x^3 - x - 2$.

n	p_n	$f(p_n)$
0	1.0	−2.0
1	1.2	−1.472
2	1.4	−0.656
3	1.52495614	0.02131598
4	1.52135609	−0.00014040
5	1.52137971	−0.00000001

SECTION 2.6 ITERATION FOR NONLINEAR SYSTEMS: page 108

1. $J(x, y) = \begin{pmatrix} 1 - x & y/4 \\ (1 - x)/2 & (2 - y)/2 \end{pmatrix}$, $J(1.1, 2.0) = \begin{pmatrix} -0.1 & 0.5 \\ -0.05 & 0.0 \end{pmatrix}$

	Fixed-point iteration		Seidel iteration	
k	p_k	q_k	p_k	q_k
0	1.1	2.0	1.1	2.0
1	1.12	1.9975	1.12	1.9964
2	1.1165508	1.9963984	1.1160016	1.9966327
∞	1.1165151	1.9966032	1.1165151	1.9966032

3. $J(x, y) = \begin{pmatrix} 1 - x & 1/2 \\ 2(1 - x)/9 & (2 - y)/2 \end{pmatrix}$, $J(1.4, 2.0) = \begin{pmatrix} -0.4 & 0.5 \\ -0.088889 & 0.0 \end{pmatrix}$

	Fixed-point iteration		Seidel iteration	
k	p_k	q_k	p_k	q_k
0	1.4	2.0	1.4	2.0
1	1.42	1.9822222	1.42	1.9804
2	1.4029111	1.9803210	1.402	1.9819480
∞	1.4076401	1.9814506	1.4076401	1.9814506

9. Fixed-point iteration will find the solution near the origin $(p, q) \approx (-0.090533, -0.099864)$. Eight other solutions are near: $(1, 0)$, $(0, 1)$, $(-1, 0)$, $(0, -1)$, $(1, 1)$, $(1, -1)$ $(-1, 1)$, $(-1, -1)$.

SECTION 2.7 NEWTON'S METHOD FOR SYSTEMS: page 117

1. $0 = x^2 - y - 0.2$, $0 = y^2 - x - 0.3$

$\mathbf{P}_k$	Solution of the linear system: $J(\mathbf{P}_k)\,d\mathbf{P} = -\mathbf{F}(\mathbf{P}_k)$	$\mathbf{P}_k + d\mathbf{P}$
$\begin{pmatrix} 1.2 \\ 1.2 \end{pmatrix}$	$\begin{pmatrix} 2.4 & -1.0 \\ -1.0 & 2.4 \end{pmatrix}\begin{pmatrix} -0.0075630 \\ 0.0218487 \end{pmatrix} = -\begin{pmatrix} 0.04 \\ -0.06 \end{pmatrix}$	$\begin{pmatrix} 1.192437 \\ 1.221849 \end{pmatrix}$
$\begin{pmatrix} 1.192437 \\ 1.221849 \end{pmatrix}$	$\begin{pmatrix} 2.384874 & -1.0 \\ -1.0 & 2.443697 \end{pmatrix}\begin{pmatrix} -0.0001278 \\ -0.0002476 \end{pmatrix} = -\begin{pmatrix} 0.0000572 \\ 0.0004774 \end{pmatrix}$	$\begin{pmatrix} 1.192309 \\ 1.221601 \end{pmatrix}$

(a) Therefore, $(p_1, q_1) = (1.192437, 1.221849)$ and $(p_2, q_2) = (1.192309, 1.221601)$.

$\mathbf{P}_k$	Solution of the linear system: $J(\mathbf{P}_k)\,d\mathbf{P} = -\mathbf{F}(\mathbf{P}_k)$	$\mathbf{P}_k + d\mathbf{P}$
$\begin{pmatrix} -0.2 \\ -0.2 \end{pmatrix}$	$\begin{pmatrix} -0.4 & -1.0 \\ -1.0 & -0.4 \end{pmatrix}\begin{pmatrix} -0.0904762 \\ 0.0761905 \end{pmatrix} = -\begin{pmatrix} 0.04 \\ -0.06 \end{pmatrix}$	$\begin{pmatrix} -0.2904762 \\ -0.1238095 \end{pmatrix}$
$\begin{pmatrix} -0.2904762 \\ -0.1238095 \end{pmatrix}$	$\begin{pmatrix} -0.5809524 & -1.0 \\ -1.0 & -0.2476190 \end{pmatrix}\begin{pmatrix} 0.0044128 \\ 0.0056223 \end{pmatrix} = -\begin{pmatrix} 0.0081859 \\ 0.0058050 \end{pmatrix}$	$\begin{pmatrix} -0.2860634 \\ -0.1181872 \end{pmatrix}$

(b) Therefore, $(p_1, q_1) = (-0.2904762, -0.1238095)$ and $(p_2, q_2) = (-0.2860634, -0.1181872)$.

2. $0 = x^2 + y^2 - 2$, $0 = x^2 - y - 0.5x + 0.1$

(a) $(p_1, q_1) = (1.144118, 0.833824)$ and $(p_2, q_2) = (1.142320, 0.833732)$.

(b) $(p_1, q_1) = (-0.809639, 1.160241)$ and $(p_2, q_2) = (-0.809350, 1.159721)$.

3. (b) The values of the Jacobian determinant at the solution points are $|J(1, 1)| = 0$ and $|J(-1, -1)| = 0$. Newton's method depends on being able to solve a linear system where the matrix is $J(p_n, q_n)$ and (p_n, q_n) is near a solution. For this example, the system equations are ill-conditioned and thus hard to solve with precision. In fact, for some values near a solution we have $J(x_0, y_0) = 0$, for example, $J(1.0001, 1.0001) = 0$.

6. $0 = 2xy - 3$, $0 = x^2 - y - 2$

$(p_0, q_0) = (1.5, 0.9)$, $(p_1, q_1) = (1.7083333, 0.875)$

$(p_2, q_2) = (1.6980622, 0.8833096)$, $(p, q) = (1.6980481, 0.8833672)$

8. $0 = 3x^2 - 2y^2 - 1$, $0 = x^2 - 2x + y^2 + 2y - 8$

I. $(p_0, q_0) = (-1.0, 1.0)$, $(p_1, q_1) = (-1.2, 1.3)$

$(p_2, q_2) = (-1.1928571, 1.2785714)$, $(p, q) = (-1.1928731, 1.2784441)$

II. $(p_0, q_0) = (3.0, -3.4)$, $(p_1, q_1) = (2.925, -3.5125)$

$(p_2, q_2) = (2.9234928, -3.5100168)$, $(p, q) = (2.9234921, -3.5100156)$

10. $0 = 2x^3 - 12x - y - 1$, $0 = 3y^2 - 6y - x - 3$

I. $(p, q) = (2.5908586, 2.6922233)$ II. $(p, q) = (2.4627339, -0.6795569)$

III. $(p, q) = (-0.0493209, -0.4083890)$ IV. $(p, q) = (-2.3078910, 2.1093705)$

V. $(p, q) = (-2.4108204, -0.0937976)$ VI. $(p, q) = (-0.2855601, 2.3801497)$

12. $0 = 7x^3 - 10x - y - 1$, $0 = 8y^3 - 11y + x - 1$

I. $(-0.0905331, -0.0998637)$ II. $(1.2433858, 0.0221339)$

III. $(-0.2311448, 1.2250008)$ IV. $(-1.1531142, -0.2017060)$

V. $(0.0124852, -1.1248379)$ VI. $(1.2912933, 1.1591321)$

VII. $(1.1860751, -1.1809721)$ VIII. $(-1.0604370, 1.2569430)$

IX. $(-1.1980104, -1.0558300)$

SECTION 3.1 INTRODUCTION TO VECTORS AND MATRICES: page 129

1. **(i).** **(a)** $(1, 4)$ **(b)** $(5, -12)$ **(c)** $(9, -12)$ **(d)** 5 **(e)** -38 **(f)** $(-5, 12)$ **(g)** 13
(iii). **(a)** $(5, -20, -10)$ **(b)** $(3, 4, 12)$ **(c)** $(12, -24, 3)$ **(d)** 9 **(e)** 89
(f) $(-3, -4, -12)$ **(g)** 13
2. **(a)** $\theta = \arccos(-16/21) \approx 2.437045$ radians

SECTION 3.2 PROPERTIES OF VECTORS AND MATRICES: page 140

1. $AB = \begin{pmatrix} -11 & -12 \\ 13 & -24 \end{pmatrix}$, $\quad BA = \begin{pmatrix} -15 & 10 \\ -12 & -20 \end{pmatrix}$

3. **(a)** $(AB)C = A(BC) = \begin{pmatrix} 2 & -5 \\ -88 & -56 \end{pmatrix}$

5. **(a)** 33 **(c)** The determinant does not exist because the matrix is not square.

6.

U	$V = R_y\left(\dfrac{\pi}{6}\right) U$	$W = R_z\left(\dfrac{\pi}{4}\right) R_y\left(\dfrac{\pi}{6}\right) U$
$(0, 0, 0)$	$(0.000000, 0, \quad 0.000000)$	$(\ \ 0.000000, 0.000000, \quad 0.000000)$
$(1, 0, 0)$	$(0.866025, 0, -0.500000)$	$(\ \ 0.612372, 0.612372, -0.500000)$
$(0, 1, 0)$	$(0.000000, 1, \quad 0.000000)$	$(-0.707107, 0.707107, \quad 0.000000)$
$(0, 0, 1)$	$(0.500000, 0, \quad 0.866025)$	$(\ \ 0.353553, 0.353553, \quad 0.866025)$
$(1, 1, 0)$	$(0.866025, 1, -0.500000)$	$(-0.094734, 1.319479, -0.500000)$
$(1, 0, 1)$	$(1.366025, 0, \quad 0.366025)$	$(\ \ 0.965926, 0.965926, \quad 0.366025)$
$(0, 1, 1)$	$(0.500000, 1, \quad 0.866025)$	$(-0.353553, 1.060660, \quad 0.866025)$
$(1, 1, 1)$	$(1.366025, 1, \quad 0.366025)$	$(\ \ 0.258819, 1.673033, \quad 0.366025)$

11. **(a)** MN **(b)** $M(N - 1)$

15. $XX^T = (6)$, $\quad X^T X = \begin{pmatrix} 1 & -1 & 2 \\ -1 & 1 & -2 \\ 2 & -2 & 4 \end{pmatrix}$

SECTION 3.3 UPPER-TRIANGULAR LINEAR SYSTEMS: page 146

1. $x_1 = 2$, $x_2 = -2$, $x_3 = 1$, $x_4 = 3$, and $\det(A) = 120$
6. $x_1 = 3$, $x_2 = 2$, $x_3 = 1$, $x_4 = -1$, and $\det(A) = -24$
8. The forward substitution algorithm. $x_1 = b_1/a_{1,1}$. Then compute

$$x_k = \frac{b_k - \displaystyle\sum_{j=1}^{k-1} a_{k,j} x_j}{a_{k,k}} \quad \text{for } k = 2, 3, \ldots, N.$$

SECTION 3.4 GAUSSIAN ELIMINATION AND PIVOTING: page 157

1. $x_1 = -3$, $x_2 = 2$, $x_3 = 1$
5. $y = 5 - 3x + 2x^2$
10. $x_1 = 1$, $x_2 = 3$, $x_3 = 2$, $x_4 = -?$
15. Algorithm for the solution of a tridiagonal linear system.

```
FOR  R = 2  TO  N  DO                              {Start the upper-triangularization}
  |      T := A(R−1)/D(R−1)
  |      D(R) := D(R) − T*C(R−1)
  |      B(R) := B(R) − T*B(R−1)
END                                                {End upper-triangularization}

         X(N) := B(N)/D(N)                         {Start the back substitution}
FOR  R = N−1  DOWNTO  1  DO
  |_____ X(R) := [B(R) − C(R)*X(R+1)]/D(R)         {End back substitution}
```

22. (a) Solution for Hilbert matrix A:

$x_1 = 25, x_2 = -300, x_3 = 1050, x_4 = -1400, x_5 = 630$

(b) Solution for the other matrix A:

$x_1 = 28.02304, x_2 = -348.5887, x_3 = 1239.781$
$x_4 = -1666.785, x_5 = 753.5564$

SECTION 3.5 MATRIX INVERSION: page 164

1. (a) $\begin{pmatrix} 2 & -1 & 0 \\ -1 & -3 & 2 \\ 0 & 2 & -1 \end{pmatrix}$ **3. (a)** $\begin{pmatrix} -\frac{13}{3} & -3 & -\frac{2}{3} \\ \frac{1}{3} & 0 & -\frac{1}{3} \\ 2 & 1 & 0 \end{pmatrix}$

5. (a) $\begin{pmatrix} -0.20 & -0.76 & -2.36 & 1.08 \\ 0.40 & 0.72 & 2.92 & -0.76 \\ -0.20 & -0.16 & -0.76 & 0.28 \\ 0.80 & 1.44 & 4.84 & -1.52 \end{pmatrix}$

12. The inverse of the Hilbert matrix of dimension 5×5 is

$$\begin{pmatrix} 25 & -300 & 1{,}050 & -1{,}400 & 630 \\ -300 & 4{,}800 & -18{,}900 & 26{,}880 & -12{,}600 \\ 1{,}050 & -18{,}900 & 79{,}380 & -117{,}600 & 56{,}700 \\ -1{,}400 & 26{,}880 & -117{,}600 & 179{,}200 & -88{,}200 \\ 630 & -12{,}600 & 56{,}700 & -88{,}200 & 44{,}100 \end{pmatrix}$$

SECTION 3.6 TRIANGULAR FACTORIZATION: page 177

1. (a) $\mathbf{Y}^T = (-4, 12, 3)$, $\mathbf{X}^T = (-3, 2, 1)$ **(b)** $\mathbf{Y}^T = (20, 39, 9)$, $\mathbf{X}^T = (5, 7, 3)$

4. (a) $\begin{pmatrix} -5 & 2 & -1 \\ 1 & 0 & 3 \\ 3 & 1 & 6 \end{pmatrix} = \begin{pmatrix} 1 & 0 & 0 \\ -0.2 & 1 & 0 \\ -0.6 & 5.5 & 1 \end{pmatrix} \begin{pmatrix} -5 & 2 & -1 \\ 0 & 0.4 & 2.8 \\ 0 & 0 & -10 \end{pmatrix}$

6. (a) $\mathbf{Y}^T = (8, -6, 12, 2)$, $\mathbf{X}^T = (3, -1, 1, 2)$ **(b)** $\mathbf{Y}^T = (28, 6, 12, 1)$, $\mathbf{X}^T = (3, 1, 2, 1)$

9. The triangular factorization $A = LU$ is

$$\begin{pmatrix} 1 & 1 & 0 & 4 \\ 2 & -1 & 5 & 0 \\ 5 & 2 & 1 & 2 \\ -3 & 0 & 2 & 6 \end{pmatrix} = \begin{pmatrix} 1 & 0 & 0 & 0 \\ 2 & 1 & 0 & 0 \\ 5 & 1 & 1 & 0 \\ -3 & -1 & -1.75 & 1 \end{pmatrix} \begin{pmatrix} 1 & 1 & 0 & 4 \\ 0 & -3 & 5 & -8 \\ 0 & 0 & -4 & -10 \\ 0 & 0 & 0 & -7.5 \end{pmatrix}$$

17. (a) $I_1 = 3, I_2 = 5, I_3 = 1$ **(c)** $I_1 = 4, I_2 = 3, I_3 = -1$

SECTION 3.7 ITERATIVE METHODS FOR LINEAR SYSTEMS: page 188

1. (a) Jacobi iteration
$P_1 = (3.75, 1.8)$
$P_2 = (4.2, 1.05)$
$P_3 = (4.0125, 0.96)$
Iteration will converge to
$P = (4, 1)$.

(b) Gauss–Seidel iteration
$P_1 = (3.75, 1.05)$
$P_2 = (4.0125, 0.9975)$
$P_3 = (3.999375, 1.000125)$
Iteration will converge to
$P = (4, 1)$.

3. (a) Jacobi iteration
$P_1 (-1, -1)$
$P_2 = (-4, -4)$
$P_3 = (-13, -13)$
The iteration diverges
away from the solution
$P = (0.5, 0.5)$.

(b) Gauss–Seidel iteration
$P_1 = (-1, -4)$
$P_2 = (-13, -40)$
$P_3 = (-121, -364)$
The iteration diverges
away from the solution
$P = (0.5, 0.5)$.

5. (a) Jacobi iteration
$P_1 = (2, 1.375, 0.75)$
$P_2 = (2.125, 0.96875, 0.90625)$
$P_3 = (2.0125, 0.95703125, 1.0390625)$

Iteration will converge to
$P = (2, 1, 1)$.

(b) Gauss–Seidel iteration
$P_1 = (2, 0.875, 1.03125)$
$P_2 = (1.96875, 1.01171875, 0.989257813)$
$P_3 = (2.00449219, 0.99753418, 1.0017395)$

Iteration will converge to
$P = (2, 1, 1)$.

9. No. The matrix is not diagonally dominant, and interchanging the rows will not produce a diagonally dominant matrix. However, we can use transformations to get an equivalent linear system such as

$$5x + 3y = 6$$
$$-6x - 8y = -4$$

which can be solved with iteration.

12. (a) $M_1 = M_{50} = 0.633974596, \quad M_2 = M_{49} = 0.464101615$
$M_3 = M_{48} = 0.509618943, \quad M_4 = M_{47} = 0.497422612$
$M_5 = M_{46} = 0.500690609, \quad M_6 = M_{45} = 0.499814952$
$M_7 = M_{44} = 0.500049584, \quad M_8 = M_{43} = 0.499986714$
$M_9 = M_{42} = 0.500003560, \quad M_{10} = M_{41} = 0.499999046$
$M_{11} = M_{40} = 0.500000255, \quad M_{12} = M_{39} = 0.499999932$
$M_{13} = M_{38} = 0.500000009, \quad M_{14} = M_{37} = 0.499999995$
$M_{15} = M_{36} = 0.500000001$
$M_k = 0.500000000 \qquad$ for $k = 16$ TO 35

SECTION 4.1 TAYLOR SERIES AND CALCULATION OF FUNCTIONS: page 204

1. (a) $P_5(x) = x - x^3/3! + x^5/5!$
$P_7(x) = x - x^3/3! + x^5/5! - x^7/7!$
$P_9(x) = x - x^3/3! + x^5/5! - x^7/7! + x^9/9!$
(b) $P_5(0.5) = 0.4794270834, E_5(0.5) = -0.0000015448$
$P_7(0.5) = 0.4794255333, E_7(0.5) = 0.0000000053$

$P_9(0.5) = 0.4794255387, E_9(0.5) = -0.0000000001$
$\sin(0.5) = 0.4794255386$
$P_5(1.0) = 0.8416666667, E_5(1.0) = -0.0001956818$
$P_7(1.0) = 0.8414682539, E_7(1.0) = 0.0000027309$
$P_9(1.0) = 0.8414710096, E_9(1.0) = -0.0000000248$
$\sin(1.0) = 0.8414709848$

(c) $|E_9(x)| = |\sin(c)x^{10}/10!|, 1 \times 1^{10}/10! = 0.0000002755 \ldots$

(d) $P_5(x) = 2^{-1/2}[1 + (x - \pi/4) - (x - \pi/4)^2/2 - (x - \pi/4)^3/6 + (x - \pi/4)^4/24 + (x - \pi/4)^5/120]$

(e) $P_5(0.5) = 0.4794260467, E_5(0.5) = -0.0000005081$
$\sin(0.5) = 0.4794255386$
$P_5(1.0) = 0.8414710840, E_5(1.0) = -0.0000000992$
$\sin(1.0) = 0.8414709848$

3. No. Because the derivatives of $f(x)$ are undefined at $x_0 = 0$.

5. $P_3(x) = 1 + 0x - x^2/2 + 0x^3$

6. (a) $f(2) = 2, f'(2) = \frac{1}{4}, f^{(2)}(2) = -\frac{1}{32}, f^{(3)}(2) = \frac{3}{256}$

$P_3(x) = 2 + (x - 2)/4 - (x - 2)^2/64 + (x - 2)^3/512$

(b) $P_3(1) = 1.732421875$; compare with $3^{1/2} = 1.732050808$

(c) $f^{(4)}(x) = -15(2 + x)^{-7/2}/16$; the maximum of $|f^{(4)}(x)|$ on the interval $1 \le x \le 3$ occurs when $x = 1$ and $|f^{(4)}(x)| \le |f^{(4)}(1)| \le 3^{-7/2} \times 15/16 \approx 0.020046. \ldots$ Therefore,

$$|E_3(x)| \le \frac{(0.020046. \ldots) \times 1^4}{4!} = 0.00083529. \ldots$$

8. (d) $P_3(0.5) = 0.41666667$ **10. (d)** $P_2(0.5) = 1.21875000$
$P_6(0.5) = 0.40468750$ $P_4(0.5) = 1.22607422$
$P_9(0.5) = 0.40553230$ $P_6(0.5) = 1.22660828$
$\ln(1.5) = 0.40546511$ $(1.5)^{1/2} = 1.22474487$

11. $f(x) = x^3 - 2x^2 + 2, f'(x) = 3x^2 - 4x, f^{(2)}(x) = 6x - 4, f^{(3)}(x) = 6$
$f(1) = 1, f'(1) = 1, f^{(2)}(1) = 2, f^{(3)}(1) = 6$
$P_3(x) = 1 + (x - 1) + (x - 1)^2 + (x - 1)^3$

13. (a) $P_3(2.1) = 2.254, P_3(1.9) = 1.646$ **(b)** $[P_3(2.1) - P_3(1.9)]/0.2 = 3.04$
(c) $f(2) = P_3(2) = 2, f'(2) = P_3'(2) = 3$

18. (b) $P_2(1.05, 1.1) = 1 - (0.05) + (0.1) + (0.05)^2 - (0.05)(0.1) = 1.0475$. Compare with the function value $f(1.05, 1.1) = 1.047619048$.

SECTION 4.2 INTRODUCTION TO INTERPOLATION: page 213

1. $P(x) = -0.02x^3 + 0.1x^2 - 0.2x + 1.66$
(a) Use $x = 4$ and get $b_3 = -0.02, b_2 = 0.02, b_1 = -0.12, b_0 = 1.18$. Hence $P(4) = 1.18$.
(b) Use $x = 4$ and get $d_2 = -0.06, d_1 = -0.04, d_0 = -0.36$. Hence $P'(4) = -0.36$.
(c) Use $x = 4$ and get $i_4 = -0.005, i_3 = 0.01333333, i_2 = -0.04666667, i_1 = 1.47333333, i_0 = 5.89333333$. Hence $I(4) = 5.89333333$. Similarly, use $x = 1$ and get $I(1) = 1.58833333$.

$$\int_1^4 P(x) \, dx = I(4) - I(1) = 5.89333333 - 1.58833333 = 4.305$$

(d) Use $x = 5.5$ and get $b_3 = -0.02, b_2 = -0.01, b_1 = -0.255, b_0 = 0.2575$. Hence $P(5.5) = 0.2575$.

4. Approximations for $f(x) = \exp(x)$

		$P(x)$	$\exp(x)$	$\exp(x) - P(x)$
(a)	$x = 0.3$	1.34985810	1.34985881	0.00000071
	$x = 0.4$	1.49182470	1.49182470	0.00000000
	$x = 0.5$	1.64872179	1.64872127	-0.00000052
(b)	$x = -0.1$	0.90481537	0.90483742	0.00002205
	$x = 1.1$	3.00413986	3.00416602	0.00002616

SECTION 4.3 LAGRANGE APPROXIMATION: page 225

1. **(a)** $P_1(x) = -1(x - 0)/(-1 - 0) + 0 = x + 0 = x$
  (b) $P_2(x) = -1(x - 0)(x - 1)/(-1 - 0)(-1 - 1) + 0 + 1(x + 1)(x - 0)/(1 + 1)(1 - 0)$
    $= -0.5(x)(x - 1) + 0.5(x)(x + 1) = 0x^2 + x + 0 = x$
  (c) $P_3(x) = -1(x)(x - 1)(x - 2)/(-1)(-2)(-3) + 0$
      $+ 1(x + 1)(x)(x - 2)/(2)(1)(-1) + 8(x + 1)(x)(x - 1)/(3)(2)(1)$
    $P_3(x) = x^3 + 0x^2 + 0x + 0 = x^3$
  (d) $P_1(x) = 1(x - 2)/(1 - 2) + 8(x - 1)/(2 - 1)$
    $P_1(x) = -(x - 2) + 8(x - 1) = 7x - 6$
  (e) $P_2(x) = 0 + (x)(x - 2)/(1)(-1) + 8(x)(x - 1)/(2)(1)$
    $P_2(x) = -(x)(x - 2) + 4(x)(x - 1) = 3x^2 - 2x + 0$
3. **(a)** $P_2(x) = -2x^2 + 6x + 0$, $P_2(1.5) = 4.5$
  (b) $P_3(x) = 0.5x^3 - 3.5x^2 + 7x + 0$, $P_3(1.5) = 4.3125$
6. **(c)** $f^{(4)}(c) = 120(c - 1)$ for all c, so that
    $E_3(x) = 5(x + 1)(x)(x - 3)(x - 4)(c - 1)$
10. $|f^{(2)}(c)| \le |-\sin(1)| = 0.84147098 = M_2$
  (a) $h^2 M_2/8 = h^2 \times 0.84147098/8 < 5 \times 10^{-7}$
    $h^2 < 4.753580 \times 10^{-6}$ implies that $h < 0.00218027$
12. **(a)** $z = 3 - 2x + 4y$

SECTION 4.4 NEWTON POLYNOMIALS: page 235

1. $P_1(x) = 4 - (x - 1)$
  $P_2(x) = 4 - (x - 1) + 0.4(x - 1)(x - 3)$
  $P_3(x) = 4 - (x - 1) + 0.4(x - 1)(x - 3) + 0.01(x - 1)(x - 3)(x - 4)$
  $P_4(x) = P_3(x) - 0.002(x - 1)(x - 3)(x - 4)(x - 4.5)$
  $P_1(2.5) = 2.5$, $P_2(2.5) = 2.2$, $P_3(2.5) = 2.21125$, $P_4(2.5) = 2.21575$
5. $f(x) = 3 \times 2^x$
  $P_4(x) = 1.5 + 1.5(x + 1) + 0.75(x + 1)(x) + 0.25(x + 1)(x)(x - 1)$
      $+ 0.0625(x + 1)(x)(x - 1)(x - 2)$
  $P_1(1.5) = 5.25$, $P_2(1.5) = 8.0625$
  $P_3(1.5) = 8.53125$, $P_4(1.5) = 8.47265625$
7. $f(x) = 3.6/x$
  $P_4(x) = 3.6 - 1.8(x - 1) + 0.6(x - 1)(x - 2) - 0.15(x - 1)(x - 2)(x - 3)$
      $+ 0.03(x - 1)(x - 2)(x - 3)(x - 4)$

$P_1(2.5) = 0.9, \qquad P_2(2.5) = 1.35$
$P_3(2.5) = 1.40625, \ P_4(2.5) = 1.423125$

11. $f(x) = x^{1/2}$

$P_4(x) = 2.0 + 0.23607(x - 4) - 0.01132(x - 4)(x - 5)$
$\qquad\qquad + 0.00091(x - 4)(x - 5)(x - 6) - 0.00008(x - 4)(x - 5)(x - 6)(x - 7)$

$P_1(4.5) = 2.11804, \ P_2(4.5) = 2.12086$
$P_3(4.5) = 2.12121, \ P_4(4.5) = 2.12128$

15. $P_2(x) = -2 + 4(x) - (x)(x - 1) = -x^2 + 5x - 2$

17. $P_3(x) = 5 - (x)(x - 1) + (x)(x - 1)(x - 2) = x^3 - 4x^2 + 3x + 5$

SECTION 4.5 CHEBYSHEV POLYNOMIALS: page 247

9. (a) $\ln(x + 2) \approx 0.69549038 + 0.49905042x - 0.14334605x^2 + 0.04909073x^3$

    (b) $|f^{(4)}(x)|/(2^3 \times 4!) \leq |-6|/(2^3 \times 4!) = 0.03125000$

11. (a) $\cos(x) \approx 1 - 0.46952087x^2$

    (b) $|f^{(3)}(x)|/(2^2 \times 3!) \leq |\sin(1)|/(2^2 \times 3!) = 0.03506129$

13. The error bound for Taylor's polynomial is

$$\frac{|f^{(8)}(x)|}{8!} \leq \frac{|\sin(1)|}{8!} = 0.00002087$$

The error bound for the minimax approximation is

$$\frac{|f^{(8)}(x)|}{2^7 \times 8!} \leq \frac{|\sin(1)|}{2^7 \times 8!} = 0.00000016$$

17. $f(x) = \sin(x)$, Chebyshev coefficients:

    (a) $C(0) = 0, \ C(1) = 0.88010417, \ C(2) = 0, \ C(3) = -0.03962622$

    (b) $C(0) = 0, \ C(1) = 0.88010116, \ C(2) = 0, \ C(3) = -0.03912370$
        $C(4) = 0$

    (c) $C(0) = 0, \ C(1) = 0.88010117, \ C(2) = 0, \ C(3) = -0.03912672$
        $C(4) = 0, \ C(5) = 0.00050252$

    (d) $C(0) = 0, \ C(1) = 0.88010117, \ C(2) = 0, \ C(3) = -0.03912671$
        $C(4) = 0, \ C(5) = 0.00049951, \ C(6) = 0$

SECTION 4.6 PADÉ APPROXIMATIONS: page 254

1. $1 = p_0, \ 1 + q_1 = p_1, \ \frac{1}{2} + q_1 = 0, \ q_1 = -\frac{1}{2}, \ p_1 = \frac{1}{2}$

    $\exp(x) \approx R_{1,1}(x) = (2 + x)/(2 - x)$

3. $1 = p_0, \ \frac{1}{3} + 2q_1/15 = p_1, \ \frac{2}{15} + q_1/3 = 0, \ q_1 = -\frac{2}{5}, \ p_1 = -\frac{1}{15}$

5. $1 = p_0, \ 1 + q_1 = p_1, \ \frac{1}{2} + q_1 + q_2 = p_2$

$$\left.\begin{array}{l} \dfrac{1}{6} + \dfrac{q_1}{2} + q_2 = 0 \\[2em] \dfrac{1}{24} + \dfrac{q_1}{6} + \dfrac{q_2}{2} = 0 \end{array}\right\} \quad \text{First solve this system.}$$

$q_1 = -\frac{1}{2}, q_2 = \frac{1}{12}, p_1 = \frac{1}{2}, p_2 = \frac{1}{12}$

7. $1 = p_0, \frac{1}{3} + q_1 = p_1, \frac{2}{15} + q_1/3 + q_2 = p_2$

$$\left.\begin{array}{c} \dfrac{17}{315} + \dfrac{2q_1}{15} + \dfrac{q_2}{3} = 0 \\[3mm] \dfrac{62}{2835} + \dfrac{17q_1}{315} + \dfrac{2q_2}{15} = 0 \end{array}\right\}\quad \text{First solve this system.}$$

$q_1 = -\frac{4}{9}, q_2 = \frac{1}{63}, p_1 = -\frac{1}{9}, p_2 = \frac{1}{945}$

11.

x_k	Taylor polynomial approximation	Padé approximation	Exact value, $f(x) = \exp(x)$
−1.6	0.270400	0.205298	0.201897
−1.2	0.318400	0.302326	0.301194
−0.8	0.451733	0.449541	0.449329
−0.4	0.670400	0.670330	0.670320
0.0	1.000000	1.000000	1.000000
0.4	1.491733	1.491803	1.491825
0.8	2.222400	2.224490	2.225541
1.2	3.294400	3.307692	3.320117
1.6	4.835733	4.870968	4.953032

13.

x_k	Taylor polynomial approximation	Padé approximation	Exact value, $f(x) = \tan(x)$
0.0	0.000000	0.000000	0.000000
0.2	0.202710	0.202710	0.202710
0.4	0.422793	0.422793	0.422793
0.6	0.684099	0.684137	0.684137
0.8	1.028611	1.029639	1.029639
1.0	1.542504	1.557407	1.557408
1.2	2.413996	2.572147	2.572152
1.4	4.052510	5.797765	5.797884

SECTION 5.1 LEAST-SQUARES LINE: page 265

1. (a) $10A + 0B = 7$
$0A + 5B = 13$
$y = 0.70x + 2.60, E_2(f) \approx 0.2449$

2. (a) $40A + 0B = 58$
$0A + 5B = 31.2$
$y = 1.45x + 6.24, E_2(f) \approx 0.8958$

3. (a) $60A + 10B = 120$
$10A + 5B = 31$
$y = 1.45x + 3.3, E_2(f) \approx 0.7348$

4. (a) $\sum_{k=1}^{5} x_k y_k \Big/ \sum_{k=1}^{5} x_k^2 = 86.9/55 = 1.58$

$y = 1.58x, \; E_2(f) \approx 0.1720$

11. (a) $y = 1.6866x^2, \; E_2(f) \approx 1.3$

$y = 0.5902x^3, \; E_2(f) \approx 0.29.$ This is the best fit.

13. (a) $k = 39.98/2.2 \approx 18.1727$

SECTION 5.2 CURVE FITTING: page 281

1. (a) $164A \;\;\;\; + 20C = \;\; 186$

$ 20B = -34$

$20A \;\;\; + \;\; 4C = \;\;\; 26$

$y = 0.875x^2 - 1.70x + 2.125 = 7/8x^2 - 17/10x + 17/8$

3. (a) $y = 0.5102x^2, \; E_1(f) \approx 0.20$ is the best fit.

$y = 0.4304 \; 2^x, \; E_1(f) \approx 0.62$

4. (a) $55A + 15B = 25.9297$

$15A + \;\; 5B = \;\; 6.1515$

$A = 0.7475, \; B = -1.0123, \; C = \exp(B) = 0.3634$

$y = 0.3634e^{0.7475x}, \; E_1(f) \approx 0.87$

(b) $6.1993A + 4.7874B = 8.9366$

$4.7874A + 5B = 6.1515$

$A = 1.8859, \; B = -0.5755, \; C = \exp(B) = 0.5625$

$y = 0.5625x^{1.8859}, \; E_1(f) \approx 0.27$ is the best fit.

6. (a) $15A + 5B = -0.8647$

$5A + 5B = 4.2196$

$A = -0.5084, \; B = 1.3524, \; C = \exp(B) = 3.8665$

$y = 3.8665e^{-0.5084x}, \; E_1(f) \approx 0.10$

9.

	Using linearization	Minimizing least squares
(a)	$\dfrac{1000}{1 + 4.3018 \exp(-1.0802t)}$	$\dfrac{1000}{1 + 4.2131 \exp(-1.0456t)}$
(b)	$\dfrac{5000}{1 + 8.9991 \exp(-0.81138t)}$	$\dfrac{5000}{1 + 8.9987 \exp(-0.81157t)}$

12. (a) $14A + 15B + 8C = 82,$ $\quad A = 2.4, \; B = 1.2, \; C = 3.8$

$15A + 19B + 9C = 93$ $\quad\;$ Yields $z = 2.4x + 1.2y + 3.8$

$8A + 9B + 5C = 49$

SECTION 5.3 INTERPOLATION BY SPLINE FUNCTIONS: page 298

6. (i) $m_0 = 10, \; m_1 = -14, \; m_2 = 4, \; m_3 = 10$

$S_0(x) = -4x^3 + 5x^2 + 2x + 1$

$S_1(w) = 3w^3 - 7w^2 + 4, \; w = x - 1$

$S_2(w) = w^3 + 2w^2 - 5w, \; w = x - 2$

(ii) $m_0 = 0, \; m_1 = -12, \; m_2 = 6, \; m_3 = 0$

$S_0(x) = -2x^3 + 5x + 1$

$S_1(w) = 3w^3 - 6w^2 - w + 4, \ w = x - 1$
$S_2(w) = -w^3 + 3w^2 - 4w, \ w = x - 2$

(iii) $m_0 = -16, \ m_1 = -7, \ m_2 = 2, \ m_3 = 11$
$S_0(x) = 1.5x^3 - 8x^2 + 9.5x + 1$
$S_1(w) = 1.5w^3 - 3.5w^2 - 2w + 4, \ w = x - 1$
$S_2(w) = 1.5w^3 + w^2 - 4.5w, \ w = x - 2$

(iv) $m_0 = -9.25, \ m_1 = -9.25, \ m_2 = 4.25, \ m_3 = 4.25$
$S_0(x) = -4.625x^2 + 7.625x + 1$
$S_1(w) = 2.25w^3 - 4.625w^2 - 1.625w + 4, \ w = x - 1$
$S_2(w) = 2.125w^2 - 4.125w, \ w = x - 2$

(v) $m_0 = -1.5, \ m_1 = -11.4, \ m_2 = 5.1, \ m_3 = 3$
$S_0(x) = -1.65x^3 - 0.75x^2 + 5.4x + 1$
$S_1(w) = 2.75w^3 - 5.7w^2 - 1.05w + 4, \ w = x - 1$
$S_2(w) = -0.35w^3 + 2.55w^2 - 4.2w, \ w = x - 2$

10. (i) $m_0 = -4.25, \ m_1 = 2.5, \ m_2 = 6.25, \ m_3 = -9.5, \ m_4 = 7.75$
$S_0(x) = 1.125x^3 - 2.125x^2 - 2x + 5$
$S_1(w) = 0.625w^3 + 1.25w^2 - 2.875w + 2, \ w = x - 1$
$S_2(w) = -2.625w^3 + 3.125w^2 + 1.5w + 1, \ w = x - 2$
$S_3(w) = 2.875w^3 - 4.75w^2 - 0.125w + 3, \ w = x - 3$

(ii) $m_0 = 0, \ m_1 = 1.5, \ m_2 = 6, \ m_3 = -7.5, \ m_4 = 0$
$S_0(x) = 0.25x^3 - 3.25x + 5$
$S_1(w) = 0.75w^3 + 0.75w^2 - 2.5w + 2, \ w = x - 1$
$S_2(w) = -2.25w^3 + 3w^2 + 1.25w + 1, \ w = x - 2$
$S_3(w) = 1.25w^3 - 3.75w^2 + 0.5w + 3, \ w = x - 3$

(iii) $m_0 = -1, \ m_1 = 2, \ m_2 = 5, \ m_3 = -4, \ m_4 = -13$
$S_0(x) = 0.5x^3 - 0.5x^2 - 3x + 5$
$S_1(w) = 0.5w^3 + w^2 - 2.5w + 2, \ w = x - 1$
$S_2(w) = -1.5w^3 + 2.5w^2 + w + 1, \ w = x - 2$
$S_3(w) = -1.5w^3 - 2w^2 + 1.5w + 3, \ w = x - 3$

(iv) $m_0 = \frac{19}{15}, \ m_1 = \frac{19}{15}, \ m_2 = \frac{17}{3}, \ m_3 = -\frac{89}{15}, \ m_4 = -\frac{89}{15}$
$S_0(x) = 0.63333x^2 - 3.63333x + 5$
$S_1(w) = 0.73333w^3 + 0.63333w^2 - 2.36666w + 2, \ w = x - 1$
$S_2(w) = -1.93333w^3 + 2.83333w^2 + 1.1w + 1, \ w = x - 2$
$S_3(w) = -2.96666w^2 + 0.96666w + 3, \ w = x - 3$

(v) $m_0 = 0.5, \ m_1 = 1.4, \ m_2 = 5.9, \ m_3 = -7, \ m_4 = -1.9$
$S_0(x) = 0.15x^3 + 0.25x^2 - 3.4x + 5$
$S_1(w) = 0.75w^3 + 0.7w^2 - 2.45w + 2, \ w = x - 1$
$S_2(w) = -2.15w^3 + 2.95w^2 + 1.2w + 1, \ w = x - 2$
$S_3(w) = 0.85w^3 - 3.5w^2 + 0.65w + 3, \ w = x - 3$

15. (i) $m_0 = -2.08, \ m_1 = 1.76, \ m_2 = 1.04, \ m_3 = 0.08$
$m_4 = -1.36, \ m_5 = -0.64, \ m_6 = -2.08$
$S_0(x) = 0.64x^3 - 1.04x^2 - 0.6x + 1$
$S_1(w) = -0.12w^3 + 0.88w^2 - 0.76w, \ w = x - 1$
$S_2(w) = -0.16w^3 + 0.52w^2 + 0.64w, \ w = x - 2$
$S_3(w) = -0.24w^3 + 0.04w^2 + 1.2w + 1, \ w = x - 3$
$S_4(w) = 0.12w^3 - 0.68w^2 + 0.56w + 2, \ w = x - 4$
$S_5(w) = -0.24w^3 - 0.32w^2 - 0.44w + 2, \ w = x - 5$

(ii) $m_0 = 0$, $m_1 = 1.2$, $m_2 = 1.2$, $m_3 = 0$
   $m_4 = -1.2$, $m_5 = -1.2$, $m_6 = 0$
   $S_0(x) = 0.2x^3 - 1.2x + 1$
   $S_1(w) = 0.6w^2 - 0.6w$, $w = x - 1$
   $S_2(w) = -0.2w^3 + 0.6w^2 + 0.6w$, $w = x - 2$
   $S_3(w) = -0.2w^3 + 1.2w + 1$, $w = x - 3$
   $S_4(w) = -0.6w^2 + 0.6w + 2$, $w = x - 4$
   $S_5(w) = 0.2w^3 - 0.6w^2 - 0.6w + 2$, $w = x - 5$

(iii) $m_0 = 0.75$, $m_1 = 1$, $m_2 = 1.25$, $m_3 = 0$
   $m_4 = -1.25$, $m_5 = -1$, $m_6 = -0.75$
   $S_0(x) = 0.041\overline{6}x^3 + 0.375x^2 - 1.41\overline{6}x + 1$
   $S_1(w) = 0.041\overline{6}w^3 + 0.5w^2 - 0.541\overline{6}w$, $w = x - 1$
   $S_2(w) = -0.2083\overline{3}w^3 + 0.625w^2 + 0.5833\overline{3}w$, $w = x - 2$
   $S_3(w) = -0.2083\overline{3}w^3 + 1.208\overline{3}w + 1$, $w = x - 3$
   $S_4(w) = 0.041\overline{6}w^3 - 0.625w^2 + 0.5833\overline{3}w + 2$, $w = x - 4$
   $S_5(w) = -0.041\overline{6}w^3 - 0.5w^2 - 0.541\overline{6}w + 2$, $w = x - 5$

SECTION 5.4 FOURIER SERIES AND TRIGONOMETRIC POLYNOMIALS: page 312

1. $f(x) = \dfrac{4}{\pi}\left[\sin(x) + \dfrac{\sin(3x)}{3} + \dfrac{\sin(5x)}{5} + \dfrac{\sin(7x)}{7} + \cdots\right]$

5. $f(x) = \dfrac{\pi}{4} + \displaystyle\sum_{j=1}^{\infty}\left[\dfrac{(-1)^j - 1}{\pi j^2}\right]\cos(jx) - \sum_{j=1}^{\infty}\left[\dfrac{(-1)^j}{j}\right]\cos(jx)$

7. $f(x) = \dfrac{4}{\pi}\left[\sin(x) - \dfrac{\sin(3x)}{3^2} + \dfrac{\sin(5x)}{5^2} - \dfrac{\sin(7x)}{7^2} + \cdots\right]$

13. $a_0 = 0.82246703$, $a_1 = -1$, $a_2 = 0.25$, $a_3 = -0.11111111$
   $a_4 = 0.0625$, $a_5 = -0.04$

15. (a) *Remark.* Use $f(-\pi) = 0$, $f(0) = 0$, $f(\pi) = 0$.
$$P(x) = 1.244017\sin(x) + 0.333333\sin(3x) + 0.089316\sin(5x)$$

   (c) *Remark.* Use $f(-\pi) = \pi/2$, $f(\pi) = \pi/2$.
$$P(x) = 0.785398 - 0.651366\cos(x) + 0.977049\sin(x)$$
$$-0.453450\sin(2x) - 0.087266\cos(3x) + 0.261799\sin(3x)$$
$$-0.151150\sin(4x) - 0.046766\cos(5x) + 0.070149\sin(5x)$$

17. Average temperature in Fairbanks, Alaska (4-week intervals)
$P(x) = 27.3077 - 37.9050 \cos(2\pi x/13) - 7.260548 \sin(2\pi x/13)$

Date	x	y	$P(x)$
Jan. 1	0	-14	-10.6
Jan. 29	1	-9	-9.6
Feb. 26	2	2	-0.2
Mar. 26	3	15	15.5
Apr. 23	4	35	34.0
May 21	5	52	50.9
June 18	6	62	62.4
July 16	7	63	65.8
Aug. 13	8	58	60.5
Sept. 10	9	50	47.5
Oct. 8	10	34	30.0
Nov. 5	11	12	11.8
Dec. 3	12	-5	-2.9
Jan. 1	13	-14	-10.6

SECTION 6.1 APPROXIMATING THE DERIVATIVE: page 328

1. $f(x) = \sin(x)$

h	Approximate $f'(x)$ formula (3)	Error in the approximation	Bound for the truncation error
0.1	0.695546112	0.001160597	0.001274737
0.01	0.696695100	0.000011609	0.000012747
0.001	0.696706600	0.000000109	0.000000127

3. $f(x) = \sin(x)$

h	Approximate $f'(x)$ formula (10)	Error in the approximation	Bound for the truncation error
0.1	0.696704390	0.000002320	0.000002322
0.01	0.696706710	-0.000000001	0.000000000

5. $f(x) = x^3$ **(a)** $f'(2) \approx 12.0025000$ **(b)** $f'(2) \approx 12.0000000$
(c) For part (a): $O(h^2) = -(0.05)^2 f^{(3)}(c)/6 = -0.0025000$
For part (b): $O(h^4) = -(0.05)^4 f^{(5)}(c)/30 = -0.0000000$

7. $f(x, y) = xy/(x + y)$

(a) $f_x(x, y) = [y/(x + y)]^2, f_x(2, 3) = 0.36$

h	Approximation to $f_x(2, 3)$	Error in the approximation
0.1	0.360144060	−0.000144060
0.01	0.360001400	−0.000001400
0.001	0.360000000	0.000000000

$f_y(x, y) = [x/(x + y)]^2, f_y(2, 3) = 0.16$

h	Approximation to $f_y(2, 3)$	Error in the approximation
0.1	0.160064030	−0.000064030
0.01	0.160000600	−0.000000600
0.001	0.160000000	0.000000000

9. (a) Formula (3) gives $I'(1.2) \approx -13.5840$ and $E(1.2) \approx 11.3024$.
   Formula (10) gives $I'(1.2) \approx -13.6824$ and $E(1.2) \approx 11.2975$.
(b) Using differentiation rules from calculus, we obtain
   $I'(1.2) \approx -13.6793$ and $E(1.2) \approx 11.2976$.

11. $f(x) = \cos(x), f^{(3)}(x) = \sin(x)$
Use the bound $|f^{(3)}(x)| \leq \sin(1.3) \approx 0.96356$.

h	App. $f'(x)$ Equation (17)	Error in the approximation	Equation (19), total error bound |round-off| + |trunc.|
0.1	−0.93050	−0.00154	0.00005 + 0.00161 = 0.00166
0.01	−0.93200	−0.00004	0.00050 + 0.00002 = 0.00052
0.001	−0.93000	−0.00204	0.00500 + 0.00000 = 0.00500

15. $f(x) = \cos(x), f^{(5)}(x) = -\sin(x)$
Use the bound $|f^{(5)}(x)| \leq \sin(1.4) \approx 0.98545$.

h	App. $f'(x)$ equation (22)	Error in the approximation	Equation (24), total error bound |round-off| + |trunc.|
0.1	−0.93206	0.00002	0.00008 + 0.00000 = 0.00008
0.01	−0.93208	0.00004	0.00075 + 0.00000 = 0.00075
0.001	−0.92917	−0.00287	0.00750 + 0.00000 = 0.00750

SECTION 6.2 NUMERICAL DIFFERENTIATION FORMULAS: page 342

1. $f(x) = \ln(x)$ (a) $f''(5) \approx -0.040001600$
(b) $f''(5) \approx -0.040007900$ (c) $f''(5) \approx -0.039999833$
(d) $f''(5) = -0.04000000 = -1/5^2$
   The answer in part (b) is more accurate.
3. $f(x) = \ln(x)$ (a) $f''(5) \approx 0.0000$
(b) $f''(5) \approx -0.0400$ (c) $f''(5) \approx 0.0133$
(d) $f''(5) = -0.0400 = -1/5^2$
   The answer in part (b) is more accurate.

5. (a) $f(x) = x^2, f''(1) \approx 2.0000$
   (b) $f(x) = x^4, f''(1) \approx 12.0002$
9. (a)

x	$f'(x)$
0.0	0.141345
0.1	0.041515
0.2	−0.058275
0.3	−0.158025

SECTION 7.1 INTRODUCTION TO QUADRATURE: page 355

1. (a) $f(x) = \sin(\pi x)$

Trapezoidal rule	0.0
Simpson's rule	0.666667
Simpson's $\frac{3}{8}$ rule	0.649519
Boole's rule	0.636165

(c) $f(x) = \sin(\sqrt{x})$

Trapezoidal rule	0.420735
Simpson's rule	0.573336
Simpson's $\frac{3}{8}$ rule	0.583143
Boole's rule	0.593376

2. (a) $f(x) = \sin(\pi x)$

Composite trapezoidal rule	0.603553
Composite Simpson rule	0.638071
Boole's rule	0.636165

(c) $f(x) = \sin(\sqrt{x})$

Composite trapezoidal rule	0.577889
Composite Simpson rule	0.592124
Boole's rule	0.593376

SECTION 7.2 COMPOSITE TRAPEZOIDAL AND SIMPSON'S RULE: page 366

1. (a) $F(x) = \arctan(x), F(1) - F(-1) = \pi/2 \approx 1.57079632679$

M	h	$T(f, h)$	$E_T(f, h)$	$S(f, h)$	$E_S(f, h)$
10	0.2	1.56746305691	0.00333326989	1.57079538809	0.00000093870
20	0.1	1.56996299445	0.00083333234	1.57079630697	0.00000001983
40	0.05	1.57058799348	0.00020833332	1.57079632648	0.00000000031
80	0.025	1.57074424346	0.00005208333	1.57079632679	0.00000000000
160	0.0125	1.57078330596	0.00001302083	1.57079632679	0.00000000000

(c) $F(x) = 2\sqrt{x}, F(4) - F\left(\frac{1}{4}\right) = 3$

M	h	$T(f, h)$	$E_T(f, h)$	$S(f, h)$	$E_S(f, h)$
10	0.375	3.04191993765	−0.04191993765	3.00762208163	−0.00762208163
20	0.1875	3.01118533878	−0.01118533878	3.00094047249	−0.00094047249
40	3.75/40	3.00285937995	−0.00285937995	3.00008406033	−0.00008406033
80	3.75/80	3.00071938914	−0.00071938914	3.00000605887	−0.00000605887
160	3.75/160	3.00018014420	−0.00018014420	3.00000039588	−0.00000039588

2. (a) $\int_0^1 \sqrt{1 + 9x^4}\, dx = 1.54786565469019$

$M = 10$, $T(f, 1/10) = 1.55260945$
$M = 5$, $S(f, 1/10) = 1.54786419$

3. Trapezoidal rule approximation:

$\frac{1}{2}(19 + 2 \times 13 + 10) + \frac{2}{2}(10 + 2 \times 7 + 5) + \frac{4}{2}(5 + 2 \times 2 + 1) = \dfrac{153}{2} = 76.5$

4. Simpson's rule approximation:

$\frac{1}{3}(19 + 4 \times 13 + 10) + \frac{2}{3}(10 + 4 \times 7 + 5) + \frac{4}{3}(5 + 4 \times 2 + 1) = \dfrac{223}{3} = 74.333333$

5. (a) $2\pi \int_0^1 x^3 \sqrt{1 + 9x^4}\, dx = 3.5631218520124$

$M = 10$, $T(f, 1/10) = 3.64244664$
$M = 5$, $S(f, 1/10) = 3.56372816$

10. (a) Use the bound $|f^{(2)}(x)| = |-\cos(x)| \le |\cos(0)| = 1$, and obtain $\dfrac{(\pi/3 - 0)1h^2}{12} \le 5 \times 10^{-9}$, then substitute $h = \pi/(3M)$ and get $\pi^3/162 \times 10^8 \le M^2$. Solve and get $4374.89 \le M$, since M must be an integer, $M = 4375$ and $h = 0.000239359$.

11. (a) Use the bound $|f^{(4)}(x)| = |\cos(x)| \le |\cos(0)| = 1$, and obtain $\dfrac{(\pi/3 - 0)1h^4}{180} \le 5 \times 10^{-9}$, then substitute $h = \pi/(6M)$ and get $\pi^5/34{,}992 \times 10^7 \le M^4$. Solve and get $17.1967 \le M$, since M must be an integer, $M = 18$ and $h = 0.029088821$.

12.

M	h	$T(f, h)$	$E_T(f, h) = O(h^2)$
1	0.2	0.1990008	0.0006660
2	0.1	0.1995004	0.0001664
4	0.05	0.1996252	0.0000416
8	0.025	0.1996564	0.0000104
16	0.0125	0.1996642	0.0000026

SECTION 7.3 RECURSIVE RULES AND ROMBERG INTEGRATION: page 380

1. (a)

J	$R(J, 0)$	$R(J, 1)$	$R(J, 2)$	$R(J, 3)$	$R(J, 4)$
0	−0.00171772				
1	0.02377300	0.03226990			
2	0.60402717	0.79744521	0.84845691		
3	0.64844713	0.66325379	0.65430770	0.65122596	
4	0.66591329	0.67173534	0.67230077	0.67258638	0.67267015
5	0.67029931	0.67176132	0.67176306	0.67175452	0.67175126
6	0.67139339	0.67175808	0.67175786	0.67175778	0.67175779
7	0.67166676	0.67175788	0.67175786	0.67175786	0.67175786
8	0.67173509	0.67175787	0.67175786	0.67175786	0.67175786

$R(5, 5) = 0.67175036$, $R(6, 6) = 0.67175780$, $R(7, 7) = 0.67175787$, $R(8, 8) = 0.67175786$

(c)

J	$R(J, 0)$	$R(J, 1)$	$R(J, 2)$	$R(J, 3)$	$R(J, 4)$
0	2.88				
1	2.10564024	1.84752031			
2	1.78167637	1.67368841	1.66209962		
3	1.65849527	1.61743491	1.61368467	1.61291618	
4	1.61691082	1.60304933	1.60209029	1.60190626	1.60186308
5	1.60450987	1.60037622	1.60019801	1.60016798	1.60016116
6	1.60115269	1.60003363	1.60001079	1.60000781	1.60000719
7	1.60028999	1.60000242	1.60000034	1.60000018	1.60000015
8	1.60007262	1.60000016	1.60000001	1.60000000	1.60000000

$R(5, 5) = 1.60015950$, $R(6, 6) = 1.60000700$, $R(7, 7) = 1.60000014$, $R(8, 8) = 1.60000000$

2. $R(1, 1) = 3.133333333333333333$, $R(2, 2) = 3.142117647058823529$,
$R(3, 3) = 3.141585783761873844$, $R(4, 4) = 3.141592665277717401$,
$R(5, 5) = 3.141592653638243501$, $R(6, 6) = 3.141592653589722283$,
$R(7, 7) = 3.141592653589793258$, $R(8, 8) = 3.141592653589793238$

3. $F(0.5) = 0.69146246$, $F(1.0) = 0.84134475$, $F(1.5) = 0.93319280$, $F(2.0) = 0.97724987$,
$F(2.5) = 0.99379033$, $F(3.0) = 0.99865010$, $F(3.5) = 0.99976737$, $F(4.0) = 0.99996833$

6. $w_0 = \dfrac{14}{45}$, $w_1 = \dfrac{64}{45}$, $w_2 = \dfrac{24}{45}$, $w_3 = \dfrac{64}{45}$, and $w_4 = \dfrac{14}{45}$

13. (a) $N = 3$ **(b)** $N = 5$

15. (ii) For $\displaystyle\int_0^1 x^{1/2}\, dx$ Romberg integration converges slowly because the higher derivatives of the integrand $f(x) = x^{1/2}$ are not bounded near $x = 0$.

SECTION 7.4 ADAPTIVE QUADRATURE: page 390

1. The true value of the integral $I[f]$, the numerical quadrature value $Q[f]$, and the list of Simpson approximations $\{(a_k, b_k, S(a_{k1}, b_{k1}) + S(a_{k2}, b_{k2})\}$ is
(a) $I[f] = 0.6717578646...$, $Q[f] = 0.6717603$,
$\{(0, 0.1875, 0.0347439), (0.1875, 0.375, 0.0991360), (0.375, 0.5625, 0.1463674),$
$(0.5625, 0.65625, 0.0809331), (0.65625, 0.75, 0.0786779), (0.75, 0.9375, 0.1294425),$
$(0.9375, 1.03125, 0.0449459), (1.03125, 1.125, 0.0319015), (1.125, 1.3125, 0.0337709),$
$(1.3125, 1.5, 0.0098815), (1.5, 1.6875, -0.0004885), (1.6875, 1.875, -0.0039295),$
$(1.875, 2.25, -0.0079563), (2.25, 3, -0.0056658)\}$
(c) $I[f] = 1.6$, $Q[f] = 1.6000023$
$\{(0.04, 0.055, 0.0690417), (0.055, 0.070, 0.0601087), (0.070, 0.085, 0.0539449),$
$(0.085, 0.10, 0.0493603), (0.10, 0.13, 0.0886548), (0.13, 0.16, 0.0788898),$
$(0.16, 0.22, 0.1380834), (0.22, 0.28, 0.1202174), (0.28, 0.4, 0.2066110),$
$(0.4, 0.52, 0.1773096), (0.52, 0.76, 0.3013400), (0.76, 1, 0.2564406)\}$

3. The true value of the integral $I[f]$, the numerical quadrature value $Q_B[f]$, and the list of Simpson approximations $\left\{\left(a_k, b_k, \dfrac{16(S(a_{k1},b_{k1}) + S(a_{k2},b_{k2})) - S(a_k,b_k)}{15}\right)\right\}$ is
(a) $I[f] = 0.6717578646...$, $Q_B[f] = 0.6717584$ **(c)** $I[f] = 1.6$, $Q_B[f] = 1.6000001$

SECTION 7.5 GAUSS–LEGENDRE INTEGRATION: page 397

1. $\int_0^2 6t^5 \, dt = 64$

    (b) $G(f, 2) = 58.6666667$ **(c)** $G(f, 3) = 64$ **(d)** $G(f, 4) = 64$

3. $\int_0^1 \sin(t)/t \, dt \approx 0.9460831$

    (b) $G(f, 2) = 0.9460411$ **(c)** $G(f, 3) = 0.9460831$ **(d)** $G(f, 4) = 0.9460831$

6. (a) $N = 4$ **(b)** $N = 6$

8. If the fourth derivative does not change too much, then

$$\left| \frac{f^{(4)}(c_1)}{135} \right| < \left| \frac{-f^{(4)}(c_2)}{90} \right|.$$

The truncation error term for the Gauss–Legendre rule will be less than the truncation error term for Simpson's rule.

SECTION 8.1 MINIMIZATION OF A FUNCTION: page 420

4. $f(x) = 3x^2 - 2x + 5$
    $f'(x) = 6x - 2$. Local minimum at $\frac{1}{3}$.

6. $f(x) = 4x^3 - 8x^2 - 11x + 5$
    $f'(x) = 12x^2 - 16x - 11$
    Critical points $\frac{11}{6}$, $-\frac{1}{2}$. Local minimum at $\frac{11}{6}$.

8. $f(x) = (x + 2.5)/(4 - x^2)$
    $f'(x) = (x^2 + 5x + 4)/(4 - x^2)^2$
    Critical points $-1, -4$. Local minimum at -1.

10. $f(x) = -\sin(x) - \sin(3x)/3$ on $[0, 2]$
    $f'(x) = -[\cos(x) + \cos(3x)]$. Local minimum at 0.785398163.

14. $f(x, y) = x^3 + y^3 - 3x - 3y + 5$
    $f_x(x, y) = 3(x^2 - 1), f_y(x, y) = 3(y^2 - 1)$
    Critical points $(1, 1), (1, -1), (-1, 1), (-1, -1)$. Local minimum at $(1, 1)$.

16. $f(x, y) = x^2 y + xy^2 - 3xy$
    $f_x(x, y) = y(2x + y - 3), f_y(x, y) = x(x + 2y - 3)$
    Critical points $(0, 0), (0, 3), (3, 0), (1, 1)$. Local minimum at $(1, 1)$.

18. $f(x, y) = (x - y)/(2 + x^2 + y^2)$
    $f_x(x, y) = (y^2 + 2xy + 2 - x^2)/(2 + x^2 + y^2)^2$
    $f_y(x, y) = (y^2 - 2xy - 2 - x^2)/(2 + x^2 + y^2)^2$
    Critical points $(1, -1), (-1, 1)$. Local minimum at $(-1, 1)$.

20. $f(x, y) = (x - y)^4 + (x + y - 2)^2$
    $f_x(x, y) = 4(x - y)^3 + 2(x + y - 2), f_y(x, y) = -4(x - y)^3 + 2(x + y - 2)$
    Local minimum at $(1, 1)$.

32. $f(x, y, z) = 2x^2 + 2y^2 + z^2 - 2xy + yz - 7y - 4z$
    $f_x = 0$: $4x - 2y \qquad = 0$
    $f_y = 0$: $-2x + 4y + z - 7 = 0$
    $f_z = 0$: $y + 2z - 4 = 0$ Local minimum at $(1, 2, 1)$.

34. $f(x, y, z) = x^4 + y^4 + z^4 - 4xyz$
$f_x(x, y, z) = 4(x^3 - yz), f_y(x, y, z) = 4[y^3 - xz]$
$f_z(x, y, z) = 4(z^3 - xy)$.   Local minimum at $(1, 1, 1)$.
36. $f(x, y, z, u) = 2(x^2 + y^2 + z^2 + u^2) + x(y + z - u) + yz - 5x - 6y - 6z - 3u$
$f_x = 0$: $4x + y + z - u - 5 = 0$
$f_y = 0$: $x + 4y + z \qquad - 6 = 0$
$f_z = 0$: $x + y + 4z \qquad - 6 = 0$
$f_u = 0$: $-x \qquad\qquad + 4u - 3 = 0$
Local minimum at $(1, 1, 1, 1)$.

SECTION 9.1 INTRODUCTION TO DIFFERENT EQUATIONS: page 427

1. (b) $L = 1$
3. (b) $L = 3$
5. (b) $L = 60$

10. (c) No because $f_y(t, y) = \frac{1}{2} y^{-2/3}$ is not continuous when $t = 0$, and $\lim_{y \to 0} f_y(t, y) = \infty$.

13. $y(t) = t^3 - \cos(t) + 3$

15. $y(t) = \displaystyle\int_0^t \exp(-s^2/2) \, ds$

17. (b) $y(t) = y_0 \exp(-0.000120968t)$ **(c)** 2808 years   **(d)** 6.9237 sec

SECTION 9.2 EULER'S METHOD: page 435

1. $y' = t^2 - y$ with $y(0) = 1$. The Euler solutions are

	y_k			Exact
t_k	$h = 0.2$	$h = 0.1$	$h = 0.05$	$y(t_k)$
0.05			0.95	0.951271
0.1		0.9	0.902625	0.905163
0.15			0.857994	0.861792
0.2	0.8	0.811	0.816219	0.821269
1.0	0.537856	0.586189	0.609438	0.632121
2.0	1.714101	1.790581	1.827913	1.864665

3. $y' = -ty$ with $y(0) = 1$. The Euler solutions are

	y_k			Exact
t_k	$h = 0.2$	$h = 0.1$	$h = 0.05$	$y(t_k)$
0.05			1.0	0.998751
0.1		1.0	0.9975	0.995012
0.15			0.992513	0.988813
0.2	1.0	0.99	0.985069	0.980199
1.0	0.652861	0.628157	0.616984	0.606531
2.0	0.124379	0.130400	0.132980	0.135335

6. $y_5 = 1000(1 + 0.12)^5 = 1762.3417$

$y_{60} = 1000(1 + 0.01)^{60} = 1816.6967$

$y_{1800} = 1000\left(1 + \dfrac{0.12}{360}\right)^{1800} = 1821.9355$

8. $P_{k+1} = P_k + (0.02P_k - 0.00004P_k^2)10$ for $k = 1, 2, \ldots, 8.$

Year	t_k	Actual population at t_k, $P(t_k)$	P_k Euler rounded at each step	P_k Euler with more digits carried at each step
1900	0.0	76.1	76.1	76.1
1910	10.0	92.4	89.0	89.0035
1920	20.0	106.5	103.6	103.6356
1930	30.0	123.1	120.0	120.0666
1940	40.0	132.6	138.2	138.3135
1950	50.0	152.3	158.2	158.3239
1960	60.0	180.7	179.8	179.9621
1970	70.0	204.9	202.8	203.0000
1980	80.0	226.5	226.9	227.1164

13. No. For any M, Euler's method produces $0 < y_1 < y_2 < \cdots < y_M$. The mathematical solution is $y(t) = \tan(t)$ and $y(3) < 0$.

SECTION 9.3 HEUN'S METHOD: page 442

1. $y' = t^2 - y$ with $y(0) = 1$. The Heun solutions are

t_k	y_k $h = 0.2$	y_k $h = 0.1$	y_k $h = 0.05$	Exact $y(t_k)$
0.05			0.951313	0.951271
0.1		0.9055	0.905245	0.905163
0.15			0.861915	0.861792
0.2	0.824	0.821928	0.821431	0.821269
1.0	0.643244	0.634782	0.632772	0.632121
2.0	1.881720	1.868726	1.865656	1.864665

3. $y' = -ty$ with $y(0) = 1$. The Heun solutions are

t_k	y_k $h = 0.2$	y_k $h = 0.1$	y_k $h = 0.05$	Exact $y(t_k)$
0.05			0.99875	0.998751
0.1		0.995	0.995011	0.995012
0.15			0.988811	0.988813
0.2	0.98	0.980175	0.980196	0.980199
1.0	0.606975	0.606718	0.606586	0.606531
2.0	0.139628	0.136318	0.135571	0.135335

6. Solution of $v' = -32 - 0.1v$ over [0, 30] with $v(0) = 160$.

t_k	v_k
1.00	114.341
2.00	73.025
3.00	35.639
4.00	1.809
5.00	-28.802
10.00	-143.341
20.00	-254.983
30.00	-296.071

9. Richardson improvement for solving $y' = (t - y)/2$ over [0, 3] with $y(0) = 1$. The table entries are approximations to $y(3)$.

h	y_h	$[4y_h - y_{2h}]/3$
1	1.732422	
$\frac{1}{2}$	1.682121	1.665354
$\frac{1}{4}$	1.672269	1.668985
$\frac{1}{8}$	1.670076	1.669345
$\frac{1}{16}$	1.669558	1.669385
$\frac{1}{32}$	1.669432	1.669390
$\frac{1}{64}$	1.669401	1.669391

10. $y' = 1.5y^{1/3}$, $f(t, y) = 1.5y^{1/3}$, $f_y(t, y) = 0.5y^{-2/3}$.

$f_y(0, 0)$ does not exist. The I.V.P. is not well-posed on any rectangle that contains (0, 0).

SECTION 9.4 TAYLOR METHODS: page 448

1. $y' = t^2 - y$, $y'' = -t^2 + 2t + y$, $y''' = t^2 - 2t + 2 - y$
$y^{(4)} = -t^2 + 2t - 2 + y$ with $y(0) = 1$. Taylor's solutions:

t_k	y_k $h = 0.2$	y_k $h = 0.1$	y_k $h = 0.05$	Exact $y(t_k)$
0.05			0.9512706	0.9512706
0.1		0.9051625	0.9051626	0.9051626
0.15			0.8617920	0.8617920
0.2	0.8212667	0.8212691	0.8212692	0.8212692
1.0	0.6321148	0.6321202	0.6321205	0.6321206
2.0	1.8646605	1.8646645	1.8646647	1.8646647

3. $y' = -ty$, $y'' = t^2y - y$, $y''' = -t^3y + 3ty$
$y^{(4)} = t^4y - 6t^2y + 3y$ with $y(0) = 1$. Taylor solutions:

t_k	y_k			Exact $y(t_k)$
	$h = 0.2$	$h = 0.1$	$h = 0.05$	
0.05			0.9987508	0.9987508
0.1		0.9950125	0.9950125	0.9950125
0.15			0.9888131	0.9888130
0.2	0.9802	0.9801988	0.9801987	0.9801987
1.0	0.6065735	0.6065333	0.6065308	0.6065307
2.0	0.1353374	0.1353354	0.1353353	0.1353353

6. Richardson improvement for the Taylor solution to $y' = (t - y)/2$ over $[0, 3]$ with $y(0) = 1$. The table entries are approximations to $y(3)$.

h	y_h	$[16y_h - y_{2h}]/15$
1	1.6701860	
1/2	1.6694308	1.6693805
1/4	1.6693928	1.6693903
1/8	1.6693906	1.6693905

SECTION 9.5 RUNGE–KUTTA METHODS: page 462

1. $y' = t^2 - y$ with $y(0) = 1$. The RK4 solutions are

t_k	y_k			Exact $y(t_k)$
	$h = 0.2$	$h = 0.1$	$h = 0.05$	
0.05			0.9512706	0.9512706
0.1		0.9051627	0.9051626	0.9051626
0.15			0.8617920	0.8617920
0.2	0.8212733	0.8212695	0.8212693	0.8212692
1.0	0.6321380	0.6321216	0.6321206	0.6321206
2.0	1.8646923	1.8646664	1.8646648	1.8646647

3. $y' = -ty$ with $y(0) = 1$. The RK4 solutions are

t_k	y_k			Exact $y(t_k)$
	$h = 0.2$	$h = 0.1$	$h = 0.05$	
0.05			0.9987508	0.9987508
0.1		0.9950125	0.9950125	0.9950125
0.15			0.9888130	0.9888130
0.2	0.9801987	0.9801987	0.9801987	0.9801987
1.0	0.6065314	0.6065307	0.6065307	0.6065307
2.0	0.1353590	0.1353366	0.1353354	0.1353353

6. Solution of $y' = 0.01(70 - y)(50 - y)$ over $[0, 20]$ with $y(0) = 0$.

t_k	v_k
0.50	13.45109
1.00	21.82776
2.00	31.62582
3.00	37.10403
4.00	40.54658
5.00	42.87100
10.00	47.85965
20.00	49.73487

SECTION 9.6 PREDICTOR–CORRECTOR METHODS: page 474

1. $y' = t^2 - y$ with $y(0) = 1$

	y_k		
t_k	Adams–Bashforth–Moulton Solution	Milne–Simpson Solution	Hamming's Solution
0.2	0.8212693	0.8212692	0.8212692
0.25	0.7836992	0.7836992	0.7836992
0.3	0.7491818	0.7491818	0.7491818
0.4	0.6896800	0.6896800	0.6896800
0.5	0.6434694	0.6434693	0.6434693

3. $y' = -t/y$ with $y(1) = 1$

	y_k		
t_k	Adams–Bashforth–Moulton Solution	Milne–Simpson Solution	Hamming's Solution
1.2	0.7483205	0.7483273	0.7483296
1.25	0.6613998	0.6614224	0.6614250
1.3	0.5566583	0.5567251	0.5567253
1.35	0.4208572	0.4210904	0.4210644
1.4	0.1974740	0.1988783	0.1982560

5. $y' = 2ty^2$ with $y(0) = 1$

	y_k		
t_k	Adams–Bashforth–Moulton Solution	Milne–Simpson Solution	Hamming's Solution
0.2	1.0416675	1.0416670	1.0416668
0.25	1.0666688	1.0666673	1.0666671
0.3	1.0989052	1.0989025	1.0989020
0.4	1.1904878	1.1904801	1.1904788
0.5	1.3333631	1.3333439	1.3333404

7. $y' = 2y - y^2$ with $y(0) = 1$

	y_k		
t_k	Adams–Bashforth–Moulton Solution	Milne–Simpson Solution	Hamming's Solution
0.2	1.1972300	1.1973754	1.1973753
0.25	1.2447770	1.2449187	1.2449186
0.3	1.2911748	1.2913127	1.2913126
0.4	1.3798202	1.3799491	1.3799489
0.5	1.4619988	1.4621172	1.4621170

9. $y' = y^2 \sin(t)$ with $y(0) = 1$

	y_k		
t_k	Adams–Bashforth–Moulton Solution	Milne–Simpson Solution	Hamming's Solution
0.2	1.0203389	1.0203389	1.0203388
0.25	1.0320852	1.0320850	1.0320850
0.3	1.0467519	1.0467517	1.0467516
0.4	1.0857051	1.0857046	1.0857045
0.5	1.1394953	1.1394943	1.1394941

SECTION 9.7 EXERCISES FOR SYSTEMS OF DIFFERENTIAL EQUATIONS: page 480

1.

t_k	Euler's Solution		Runge–Kutta Solution	
	x_k	y_k	x_k	y_k
0.00	-2.7000000	2.8000000	-2.7000000	2.8000000
0.05	-2.5500000	2.6700000	-2.5521092	2.6742492
0.10	-2.4040735	2.5485015	-2.4078422	2.5570240
0.15	-2.2613042	2.4355685	-2.2662276	2.4484383
0.20	-2.1206612	2.3313677	-2.1261657	2.3487177
0.25	-1.9809701	2.2361870	-1.9863951	2.2582202
0.30	-1.8408766	2.1504585	-1.8454540	2.1774612
0.35	-1.6988028	2.0747861	-1.7016310	2.1071446
0.40	-1.5528942	2.0099803	-1.5529068	2.0482011
0.45	-1.4009547	1.9570997	-1.3968823	2.0018347
0.50	-1.2403677	1.9175032	-1.2306913	1.9695802
0.55	-1.0679998	1.8929130	-1.0508942	1.9533731
0.60	-0.8800833	1.8854923	-0.8533475	1.9556353
0.65	-0.6720727	1.8979393	-0.6330458	1.9793801
0.70	-0.4384703	1.9336037	-0.3839279	2.0283406
0.75	-0.1726120	1.9966280	-0.0986403	2.1071260
0.80	0.1335929	2.0921202	0.2317506	2.2214137
0.85	0.4899800	2.2263660	0.6181117	2.3781833
0.90	0.9085186	2.4070863	1.0736725	2.5860022
0.95	1.4037862	2.6437537	1.6145507	2.8553759
1.00	1.9935477	2.9479775	2.2603937	3.1991747

5. $2x'' - 5x' - 3x = 45 \exp(2t)$ with $x(0) = 2$ and $x'(0) = 1$
$x' = y$, $y' = 1.5x + 2.5y + 22.5 \exp(2t)$

t_k	Euler's Solution		Runge–Kutta Order 4	
	x_k	y_k	x_k	y_k
0.05	2.05	2.4	2.0875384	2.5548149
0.1	2.17	4.0970673	2.2612983	4.4593080
0.2	2.6821548	8.6109876	2.9477416	9.6019238
0.3	3.6910454	15.1009428	4.2609483	17.1320033
0.4	5.4118562	24.3286106	6.4858063	28.0250579
0.5	8.1422522	37.3283984	10.0223747	43.6284511

SECTION 9.8 BOUNDARY VALUE PROBLEMS: page 489

2.

t_k	x_k
0.0	0.95
0.1	0.60878730
0.2	0.40271388
0.3	0.28465173
0.4	0.21949736
0.5	0.18169660
0.6	0.15343074
0.7	0.12322870
0.8	0.08483430
0.9	0.03621082
1.0	−0.02138964
1.1	−0.08432878
1.2	−0.14746376
1.3	−0.20499428
1.4	−0.25124697
1.5	−0.28136156
1.6	−0.29184300
1.7	−0.28094858
1.8	−0.24888884
1.9	−0.19783359
2.0	−0.13172934
2.1	−0.05594944
2.2	0.02318879
2.3	0.09899431
2.4	0.16500770
2.5	0.21558615
2.6	0.24641578
2.7	0.25490439
2.8	0.24042007
2.9	0.20435366
3.0	0.15

7.

t_k	x_k
1.0	1.0
1.2	1.0546381
1.4	1.0831232
1.6	1.0798710
1.8	1.0388296
2.0	0.9542127
2.2	0.8212266
2.4	0.6367661
2.6	0.4000418
2.8	0.1130972
3.0	−0.2188219
3.2	−0.5870815
3.4	−0.9796809
3.6	−1.3814962
3.8	−1.7747499
4.0	−2.1396926
4.2	−2.4554759
4.4	−2.7011783
4.6	−2.8569364
4.8	−2.9051254
5.0	−2.8315261
5.2	−2.6264119
5.4	−2.2854900
5.6	−1.8106358
5.8	−1.2103650
6.0	−0.5

SECTION 9.9 FINITE-DIFFERENCE METHOD: page 497

2.

t_k	x_k
0.0	0.95
0.1	0.5317266
0.2	0.2857011
0.3	0.1511952
0.4	0.0839456
0.5	0.0523099
0.6	0.0344961
0.7	0.0165091
0.8	−0.0094459
0.9	−0.0462800
1.0	−0.0931759
1.1	−0.1466461
1.2	−0.2015235
1.3	−0.2518848
1.4	−0.2918877
1.5	−0.3164939
1.6	−0.3220454
1.7	−0.3066663
1.8	−0.2704721
1.9	−0.2155770
2.0	−0.1459092
2.1	−0.0668545
2.2	0.0152348
2.3	0.0936205
2.4	0.1617937
2.5	0.2140701
2.6	0.2461103
2.7	0.2553193
2.8	0.2410884
2.9	0.2048586
3.0	0.15

7.

t_k	x_k
1.0	1.0
1.2	1.0567979
1.4	1.0872681
1.6	1.0857698
1.8	1.0461730
2.0	0.9626076
2.2	0.8302022
2.4	0.6457893
2.6	0.4085406
2.8	0.1204907
3.0	−0.2130896
3.2	−0.5835053
3.4	−0.9786586
3.6	−1.3832965
3.8	−1.7794839
4.0	−2.1472943
4.2	−2.4656913
4.4	−2.7135651
4.6	−2.8708760
4.8	−2.9198458
5.0	−2.8461364
5.2	−2.6399466
5.4	−2.2969609
5.6	−1.8190898
5.8	−1.2149440
6.0	−0.5

SECTION 10.1 HYPERBOLIC EQUATIONS: page 507

4. (a)

t_j	x_2	x_3	x_4	x_5
0.0	0.587785	0.951057	0.951057	0.587785
0.1	0.475528	0.769421	0.769421	0.475528
0.2	0.181636	0.293893	0.293893	0.181636
0.3	−0.181636	−0.293893	−0.293893	−0.181636
0.4	−0.475528	−0.769421	−0.769421	−0.475528
0.5	−0.587785	−0.951057	−0.951057	−0.587785

5. (a)

t_j	x_2	x_3	x_4	x_5
0.0	0.500	1.000	1.500	0.750
0.1	0.500	1.000	0.875	0.800
0.2	0.500	0.375	0.300	0.125
0.3	−0.125	−0.200	−0.375	−0.500
0.4	−0.700	−0.875	−1.000	−0.500
0.5	−0.750	−1.500	−1.000	−0.500

SECTION 10.2 PARABOLIC EQUATIONS: page 518

3.

$x_1 = 0.0$	$x_2 = 0.2$	$x_3 = 0.4$	$x_4 = 0.6$	$x_5 = 0.8$	$x_6 = 1.0$
0.	0.587785	0.951057	0.951057	0.587785	0.
0.	0.475528	0.769421	0.769421	0.475528	0.
0.	0.384710	0.622475	0.622475	0.384710	0.
0.	0.311237	0.503593	0.503593	0.311237	0.
0.	0.251796	0.407415	0.407415	0.251796	0.
0.	0.203707	0.329606	0.329606	0.203707	0.
0.	0.164803	0.266657	0.266657	0.164803	0.
0.	0.133328	0.215730	0.215730	0.133328	0.
0.	0.107865	0.174529	0.174529	0.107865	0.
0.	0.087264	0.141197	0.141197	0.087264	0.
0.	0.070598	0.114231	0.114231	0.070598	0.

5.

$x_2 = 0.1$	$x_3 = 0.2$	$x_4 = 0.3$	$x_5 = 0.4$	$x_6 = 0.5$	$x_7 = 0.6$	$x_8 = 0.7$	$x_9 = 0.8$	$x_{10} = 0.9$
0.896802	1.538842	1.760074	1.538842	1.000000	0.363271	−0.142039	−0.363271	−0.278768
0.679453	1.178972	1.379558	1.261578	0.906680	0.463031	0.087481	−0.113105	−0.119094
0.525254	0.922045	1.103912	1.053055	0.822069	0.510614	0.226224	0.044356	−0.017187
0.414563	0.736209	0.901105	0.893110	0.745354	0.524638	0.304904	0.140008	0.046091
0.333982	0.599719	0.749227	0.767871	0.675798	0.517573	0.344237	0.194729	0.083684
0.274357	0.497707	0.633263	0.667755	0.612733	0.497732	0.358159	0.222603	0.104333
0.229422	0.419983	0.542889	0.586109	0.555553	0.470615	0.356015	0.233109	0.113928
0.194882	0.359543	0.470979	0.518282	0.503709	0.439829	0.344039	0.232602	0.116428
0.167775	0.311558	0.412595	0.460997	0.456703	0.407704	0.326366	0.225329	0.114483
0.146059	0.272679	0.364288	0.411917	0.414084	0.375717	0.305714	0.214105	0.109858
0.128313	0.240573	0.323633	0.369362	0.375442	0.344771	0.283844	0.200785	0.103722

SECTION 10.3 ELLIPTIC EQUATIONS: page 532

1. (a)
$$-4p_1 + p_2 + p_3 \qquad\quad = -80$$
$$p_1 - 4p_2 \qquad + p_4 = -10$$
$$p_1 \qquad - 4p_3 + p_4 = -160$$
$$p_2 + p_3 - 4p_4 = -90$$
(b) $p_1 = 41.25$, $p_2 = 23.75$, $p_3 = 61.25$, $p_4 = 43.75$

5. (b) Use $B = (-100, -10, -50, -90, 0, -40, -210, -120, -160)^T$ and get:
$P = (p_1, p_2, p_3, p_4, p_5, p_6, p_7, p_8, p_9)^T$
$= (54.2857, 41.4286, 36.4286, 75.7143, 65., 54.2857, 93.5714, 88.5714, 75.7143)^T$

(d) Start with $u_{i,j} = 65$ at each interior grid point. Use the SOR method with $w = 1.44646$. After 19 iterations the approximate solution is

	x_1	x_2	x_3	x_4	x_5	x_6	x_7	x_8	x_9
y_9	105.	120.	120.	120.	120.	120.	120.	120.	80.
y_8	90.	102.214	105.39	105.377	103.746	100.417	93.809	78.9553	40.
y_7	90.	93.4666	93.9694	92.3712	89.19	84.1128	75.8634	62.0122	40.
y_6	90.	87.6821	84.6489	80.9478	76.5297	70.9805	63.5197	53.2301	40.
y_5	90.	82.6121	75.9958	70.2411	65.0004	59.7598	54.005	47.3884	40.
y_4	90.	76.7703	66.4809	59.0203	53.4712	49.0532	45.3521	42.3186	40.
y_3	90.	67.9881	54.1371	45.8879	40.8108	37.6298	36.0316	36.5341	40.
y_2	90.	51.0449	36.1913	29.5834	26.2545	24.6236	24.6103	27.7861	40.
y_1	50.	10.	10.	10.	10.	10.	10.	10.	25.

SECTION 11.1 HOMOGENEOUS SYSTEMS: THE EIGENVALUE PROBLEM: page 548

6. (a) $p(\lambda) = \lambda^2 - 3\lambda - 4$, and $\lambda_1 = -1$, $V_1 = (-1, 1)$; $\lambda_2 = 4$, $V_2 = (2, 3)$
   (c) $p(\lambda) = \lambda^2 - 2\lambda - 8$, and $\lambda_1 = -2$, $V_1 = (-1, 1)$; $\lambda_2 = 4$, $V_2 = (1, 1)$
10. (a) $p(\lambda) = -\lambda^3 + 7\lambda^2 - 14\lambda + 8$, and $\lambda_1 = 4$, $V_1 = (1, 1, 2)$; $\lambda_2 = 2$, $V_2 = (2, 3, 5)$; and $\lambda_3 = 1$, $V_3 = (1, 2, 4)$

SECTION 11.2 THE POWER METHOD: page 560

4. $p(\lambda) = -\lambda^3 + 3\lambda^2 + 6\lambda - 8$, and
   $\lambda_1 = -2$, $X_1 = (-1, 2, 1)$; $\lambda_2 = 4$, $X_2 = (-3, 4, 2)$; and $\lambda_3 = 1$, $X_3 = (-2, 1, 0)$
6. $p(\lambda) = -\lambda^3 + 3\lambda^2 + 6\lambda - 8$, and
   $\lambda_1 = -2$, $X_1 = (-1, 2, 1)$; $\lambda_2 = 4$, $X_2 = (-2, 1, 0)$; and $\lambda_3 = 1$, $X_3 = (-3, 4, 2)$
8. $p(\lambda) = -\lambda^3 + 8\lambda^2 - 17\lambda + 10$, and
   $\lambda_1 = 6$, $X_1 = (0, -1, -1, 1)$; $\lambda_2 = 5$, $X_2 = (1, 1, 1, 0)$;
   $\lambda_3 = 3$, $X_3 = (-1, 1, 0, 0)$; $\lambda_4 = 1$, $X_4 = (1, 0, 1, 1)$
10. The sequence of vectors is

```
(2000,    6000,     4000   )
(3000.,   4800.,    4200.   )
(3060.,   4560.,    4380.   )
(3030.,   4512.,    4458.   )
(3011.4,  4502.4,   4486.2  )
(3003.9,  4500.48,  4495.62 )
(3001.27, 4500.10,  4498.64 )
(3000.40, 4500.02,  4499.58 )
(3000.12, 4500.,    4499.87 )
(3000.04, 4500.,    4499.96 )
(3000.01, 4500.,    4499.99 )
(3000.  , 4500.,    4500.   )
(3000.  , 4500.,    4500.   )
(3000.  , 4500.,    4500,   )
```

SECTION 11.3 JACOBI'S METHOD: page 572

1. $\lambda_1 = \dfrac{7 + \sqrt{33}}{2} = 6.37228132, \lambda_2 = 2, \lambda_3 = \dfrac{7 - \sqrt{33}}{2} = 0.627718677$

$\mathbf{X}_1 = (0.843070331, 1.000000000, 0.843070331)$
$\mathbf{X}_2 = (1.000000000, 0.000000000, -1.000000000)$
$\mathbf{X}_3 = (-0.593070331, 1.000000000, -0.593070331)$

3. $\lambda_1 = -9, \lambda_2 = 6, \lambda_3 = 3$ and $\mathbf{X}_1 = (-2, -1, 2), \mathbf{X}_2 = (2, -2, 1), \mathbf{X}_3 = (1, 2, 2)$

5. $\lambda_1 = 6 + \sqrt{26} = 11.0990195, \quad \lambda_2 = 2 + \sqrt{2} = 3.41421356,$
$\lambda_3 = 6 - \sqrt{26} = 0.900980486, \quad \lambda_4 = 2 - \sqrt{2} = 0.585786438$

$\mathbf{X}_1 = (0.819803903, 1.000000000, 1.000000000, 0.819803903)$
$\mathbf{X}_2 = (1.000000000, 0.414213562, -0.414213562, -1.000000000)$
$\mathbf{X}_3 = (1.000000000, -0.819803903, -0.819803903, 1.000000000)$
$\mathbf{X}_4 = (-0.414213562, 1.000000000, -1.000000000, 0.414213562)$

7. $\lambda_1 = -2, \lambda_2 = 6, \lambda_3 = 4, \lambda_4 = 1$
$\mathbf{X}_1 = (-1, -1, 1, 1), \mathbf{X}_2 = (-1, 1, -1, 1), \mathbf{X}_3 = (1, -1, -1, 1), \mathbf{X}_4 = (1, 1, 1, 1)$

9. $\lambda_1 = 10, \lambda_2 = 5, \lambda_3 = -4, \lambda_4 = 1, \lambda_5 = -2$
$\mathbf{X}_1 = (1.5, 1.0, 1.0, -1.0, -1.0), \mathbf{X}_2 = (1.0, 1.5, -1.0, 1.0, 1.0),$
$\mathbf{X}_3 = (1.0, -1.0, 1.5, 1.0, 1.0), \mathbf{X}_4 = (-1.0, 1.0, 1.0, -1.0, 1.5),$
$\mathbf{X}_5 = (-1.0, 1.0, 1.0, 1.5, -1.0)$

11. (a) $\lambda_1 = 6.28995, \qquad \lambda_2 = 2.29428, \qquad \lambda_3 = 0.415775$
$\omega_1 = 2.50798, \qquad \omega_2 = 1.51469, \qquad \omega_3 = 0.644806$
$\mathbf{V}_1 = (0.835992, -0.539192, 0.101928)$
$\mathbf{V}_2 = (-0.504896, -0.683054, 0.527748)$
$\mathbf{V}_3 = (0.214935, 0.492656, 0.843263)$

SECTION 11.4 EIGENVALUES OF SYMMETRIC MATRICES: page 589

3. (a) Solution:
$\lambda_1 = 2 - \sqrt{2} = 0.585786438$
$\lambda_2 = 2$
$\lambda_3 = 2 + \sqrt{2} = 3.41421356$
$\mathbf{X}_1 = \left(\dfrac{-\sqrt{2}}{2}, 1, \dfrac{-\sqrt{2}}{2}\right)$
$\mathbf{X}_2 = (1, 0, -1)$
$\mathbf{X}_3 = \left(\dfrac{\sqrt{2}}{2}, 1, \dfrac{\sqrt{2}}{2}\right)$

(b) Solution:
$\lambda_1 = 4 - \sqrt{2} = 2.58578644$
$\lambda_2 = 4$
$\lambda_3 = 4 + \sqrt{2} = 5.41421356$
$\mathbf{X}_1 = \left(\dfrac{-\sqrt{2}}{2}, 1, \dfrac{-\sqrt{2}}{2}\right)$
$\mathbf{X}_2 = (1, 0, -1)$
$\mathbf{X}_3 = \left(\dfrac{\sqrt{2}}{2}, 1, \dfrac{\sqrt{2}}{2}\right)$

5. (a) Solution:

$$\lambda_1 = 4 - \frac{\sqrt{6 + 2\sqrt{5}}}{2} = 2.381966011$$

$$\lambda_2 = 4 + \frac{\sqrt{6 + 2\sqrt{5}}}{2} = 5.61803399$$

$$\lambda_3 = 4 - \frac{\sqrt{6 - 2\sqrt{5}}}{2} = 3.38196601$$

$$\lambda_4 = 4 + \frac{\sqrt{6 - 2\sqrt{5}}}{2} = 4.61803399$$

$\mathbf{X}_1 = (-0.618033989, 1., -1., 0.618033989)$
$\mathbf{X}_2 = (0.618033989, 1., 1., 0.618033989)$
$\mathbf{X}_3 = (1., -0.618033989, -0.618033989, 1.)$
$\mathbf{X}_4 = (1., 0.618033989, -0.618033989, -1.)$

(b) Solution:

$$\lambda_1 = 4 - \frac{\sqrt{6 + 2\sqrt{5}}}{2} = 2.381966011$$

$$\lambda_2 = 4 + \frac{\sqrt{6 + 2\sqrt{5}}}{2} = 5.61803399$$

$$\lambda_3 = 4 - \frac{\sqrt{6 - 2\sqrt{5}}}{2} = 3.38196601$$

$$\lambda_4 = 4 + \frac{\sqrt{6 - 2\sqrt{5}}}{2} = 4.61803399$$

$\mathbf{X}_1 = (0.618033989, 1., 1., 0.618033989)$
$\mathbf{X}_2 = (-0.618033989, 1., -1., 0.618033989)$
$\mathbf{X}_3 = (1., 0.618033989, -0.618033989, -1.)$
$\mathbf{X}_4 = (1., -0.618033989, -0.618033989, 1.)$

Index

TENS OF THOUSANDS OF STUDENTS NATIONWIDE ARE ALREADY USING
THE STUDENT EDITION OF MATLAB®.
YOU CAN TOO!

VERSION	TITLE CODE	ISBN NUMBER
MS/PC DOS (IBM® PCs and Compatibles)		
☐ THE STUDENT EDITION OF MATLAB FOR MS-DOS PERSONAL COMPUTERS WITH 5-1/4″ DISKS $50.00	85598-1	0-13-855982-1
☐ THE STUDENT EDITION OF MATLAB FOR MS-DOS PERSONAL COMPUTERS WITH 3-1/2″ DISKS $50.00	85597-3	0-13-855974-0
Macintosh® Personal Computers		
☐ THE STUDENT EDITION OF MATLAB FOR MACINTOSH COMPUTERS $50.00	85599-9	0-13-855990-2
Book Alone		
☐ THE STUDENT EDITION OF MATLAB STUDENT USER GUIDE $30.00	85600-5	0-13-856006-1

The Student Edition of MATLAB is solely for use by students working toward a degree at a degree-granting educational institution. I certify that I am such a student.

Signature _____ School _____

SHIPPING ADDRESS:

Name _____

Address _____

City _____ State _____ Zip _____

If payment accompanies order, Prentice Hall pays postage and handling charges. Same return privilege refund guaranteed.

☐ PAYMENT ENCLOSED—shipping and handling to be paid by publisher.

☐ I prefer to charge my   ☐ VISA   ☐ MasterCard

Card Number _____

Expiration Date _____

Signature _____

☛ MAIL TO: Prentice Hall/Neodata   PHONE: (515) 284-6761
P.O. Box 11073   FAX: (515) 284-2607
Des Moines, Iowa 50336-1073

Prices slightly higher outside the U.S. and subject to change without notice.
Available at college bookstores or direct from Prentice Hall.

D-1314-AD(7)

List of Theorems